UNDERSTANDING
Social Problems

10e

Linda A. Mooney

David Knox

Caroline Schacht

EAST CAROLINA UNIVERSITY

CENGAGE
Learning

Australia • Brazil • Mexico • Singapore • United Kingdom • United States

Understanding Social Problems,
Tenth Edition
Linda A. Mooney, David Knox,
and Caroline Schacht

Product Director: Marta Lee-Perriard

Product Manager: Libby Beiting-Lipps

Content Developer: Jessica Alderman

Product Assistant: Chelsea Meredith

Marketing Director: Jennifer Levanduski

Content Project Manager: Cheri Palmer

Art Director: Vernon Boes

Manufacturing Planner: Judy Inouye

Production Service: Jill Traut, MPS Limited

Photo and Text Researchers: Lumina
Datamatics

Copy Editor: Laura Larson

Illustrator/Compositor: MPS Limited

Text Designer: Lisa Buckley

Cover Designer: Larry Didona

Cover Images: Alcoholic—Edward/Yay Micro/
Age Fotostock. USA, Florida, industrial
smokestacks—Dkar Images/Tetra Images/
Corbis. Door with graffiti—Kenneth O'Quinn/
iStock/Getty Images Plus/Getty Images.
Broken glass and reflections—Barbara Fischer,
Australia/Moment/Getty Images. A polar bear
floating in Arctic sea—Jan Will/iStock/Getty
Images Plus/Getty Images. Two siblings—
Marcus Lindstrom/E+/Getty Images. Little girl
sitting on curb in dirty dress—Tressie Davis/
iStock/Getty Images Plus/Getty Images. Young
man with rainbow flag painted on arm—Sami
Sert/E+/Getty Images. Black power/civil
rights—PJPhoto69/E+/Getty Images. Military
helicopter landing in desert—Frank Rossoto
Stocktrek/DigitalVision/Getty Images

For product information and technology assistance, contact us at
Cengage Learning Customer & Sales Support, 1-800-354-9706.

For permission to use material from this text or product,
submit all requests online at **www.cengage.com/permissions**
Further permissions questions can be e-mailed to
permissionrequest@cengage.com

Library of Congress Control Number: 2015951382

Student Edition:
ISBN: 978-1-305-57651-3

Loose-leaf Edition:
ISBN: 978-1-305-85962-3

Cengage Learning
20 Channel Center Street
Boston, MA 02210
USA

Cengage Learning is a leading provider of customized learning solutions
with employees residing in nearly 40 different countries and sales in more
than 125 countries around the world. Find your local representative at
www.cengage.com.

Cengage Learning products are represented in Canada by Nelson Education,
Ltd.

To learn more about Cengage Learning Solutions, visit **www.cengage.com**

Purchase any of our products at your local college store or at our preferred
online store **www.cengagebrain.com**

Printed in the United States of America
Print Number: 01   Print Year: 2016

For our grandchildren: Lana, Juno, and Sky

They give us hope.

Brief Contents

Contents

7 Work and Unemployment 213

8 Problems in Education 245

Features

Animals and Society

Preface

U nderstanding Social Problems is intended for use in college-level sociology courses. We recognize that many students enrolled in undergraduate sociology classes are not sociology majors. Thus, we have designed our text with the aim of inspiring students—no matter what their academic major or future life path may be—to care about social problems. In addition to providing a sound theoretical and research basis for sociology majors, *Understanding Social Problems* also speaks to students who are headed for careers in business, psychology, health care, social work, criminal justice, and the nonprofit sector, as well as to those pursuing degrees in education, fine arts and the humanities, or to those who are "undecided." Social problems, after all, affect each and every one of us, directly or indirectly. And everyone—whether a leader in business or politics, a stay-at-home parent, or a student—can become more mindful of how his or her actions, or inactions, perpetuate or alleviate social problems. We hope that *Understanding Social Problems* plants seeds of social awareness that will grow no matter what academic, occupational, and life path students choose.

New to This Edition

The tenth edition of *Understanding Social Problems* features expanded coverage of Native Americans, women in the military, LGBT issues, prescription drug abuse, "fracking," climate deniers, terrorism, and human rights issues. Learning objectives are now presented at the beginning of each chapter to guide student learning. Other pedagogical features that students and professors have found useful have been retained, including a running glossary, list of key terms, chapter reviews, and *Test Yourself* sections. Most of the opening vignettes in the tenth edition are new, as are many of the *What Do You Think?* questions, which are designed to engage students in critical thinking and stimulate classroom discussion. Many of the boxed chapter features (*The Human Side*, *Self and Society*, *Social Problems Research Up Close*, and *Animals and Society*) have been updated or replaced with new content. Finally, the tenth edition has new or updated research, data, tables, figures, and photos in each chapter, as well as new and revised material, detailed as follows.

Chapter 1 ("Thinking about Social Problems") now includes the results of a global survey on social problems around the world, including a table with rankings of the "greatest problems in the world" by region. This revised chapter also features an updated *Self and Society* and *Social Problems Research Up Close*, as well as new data from Gallup Polls, the Pew Research Center, and the Centers for Disease Control and Prevention.

Chapter 2 ("Physical and Mental Health and Health Care") begins with a new opening vignette about the Ebola epidemic. A new *Social Problems Research Up Close* feature asks, "Are Americans the Healthiest Population in the World?" There is a new table on "Successful People with Mental Illness." New topics include peer-to-peer mental health support groups on campus, food deserts, Indian Health Service, Military Health Service, and the male health disadvantage. Updated topics include the Affordable Health Care Act, medical debt, and complementary and alternative health care. A new *What Do You Think?* question asks readers if they agree with the 2014 Supreme Court ruling that craft store chain Hobby Lobby and other closely held for-profit companies may choose not to pay for coverage of birth control in their workers' health plans if the company's owner has religious objections.

Chapter 3 ("Alcohol and Other Drugs") begins with a new opening vignette, followed by a new *Social Problems Research Up Close* feature on the portrayal of cigarette smoking in popular movies. The chapter has been reorganized for this edition: Misuse of prescription drugs has been added to the section on "Frequently Used Legal Drugs"; and sections on "Tobacco Advertising," "Alternative Nicotine Delivery Systems," "Prevention" of alcohol, tobacco and prescription drug abuse, and "Synthetic Marijuana" have also been added. The "Health Costs of Drug Use" section is now divided into legal and illegal drugs, as are the strategies for action.

There are numerous new topics including the dramatic increase in teenagers' use of heroin and prescription drugs, the modification of the D.A.R.E. curriculum, e-cigarette use, the impact of heavy drinking on others in the home, the impact of alcohol consumption combined with other drugs on driving, drug overdoses, a cost–benefit analysis of drug courts, Internet drug sales and e-pharmacies, the sociological risk factors in drug use, the MPOWER program of the World Health Organization, and pro-tobacco and anti-tobacco social forces on tobacco use.

Chapter 4 ("Crime and Social Control") contains a new opening vignette, a new section on technology and crime, and two new features. The *Social Problems Research Up Close* examines the role of race in criminal justice policies, and the *Self and Society* assesses students' fear of criminal victimization. New crime and social control topics include a discussion of Agnew's General Strain Theory; incarceration as racism; public perceptions of black criminals; General Motors, Honda, and Takata as corporate criminals; gangs and schools; police shootings of unarmed minorities; the safety gender gap; the socioemotional impact of violent crime; aging and crime; the difficulty in assessing crime prevention and recidivism; recent crime legislation; social forces leading to and away from "get tough" crime policies; and federal reforms and the "Smart on Crime Initiative."

Chapter 5 ("Family Problems") opens with a new vignette about the domestic violence case of Janay and Ray Rice. We added several new topics, including polyamory and poly families, grandfamilies, gray divorces, relationship literacy education, intentional communities, and the "Period of Purple Crying." The updated and reorganized section on "Strategies for Action" includes a new section on "Strategies to Strengthen Families" and focuses on expanding definitions of family. There is also a new discussion of Child Protective Services. The revised chapter includes updated global data on child abuse, updated statistics on domestic violence, new Census Bureau data on interethnic and interracial marriages and relationships, and new Pew Research data on U.S. marriage and family patterns and values. A new *Social Problems Research Up Close* feature presents research on "The Polyamorists Next Door."

Chapter 6 ("Economic Inequality, Wealth, and Poverty") opens with a new vignette about a dog, Cosmo, who enjoys a higher standard of living than many poor people. This revised chapter includes new data on inequality within the top 1 percent, inequality in the global distribution of household wealth, inequality in the United States, and updated census data on poverty and poverty thresholds. New topics include wage theft, corporate tax inversion, plutocracy, and the marriage opportunity gap. New figures display average U.S. family wealth and distribution of U.S. wealth, and a new table presents the United Nations' 17 sustainable development goals. A new *What Do You Think?* question asks why wage theft gets so little attention in the media compared with other types of theft.

Chapter 7 ("Work and Unemployment") opens with a new vignette about unsafe working conditions in Bangladesh's garment industry. New topics in this revised chapter include communism, full employment, frictional unemployment, and "right-to-work" laws. We have added new research on work-related stressors and health and a new table on common work-related stressors, and new research on life satisfaction among union members compared with nonunion members. This chapter frames employment-related concerns as human rights issues and presents examples of how and why these human rights are being violated in workplaces around the world.

Chapter 8 ("Problems in Education") has been significantly revised with a new opening vignette, and all new chapter features. The *Self and Society* asks students to assess the importance of various aspects of their high school experiences in securing a job, and

The Human Side recounts the story of a former teacher who describes why she left the profession. The *Social Problems Research Up Close,* using a national sample, examines the process that leads students to drop out of high school. New *What Do You Think?* topics include the worldwide availability of *Sesame Street,* the desirability of same-sex classrooms, the case of a black college student suing his white roommates, and the future of teacher tenure. The section on "Crime, Violence, and School Discipline" is now reorganized into four areas: "Crime and Violence against Students," "Crime and Violence against Teachers," "School Discipline," and "Bullying."

New topics include comparisons of student outcomes by socioeconomic status in China and the United States, a longitudinal study of students from first grade to young adulthood, the social costs of dropouts, the impact of a disadvantaged school environment on teacher effectiveness, the diversity gap, "degrees of inequality" in higher education, for-profit online colleges and universities, merit-based versus need-based financial aid, "separate but unequal" college admissions, Parents' Revolution and the Network for Public Education, and the UNC scandal.

Chapter 9 ("Race, Ethnicity, and Immigration") opens with a new vignette about the anti–Columbus Day movement. This chapter features a new *The Human Side* feature: "A Cherokee Citizen's View of Andrew Jackson." There is a new section and table on racial microaggressions, and a new section on implicit prejudice. Other new topics include colorism, "sundown towns," and state laws banning Sharia law. The revised chapter includes updated U.S. Census data on Hispanic, racial, and foreign-born U.S. populations, new FBI data on hate crimes, and an updated section on white power music. New *What Do You Think?* questions ask (1) about views toward Rachel Dolezal's choice to identify as black, (2) if black racism toward whites is equivalent to white racism toward blacks, (3) what students think about the phrase "Black lives matter," and (4) if Barack Obama would have been elected president if he had darker skin color.

Chapter 10 ("Gender Inequality") features a new *The Human Side*—a suicide note written by a transgender teen—and a new *Self and Society* on whether men, women, or both make strong financial and political leaders. There is also a new subsection on same-sex education under "The School Experience and Cultural Sexisms" heading.

New topics and terms include the "missing girls" of China, vulnerable employment, family well-being and the gender pay gap, attributional gender bias, reinforcement of gender stereotypes in same sex classrooms, gender role content analyses of the 10 most popular programs on Cartoon Network and of 120 children's films from 11 countries, income differentials, Freeman and Freeman's *Stressed Sex: Uncovering the Truth about Men, Women, and Mental Health*, human trafficking of women and girls, A Voice for Men, the Paycheck Fairness Act, and the Workplace Advancement Act.

Chapter 11 ("Sexual Orientation and the Struggle for Equality") has been significantly revised in light of the recent Supreme Court decision legalizing gay marriage in the United States. A new section, "The Consequences of Anti-LGBT Bias," has also been added that includes topics on the relationship between LGBT status and (1) physical and mental health; (2) substance abuse; (3) economic inequality, poverty, and homelessness; and (4) aging and retirement. There is also a new opening vignette, and two new features. The *Self and Society* assess student attitudes toward gay and lesbian issues and *The Human Side*, "I Needed to Do Something . . .," an essay by Tim Cook, CEO of Apple, on why he felt he had to come out to the public.

New topics and terms include estimates of the U.S. LGBT population and the number of LGBT married couples; results of public opinion polls; the *Obergefell v. Hodges* decision and reactions to it; the backlash against LGBT rights successes; religious freedom laws and the corporate response; societal beliefs about the origin of sexual orientation; banning sexual orientation change efforts; social forces that led to greater social support for LGBT individuals; demographic differences between same- and different-sex couples; gay fathers' connectivity between emotional and cognitive parts of their brains; the American Sociological Association's amicus curiae brief; "corrective rape" of lesbians; LGBT higher rates of depression, suicide, poverty, homelessness, and physical illness; dissenting opinions in *Obergefell*; the social costs of continued homonegativity; the "gay stimulus package"; the First Amendment Defense Act; children's response to learning a parent is gay; and the self-fulfilling prophecy of stereotypical gay men and women's appearances.

Chapter 12 ("Population Growth and Aging") includes a new *The Human Side* feature describing one woman's decision (with her husband) to remain childfree. This revised chapter includes updated information about Social Security and updated figures, tables, and data from the Population Reference Bureau. A new *What Do You Think?* question asks if professors should retire after a certain age.

Chapter 13 ("Environmental Problems") begins with a new vignette about the 2015 Indian heat wave. A new *Social Problems Research Up Close* feature, "The Climate Deception Dossiers," presents documentation of how the fossil fuel industry has deceived the public on issues related to global warming and climate change. The revised chapter also includes new sections on climate deniers, fracking, environmental migrants, and hunters and anglers as environmentalists. A new *The Human Side* feature, written specifically for this text, is titled, "Fracking Stories Told by Someone Who Isn't Gagged." There is also a new figure on "The Cycle of Fracking Denial." New topics include charismatic megafauna, NIMBY ("Not in My Backyard"), the landmark Dutch court ruling that orders the government to step up efforts to reduce greenhouse gas emissions, and the 2015 Senate vote that climate change is real. The section on religion and environmentalism has been updated, including the addition of Pope Francis's call to action on climate change and environmental protection. This revised chapter also mentions the Permanent People's Tribunal consideration of whether fracking violates human rights, and also discusses how state officials and employees in Florida were ordered to not use the terms *global warming* or *climate change*. A new *What Do You Think?* feature asks students what Pope Francis's statement "There can be no renewal of our relationship with nature without a renewal of humanity itself" means to them.

Chapter 14 ("Science and Technology") begins with a new opening vignette on medical technology. There are also two new features and two updated features. *The Human Side* concerns the consequences, for one young woman, of being denied an abortion, and the *Social Problems Research Up Close* examines gender differences, or lack thereof, in Internet use. Both the *Self and Society* and *Animals and Society* features have been updated. New topics or terms include the increased use of "digital agents" for blue- and white-collar jobs, technology corporate lawsuits, the Internet of Things (IoT), use of social media in political unrest, the growth of genetically modified (GM) organisms, GM foods and their consequences, wearable technology, the use of nanotechnology in "nanofoods," new state restrictions on abortion, genetic cloning, reproductive cloning, and therapeutic cloning, the expansion of cybersecurity breaches and Internet vulnerability, cyberattack threats to global security, automation of language and reasoning skills, the FCC's right to regulate net neutrality (no blocking, no throttling, no paid prioritization), slowness rage, the deep or dark web, Silk Road, Acxiom—the "cookie" collecting company, reform of the NSA's surveillance program, and the Marketplace Fairness Act. New *What Do You Think?* topics include the impact of self-driving cars, restrictions on privately owned drones, "right to forget" Internet laws, social groups on Mars, and sexism, racism, and diversity in the gaming community.

Chapter 15 ("Conflict, War, and Terrorism") begins with a new opening vignette. There is a new *The Human Side* on the refugee crisis, and a new *Self and Society* feature on "National Defense and the U.S. Military." The "Economics of Military Spending" has new subsections on weapons sales and the cost of war. Feminist theories of war and an expanded section on women in the military are now standalone headings. Reorganization of the chapter also includes adding two new sections ("Guantánamo Detention Center" and "Weapons of Mass Destruction") under the "America's Response to Terrorism" heading.

New topics of discussion include the use of unoccupied aerial vehicles (drones), the direct and indirect costs of violence, the devastation of Afghanistan and Syrian society, refugees and asylum seekers, women in the military, "rally around the flag," gender norming, occupationally specific standards validation, conflict minerals, civil war in Yemen, Charlie Hebdo and the terrorist attacks in Paris and Garland, Texas, Boko Haram, the evolution and funding of the Islamic State in Iraq and Syria (ISIS), public attitudes toward the use of ground troops in the Middle East, the Charleston, South Carolina, killings, the grievance models of terrorism, micro-aggression psychological models of terrorists, ecocide, the nuclear weapons agreement with Iran, and the public's national priorities by political party.

Features and Pedagogical Aids

We have integrated a number of features and pedagogical aids into the text to help students learn to think about social problems from a sociological perspective. Our mission is to help students think critically about social problems and their implications, and to increase their awareness of how social problems relate to their personal lives.

Boxed Features

Animals and Society. Several chapters contain a feature called *Animals and Society,* which examines issues, problems, policies, and/or programs concerning animals within the context of the social problem discussed in that chapter. For example, Chapter 5 ("Family Problems") includes an *Animals and Society* feature that examines "Pets and Domestic Violence," and in Chapter 14 ("Science and Technology"), the *Animals and Society* feature discusses "The Use of Animals in Scientific Research."

Self and Society. Each chapter includes a *Self and Society* feature designed to help students assess their own attitudes, beliefs, knowledge, or behaviors regarding some aspect of the social problem under discussion. In Chapter 5 ("Family Problems"), for example, the "Abusive Behavior Inventory" invites students to assess the frequency of various abusive behaviors in their own relationships. The *Self and Society* feature in Chapter 3 ("Alcohol and Other Drugs") allows students to measure the consequences of their own drinking behavior and compare it to respondents in a national sample, and students can assess their fear of criminal victimization in Chapter 4 ("Crime and Social Control").

The Human Side. Each chapter includes a boxed feature that describes personal experiences and views of individuals who have been directly affected by social problems. *The Human Side* feature in Chapter 4 ("Crime and Social Control"), for example, describes the horrific consequences of being a victim of rape, and *The Human Side* feature in Chapter 9 ("Race, Ethnicity, and Immigration") presents a Cherokee citizen's view of Andrew Jackson. In Chapter 10 ("Gender Inequality"), *The Human Side* features a suicide note from a transgender teenager.

Social Problems Research Up Close. This feature, found in every chapter, presents examples of social science research, summarizing the sampling and methods involved in data collection, and presenting findings and conclusions of the research study. Examples of *Social Problems Research Up Close* topics include job loss in midlife, polyamorists and poly families, gender and Internet use, tactics used by the fossil fuel industry to deceive the public about global warming and climate change, two-faced racism, and mental illness and suicide among U.S. veterans.

In-Text Learning Aids

Learning Objectives. We have developed a set of learning objectives that are presented at the beginning of each chapter. The learning objectives are designed to help students focus on key concepts, theories, and terms as they read each chapter.

Vignettes. Each chapter begins with a vignette designed to engage students and draw them into the chapter by illustrating the current relevance of the topic under discussion. For example, Chapter 5 ("Family Problems") begins with the domestic violence incident involving football player Ray Rice and his wife Janay Rice. Chapter 9 ("Race, Ethnicity, and Immigration") opens with details concerning the anti–Columbus Day movement and Chapter 15 ("Conflict, War, and Terrorism") describes an ISIS training camp for young boys.

Key Terms and Glossary. Important terms and concepts are highlighted in the text where they first appear. To reemphasize the importance of these words, they are listed at the end of every chapter and are included in the glossary at the end of the text.

Running Glossary. This tenth edition continues the running glossary that highlights the key terms in every chapter by putting the key terms and their definitions in the text margins.

What Do You Think? Sections. Each chapter contains multiple sections called *What Do You Think?* These sections invite students to use critical thinking skills to answer questions about issues related to the chapter content. For example, one *What Do You Think?* question in Chapter 4 ("Crime and Social Control") asks students, "What perpetuates the myth of the male-only serial killer?" and a *What Do You Think?* question in Chapter 11 ("Sexual Orientation and the Struggle for Equality") asks, "Should gay men and women who are subjected to violence in their home country be eligible for political asylum in the United States?"

Understanding [Specific Social Problem] Sections. All too often, students, faced with contradictory theories and research results walk away from social problems courses without any real understanding of their causes and consequences. To address this problem, chapter sections titled "Understanding [specific social problem]" cap the body of each chapter just before the chapter summaries. Unlike the chapter summaries, these sections sum up the present state of knowledge and theory on the chapter topic and convey the urgency for rectifying the problems discussed in the chapter.

Supplements

The tenth edition of *Understanding Social Problems* comes with a full complement of supplements designed for both faculty and students.

Supplements for Instructors

Online Instructor's Resource Manual. This supplement offers instructors learning objectives, key terms, lecture outlines, student projects, classroom activities, Internet exercises, and video suggestions.

Online Test Bank. Test items include multiple-choice and true-false questions with answers and text references, as well as short-answer and essay questions for each chapter.

Cengage Learning Testing Powered by Cognero. The Test Bank is also available through Cognero, a flexible, online system that allows instructors to author, edit, and manage test bank content as well as create multiple test versions in an instant. Instructors can deliver tests from their school's learning management system, classroom, office, or home.

Online PowerPoints. These vibrant, Microsoft® PowerPoint® lecture slides for each chapter assist instructors with lectures by providing concept coverage using images, figures, and tables directly from the textbook.

Supplements for Students

MindTap Sociology for *Understanding Social Problems*. With MindTap™ Sociology for *Understanding Social Problems* students have the tools to better manage their time, allowing them flexibility in when and where they complete assignments. Course material that is specially customized by the instructor in an easy-to-use interface keeps students engaged and active in the course. MindTap helps students achieve better grades by cultivating a true understanding of course concepts, and includes a mobile app to help keep students on track. With a wide array of course specific tools and apps—from note taking to flashcards—MindTap is a worthwhile and valuable investment in students' education.

Students will stay engaged with MindTap's interactive activities and remain motivated by information that shows where they stand at all times—both individually and compared to the highest performers in class. MindTap eliminates the guesswork, focusing on what's most important with a learning path designed specifically by the instructor. Students can master the most important information with built-in study tools such as visual chapter summaries that help students stay organized and use time efficiently.

Acknowledgments

This text reflects the work of many people. We would like to thank the following for their contributions to the development of this text: Libby Beiting-Lipps, Product Manager; Jessica Alderman, Associate Content Developer; Cheri Palmer, Senior Content Project Manager; Jill Traut, Project Manager at MPS Limited; Vernon Boes, Senior Art Director; Deanna Ettinger, Intellectual Property Analyst; and Nick Barrows, Intellectual Property Project Manager. We would also like to acknowledge the support and assistance of Carol L. Jenkins, Ben Lewis, Sharon Wilson, Molly Clever, Marieke Van Willigen, James and Mabelle Miller, and Don and Jean Fowler. To each, we send our heartfelt thanks. Special thanks also to George Glann, whose valuable contributions have assisted in achieving the book's high standard of quality from edition to edition.

Additionally, we are indebted to those who read the manuscript in its various drafts and provided valuable insights and suggestions, many of which have been incorporated into the final manuscript:

Annette Allen, Troy University

Sandra Alvarez, American International College

Christie Barcelos, Greenfield Community College

James Botts, Belmont Abbey College

Alondo Campbell, Santa Ana College

Margaret Choka, Pellissippi State Community College

Mirelle Cohen, Olympic College

Cynthia Coleman, Tompkins Cortland and Empire State College

Tonja Conerly, San Jacinto College—South

Gayle D'Andrea, Reynolds Community College

We are also grateful to all reviewers of the previous editions.

Finally, we are interested in ways to improve the text and invite your feedback and suggestions for new ideas and material to be included in subsequent editions. You can contact us at mooneyl@ecu.edu, knoxd@ecu.edu, or cschacht@suddenlink.net.

Andrew Rich/Getty Images

Unless someone like you cares a whole awful lot, nothing is going to get better. It's not."

DR. SEUSS
The Lorax

1

Thinking about Social Problems

Learning Objectives

After studying this chapter, you will be able to . . .

1 Define a social problem.

2 Discuss the elements of the social structure and culture of society.

3 Understand the connections between private troubles and public issues, and how they relate to the sociological imagination.

4 Summarize structural functionalism, conflict theory, and symbolic interactionism and their respective theories of social problems.

5 Describe the stages in conducting a research study.

After the economic turndown of 2008, the U.S. Congress passed the American Recovery and Reinvestment Act of 2009. The stimulus package was designed to help failing industries, create jobs, promote consumer spending, rescue the failed housing market, and encourage energy-related investments. To date, the distribution of stimulus funds amounts to over $840 billion (Recovery.gov 2015).

Rubberball/Fotosearch

IN A JUNE 2015 Gallup Poll, a random sample of Americans was asked, "What do you think is the most important problem facing this country today?" Leading problems included economic issues (i.e., wages, corporate corruption, the gap between the rich and poor, etc.), which were the clear majority of responses, and noneconomic issues such as immigration, distrust of government, health care, the family, education, and poverty (Gallup 2015a). Moreover, a recent survey indicates that just 28 percent of Americans are satisfied "with the way things are going in the United States"— a number significantly lower than a decade ago when 42 percent of Americans were satisfied with the direction of the country (Gallup 2015b).

We should not, however, confine our concerns to social problems in the United States. Globalization requires an understanding of the interrelationship between countries and regions around the world. Although some social problems are clearly global in nature, others appear to only impact the nation in which they occur. The economy, for example, is often discussed in terms of the U.S. job growth, the U.S. inflation rate, or American's consumer confidence. And yet, in 2015, when China's stock market plunged, the NYSE recorded its steepest losses since the recession of 2008. Nonetheless, whether measured by travel patterns, languages spoken, or student study abroad, Americans have shown little interest in other countries. Calling this "unfamiliarity with the world" a crisis, Ungar (2015) comments that the "continued ignorance of, or indifference toward, how other people see the world is a concrete threat to our own security and safety (p. 1).

Problems related to poverty, inadequate education, crime and violence, oppression of minorities, environmental destruction, and war and terrorism as well as many other social issues are both national and international concerns. Such problems present both a threat and a challenge to our national and global society. The primary goal of this textbook is to facilitate increased awareness and understanding of problematic social conditions in U.S. society and throughout the world.

Although the topics covered in this book vary widely, all chapters share common objectives: to explain how social problems are created and maintained; to indicate how they affect individuals, social groups, and societies as a whole; and to examine programs and policies for change. We begin by looking at the nature of social problems.

What Is a Social Problem?

There is no universal, constant, or absolute definition of what constitutes a social problem. Rather, social problems are defined by a combination of objective and subjective criteria that vary across societies, among individuals and groups within a society, and across historical time periods.

Objective and Subjective Elements of Social Problems

Although social problems take many forms, they all share two important elements: an objective social condition and a subjective interpretation of that social condition. The **objective element of a social problem** refers to the existence of a social condition. We become aware of social conditions through our own life experience, through the media, and through education. We see the homeless, hear gunfire in the streets, and see battered women in hospital emergency rooms. We read about employees losing their jobs as businesses downsize and factories close. In television news reports, we see the anguished faces of parents whose children have been killed by violent youths.

> For a condition to be defined as a social problem, there must be public awareness of the condition. How do you think the widespread use of communication technology—such as smartphones, Facebook, Twitter, and YouTube—has affected public awareness of problematic social conditions? Can you think of social problems that you became aware of through communication technology that you probably would not have been aware of if such technology were not accessible?

The **subjective element of a social problem** refers to the belief that a particular social condition is harmful to society or to a segment of society and that it should and can be changed. We know that crime, drug addiction, poverty, racism, violence, and pollution exist. These social conditions are not considered social problems, however, unless at least a segment of society believes that these conditions diminish the quality of human life.

By combining these objective and subjective elements, we arrive at the following definition: A **social problem** is a social condition that a segment of society views as harmful to members of society, and is in need of remedy.

Variability in Definitions of Social Problems

Individuals and groups frequently disagree about what constitutes a social problem. For example, some Americans view gun control as a necessary means of reducing gun violence whereas others believe that gun control is a threat to civil rights and individual liberties. Similarly, some Americans view the availability of abortion as a social problem, whereas others view restrictions on abortion as a social problem.

Definitions of social problems vary not only within societies but also across societies and geographic regions. Table 1.1 graphically portrays responses to a global survey (44 countries, $N = 48,643$) concerning the most pressing social problems in the world. Note that in the more advanced regions (Europe and the United States), social inequality is considered the most dangerous world problem, while in Africa, AIDS and other infectious diseases are viewed as the most important issues facing the world. Not surprisingly, religious and ethnic hatred was the top response for Middle Easterners (Pew 2014).

TABLE 1.1 **The Greatest Problems in the World, 2014 (44 Countries, _N_ = 48,643)**

	Religious & Ethnic Hatred	Inequality	Pollution & Environment	Nuclear Weapons	AIDS & Other Diseases
Middle East	34%	18%	9%	20%	10%
Europe	15	32	14	19	5
Asia	13	18	22	21	12
Latin America	9	18	25	26	19
Africa	24	18	7	22	29
U.S.	24	27	15	23	7

■ Top Choice
NOTE: Regional medians, Russia and Ukraine not included in Europe median.
SOURCE: Pew 2014.

WHAT do you THINK?

Some Americans view gun control as a necessary means of reducing gun violence whereas others believe that gun control is a threat to civil rights and individual liberties.

objective element of a social problem Awareness of social conditions through one's own life experiences and through reports in the media.

subjective element of a social problem The belief that a particular social condition is harmful to society, or to a segment of society, and that it should and can be changed.

social problem A social condition that a segment of society views as harmful to members of society and in need of remedy.

What constitutes a social problem also varies by historical time periods. For example, before the 19th century, a husband's legal right and marital obligation was to discipline and control his wife through the use of physical force. Today, the use of physical force is regarded as a social problem rather than a marital right.

Lastly, social problems change over time not only because *definitions* of conditions change, as in the example of the use of force in marriage, but also because the *conditions* themselves change. The use of cell phones while driving was not considered a social problem in the 1990s, as cell phone technology was just beginning to become popular. Now, with most U.S. adults having a cell phone, the issue of "distracted driving" has become a national problem. According to the National Highway Traffic Safety Administration (NHTSA 2014), in 2012, over 650,000 daytime drivers used their cell phones or operated some other kind of electronic device while driving. In the same year, an estimated 421,000 were injured in distracted driving automobile crashes.

WHAT do you THINK?

Many drivers see using mobile phones while driving as risky when other drivers do it, but view their own mobile phone use while driving as safe (NHTSA 2013). Why do you think this is so? Do you think using mobile phones or other electronic devices while driving is safe?

The Washington Post/Getty Images

A young girl and her family celebrate President Obama's 2014 announcement of immigration reform which, in part, will allow nearly 5 million undocumented immigrants to remain in the United States (Holland and Rampton 2014). Both officials and the public often disagree on the best strategy to remedy a social problem. Shortly after President Obama's address to the nation, Republican opponents heavily criticized the proposal.

Because social problems can be highly complex, it is helpful to have a framework within which to view them. Sociology provides such a framework. Using a sociological perspective to examine social problems requires knowledge of the basic concepts and tools of sociology. In the remainder of this chapter, we discuss some of these concepts and tools: social structure, culture, the "sociological imagination," major theoretical perspectives, and types of research methods.

Elements of Social Structure and Culture

Although society surrounds us and permeates our lives, it is difficult to "see" society. By thinking of society in terms of a picture or image, however, we can visualize society and therefore better understand it. Imagine that society is a coin with two sides: On one side is the structure of society, and on the other is the culture of society. Although each side is distinct, both are inseparable from the whole. By looking at the various elements of social structure and culture, we can better understand the root causes of social problems.

Elements of Social Structure

structure The way society is organized including institutions, social groups, statuses, and roles.

institution An established and enduring pattern of social relationships.

The **structure** of a society refers to the way society is organized. Society is organized into different parts: institutions, social groups, statuses, and roles.

Institutions. An **institution** is an established and enduring pattern of social relationships. The five traditional institutions are family, religion, politics, economics, and education, but some sociologists argue that other social institutions—such as science and technology, mass media, medicine, sports, and the military—also play important roles in modern society. Many social problems are generated by inadequacies in various institutions. For example, unemployment may be influenced by the educational institution's failure to prepare individuals for the job market and by alterations in the structure of the economic institution.

Social Groups. Institutions are made up of social groups. A **social group** is defined as two or more people who have a common identity, interact, and form a social relationship. For example, the family in which you were reared is a social group that is part of the family institution. The religious association to which you may belong is a social group that is part of the religious institution.

Social groups can be categorized as primary or secondary. **Primary groups**, which tend to involve small numbers of individuals, are characterized by intimate and informal interaction. Families and friends are examples of primary groups. **Secondary groups**, which may involve small or large numbers of individuals, are task oriented and characterized by impersonal and formal interaction. Examples of secondary groups include employers and their employees and clerks and their customers.

Statuses. Just as institutions consist of social groups, social groups consist of statuses. A **status** is a position that a person occupies within a social group. The statuses we occupy largely define our social identity. The statuses in a family may consist of mother, father, stepmother, stepfather, wife, husband, partner, child, and so on. Statuses can be either ascribed or achieved. An **ascribed status** is one that society assigns to an individual on the basis of factors over which the individual has no control. For example, we have no control over the sex, race, ethnic background, and socioeconomic status into which we are born. Similarly, we are assigned the status of child, teenager, adult, or senior citizen on the basis of our age—something we do not choose or control.

An **achieved status** is assigned on the basis of some characteristic or behavior over which the individual has some control. Whether you achieve the status of college graduate, spouse, parent, bank president, or prison inmate depends largely on your own efforts, behavior, and choices. One's ascribed statuses may affect the likelihood of achieving other statuses, however. For example, if you are born into a poor socioeconomic status, you may find it more difficult to achieve the status of college graduate because of the high cost of a college education.

Every individual has numerous statuses simultaneously. You may be a student, parent, tutor, volunteer fund-raiser, female, and Hispanic. A person's *master status* is the status that is considered the most significant in a person's social identity. In the United States, a person's occupational status is typically regarded as a master status. If you are a full-time student, your master status is likely to be student.

Roles. Every status is associated with many **roles**, or the set of rights, obligations, and expectations associated with a status. Roles guide our behavior and allow us to predict the behavior of others. As students, you are expected to attend class, listen and take notes, study for tests, and complete assignments. Because you know what the role of teacher involves, you can predict that your teachers will lecture, give exams, and assign grades based on your performance on tests.

A single status involves more than one role. The status of prison inmate includes one role for interacting with prison guards and another role for interacting with other prison inmates. Similarly, the status of nurse involves different roles for interacting with physicians and with patients.

Elements of Culture

Whereas the social structure refers to the organization of society, the **culture** refers to the meanings and ways of life that characterize a society. The elements of culture include beliefs, values, norms, sanctions, and symbols.

Whereas the social structure refers to the organization of society, the culture refers to the meanings and ways of life that characterize a society.

social group Two or more people who have a common identity, interact, and form a social relationship.

primary groups Usually small numbers of individuals characterized by intimate and informal interaction.

secondary groups Involving small or large numbers of individuals, groups that are task oriented and are characterized by impersonal and formal interaction.

status A position that a person occupies within a social group.

ascribed status A status that society assigns to an individual on the basis of factors over which the individual has no control.

achieved status A status that society assigns to an individual on the basis of factors over which the individual has some control.

roles The set of rights, obligations, and expectations associated with a status.

culture The meanings and ways of life that characterize a society, including beliefs, values, norms, sanctions, and symbols.

Beliefs. **Beliefs** refer to definitions and explanations about what is assumed to be true. The beliefs of an individual or group influence whether that individual or group views a particular social condition as a social problem. Does secondhand smoke harm nonsmokers? Are nuclear power plants safe? Does violence in movies and on television lead to increased aggression in children? Our beliefs regarding these issues influence whether we view the issues as social problems. Beliefs influence not only how a social condition is interpreted but also the existence of the condition itself.

Values. **Values** are social agreements about what is considered good and bad, right and wrong, desirable and undesirable. Frequently, social conditions are viewed as social problems when the conditions are incompatible with or contradict closely held values. For example, poverty and homelessness violate the value of human welfare; crime contradicts the values of honesty, private property, and nonviolence; racism, sexism, and heterosexism violate the values of equality and fairness. Often responses to opinion surveys (see this chapter's *Self and Society* feature) reveal an individual's values. For example, agreeing with the statement "a chief benefit of a college education is that it increases one's earning power" reflects the American value of economic well-being.

Values play an important role not only in the interpretation of a condition as a social problem but also in the development of the social condition itself. For example, most Americans view capitalism, characterized by free enterprise and the private accumulation of wealth, positively (Newport 2012). Nonetheless, a capitalist system, in part, is responsible for the inequality in American society as people compete for limited resources.

Norms and Sanctions. **Norms** are socially defined rules of behavior. Norms serve as guidelines for our behavior and for our expectations of the behavior of others.

There are three types of norms: folkways, laws, and mores. *Folkways* refer to the customs, habits, and manners of society—the ways of life that characterize a group or society. In many segments of our society, it is customary to shake hands when being introduced to a new acquaintance, to say "excuse me" after sneezing, and to give presents to family and friends on their birthdays. Although no laws require us to do these things, we are expected to do them because they are part of the cultural tradition, or folkways, of the society in which we live.

Laws are norms that are formalized and backed by political authority. It is normative for a Sikh to where a turban, and to have long hair and a beard. However, when a Hofstra University student who was also a Sikh sought to enlist in his school's ROTC program, he was denied a religious exemption from the army's "grooming policies." The army later argued that Mr. Singh could not request a religious exemption unless he was an ROTC cadet. A newly filed lawsuit against the United States Army notes the catch-22: Mr. Singh cannot become an ROTC cadet unless he is granted a religious exemption and cannot request a religious exemption unless he is an ROTC cadet (Shortell 2014).

Mores are norms with a moral basis. Both littering and child sexual abuse are violations of law, but child sexual abuse is also a violation of our mores because we view such behavior as immoral.

All norms are associated with **sanctions**, or social consequences for conforming to or violating norms. When we conform to a social norm, we may be rewarded by a positive sanction. These may range from an approving smile to a public ceremony in our honor. When we violate a social norm, we may be punished by a negative sanction, which may range from a disapproving look to the death penalty or life in prison. Most sanctions are spontaneous expressions of approval or disapproval by groups or individuals—these are referred to as informal sanctions. Sanctions that are carried out according to some recognized or formal procedure are referred to as formal sanctions. Types of sanctions, then, include positive informal sanctions, positive formal sanctions, negative informal sanctions, and negative formal sanctions (see Table 1.2).

TABLE 1.2 Types and Examples of Sanctions

	Positive	Negative
Informal	Being praised by one's neighbors for organizing a neighborhood recycling program	Being criticized by one's neighbors for refusing to participate in the neighborhood recycling program
Formal	Being granted a citizen's award for organizing a neighborhood recycling program	Being fined by the city for failing to dispose of trash properly

© Cengage Learning

beliefs Definitions and explanations about what is assumed to be true.

values Social agreements about what is considered good and bad, right and wrong, desirable and undesirable.

norms Socially defined rules of behavior, including folkways, laws, and mores.

sanctions Social consequences for conforming to or violating norms.

Social Opinion Survey

Indicate with a check mark the items you "strongly agree" or "somewhat agree" with.

	Strongly Agree or Somewhat Agree
1. Wealthy people should pay a larger share of taxes than they do now.	_____
2. Affirmative action in college admissions should be abolished.	_____
3. The federal government should do more to control the sale of handguns.	_____
4. A national health care plan is needed to cover everybody's medical costs.	_____
5. The federal government should raise taxes to reduce the deficit.	_____
6. Addressing global warming should be a federal priority.	_____
7. The chief benefit of a college education is that it increases one's earning power.	_____
8. Gays and lesbians should have the legal right to adopt a child.	_____
9. Undocumented immigrants should be denied access to public education.	_____
10. How would you characterize your political views?	
Far left	
Liberal	_____
Middle of the road	_____
Conservative	_____
Far right	_____

Percentage of first-year college students at bachelor's institutions who "strongly agree" or "somewhat agree" with the following statements*

	Strongly Agree or Somewhat Agree
1. Wealthy people should pay a larger share of taxes than they do now.	68.1
2. Affirmative action in college admissions should be abolished.	52.0
3. The federal government should do more to control the sale of handguns.	63.8
4. A national health care plan is needed to cover everybody's medical costs.	61.3
5. The federal government should raise taxes to reduce the deficit.	36.9
6. Addressing global warming should be a federal priority.	60.8
7. The chief benefit of a college education is that it increases one's earning power.	72.0
8. Gays and lesbians should have the legal right to adopt a child.	83.3
9. Undocumented immigrants should be denied access to public education.	40.7
10. How would you characterize your political views?	
Far left	2.8
Liberal	27.7
Middle of the road	46.3
Conservative	21.2
Far right	1.9

*Percentages are rounded.
SOURCE: Eagan et al. 2014.

Symbols. A **symbol** is something that represents something else. Without symbols, we could not communicate with one another or live as social beings.

The symbols of a culture include language, gestures, and objects whose meanings the members of a society commonly understand. In our society, a red ribbon tied around a car antenna symbolizes Mothers against Drunk Driving, a peace sign symbolizes the value of nonviolence, and a white-hooded robe symbolizes the Ku Klux Klan. Sometimes people attach different meanings to the same symbol. The Confederate flag is a symbol of southern pride to some and a symbol of racial bigotry to others.

The elements of the social structure and culture just discussed play a central role in the creation, maintenance, and social responses to various social problems. One of the

symbol Something that represents something else.

goals of taking a course in social problems is to develop an awareness of how the elements of social structure and culture contribute to social problems. Sociologists refer to this awareness as the "sociological imagination."

The Sociological Imagination

The **sociological imagination**, a term C. Wright Mills (1959) developed, refers to the ability to see the connections between our personal lives and the social world in which we live. When we use our sociological imagination, we are able to distinguish between "private troubles" and "public issues" and to see connections between the events and conditions of our lives and the social and historical context in which we live.

For example, that one person is unemployed constitutes a private trouble. That millions of people are unemployed in the United States constitutes a public issue. Once we understand that other segments of society share personal troubles such as intimate partner abuse, drug addiction, criminal victimization, and poverty, we can look for the elements of social structure and culture that contribute to these public issues and private troubles. If the various elements of social structure and culture contribute to private troubles and public issues, then society's social structure and culture must be changed if these concerns are to be resolved.

Rather than viewing the private trouble of obesity and all of its attending health concerns as a result of an individual's faulty character, lack of self-discipline, or poor choices regarding food and exercise, we may understand the obesity epidemic as a public issue that results from various social and cultural forces, including government policies that make high-calorie foods more affordable than healthier, fresh produce; powerful food lobbies that fight against proposals to restrict food advertising to children; and technological developments that have eliminated many types of manual labor and replaced them with sedentary "desk jobs" (see Chapter 2).

Theoretical Perspectives

Theories in sociology provide us with different perspectives with which to view our social world. A perspective is simply a way of looking at the world. A **theory** is a set of interrelated propositions or principles designed to answer a question or explain a particular phenomenon; it provides us with a perspective. Sociological theories help us to explain and predict the social world in which we live.

Sociology includes three major theoretical perspectives: the structural-functionalist perspective, the conflict perspective, and the symbolic interactionist perspective. Each perspective offers a variety of explanations about the causes of and possible solutions to social problems.

Structural-Functionalist Perspective

The structural-functionalist perspective is based largely on the works of Herbert Spencer, Emile Durkheim, Talcott Parsons, and Robert Merton. According to structural functionalism, society is a system of interconnected parts that work together in harmony to maintain a state of balance and social equilibrium for the whole. For example, each of the social institutions contributes important functions for society: Family provides a context for reproducing, nurturing, and socializing children; education offers a way to transmit a society's skills, knowledge, and culture to its youth; politics provides a means of governing members of society; economics provides for the production, distribution, and consumption of goods and services; and religion provides moral guidance and an outlet for worship of a higher power.

The structural-functionalist perspective emphasizes the interconnectedness of society by focusing on how each part influences and is influenced by other parts. For example, the increase in single-parent and dual-earner families has contributed to the number of children who are failing in school because parents have become less available to supervise their children's homework. As a result of changes in technology, colleges are offering more

---

When we use our sociological imagination, we are able to distinguish between "private troubles" and "public issues" and to see connections between the events and conditions of our lives and the social and historical context in which we live.

The structural-functionalist perspective emphasizes the interconnectedness of society by focusing on how each part influences and is influenced by other parts.

sociological imagination The ability to see the connections between our personal lives and the social world in which we live.

theory A set of interrelated propositions or principles designed to answer a question or explain a particular phenomenon.

technical programs, and many adults are returning to school to learn new skills that are required in the workplace. The increasing number of women in the workforce has contributed to the formulation of policies against sexual harassment and job discrimination.

Structural functionalists use the terms *functional* and *dysfunctional* to describe the effects of social elements on society. Elements of society are functional if they contribute to social stability and dysfunctional if they disrupt social stability. Some aspects of society can be both functional and dysfunctional. For example, crime is dysfunctional in that it is associated with physical violence, loss of property, and fear. But according to Durkheim and other functionalists, crime is also functional for society because it leads to heightened awareness of shared moral bonds and increased social cohesion.

Sociologists have identified two types of functions: manifest and latent (Merton 1968). **Manifest functions** are consequences that are intended and commonly recognized. **Latent functions** are consequences that are unintended and often hidden. For example, the manifest function of education is to transmit knowledge and skills to society's youth. But public elementary schools also serve as babysitters for employed parents, and colleges offer a place for young adults to meet potential mates. The babysitting and mate selection functions are not the intended or commonly recognized functions of education; hence, they are latent functions.

> In viewing society as a set of interrelated parts, structural functionalists argue that proposed solutions to social problems may lead to other social problems. For example, urban renewal projects displace residents and break up community cohesion. Racial imbalance in schools led to forced integration, which in turn generated violence and increased hostility between the races. What are some other "solutions" that have led to social problems? Do all solutions come with a price to pay? Can you think of a solution to a social problem that has no negative consequences?

Structural-Functionalist Theories of Social Problems

Two dominant theories of social problems grew out of the structural-functionalist perspective: social pathology and social disorganization.

Social Pathology. According to the social pathology model, social problems result from some "sickness" in society. Just as the human body becomes ill when our systems, organs, and cells do not function normally, society becomes "ill" when its parts (i.e., elements of the structure and culture) no longer perform properly. For example, problems such as crime, violence, poverty, and juvenile delinquency are often attributed to the breakdown of the family institution; the decline of the religious institution; and inadequacies in our economic, educational, and political institutions.

Social "illness" also results when members of a society are not adequately socialized to adopt its norms and values. People who do not value honesty, for example, are prone to dishonesties of all sorts. Early theorists attributed the failure in socialization to "sick" people who could not be socialized. Later theorists recognized that failure in the socialization process stemmed from "sick" social conditions, not "sick" people. To prevent or solve social problems, members of society must receive proper socialization and moral education, which may be accomplished in the family, schools, places of worship, and/or through the media.

Social Disorganization. According to the social disorganization view of social problems, rapid social change (e.g., the cultural revolution of the 1960s) disrupts the norms in a society. When norms become weak or are in conflict with each other, society is in a state of **anomie**, or *normlessness*. Hence, people may steal, physically abuse their spouses or children, abuse drugs, commit rape, or engage in other deviant behavior because the norms regarding these behaviors are weak or conflicting. According to this view, the solution to social problems lies in slowing the pace of social change and strengthening social norms. For example, although the use of alcohol by teenagers is considered a violation of a social norm in our society, this norm is weak. The media portray young people drinking alcohol, teenagers teach each other to drink alcohol and buy fake identification cards (IDs) to purchase alcohol, and parents model drinking behavior by having a few drinks

manifest functions Consequences that are intended and commonly recognized.

latent functions Consequences that are unintended and often hidden.

anomie A state of normlessness in which norms and values are weak or unclear.

after work or at a social event. Solutions to teenage drinking may involve strengthening norms against it through public education, restricting media depictions of youth and alcohol, imposing stronger sanctions against the use of fake IDs to purchase alcohol, and educating parents to model moderate and responsible drinking behavior.

Conflict Perspective

Contrary to the structural-functionalist perspective, the conflict perspective views society as composed of different groups and interests competing for power and resources. The conflict perspective explains various aspects of our social world by looking at which groups have power and benefit from a particular social arrangement. For example, feminist theory argues that we live in a patriarchal society—a hierarchical system of organization controlled by men. Although there are many varieties of feminist theory, most would hold that feminism "demands that existing economic, political, and social structures be changed" (Weir and Faulkner 2004, p. xii).

The origins of the conflict perspective can be traced to the classic works of Karl Marx. Marx suggested that all societies go through stages of economic development. As societies evolve from agricultural to industrial, concern over meeting survival needs is replaced by concern over making a profit, the hallmark of a capitalist system. Industrialization leads to the development of two classes of people: the bourgeoisie, or the owners of the means of production (e.g., factories, farms, businesses), and the proletariat, or the workers who earn wages.

The division of society into two broad classes of people—the "haves" and the "have-nots"—is beneficial to the owners of the means of production. The workers, who may earn only subsistence wages, are denied access to the many resources available to the wealthy owners. According to Marx, the bourgeoisie use their power to control the institutions of society to their advantage. For example, Marx suggested that religion serves as an "opiate of the masses" in that it soothes the distress and suffering associated with the working-class lifestyle and focuses the workers' attention on spirituality, God, and the afterlife rather than on worldly concerns such as living conditions. In essence, religion diverts the workers so that they concentrate on being rewarded in heaven for living a moral life rather than on questioning their exploitation.

Conflict Theories of Social Problems

There are two general types of conflict theories of social problems: Marxist and non-Marxist. Marxist theories focus on social conflict that results from economic inequalities; non-Marxist theories focus on social conflict that results from competing values and interests among social groups.

Marxist Conflict Theories. According to contemporary Marxist theorists, social problems result from class inequality inherent in a capitalistic system. A system of haves and have-nots may be beneficial to the haves but often translates into poverty for the have-nots. For example, in 2013, the typical pay for a CEO of a Standard & Poor's company increased 9 percent to $10.5 million—257 times the national average. During the same time period, the wages of a U.S. worker increased just 1.3 percent (Boak 2014). As we will explore later in this textbook, many social problems, including physical and mental illness, low educational achievement, and crime, are linked to poverty.

In addition to creating an impoverished class of people, capitalism also encourages "corporate violence." *Corporate violence* can be defined as actual harm and/or risk of harm inflicted on consumers, workers, and the general public as a result of decisions by corporate executives or managers. Corporate violence can also result from corporate negligence; the quest for profits at any cost; and willful violations of health, safety, and environmental laws (Reiman and Leighton 2013). Our profit-motivated economy encourages individuals who are otherwise good, kind, and law abiding to knowingly participate in the manufacturing and marketing of defective products, such as brakes on American jets, fuel tanks on automobiles, and salmonella-contaminated peanut butter (Basu 2014).

In 2010, a British Petroleum (BP) oil well off the coast of Louisiana ruptured, killing 11 people and spewing millions of gallons of oil into the Gulf of Mexico (see Chapter 13).

Evidence suggests that BP officials knew of the unstable cement seals on the rigs long before what has been called the worst offshore disaster in U.S. history (Pope 2011). Since May 2010, BP has paid over $28 billion in claims, advances, settlements, and other related costs (BP 2015).

Marxist conflict theories also focus on the problem of **alienation**, or powerlessness and meaninglessness in people's lives. In industrialized societies, workers often have little power or control over their jobs, a condition that fosters in them a sense of powerlessness. The specialized nature of work requires employees to perform limited and repetitive tasks; as a result, workers may come to feel that their lives are meaningless.

Alienation is bred not only in the workplace but also in the classroom. Students have little power over their education and often find that the curriculum is not meaningful to their lives. Like poverty, alienation is linked to other social problems, such as low educational achievement, violence, and suicide.

Marxist explanations of social problems imply that the solution lies in eliminating inequality among classes of people by creating a classless society. The nature of work must also change to avoid alienation. Finally, stronger controls must be applied to corporations to ensure that corporate decisions and practices are based on safety rather than on profit considerations.

Mark Wilson/Getty Images

Preschooler Jacob Hurley, who became seriously ill after eating peanut butter manufactured by the Peanut Corporation of America, is shown sitting with his father Peter Hurley, who is testifying before a House Energy and Commerce Committee hearing on Capitol Hill in Washington, DC, in January 2009. Nine deaths and over 700 illnesses resulted from the salmonella-tainted peanuts and, in 2013, former officials of the company were indicted on over 76 criminal counts (Schoenberg & Mattingly 2013).

Non-Marxist Conflict Theories. Non-Marxist conflict theorists, such as Ralf Dahrendorf, are concerned with conflict that arises when groups have opposing values and interests. For example, antiabortion activists value the life of unborn embryos and fetuses; pro-choice activists value the right of women to control their own bodies and reproductive decisions. These different value positions reflect different subjective interpretations of what constitutes a social problem. For anti-abortionists, the availability of abortion is the social problem; for pro-choice advocates, the restrictions on abortion are the social problem. Sometimes the social problem is not the conflict itself but rather the way that conflict is expressed. Even most pro-life advocates agree that shooting doctors who perform abortions and blowing up abortion clinics constitute unnecessary violence and lack of respect for life. Value conflicts may occur between diverse categories of people, including nonwhites versus whites, heterosexuals versus homosexuals, young versus old, Democrats versus Republicans, and environmentalists versus industrialists.

Solving the problems that are generated by competing values may involve ensuring that conflicting groups understand each other's views, resolving differences through negotiation or mediation, or agreeing to disagree. Ideally, solutions should be win-win, with both conflicting groups satisfied with the solution. However, outcomes of value conflicts are often influenced by power; the group with the most power may use its position to influence the outcome of value conflicts. For example, when Congress could not get all states to voluntarily increase the legal drinking age to 21, it threatened to withdraw federal highway funds from those that would not comply.

Symbolic Interactionist Perspective

Both the structural-functionalist and the conflict perspectives are concerned with how broad aspects of society, such as institutions and large social groups, influence the social world. This level of sociological analysis is called *macro-sociology*: It looks at the big picture of society and suggests how social problems are affected at the institutional level.

alienation A sense of powerlessness and meaninglessness in people's lives.

Micro-sociology, another level of sociological analysis, is concerned with the social-psychological dynamics of individuals interacting in small groups. Symbolic interactionism reflects the micro-sociological perspective and was largely influenced by the work of early sociologists and philosophers such as Max Weber, Georg Simmel, Charles Horton Cooley, G. H. Mead, W. I. Thomas, Erving Goffman, and Howard Becker. Symbolic interactionism emphasizes that human behavior is influenced by definitions and meanings that are created and maintained through symbolic interaction with others.

Sociologist W. I. Thomas (1931/1966) emphasized the importance of definitions and meanings in social behavior and its consequences. He suggested that humans respond to their definition of a situation rather than to the objective situation itself. Hence, Thomas noted that situations that we define as real become real in their consequences.

Symbolic interactionism also suggests that social interaction shapes our identity or sense of self. We develop our self-concept by observing how others interact with us and label us. By observing how others view us, we see a reflection of ourselves that Cooley calls the "looking-glass self."

Last, the symbolic interactionist perspective has important implications for how social scientists conduct research. German sociologist Max Weber argued that, to understand individual and group behavior, social scientists must see the world through the eyes of that individual or group. Weber called this approach *verstehen,* which in German means "to understand." *Verstehen* implies that, in conducting research, social scientists must try to understand others' views of reality and the subjective aspects of their experiences, including their symbols, values, attitudes, and beliefs.

Symbolic Interactionist Theories of Social Problems

A basic premise of symbolic interactionist theories of social problems is that a condition must be *defined or recognized* as a social problem for it to *be* a social problem. Three symbolic interactionist theories of social problems are based on this general premise.

Blumer's Stages of a Social Problem. Herbert Blumer (1971) suggested that social problems develop in stages. First, social problems pass through the stage of *societal recognition*—the process by which a social problem, for example, drunk driving, is "born." Drunk driving wasn't illegal until 1939, when Indiana passed the first state law regulating alcohol consumption and driving (Indiana State Government 2013). Second, *social legitimation* takes place when the social problem achieves recognition by the larger community, including the media, schools, and churches. As the visibility of traffic fatalities associated with alcohol increased, so did the legitimation of drunk driving as a social problem. The next stage in the development of a social problem involves *mobilization for action*, which occurs when individuals and groups, such as Mothers against Drunk Driving, become concerned about how to respond to the social condition. This mobilization leads to the *development and implementation of an official plan* for dealing with the problem, involving, for example, highway checkpoints, lower legal blood-alcohol levels, and tougher regulations for driving drunk.

Blumer's stage development view of social problems is helpful in tracing the development of social problems. For example, although sexual harassment and date rape occurred throughout the 20th century, these issues did not begin to receive recognition as social problems until the 1970s. Social legitimation of these problems was achieved when high schools, colleges, churches, employers, and the media recognized their existence. Organized social groups mobilized to develop and implement plans to deal with these problems. Groups successfully lobbied for the enactment of laws against sexual harassment and the enforcement of sanctions against violators of these laws. Groups also mobilized to provide educational seminars on date rape for high school and college students and to offer support services to victims of date rape.

Some disagree with the symbolic interactionist view that social problems exist only if they are recognized. According to this view, individuals who were victims of date rape in the 1960s may be considered victims of a problem, even though date rape was not recognized as a social problem at that time.

Labeling Theory. Labeling theory, a major symbolic interactionist theory of social problems, suggests that a social condition or group is viewed as problematic if it is labeled as such. According to labeling theory, resolving social problems sometimes involves changing the meanings and definitions that are attributed to people and situations. For example, so long as teenagers define drinking alcohol as "cool" and "fun," they will continue to abuse alcohol. So long as our society defines providing sex education and contraceptives to teenagers as inappropriate or immoral, the teenage pregnancy rate in the United States will continue to be higher than that in other industrialized nations. Individuals who label their own cell phone use while driving as safe will continue to use their cell phones as they drive, endangering their own lives and the lives of others.

Social Constructionism. Social constructionism is another symbolic interactionist theory of social problems. Similar to labeling theorists and symbolic interactionism in general, social constructionists argue that individuals who interpret the social world around them socially construct reality. Society, therefore, is a social creation rather than an objective given. As such, social constructionists often question the origin and evolution of social problems. For example, social constructionist theory has been used to

> analyze the history of the temperance and prohibition movements[,] ... the rise of alcoholism as a disease movement in the post-prohibition era[,] ... and the crusade against drinking and driving in the 1980s in the United States.... These studies [each] analyzed the shifts in social meanings attributed to alcohol beverage use and to problems within the changing landscapes of social, economic, and political power relationships in American society. (Herd 2011, p. 7)

Central to this idea of the social construction of social problems are the media, universities, research institutes, and government agencies, which are often responsible for the public's initial "take" on the problem under discussion.

Table 1.3 summarizes and compares the major theoretical perspectives, their criticisms, and social policy recommendations as they relate to social problems. The study of social

TABLE 1.3 Comparison of Theoretical Perspectives

	Structural Functionalism	Conflict Theory	Symbolic Interactionism
Representative theorists	Emile Durkheim, Talcott Parsons, Robert Merton	Karl Marx, Ralf Dahrendorf	George H. Mead, Charles Cooley, Erving Goffman
Society	Society is a set of interrelated parts; cultural consensus exists and leads to social order; natural state of society—balance and harmony.	Society is marked by power struggles over scarce resources; inequities result in conflict; social change is inevitable; natural state of society—imbalance.	Society is a network of interlocking roles; social order is constructed through interaction as individuals, through shared meaning, making sense out of their social world.
Individuals	Individuals are socialized by society's institutions; socialization is the process by which social control is exerted; people need society and its institutions.	People are inherently good but are corrupted by society and its economic structure; institutions are controlled by groups with power; "order" is part of the illusion.	Humans are interpretive and interactive; they are constantly changing as their "social beings" emerge and are molded by changing circumstances.
Cause of social problems?	Rapid social change; social disorganization that disrupts the harmony and balance; inadequate socialization and/or weak institutions.	Inequality; the dominance of groups of people over other groups of people; oppression and exploitation; competition between groups.	Different interpretations of roles; labeling of individuals, groups, or behaviors as deviant; definition of an objective condition as a social problem.
Social policy/ solutions	Repair weak institutions; assure proper socialization; cultivate a strong collective sense of right and wrong.	Minimize competition; create an equitable system for the distribution of resources.	Reduce impact of labeling and associated stigmatization; alter definitions of what is defined as a social problem.
Criticisms	Called "sunshine sociology"; supports the maintenance of the status quo; needs to ask "functional for whom?"; does not deal with issues of power and conflict; incorrectly assumes a consensus.	Utopian model; Marxist states have failed; denies existence of cooperation and equitable exchange; cannot explain cohesion and harmony.	Concentrates on micro issues only; fails to link micro issues to macro-level concerns; too psychological in its approach; assumes label amplifies problem.

Each chapter in this book contains a *Social Problems Research Up Close* box that describes a research study that examines some aspect of a social problem and is presented in a report, book, or journal. Academic sociologists, those teaching at community colleges, colleges, or universities, as well as other social scientists, primarily rely on journal articles as the means to exchange ideas and information. Some examples of the more prestigious journals in sociology include the *American Sociological Review*, the *American Journal of Sociology*, and *Social Forces*. Most journal articles begin with *an introduction and review of the literature*. Here, the investigator examines previous research on the topic, identifies specific research areas, and otherwise "sets the stage" for the reader. Often in this section, research hypotheses are set forth, if applicable. A researcher, for example, might hypothesize that the sexual behavior of adolescents has changed over the years as a consequence of increased fear of sexually transmitted diseases and that such changes vary on the basis of sex.

The next major section of a journal article is *sample and methods*. In this section, an investigator describes how the research sample was selected, the characteristics of the research sample, the details of how the research was conducted, and how the data were analyzed (see Appendix). Using the sample research question, a sociologist might obtain data from the Youth Risk Behavior Surveillance Survey collected by the Centers for Disease Control and Prevention. This self-administered questionnaire is distributed biennially to more than 10,000 high school students across the United States.

The final section of a journal article includes the *findings and conclusions*. The findings of a study describe the results, that is, what the researcher found as a result of the investigation. Findings are then discussed within the context of the hypotheses and the conclusions that can be drawn. Often, research results are presented in tabular form. Reading tables carefully is an important part of drawing accurate conclusions about the research hypotheses. In reading a table, you should follow the steps listed here (see the table within this box):

1. *Read the title of the table and make sure that you understand what the table contains.* The title of the table indicates the unit of analysis (high school students), the dependent variable (sexual risk behaviors), the independent variables (sex and year), and what the numbers represent (percentages).

2. *Read the information contained at the bottom of the table, including the source and any other explanatory information.* For example, the information at the bottom of this table indicates that the data are from the Centers for Disease Control and Prevention, that "sexually active" was defined as having intercourse in the last three months, and that data on condom use were only from those students who were defined as being currently sexually active.

3. *Examine the row and column headings.* This table looks at the percentage of males and females, over four years, who reported ever having sexual intercourse, having four or more sex partners in a lifetime, being currently sexually active, and using condoms during the last sexual intercourse.

4. *Thoroughly and carefully examine the data in the table, looking for patterns between variables.* As indicated in the table, compared to 2011, fewer high school males engaged in the risky behaviors in 2013 with the exception of condom use, which declined over the time period. In the same year females report the highest rate of ever engaged in sexual intercourse

problems is based on research as well as on theory, however. Indeed, research and theory are intricately related. As Wilson (1983) stated:

> Most of us think of theorizing as quite divorced from the business of gathering facts. It seems to require an abstractness of thought remote from the practical activity of empirical research. But theory building is not a separate activity within sociology. Without theory, the empirical researcher would find it impossible to decide what to observe, how to observe it, or what to make of the observations. (p. 1)

Social Problems Research

Most students taking a course in social problems will not become researchers or conduct research on social problems. Nevertheless, we are all consumers of research that is reported in the media. Politicians, social activist groups, and organizations attempt to justify their decisions, actions, and positions by citing research results. As consumers of research, we need to understand that our personal experiences and casual observations are less reliable

although only a fraction higher than in 2007. However, in 2013 females reported the highest rate of four or more lifetime sex partners. Comparing 2007 to 2013, we note very little difference in the percent of females reporting being sexually active, although condom use during last intercourse was the lowest in the four years measured.

5. *Use the information you have gathered in step 4 to address the hypotheses.* Clearly, sexual practices, as hypothesized, have changed over time. For example, contrary to expectations, both males and females, when we compare data from 2007 to 2013, report a general decrease in condom use during sexual intercourse. Furthermore, the percentage of males and females reporting four or more sex partners has also increased during the same time period. Look at the table and see what patterns you detect, including how these patterns address the hypothesis.

6. *Draw conclusions consistent with the information presented.* From the table, can we conclude that sexual practices have changed over time? The answer is probably yes, although the limitations of the survey, the sample, and the measurement techniques used always should be considered. Can we conclude that the observed changes are a consequence of the fear of sexually transmitted diseases? The answer is *no*, and not just because of the results. Having no measure of fear of sexually transmitted diseases over the time period studied, we are unable to come to such a conclusion. More information, from a variety of sources, is needed. The use of multiple methods and approaches to study a social phenomenon is called *triangulation*.

Percentages of High School Students Reporting Sexually Risky Behaviors, by Sex and Survey Year

Survey Year	Ever Had Sexual Intercourse	Four or More Sex Partners during Lifetime	Currently Sexually Active*	Condom Used during Last Intercourse[†]
Males				
2007	49.8	17.9	34.3	68.5
2009	46.1	16.2	32.6	70.4
2011	49.2	17.8	34.2	67.0
2013	47.5	16.8	32.7	65.8
Females				
2007	45.9	11.8	35.6	54.9
2009	45.7	11.2	35.7	57.0
2011	45.6	12.6	34.2	53.6
2013	46.0	13.2	35.2	53.1

*Sexual intercourse during the three months preceding the survey.
[†]Among currently sexually active students.
SOURCES: Centers for Disease Control and Prevention 2008, 2010, 2012, 2014.

than generalizations based on systematic research. One strength of scientific research is that it is subjected to critical examination by other researchers (see this chapter's *Social Problems Research Up Close* feature). The more you understand how research is done, the better able you will be to critically examine and question research rather than to passively consume research findings. In the remainder of this section, we discuss the stages of conducting a research study and the various methods of research that sociologists use.

Stages of Conducting a Research Study

Sociologists progress through various stages in conducting research on a social problem. In this section, we describe the first four stages: (1) formulating a research question, (2) reviewing the literature, (3) defining variables, and (4) formulating a hypothesis.

Formulating a Research Question. A research study usually begins with a research question. Where do research questions originate? How does a particular researcher come to ask a particular research question? In some cases, researchers have a personal

> The more you understand how research is done, the better able you will be to critically examine and question research rather than to passively consume research findings.

interest in a specific topic because of their own life experiences. For example, a researcher who has experienced spouse abuse may wish to do research on such questions as "What factors are associated with domestic violence?" and "How helpful are battered women's shelters in helping abused women break the cycle of abuse in their lives?" Other researchers may ask a particular research question because of their personal values—their concern for humanity and the desire to improve human life. Researchers may also want to test a particular sociological theory, or some aspect of it, to establish its validity or conduct studies to evaluate the effect of a social policy or program. Research questions may also be formulated by the concerns of community groups and social activist organizations in collaboration with academic researchers. Government and industry also hire researchers to answer questions such as "How many vehicle crashes are caused by 'distracted driving' involving the use of cell phones?" and "What types of cell phone technologies can prevent the use of cell phones while driving?"

Reviewing the Literature.   After a research question is formulated, researchers review the published material on the topic to find out what is already known about it. Reviewing the literature also provides researchers with ideas about how to conduct their research and helps them formulate new research questions. A literature review serves as an evaluation tool, allowing a comparison of research findings and other sources of information, such as expert opinions, political claims, and journalistic reports.

In a free society, there must be freedom of information. That is why the U.S. Constitution and, more specifically, the First Amendment protect journalists' sources. If journalists are compelled to reveal their sources, their sources may be unwilling to share information, which would jeopardize the public's right to know. A journalist cannot reveal information given in confidence without permission from the source or a court order. Do you think sociologists should be granted the same protections as journalists? If a reporter at your school newspaper uncovered a scandal at your university, should he or she be protected by the First Amendment?

Defining Variables.   A **variable** is any measurable event, characteristic, or property that varies or is subject to change. Researchers must operationally define the variables they study. An *operational definition* specifies how a variable is to be measured. For example, an operational definition of the variable "religiosity" might be the number of times the respondent reports going to church or synagogue. Another operational definition of "religiosity" might be the respondent's answer to the question "How important is religion in your life?" (for example, 1 is not important; 2 is somewhat important; 3 is very important).

Operational definitions are particularly important for defining variables that cannot be directly observed. For example, researchers cannot directly observe concepts such as "mental illness," "sexual harassment," "child neglect," "job satisfaction," and "drug abuse." Nor can researchers directly observe perceptions, values, and attitudes.

variable Any measurable event, characteristic, or property that varies or is subject to change.

hypothesis A prediction or educated guess about how one variable is related to another variable.

dependent variable The variable that researchers want to explain; that is, it is the variable of interest.

independent variable The variable that is expected to explain change in the dependent variable.

Formulating a Hypothesis.   After defining the research variables, researchers may formulate a **hypothesis,** which is a prediction or educated guess about how one variable is related to another variable. The **dependent variable** is the variable that researchers want to explain; that is, it is the variable of interest. The **independent variable** is the variable that is expected to explain change in the dependent variable. In formulating a hypothesis, researchers predict how the independent variable affects the dependent variable. For example, Kmec (2003) investigated the impact of segregated work environments on minority wages, concluding that "minority concentration in different jobs, occupations, and establishments is a considerable social problem because it perpetuates racial wage inequality" (p. 55). In this example, the independent variable is workplace segregation, and the dependent variable is wages.

Arctic Circle

Methods of Data Collection

After identifying a research topic, reviewing the literature, defining the variables, and developing hypotheses, researchers decide which method of data collection to use. Alternatives include experiments, surveys, field research, and secondary data.

Experiments. **Experiments** involve manipulating the independent variable to determine how it affects the dependent variable. Experiments require one or more experimental groups that are exposed to the experimental treatment(s) and a control group that is not exposed. After a researcher randomly assigns participants to either an experimental group or a control group, the researcher measures the dependent variable. After the experimental groups are exposed to the treatment, the researcher measures the dependent variable again. If participants have been randomly assigned to the different groups, the researcher may conclude that any difference in the dependent variable among the groups is due to the effect of the independent variable.

An example of a "social problems" experiment on poverty would be to provide welfare payments to one group of unemployed single mothers (experimental group) and no such payments to another group of unemployed single mothers (control group). The independent variable would be welfare payments; the dependent variable would be employment. The researcher's hypothesis could be that mothers in the experimental group would be less likely to have a job after 12 months than mothers in the control group.

The major strength of the experimental method is that it provides evidence for causal relationships, that is, how one variable affects another. A primary weakness is that experiments are often conducted on small samples, often in artificial laboratory settings; thus, the findings may not be generalized to other people in natural settings.

Surveys. **Survey research** involves eliciting information from respondents through questions. An important part of survey research is selecting a sample of those to be questioned. A **sample** is a portion of the population, selected to be representative so that the information from the sample can be generalized to a larger population. For example, instead of asking all middle school children about their delinquent activity, the researcher would ask a representative sample of them and assume that those who were not questioned would give similar responses. After selecting a representative sample, survey researchers either interview people, ask them to complete written questionnaires, or elicit responses to research questions through web-based surveys. Some surveys are conducted annually or every other year so that researchers can observe changes in responses over time. This chapter's *Self and Society* feature allows you to voice your opinion on various social issues through the use of a written questionnaire. After completing the survey you can compare your responses to a national sample of first-year college students.

Interviews. In interview survey research, trained interviewers ask respondents a series of questions and make written notes about or tape-record the respondents' answers. Interviews may be conducted over the phone or face-to-face.

experiments Manipulating the independent variable to determine how it affects the dependent variable. Experiments require one or more experimental groups that are exposed to the experimental treatment(s) and a control group that is not exposed.

survey research Eliciting information from respondents through questions.

sample A portion of the population, selected to be representative so that the information from the sample can be generalized to a larger population.

A Sociologist's "Human Side"

Some of us know early in life exactly what we want to be, or at least what we think we want to be, when we "grow up." Starting in high school and continuing in college, students are often frantic about what to "do." I see it all the time as a sociology student adviser. "I really love sociology but what can I do with a sociology degree?" Actually, I don't find that question particularly surprising. Few of us grow up hearing about sociology or knowing what it is. Even I didn't consciously choose to be a sociologist as one might choose a career in law or nursing, or increasingly, in business and computer science.

There is a theory called "drift theory," which argues that delinquency is a "relatively inarticulate oral tradition" (Matza 1990, p. 52), in which youths drift back and forth between conforming and nonconforming behavior.

Although believing strongly and hoping fervently that sociology is more than a "relatively inarticulate oral tradition," I think the concept of drift applied to me. I had no vision, no calling, no great quest to be a sociologist. I don't even remember, although this was many years ago, the first time I was aware of the fact that such creatures existed.

But as a child of 13, I left the safety and security of suburbia to take a train to downtown Cleveland, Ohio, to an area called Hough to investigate why some people, who were very, very different than me, were burning down a town. It was, as you may know, part of the urban riots and civil rights protests of the 1960s. I suppose, looking back, it was my first field research.

Throughout high school, I consciously drifted in and out of various cliques, fascinated by each, and at Kent State University, where I did my undergraduate work, I was faced with the reality of anti–Vietnam War demonstrations not unlike the many student demonstrations of today (e.g., Occupy Wall Street). There was one difference, though. At Kent State, four students were killed and nine injured by Ohio National Guardsmen called in to quiet the protests.

I think these events, as various events you could isolate in your own lives, were instrumental in molding me as a person and, eventually, as a sociologist. Although, as I said, I had no burning desire to be a sociologist growing up, today I am possessed by a "very special kind of passion" (Berger 1963, p. 12), driven by a demon, perhaps the one that led me to Hough so many years ago.

I feel privileged and honored to be a sociologist and to be able to do what I want to do—and to get paid for it.

The beauty of sociology, among other things, is that it provides a framework—a lens, if you will—to problem-solve within a variety of venues: social service departments; consulting firms; hospitals; federal, county, or local government agencies; nursing homes and rehabilitation facilities; law offices; and so forth. Or, as I did, you might want to become an academic sociologist.

Not only will you find sociologists working in almost every imaginable location, the list of what they do is endless. As you may know, you can find sociology courses on race, class, and gender; social movements; family; criminology; sexuality; mass media; religion; the environment; education; health; social psychology; aging;

immigration;…and yes, social problems. We have such diversity of topics because we do such a variety of jobs. Would you like to be a Foreign Affairs Officer and work in the State Department? Among other degrees listed for this entry-level position is a bachelor's degree in sociology (U.S. Department of State 2013).

But then there's that other pesky problem—what about pay? I know you're all smart enough to know that income in itself is not predictive of job satisfaction. You could earn $249,999 a year, a lot of money to you and me, but if you never saw your spouse/partner, developed ulcers from the stress, worked in a life-threatening environment, and had little job security, I hope most of you would run the other way. So let's take a more balanced approach.

As discussed in the *Wall Street Journal* (2013) and *Forbes* (Smith 2013), an annual listing of the 200 "Best and Worst Jobs" indicate that sociologists are near the top, checking in at number 19. Imagine that you are number 19 out of 200 students graduating from your high school class. Not bad. And the best part is that these calculations were based on official data (e.g., from the U.S. Department of Labor) in *five areas*: (1) environmental factors (e.g., stamina required, competitiveness), (2) income, (3) outlook (e.g., expected employment growth), (4) physical demands (e.g., requires lifting), and (5) stress (e.g., deadlines, travel) (Career Cast 2013).

I admit it would be nice to be number one but right now, I couldn't be happier with number 19.

SOURCE: Mooney 2015.

One advantage of interview research is that researchers are able to clarify questions for respondents and follow up on answers to particular questions. Researchers often conduct face-to-face interviews with groups of individuals who might otherwise be inaccessible. For example, some AIDS-related research attempts to assess the degree to which individuals engage in behavior that places them at high risk for transmitting or contracting HIV. Street youth and intravenous drug users, both high-risk groups for HIV infection, may not have a telephone or address because of their transient lifestyle. These groups may be accessible, however, if the researcher locates their hangouts and conducts face-to-face interviews.

The most serious disadvantages of interview research are cost and the lack of privacy and anonymity. Respondents may feel embarrassed or threatened when asked questions that relate to personal issues such as drug use, domestic violence, and sexual behavior. As a result, some respondents may choose not to participate in interview research on sensitive topics. Those who do participate may conceal or alter information or give socially desirable answers to the interviewer's questions (e.g., "No, I do not use drugs" or "No, I do not text while driving.").

Questionnaires. Instead of conducting personal or phone interviews, researchers may develop questionnaires that they either mail, post online, or give to a sample of respondents. Questionnaire research offers the advantages of being less expensive and less time-consuming than face-to-face or telephone surveys. Questionnaire research also provides privacy and anonymity to the research participants, thus increasing the likelihood that respondents will provide truthful answers.

The major disadvantage of mail or online questionnaires is that it is difficult to obtain an adequate response rate. Many people do not want to take the time or make the effort to complete a questionnaire. Others may be unable to read and understand the questionnaire.

Web-based surveys. In recent years, technological know-how and the expansion of the Internet have facilitated the use of online surveys. Web-based surveys, although still less common than interviews and questionnaires, are growing in popularity and are thought by some to reduce many of the problems associated with traditional survey research (Farrell and Petersen 2010). For example, the response rate of telephone surveys has been declining as potential respondents have caller ID, unlisted telephone numbers, answering machines, or no home (i.e., landline) telephone (Farrell and Petersen 2010). On the other hand, the use of and access to the Internet continues to grow. In 2015, the number of Americans connected to the Internet was higher than in any other single year (Internet Live 2015).

Field Research. **Field research** involves observing and studying social behavior in settings in which it occurs naturally. Two types of field research are participant observation and nonparticipant observation.

In *participant observation research*, researchers participate in the phenomenon being studied so as to obtain an insider's perspective on the people and/or behavior being observed. Palacios and Fenwick (2003), two criminologists, attended dozens of raves over a 15-month period to investigate the South Florida drug culture. In *nonparticipant observation research*, researchers observe the phenomenon being studied without actively participating in the group or the activity. For example, Simi and Futrell (2009) studied white power activists by observing and talking to organizational members but did not participate in any of their organized activities.

Sometimes sociologists conduct in-depth detailed analyses or case studies of an individual, group, or event. For example, Fleming (2003) conducted a case study of young auto thieves in British Columbia. He found that, unlike professional thieves, the teenagers' behavior was primarily motivated by thrill seeking—driving fast, the rush of a possible police pursuit, and the prospect of getting caught.

The main advantage of field research on social problems is that it provides detailed information about the values, rituals, norms, behaviors, symbols, beliefs, and emotions of those being studied. A potential problem with field research is that the researcher's observations may be biased (e.g., the researcher becomes too involved in the group to be objective). In addition, because field research is usually based on small samples, the findings may not be generalizable.

Secondary Data Research. Sometimes researchers analyze secondary data, which are data that other researchers or government agencies have already collected or that exist in forms such as historical documents, police reports, school records, and official records

field research Observing and studying social behavior in settings in which it occurs naturally.

of marriages, births, and deaths. A major advantage of using secondary data in studying social problems is that the data are readily accessible, so researchers avoid the time and expense of collecting their own data. Secondary data are also often based on large representative samples. The disadvantage of secondary data is that researchers are limited to the data already collected.

Ten Good Reasons to Read This Book

Most students reading this book are not majoring in sociology and do not plan to pursue sociology as a profession. So, why should students take a course on social problems? How can reading this textbook about social problems benefit you?

1. *Understanding that the social world is too complex to be explained by just one theory will expand your thinking about how the world operates.* For example, juvenile delinquency doesn't have just one cause—it is linked to (1) an increased number of youths living in inner-city neighborhoods with little or no parental supervision (social disorganization theory); (2) young people having no legitimate means of acquiring material wealth (anomie theory); (3) youths being angry and frustrated at the inequality and racism in our society (conflict theory); and (4) teachers regarding youths as "no good" and treating them accordingly (labeling theory).

2. *Developing a sociological imagination will help you see the link between your personal life and the social world in which you live.* In a society that values personal responsibility, there is a tendency to define failure and success as consequences of individual free will. The sociological imagination enables us to understand how social forces influence our personal misfortunes and failures, and contribute to personal successes and achievements.

3. *Understanding globalization can help you become a safe, successful, and productive world citizen.* Social problems cross national boundaries. Problems such as obesity, war, climate change, human trafficking, and overpopulation are global problems. Problems that originate in one part of the world may affect other parts of the world, and may be caused by social policies in other nations. Thus, understanding social problems requires consideration of the global interconnectedness of the world. And solving today's social problems requires collective action among citizens across the globe. To better prepare students for a globalized world, many colleges and universities have made changes to the curriculum such as adding new general education or core curriculum courses on global concerns and perspectives, revamping existing courses to increase emphasis on global issues, and offering a "global certificate" that students can earn by completing a certain number of courses with an international focus (Wilhelm 2012).

WHAT do you THINK?

Some colleges and universities have instituted policies that require students to take one or more global courses—courses with a global or international focus—in order to graduate. Do you think colleges and universities should require some minimum number of global courses for undergraduates? Why or why not?

4. *Understanding the difficulty involved in "fixing" social problems will help you make decisions about your own actions, for example, who you vote for or what charity you donate money to.* It is important to recognize that "fixing" social problems is a very difficult and complex enterprise. One source of this difficulty is that we don't all agree on what the problems are. We also don't agree on what the root causes are of social problems. Is the problem of gun violence in the United States a problem caused by gun availability? Violence in the media? A broken mental health care system? Masculine gender norms? If we socialized boys to be more nurturing and gentle, rather than aggressive and competitive, we might reduce gun violence, but we would also potentially create a generation of boys who would not want to sign up for combat duties in the military, and our armed forces would not have enough

recruits. Thus, solving one social problem (gun violence) may create another social problem (too few military recruits). It should also be noted that although some would see low military recruitment as a problem, others would see it as a positive step toward a less militaristic society.

5. *Although this is a social problems book, it may actually make you more, rather than less, optimistic.* Yes, all the problems discussed in the book are real, and they may seem insurmountable, but they aren't. You'll read about positive social change (for example, the number of people who smoke cigarettes in the United States has dramatically dropped, as have rates of homophobia, racism, and sexism). Life expectancy has increased, and more people go to college than ever before. Change for the better can and does happen.

6. *Knowledge is empowering.* Social problems can be frightening, in part, because most people know very little about them beyond what they hear on the news or from their friends. Misinformation can make problems seem worse than they are. The more accurate the information you have, the more you will realize that we, as a society, have the power to solve the problems, and the less alienated you will feel.

7. *The* Self and Society *exercises increase self-awareness and allow you to position yourself within the social landscape.* For example, earlier in this chapter, you had the opportunity to assess your personal values and to compare your responses to a national sample of first-year college students.

8. The Human Side *features make you a more empathetic and compassionate human being by personalizing the topic at hand.* The study of social problems is always about the quality of life of individuals. By conveying the private pain and personal triumphs associated with social problems, we hope to elicit a level of understanding that may not be attained through the academic study of social problems alone. The Human Side in this chapter highlights one of the author's paths to becoming a sociologist.

9. *The* Social Problems Research Up Close *features teach you the basics of scientific inquiry, making you a smarter consumer of "pop" sociology, psychology, anthropology, and the like.* These boxes demonstrate the scientific enterprise, from theory and data collection to findings and conclusions. Examples of research topics featured in later chapters of this book include the portrayal of smoking in popular films, polyamory and poly families, gender and the Internet, how the fossil fuel industry has deceived the public about climate change, responses to masculinity threats, and military suicides.

10. *Learning about social problems and their structural and cultural origins helps you— individually or collectively—make a difference in the world.* Individuals can make a difference in society through the choices they make. You may choose to vote for one candidate over another, demand the right to reproductive choice or protest government policies that permit it, drive drunk or stop a friend from driving drunk, repeat a homophobic or racist joke or chastise the person who tells it, and practice safe sex or risk the transmission of sexually transmitted diseases.

Collective social action is another, often more powerful way to make a difference. You may choose to create change by participating in a **social movement**—an organized group of individuals with a common purpose of promoting or resisting social change through collective action. Some people believe that, to promote social change, one must be in a position of political power and/or have large financial resources. However, the most important prerequisite for becoming actively involved in improving levels of social well-being may be genuine concern and dedication to a social "cause."

Understanding Social Problems

At the end of each chapter, we offer a section with a title that begins with "Understanding…," in which we reemphasize the social origin of the problem being discussed, the consequences, and the alternative social solutions. Our hope is that readers will end each chapter with a "sociological imagination" view of the problem and with an idea of how, as a society, we might approach a solution.

social movement An organized group of individuals with a common purpose of promoting or resisting social change through collective action.

Sociologists have been studying social problems since the Industrial Revolution. Industrialization brought about massive social changes: The influence of religion declined, and families became smaller and moved from traditional, rural communities to urban settings. These and other changes have been associated with increases in crime, pollution, divorce, and juvenile delinquency. As these social problems became more widespread, the need to understand their origins and possible solutions became more urgent. The field of sociology developed in response to this urgency. Social problems provided the initial impetus for the development of the field of sociology and continue to be a major focus of sociology.

There is no single agreed-on definition of what constitutes a social problem. Most sociologists agree, however, that all social problems share two important elements: an objective social condition and a subjective interpretation of that condition. Each of the three major theoretical perspectives in sociology—structural-functionalist, conflict, and symbolic interactionist—has its own notion of the causes, consequences, and solutions of social problems.

Chapter Review

- **What is a social problem?**
Social problems are defined by a combination of objective and subjective criteria. The objective element of a social problem refers to the existence of a social condition; the subjective element of a social problem refers to the belief that a particular social condition is harmful to society or to a segment of society and that it should and can be changed. By combining these objective and subjective elements, we arrive at the following definition: A social problem is a social condition that a segment of society views as harmful to members of society and in need of remedy.

- **What is meant by the structure of society?**
The structure of a society refers to the way society is organized.

- **What are the components of the structure of society?**
The components are institutions, social groups, statuses, and roles. Institutions are an established and enduring pattern of social relationships and include family, religion, politics, economics, and education. Social groups are defined as two or more people who have a common identity, interact, and form a social relationship. A status is a position that a person occupies within a social group and that can be achieved or ascribed. Every status is associated with many roles, or the set of rights, obligations, and expectations associated with a status.

- **What is meant by the culture of society?**
Whereas *social structure* refers to the organization of society, *culture* refers to the meanings and ways of life that characterize a society.

- **What are the components of the culture of society?**
The components are beliefs, values, norms, and symbols. *Beliefs* refer to definitions and explanations about what is assumed to be true. *Values* are social agreements about what is considered good and bad, right and wrong, desirable and undesirable. *Norms* are socially defined rules of behavior. Norms serve as guidelines for our behavior and for our expectations of the behavior of others. Finally, a *symbol* is something that represents something else.

- **What is the sociological imagination, and why is it important?**
The *sociological imagination*, a term that C. Wright Mills (1959) developed, refers to the ability to see the connections between our personal lives and the social world in which we live. It is important because, when we use our sociological imagination, we are able to distinguish between "private troubles" and "public issues" and to see connections between the events and conditions of our lives and the social and historical context in which we live.

- **What are the differences between the three sociological perspectives?**
According to structural functionalism, society is a system of interconnected parts that work together in harmony to maintain a state of balance and social equilibrium for the whole. The conflict perspective views society as composed of different groups and interests competing for power and resources. Symbolic interactionism reflects the microsociological perspective and emphasizes that human behavior is influenced by definitions and meanings that are created and maintained through symbolic interaction with others.

- **What are the first four stages of a research study?**
The first four stages of a research study are formulating a research question, reviewing the literature, defining variables, and formulating a hypothesis.

- **How do the various research methods differ from one another?**
Experiments involve manipulating the independent variable to determine how it affects the dependent variable. Survey research involves eliciting information from respondents through questions. Field research involves observing and studying social behavior in settings in which

it occurs naturally. Secondary data are data that other researchers or government agencies have already collected or that exist in forms such as historical documents, police reports, school records, and official records of marriages, births, and deaths.

• **What is a social movement?**
Social movements are one means by which social change is realized. A social movement is an organized group of individuals with a common purpose to either promote or resist social change through collective action.

Test Yourself

1. Definitions of social problems are clear and unambiguous.
   a. True
   b. False
2. The social structure of society contains
   a. statuses and roles.
   b. institutions and norms.
   c. sanctions and social groups.
   d. values and beliefs.
3. The culture of society refers to its meaning and the ways of life of its members.
   a. True
   b. False
4. Alienation
   a. refers to a sense of normlessness.
   b. is focused on by symbolic interactionists.
   c. can be defined as the powerlessness and meaninglessness in people's lives.
   d. is a manifest function of society.
5. Blumer's stages of social problems begin with
   a. mobilization for action.
   b. societal recognition.
   c. social legitimation.
   d. development and implementation of a plan.

6. The independent variable comes first in time (i.e., it precedes the dependent variable).
   a. True
   b. False
7. The third stage in defining a research study is
   a. formulating a hypothesis.
   b. reviewing the literature.
   c. defining the variables.
   d. formulating a research question.
8. A sample is a subgroup of the population—the group to whom you actually give the questionnaire.
   a. True
   b. False
9. Studying police behavior by riding along with patrol officers would be an example of
   a. participant observation.
   b. nonparticipant observation.
   c. field research.
   d. both a and c.
10. Students benefit from reading this book because it
    a. provides global coverage of social problems.
    b. highlights social problems research.
    c. encourages students to take pro-social action.
    d. all of the above

Answers: 1. B; 2. A; 3. A; 4. C; 5. B; 6. A; 7. C; 8. A; 9. D; 10. D.

Key Terms

achieved status 7	institution 6	social movement 23
alienation 13	latent functions 11	social problem 5
anomie 11	manifest functions 11	sociological imagination 10
ascribed status 7	norms 8	status 7
beliefs 8	objective element of a social problem 5	structure 6
culture 7	primary groups 7	subjective element of a social problem 5
dependent variable 18	roles 7	survey research 19
experiments 19	sample 19	symbol 9
field research 21	sanctions 8	theory 10
hypothesis 18	secondary groups 7	values 8
independent variable 18	social group 7	variable 18

America's health care system is neither healthy, caring, nor a system."

WALTER CRONKITE

2

Physical and Mental Health and Health Care

Learning Objectives

After studying this chapter, you will be able to…

1 Compare life expectancy and mortality in low, middle, and high-income countries, and identify ways in which globalization affects health and health care.

2 Describe the prevalence, impact, and causes of mental illness.

3 Explain how conflict theory, structural-functionalism, and symbolic interactionism help us understand illness and health care.

4 Identify five lifestyle behaviors that influence health and give examples of how socioeconomic status, gender, race, and ethnicity affect health.

5 Identify and describe the various types of private insurance plans and public health care insurance programs in the United States and differentiate between allopathic medicine and complementary and alternative medicine.

6 Critically evaluate health care in the United States on the dimensions of health insurance coverage, cost of health care, and adequacy of mental health care.

7 Describe efforts to improve health in low- and middle-income countries, fight the growing problem of obesity, improve mental health care, and increase access to affordable health care in the United States.

8 Discuss the complexity of factors that affect health and that must be addressed in order to improve the health of a society.

ZOOM DOSSO/Getty Images

Poverty is an underlying social condition that contributed to the 2014 Ebola outbreak in West Africa.

ONE OF THE MOST widely covered media topics of 2014 was the West African outbreak of Ebola—a virus that is spread through contact with infected bodily fluids or surfaces contaminated with such fluids. The 2014 Ebola outbreak in West Africa was the largest Ebola outbreak since the virus was first discovered in 1976, with more Ebola cases and deaths in 2014 than all previous outbreaks combined (World Health Organization 2015a). Containing an Ebola outbreak requires identifying, treating, and isolating people who are infected with the virus, and also finding everyone who had close contact with infected individuals and tracking them for 21 days. Any of these contacts who come down with the disease are isolated from the community and then their contacts are traced, repeating the contact-tracing process.

Public health worker Rebecca Levine flew to Sierra Leone to help implement contact tracing to contain the Ebola outbreak. When Levine arrived at a Ministry of Health office in Sierra Leone, she found that the database she needed was "pretty much in shambles," with many contacts' addresses either missing or vague, like "down by the farm road" (quoted in Cohen and Bonifield 2014). Less than a third of all the contacts in the database had a usable address, making tracking down people who may have come into contact with Ebola-infected individuals a daunting task. According to public health experts, this inability to do complete contact tracing is a major reason that the Ebola outbreak of 2014 spiraled out of control (Cohen and Bonifield 2014).

> One could argue that the study of social problems is, essentially, the study of health problems, as each social problem affects the physical, mental, and social well-being of humans and the social groups of which they are a part.

In this chapter, we emphasize that health problems are not individual problems—they are social problems that have social causes and social outcomes. Social conditions that are responsible for much of the spread of Ebola include poverty, inequality, and inadequate provision of healthcare (Jones 2014). Social outcomes of the Ebola outbreak in Ebola-stricken West African nations included skyrocketing prices of consumer goods and services, school closings, and mental distress associated with loss, fear, anxiety and stress. In the United States, where the risk of contracting Ebola is small, the 2014 West African Ebola outbreak lead to fear, mandatory quarantines for some travelers from West African nations, and proposals to ban some travelers from entering the United States.

In this chapter, we use a sociological approach to physical and mental health issues, examining why some social groups experience more health problems than others and how social forces affect and are affected by health and illness. We also address problems in health care, focusing on issues related to access, cost, and quality of health care.

The World Health Organization (1946) defined **health** as "a state of complete physical, mental, and social well-being" (p. 3). One could argue that the study of social problems is, essentially, the study of health problems, as each social problem affects the physical, mental, and social well-being of humans and the social groups of which they are a part.

The Global Context: Health and Illness around the World

How long can you expect to live? What kinds of health problems might you experience in your lifetime? What are the causes of death that are likely to affect you, and your loved ones? The answers to these questions vary from country to country. In making international comparisons, social scientists commonly classify countries according to their level of economic development. (1) **Developed countries**, also known as *high-income countries*, have relatively high gross national income per capita; (2) **less developed** or **developing countries**, also known as *middle-income countries*, have relatively low gross national income per capita; and (3) **least developed countries** (known as *low-income countries*) are the poorest countries of the world. As we discuss in the following section, how long people live and what causes their death varies across the globe.

Life Expectancy and Mortality in Low-, Middle-, and High-Income Countries

Life expectancy—the average number of years that individuals born in a given year can expect to live—is significantly greater in high-income countries than in low-income countries (see Table 2.1). Life expectancy ranges from a high of 84 (in Japan) to a low of 46 (in Sierra Leone) (World Health Organization 2015d).

TABLE 2.1 Life Expectancy by Country Income Level, 2013

Country Income Level	Life Expectancy
High	79
Upper middle	74
Lower middle	66
Low	62
WORLD	71

SOURCE: World Health Organization 2015d.

The leading causes of death, or **mortality**, also vary around the world (see Table 2.2). Deaths caused by parasitic and infectious diseases, such as HIV/AIDS, tuberculosis, diarrheal diseases, and malaria are much more common in less developed countries compared with the more developed countries. Parasitic and infectious diseases spread more easily in poor and overcrowded housing conditions, and in areas with lack of clean water and sanitation (see also Chapter 6).

Worldwide, nearly two-thirds of deaths are due to noncommunicable diseases, primarily heart disease, stroke, cancer, and respiratory diseases. These noninfectious, nontransmissible diseases are the leading causes of death in wealthy countries such as the United States and are related to tobacco use, physical inactivity, unhealthy diet, and alcohol abuse. In recent decades, noncommunicable diseases—particularly heart disease—have also become leading causes of death in low and middle-income countries, as rising incomes and emerging middle classes in countries such as China and India have led to (1) increased use of tobacco (linked to cancer and respiratory diseases); (2) increased access to automobiles, televisions, and other technologies that contribute to a sedentary lifestyle, and (3) increased consumption of processed foods high in sugar and fat (linked to obesity).

TABLE 2.2 Leading Causes of Death, by Country Income Level

Low Income	High Income
1. Respiratory infections	Heart disease
2. HIV/AIDS	Stroke and other cerebrovascular disease
3. Diarrheal diseases	Trachea, bronchus, lung cancers
4. Stroke	Alzheimer's and other dementias
5. Heart disease	Chronic obstructive pulmonary disease
6. Malaria	Respiratory infections
7. Preterm birth complications	Colon and rectum cancer
8. Tuberculosis	Diabetes

SOURCE: World Health Organization 2014b.

health According to the World Health Organization, "a state of complete physical, mental, and social well-being."

developed countries Countries that have relatively high gross national income per capita, also known as high-income countries.

developing countries Countries that have relatively low gross national income per capita, also known as less developed or middle-income countries.

least developed countries The poorest countries of the world.

life expectancy The average number of years that individuals born in a given year can expect to live.

mortality Death.

Data on deaths from international terrorism and tobacco-related deaths in 37 developed and eastern European countries revealed that tobacco-related deaths outnumbered terrorist deaths by about a whopping 5,700 times (Thomson and Wilson 2005). The number of tobacco deaths was equivalent to the impact of a September 11, 2001–type terrorist attack every 14 hours! Given that tobacco-related deaths grossly outnumber terrorism-related deaths, why hasn't the U.S. government waged a "war on tobacco" on a scale similar to its "war on terrorism"?

Mortality among Infants and Children. The rates of **infant mortality** (death of live-born infants under 12 months of age), and **under-5 mortality** (death of children under age 5) provide powerful indicators of the health of a population. The *infant mortality rate*, the number of deaths of live-born infants under 1 year of age per 1,000 live births (in any given year), ranges from an average of 5 in high-income nations to 53 in low-income nations (World Health Organization 2015d). The *under-5 mortality rate*, or death rate of children under age 5, similarly is much lower in high-income countries than in low-income countries. Diarrhea, which can lead to life-threatening dehydration, often results from contaminated drinking water and lack of sanitation, or the unavailability of toilets or other hygienic means of disposing of human waste. One-third of the world's population does not have access to adequate sanitation facilities, and one in 10 people on the planet don't have access to safe drinking water (World Health Organization 2015d) (see also Chapter 6).

> When Tanzanian mothers are in labor, they often say to their older children, "I'm going to go and fetch the new baby; it is a dangerous journey and I may not return."

Maternal Mortality. Women in the United States and other developed countries generally do not experience pregnancy and childbirth as life-threatening. But for women age 15 to 49 in developing countries, **maternal mortality**—death that results from complications associated with pregnancy and childbirth—is a leading cause of death. When Tanzanian mothers are in labor, they often say to their older children, "I'm going to go and fetch the new baby; it is a dangerous journey and I may not return" (Grossman 2009). The top causes of maternal mortality are hemorrhage (severe loss of blood), infection, high blood pressure during pregnancy, and unsafe abortion.

WHAT
do you
THINK?

Suppose that you or your partner had a 1 in 52 chance of dying from a pregnancy or childbirth-related cause—the same risk of maternal death that women in the least developed countries face (see Table 2.3). Would that knowledge affect your views about (1) having children? (2) using contraception? (3) policies to ensure access to safe abortion?

Rates of maternal mortality show a greater disparity between rich and poor countries than any of the other societal health measures. Nearly all (99 percent) maternal deaths occur in low-income countries (World Health Organization 2014a). High maternal mortality rates in less developed countries are related to poor-quality and inaccessible health care; most women give birth without the assistance of trained personnel. High maternal mortality rates are also linked to malnutrition and poor sanitation and to pregnancy and childbearing at early ages. Women in many countries also lack access to family planning services and/or do not have the support of their male partners to use contraceptive methods such as condoms. Consequently, many women resort to abortion to limit their childbearing, even in countries where abortion is illegal and unsafe.

infant mortality Deaths of live-born infants under 1 year of age.

under-5 mortality Deaths of children under age 5.

maternal mortality Deaths that result from complications associated with pregnancy, childbirth, and unsafe abortion.

TABLE 2.3 Lifetime Risk of Maternal Mortality by Development Level

	Lifetime Risk of Maternal Mortality
Least developed	1 in 52
Less developed	1 in 150
More developed	1 in 3,800
United States	1 in 2,400

SOURCE: Population Reference Bureau 2013.

Compared with people who live in low-income countries, people in the United States have longer lives and better health. But how does the United States compare with other high-income countries with regard to longevity and health? This chapter's *Social Problems Research Up Close* feature reveals that the United States is one of the wealthiest countries in the world, but it is not one of the healthiest.

Globalization, Health, and Medical Care

Globalization, broadly defined as the growing economic, political, and social interconnectedness among societies throughout the world, has had both positive and negative effects on health and medical care. On the positive side, globalized communications technology is helpful in monitoring and reporting outbreaks of disease, disseminating guidelines for controlling and treating disease, and sharing medical knowledge and research findings (Lee 2003). Global travel and global trade agreements are aspects of globalization that can impact health negatively.

Effects of Global Travel on Health. International travel has become commonplace, with more than 90 million international passengers a year flying into the United States (Berman et al. 2014). Global travel can result in the spread of infectious disease, such as when Thomas Duncan flew to Texas from Liberia—one of three West African countries where a deadly Ebola outbreak in 2014 caused thousands of deaths. Duncan was diagnosed with Ebola and soon after died in a Dallas hospital, where two nurses also came in contact with and later tested positive for the virus. To contain the spread of Ebola, U.S. federal officials ordered health screening for travelers from West Africa arriving in U.S. airports. Passengers who showed symptoms of Ebola (such as an elevated body temperature) and/or who reported possible exposure to Ebola were isolated and monitored to avoid further spread of the Ebola virus.

Effects of Global Trade Agreements on Health. Global trade agreements (see Chapter 7) have expanded the range of goods available to consumers, but at a cost to global health. The international trade of tobacco, alcohol, and sugary drinks and high-calorie processed foods, and the expansion of fast-food chains across the globe, is associated with a worldwide rise in cancer, heart disease, stroke, obesity, and diabetes (Hawkes 2006). Globalization has resulted in rising incomes in the developing world, and although it has improved quality of life for many people, it has also increased access to unhealthy foods and beverages and decreased levels of physical activity. As poor populations move toward the middle class, they can afford to buy televisions, computers, automobiles, and processed foods—products that increase caloric intake and decrease physical activity, leading to increased rates of obesity around the world. Indeed, a new word has emerged to refer to the high prevalence of obesity around the world: **globesity**.

Globesity. Until recently, obesity was a public health problem only in Western industrialized countries. But over the last couple of decades, obesity has become a global problem affecting countries of every income and development level. Since 1980, worldwide obesity has nearly doubled. In 2014, more than half of the world's population was either overweight (39 percent) or obese (13 percent) (based on Body Mass Index calculations determined by height and weight) (World Health Organization 2015c). Globesity includes children, too: 42 million children under age 5 are overweight or obese (ibid.). Indeed, for the first time in human history, the world has more overweight than underweight people (Harvard School of Public Health 2013).

In general, as countries move up the income scale, rates of obesity increase as well. But in low-income countries, wealthier people are more likely to be overweight, whereas rates of obesity in high-income countries are higher among the poor (Harvard School of Public Health 2013). Why do you think this is so?

> The United States is one of the wealthiest countries in the world, but it is not one of the healthiest.

globalization The growing economic, political, and social interconnectedness among societies throughout the world.

globesity The high prevalence of obesity around the world.

WHAT do you THINK?

social problems
RESEARCH
UP CLOSE

Are Americans the Healthiest Population in the World?

Many Americans view the United States as the best country in the world, and assume that this means that Americans have the best health of any population in the world. Is this true? The National Institutes of Health asked the National Health Research Council and the Institute of Medicine (2013) to form a panel of experts to study the health of Americans compared with that of populations in other wealthy countries. This Panel on Understanding Cross-National Health Differences Among High-Income Countries included experts from public health, medicine, economics, sociology, and other disciplines. The panel's task was to examine the health of Americans across the lifespan, comparing it with that of other high-income countries, and investigate explanations for why Americans have poorer health outcomes compared with other populations.

Sample and Methods
The panel selected 16 comparable high income, or "peer" countries to compare with the United States on

various measures of health and longevity. These countries include Austria, Australia, Canada, Denmark, Finland, France, Germany, Italy, Japan, Norway, Portugal, Spain, Sweden, Switzerland, the Netherlands, and the United Kingdom. The panel relied on secondary data analysis—the analysis of data that have already been collected by other researchers and organizations, including the World Health Organization and the Organisation for Economic Co-operation and Development (OECD).

Selected Results
How did the United States compare with its peer countries on measures of health? Let's start with the good news: Compared with people in other industrialized countries, Americans are less likely to smoke and drink heavy alcohol, and they have better control over their cholesterol levels. The United States also has higher rates of cancer screening and survival, and has higher survival after age 75.

But the report's main finding was that despite the fact that the United

States spends more on health care per person than any other industrialized country, Americans die sooner and have higher rates of disease or injury. Life expectancy of U.S. men is shorter than for men in any of the other 16 countries, and only one country (Denmark) has a lower life expectancy for women than that of U.S. women. The United States ranked last or near the bottom in nine key areas of health: infant mortality and low birth weight; injuries and homicide; teenage pregnancy and sexually transmitted infections; HIV/AIDS prevalence; drug-related deaths; obesity and diabetes; heart disease; lung disease; and disability. This "U.S. health disadvantage" has been getting worse for three decades, especially among women. And although health outcomes are generally worse among socially disadvantaged members of a population, even advantaged Americans—those who are college educated, upper income, or insured—have poorer health than similar individuals in other industrialized countries.

medical tourism A global industry that involves traveling, primarily across international borders, for the purpose of obtaining medical care.

mental health Psychological, emotional, and social well-being.

mental illness Refers collectively to all mental disorders, which are characterized by sustained patterns of abnormal thinking, mood (emotions), or behaviors that are accompanied by significant distress and/or impairment in daily functioning.

Medical Tourism. The globalization of medical care involves increased international trade in health products and services. **Medical tourism**—a growing multibillion-dollar global industry—involves traveling across international borders to obtain medical care. Health care consumers travel to other countries for medical care for three primary reasons: (1) to obtain medical treatment that is not available in their home country; (2) to avoid waiting periods for treatment; and/or (3) to save money on the cost of medical treatment. Steve Jobs reportedly traveled to Switzerland for special cancer treatment, and National Football League quarterback Peyton Manning flew to Europe for a stem cell procedure to treat his injured neck (Turner and Hodges 2012). Popular medical tourism destinations that lure health care consumers with competitively priced medical care include Mexico, Singapore, Thailand, South Korea, and India, among other countries. Medical tourism companies offer packages that bundle air travel, ground transportation, hotel accommodations, and guided tours, along with arranging medical treatment, such as organ transplants, dental work, stem cell therapies, cosmetic surgery, reproductive assistance, weight loss surgery, cardiac surgery, and many other medical treatments and services.

Although medical tourism can benefit some patients in providing timely, reduced-cost, quality medical care, there are a number of risks and problems involved. Unlike the highly regulated health care industry in the United States, medical services, products, and facilities in other countries may not be regulated, so quality control is a concern.

The research report offered the following explanations for the U.S. health disadvantage:

- **Health systems.** Americans are more likely to find health care inaccessible or unaffordable. Unlike other industrialized countries, the United States has a relatively large uninsured population and more limited access to health care.
- **Unhealthy behaviors.** Compared with other industrialized populations, Americans have higher rates of prescription and illegal drug abuse, are more likely to use firearms in acts of violence, are less likely to use seat belts, and are more likely to be involved in traffic accidents that involve alcohol. Americans also consume the most calories per person and have the highest obesity rates.
- **Social and economic conditions.** Although the income of Americans is higher on average than in other countries, the United States has higher rates of poverty (especially child poverty), more income inequality, and less social mobility (see also Chapter 6). The United States also lags behind other countries

in the education of youth, which also negatively affects health. And compared with other industrialized populations, Americans benefit less from social safety programs that help buffer the adverse health effects of poverty and low educational attainment.

- **Physical and social environment.** The physical environment in most U.S. communities discourages physical activity, as the environment is designed for automobiles rather than pedestrians. Indeed, U.S. adults take the fewest steps of any industrialized nation, averaging slightly over 5,000 steps a day compared with adults in Australia and Switzerland who average nearly 10,000 steps a day (Trust for America's Health 2012). And in the absence of other transportation options, greater reliance on automobiles in the United States contributes to higher traffic fatalities (National Research Council and Institute of Medicine 2013). The social environment adversely affects Americans' health in a number of ways: (1) Americans' unhealthy patterns of food

consumption are shaped by the agricultural and food industries, grocery store and restaurant offerings, and marketing; (2) the higher rate of firearm-related deaths in the United States is at least in part due to the fact that firearms are more available in the United States than in peer countries; (3) Americans' higher rates of substance abuse, physical illness, and family violence may be related to the higher-stress lifestyle in the United States. For example, Americans tend to work more hours and have less vacation time compared to workers in other industrialized countries (see also Chapter 7).

In sum, the U.S. health disadvantage has multiple causes involving inadequate health care, unhealthy behaviors, adverse social and economic conditions, physical and social environmental factors, and the cultural values and public policies that shape these factors. Unless these conditions change, Americans will continue to have shorter lives and poorer health than people in other industrialized countries.

Medical travel may contribute to the spread of infectious disease, and the medical tourism industry may encourage the illegal market for human organs, as the very poor are vulnerable to being coerced into selling one of their kidneys for transplantation. Medical tourism raises ethical concerns about health equity, as health services in popular medical tourism destinations flow not to the local population, but to foreigners.

Mental Illness: The Hidden Epidemic

Although what it means to be mentally healthy varies across cultures, in the United States **mental health** is understood to include our psychological, emotional, and social well-being. **Mental illness** refers collectively to all mental disorders, which are

Chicago Tribune/Getty Images

Many Americans cross the border into Mexico to obtain less expensive dental care.

characterized by sustained patterns of abnormal thinking, mood (emotions), or behaviors that are accompanied by significant distress and/or impairment in daily functioning (see Table 2.4).

It is important to recognize that physical and mental health are connected and affect each other. For example, people with type 2 diabetes are twice as likely to experience depression as the general population, and people with diabetes who are depressed have more difficulty with self-care. Up to half of cancer patients have a mental illness, especially anxiety and depression, and some evidence suggests that treating depression in cancer patients may improve survival time. People with mental illness are twice as likely to smoke cigarettes as other people. Depression also increases the risk for having a heart attack, but treating the symptoms of depression in people who have had a heart attack improves their survival (Kolappa et al. 2013).

Mental illness is a "hidden epidemic" because the shame and embarrassment associated with mental problems discourage people from acknowledging and talking about them. Being labeled as "mentally ill" is associated with a **stigma**—a discrediting label that can negatively affect an individual's self-concept and disqualify that person from full social acceptance. Stigma associated with mental illness is perpetuated by negative stereotypes of people with mental illness. One of the most common stereotypes of people with mental illness is that they are dangerous and violent. Although untreated mental illness can result in violent behavior, the vast majority of people with severe mental illness are not violent and they are involved in only about 4 percent of violent crimes. And people with mental illness are 11 or more times as likely to be victims of violence than members of the general population (Goode and Healy 2013).

> Although untreated mental illness can result in violent behavior, the vast majority of people with mental illness are not violent and they are involved in only about 4 percent of violent crimes.

TABLE 2.4 Mental Disorders Classified by the American Psychiatric Association

Classification	Description
Anxiety disorders	Disorders characterized by anxiety that is manifest in phobias, panic attacks, or obsessive-compulsive disorder
Dissociative disorders	Problems involving a splitting or dissociation of normal consciousness, such as amnesia and multiple personality
Disorders first evident in infancy, childhood, or adolescence	Disorders including mental retardation, attention deficit/hyperactivity disorder, and stuttering
Eating or sleeping disorders	Disorders including anorexia, bulimia, and insomnia
Impulse control disorders	Problems involving the inability to control undesirable impulses, such as kleptomania, pyromania, and pathological gambling
Mood disorders	Emotional disorders such as major depression and bipolar (manic-depressive) disorder
Organic mental disorders	Psychological or behavioral disorders associated with dysfunctions of the brain caused by aging, disease, or brain damage (such as Alzheimer's disease)
Personality disorders	Maladaptive personality traits that are generally resistant to treatment, such as paranoid and antisocial personality types
Schizophrenia and other psychotic disorders	Disorders with symptoms such as delusions or hallucinations
Somatoform disorders	Psychological problems that present themselves as symptoms of physical disease, such as hypochondria
Substance-related disorders	Disorders resulting from abuse of alcohol and/or drugs, such as barbiturates, cocaine, or amphetamines

© Cengage Learning

stigma A discrediting label that can negatively affect an individual's self-concept and disqualify that person from full social acceptance.

Extent and Impact of Mental Illness

Of 17 nations included in the World Mental Health Survey, the United States has the highest rate of mental illness (Shern and Lindstrom 2013). The Substance Abuse and Mental Health Services Administration (2014) found that in 2013, 43.8 million U.S. adults—nearly one in five adults—had a mental illness in the past year. About a fourth of these adults had serious mental illness that caused impairment in daily functioning. One in 10 U.S. adolescents (aged 12 to 17) experienced a major depressive episode (lasting two weeks or more) in the past year. About half of all Americans will experience some form of mental disorder in their lifetime, with first onset usually occurring in childhood or adolescence (Shally-Jensen 2013).

Untreated mental illness can lead to poor educational achievement, lost productivity, unsuccessful relationships, significant distress, violence and abuse, incarceration, unemployment, homelessness, and poverty. Suicide is the tenth leading cause of death in the United States and the third leading cause of death among persons aged 15 to 24. For every person who commits suicide, 10 more try but fail. Most people who commit suicide are suffering with a mental disorder—most commonly depression or substance abuse—at the time of their death (Shally-Jensen 2013). One population at high risk for suicide is veterans. In recent years, about 18 to 22 U.S. veterans have committed suicide each day (Smith-McDowell 2013).

Causes of Mental Illness

There are many misconceptions about the causes of mental illness, such as the misconception that mental illness is caused by personal weakness, or results from engaging in immoral behavior. In some cultures, people with mental illness are viewed as being possessed by evil spirits or supernatural forces.

Biomedical explanations of mental illness focus on genetic, neurological conditions, and hormonal factors that can cause mental illness. Social and environmental influences that can trigger mental illness include physical, emotional, and sexual abuse; poverty and homelessness; job loss; divorce; the death of a loved one; devastation from a natural disaster such as flood or earthquake; the onset of illness or disabling injury; and the trauma of war. Broadly speaking, "mental health is impacted detrimentally when civil, cultural, economic, political, and social rights are infringed" (World Health Organization 2010, p. xxvi).

> More than one in four college students has been diagnosed or treated by a professional for a mental health problem within the past year.

Mental Illness among College Students

Mental health problems are not uncommon among college students (see Table 2.5). Nearly one in four college students has been diagnosed or treated by a professional for a mental health problem within the past year—14 percent of college students were diagnosed or treated for anxiety, 12 percent for depression, and 6 percent for panic attacks (American College Health Association 2014). The National Alliance on Mental Illness (2012) (NAMI) surveyed 765 individuals with a mental health condition who were enrolled in college currently or within the past five years, and found that only half had disclosed their diagnosis to their college. Table 2.6 lists the five top reasons why students chose either to disclose or not to disclose their diagnosis.

TABLE 2.5 Percentage of College Students Experiencing Selected Mental Health Difficulties Anytime in the Past 12 Months

Mental Health Difficulty	Percentage
Felt so depressed it was difficult to function	33
Felt overwhelming anxiety	54
Felt very lonely	59
Felt things were hopeless	46
Seriously considered suicide	8
Intentionally cut, burned, bruised, or otherwise hurt yourself	6

NOTE: Percentages are rounded.
SOURCE: Adapted from American College Health Association (2014), *American College Health Association National College Health Assessment II: Reference Groups Executive Summary Spring 2014*. Hanover, MD: American College Health Association.

SELF and society

Warning Signs for Mental Illness

Do you or someone you know, such as a roommate, friend, or family member have a mental illness and not realize it? Read each of the following warning signs for mental illness, and put a check mark next to each one that applies to yourself or to someone you are concerned about. If you or someone you know shows any of these signs, consider seeking help from a qualified health professional. To find out what mental health help is in your area, you can call the NAMI hotline (1-800-950-6264; Monday through Friday, 10 a.m. to 6 p.m.). If there is a risk of suicide, the National Suicide Prevention Lifeline (1-800-273-8255) is available 24 hours a day.

Warning Sign	You	Someone You Are Concerned About
1. Excessive worrying or fear	_____	_____
2. Feeling excessively sad or low	_____	_____
3. Confused thinking or problems concentrating or learning	_____	_____
4. Extreme mood changes, including uncontrollable "highs" or feelings of euphoria	_____	_____
5. Prolonged or strong feelings of irritability or anger	_____	_____
6. Avoiding friends and social activities	_____	_____
7. Difficulties understanding or relating to other people	_____	_____
8. Changes in sleeping habits or feeling tired and low energy	_____	_____
9. Changes in eating habits such as increased hunger or lack of appetite	_____	_____
10. Abuse of alcohol or drugs	_____	_____
11. Changes in sex drive	_____	_____
12. Difficulty perceiving reality (delusions or hallucinations)	_____	_____
13. Multiple physical ailments without obvious causes (such as headaches, stomachaches)	_____	_____
14. Inability to carry out daily activities or handle daily stress or problems	_____	_____
15. An intense fear of weight gain or concern with appearance	_____	_____
16. Thinking about suicide	_____	_____

SOURCE: Adapted from Mental Health: Know the Warning Signs (AKA-NAMI Partnership 2015).

More than half of respondents in the NAMI survey did not access their college or university's Disabilities Resource Center to request accommodations such as excused absences for treatment and adjustments in test settings and test times. The top reason? Students were unaware that they qualified for and had a right to receive accommodations. Students also cited fear of stigma as a reason for not requesting accommodations. The NAMI survey found that 40 percent of students (both currently and previously enrolled) with mental health conditions did not seek mental health services and supports on campus; the number one reason students did not seek clinical services is concern about the stigma associated with mental illness. Students also cited busy schedules as a barrier to seeking services. Students with mental health problems may also not seek help because they do not view themselves as having mental illness (see this chapter's *Self and Society* feature).

TABLE 2.6 **Top Reasons for Mental Health Diagnosis Disclosure or Nondisclosure among College Students**

Top Five Reasons Why Students Disclose
To receive accommodations
To receive clinical services and supports on campus
To be a role model and to reduce stigma
To educate students, staff, and faculty about mental health
To avoid disciplinary action by the school and to avoid losing financial aid

Top Five Reasons Why Students Do Not Disclose
Fear or concern for the impact disclosure would have on how students, faculty, and staff perceive them, especially in mental health degree programs
No opportunity to disclose
Diagnosis does not impact academic performance
Lack of knowledge that disclosing could help secure accommodations
Mistrust that medical information will remain confidential

SOURCE: Based on National Alliance on Mental Illness (2012), *College Students Speak: A Survey Report on Mental Health.* Available at www.nami.org

Sociological Theories of Illness and Health Care

The three major sociological theories—structural functionalism, conflict theory, and symbolic interactionism—each contribute to our understanding of illness and health care.

Structural-Functionalist Perspective

According to the structural-functionalist perspective, health care is a social institution that functions to maintain the well-being of societal members and, consequently, of the social system as a whole. Thus, this perspective points to how failures in the health care system affect not only the well-being of individuals, but also the health of other social institutions, such as the economy and the family.

The structural-functionalist perspective also examines how changes in society affect health. As societies develop and provide better living conditions, life expectancy increases and birthrates decrease (Weitz 2013). At the same time, the main causes of death and disability shift from infectious disease, and infant, child, and maternal mortality to chronic, noninfectious illness and disease such as cancer, heart disease, Alzheimer's disease, and arthritis.

Just as social change affects health, health concerns may lead to social change. The emergence of HIV and AIDS in the U.S. gay male population helped unite and mobilize gay rights activists. Concern over the effects of exposure to tobacco smoke—the greatest cause of disease and death in the United States and other developed countries—led to legislation banning smoking in public places.

Finally, the structural-functionalist perspective draws attention to latent dysfunctions, or unintended and often unrecognized negative consequences of social patterns or behavior. For example, the pervasive use of antibiotics in factory farm animal feed has a serious unintended consequence: the growth of antibiotic-resistant bacteria, or "superbugs" that make treating infections more difficult and costly. Annually, at least 23,000 people in the United States die from infections that are resistant to antibiotics (Grow et al. 2014).

Another example of a latent function is a New York law, passed just weeks after the Sandy Hook shooting in Newtown, Connecticut, requiring mental health practitioners to inform authorities about potentially dangerous patients, enabling law enforcement officials to confiscate any firearm owned by such a patient. New York's law also allows guns to

Most antibiotics sold in the United States are used in meat and poultry production. Human consumption of animal products that contain antibiotics has contributed to the rise of "super bugs": infections that are resistant to antibiotic drug treatment.

Visual Mozart/Getty images

be taken from people who voluntarily commit themselves to hospitalization for mental health treatment. Critics of the law argue that it will have unintended consequences—that it will deter people from seeking treatment and prevent people in treatment from talking about violence (Goode and Healy 2013).

Conflict Perspective

The conflict perspective focuses on how socioeconomic status, power, and the profit motive influence illness and health care. As we discuss later in this chapter, socioeconomic status affects access to quality healthcare, and influences living and working conditions that affect our health.

The conflict perspective points to ways in which powerful groups and wealthy corporations influence health-related policies and laws through lobbying and financial contributions to politicians and political candidates. Private health insurance companies have much to lose if the United States adopts a national public health insurance program or even a public insurance option, and have spent millions of dollars opposing such proposals (Mayer 2009). The "health care industrial complex," which includes pharmaceutical and health care product industries, and organizations representing doctors, hospitals, nursing homes, and other health services industries, spends more than three times what the military-industrial complex spends on lobbying in Washington, DC (Brill 2013). Corporations also hire public relations (PR) companies to influence public opinion about health care issues. In his book *Deadly Spin* (2010), insurance industry insider Wendell Potter describes how the insurance industry hired a PR firm to manipulate public opinion on health care reform in part by discrediting Michael Moore's 2007 documentary *Sicko*.

The conflict perspective criticizes the pharmaceutical and health care industry for placing profits above people. "Drugmakers, device makers, and insurers decide which products to develop based not on what patients need, but on what their marketers tell them will sell—and produce the highest profit" (Mahar 2006, p. xviii). For example, not enough drugs are being developed to combat the growing public health threat of antibiotic-resistant infections, in part because pharmaceutical companies do not have a financial incentive. "Antibiotics . . . have a poor return on investment because they are taken for a short period of time and cure their target disease. In contrast, drugs that treat chronic illness, such as high blood pressure, are taken daily for the rest of a patient's life" (Braine 2011).

Many industries place profit above health considerations of workers and consumers. Chapter 7, "Work and Unemployment," discusses how employers often cut costs by neglecting to provide adequate safety measures for their employees. Chapter 13, "Environmental Problems," looks at how corporations often ignore environmental laws and policies, exposing the public to harmful pollution. The food industry is more concerned about profits than about public health. For example, most meat and dairy producers routinely feed antibiotics to animals, which humans then consume. Antibiotics in animal feed have contributed to the development of strains of antibiotic-resistant bacteria in humans. Proposals to limit the use of antibiotics in animal feed have been blocked by the livestock industry lobbies, whose profits would be threatened if antibiotic use in food animals was limited (Katel 2010).

Symbolic Interactionist Perspective

Symbolic interactionists focus on (1) how meanings, definitions, and labels influence health, illness, and health care; and (2) how such meanings are learned through interaction with others and through media messages and portrayals. According to the symbolic interactionist perspective of illness, "there are no illnesses or diseases in nature. There are only conditions that society, or groups within it, has come to define as illness or disease" (Goldstein 1999, p. 31). Psychiatrist Thomas Szasz (1961/1970) argued that what we call "mental illness" is no more than a label conferred on those individuals who are "different," that is, those who do not conform to society's definitions of appropriate behavior.

Defining or labeling behaviors and conditions as medical problems is part of a trend known as **medicalization**. Behaviors and conditions that have undergone medicalization include post-traumatic stress disorder, premenstrual syndrome, menopause, childbirth, attention deficit/hyperactivity disorder, and even the natural process of dying. Conflict theorists view medicalization as resulting from the medical profession's domination and pursuit of profits. A symbolic interactionist perspective suggests that medicalization results from the efforts of sufferers to "translate their individual experiences of distress into shared experiences of illness" (Barker 2002, p. 295).

According to symbolic interactionism, conceptions of health and illness are socially constructed. It follows, then, that definitions of health and illness vary over time and from society to society. In some countries, being fat is a sign of health and wellness; in others, it is an indication of mental illness or a lack of self-control. Among some cultural groups, perceiving visions or voices of religious figures is considered a normal religious experience, whereas such "hallucinations" would be indicative of mental illness in other cultures. In 18th- and 19th-century America, masturbation was considered an unhealthy act that caused a range of physical and mental health problems (Allen 2000). Today, most health professionals agree that masturbation is a normal, healthy aspect of sexual expression.

Symbolic interactionism draws attention to the effects that meanings and labels have on health and health risk behaviors. For example, among white Americans, having a "tan" is culturally defined as youthful and attractive, and so many white Americans sunbathe and use tanning beds—behaviors that increase one's risk of developing skin cancer. Meanings and labels also affect health policies. After the International Agency for Research on Cancer issued a 2009 report labeling tanning beds as "carcinogenic to humans," many states proposed and/or enacted legislation prohibiting the use of tanning beds among minors, or requiring parental permission (National Conference of State Legislators 2015a; Reinberg 2009).

> People who have used indoor tanning devices, such as tanning beds, have a 59 percent increased risk of developing melanoma—a potentially fatal form of skin cancer. In 2014, the Food and Drug Administration changed its label for tanning devices from "low-risk" to "moderate-risk" and ruled that such devices must carry a "black box" warning label, visible to consumers, stating that the device should not be used by people under age 18 (Willingham 2014). Do you think that seeing a warning label on tanning devices will change people's behavior, dissuading them from using tanning devices? Why or why not?

WHAT do you THINK?

Symbolic interactionists also focus on the stigmas associated with certain health conditions. Individuals with mental illness, drug addiction, physical deformities and impairments, missing or decayed teeth, obesity, and HIV infection and AIDS are often stigmatized, and consequently, discriminated against. For example, the HIV/AIDS-related stigma, which stems from societal views that people with HIV/AIDS are immoral and shameful, results in discrimination in employment, housing, social relationships, and medical care. The stigma associated with health problems implies that individuals—rather than society—are responsible for their health. In U.S. culture, "sickness increasingly seems to be construed as a personal failure—a failure of ethical virtue, a failure to take care of oneself 'properly' by eating the 'right' foods or getting 'enough' exercise, a failure

medicalization Defining or labeling behaviors and conditions as medical problems.

to get a Pap smear, a failure to control sexual promiscuity, genetic failure, a failure of will, or a failure of commitment—rather than society's failure to provide basic services to all of its citizens" (Sered and Fernandopulle 2005, p. 16).

Social Factors and Lifestyle Behaviors Associated with Health and Illness

Health problems are linked to lifestyle behaviors such as excessive alcohol consumption and cigarette smoking (see Chapter 3), unprotected sexual intercourse, physical inactivity, and unhealthy diet. For example, most college students do not eat the recommended five or more servings of fruits and vegetables daily, or exercise regularly. These lifestyle behaviors help to explain why one-third of college students are either overweight or obese (American College Health Association 2014). Obesity and other health problems are affected not only by lifestyle behaviors but also by social factors such as socioeconomic status, race/ethnicity, and gender.

Socioeconomic Status and Health

Socioeconomic status refers to a person's position in society based on that person's level of educational attainment, occupation, and household income (see also Chapter 6). One's socioeconomic status—one's income level, education, and occupation—greatly influences one's health. Low socioeconomic status is associated with less access to quality healthcare, increased likelihood of having an unhealthy lifestyle, and higher exposure to adverse living conditions, injury, and disease (Cockerham 2013). People living in poverty are more likely to suffer from malnutrition; hazardous environmental, housing, and working conditions; lack of clean water and sanitation; (see and inadequate medical care. In low-income countries, for example, people with cancer lack access to medical treatment and therefore have lower survival rates compared with cancer patients in high-income countries (Farmer et al. 2010). Most pain medication (about 80 percent) is used by people living in developed countries; few people in less developed countries can afford pain medication or access it when needed. About 4.8 million people with severe cancer pain go untreated annually (Bafana 2013).

In the United States, low socioeconomic status is associated with higher incidence and prevalence of health problems, and lower life expectancy. A study of mortality rates in 3,140 U.S. counties found that the factors most strongly associated with higher mortality were poverty and lack of college education (Kindig and Cheng 2013). In the United States, rates of overweight and obesity are higher among people living in poverty. This is, in part, because high-calorie processed foods tend to be more affordable than fresh vegetables, fruits, and lean meats or fish. And residents of low-income areas often live in **food deserts**, which are areas where residents lack access to grocery stores that sell fresh

In less developed countries, few people can afford or access pain medication.

Nikos Pilos/Kiriakatiki-E/ZUMAPRESS/Newscom

socioeconomic status
A person's position in society based on the level of educational attainment, occupation, and income of that person or that person's household.

fruits and vegetables, and instead rely on convenience stores and fast-food chains that sell mostly high-calorie processed food.

In addition, members of the lower class are subjected to the most stress and have the fewest resources to cope with it (Cockerham 2007). U.S. adults living below the poverty threshold are nearly ten times more likely to report experiencing serious psychological distress as adults in families with an income at least four times the poverty level (National Center for Health Statistics 2015). Stress has been linked to a variety of physical and mental health problems, including high blood pressure, cancer, chronic fatigue, and substance abuse.

U.S. adults who have completed college live an average of ten more years than adults who do not have a high school diploma (Hummer and Hernandez 2013). Why is higher educational attainment linked to better health? First, higher education can lead to higher-paying jobs that provide income to afford better health care, a safer living environment, physically active recreation, and a diet with more fresh (and healthy) foods. A second explanation is that higher levels of education lead to greater knowledge about health issues, and encourage the development of cognitive skills that enable individuals to make better health-related choices, such as the choice to exercise, avoid smoking and heavy drinking, use contraceptives and condoms to avoid sexually transmissible infections, seek prenatal care, and follow doctors' recommendations for managing health problems.

Just as socioeconomic status affects health, health affects socioeconomic status. Physical and mental health problems can limit one's ability to pursue education or vocational training and to find or keep employment. The high cost of health care not only deepens the poverty of people who are already barely getting by but also can financially devastate middle-class families. Based on data from 89 countries around the world, an estimated 100 million people each year are pushed into poverty as a result of paying for needed health services (World Health Organization 2012). Later in this chapter, we look more closely at the high cost of health care and its consequences for individuals and families.

Gender and Health

In many societies, men have more access to social power, privileges, resources, and opportunities. Yet, these advantages do not translate into living longer lives. Throughout the world, women live longer than men. Globally, the life expectancy gender gap was 5 years in 2013 (World Health Organization 2015d). In the United States, average life expectancy in 2013 was 81 for women and 76 for men.

Lower life expectancy in males is related to males' greater exposure to occupational hazards (see also Chapter 7), and social norms that encourage male risk-taking behavior, such as alcohol and drug use, dangerous sports, violence, and fast driving. Men are also less likely than women to seek health care when they are ill, and when they see a doctor, men are less likely to disclose any symptoms of disease or illness they may be experiencing (Baker et al. 2014).

Although women tend to have longer life expectancies than men, they suffer other health-related disadvantages. In many societies, women and girls are viewed and treated as socially inferior, and are denied equal access to nutrition and health care, particularly reproductive health care. Traditional family responsibilities also affect women's health. Women do most of the food preparation and, in many areas of the world where solid fuels are used for cooking indoors, women are more likely than men to suffer from respiratory problems due to exposure to cooking fumes. Gender inequality also exposes women to domestic and sexual exploitation, increasing women's risk of physical injury and of acquiring HIV and other sexually transmissible infections. Globally, 30 percent of women aged 15 and over have experienced physical and/or sexual violence perpetrated by an intimate partner (Devries et al. 2013). "Although neither health care workers nor the general public typically thinks of battering as a health problem, it is a major cause of injury, disability, and death among American women, as among women worldwide" (Weitz 2013, p. 66).

food deserts Areas where residents lack access to grocery stores that sell fresh fruits and vegetables, and instead rely on convenience stores and fast-food chains that sell mostly high-calorie processed food.

Regarding mental health, women are more likely than men to experience mental health problems (Freeman and Freeman 2013b). Biological differences can account for some gender differences in mental health. For example, hormonal changes after childbirth can result in some women suffering from postpartum depression—one study found that one in seven women experienced postpartum depression (Wisner et al. 2013). Gender differences in mental health can also be attributed to gender roles and the unequal status of women.

> Being judged on one's appearance and the degree to which one conforms to a largely unattainable physical "ideal," shouldering the burden of responsibility for family, home and career, growing up in a society that routinely valorises masculinity while belittling femininity . . . all of these factors are likely to help lower women's self-esteem, increase their level of stress and leave them vulnerable to mental health problems. . . . And that's without taking account of the effects of sexual abuse, a trauma that's frequently implicated in later psychological illness. (Freeman and Freeman 2013a, n.p.)

Race, Ethnicity, and Health

Many racial and ethnic minorities get sick at younger ages and die sooner than non-Hispanic whites. U.S. black women and men have a lower life expectancy compared with their white counterparts. Black males are significantly more likely to die from homicide (see Chapter 4), and black men and women are much more likely than whites to die of heart disease and stroke than are those from other racial and ethnic groups. Black women have the highest rate of U.S. infant mortality, followed by American Indian/Alaskan Native, with Asian/Pacific Islander women having the lowest infant mortality rate. These are just a few examples of the health disparities among U.S. racial/ethnic groups.

As shown in Table 2.7, Hispanic women and men live longer than either blacks or whites, and Hispanics have higher survival rates for conditions such as cancer, heart disease, HIV/AIDS, kidney disease, and stroke (Ruiz et al. 2013). The Hispanic health advantage—known as the "Hispanic paradox"—is puzzling because Hispanics share some of the same risk factors that contribute to poorer health among blacks—higher rates of poverty and obesity, and lower educational attainment compared with non-Hispanic whites (Kindig and Cheng 2013). One theory for this "Hispanic paradox" is that Hispanics who immigrate to the United States are among the healthiest from their countries. Another reason for Hispanic longevity is that Hispanic cultural values promote close and supportive family and community relationships and build strong social support, which is associated with better health outcomes.

Health disparities are largely due to substantial racial/ethnic differences in income, education, housing, and access to health care. Racial and ethnic minorities are less likely than whites to have health insurance and so are less likely to receive preventive services (such as colon cancer screening), medical treatment for chronic conditions, and prenatal care. Minorities are also more likely than whites to live in environments where they are exposed to hazards such as toxic chemicals and other environmental hazards (see also Chapter 13). However, although high-income blacks and whites live longer than their low-income counterparts, whites at every income level live at least three years longer than blacks. And the infant mortality rate of college-educated African American women is more than 2.5 times as high as for college-educated whites and Hispanics. In fact, black female college graduates have a higher infant mortality rate than Hispanic and white women who have not completed high school (Williams 2012). One explanation for the effect of race on health, independent of socioeconomic status, is that health is affected not only by one's current socioeconomic status, but also by social and economic circumstances experienced

TABLE 2.7 Life Expectancy at Birth by Race/Hispanic Origin and Sex: United States, 2013

	Non-Hispanic White	Black	Hispanic
Female	81.2	78.4	83.8
Male	76.5	72.3	79.1

SOURCE: Centers for Disease Control and Prevention (2015), "Deaths: Final Data for 2013." *National Vital Statistics Report* 64(2).

over the life course. Minorities are more likely than whites to have experienced social and economic adversity in childhood that can affect their health in adulthood (Williams 2012).

Health disparities are sometimes explained by differences in lifestyle behaviors. Compared with other racial and ethnic populations, Native Americans/Alaskan Natives have the highest death rate from motor vehicle crashes, because they have the highest rates of alcohol-impaired driving as well as seatbelt nonuse (Centers for Disease Control and Prevention 2013). Black Americans have the highest rate of obesity in part because of racial differences in eating behavior. But lifestyle behaviors are often influenced by social factors. Blacks, on average, have lower incomes than whites, and so are more likely to choose foods that are more affordable, and junk foods and fast foods tend to be cheaper than fruits, vegetables, and lean sources of protein (Centers for Disease Control and Prevention 2013).

Another factor that contributes to racial/ethnic health disparities is prejudice and discrimination. Stress associated with prejudice and discrimination can raise blood pressure, which may help to explain why African Americans have higher blood pressure and higher rates of cardiovascular disease than whites (Fischman 2010; Lewis et al. 2014). Prejudice and discrimination can adversely affect health by reducing minorities' access to jobs and safe housing (see also Chapter 9) (Williams 2012). Finally, research shows that even after controlling for socioeconomic status, health insurance coverage, and other factors, racial and ethnic minorities tend to receive fewer medical procedures and poorer quality medical care than whites (Smedley et al. 2003).

Regarding mental health, research finds no significant difference among races in their overall rates of mental illness (Cockerham 2007). Differences that do exist are often associated more with social class than with race or ethnicity. However, some studies suggest that minorities have a higher risk for mental disorders, such as anxiety and depression, in part because of racism and discrimination, which adversely affect physical and mental health. Minorities also have less access to or are less likely to receive needed mental health services, often receive lower-quality mental health care, and are underrepresented in mental health research.

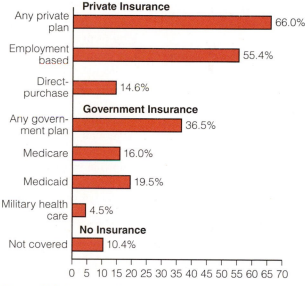

Figure 2.1 Coverage by Type of Health Insurance, 2014
SOURCE: Smith and Medalia 2015.

U.S. Health Care: An Overview

In the United States, there is no one health care system; rather, health care is offered through various private and public means (see Figure 2.1). In traditional health insurance plans, the insured choose their health care providers, whose fees are reimbursed by the insurance company. Insured individuals typically pay an out-of-pocket "deductible" (usually ranging from a few hundred to a thousand dollars or more per year per person) as well as a percentage of medical expenses (e.g., 20 percent) until a maximum out-of-pocket expense amount is reached (after which insurance will cover 100 percent of medical costs). Most insurance companies control costs through **managed care**, which involves monitoring and controlling the decisions of health care providers. The insurance company may, for example, require doctors to receive approval before they can hospitalize a patient, perform surgery, or order an expensive diagnostic test.

Public Health Insurance Programs

Medicare is funded by the federal government and reimburses the elderly and people with certain disabilities for their health care. Individuals contribute payroll taxes to Medicare throughout their working lives and generally become eligible for Medicare when they

managed care Any medical insurance plan that controls costs through monitoring and controlling the decisions of health care providers.

Medicare A federally funded program that provides health insurance benefits to the elderly, disabled, and those with advanced kidney disease.

Medicaid A public health insurance program, jointly funded by the federal and state governments, that provides health insurance coverage for the poor who meet eligibility requirements.

State Children's Health Insurance Program (SCHIP) A public health insurance program, jointly funded by the federal and state governments, that provides health insurance coverage for children whose families meet income eligibility standards.

Military Health System (MHS) The federal entity that provides medical care in military hospitals and clinics, and in combat zones and at bases overseas and on ships, and that provides health insurance known as Tricare to active duty service members, military retirees, their eligible family members, and survivors.

Veterans Health Administration (VHA) A system of hospitals, clinics, counseling centers and long-term care facilities that provides care to military veterans.

Indian Health Service A federal agency that provides health services to members of 566 federally recognized American Indian and Alaska Native tribes and their descendants.

allopathic medicine The conventional or mainstream practice of medicine; also known as Western medicine.

complementary and alternative medicine Refers to a broad range of health care approaches, practices, and products that are not considered part of conventional medicine.

integrative medicine A practice of medicine that combines mainstream medicine with complementary health care approaches.

reach 65, regardless of their income or health status. Medicare consists of four separate programs: Part A is hospital insurance for inpatient care, which is free, but enrollees may pay a deductible and a co-payment. Part B is a supplementary medical insurance program, which helps pay for physician, outpatient, and other services. Part B is voluntary and is not free; enrollees must pay a monthly premium as well as a co-payment for services. Medicare does not cover long-term nursing home care, dental care, eyeglasses, and other types of services, which is why many individuals who receive Medicare also enroll in Part C, which allows beneficiaries to purchase private supplementary insurance that receives payments from Medicare. Part D is an outpatient drug benefit that is also voluntary and requires enrollees to pay a monthly premium, meet an annual deductible, and pay coinsurance for their prescriptions.

Medicaid, which provides health care coverage for the poor, is jointly funded by the federal and state governments. Eligibility rules and benefits vary from state to state; and, in many states, Medicaid provides health care only for those who are well below the federal poverty level. The **State Children's Health Insurance Program (SCHIP)** provides health coverage to children without insurance, many of whom come from families with income too high to qualify for Medicaid but too low to afford private health insurance. Under this initiative, states receive matching federal funds to provide medical insurance to children without insurance.

Military health care includes the **Military Health System (MHS)**, which provides medical care in military hospitals and clinics, in combat zones, and at bases overseas and on ships. The MHS also has a health insurance system known as Tricare that provides health services to millions of active duty service members, military retirees, their eligible family members, and survivors. Military health care also includes the **Veterans Health Administration (VHA)**, which is a system of hospitals, clinics, counseling centers, and long-term care facilities that provides care to military veterans. In recent years, the VHA has been criticized for long delays in providing needed care, and the MHS also came under scrutiny for lapses in care.

The **Indian Health Service** is a federal agency that provides health services to members of 566 federally recognized American Indian and Alaska Native tribes and their descendants (Indian Health Service 2014). Most IHS funds go to providing services to American Indians and Alaska Natives who live on or near reservations or Alaska villages, but some funding supports healthcare programs for American Indians and Alaska Natives who live in urban areas. The IHS funding, which is appropriated by Congress each year, has increased over time but is not sufficient to meet health care needs, leaving many American Indians and Alaska Natives who rely on IHS for care without access to needed care (Artiga, Arguello, and Duckett 2013).

Complementary and Alternative Medicine (CAM)

In Western nations such as the United States, the conventional or mainstream practice of medicine is known as **allopathic**, or Western, medicine. **Complementary and alternative medicine (CAM)** refers to a broad range of health care approaches, practices, and products that are not considered part of conventional medicine. Types of CAM include herbal and homeopathic remedies, vitamins, meditation, Pilates, yoga, tai chi, acupuncture, chiropractic care, massage therapy, Reiki and other energy work, and the use of traditional healers. Some people consider prayer a form of traditional healing. **Integrative medicine** is an approach that combines mainstream medicine with complementary health care approaches. For example, an integrative cancer treatment center may offer mainstream radiation and chemotherapy as well as meditation and acupuncture to manage pain.

More than 30 percent of U.S. adults and about 12 percent of U.S. children use alternative or complementary health care (National Center for Complementary and Alternative Medicine 2015). Americans spend $9 billion on CAM each year (Davis et al. 2013). The cost of CAM is paid by the consumer primarily out of pocket, although some services, such as chiropractic care, are covered by most health insurance.

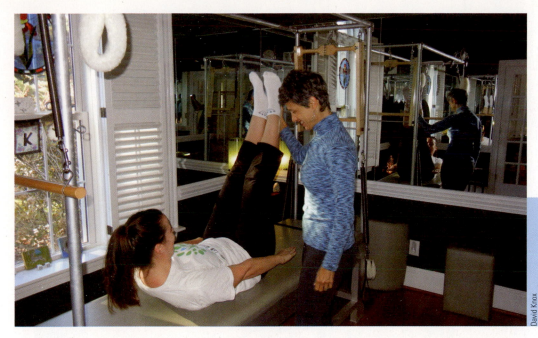

Pilates is a form of exercise that can help relieve low back pain—a condition that most people will experience at some point in their lives.

David Knox

A controversial form of CAM is the medical use of cannabis, or marijuana. Marijuana was first used for medicinal purposes in 2737 B.C. In 1851, marijuana was classified as a legitimate medical compound in the United States, but it was criminalized in 1937, against the advice of the American Medical Association. In 1996, California became the first state to legalize the medical use of marijuana, and as of June 2015, 23 states and the District of Columbia have legalized medical marijuana (see Table 2.8).

Medical marijuana can be smoked in cigarettes and can also be delivered in a variety of nonsmoked preparations, including patches, suppositories, vaporizations, tinctures, nasal sprays, and edible preparations. Medical marijuana is most frequently used to alleviate pain and muscle spasms, but it has also been effective for a variety of other conditions, including nausea and vomiting (a common side effect of cancer chemotherapy), and anorexia and weight loss in patients with AIDS and other conditions that reduce appetite. Adverse side effects of medical marijuana are typically not serious, with the most common being dizziness, and some users experience anxiety and paranoia (Borgelt et al. 2013). Safety concerns regarding medical marijuana include impairment in memory and cognition; and, among frequent adolescent users of marijuana, there is an increased risk of developing schizophrenia. Another concern is the accidental consumption of food products containing marijuana (e.g., cookies or brownies) among children. Unlike many conventional prescription drugs, no deaths from overdoses of marijuana have been reported. This chapter's *The Human Side* features testimonies from people with chronic health problems who found enormous relief through the use of medical marijuana.

TABLE 2.8 States with Legalized Medical Marijuana (as of June 2015)

State	Year Legalized	State	Year Legalized	State	Year Legalized
Alaska	1998	Illinois	2013	New Hampshire	2013
Arizona	2010	Maine	1999	New Jersey	2010
California	1996	Maryland	2014	New Mexico	2007
Colorado	2000	Massachusetts	2012	New York	2014
Connecticut	2012	Michigan	2008	Oregon	1998
DC	2010	Minnesota	2014	Rhode Island	2006
Delaware	2011	Montana	2004	Vermont	2004
Hawaii	2000	Nevada	2000	Washington	1998

SOURCE: National Conference of State Legislators 2015b.

Bill Delany has Crohn's disease, an inflammatory bowel disease that causes abdominal pain, chronic diarrhea and malnutrition, and can result in death (Delany 2012).

> My Crohn's reached a level that I was visiting my toilet about 30 times a day/night, for about a year. . . . I had lost 50 to 60 pounds in a matter of weeks. . . . I had to apply for Social Security Disability in 2008—after dealing with a severe case of Crohn's (four surgeries) since 1999—it was quickly granted. . . . I had lived (and suffered) in Colorado for three and a half years before I heard of the medical marijuana law. . . . I was desperate. I called a doctor and he was willing to sign for me to get a medical marijuana card. At the time the only source I knew of for the medicine was a shadowy guy in Durango, literally in an alleyway apartment. I put on my Depends and drove over there. . . . Medical marijuana was supposed to be legal, but it sure didn't feel like it to me back then before Colorado dispensaries became actual stores instead of alleyways. I went home and medicated. The vapors immediately started to open some of the blockages in my intestinal tract and I knew there was some hope, for the first time in 11 years. The VA recommended removing my colon and rectum, leaving a pouch, just months earlier—I'm certainly glad that I didn't allow it. Within three months I was able to wean myself off of prescription drugs and I knew that I was going to reclaim my life from this disease. My VA doctors were initially skeptical about my new therapy, but they had no better suggestions—now they are pleased that I'm no longer dying a slow death, at taxpayer expense. (Delany 2012)

Medical marijuana dispensary.

AP Images/SIPA/Kennell Krista

Sherry Smith suffers from multiple sclerosis, a progressive autoimmune disease that affects the central nervous system and produces symptoms such as numbness, impaired muscular coordination and speech, blurred vision, and severe fatigue.

> I've been disabled with MS for so long that I got to the point where I didn't know which was worse, the dozens of medications prescribed by my doctors or the disease itself. . . . My prescription drugs were making me depressed and sick. . . . Now I take hemp oil capsules three times a day, which completely controls my muscle spasticity and pain, and then supplement, when necessary every once in a while, with smokable herb. (Quoted in Mannix 2009)

Vicki Burk, a grandmother and member of Idaho Moms for Marijuana and the Southern Idaho Cannabis Coalition describes how marijuana has provided relief from stomach distress and chronic pain ("Testimonials" 2013).

> My story began back in 1997–1998. I was having severe stomachaches after I ate. It didn't matter what it was. I went through a barrage of tests and no cause was ever found. In 2005, my liver enzymes were increased. I was told I had a faulty liver. In 2006 I was diagnosed with PBC (primary biliary cirrhosis). I was so sick I threw up everything. My boyfriend at the time told me to take a few hits and it would calm my stomach down so that I could eat and drink. That way I would not become dehydrated. I did as he suggested and it was like night & day. What a relief! I now use it to relieve my chronic pain from arthritis and fibromyalgia as well. It works better than the morphine pills like Kadian, which I take. I also take oxycodone. The marijuana works better than both of them together.

Problems in U.S. Health Care

In a comparison of health care in 11 wealthy nations, the United States ranked *last*, despite the fact that health care spending—both per person and as a percentage of gross domestic product—is considerably higher in the United States than in the other countries (Australia, Canada, France, Germany, the Netherlands, New Zealand, Norway, Sweden, Switzerland, and the United Kingdom) (Davis et al. 2014). In the following sections we discuss three main problems in U.S. health care: inadequate health insurance coverage, the high cost of medical care, and inadequacies in mental health care.

Inadequate Health Insurance Coverage

Many other countries, including 31 European countries and Canada, have national health insurance systems that provide **universal health care**—health care to all citizens. National health insurance is typically administered and paid for by government and funded by taxes or Social Security contributions. Despite differences in how national health insurance works in various countries, typically, the government (1) directly controls the financing and organization of health services, (2) directly pays providers, (3) owns most of the medical facilities (Canada is an exception), (4) guarantees universal access to health care, (5) allows private care for individuals who are willing to pay for their medical expenses, and (6) allows individuals to supplement their national health care with private insurance as an upgrade to a higher class of service and a larger range of services (Cockerham 2007; Quadagno 2004). Any rationing of health care in countries with national health insurance is done on the basis of medical need, not ability to pay.

In 2014, 10.4 percent of Americans (33 million people) did not have health insurance coverage for the entire year (Smith and Medalia 2015). Non-Hispanic whites are more likely than racial and ethnic minorities to have health insurance. Hispanics are the least likely to have insurance, with one in five Hispanics lacking health insurance in 2014 (Smith and Medalia 2015). Of all age groups, young adults aged 19 to 34 are the least likely to have health insurance. Nearly all seniors aged 65 and older have health insurance through Medicare.

Employed individuals and individuals with higher incomes are more likely to have health insurance. However, employment is no guarantee of health care coverage; in 2014, 11.2 percent of full-time workers were uninsured (Smith and Medalia 2015). Some businesses do not offer health benefits to their employees, some employees are not eligible for health benefits because of waiting periods or part-time status, and some employees who are eligible may not enroll in employer-provided health insurance because they cannot afford their share of the premiums.

Consequences of Inadequate Health Insurance. An estimated 45,000 deaths per year in the United States are attributable to lack of health insurance (Park 2009). Individuals who lack health insurance are less likely to receive preventive care, are more likely to be hospitalized for avoidable health problems, and are more likely to have disease diagnosed in the late stages.

Because most health care providers do not accept patients who do not have insurance, many individuals without insurance resort to using the local hospital emergency room (Scal and Town 2007). The federal Emergency Medical Treatment and Active Labor Act requires hospitals to assess all patients who come to their emergency rooms to determine whether an emergency medical condition exists and, if it does, to stabilize patients before transferring them to another facility. Hospital patients without insurance are almost always billed at a much higher cost than the prices negotiated by insurance companies.

> An estimated 45,000 deaths per year in the United States are attributable to lack of health insurance.

universal health care A system of health care, typically financed by the government, that ensures health care coverage for all citizens.

Individuals who lack dental insurance commonly have untreated dental problems, which can lead to or exacerbate other health problems:

> Because they affect the ability to chew, untreated dental problems tend to exacerbate conditions such as diabetes or heart disease. . . . Missing and rotten teeth make it painful if not impossible to chew fruits, whole grain foods, salads, or many of the fiber-rich foods recommended by doctors and nutrition experts. (Sered and Fernandopulle 2005, pp. 166–167)

In their book *Uninsured in America*, Sered and Fernandopulle (2005) described one interviewee who "covered her mouth with her hand during our entire interview because she was embarrassed about her rotting teeth," and another interviewee "used his pliers to yank out decayed and aching teeth" (p. 166). The authors note that "almost every time we asked interviewees what their first priority would be if the president established universal health coverage tomorrow, the immediate answer was 'my teeth'" (p. 166).

The High Cost of Health Care

Health care spending in the United States is far greater than in other industrialized countries. Yet nearly every other wealthy nation has better health outcomes, as measured by life expectancy and infant mortality.

It is widely believed that U.S. health costs have gone up due to the aging of the population; people are living longer today than in previous generations, and older people have greater health care needs. However, the United States has a relatively young population compared with many other high-income countries that spend much less on health care than does the United States (Squires 2012). Compared to other OECD countries, hospitalizations in the United States are less frequent and shorter. Yet, spending per hospital discharge in the United States (more than $18,000) is nearly three times higher than the OECD median ($6,222) (Squires 2012). The United States has high rates of performing medical tests and procedures involving advanced medical technology, such as diagnostic imaging, coronary procedures (angioplasty, stenting, and cardiac catheterizations), knee replacements, and dialysis; and charges for these medical tests and procedures are much higher in the United States than in other countries. Consider the advancements in the treatment of preterm babies, for which very little could be done in 1950. By 1990, special ventilators and neonatal intensive care became standard treatment for preterm babies in the United States (Kaiser Family Foundation 2007). The high rate of obesity in the United States—about a third of the adult population—accounts for nearly 10 percent of medical spending (reported in Squire 2012).

Prescription drug prices in the United States are 50 percent more than comparable drugs sold in other developed countries (Brill 2013). Drug prices are not regulated by the U.S. government; they are regulated in other countries. The pharmaceutical industry, which is among the most profitable industries in the United States, argues that U.S. drug prices are high because of the high cost of researching and developing (R & D) new drugs. But a critical analysis reveals that the industry purposely overestimates R & D costs to justify their high drug prices (Light and Warburton 2011). Furthermore, most large drug companies pay substantially more for marketing, advertising, and administration than for research and development (Families USA 2007).

Health insurance in the United States is another health care expense. In 2014, the average annual premiums for employer-sponsored coverage were $6,025 for an individual (worker's share was $1,081) and $16,834 for a family (worker's share was $4,823). From 2004 to 2014, average premiums for family coverage rose 69 percent, outpacing increases in both workers' wages and inflation (Kaiser Family Foundation/HRET 2014). U.S. health

TABLE 2.9 Cutting Back on Medical Care Due to Cost

In the past 12 months, because of cost, have you or another family member living in your household . . . ?	Percent Saying "Yes"
Relied on home remedies or over-the-counter (OTC) drugs instead of seeing a doctor	38%
Skipped dental care or check-ups	35%
Put off or postponed getting health care you needed	29%
Skipped recommended medical test or treatment	25%
Not filled a prescription for medicine	24%
Cut pills in half or skipped doses of medicine	16%
Had problems getting mental health care	8%
Did *any* of the above	58%

SOURCE: Adapted from Kaiser Family Foundation 2012.

> Having insurance does not guarantee that one is protected against financial devastation resulting from illness or injury, because even the insured typically must pay co-payments, deductibles, and exclusions.

insurance is costly largely because of high administrative expenses, which per capita are six times higher than in western European nations (National Coalition on Health Care 2009). Harrison (2008) explains the reason:

> The United States has the most bureaucratic health care system in the world, including over 1,500 different companies, each offering multiple plans, each with its own marketing program and enrollment procedures, its own paperwork and policies, its CEO salaries, sales commissions, and other nonclinical costs—and, of course, if it is a for-profit company, its profits.

Consequences of the High Cost of Health Care for Individuals and Families. One in three nonelderly U.S. adults struggle to pay medical bills: they have had problems affording medical bills in the last year, are making medical bill payments over time, or have medical bills they can't afford to pay at all (Pollitz et al. 2014). **Medical debt** is debt that results when people cannot afford to pay their medical bills. In 2014, nearly one in five U.S. adults were contacted by a debt collector about medical debt (LaMontagne 2014). More than one-third of debt collected by debt collection agencies is for medical debt (see Figure 2.2).

Medical debt can lead to bankruptcy, depletion of retirement or college savings, home foreclosure, damaged credit rating, reduced standard of living, and inability to receive needed medical care (Pollitz et al. 2014). Forgoing medicine or medical care (see Table 2.9) often exacerbates a medical condition, leading to even higher medical costs, or tragically, leading to death. Having insurance does not guarantee that one is protected against financial devastation resulting from illness or injury, because even the insured typically must pay co-payments, deductibles, and exclusions. The majority (70 percent) of people with medical debt are insured (Pollitz et al. 2014).

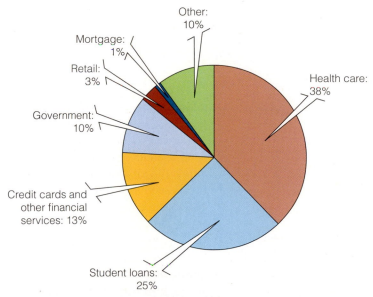

Other: 10%
Mortgage: 1%
Retail: 3%
Government: 10%
Credit cards and other financial services: 13%
Student loans: 25%
Health care: 38%

Figure 2.2 Types of Debt Collected from U.S. Consumers
SOURCE: Adapted from Montagne 2014.

medical debt Debt that results when people cannot afford to pay their medical bills.

Inadequate Mental Health Care

In the 1960s, the U.S. model for psychiatric care shifted from long-term inpatient care in institutions to drug therapy and community-based mental health centers. This transition, known as **deinstitutionalization**, has resulted in a significant decrease in the number of mental health facilities with 24-hour or residential treatment and the number of psychiatric treatment beds available. Deinstitutionalization removed patients from facilities where they were sometimes treated in a neglectful or inhumane manner, and restored freedom of choice to mental health consumers, including the right to refuse treatment. During the deinstitutionalization era, a variety of laws were passed making it illegal to commit psychiatric patients against their will unless they posed an immediate threat to themselves or to others. However, community mental health programs have not adequately met the need for care, and millions of Americans with mental disabilities go without needed care or rely on hospital emergency room care when their condition deteriorates into a major mental health breakdown.

The most frequently used source of care for mental health problems has become primary care and general doctors and nurses. Other "nonspecialty" care providers include community health centers, schools, nursing homes, correctional institutions, and emergency rooms. This fragmented system of mental health care leaves many people with mental health problems to fall through the cracks. Nearly one-third of the 10 million U.S. adults with serious mental illness in 2013 did not receive any mental health services in the past year (Substance Abuse and Mental Health Services Administration 2014).

Mental health services are often inaccessible, especially in rural areas. In most states, services are available from "9 to 5"; the system is "closed" in the evenings and on weekends when many people with mental illness experience the greatest need. Across the nation, people with severe mental illness end up in jails and prisons (see also Chapter 4), homeless shelters, and hospital emergency rooms. Many children with untreated mental disorders drop out of school or end up in foster care or the juvenile justice system. Given the increasing growth of minority populations, another deficit in the mental health system is the inadequate number of mental health clinicians who speak the client's language and who are aware of cultural norms and values of minority populations.

Strategies for Action: Improving Health and Health Care

Some strategies for improving health focus on interventions that target specific health problems, such as promoting condom use to prevent HIV infections and encouraging physical activity to reduce obesity. More comprehensive interventions focus on the broader social determinants of health, such as poverty and economic inequality, gender inequality, racial/ethnic discrimination, and environmental pollution. In addition to the strategies discussed in this section, efforts to alleviate social problems discussed in other chapters of this book are also essential elements to improving health.

Improving Health in Low- and Middle-Income Countries

Efforts to improve health in low- and middle-income countries include improving access to adequate nutrition, clean water and sanitation, and medical care. More targeted interventions include increasing immunizations for diseases such as measles, distributing mosquito nets to prevent malaria, and promoting the use of condoms to prevent the spread of HIV/AIDS.

Efforts to reduce maternal mortality—a major cause of death among women in the poorest countries of the world—have focused largely on providing access to good-quality reproductive care and family planning services. There are 225 million women in developing countries who want to postpone or avoid childbearing, but are not using contraception (World Health Organization 2015b). Some women's health advocates are fighting to pass legislation aimed at preventing "child marriage," as pregnant girls under age 18 are much more likely

deinstitutionalization The removal of individuals with psychiatric disorders from mental hospitals and large residential institutions to outpatient community mental health centers.

to die during pregnancy or childbirth than women in their 20s (Biset 2013). Globally, about one in four women aged 20 to 24 were married before age 18; about 8 percent of young women today were married before age 15 (UNICEF 2014).

Another strategy to improve the health of women and children in low-income countries is to provide women with education and income-producing opportunities. Promoting women's education increases the status and power of women to control their reproductive lives, exposes women to information about health issues, and also delays marriage and childbearing. In many developing countries, women's lack of power and status means that they have little control over health-related decisions. Men make the decisions about whether or when their wives (or partners) will have sexual relations, use contraception, or use health services.

Childbearing at an early age involves higher health risks for women and infants.

These and other efforts have had some success, as child and maternal death rates have declined over the last decade. However, a number of countries with the highest rates of maternal mortality have made little or no progress.

Essential health services in low-income countries—those focused on HIV, tuberculosis, malaria, maternal and child health, and prevention of noncommunicable diseases—are estimated to cost only $44 per person per year. Less developed countries need financial assistance to improve the health their populations. However, "higher levels of funding might not translate into better service coverage or improved health outcomes if the resources are not used efficiently or equitably" (World Health Organization 2012, p. 40).

Fighting the Growing Problem of Obesity

Sociological strategies to reducing and preventing obesity involve enacting programs and policies that encourage people (1) to eat a diet with sensible portions, with lots of high-fiber fruits and vegetables, and with minimal sugar and fat, and (2) to engage in regular physical activity.

Federal, State, and Local Antiobesity Policies. In 2015, the Food and Drug Administration announced a ruling that will effectively ban partially hydrogenated oils—the main source of trans fats in processed foods. Trans fats, which were banned in New York City restaurants in 2006, are believed to contribute to heart disease. Another FDA rule announced in 2014 requires chain restaurants, movie theaters, and pizza parlors to post calorie counts on their menus. About 18 states and cities already had menu-labeling rules in effect. Berkeley, California, became the first U.S. city to pass a law (in 2014) placing a 1-cent-an-ounce tax on sodas and other sugary drinks. In 2012, New York mayor Mike Bloomberg passed a law to ban large-size sugary beverages, but the law was struck down by a New York state judge. The 2012 federal Healthy Hunger-Free Kids Act established new standards for the National School Lunch and School Breakfast programs that require more fruits and vegetables and whole-grain foods; only fat-free or low-fat milk; and less saturated fats, trans fats, and sodium. Some states have enacted policies to improve school nutrition, increase school-based physical activity, and screen students' body mass index.

To encourage physical activity as well as green transportation, many state and local governments are adopting "Complete Street"

15 teaspoons of sugar per bottle

Researchers at Tufts University found that sugary drinks are responsible for 184,000 deaths each year, including more than 25,000 deaths in the United States.

policies. Complete Streets are roads designed with sidewalks, bike lanes, and other features that encourage walking, bicycling, and use of public transportation. Some cities and states are also addressing the problem of food deserts—areas where residents lack access to fresh fruits and vegetables—by such strategies as establishing urban and community gardens, farmers markets, and incentives to bring grocery stores into food desert areas.

Regulation of Food Marketing to Youth. Food marketing influences children's food preferences, and "children . . . play an important role in which products their parents purchase at the store, and which restaurants they frequent" (Federal Trade Commission 2012, p. ES-9). Recognizing how powerful advertising is in influencing the food and beverage choices of youth, the federal government has proposed guidelines for food marketing to children. But the powerful food industry has fought government regulations on advertising and, instead, self-regulates its food marketing to youth according to guidelines set out by the Children's Food and Beverage Advertising Initiative (CFBAI). But critics of the program claim that self-regulation has led to minimal improvement and that "the overwhelming majority of foods advertised to kids is still of poor nutritional quality. Under the Children's Food and Beverage Advertising Initiative nutritional standards, Cocoa Puffs, Popsicles, SpaghettiOs, and Fruit Roll-Ups are considered nutritious foods" (Wootan 2012).

Workplace Wellness Programs. Some workplaces have employee wellness programs that encourage employees to exercise, make healthy food choices, and engage in other health promotion behaviors such as quitting smoking. Some workplaces have onsite gyms, reimburse employees for gym membership, or organize lunchtime walking or jogging activities. Some employers have contests, awarding prizes to employees who lose the most weight or spend the most time exercising. Research suggests that for every dollar a company spends on wellness programs, it saves $3.27 in medical costs and $2.75 in absentee costs (Trust for America's Health 2012).

Soldiers and veterans suffering from mental health issues often need encouragement to seek professional mental health care.

Strategies to Improve Mental Health Care

Some of the strategies for improving mental health care in the United States include eliminating the stigma associated with mental illness, improving access to mental health services, and supporting mental health needs of college students.

Eliminating the Stigma of Mental Illness. Eliminating the stigma of mental illness is important because the negative label of "mental illness" and the social rejection and stigmatization associated with mental illness discourages individuals from seeking mental health treatment. The National Alliance on Mental Illness (NAMI) has fought against negative portrayals of mental illness in movies and television, and stigmatizing and inaccurate language in the news media. A major victory in the fight against the stigma of mental illness occurred in 2013, when the Associated Press—a global news network seen or heard by more than half the world's population—adopted new rules on how editors and reporters report about mental illness. The new rules include the following:

- Mental illness is a general condition. Specific disorders or types of mental illness should be used whenever possible.
- Do not use derogatory terms, such as insane, crazy/crazed, nuts, or deranged, unless they are part of a quotation that is essential to the story.
- Whenever possible, rely on people with mental illness to talk about their own diagnoses.
- Avoid using mental health terms to describe non–health issues. For example, don't say that an awards show was schizophrenic.
- Do not assume that mental illness is a factor in a violent crime, and verify statements to that effect. Research has shown that the vast majority of people with mental illness are not violent, and most people who are violent do not suffer from mental illness. (Carolla 2013)

The Department of Defense has launched an anti-stigma campaign called "Real Warriors, Real Battles, Real Strength" designed to assure military personnel that seeking mental health treatment will not harm their career and to publicize stories of military personnel who have been successfully treated for mental health problems (Dingfelder 2009). Effective anti-stigma campaigns not only focus on eradicating negative stereotypes of people with mental illness, but also emphasize the positive accomplishments and contributions of people with mental illness (see Table 2.10).

WHAT do you THINK?

Military service men and women may be awarded the Purple Heart medal if they are wounded or killed in military action. Those wounds must be physical—emotional wounds such as post-traumatic stress disorder (PTSD) do not qualify. The U.S. Department of Defense (2014, n.p.) explains:

> PTSD is defined as an anxiety disorder caused by witnessing or experiencing a traumatic event; it is not a wound intentionally caused by the enemy from an "outside force or agent," but is a secondary effect caused by witnessing or experiencing a traumatic event.

Do you agree with the Department of Defense? Or do you think veterans with PTSD should be eligible to receive the Purple Heart medal?

Improving Access to Mental Health Care. Improving access to mental health services involves (1) recruiting more mental health professionals, especially those willing to serve in rural and impoverished communities and who have cultural competency to work with clients from diverse cultural backgrounds; (2) improving health insurance coverage for mental health problems; and (3) expanding mental health screening.

TABLE 2.10 Successful People with Mental Illness

DEPRESSION	BIPOLAR DISORDER	ANXIETY DISORDER
Harrison Ford	Ludwig von Beethoven	Oprah Winfrey
Abraham Lincoln	Catherine Zeta Jones	Eric Clapton
Ernest Hemingway	Vincent Van Gogh	Johnny Depp
ADHD*	**SCHIZOPHRENIA**	**OCD†**
Benjamin Franklin	John Nash (mathematician)	Albert Einstein
Malcolm Forbes (NJ senator, magazine publisher)	Vaclav Nijinsky (dancer)	Howie Mandel
Michael Phelps (Olympic swimmer)	Peter Green (guitarist)	Cameron Diaz

*Attention deficit/hyperactivity disorder
†Obsessive compulsive disorder
SOURCES: Based on Holmes 2014; Mental Health Advocacy, n. d.

Many people living with mental illness attain success in their lives, such as Catherine-Zeta Jones, who has bipolar disorder.

Helga Esteb/Shutterstock.com

The 2010 Affordable Care Act (ACA) included a new program—the Mental and Behavioral Health Education and Training Grant program—that provides funds to institutions of higher education to recruit and train students pursuing graduate degrees in clinical mental and behavioral health. A 2012 executive order signed by President Obama increased mental health staffing at the Department of Veterans Affairs trained mental health peer specialists, and expanded the capacity of the Veterans Crisis Line, as well as suicide prevention awareness campaigns. In 2014, President Obama announced further executive orders to improve mental health care for military service members (White House 2014).

The Affordable Care Act has improved mental health care by greatly expanding mental health and substance use disorder insurance coverage. Under the ACA, all new small-group and individual market insurance plans are required to cover 10 Essential Health Benefit categories, including mental health and substance use disorder services, and must cover them with the same benefits as medical and surgical benefits—a concept known as **parity**. The Affordable Care Act has expanded mental health and substance use insurance coverage and federal parity protections to 62 million Americans (Beronio et al. 2013). In 2014, the VA expanded its mental health care services to cover veterans who have suffered from sexual assault or harassment during their military service.

Another strategy to improve access to mental health care involves making mental health screening a standard practice reimbursed by insurance, just like mammograms and other screening tests are reimbursed. And although most schools screen children and adolescents for hearing and vision problems, some schools also screen for mental health problems such as depression.

Mental Health Support for College Students. College students with mental health problems tend to turn to their friends for support instead of or in addition to seeking professional help (Kirsch et al. 2014). Consider the case of John, an average college student and fraternity member who suffered from anxiety and who had difficulty making friends.

> In his senior year, he suffered an acute manic psychosis, requiring a medical leave of absence from school. . . . His fraternity brothers had visited him in the hospital, kept in touch during his leave, and enthusiastically welcomed him back. They provided tutoring and study help, monitored his use of alcohol and drugs, assisted with medication adherence, and helped him get to class and appointments. The fraternity played a major role in John's recovery. (Kirsch et al. 2014, p. 524)

Recognizing the important role that students can play in supporting each other, many colleges and universities have peer-to-peer intervention programs that train students how to recognize and respond to individuals experiencing distress, and how to refer these individuals, when appropriate, to professional resources.

Most colleges and universities offer mental health services to students and provide accommodations for students with documented mental health conditions (e.g., adjustments in test setting and times and excused absences for treatment). Colleges can improve these services in a variety of ways, such as making sure that students know these services exist, employing more mental health professionals, and offering extended and flexible hours of service. In a survey of college students diagnosed with a mental health condition, respondents were asked, on average, how long they had to wait for an appointment to access campus mental health services; nearly 4 in 10 waited more than five days for an appointment (National Alliance on Mental Illness 2012).

As discussed earlier in this chapter, many college students with mental health conditions do not disclose their condition and do not seek accommodations or services

parity In health care, a concept requiring equality between mental health care insurance coverage and other health care coverage.

largely because of the stigma involved in being identified as having a mental illness. To reduce the stigma of mental illness on campuses, it is critical to provide information to the campus community on how common mental health conditions are and to emphasize the importance of getting help.

On more than 25 college campuses throughout the United States, students are getting involved in clubs that offer support and advocacy for students with mental illness. NAMI on Campus clubs, which are affiliated with the National Alliance on Mental Illness, are student-led clubs that raise awareness of mental health issues, educate the campus community, and support students (Crudo 2013).

Stress can exacerbate mental health problems, and even trigger their onset. To help students cope with the stress of exam week, some colleges and universities provide students with access to "therapy dogs" as a way to ease stress. The University of Connecticut provides therapy dogs at the library during exam week in a program called "Paws to Relax." At Harvard Medical School and Yale Law School, students can borrow therapy dogs through the library's card catalogue, just like they borrow books. This chapter's *Animals and Society* feature describes how dogs and other animals are used in mental health counseling.

The Affordable Care Act of 2010

Following much heated debate, in 2010, the Patient Protection and Affordable Care Act, commonly referred to as the **Affordable Care Act (ACA)** or "Obamacare," was signed into law by President Obama, with the overarching goal of increasing health insurance coverage to Americans. Just a few of the many provisions of ACA include the following:

- Establishes an "individual mandate" that requires U.S. citizens and legal residents to have health insurance or pay a penalty (waivers granted for financial hardship)
- Creates health insurance exchanges: online marketplaces where consumers can shop for, compare, and enroll in insurance plans
- Provides tax credits to businesses that provide insurance to their employees
- Requires health insurance plans to provide dependent coverage for children up to age 26
- Prohibits health insurance plans from placing lifetime limits on the dollar value of coverage; restricting annual limits on coverage; prohibiting insurers from canceling coverage except in cases of fraud; and prohibiting denial of insurance due to pre-existing conditions
- Requires insurance companies to use a certain percentage of the premiums they collect on medical care, as opposed to administrative expenses and profits
- Expands Medicaid to cover more low-income individuals/families
- Provides discounts on brand-name prescription drugs and free preventive services and annual wellness exams for Medicare enrollees, and raising Medicare premiums for some higher-income seniors

Affordable Care Act (ACA) Health care reform legislation that President Obama signed into law in 2010, with the goal of expanding health insurance coverage to more Americans; also known as the Patient Protection and Affordable Care Act, or "Obamacare."

One provision of the ACA requires insurance plans to cover the cost of contraceptives. However, in 2014, the Supreme Court ruled that craft store chain Hobby Lobby and other closely held for-profit companies may choose not to pay for coverage of birth control in their workers' health plans if the company's owner has religious objections. Public opinion on the Court's decision is evenly divided, with 47 percent of U.S. adults saying they approve and 49 percent disapproving (Hamel et al. 2014). Do you agree with the Court's ruling? What if the company's owner had a religious objection to other health services, such as blood transfusions or vaccinations?

Animal-assisted therapy involves using animals such as dogs, cats, and horses in the treatment of anxiety, depression, family problems, autism, eating disorders, post-traumatic stress, attention deficit disorders, and other mental health problems (Fine 2010; Peters 2011). Animals contribute to therapy in a variety of ways, including (1) reducing anxiety, (2) helping the trust and rapport-building process between the client and the therapist, (3) increasing motivation to attend and participate in therapy because of the desire to spend time with the therapy animal, and (4) stimulating conversation about difficult topics. One clinician reported an experience working with a child who had been traumatized by sexual abuse:

> I told one child that Buster [a dog] had a nightmare. I then asked the child, "What do you think Buster's nightmare was about?" The child said, "The nightmare was about being afraid of getting hurt again by someone mean." (Cited in Kruger and Serpell 2010, p. 39)

AAT is also used with individuals with severe mental disorders. Marsha was a 23-year-old woman who was diagnosed with catatonic schizophrenia. She was treated with medication and electroshock therapy, without improvement. She was withdrawn, frozen, and nearly mute. A therapy dog was introduced into her treatment:

> At first there was no improvement in Marsha's behavior. . . . She remained very withdrawn and the only signs of communication were when she was with the dog. When the dog was taken away, she would get off her chair and go after it. She began to walk the dog a little and . . . was given a written schedule of the hours when the dog would come and visit her; she began to look forward to the visits and to talk about the dog with the other patients. Six days after the introduction of the dog Marsha suddenly showed marked improvement and shortly thereafter she was discharged. (cited in Urichuk with Anderson 2003, pp. 107–108)

The most common use of AAT in mental health services involves therapists working in partnership with their own pet that has been evaluated and certified as appropriate for therapy work (Chandler 2005). Dogs used as therapy animals must meet rigorous requirements through organizations such as Therapy Dogs Inc., Delta Society®, and Therapy Dogs International. For the dogs, this includes a temperament test, obedience class training, and additional AAT training in which the dogs learn things like not reacting to loud noises, how to ride on elevators, and being comfortable around patients who use wheelchairs or walkers. In addition, the dogs must maintain good health and remain current on vaccinations.

Not all AAT involves using specially trained or certified animals. Some individuals achieve improvements in mental health functioning and well-being as a result of interacting with and taking care of farm animals (Arehart-Treichel 2008). One case study describes how Mark, a young teenager with autism, benefited through his interactions with donkeys:

> At first Mark could not get near the donkeys. Naturally wary of people, the donkeys ran away from Mark as he marched after them. . . . But Mark was motivated, and with guidance and patience has learnt to approach the donkeys slowly and gently. He has become aware of the donkeys' feelings and of how his actions impact them; he has built a relationship with the donkeys based upon mutual trust and respect. He is now rewarded each week by Ceilidh running up to him to have her face rubbed. This is a new experience for Mark that we are working on transferring to his human relationships. (cited in Urichuk with Anderson 2003, p. 80)

Small animals such as rabbits, guinea pigs, and birds can also be used as therapy animals:

> Marta was an 8-year-old diagnosed as an emotionally disturbed child of a strict and abusive mother. She was aggressive and hyperactive, sexually precocious, and had temper

Therapy dogs are used in a variety of settings, including clinics, private medical offices, nursing homes, schools, and hospitals.

AP Images/Tim Roske

> tantrums. In her first few months at the residential school, no one could get her to talk about her relationship with her mother. In the first session with a small furry rabbit, she held him in her lap and stroked him, telling the therapist that the rabbit's ears had been chewed by the mother rabbit. (The rabbit's ears were normal.) The therapist asked her why this was so. Marta responded, "The mother rabbit chewed the baby rabbit's ears all up. She wanted the baby to leave home." The therapist then asked, "How did the baby rabbit feel?" In answering, Marta said "Sad. The baby rabbit loves the mother rabbit but the mother rabbit no longer loves the baby." This dialogue about the rabbit was an opener for Marta to then talk of her own feelings about the mother who badly beat her. (cited in Urichuk with Anderson 2003, pp. 64–65)

It is important to note that AAT is not appropriate for individuals who are fearful of or allergic to animals. Even with careful selection and training of therapy animals, there is some risk that a therapy animal could injure a client, and also risk that a client could hurt the animal. But with close supervision and careful management, animal-assisted therapy provides opportunities for improving the well-being and functioning of children and adults with a variety of mental health problems.

Political opposition to the Affordable Care Act, primarily from Republicans and Tea Party members, resulted in numerous efforts in Congress to repeal the law. Challenges to the constitutionality of the ACA eventually led to the Supreme Court, which, in a 5–4 vote, ruled in 2012 that the individual mandate is a constitutional exercise of Congress's power to levy taxes (Musumeci 2012). But the Court also ruled that states may opt out of the Medicaid expansion provision of the ACA; and as this book goes to press nearly half the states have not expanded Medicaid, leaving millions of low-income Americans uninsured. Researchers estimate that states' rejection of Medicaid expansion will result in as many as 17,100 deaths annually (Dickman et al. 2014).

A year after the implementation of the ACA, an estimated 8 to 11 million Americans had newly enrolled in health insurance and the number of Americans without health insurance declined by about 25 percent (Sanger-Katz 2014). More than half of the new enrollees signed up for Medicaid, especially in states that chose to expand Medicaid eligibility. The majority of people (85 percent) who signed up for private insurance through online exchanges during the first enrollment period qualified for federal subsidies that lowered their premiums (Goodnough et al. 2014). Because the ACA requires insurers to cover people with preexisting conditions and to provide a broader array of benefits, some people who already had insurance saw their premiums rise.

Five years after the ACA was implemented the percentage of people lacking health insurance had declined to its lowest level in more than a decade, with the greatest gains in insurance coverage found among young adults, Hispanics, blacks, and those with low incomes (Blumenthal et al. 2015; Collins et al. 2015). The ACA is also credited with reducing the number of U.S. adults who (1) reported not getting needed care because of cost, and (2) had problems paying their medical bills (Collins et al. 2015).

The public continues to be divided over the ACA, with about half of Americans having an unfavorable view of the law, although the majority wants Congress to improve the law rather than to repeal it (Hamel, Firth, and Brodie 2014). People who favor the ACA tend to approve of the law's goal of expanding access to health care and insurance; those who oppose the law often cite concern over the cost, and are opposed to the individual mandate and government involvement in health care. Public opinion is often based on misperceptions: fewer than four in ten Americans are aware that enrollees under the ACA have a choice between private plans; a quarter mistakenly believe that enrollees under the ACA are covered by a single government health insurance plan (Hamel et al. 2014). Some Americans criticize the ACA for not going far enough to ensure access to health care and call for further health care reform to create a single-payer health care system.

The Debate over Single-Payer Health Care

In a **single-payer health care** system, a single tax-financed public insurance program replaces private insurance companies. Advocates of single-payer health care financing point out that administration costs of private insurance consume nearly a third of Americans' health dollars. These costs include insurance company overhead, underwriting, billing, sales and marketing departments, exorbitant executive pay, and profits. In addition, hospitals and doctors must pay administrative staff to deal with the various billing policies and procedures of different insurers. Replacing private insurance companies with single-payer health care would save more than $400 billion per year—enough to provide universal coverage to every U.S. resident without copayments, deductibles, or increase in health expenditures (Himmelstein and Woolhandler 2014).

A bill to create a single-payer health care system, the Expanded and Improved Medicare for All Act, would replace private insurance companies with one public agency that would pay for medical care for all Americans, much like Medicare works for seniors. In 2011, Vermont became the first state to pass legislation to establish state-level single-payer health care, which is expected to be implemented in 2017.

The insurance industry opposes the adoption of a single-payer health care system and spends a great deal of money on lobbying, political contributions, and public relations to influence the health reform debate. Opponents argue that a single-payer national health insurance program would amount to a "government takeover" of health care and would

single-payer health care A health care system in which a single tax-financed public insurance program replaces private insurance companies.

Supporters of a single-payer health care system argue that access to health care is a basic human right that, like public education, should be provided with public funds.

result in higher costs, less choice, rationing, and excessive bureaucracy—the very outcomes that have resulted from corporatized medicine (Nader 2009). The rise of the grassroots Tea Party political movement has fueled opposition to "big government," viewing government "takeover" of health care as an intrusion into individual freedoms.

Supporters of a single-payer health care system argue that access to health care is a human right—a public responsibility that, like education, should be paid for with public monies rather than provided through a market-based system. Himmelstein and Woolhandler (2014) note:

> Economic texts preach that markets breed efficiency, but the most market-oriented health systems are the least efficient. . . . A simple national health insurance program is the best way to meet both the moral imperative to care for the sick and the economic imperative to do so efficiently. (p. 2081)

Understanding Problems of Illness and Health Care

Although human health has probably improved more over the past half-century than over the previous three millennia, the gap in health between rich and poor remains wide. Poor countries need economic and material assistance to alleviate problems such as Ebola, HIV/AIDS, high maternal and infant mortality rates, and malaria. Cancer, once viewed as a disease that affects primarily wealthy countries, has now become prevalent in low-income countries where treatment is either not available or not affordable (Farmer et al. 2010). Obesity and its associated health problems have also spread throughout the world, adding to the burden of infectious diseases that already plague low-income countries.

Although poverty may be the most powerful social factor affecting health, other social factors that affect health include globalization, increased longevity, family structure, gender, education, and race or ethnicity. Although individuals make choices that affect their health—choices such as whether to smoke, exercise, eat a healthy diet, engage in risky sexual activity, wear a seat belt, and so on—those choices are also influenced by social, economic, and political forces that must be addressed if the goal is to improve the health not only of individuals but also of entire populations. By focusing on individual behaviors that affect health and illness, we often overlook social causes of health problems (Link and Phelan 2001). A sociological view of illness and health care looks not only at the social causes, but also at the social *consequences* of health problems—consequences that potentially affect us all. In *Uninsured in America* (2005), Sered and Fernandopulle explain:

> If millions of American children do not have reliable, basic health care, all children who attend American schools are at risk through daily exposure to untreated disease. If millions of restaurant and food industry workers do not have health insurance, people preparing food and waiting tables are sharing their health problems with everyone they serve. . . . If tens of millions of Americans go without basic and preventive care, we all pay the bill when their health problems turn into complex medical emergencies necessitating expensive . . . treatment. (p. 20)

A sociological approach to illness and health care also looks at social solutions such as federal, state, and local government policies and laws designed to improve public health, and examines the conflicts between public health initiatives and industries whose profits might be threatened by such initiatives.

Improving public health is a complex endeavor. A comprehensive approach to health is informed by the fact that "there is no single silver bullet for population health improvement. Investments in all determinants of health—including health care, public health, health behaviors, and residents' social and physical environments—will be required" (Kindig and Cheng 2013, p. 456). A comprehensive approach to improving the health of a society requires addressing diverse issues such as poverty and economic inequality, gender inequality, population growth, environmental issues, education, housing, energy, water and sanitation, agriculture, and workplace safety. Despite the significance of recent health care reform efforts to improve Americans' health insurance coverage, access to health care is only one piece of the puzzle: "Health and longevity are also profoundly influenced by where and how Americans live, learn, work, and play" (Williams, McClellan, and Rivlin 2010, p. 1481).

Improving the health of the world also means seeking nonmilitary solutions to international conflicts. In addition to the deaths, injuries, and illnesses that result from combat, war diverts economic resources from health programs, leads to hunger and disease caused by the destruction of infrastructure, causes psychological trauma, and contributes to environmental pollution (Sidel and Levy 2002). Thus, "the prevention of war . . . is surely one of the most critical steps mankind can make to protect public health" (White 2003, p. 228).

The tragic and senseless shooting deaths at Sandy Hook Elementary School in 2012 renewed public concern for affordable and accessible mental health care—another critical but often neglected aspect of public health. The Sandy Hook tragedy also ignited debates over gun control in the United States (see also Chapter 4), raising the question of whether widespread availability of guns is a public health problem. The rate of firearm-related deaths in the United States is nearly 20 times that in other developed nations (Shern and Lindstrom 2013). In other developed and civilized countries that grant their citizens the right to universal health care, owning a gun is a privilege and not a right. In the United States, Americans have a constitutional right to own a gun, but no similar right to health care. Until the United States joins the rest of the developed world as well as the United Nations in declaring access to health care to be a basic human right, it may be easier for many Americans to access a gun than it is to access health care.

Chapter Review

- **How do the authors argue that the study of social problems is, essentially, the study of health problems?**
  According to the World Health Organization, health is "a state of complete physical, mental, and social well-being." Based on this definition, the authors suggest that the study of social problems is, essentially, the study of health problems, because each social problem affects the physical, mental, and social well-being of humans and the social groups of which they are a part.

- **What are some major differences in the health of populations living in high-income countries compared with the health of populations living in low-income countries?**
  Life expectancy is significantly greater in high-income countries compared with low-income countries. Although the majority of deaths worldwide are caused by noncommunicable diseases such as heart disease, stroke, cancer, and respiratory disease, low-income countries have a

comparatively higher rate of infectious and parasitic diseases, infant and child deaths, and maternal mortality.

- **How has globalization affected health worldwide?**
Increased global transportation and travel contribute to the spread of infectious disease. Globalization is linked to the rise in obesity worldwide due to increased access to unhealthy foods and beverages, and to televisions, computers, and motor vehicles, which are associated with increased sedentary behavior. These factors have contributed to *globesity*—a worldwide increase in overweight and obesity. On the positive side, globalized communications technology is helpful in monitoring and reporting on outbreaks of disease, disseminating guidelines for controlling and treating disease, and sharing medical knowledge and research findings. Another aspect of globalization and health is the growth of medical tourism—a multibillion-dollar global industry that involves traveling, primarily across international borders, for the purpose of obtaining medical care.

- **Are Americans the healthiest population in the world?**
The United States is one of the wealthiest countries in the world, but it is not one of the healthiest. In a comparison of health outcomes in the United States with those of 16 other high-income, industrialized countries, Americans are less likely to smoke and drink heavy alcohol, and they have better control over their cholesterol levels. The United States also has higher rates of cancer screening and survival, and has higher survival after age 75. But despite the fact that the United States spends more on health care per person than any other industrialized country, Americans die sooner and have higher rates of disease or injury. The United States ranked last or near the bottom in nine key areas of health: infant mortality and low birth weight; injuries and homicide; teenage pregnancy and sexually transmitted infections; HIV/AIDS prevalence; drug-related deaths; obesity and diabetes; heart disease; lung disease; and disability.

- **Why is mental illness referred to as a "hidden epidemic"?**
Mental illness is a "hidden epidemic" because the shame and embarrassment associated with mental problems discourage people from acknowledging and talking about them. The stigma of being labeled as "mentally ill" can negatively affect an individual's self-concept and disqualify that person from full social acceptance. Negative stereotypes of people with mental illness contribute to its stigma. One of the most common stereotypes of people with mental illness is that they are dangerous and violent. Although untreated mental illness can result in violent behavior, the vast majority of people with severe mental illness is not violent and is involved in only about 4 percent of violent crimes. In fact, people with mental illness are much more likely to be victims of violence than members of the general population.

- **How common is mental illness in the United States?**
In 2013, nearly one in five U.S adults had a mental illness in the past year. About half of all Americans will experience some form of mental disorder in their lifetime. Nearly one in four college students has been diagnosed or treated by a professional for a mental health problem within the past year.

- **How do structural-functionalism, conflict theory, and symbolic interactionism help us understand illness and health care?**
Structural-functionalism examines (1) how failures in the health care system affect not only the well-being of individuals, but also the health of other social institutions, such as the economy and the family; (2) how changes in society affect health, and how health concerns may lead to social change; and (3) latent dysfunctions, or unintended and often unrecognized negative consequences of health-related social patterns or behavior. The conflict perspective (1) focuses on how socioeconomic status, power, and the profit motive influence illness and health care; (2) points to ways in which powerful groups and wealthy corporations influence health-related policies and laws through lobbying and financial contributions to politicians and political candidates; and (3) criticizes the pharmaceutical and health care industry for placing profits above people. Symbolic interactionism focuses on (1) how meanings, definitions, and labels influence health, illness, and health care; (2) the process of medicalization whereby behaviors and conditions come to be labeled as medical problems have undergone medicalization; (3) how conceptions of health and illness are socially constructed, and vary over time and across societies; and (4) the stigmas associated with certain health conditions.

- **What are three main social factors that are associated with health and illness?**
Three main social factors associated with health and illness are socioeconomic status, race/ethnicity, and gender.

- **How does health care in the United States compare with that of many other high-income nations?**
Many other advanced countries have national health insurance systems—typically administered and paid for by government—that provide universal health care (health care to all citizens). The United States does not have a health care system per se, but rather has a patchwork that includes both private insurance (purchased individually or through employers or other groups), and public insurance plans such as Medicare and Medicaid.

- **What are examples of complementary and alternative medicine?**
Complementary and alternative medicine (CAM) refers to a broad range of health care approaches, practices, and products that are not considered part of conventional medicine, including herbal and homeopathic remedies, vitamins, meditation, Pilates, yoga, tai chi, acupuncture, chiropractic care, massage therapy, Reiki and other energy work, the use of traditional healers, and medical marijuana.

- **What are three problems in U.S. health care discussed in this chapter?**
Three problems in U.S. health care include inadequate health insurance, the high cost of health care, and inadequate mental health care.

- **What is the main goal of the Patient Protection and Affordable Care Act, commonly referred to as the Affordable Care Act (ACA) or "Obamacare"?**
The main goal of the Affordable Care Act is to increase health insurance coverage to Americans.

- **What is single-payer health care?**
  In a single-payer health care system, a single tax-financed public insurance program replaces private insurance companies. Advocates of single-payer health care financing point out that replacing private insurance companies with single-payer health care would save more than $400 billion per year—enough to provide universal coverage to every U.S. resident without copayments, deductibles, or increase in health expenditures.

Test Yourself

1. Deaths due to _____ are much more common in less developed countries compared with more developed countries.
   a. heart disease
   b. cancer
   c. stroke
   d. infectious diseases
2. Americans are the healthiest population in the world.
   a. True
   b. False
3. Obesity is only a problem in wealthy, developed countries.
   a. True
   b. False
4. How many Americans will experience some form of mental disorder in their lifetime?
   a. One in 20
   b. One in 10
   c. One in 5
   d. About half
5. A study of mortality rates in 3,140 U.S. counties found that the factors most strongly associated with higher mortality were poverty and
   a. gender.
   b. lack of college education.
   c. race.
   d. marital status.
6. Medical marijuana is most frequently used to
   a. lose weight.
   b. treat depression.
   c. treat attention deficit/hyperactivity disorder.
   d. treat pain and muscle spasms.
7. In the United States, who is most likely to lack health insurance?
   a. The elderly
   b. Blacks/African Americans
   c. Hispanics
   d. Children
8. Which type of debt is most commonly collected by collection agencies?
   a. Medical
   b. Credit card
   c. Student loan
   d. Mortgage
9. In the United States, the most frequently used source of care for mental health problems is
   a. psychologists.
   b. psychiatrists.
   c. primary care physicians.
   d. social workers.
10. The Affordable Care Act requires health insurance plans to provide dependent coverage for children up to age
   a. 18.
   b. 20.
   c. 24.
   d. 26.

Answers: 1. D; 2. B; 3. B; 4. D; 5. B; 6. D; 7. C; 8. A; 9. C; 10. D.

Key Terms

AP Images/Ian West

> " Substance abuse, the nation's number one preventable health problem, places an enormous burden on American society, harming health, family life, the economy, and public safety, and threatening many other aspects of life."

THE ROBERT WOOD JOHNSON FOUNDATION,
Institute for Health Policy, Brandeis University

3

Alcohol and Other Drugs

Learning Objectives

After studying this chapter, you will be able to . . .

1 Compare international drug use trends.

2 Propose recommendations to reduce drug abuse based on each of the three sociological theories.

3 Summarize the use and abuse of legal drugs.

4 Summarize the use and abuse of illegal drugs.

5 Argue which consequence of drug abuse is the most harmful for society.

6 Evaluate each of the treatment alternative types.

7 Assess the viability of each of the proposed drug-fighting strategies.

HE WAS FOUND the morning after fraternity members had dragged him into an empty bedroom when he passed out (Placko 2014). The Northern Illinois University freshman had attended a party where pledges were instructed to drink every time they failed to find their "pledge mom"—a member of an affiliated sorority. David Bogenberger's death was caused by alcohol poisoning. His blood alcohol level was five times the legal limit, having consumed an estimated 40 ounces of vodka and other spirits in 90 minutes (Walberg 2013). The 19-year-old's parents named the fraternity and 22 of its members in a wrongful death suit. All were found guilty of misdemeanor hazing or misdemeanor reckless conduct. Addressing the defendants in court, David's mother said, "The human decency that most of us would render to a sick animal, these self-proclaimed 'brothers' would not even extend to a young man they pledged a lifelong brotherhood to" (Ward 2015).

During the victim impact phase of the trial of five former Northern Illinois University Phi Kappa Alpha fraternity members, David Bogenberger's father breaks down. Addressing the defendants he stated, "You left him alone to die."

Drug-related deaths are just one of the many negative consequences of alcohol and drug abuse. The abuse of alcohol and other drugs is a social problem when it interferes with the well-being of individuals and/or the societies in which they live—when it jeopardizes health, safety, work and academic success, family, and friends. But managing the drug problem is a difficult undertaking, particularly when attitudes vary dramatically by demographic groups (see Table 3.1). In dealing with drugs, a society must balance individual rights and civil liberties against the personal and social harm that drugs promote—fetal alcohol syndrome, suicide, drunk driving, industrial accidents, mental illness, unemployment, and teenage addiction. When to regulate, what to regulate, and whether to regulate are complex social issues. Our discussion begins by looking at how drugs are used and regulated in other societies.

The Global Context: Drug Use and Abuse

Pharmacologically, a **drug** is any substance other than food that alters the structure or functioning of a living organism when it enters the bloodstream. Using this definition, everything from vitamins to aspirin is a drug. Sociologically, the term *drug* refers to any chemical substance that (1) has a direct effect on users' physical, psychological, and/or intellectual functioning; (2) has the potential to be abused; and (3) has adverse consequences for individuals and/or society. Societies vary in how they define and respond to drug use. Thus, drug use is influenced by the social context of the particular society in which it occurs.

drug Any substance other than food that alters the structure or functioning of a living organism when it enters the bloodstream.

Drug Use and Abuse around the World

Globally, 3.5 to 7.0 percent of the world's population between the ages of 15 and 64—162 to 324 million people—reported using at least one illicit drug in the previous year (World Drug Report [WDR] 2014). According to the most recent report, cannabis (i.e., marijuana and hashish) remains by far the most widely used illegal drug, followed by amphetamine-type stimulants (ATS), cocaine, and opiate-based drugs.

The *Global Status Report on Alcohol and Health* (World Health Organization [WHO] 2014a) indicates that, in the year prior to data collection, 38.8 percent of respondents had consumed alcohol. Furthermore, 16.0 percent of the world's 15 years and older population were "heavy episodic" drinkers, more commonly called binge drinking. In 2012, the most recent year for which data are available, over 3 million deaths were attributable to alcohol (WHO 2014a).

Alcohol consumption is highest in Eastern European countries such as Belarus, Moldova, Lithuania, and the Russian Federation. To understand regional variation in alcohol consumption rates, one must examine a host of social variables. For example, Russians' high alcohol consumption, although descending recently, reflects access and low cost, high unemployment and the resulting boredom, and peer pressure (Jargin 2012). Similarly, rates are low in the Middle East and Asia where religious dictates discourage alcohol use, availability is difficult, and punishment is severe.

Globally, nearly 20 percent of the adult population smokes cigarettes (Eriksen et al. 2012), and 80 percent of the people who smoke cigarettes are from low- and middle-income countries (WHO 2014b). Prevalence of smoking, however, varies dramatically by country. For example, estimated prevalence rates for men in Armenia, Russia, and Laos exceeds 50 percent but less than 10 percent in Ghana, Sudan, Antigua, and Ethiopia. Prevalence rates for women also vary by country. Five percent or more of women in Chile, France, Austria, and Hungary smoke tobacco compared to 1 percent or less in Ethiopia, Sudan, Cameroon, and Morocco (Ng et al. 2014).

Illegal drug use also varies by location. Over 80 million European adults report using an illicit drug at least once in their lifetime—about 25 percent of the population. However, the "levels of lifetime use vary considerably between countries, from around one third of adults in Denmark, France and the United Kingdom, to less than one in ten in Bulgaria, Greece, Cyprus, Hungary, Portugal, Romania and Turkey" (European Monitoring Centre for Drugs and Drug Addiction [EMCDDA] 2014a, p. 33). Furthermore, lifetime use of *any* illicit drug other than marijuana by 15- and 16-year-olds is the highest in the United States (16.0 percent) when compared to 36 European countries (Wadley 2012).

Finally, drug use varies over time. Figure 3.1, based on data from Monitoring the Future (MTF), a national in-school survey of secondary school students' drug use, indicates that alcohol use and smoking among the three age groups, in general, has decreased over the last two decades. Illicit drug use for 8th, 10th, and 12th graders, although variable over the time period, was at its highest levels in 2013. It should also be noted that Figure 3.1 compares 8th, 10th, and 12th grader drug use across drug type. In each case, drug use increases with age (MTF 2014).

TABLE 3.1 U.S. Attitudes toward Drug Abuse, 2014 (*N* = 1,821)

| | The problem of drug abuse across the country is a... | | |
	Crisis %	Serious Problem %	Minor/Not a Problem %
Total	32	55	13
Men	29	55	15
Women	34	55	11
White	32	54	13
Black	36	52	11
Hispanic	28	60	12
18–29	20	56	23
30–49	36	49	14
50–64	34	57	9
65+	36	59	3
College grad+	27	57	15
Some college	34	52	14
HS or less	33	57	10

SOURCE: Pew 2014.

> Globally, nearly 20 percent of the adult population smokes cigarettes...and 80 percent of the people who smoke cigarettes are from low- and middle-income countries.

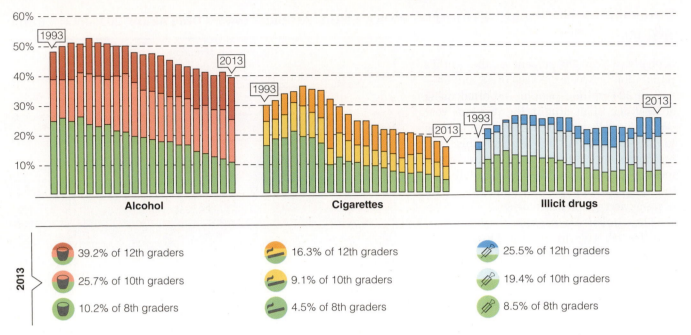

2013		
39.2% of 12th graders	16.3% of 12th graders	25.5% of 12th graders
25.7% of 10th graders	9.1% of 10th graders	19.4% of 10th graders
10.2% of 8th graders	4.5% of 8th graders	8.5% of 8th graders

Figure 3.1 Alcohol, Cigarette, and Illicit Drug Use over the Last Two Decades for 8th, 10th, and 12th Graders, 2013*
SOURCE: NIDA 2014.
*Past 30 day use.

Some have argued that differences in drug use can be attributed to variations in drug policies. The Netherlands, for example, has had an official government policy of treating the use of "soft" drugs such as marijuana and hashish as a public health issue rather than a criminal justice issue since the mid-1970s. Despite the proliferation of "cannabis cafés," sales of marijuana for personal use in the Netherlands do not significantly differ from that in other European countries (Dutch Drug Policy 2014).

Historically, Great Britain has also adopted a "medical model," particularly in regard to heroin and cocaine. As recently as the 1960s, English doctors prescribed opiates and cocaine for their drug-addicted patients who were unlikely to quit using drugs on their own and for the treatment of withdrawal symptoms. By the 1970s, however, the emphasis was on prohibition, criminalization, and incarceration and, as in the United States, it became known as the "war on drugs." Today, British officials are weighing the relative merits of returning to a medical model that embraces drug addiction as a disorder (British Medical Association [BMA] 2013; Feilding 2013). Similarly, the European Union, has adopted a 2013–2020 drug policy that for the first time includes the objective of reducing "the health and social risks and harms caused by drugs" (EMCDDA 2014b, p. 2).

WHAT do you THINK?

As of 2015, the recreational use of marijuana was legal in four states—Alaska, Colorado, Oregon, and Washington—and the District of Columbia, with likely legalization initiatives in five or more states in 2016 (Boyette and Wilson 2015). Nonetheless, to date, the drug remains illegal under federal law. Do you think the federal government should legalize marijuana? What other drugs, if any, should be legal?

In stark contrast to such health-based policies, many other countries execute drug users and/or dealers or subject them to corporal punishment that may include whipping, stoning, beating, and torture. In Iran alone, over 318 people were executed for drug-related offenses in 2014, and the number is growing yearly (Human Rights Watch [HRW] 2014). Death sentences in Pakistan have tripled (Gallahue et al. 2012). Thirty-three countries or territories have the death penalty for drug violations, including Saudi Arabia, where violators are beheaded (Ghosh 2014).

Drug Use and Abuse in the United States

In the United States, cultural definitions of drug use are contradictory—condemning it, on the one hand (e.g., heroin), yet encouraging and tolerating it, on the other (e.g., alcohol). At various times in U.S. history, many drugs that are illegal today were legal and readily available. In the 1800s and the early 1900s, opium was routinely used in medicines as a pain reliever, and morphine was taken as a treatment for dysentery and fatigue. Amphetamine-based inhalers were legally available until 1949, and cocaine was an ingredient in Coca-Cola until 1906, when it was replaced with another drug—caffeine (Abadinsky 2013; Witters et al. 1992). Not surprisingly, Americans' concerns with drugs have varied over the years. In 2000, 31 percent of Americans believed that marijuana should be legal; in 2015, the majority of Americans, 53 percent, thought that marijuana should be legal (Pew 2015).

Use of illegal drugs in the United States is a fairly common phenomenon. According to the most recent National Survey on Drug Use and Health (NSDUH 2014) available, in 2013, over 24 million Americans over the age of 12 had used an illicit drug in the month prior to the survey year, representing 9.4 percent of the 12 and older population. For purposes of the survey, illicit drugs included "marijuana/hashish, cocaine (including crack), heroin, hallucinogens, inhalants, or prescription-type psychotherapeutics (e.g., pain relievers, tranquilizers, stimulants, and sedatives) used non-medically" (NSDUH 2014, p. 2). Furthermore, 52.2 percent of the American 12 and older population reported alcohol consumption in the previous month and 25.5 percent reported tobacco use in the previous month (NSDUH 2014).

Sociological Theories of Drug Use and Abuse

Drug abuse occurs when acceptable social standards of drug use are violated, resulting in adverse physiological, psychological, and/or social consequences. When an individual's drug use leads to hospitalization, arrest, or divorce, such use is usually considered abusive. Drug abuse, however, does not always entail drug addiction. Drug addiction, or **chemical dependency**, refers to a condition in which drug use is compulsive—users are unable to stop because of their dependency. The dependency may be psychological (the individual needs the drug to achieve a feeling of well-being) and/or physical (withdrawal symptoms occur when the individual stops taking the drug). For example, withdrawal from marijuana includes mood swings, depression, anger, decreased appetite, and restlessness (Allsop et al. 2012; Zickler 2003).

In 2013, more than 21 million Americans, 8.2 percent of the population 12 or older, were defined as being dependent on or abusers of alcohol and/or other drugs. Of that number, 14.7 million were dependent on or abused alcohol only, 4.3 million were dependent on or abused illicit drugs but not alcohol, and 2.6 million were dependent on or abused both illicit drugs and alcohol. The most common illicit drug to be dependent on was marijuana followed by pain relievers, cocaine, heroin, stimulants, tranquilizers, hallucinogens, inhalants, and sedatives. Individuals who were dependent on or abuse illegal drugs and/or alcohol were disproportionately from the western part of the United States, male, American Indians or Alaska Natives, unemployed, between the ages of 18 and 25, and from nonmetropolitan areas. Among adults 18 and over, respondents with college degrees had the lowest rates of dependency, and those without a high school degree had the highest (NSDUH 2014).

Various theories provide explanations for why some people use and abuse drugs. Drug use is not simply a matter of individual choice. Theories of drug use explain how structural and cultural forces as well as biological and psychological factors influence drug use and society's responses to it.

drug abuse The violation of social standards of acceptable drug use, resulting in adverse physiological, psychological, and/or social consequences.

chemical dependency A condition in which drug use is compulsive and users are unable to stop because of physical and/or psychological dependency.

Structural-Functionalist Perspective

Structural functionalists argue that drug abuse is a response to weakening societal norms. As society becomes more complex and as rapid social change occurs, norms and values become unclear and ambiguous, resulting in anomie—a state of normlessness. Anomie may exist at the societal level, resulting in social strains and inconsistencies that lead to drug use. For example, research indicates that increased alcohol consumption in the 1830s and the 1960s was a response to rapid social change and the resulting stress (Rorabaugh 1979). Anomie produces inconsistencies in cultural norms regarding drug use. For example, although public health officials and health care professionals warn of the dangers of alcohol and tobacco use, advertisers glorify the use of alcohol and tobacco, and the U.S. government subsidizes the alcohol and tobacco industries. Furthermore, cultural traditions, such as giving away cigars to celebrate the birth of a child and toasting a bride and groom with champagne, persist.

Anomie may also exist at the individual level, as when a person suffers feelings of estrangement, isolation, and turmoil over appropriate and inappropriate behavior. An adolescent whose parents are experiencing a divorce, who is separated from friends and family as a consequence of moving, or who lacks parental supervision and discipline may be more vulnerable to drug use because of such conditions. Thus, from a structural-functionalist perspective, drug use is a response to the absence of a perceived bond between the individual and society and to the weakening of a consensus regarding what is considered acceptable.

> Anomie produces inconsistencies in cultural norms regarding drug use....[P]ublic health officials and health care professionals warn of the dangers of alcohol and tobacco use, advertisers glorify the use of alcohol and tobacco, and the U.S. government subsidizes the alcohol and tobacco industries.

WHAT do you THINK?

In recent years, there has been an increase in the use of human growth hormones and steroids, both performance-enhancing substances (PESs), among teenagers. Use of these drugs without a prescription, often obtained from online outlets, is particularly dangerous since product safety is unknown (Feliz 2014). In a survey, over half of parents reported discussing the dangers of PES with their teens, yet only 12 percent of the teens remembered PESs being discussed with their parents in the most recent "the dangers of drugs" conversation. What do you think is responsible for this misperception? What sociological theory would be most helpful in understanding the difference between what is said and what is heard?

Consistent with this perspective, a study on attitudes and drug use concluded that the dramatic increase in teenage use of prescription drugs is, in part, the result of the "lax attitudes and beliefs of parents and caregivers....Parents are not effectively communicating the dangers of Rx medicine misuse and abuse to their kids, nor are they safeguarding their medications at home and disposing of unused medications properly" (Partnership Attitude Tracking Study 2013, p. 1). Research also indicates that when risks are communicated to children, they need to be sustained through adolescence (Zehe and Colder 2015). The importance of the family in deterring drug use was highlighted in the national youth media campaign—"Parents. The Anti-Drug" (Office of National Drug Control Policy [ONDCP] 2009).

Conflict Perspective

The conflict perspective emphasizes the importance of power differentials in influencing drug use behavior and societal values concerning drug use. From a conflict perspective, drug use occurs as a response to the inequality perpetuated by a capitalist system. Societal members, alienated from work, friends, and family as well as from society and its institutions, turn to drugs as a means of escaping the oppression and frustration caused by the inequality they experience. Furthermore, conflict theorists emphasize that the most powerful members of society influence the definitions of which drugs are illegal and the penalties associated with illegal drug production, sales, and use.

For example, alcohol is legal because it is often consumed by those who have the power and influence to define its acceptability—white males (NSDUH 2014). This group also disproportionately profits from the sale and distribution of alcohol and can afford powerful lobbying groups in Washington, DC, to guard the alcohol industry's interests. Because this group also commonly uses tobacco and caffeine, societal definitions of these substances are also relatively accepting. Conversely, minority group members disproportionately use crack cocaine rather than powder cocaine (Fryer et al. 2014). Although the pharmacological properties of the two drugs are the same, possession of 5 grams of crack cocaine carried the same penalty under federal law as possession of 500 grams of powdered cocaine (Taifia 2006). In 2010, Congress voted to change the 1986 law that established the 100 to 1 ratio sentencing disparity and also eliminated the five-year mandatory minimum for first-time possession of crack cocaine. Inequities in state sentencing laws continue but are slowly changing (Crisp 2012).

The use of opium by Chinese immigrants in the 1800s provides a historical example. The Chinese, who had been brought to the United States to work on the railroads, regularly smoked opium as part of their cultural tradition. As unemployment among white workers increased, however, so did resentment of Chinese laborers. Attacking the use of opium became a convenient means of attacking the Chinese, and in 1877, Nevada became the first of many states to prohibit opium use. As Morgan (1978) observed:

> The first opium laws in California were not the result of a moral crusade against the drug itself. Instead, it represented a coercive action directed against a vice that was merely an appendage of the real menace—the Chinese—and not the Chinese per se, but the laboring "Chinamen" who threatened the economic security of the white working class. (p. 59)

The criminalization of other drugs, including cocaine, heroin, and marijuana, follows similar patterns of social control of the powerless, political opponents, and/or minorities. In the 1940s, marijuana was used primarily by minority group members, and users faced severe criminal penalties. However, after white, middle-class college students began to use marijuana in the 1970s, the government reduced the penalties associated with its use. Today, those college students are doctors, lawyers, nurses, and business professionals, and the legalization of marijuana for recreational use has occurred in four states and the District of Columbia, with other ballot initiatives yet to be decided. Although the nature and pharmacological properties of the drug have not changed, the population of users is now connected to power and influence. Thus, conflict theorists regard the regulation of certain drugs, as well as drug use itself, as a reflection of differences in the political, economic, and social power of various interest groups.

Symbolic Interactionist Perspective

Symbolic interactionism, which emphasizes the importance of definitions and labeling, concentrates on the social meanings associated with drug use. If the initial drug use experience is defined as pleasurable, it is likely to recur, and the individual may earn the label of "drug user" over time. If this definition is internalized so that the individual assumes an identity of a drug user, the behavior will probably continue and may even escalate. Conversely, Copes, Hochstetler, and Williams (2008) observed that respondents who self-identified as "hustlers" rather than "crack-heads"

Advertising Archives

Advertising for tobacco products has a long history of targeting minorities. Here, a magazine article from the early 1900s appeals to traditional female concerns about femininity, beauty, and body image.

were less likely to fall prey to the debilitating effects of the drug, for "[s]lipping into uncontrollable addiction is antithetical to the hustler identity" (p. 256).

Drug use is also learned through symbolic interaction in small groups. In a study of more than 7,000 adolescents in 231 different schools in Australia, researchers found that alcohol use is predicted by, regardless of grade or age, the number of peers who consume alcohol (Kelly et al. 2012). Furthermore, in a survey of 12- to 17-year-olds, 75 percent report that "seeing pictures on social networking sites like Facebook and MySpace of kids partying with alcohol and marijuana encourages other teens to want to party like that" (National Center on Addiction and Substance Abuse [CASA] 2012, p. 8). There is also evidence that through viewing peers, first-time users learn not only the motivations for drug use and the techniques but also what to experience. Becker (1966) explained how marijuana users learn to ingest the drug. A novice being coached by a regular user reported the experience:

> I was smoking like I did an ordinary cigarette. He said, "No, don't do it like that." He said, "Suck it, you know, draw in and hold it in your lungs...for a period of time." I said, "Is there any limit of time to hold it?" He said, "No, just till you feel that you want to let it out, let it out." So I did that three or four times. (p. 47)

Marijuana users learn not only how to ingest the smoke but also how to label the experience positively. When peers define certain drugs, behaviors, and experiences as not only acceptable but also pleasurable, drug use is likely to continue.

Interactionists also emphasize that symbols can be manipulated and used for political and economic agendas. The popular D.A.R.E. (Drug Abuse Resistance Education) program, with its antidrug emphasis fostered by local schools and police, carried a powerful symbolic value with which politicians want the public to identify. Ironically, research has consistently shown that the D.A.R.E program, as originally conceived, did not significantly prevent drug use among school-aged children. The D.A.R.E. curriculum was modified in 2009, and by 2013, both middle and elementary school students were required to participate in the newer version of the program. Evaluation of the program indicates "reduced substance abuse and maintained anti-drug attitudes over time among students in early trials" (Nordrum 2014).

Biological and Psychological Theories

Drug use and addiction are the result of a complex interplay of social, psychological, and biological forces. For example, some researchers suggest that drug use and addiction are caused by a "bio-behavioral disorder," which combines biological and psychological factors (Margolis and Zweben 2011). Biological research has primarily concentrated on the role of genetics in predisposing an individual to drug use. For example, several genes determine how quickly alcohol is metabolized and the extent to which acetaldehyde, a toxic byproduct of alcohol, is absorbed. If a person has high levels of acetaldehyde in their body, alcohol consumption will lead to negative side effects, thus reducing the likelihood of alcohol abuse (National Institute on Alcohol Abuse and Alcoholism [NIAAA] 2013).

Research also indicates that severe, early-onset alcoholism may be genetically predisposed, with some men having 10 times the risk of addiction as those without a genetic predisposition. Interestingly, other problems such as depression, chronic anxiety, and attention deficit disorder are also linked to the likelihood of addiction. Nonetheless, researchers warn "that genes alone do not determine our destiny—lifestyle choices and other environmental factors have a substantial impact" (NIAAA 2013, p. 4).

Psychological explanations focus on the tendency of certain personality types to be more susceptible to drug use. Individuals who are particularly prone to anxiety may be more likely to use drugs as a way to "self-medicate." Research indicates that child abuse or neglect, particularly among females, contributes to alcohol and drug abuse that extends into adulthood (Gilbert et al. 2009). Research also indicates that stressful life events (e.g., death of a loved one) also impact the onset of illicit drug use.

Frequently Used Legal Drugs

Social definitions regarding which drugs are legal or illegal vary over time, circumstance, and societal forces. In the United States, two of the most dangerous and widely abused drugs, alcohol and tobacco, are legal. Compare, for example, the number of people who report past-month use (i.e., current use) of various illicit drugs, 24.6 million, to the 66.9 million current tobacco users and the 136.9 million current alcohol drinkers (NSDUH 2014).

Alcohol: The Drug of Choice

Americans' relationship with alcohol has a long and varied history, much longer than once thought. New research indicates that humans have been consuming alcohol, although often in the form of fermented fruit, for well over 10 million years and not the earlier estimate of 10,000 years (Carrigan et al. 2014). Throughout much of this time, "attitudes toward drinking were characterized by a continued recognition of the positive nature of moderate consumption" (Hanson 1997, n.p.). However, in the 18th century, concerns about excessive alcohol consumption and the negative effects of drinking gave rise to the emergence of a temperance movement (Hanson 2013).

In the United States, by 1920, the federal government had prohibited the manufacture, sale, and distribution of alcohol through the passage of the Eighteenth Amendment to the U.S. Constitution. Many have argued that Prohibition, like the opium regulations of the late 1800s, was in fact a "moral crusade" (Gusfield 1963) against immigrant groups who were more likely to use alcohol. The amendment had little popular support and was repealed in 1933. Today, the U.S. population is experiencing a resurgence of concern about alcohol. What has been called a "new temperance" has manifested itself in federally mandated 21-year-old drinking age laws, warning labels on alcohol bottles, increased concern over fetal alcohol syndrome and underage drinking, stricter enforcement of drinking and driving regulations (e.g., checkpoint traffic stops), and zero-tolerance policies. Such practices may have had an effect on drinking norms, particularly for young people. Between 2002 and 2013, the rate of current alcohol use by 12- to 17-year-olds steadily declined, as did the number of 8th, 10th, and 12th graders who report ever being drunk (MTF 2014; NSDUH 2014).

Despite such restrictive policies, alcohol remains the most widely used and abused drug in the United States. According to a recent poll, 64 percent of U.S. adults drink alcohol, preferring beer to wine, and wine to liquor (Saad 2014). Although most people who drink alcohol do so moderately and experience few negative effects (see this chapter's *Self and Society* feature), people with alcoholism are psychologically and physically addicted to alcohol and suffer various degrees of physical, economic, psychological, and personal harm.

The National Survey on Drug Use and Health, conducted by the U.S. Department of Health and Human Services, reported that in 2013 about half of Americans aged 12 and older consumed alcohol at least once in the month preceding the survey; that is, they were *current users* (NSDUH 2014). Of this number, 6.3 percent reported **heavy drinking**, and 22.9 percent—60.1 million people—reported **binge drinking**.

Even more troubling were the nearly nine million current users of alcohol who were 12 to 20 years old—underage drinkers—many of whom got their alcohol for free from adults—often their parents. Just over 14.0 percent were binge drinkers, and 3.7 percent were heavy drinkers (NSDUH 2014). Demographically, white males from the metropolitan areas of the Northeast of the United States were the most likely to be underage drinkers. Drinking most often occurred in social settings—that is, in groups of two or more people in their own home or in the home of someone else (NSDUH 2014).

Although teen drinking has decreased in recent years, in part as a result of reduced perceived availability of alcohol (MTF 2015), binge drinking in college continues to attract the public's attention. The likelihood of a college student binge drinking is impacted by environmental variables including place of residence (e.g., on campus versus off campus); cost and availability of alcohol; campus, local, and state alcohol policies; age, gender, and ethnic and racial makeup of the student population; prevention strategies; and the college drinking culture (Wechsler and Nelson 2008). Moreover, research indicates that college students in fraternities and sororities drink more than non-Greek students (Chauvin 2012).

> Research indicates that the younger the age of onset, the higher the probability that an individual will develop a drinking disorder at some time in his or her life.

heavy drinking As defined by the U.S. Department of Health and Human Services, five or more drinks on the same occasion on each of five or more days in the past 30 days prior to the National Survey on Drug Use and Health.

binge drinking As defined by the U.S. Department of Health and Human Services, drinking five or more drinks on the same occasion on at least one day in the past 30 days prior to the National Survey on Drug Use and Health.

The Consequences of Alcohol Consumption

Indicate whether you have or have not experienced any of the following in the last 12 months as a consequence of your own drinking. When finished, compare your responses to those of a national sample of college students.

Consequence	Yes	No
1. Did something you later regretted	____	____
2. Forgot where you were or what you did	____	____
3. Got in trouble with the police	____	____
4. Someone had sex with me without my consent.	____	____
5. Had sex with someone without their consent	____	____
6. Had unprotected sex	____	____
7. Physically injured yourself	____	____
8. Physically injured another person	____	____
9. Seriously considered suicide	____	____
10. Reported one or more of the above	____	____

These survey items are from the American College Health Association's (ACHA) National College Health Assessment II (2014). The following data are from 2014. All students are from public or private, two- or four-year colleges and universities. The average age of respondents was 22, 75 percent of the sample was white, 64 percent of the sample was female, 40.0 percent lived in campus residence halls, and 88.2 percent were single at the time of the survey (ACHA 2014).*

Percent (%)	Male	Female	Total
1. Did something you later regretted	36.5	36.6	36.5
2. Forgot where you were or what you did	34.0	31.5	32.3
3. Got in trouble with the police	4.2	2.3	3.0
4. Someone had sex with me without my consent.	1.2	2.6	2.1
5. Had sex with someone without their consent	0.6	0.5	0.6
6. Had unprotected sex	22.3	19.4	20.4
7. Physically injured yourself	16.2	14.2	14.9
8. Physically injured another person	2.9	1.3	1.8
9. Seriously considered suicide	2.6	2.4	2.5
10. *Reported one or more of the above*	55.3	52.5	53.4

*Students responding "N/A, don't drink" were excluded from this analysis.

SOURCE: American College Health Association, American College Health Association National College Health Assessment II. Reference Group Executive Summary Spring 2014. Hanover, MD: American College Health Association.

Many binge drinkers began drinking in high school, with almost one-third having their first drink before age 13. Research indicates that the younger the age of onset, the higher the probability that an individual will develop a drinking disorder at some time in his or her life (Behrendt et al. 2009; Blomeyer et al. 2013; Hingson et al. 2006). For example, an individual's chance of becoming dependent on alcohol is 40.0 percent if the person's drinking began before the age of 13. As drinking levels increase from abstinence to heavy alcohol use, the likelihood of using illegal drugs or tobacco products also increases (see Figure 3.2).

More males than females aged 12 to 20 report binge drinking, heavy drinking, and current alcohol use. Researchers have long questioned the relationship between gender and drinking behavior, including binge drinking. After administering two measures of gender identity to a sample of college students, Peralta et al. (2010) concluded, "[M]ales, who are socialized to be masculine, may rely on heavy alcohol use to coincide with other forms of male-associated behaviors (e.g., sports, risk-taking). Women, who are socialized to be feminine, on the other hand, may not engage in heavy drinking practices because this is not a part of normative femininity expression rituals" (p. 377).

WHAT do you THINK?

It's clear that drinking norms vary between males and females, but some evidence indicates that may be changing. A report by the Centers for Disease Control and Prevention (CDC) indicates that one in five high school girls report binge drinking, and one in eight adult women report binge drinking, the latter averaging three binges a month and six drinks per binge (CDC 2013b). Why do you think binge drinking has increased among women and girls?

The Tobacco Crisis

Native Americans first cultivated tobacco and introduced it to the European settlers in the 1500s. The Europeans believed that tobacco had medicinal properties, and its use spread throughout Europe, ensuring the economic success of the colonies in the New World. Tobacco was initially used primarily through chewing and snuffing, but smoking became more popular in time, even though scientific evidence that linked tobacco smoking to lung cancer existed as early as 1859 (Feagin and Feagin 1994). However, the U.S. surgeon general did not conclude that tobacco products are addictive and that nicotine causes dependency until 1989.

Globally, over 80 percent of the nearly one billion smokers in the world live in low- or middle-income countries (WHO 2013b). Using what

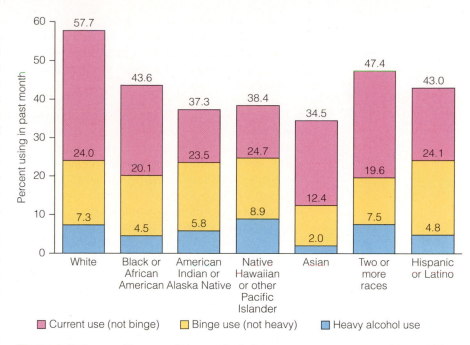

Figure 3.2 Current, Binge, and Heavy Alcohol Use among Persons Aged 12 or Older, by Race/Ethnicity, 2013
SOURCE: NSDUH 2014.

is called **meta-analysis**, researchers examined 125 scientific papers that included 31,146,096 respondents worldwide (WHO 2011b). The results are counterintuitive. Despite the obvious additional cost of using tobacco products, the analysis found strong evidence of an inverse relationship between income and smoking; that is, as income goes down, the prevalence of smoking goes up—a relationship that is stronger for younger age groups than older. The authors suggest a four-stage explanatory model:

> In earlier stages, smoking disseminates among higher-income groups who are more open to innovation. During the intermediate stages, smoking diffuses to the rest of the population. Later, smoking declines among the high-income level strata, as they are concerned with health, fitness, and the harm of smoking. Only after a long history of cigarette consumption, when all SES [socioeconomic status] groups have been similarly exposed to smoking, does the inverse social status gradient emerge. (p. 27)

Tobacco is one of the most widely used drugs in the United States. According to a U.S. Department of Health and Human Services survey, 66.9 million Americans—25.5 percent of those 12 and older—were current tobacco users in 2013 (NSDUH 2014). Current use of all tobacco products (e.g., cigarettes, smokeless tobacco, cigars) is higher for high school graduates than for college graduates, for unemployed rather than employed, and males, Americans Indians, and Alaska Natives. In general, tobacco use has steadily decreased since 2002, and cigarette smoking for 8th, 10th, and 12th graders is at an all-time low, falling by nearly 50 percent in the last five years (MTF 2015).

Alternative Nicotine Delivery Systems. While cigarette use among young people has steadily declined, the consumption of nicotine through two less traditional devices has become popular. For example, the percent of high school seniors reporting hookah use (i.e., a water pipe) in the previous year has increased dramatically, from 17.1 percent in 2010 to 22.9 percent in 2014 (MTF 2015). However, research indicates that the frequency of annual use is fairly low, with only 14 percent of 12th graders indicating use on more than two occasions during the 12 months prior to the survey (MTF 2014). Furthermore, unlike cigarette consumption, male and female hookah use is statistically the same.

E-cigarettes are battery-operated devices that produce a vapor that contains nicotine, which is then inhaled. In 2014, more 8th, 10th, and 12th graders used e-cigarettes than conventional cigarettes or any other tobacco product (MTF 2015). For

meta-analysis Meta-analysis combines the results of several studies addressing a research question—it is the analysis of analyses.

e-cigarettes Battery-operated devices that produce a vapor that contains nicotine, which can then be inhaled.

example, 16 percent of 10th graders reported using e-cigarettes compared to 7 percent using conventional cigarettes. Young people's increased concerns about the safety of conventional cigarettes is, in part, responsible for the decade-old decrease in smoking. Although the safety of e-cigarettes remains unknown, only 15 percent of 8th grade students believe they are harmful.

Tobacco Advertising. Research evidence suggests that youth develop attitudes and beliefs about tobacco products at an early age (Freeman et al. 2005). Advertising of tobacco products continues to have an influence on youth despite the 2009 Family Smoking Prevention and Tobacco Control Act, which, among other provisions, outlawed flavored cigarettes most often marketed to children (see "Government Regulation" in the "Strategies for Action" section). Tobacco company executives, however, have argued that the 2009 act only covers "cigarettes, cigarette tobacco, roll-your-own tobacco, and smokeless tobacco," and that cigars are excluded from the control of the Food and Drug Administration (Myers 2011, p. 1). Several tobacco companies are now selling small, inexpensive, sweet-flavored cigars and, between 2000 and 2012, cigar use more than doubled (Campaign for Tobacco Free Kids 2013).

There are also concerns regarding e-cigarette advertising and its impact on young people. Research commissioned by the American Legacy Foundation (2014) indicates that "teens and young adults are heavily exposed to e-cigarette messaging." Research also shows that exposure to e-cigarette advertising in television has increased dramatically in recent years—a 256 percent increase for 12- to 17-year-olds between 2011 and 2013 (Duke et al. 2014). During the same time period, the number of never smoking teens who reported using e-cigarettes tripled. The use of e-cigarettes was also found to be predictive of intentions to smoke conventional cigarettes (Bunnell et al. 2014).

There is also considerable evidence that tobacco advertisers target minorities. Primack et al. (2007) found that tobacco advertisements in African American communities were 2.6 times higher per person than in white communities. In addition, the likelihood of tobacco-related billboards was 70 percent higher in African American than in white communities. Henrisken et al. (2011) report that menthol cigarette promotions are higher and average price per pack lower in African American neighborhoods. Menthol cigarettes, often billed as a healthier alternative to nonmenthol cigarettes, increase the likelihood of smoking initiation, are more addictive, and are more difficult to quit using than nonmenthol cigarettes (Food and Drug Administration [FDA] 2013b). Menthol cigarettes are disproportionately consumed by racial and ethnic minorities, younger and less educated smokers, and women and girls.

The tobacco industry also has a long history of targeting women (Schmidt 2014). Researchers at Stanford University collected more than 1,500 cigarette advertisements from historical and contemporary magazines and newspapers. They concluded that marketing cigarettes to women and, today, girls has always been tied to body image and the evolving role of women. For example, during the second wave of feminism in the 1960s, Philip Morris's "You've Come a Long Way, Baby" campaign was developed for Virginia Slims. One researcher noted that, even today, "women-targeted cigarette brands are almost universally promoted as slender, thin, slim, lean, or light. Some brands

Diedra Laird/Charlotte Observer/Tribune News Service/Getty Images

Despite the growing popularity of e-cigarettes, many experts are concerned about the lack of research on their health impacts. Additionally, e-cigarette smokers, particularly young e-cigarette smokers, are now replacing the liquid nicotine with marijuana oil, all but eliminating telltale odor of the drug. Consequently, marijuana smoking is now taking place in locations previously unheard of including schools, malls, public walkways, and restaurants where electronic cigarette smoking is permitted.

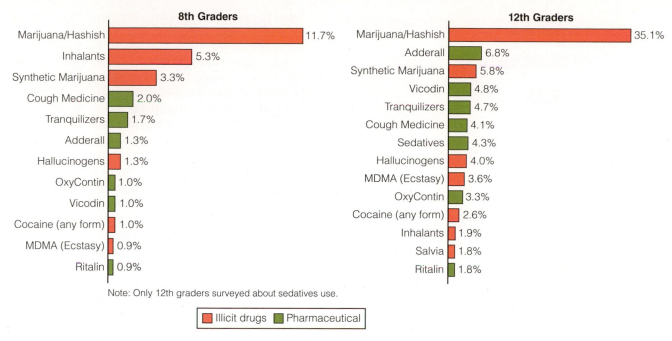

Figure 3.3 Top Drugs among 8th and 12th Graders, Past Year Use, 2014
SOURCE: MTF 2015.

have even gone so far as to recommend 'cigarette diets'" (quoted in Marine-Street 2012, p. 1). Women in developing countries are also being targeted.

Prescription Drugs

Like alcohol, prescription drugs are legal but become part of the drug abuse conversation when misused. Worldwide, the misuse of prescription drugs is a growing health problem (WDR 2014) despite the fact that lifetime use of **psychotherapeutic drugs**, sometimes called pharmaceuticals (i.e., nonmedical use of any prescription pain reliever, stimulant, sedative, or tranquilizer) has remained fairly stable since 2002 (NSDUH 2014). Over half of the U.S. population 12 and older who reported abuse of pain relievers, the largest single category of abused prescription drugs, reported receiving the drug from a friend or relative for free (NSDUH 2014). Of those who received the drug from a friend or relative for free, 83.8 percent of the friends or relatives reported obtaining the pills from a physician.

Figure 3.3 graphically portrays the seriousness of pharmaceutical use among today's teenagers. Of the top 12 drugs used by 8th and 12th graders in 2013, at least half of them are pharmaceuticals: cough medicine, tranquilizers, Adderall, OxyContin, Vicodin, and Ritalin, (MTF 2015). It is also notable that, with the exception of inhalants, the use of each of the drugs listed doubles, and in some cases triples, between 8th and 12th grade.

Frequently Used Illegal Drugs

More than 24.6 million people in the United States are current users of illegal drugs including, for measurement purposes, psychotherapeutics (see "Prescription Drugs"), representing 9.4 percent of the population of aged 12 and older (NSDUH 2014). Users of illegal drugs, although varying by type of drug used, are more likely to be male, young, and a member of a minority group (NSDUH 2014). According to the most recent Monitoring the Future survey, over the last several decades use of any illicit drug in the past year has generally declined for 8th, 10th, and 12th grade students (MTF 2015). Lifetime use of illegal drugs by European teens, about one-quarter of 15- and 16 year-olds is, as in the United States, higher among males than females, and it is highest for cannabis consumption (EMCDDA 2014a).

psychotherapeutic drugs
The nonmedical use of any prescription pain reliever, stimulant, sedative, or tranquilizer.

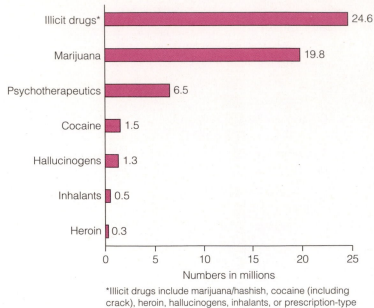

Illicit drugs*	24.6
Marijuana	19.8
Psychotherapeutics	6.5
Cocaine	1.5
Hallucinogens	1.3
Inhalants	0.5
Heroin	0.3

Numbers in millions

*Illicit drugs include marijuana/hashish, cocaine (including crack), heroin, hallucinogens, inhalants, or prescription-type psychotherapeutics used nonmedically.

Figure 3.4 Past-Month Illicit Drug Use among U.S. Residents Aged 12 or Older, 2013
SOURCE: NSDUH 2014.

Marijuana Madness

Marijuana is the most commonly used and most heavily trafficked illicit drug in the world. It is also the most commonly abused drug (see Figure 3.4). Globally, it is estimated that there are between 177 and 227 million marijuana users, representing nearly 4.0 percent of the world's 15- to 64-year-old population. Regionally, marijuana is also the most dominant illicit drug, and its consumption is particularly high in West and Central Africa, Oceania, and North America, and to a lesser extent Western and Central Europe, where its use seems to be decreasing (WDR 2014).

Marijuana's active ingredient is THC (Δ^9-tetra-hydrocannabinol), which in varying amounts can act as a sedative or a hallucinogen. When just the top of the marijuana plant is sold, it is called hashish. Hashish is much more potent than marijuana, which comes from the entire plant. Marijuana use dates back to 2737 B.C. in China, and marijuana has a long tradition of use in India, the Middle East, and Europe. In North America, hemp, as it was then called, was used to make rope and as a treatment for various ailments. Nevertheless, in 1937, Congress passed the Marijuana Tax Act that restricted the use of marijuana; the law was passed as a result of a media campaign that portrayed marijuana users as "dope fiends." In the only congressional hearing on the act, the then commissioner of narcotics testified:

> There are 100,000 total marijuana users in the U.S. and most are Negroes, Hispanics, Filipinos, and entertainers. Their Satanic music, jazz and swing, result from marijuana use. This marijuana causes white women to seek sexual relations with Negroes, entertainers, and any others. The primary reason to outlaw marijuana is its effect on degenerative races. Marijuana is an addictive drug which produces in its users insanity, criminality, and death. You smoke a joint and you're likely to kill your brother. Marijuana is the most violence-causing drug in the history of mankind. (quoted in Rabinowitz and Lurigio 2009)

More than 19.8 million current marijuana users live in the United States, representing 7.5 percent of the U.S. population aged 12 and older (NSDUH 2014). The number of Americans who use marijuana daily or almost daily has increased over the last decade from 3.1 percent in 2003 to 5.7 percent in 2013. According to the most recent Monitoring the Future survey, after five years of increases, marijuana use declined among teens in 2014 (MTF 2015).

In the present debate over the legalization of marijuana, as with cigarettes and alcohol, many express fears that it is a **gateway drug**, the use of which causes progression to other drugs. Most research, however, suggests that people who experiment with one drug are more likely to experiment with another. Indeed, most drug users use several drugs concurrently, most commonly cigarettes, alcohol, marijuana, and cocaine (Lee and Abdel-Ghany 2004).

Cocaine: From Coca-Cola to Crack

Cocaine is classified as a stimulant and, as such, produces feelings of excitation, alertness, and euphoria. Although prescription stimulants such as methamphetamine and dextroamphetamine are commonly abused, societal concern over drug abuse has focused on cocaine over the last 20 years. Its increased use, addictive qualities, physiological effects, and worldwide distribution have fueled such concerns. More than any other single substance, cocaine led to the early phases of the war on drugs.

gateway drug A drug (e.g., marijuana) that is believed to lead to the use of other drugs (e.g., cocaine).

Cocaine, which is made from the coca plant, has been used for thousands of years. Coca leaves were used in the original formula for Coca-Cola, but in the early 1900s, anti-cocaine sentiment emerged as a response to the heavy use of cocaine among urban blacks, poor whites, and criminals (Friedman-Rudovsky 2009; Thio 2007; Witters et al. 1992). Cocaine was outlawed in 1914, by the Harrison Narcotics Tax Act, but its use and effects continued to be misunderstood. For example, a 1982 *Scientific American* article suggested that cocaine was no more habit forming than potato chips (Van Dyck and Byck 1982). As demand and then supply increased, prices fell from $100 a dose to $10 a dose, and "from 1978 to 1987, the U.S. experienced the largest cocaine epidemic in history" (Witters et al. 1992, p. 256).

According to the National Survey on Drug Use and Health, 1.5 million Americans 12 years and older are current cocaine (including crack) users, representing 0.6 percent of that population (NSDUH 2014). **Crack** is a crystallized product made by boiling a mixture of baking soda, water, and cocaine. Over the last several decades, cocaine as well as crack use has declined for teenagers, as has the perceived risk of use and ease of availability (MTF 2015). In 2013, the rate of current cocaine use among 18- to 25-year-olds was just over 1 percent (NSDUH 2014).

Methamphetamine: The Meth Epidemic

Methamphetamine is a central nervous system stimulant that is highly addictive. Although the drug has only recently become popular, it is not new:

> During the Second World War, soldiers on both sides used it to reduce fatigue and enhance performance. Hitler was widely believed to be a meth addict. Later, in the 1960s, President John Kennedy also used the drug and soon it caught on among so-called "speed freaks." But, because it was extremely expensive as well as difficult to obtain, meth was never close to being as widely used as cocaine. (Thio 2007, p. 276)

In 2013, the number of persons 12 and older who used methamphetamine in the previous month was similar to that in 2012—about 0.2 percent (NSDUH 2014). Use of crystal meth (ice), the crystalline rather than powdered form of methamphetamine, is low in the general population as well as among secondary school and college students, with annual prevalence hovering around 1 percent (MTF 2015).

Because methamphetamine can be made from cold medications such as Sudafed, the U.S. Congress passed the Comprehensive Methamphetamine Control Act of 1996 that made obtaining the chemicals needed to make methamphetamine more difficult (ONDCP 2006; Thio 2007). In 2006, the Combat Methamphetamine Epidemic Act, which further articulated standards for selling over-the-counter medications used in methamphetamine production, went into effect. Since that time, several states have passed regulations that require a prescription to obtain ephedrine and pseudoephedrine, the key ingredients in methamphetamine, and with some success. The number of methamphetamine labs has declined in these states (Government Accounting Office [GAO] 2013).

> [S]everal states have passed regulations that require a prescription to obtain ephedrine and pseudoephedrine, the key ingredients in methamphetamine, and with some success. The number of methamphetamine labs has declined in these states.

Courtesy of Multnomah County Sheriff's Office

For those who use methamphetamine, the physical transformation is remarkable. The time lapse between the before (left) and after (right) pictures of this methamphetamine user is only three years, five months.

crack A crystallized illegal drug product produced by boiling a mixture of baking soda, water, and cocaine.

Heroin: The White Horse

Heroin is an analgesic—that is, a painkiller—and is one of a class of drugs known as opiates. Highly addictive, heroin can be injected, snorted, or smoked; and when used in conjunction with cocaine, it is called a "speedball." Use of the "white horse"—as it is often called—is highest in Afghanistan, a major opium-producing country, and Iran. Opium is trafficked from Afghanistan through the Balkan States and on to Western and Central Europe (WDR 2014).

In the United States, heroin is used by a relatively small proportion of the 12 and older population; varying by age, its use has remained fairly stable between 2009 and 2013 (NSDUH 2014). Compared to recent peak years, heroin use among secondary school students is down (NSDUH 2014); however, use among 18- to 24-year-olds has increased (Forliti et al. 2014).

Heroin use in the suburbs and among middle-class America has increased dramatically in recent years (Wiltz 2015), and heroin-related deaths by such celebrities as Phillip Seymour Hoffman, Peaches Geldhof, and Cory Monteith have further heightened the concerns. As OxyContin and other painkillers become more difficult to obtain, users often turn to the less expensive and more available heroin—and not by accident. Rather, it is the plan of drug lords from Mexico and Colombia, who strategically market the drug to Middle America with new, sophisticated techniques. Packets of heroin are now stamped with popular brand names like Chevrolet or Prada, or marketed using blockbuster movie names aimed at young people, like the *Twilight* series.

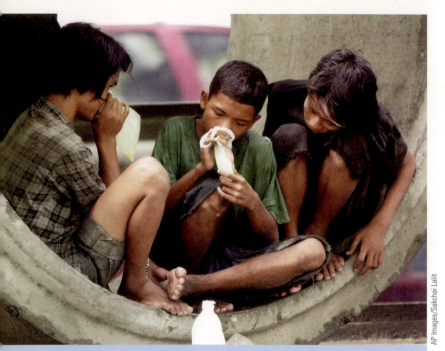

AP Images/Sakchai Lalit

Throughout much of the developing world, homeless children "huff" glue to escape hunger pains and the horrors of living on the street. Globally, UNICEF estimates that there are more than 200 million homeless children—the equivalent of two-thirds the population of the United States.

WHAT do you THINK?

In 2013, among the 12 and older population, inhalants were the third most common drug to initiate illicit drug use after marijuana and pain relievers (NSDUH 2014). Inhalants include adhesives (e.g., rubber cement), food products (e.g., vegetable cooking spray), aerosols (e.g., hair spray and air fresheners), anaesthetics (ether), gases (e.g., butane), and cleaning agents (e.g., spot remover)—more than 1,000 household products in total. "Huffing" is a serious problem and can lead to sudden sniffing death syndrome, but, nonetheless, the availability of inhalants cannot be legally controlled. What would you do to attack this potentially deadly practice?

Synthetic Drugs

Concern over **synthetic drugs**, a category of drugs that are "designed" in laboratories rather than naturally occurring in plant material, is growing and thus President Obama signed the Synthetic Drug Abuse Prevention Act of 2012. Although there are many types of synthetic drugs, some of the more popular are synthetic marijuana ("K2" or "Spice"), synthetic stimulants ("bath salts"), and synthetic hallucinogens (LSD, Ecstasy/MDMA).

Synthetic Marijuana. Synthetic marijuana is "the most common of an expanding array of [synthetic] drugs that mimic the effects of outlawed…substances but are based on manufactured and often legal compounds" (Paynter 2011). Shredded plant material is sprayed with questionable chemicals (Shuman 2014); and, because the plants and chemicals are legal, there is no oversight. The result is a global supply network enhanced by the Internet and valued at over $121 billion in North America alone.

synthetic drugs A category of drugs that are "designed" in laboratories rather than naturally occurring in plant material.

The use of synthetic marijuana is higher in the United States than in other parts of the world. For example, in a national survey in Spain, of 25,000 14- to 18-year-olds, past-year use of synthetic marijuana was a little under 1 percent (EMCDDA 2013). The percentage of U.S. 12th graders reporting the use of synthetic marijuana was 6 percent in 2014. Despite the lack of regulation and quality control, most secondary school students do not see synthetic marijuana as a dangerous drug (MTF 2015).

"Bath Salts." Often sold under such innocuous-sounding names as Bloom, Cloud Nine, Vanilla Sky, and White Lightening, bath salts are highly addictive synthetic stimulants. Use of bath salts by 8th, 10th, and 12th graders has decreased in recent years, to less than 1 percent for each grade level (MTF 2015). Calls to poison control centers about bath salts increased dramatically between 2010 and 2011 but have since declined although remaining above 2010 rates. Declines are likely to be the result of the Drug Enforcement Administration's (DEA) ban on some of the chemicals used to produce bath salts (McLaughlin 2012). Bath salts are usually snorted, orally or rectally ingested, injected, or smoked, and effects can last up eight hours (Elliott 2013).

Hallucinogens. Hallucinogen use among adults has decreased steadily since 2009 (NSDUH 2014). LSD is a synthetic hallucinogen that became popular in the 1960s and 1970s as part of the counterculture revolution. Use of LSD as well as the naturally produced hallucinogen salvia has declined among 8th, 10th, and 12th graders in recent years, as has the use of Ecstasy (MTF 2015). Recent investigations have revealed that between 1955 and 1975, U.S. military officials tested experimental drugs such as LSD on Army volunteers who were told "they would be test[ing] new Army field jackets, clothing, weapons, and things of that nature…no mention of drugs or chemicals" (quoted in Martin 2012).

Globally, over 18.8 million people between the ages of 15 and 64 use Ecstasy, the most common name for MDMA, at least once a year (WDR 2014). Although use of Ecstasy is declining worldwide, there is some evidence that its use is increasing throughout Asia (WDR 2014). In 2014, 2.3 percent of U.S. 10th graders reported past-year use of Ecstasy, the most common hallucinogen to be used (MTF 2015). "Molly," short for *molecule*, is thought to be a purer form of MDMA than Ecstasy, although as of this writing, its chemical makeup is unclear and there are no national surveys available to assess use rates. Molly users tend to be young, most often between 16 to 24 years old (Csomor 2012). There are concerns about media messages promoting the use of Molly among teenagers, including song lyrics by such artists as Kanye West and Lil Wayne, Nikki Minaj, and Miley Cyrus (Aleksander 2013; Arnowitz 2013).

Societal Consequences of Drug Use and Abuse

Drugs are a social problem not only because of their adverse effects on individuals but also because of the negative consequences their use has for society as a whole. Everyone is a victim of drug abuse. Drugs contribute to problems within the family and to escalating crime rates, are tremendously costly, and place a heavy strain on the environment. Drug abuse also has serious consequences for health at both the individual and the societal level.

The Cost to Children and Family

The cost of drug abuse to families is incalculable. In 2013, it was estimated that more than 20 million people were classified as substance abusers in the United States—8.2 percent of the 12 and older population—resulting in 1 in 10 children under the age of 18 living with a parent in need of treatment for drug or alcohol dependency (NSDUH 2014; Substance Abuse and Mental Health Services Administration [SAMHSA] 2009). Children raised in such homes are more likely to (1) live in an environment riddled with conflict, (2) have a higher probability of physical illness including injuries or death from an automobile accident, and (3) be victims of child abuse and neglect (SAMHSA 2007, 2009).

Children of alcoholics—the clear majority of children who live in homes with a drug-dependent parent(s)—are at risk for anxiety disorders, depression, and problems with cognitive and verbal skills (SAMHSA 2012). Children of alcoholics, and particularly female children of alcoholics, suffer from "significant mental health consequences...that persist far into adulthood" (Balsa et al. 2009, p. 55).

Additional harms as a result of excessive drinking in the home were identified by Berends et al. (2014) in a telephone survey of 2,649 randomly selected adults in Australia. Results indicate that about one-third of those reporting a problem drinker in the family were significantly impacted by physical (e.g., sexual coercion), social (e.g., negatively affected a social occasion), psychological (e.g., feeling threatened), and/or practical (e.g., being put at risk in a car) harms. Females were more likely to report being impacted than males, physical injury was more likely to be reported between close rather than extended family members, and "significant harm" was twice as likely to be reported by those living together in the 12 months prior to the survey than those living apart (Berends et al. 2014).

The family, however, is not the only environment in which drug use is harmful to children. A survey of more than 1,000 teens between the ages of 12 and 17 indicates that 32 percent of middle school children and 60 percent of high school students attend schools that are "drug infected," and attending "drug-infected" schools was associated with higher self-reported drug use. High levels of stress, the most common source being academic pressure, were also associated with higher student drug use. Not surprisingly, when asked, teens responded that the most important problem they face today is alcohol, tobacco, and illicit drug use (CASA 2012).

In Alabama a woman can be charged with "chemical endangerment" if she uses a controlled substance while pregnant (Alabama Code 2015). Do you think all states should have such a law? Should a pregnant woman be arrested for smoking? For drinking a glass of red wine? For refusing to take prenatal vitamins?

Crime and Drugs

An examination of the drug use among probated and paroled offenders indicates much higher use and dependence rates than in non-offender populations. In 2013, an estimated 1.7 million adults were released from prison on parole or some other type of supervised release program. Of that number, 27.4 percent were current illegal drug users and 34.3 reported being alcohol or drug dependent compared to 9.3 percent and 8.4, respectively, of the nonprison population (NSDUH 2014). Similarly, youth between the ages of 12 and 17 who reported being in a serious fight at work or at school were over twice as likely to report past month illegal drug use as their non-offending counterparts.

The relationship between crime and drug use, however, is complex. Sociologists disagree as to whether drugs actually "cause" crime or whether, instead, criminal activity leads to drug involvement. Alternatively, criminal involvement and drug use can occur at the same time; that is, someone can take drugs and commit crimes out of the desire to engage in risk-taking behaviors. Furthermore, because both crime and drug use are associated with low socioeconomic status, poverty may actually be the more powerful explanatory variable.

In addition to the hypothesized crime–drug use link, some criminal offenses are defined by the use of drugs: possession, cultivation, production, and sale of controlled substances; public intoxication; drunk and disorderly conduct; and driving while intoxicated. Driving while intoxicated is one of the most common drug-related crimes. In 2013, approximately 10.9 percent of the 12 and older population, over 28.7 million people, drove under the influence of alcohol at least once in the previous year (NSDUH 2014). In 2013, more than 10,076 Americans died in alcohol-related crashes—one-third of all traffic-related deaths (National Highway Traffic Safety Administration 2014).

Alcohol is not the only drug that impairs driving, nor are alcohol and other drugs necessarily consumed separately. Over a five-year period, researchers from DRUID (Driving Under the Influence of Drugs, Alcohols, and Medicines) studied impaired driving in Europe. Methodologically comprehensive, the researchers analyzed existing literature; conducted roadside surveys in 13 countries; sampled the general population in 15 countries; and analyzed body fluids from 50,000 randomly selected drivers, seriously injured drivers, and killed drivers (DRUID 2012). Key results indicate that (1) alcohol is the most common substance used by the general public and by those who were seriously injured or killed in a traffic accident, (2) marijuana and cocaine are the most frequently used illegal substances, (3) the majority of drivers who used illegal substances combined their use with alcohol, and (4) the combination of alcohol and illegal substances resulted in the *highest* risk factor combination followed by combinations of two or more illegal drugs.

The High Price of Alcohol and Other Drug Use

It is estimated that alcohol abuse costs the U.S. $185 billion annually, with much of that cost passed on to the public. Fifteen percent of that amount is for health and health-related expenses, and 70.0 percent is "due to reduced, lost and foregone earnings" while the "remainder is the cost of lost workforce productivity, accidents, violence, and premature death" (Research Society on Alcoholism 2011, p. 1).

Smoking-related spending by the government costs the American taxpayers over $70 billion annually—$616 per household (Campaign for Tobacco Free Kids 2013). A comprehensive report by the National Center on Addiction and Substance Abuse at Columbia University (CASA 2009) set the total annual cost of substance abuse and addiction in the United States at $467.7 billion. More importantly, the report contends that, for every dollar spent on drug abuse by federal and state governments,

> 95.6 cents went to shoveling up the wreckage and only 1.9 cents on prevention and treatment, 0.4 cents on research, 1.4 cents on taxation or regulation, and 0.7 cents on interdiction. Under any circumstances, spending more than 95 percent of taxpayer dollars on the consequences of tobacco, alcohol, and other drug abuse and addiction and less than 2 percent to relieve individuals and taxpayers of this burden would be considered a reckless misallocation of public funds. In these economic times, such upside-down-cake public policy is unconscionable. (p. i)

At the federal level, the cost of "shoveling up the wreckage" includes the cost of (1) health care due to substance abuse and addiction (the highest proportion of wreckage spending), (2) adult and juvenile crime (e.g., corrections), (3) child and family assistance programs (e.g., welfare), (4) education (e.g., Safe School Initiatives), (5) public safety (e.g., drug enforcement), (6) mental health and developmental disabilities (e.g., treatment of addiction), and (7) the federal workforce (e.g., loss of productivity). The report concludes that prevention programs must become a priority to reduce the economic costs of drug abuse.

Lastly, in a report entitled *Addressing Prescription Drug Abuse in the United States*, the Centers for Disease Control and Prevention (2013a) estimates that the health care costs of prescription drug abuse, specifically opioid drug abuse, is $72 billion annually. This expense, which is often passed on to the public, is comparable to the medical costs of asthma or HIV.

Physical and Mental Health Costs

The physical and mental health costs of alcohol and other drugs, unlike the economic costs, are incalculable—death, disease, and injury.

Alcohol, Tobacco, and Prescription Drugs. Annually, alcohol abuse is responsible for over 3.3 million deaths, nearly 6 percent of all deaths worldwide (WHO 2015), shortening the life expectancy of those who die by an average of 30 years (CDC 2014b). Alcohol kills more people than AIDS, tuberculosis, or violence and is responsible for 88,000 deaths annually in the United States (CDC 2014b; WHO 2011a). Excessive alcohol use is

TABLE 3.2 Premature Deaths Caused by Smoking and Exposure to Secondhand Smoke, 1965–2014

Cause of Death	Total
Smoking-related cancers	6,587,000
Cardiovascular and metabolic diseases	7,787,000
Pulmonary diseases	3,804,000
Conditions related to pregnancy and birth	108,000
Residential fires	86,000
Lung cancers caused by exposure to secondhand smoke	263,000
Coronary heart disease caused by exposure to secondhand smoke	2,194,000
Total	20,830,000

© Cengage Learning

SOURCE: U.S. Surgeon General 2014.

> Tobacco use is the leading preventable cause of disease and death in the world. Half of all tobacco users—6 million people—die each year from its use, and the World Health Organization warns that if something does not change soon, that number could increase to 8 million a year by 2030 (WHO 2014b).

associated with both short-term health effects such as an increased risk for violence, suicide, risky sexual behaviors, and accidents as well as long-term chronic health risks such as cirrhosis of the liver, cancer, hypertension, dementia, and psychological disorders.

Maternal prenatal alcohol use is associated with fetal alcohol spectrum disorders (FASDs) in children, a broad term used for a variety of preventable birth defects and developmental disabilities. **Fetal alcohol syndrome (FAS)** is the most serious of the FASDs and is characterized by serious mental and physical handicaps including facial deformities, growth problems, difficulty communicating, short attention spans, and hearing and vision problems, and speech and language delays (CDC 2011). In an investigation of the prevalence of FASD among a representative sample of first graders, May et al. (2014) conclude that the best predictors of FAS are (1) late recognition of pregnancy by the mother, (2) amount of alcohol consumed by the mother three months prior to pregnancy, and (3) the quantity of alcohol consumed by the father.

Tobacco use is the leading preventable cause of disease and death in the world. Half of all tobacco users—6 million people—die each year from its use, and the World Health Organization warns that if something does not change soon, that number could increase to 8 million a year by 2030 (WHO 2014b).

Americans who smoke cigarettes are more than twice as likely to develop coronary heart disease and/or to have a stroke, and both men and women have increased risks of developing lung cancer—23 times more likely and 13 times more likely, respectively (CDC 2012). It should be noted that the health impact of smoking goes beyond consumption and the effects of secondhand smoke (see Table 3.2). For example, children who work in tobacco fields are at risk for "green tobacco sickness," a disease caused by the absorption of nicotine through the skin from handling wet tobacco leaves (WHO 2014b). Additionally, smoking before and/or during pregnancy is linked to sudden infant death syndrome, pre-term delivery, congenital abnormalities, stillbirth, and spontaneous abortions (U.S. Department of Health and Human Services 2014). The health risks of e-cigarettes, as of yet, are unknown.

Prescription drug abuse also takes its toll. Drug overdoses have steadily increased over the last three decades and abuse of prescription drugs, specifically painkillers such as OxyContin, are in part responsible (CDC 2013a). Of overdose deaths in which a specific drug was identified, opioids were responsible for twice as many deaths in 2010 than in 1980. Furthermore, opioid-related "overdose deaths now outnumber overdose deaths involving all illicit drugs such as heroin and cocaine combined" (p. 3). Prescription drug abuse is also associated with increased visits to emergency rooms and higher admissions to substance abuse facilities.

Illegal Drugs. Globally, the use of illicit drugs (excluding the misuse of prescription drugs) was associated with an estimated 183,000 deaths in 2012 (WDR 2013). Of the world's 15- to 64-year-old population, illicit drug use accounts for 1 in 20 deaths in North America and Oceania, 1 in 100 deaths in Asia, 1 in 110 deaths in Europe, 1 in 150 deaths in Africa, and 1 in 200 deaths in South America (WDR 2013). Deaths from illicit drug use are disproportionately of relatively young people; 90 percent of overdose deaths in Europe are of people under the age of 25 (EMCDDA 2013). Illicit drug use is also associated with a variety of diseases. For example, injecting drug users are at a higher risk for hepatitis B, hepatitis C, and HIV (WDR 2014).

Marijuana is often considered by the public to be the least dangerous illicit drug, yet over the last decade there has been a global increase in the number of people seeking help for marijuana use, most notably in Europe, the Americas, and Oceania (WDR 2014).

fetal alcohol syndrome (FAS) A syndrome characterized by serious physical and mental handicaps as a result of maternal drinking during pregnancy.

Marijuana is addicting, as an estimated 9 percent of users become dependent, but most research suggests it is not as addicting as most commonly used drugs (Christensen and Wilson 2014). In fact, a majority of Americans believe alcohol is more dangerous to the public's health and society as a whole than marijuana (Pew 2014).

That is not to say that marijuana use is without risks. Because the tar in marijuana has a higher concentration of the chemicals linked to lung cancer, smoking marijuana is more dangerous than smoking an equivalent amount of tobacco. In addition, there has been an increase in the number of studies documenting the relationship between habitual marijuana use and structural changes of the brain (Filbey et al. 2014), psychosis in vulnerable populations (Pierre 2011), and reduced intellectual functioning (Nordqvist 2012). As a result of the debate over the legalization of marijuana, the amount of the drug supplied by the U.S. government to researchers has more than tripled (Gaffney 2015).

The Cost of Drug Use on the Environment

Although not something usually considered, the production of illegal drugs has a tremendous impact on the environment. The production of methamphetamine serves as an example. According to the Drug Enforcement Administration (DEA):

> For every pound of meth produced, 5 to 6 pounds of toxic waste are produced. Common practices by meth lab operators include dumping this waste into bath-tubs, sinks, or toilets, or outside on surrounding grounds or along roads and creeks. Some may place the waste in household or commercial trash or store it on the property. In addition to dumped waste, toxic vapors from the chemicals used and the meth-making process can permeate walls and ceilings of a home or building or the interior of a vehicle, potentially exposing unsuspecting occupants. (GAO 2013, p. 19)

The cultivation of marijuana, cocaine, and opium has an additional environmental impact. For example, the Colombian government estimates that from 1988 to 2008, nearly 5.4 million acres of rainforest, an area the size of New Jersey, were destroyed because of illegal drug production. Furthermore, studies have shown that as much as 25 percent of the deforestation that takes place in Peru is "associated with clear-cutting and burning for planting coca bushes" (DEA 2010, p. 1).

In the United States and Mexico, outdoor cannabis cultivation leads to contaminated water, clear-cutting of natural vegetation, the disposable of nonbiodegradable materials, and the diversion of natural waterways often polluted with toxic chemicals, which endanger fish and other wildlife (DEA 2014; U.S. Department of Justice 2010). Drug eradication also takes an environmental toll. Monsanto's Roundup, often used in aerial eradication of Colombia's coca fields, leads to "deforestation, contamination of water and water systems, eradication of non-coca crops and natural vegetation, and a generally negative impact on the biodiversity of the region" (Smith et al. 2014, p. 195).

Treatment Alternatives

Historically, Americans have held a punitive rather than rehabilitative attitude on how to deal with America's drug problem, particularly in reference to "hard drugs" such as heroin and cocaine. Recently, however, attitudes have shifted. Across gender, age, race, ethnicity, education, and political party affiliation, a majority of Americans now believe that the government should focus more on providing treatment for drug users than prosecution (Pew 2014).

In 2013, an estimated 21.6 million Americans aged 12 or older were classified as either dependent on or abusers of alcohol or illicit drugs (see this chapter's *The Human Side*). Of that number, 14.7 million were dependent on or abusers of alcohol only. Marijuana was the most common illicit drug to be dependent on or abuse, followed by

Real Stories, Real People

For most of us, why people begin to use drugs, why they don't stop using them, and the effects of drugs on their lives is difficult to understand. The following "real-life stories" are from those who abuse alcohol and other drugs and the people who care about them. Although not a random sample of users, their stories are not dissimilar to others you might have heard.

I didn't think I had a "drug problem"—I was buying the tablets at the chemist [drug store]. It didn't affect my work. I would feel a bit tired in the mornings, but nothing more. The fact that I had a problem came to a head when I took an overdose of about forty tablets and found myself in the hospital. I spent twelve weeks in the clinic conquering my addiction.—Alex

My friend was on drugs for four years, three of which were on hard drugs such as cocaine, LSD, morphine, and many antidepressants and painkillers. Actually anything he could get his hands on. He complained all the time of terrible pains in his body and he just got worse and worse till he finally went to see a doctor.... The doctor told him that there was nothing that could be done for him and that due to the deterioration of his body, he would not live much longer. Within days—he was dead.—Wayne, friend

My goal in life wasn't living...it was getting high. I was falling in a downward spiral towards a point of no return. Over the years, I turned to cocaine, marijuana, and alcohol under a false belief it would allow me to escape my problems. It just made things worse. I had everything, a good job, money, a loving family, yet I felt so empty inside. As if I had nothing. Over twenty years of using, I kept saying to myself, I'm going to stop permanently after using this last time. It never happened. There were even moments I had thought of giving up on life.—John

I lived with a crack addict for nearly a year. I loved that addict—who was my boyfriend—with all my heart, but I couldn't stick [with] it any more.... My ex stole incessantly and couldn't tear himself away from his pipe. I think crack is more evil than heroin—one pipe can be all it takes to turn you into an immoral monster.—Audrey, girlfriend

I was given my first joint in the playground of my school. I'm a heroin addict now, and I've just finished my eighth treatment for drug addiction.—Christian

Ecstasy made me crazy. One day I bit glass, just like I would have bitten an apple. I had to have my mouth full of pieces of glass to realize what was happening to me. Another time, I tore rags with my teeth for an hour.—Ann

When I was thirteen, friends would make fun of me if I didn't have a drink. I just gave in because it was easier to join the crowd....I was really unhappy and just drank to escape my life. I went out less and less, so started losing friends. The more lonely I got, the more I drank. I was violent and out of control. I never knew what I was doing. I was ripping my family apart....Kicked out of my home at age sixteen, I was homeless and started begging for money to buy drinks. After years of abuse, doctors told me there was irreparable harm to my health....I was only sixteen but my liver was badly damaged and I was close to killing myself from everything I was drinking.—Samantha

I started peeing blood. I felt sick....My body felt weak.... I gave up everything because I was obsessed with using....All I cared about was getting high.... I thought I could just use Coricidin for fun, that it didn't matter. I never expected to get hooked....I'll never be able to get that time back. If I could erase it and make it go away, I would.—Charlie

Jason had been at a friend's house, sniffing glue or lighter fluid, maybe both. On the way back to school, Jason kept blacking out. Finally, he fell and never got up. By the time we were able to get him to the hospital, it was too late.—Cathy, parent

SOURCE: Foundation for a Drug Free World 2013.

pain relievers and cocaine. Those with alcohol and/or drug dependence were more likely to be males, American Indians or Alaska natives, high school dropouts, unemployed, and with a history of criminal justice system involvement (NSDUH 2014). In 2013, 4.1 million Americans 12 and older received treatment for alcohol or illicit drug use.

Individuals who are interested in overcoming problem drug use have a number of available treatment alternatives (see Figure 3.5). Some options include family therapy, counseling, private and state treatment facilities, community care programs, behavior modification, drug maintenance programs, and employee assistance programs. Some argue that pharmacotherapy (i.e., use of treatment medications) holds the most promising solutions to drug abuse. For example, an anti-seizure drug, ezogabine, may help alcoholics stop drinking by blocking the rewarding effects of alcohol consumption (Knapp et al. 2014). Similarly, scientists are working on "vaccines" that, although not preventing addiction, would block the effect of a drug by preventing it from reaching the brain (Sifferlin 2015). Three commonly used techniques are inpatient or outpatient treatment, peer support groups, and drug courts.

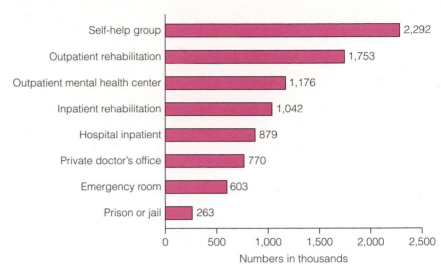

Figure 3.5 Locations Where Past Year Substance Abuse Treatment Was Received among Persons Aged 12 or Older, 2013
SOURCE: NSDUH 2014.

Inpatient and Outpatient Treatment

Inpatient treatment refers to treatment of drug dependence in a hospital and, most importantly, includes medical supervision of detoxification. Most inpatient programs last between 30 and 90 days and target individuals whose withdrawal symptoms require close monitoring (e.g., alcoholics, cocaine addicts). Some drug-dependent patients, however, can be safely treated as outpatients. Outpatient treatment allows individuals to remain in their home and work environments and is often less expensive. In outpatient treatment, patients are under the care of a physician who evaluates their progress regularly, prescribes needed medication, and watches for signs of a relapse.

The extent to which treatment is successful depends on a number of variables. A study of the effectiveness of inpatient and outpatient treatment of nearly 400 parole violators who opted for a year-long substance abuse program rather than returning to prison found that five variables predicted failure to complete the program: The participants (1) had a history of significant problems with their mothers, (2) had problems with their sexual partners in the 30 days prior to admission to the program, (3) had longer periods of incarceration, (4) had used heroin in the 30 days prior to admission to the program, and (5) were younger in age than those who successfully completed the program. Sadly, less than 33 percent of the participants completed the 12-month program (National Institute on Drug Abuse [NIDA] 2006).

Saloner and LeCook (2013) report that race is also a significant predictor of treatment completion. Using national data, they found that blacks and Hispanics are less likely to complete alcohol and drug treatment, and that Native Americans are less likely to complete alcohol treatment. Only Asians were more likely to complete both treatments than whites. Because the failure of blacks and Hispanics to complete treatment was largely due to socioeconomic variables, the authors suggest that the Affordable Care Act may mediate differences in completion rates (see Chapter 2).

Twelve-Step Programs

Both Alcoholics Anonymous (AA) and Narcotics Anonymous (NA) are voluntary associations whose only membership requirement is the desire to stop drinking or taking drugs. AA and NA are self-help groups in that nonprofessionals operate them; they offer "sponsors" to each new member and proceed along a continuum of 12 steps to recovery. Members are immediately immersed in a fellowship of caring individuals with whom they meet daily or weekly to affirm their commitment. Some have argued that AA and NA members trade their addiction to drugs for feelings of interpersonal connectedness by bonding with other group members. AA boasts more than 60,000 groups with more than a million Americans regularly attending (Flanagin 2014).

Symbolic interactionists emphasize that AA and NA provide social contexts in which people develop new meanings. Others who offer positive labels, encouragement, and social support for sobriety surround abusers. Sponsors tell the new members that they can be successful in controlling alcohol and/or drugs "one day at a time" and provide regular interpersonal reinforcement for doing so. Some research indicates that mutual support programs work. For example, Kelly and Heppner (2013) found that AA has some impact on the number of days a respondent was abstinent and the number of drinks per drinking days. Others, however, are highly critical of 12-step programs in general and, in particular, Alcoholics Anonymous. In *The Sober Truth: Debunking the Bad Science behind Twelve-Step Programs and the Rehab Industry* (2014), Dodes and Dodes state that "peer-reviewed studies peg the success rate of AA somewhere between 5 and 10 percent. That is, about one of every 15 people who enter these programs is able to become and stay sober."

Drug Courts

Concern over the punitive treatment of drug offenders and the failure of the criminal justice system to reduce recidivism rates led to the development of **drug courts**—nearly 3,000 of them in the U.S. alone (Botticelli 2015). In general, after meeting eligibility requirements, defendants must complete a specialized program designed to divert nonviolent offenders with substance abuse problems from incarceration. Emphasizing collaboration between service providers (e.g., physicians, social workers) and court personnel (e.g., judges, prosecutors), drug courts "use a system of graduated rewards and sanctions to help substance abusers attain—and maintain—a drug-free life" (Center 2015, p.1).

A cost–benefit analysis sponsored by the National Institute of Justice examined 23 drug courts and 1,787 drug offenders. Offenders were interviewed at the time they enrolled in a drug court program (treatment group) or when they would have enrolled in a drug court program (control group), and again 18 months later. Results of the study indicate that drug court participants, when compared to control group members, (1) had improved employment outcomes; (2) had lower rearrest and incarceration rates, thereby saving the criminal justice system money; and (3) were less likely to be involved in criminal activities. The authors conclude that "drug courts are doing exactly what they are supposed to do" (Downey and Roman 2014, p. 24).

Strategies for Action: America Responds

Drug use is a complex social issue that is exacerbated by the structural and cultural forces of society that contribute to its existence. Although the structure of society perpetuates a system of inequality, creating in some the need to escape, the culture of society, through the media and normative contradictions, sends mixed messages about the acceptability of drug use. Thus, trying to end drug use by developing programs, laws, or initiatives may be unrealistic. Nevertheless, numerous social policies have been implemented or proposed to help control drug use and its negative consequences with various levels of success.

Alcohol, Tobacco, and Prescription Drugs

Although there may be some overlap (e.g., education), strategies to deal with alcohol, tobacco, and prescription drug abuse are often different from those initiated to deal with illegal drugs. Prohibition, the largest social policy attempt to control a drug in the United States, was a failure, and criminalizing tobacco is likely to be just as successful. By definition prescription drugs are legal—it is access to and misuse of that needs to be controlled. Research has identified several promising strategies in reducing alcohol, tobacco, and/or prescription drug misuse including economic incentives, government regulations, legal sanctions, and prevention initiatives.

Economic Incentives. One method of reducing alcohol and tobacco use is to increase the cost of the product. After examining the cost of cigarettes by state, Dinno and Glantz (2009) conclude that increased cigarette prices are associated with fewer people smoking

> Drug use is a complex social issue that is exacerbated by the structural and cultural forces of society.... [T]he structure perpetuates a system of inequality, creating in some the need to escape, [and] the culture of society, through the media and normative contradictions, sends mixed messages about the acceptability of drug use.

drug courts Special courts that divert drug offenders to treatment programs in lieu of probation or incarceration.

and, if they smoke, fewer cigarettes are consumed. In fact, the World Health Organization (2014c) calculates that "if all countries increased taxes on cigarette packs by 50 percent, there would be 49 million fewer smokers and 11 million fewer deaths" (p. 7).

Other incentives include covering the cost of treatment and rewarding success. For example, in a study of smoking cessation, 828 employees of a large corporation were randomly assigned to one of two groups. Both groups were provided with information on smoking cessation programs but one group also received financial incentives including "$100 for completion of a smoking cessation program, $250 for cessation of smoking within 6 months after study enrollment as confirmed by a biochemical test, and $400 for abstinence for an additional 6 months after the initial cessation as confirmed by a biochemical test" (Volpp et al. 2009, p. 1). The results indicated that financial incentives reduced employees' smoking rates more than information alone. Similarly, Williams et al. (2005) note that "increasing the price of alcohol which can be achieved by eliminating price specials and promotions, or by raising price excise taxes, would lead to a reduction in both moderate and heavy drinking by college students" (p. 88). Other examples of economic incentives include reduced health insurance premiums for nonsmokers, decreased car insurance premiums for nondrinkers, advertising and promotion bans, and membership fee discounts at fitness centers.

Government Regulation. Federal, state, and local governmental regulations have each had some success in reducing tobacco and alcohol use and the problems associated with them. In 1984, Congress passed legislation that required states to raise the legal drinking age to 21 under threat of losing federal highway funds; and by 1987, all states had done so. Despite some controversy over the effectiveness of the law, a review of the research indicates that the 21 minimum drinking age requirement "has served the nation well by reducing alcohol-related traffic crashes and alcohol consumption among youths while also protecting drinkers from long-term negative outcomes they might experience in adulthood, including alcohol and other drug dependence, adverse birth outcomes, and suicide and homicide" (DeJong and Blanchette 2014, p. 113). The success of the law has inspired some to call for raising the minimum legal age for purchasing tobacco products to 21 (Knox 2015).

One of the most important pieces of legislation in recent years was the Family Smoking Prevention and Tobacco Control Act of 2009 (see the preceding "The Tobacco Crisis" section). The law gives authority to the Food and Drug Administration to regulate the manufacturing (e.g., tobacco companies must now disclose ingredients in their products), marketing (e.g., tobacco names or logos may no longer be used to sponsor sporting or entertainment events), and sale of tobacco products (e.g., terms like *light, mild,* and *low tar* may no longer be used). As a result of the FDA's regulations, the tobacco industry has shifted its attention to products not previously covered by the 2009 act (Sifferlin 2013). In reaction to such posturing, the FDA has proposed extending their authority to other tobacco products including pipe tobacco, waterpipe (i.e., hookah) tobacco, cigars, nicotine gel and dissolvables, and e-cigarettes (FDA 2014a; Nelson 2014).

World Health Organization

Higher tobacco taxes = fewer smokers, less death and healthier communities.

World Health Organization

WORLD NO TOBACCO DAY, 31 MAY
www.who.int/world-no-tobacco-day

Raising the costs of tobacco by 10 percent is associated with a 4 percent decrease in tobacco consumption in high income countries and a 5 percent decrease in tobacco consumption in low and middle income countries (WHO 2014c).

WHAT do you THINK?

The FDA's "deeming" proposal would require that (1) new tobacco products be registered with and reviewed by the FDA prior to marketing, (2) claims of reduced risk be supported by the scientific community, (3) free samples of the product not be distributed, and (4) minimum age, and identification restrictions and health warnings must be included (FDA 2014b; Nelson 2014). Should the FDA be able to extend its authority to all tobacco products? Should they be able to ban fruit- and candy-flavored tobacco products?

Clean air laws have helped reduce smoking in the workplace, bars, restaurants, and the like, and reduce consumption rates as well as secondhand smoke exposure. Smoke-free laws have also been credited with an increase in cessation rates of employees and the general population, as well as reducing the number of young new smokers (CDC 2014b). Russia is the most recent country to institute a nationwide tobacco ban and with positive results. Government officials report that the new law has resulted in a 21 percent drop in cigarette production and, if estimates are correct, will reduce the number of Russian smokers from 40 percent to 25 percent by 2020 (Kravchenko 2014).

Recent concerns over the prescription drug crisis have led to several initiatives by government regulatory agencies. For example, the FDA is encouraging research on "the development of abuse-deterrent forms of opiate medicines." The FDA, along with other stakeholders, has also been investigating the best way to treat overdoses of FDA-approved medications (Lurie 2015). In addition to these FDA's initiatives, in 2010, President Obama signed into law The Secure and Responsible Drug Disposal Act and, in 2011, outlined the administration's plan for prescription drug abuse prevention, which emphasizes education, prescription drug monitoring programs, proper disposal of prescription drugs, and law enforcement (White House 2014).

Finally, there are concerns about the sale of prescription drugs over the Internet. Illegal e-pharmacies are often operated by criminals seeking huge profits from selling illegal or counterfeit medications—half the total sales—and with much less risks (Sample 2014). Social media often facilitate such sales. In 2013, British authorities removed more than 18,000 YouTube videos directing viewers to websites selling tranquilizers, antidepressants, sleeping pills, and the like. In 2015, the mastermind behind the website Silk Road, an anonymous site in the "darkweb" of the Internet, appeared in federal court and was sentenced to life in prison for running the online drug store that, among other things, sold heroin, LSD, methamphetamine, and cocaine (McCoy 2015).

Legal Action. Suits against retailers, distributors, and manufacturers of alcohol are more recent than and often modeled after tobacco litigation. These suits primarily concern accusations of unlawful marketing, sales to underage drinkers, and failure to adequately warn of the risks of alcohol. In 2014, suits were filed against, among others, Coors, Heineken, Bacardi, the makers of Zima, and the makers of Mike's Hard Lemonade, accusing the companies of using a "long-running, sophisticated and deceptive scheme...to market alcoholic beverages to children and other underage consumers" (Willing 2014, p.1).

Federal and state governments—as well as smokers, ex-smokers, and the families of victims of smoking—have taken legal action against tobacco companies (see Figure 3.6). The 1990s brought billion-dollar judgments against the tobacco industry on behalf of states suing for reimbursement of smoking-related health care costs and families seeking punitive damages for the deaths of loved ones (Timeline 2001). In 1998, tobacco manufacturers reached a settlement with 46 states, agreeing to pay billions of dollars for reimbursement of state smoking-related health costs. The settlement also restricted the marketing, promotion, and advertising of tobacco products directed toward minors.

After years of litigation, the District of Columbia U.S. Court of Appeals held that "the tobacco industry had and continues to engage in a massive, decades-long campaign to defraud the American public...including falsely denying that nicotine is addictive, falsely representing that 'light,' and 'low tar' cigarettes present fewer health risks, falsely denying that they market to kids, and falsely denying that secondhand smoke causes disease" (Myers 2009, p. 1).

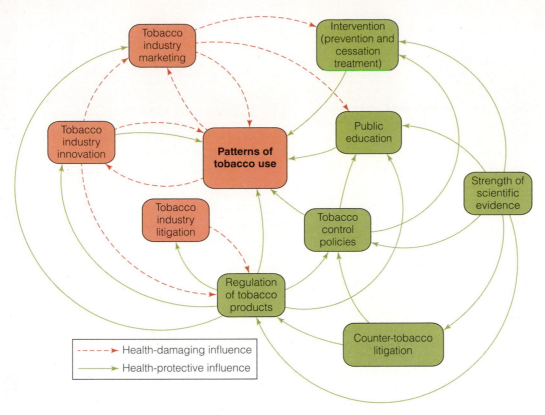

Figure 3.6 Model of Pro-Tobacco and Anti-Tobacco Social Forces on Tobacco Use
SOURCE: U.S. Surgeon General 2014.

In 2012, tobacco companies were ordered to publically admit their deception. Among other required statements to be placed in advertisements, some tobacco companies must state that "[a] Federal Court has ruled that the Defendant tobacco companies deliberately deceived the American public about designing cigarettes to enhance the delivery of nicotine, and has ordered those companies to make this statement. Here is the truth: Smoking kills, on average, 1,200 Americans. Every day" (Mears 2012). More recently, Reynolds American, Inc., was ordered to pay a Florida widow, Cynthia Robinson, $23.6 billion in damages for acting negligently in marketing its tobacco products. Robinson's husband died of lung cancer in 1996 at the age of 36 (Palazzolo 2014). The award was reduced to $16.9 million in 2015.

Although relatively recent, there have also been prescription drug-related lawsuits, most frequently involving the pain reliever OxyContin. In general, lawsuits against Purdue Pharma, the drug's manufacturer, have accused the company of failing to reveal to prescribers its addictive nature and, in some cases, asserting that it was *less* addictive than other pain medications. Criminal prosecution of physicians, pharmacists, and other health care providers, primarily for dispensing drugs for nonmedical reasons, has also been used to fight prescription drug misuse (Ausness 2014).

Prevention. Prevention of alcohol, tobacco, and prescription drug use requires reducing variables that put an individual at risk or, alternatively, increasing variables that protect the individual. Some variables which put a person at risk, for example, ethnicity or having an alcoholic parent, are ascribed and although they cannot be altered, prevention efforts can be directed toward specific at-risk populations. Other risk factors are micro-sociological (e.g., negative peer pressure) and/or macro-sociological (e.g., availability of alcohol to underage drinkers) and addressing such variables requires maximizing both micro-sociological protective variables (e.g., counteracting negative peer pressure) and macro-sociological protective variables (e.g., restricting the availability of alcohol to underage drinkers). In various ways, the prevention methods discussed next do some or all of these.

In 2012, a report by the U.S. Surgeon General concluded that "there is a causal relationship between depictions of smoking in the movies and the initiation of smoking among young people" (U.S. Surgeon General 2012, p. 6). In fact, a 2014 report by the Surgeon General estimates that if smoking or other tobacco use in films required a Motion Picture Association of America (MPAA) R rating, teen smoking would be reduced by 18 percent (U.S. Surgeon General 2014). The present research investigates smoking in the top-grossing films in the United States by answering three research questions.

Sample and Methods

Polansky et al. (2014) analyzed the top 10–grossing movies in any week of their first-run theatrical release. The investigation concentrated on three research questions (RQs): (1) what percentage of films feature tobacco-related incidents, (2) how many tobacco-related incidents occur in films, and (3) do film presentations of tobacco-related incidents vary by studio?

Tobacco incidents were operationalized as an occurrence of smoking or other tobacco use in a film. The study examined films with several Motion Picture Association of America (MPAA) ratings: G: All ages admitted, PG: Parental Guidance Suggested (some material may not be suitable for children), PG-13: Parents Strongly Cautioned (some material may be inappropriate for children under 13), and R: Restricted (under 17 requires accompanying parent or adult guardian).

Findings and Conclusions

RQ1. In every year between 2002 and 2013, R-rated films had higher rates of tobacco-related incidents than PG-13 films, and PG-13 films had higher rates of tobacco-related incidents than G/PG films. However, although the percentage of G/PG and PG-13 films with tobacco-related incidents fell by half over the time period studied, nearly 40 percent of PG-13 films contained tobacco-related incidents in 2013.

RQ2. In youth-oriented films, G/PG and PG-13, although fluctuating, there has been a general decline in the number of tobacco-related incidents per film between 2002 and 2013. However, the number of tobacco-related incidents per PG-13 film was higher in 2012 and 2013 than in 2009 through 2011. The number of incidents per R-rated films increased between 2009 and 2012.

When examining films in which smoking occurred at least once, Polansky et al. (2014) report that the number of tobacco-related incidents increased between 2010 and 2013. Furthermore, for the first time, the number of tobacco-related incidents in PG-13 films was as high as the number of tobacco-related incidents in R-rated films. The number of PG-13 films with more than 50 tobacco-related incidents per film increased from 17.0 percent in 2010 to nearly 30.0 percent in 2013.

RQ3. Film presentations of tobacco use do vary by studio. For example,

Disney all but eliminated tobacco-related incidents in its PG-13 films between 2008 and 2010; but by 2011, 60 percent of its PG-13 films contained smoking or other tobacco incidents. In 2013, the number of smoking incidents per PG-13 films was highest in Time Warner films (28.1 percent) and lowest in Fox PG-13 films (0.5 percent). Between 2010 and 2013, across the seven studios examined, the *number* of tobacco-related incidences in PG-13 films increased by 21 percent.

Polansky et al. (2014) state that although the American film industry has known for decades that smoking in films impacts the likelihood of youth initiation, their response has been inadequate. Despite the claims by the MPAA in 2007 that a "smoking" label would be included in ratings, over 80 percent of top-grossing youth-oriented films with smoking were released without the label. There has been progress, but "moderate at best and has frequently reversed. As the share of PG-13 films with smoking has declined, the amount of smoking in PG-13 films with any smoking has increased" (p. 19). The authors recommend updating the MPAA rating system whereby tobacco imagery receives a restricted (R) rating. They argue that it is "the only evidence-based method to set a transparent, enforceable, uniform standard that protects young people from toxic tobacco exposure on screen" (p. 20).

SOURCE: Polansky et al. 2014.

The World Health Organization's MPOWER initiative, one of the most successful initiatives in the global fight against tobacco use (WHO 2013a), is a good example of decreasing risk factors (e.g., banning advertising) and increasing protective factors (e.g., drug education). According to the World Health Organization, the initiative has had a significant impact in reducing the number of deaths from tobacco use and the number of people exposed to secondhand smoke. Additional alcohol, tobacco, and prescription drug misuse prevention initiatives include public education campaigns, warning labels, and school- and community-based prevention programs.

Public Education Campaign. Educating the public about the risks of alcohol, tobacco, and prescription drug use is an important component of substance abuse prevention.

The Office of National Drug Policy campaign *Above the Influence* boasts an 88 percent recognition rate among teens and has 1.7 million "likes" on its Facebook page (ONDCP 2013). The campaign delivers anti-alcohol and anti-drug messages by encouraging youth to "stay original, be yourself," and "live above" the influence—of friends and peers, what others think of you, relationships, the media, and, of course, drugs and alcohol (ONDCP 2014).

The FDA's *The Real Cost*, launched in 2014, focuses on educating 12- to 17-year-old at-risk youth about the negative consequences of tobacco use in the hope of reducing initiation rates and habitual use by those already experimenting with tobacco (FDA 2014c). The campaign uses print media, television ads, radio public service announcements, as well as social media—Facebook, YouTube, and Twitter—to reach the targeted population.

Lastly, the Partnership for *Drug-Free Kids' Medicine Abuse Project* (MAP) is a good example of a multifocused campaign that provides resources for caretakers, health care providers, educators, and community and law enforcement officials (MAP 2015). It is a five-year project directed toward preventing a half a million teenagers from abusing prescription drugs by 2017. The website includes tips for "minding your meds," a short documentary on the use of prescription drugs, a guide to treatment resources, a medicine abuse quiz, and a toll-free parent helpline.

Warning Labels. The use of warning labels on alcoholic beverages in the United States was mandated as part of the Anti-Drug Abuse Act of 1988. It is not, however, required in many other countries, including England and Canada (Taylor 2014). The impact of warning labels on drinking behavior is mixed. Although a review of the literature concludes that alcohol warning labels are "popular with the public, their effectiveness for changing drinking behavior is limited" (Thomas et al. 2014, p. 91).

In 2009, Congress passed a law requiring specific warning labels on cigarette packaging. However, after an appeals court held for the tobacco industry that the cigarette labels were a violation of free speech protections, the FDA is now in the process of creating new labels (Almasy 2013). The current warnings, now 25 years old, are largely seen as ineffective and, according to a spokesperson for the American Cancer Society, "every day that the current warnings remain in place is another day in which the tobacco industry misleads children and adults about the hazards of smoking" (Hansen 2013).

Family-, School-, and Community-Based Prevention Programs. A body of research suggests that school-based interventions may reduce alcohol and/or tobacco use. Champion et al. (2013), in a review of computer- and Internet-based prevention programs housed in schools, reported that of the seven studies reviewed, six "achieved a reduction in alcohol or drug use at post-intervention and/or follow-up" (p. 120). The authors are quick to note that technology-based interventions avoid a number of problems associated with traditional school delivery systems (e.g., cost, limited teacher time) while assuring a high degree of consistency over time, location, and student population.

There is also strong evidence that school- and home-based prevention programs in combination reduce the likelihood of nonmedical prescription drug use (Crowley et al. 2014). A sample of students were asked in the 6th grade and at the end of each year through 12th grade, "Have you ever used Vicodin, codeine, Percocet or OxyContin not prescribed by a doctor?" Students and their families were assigned to either a treatment group or control group. Members of the treatment group received one of three school-based intervention programs with or without a family-based intervention program. In the absence of family intervention, only one of the three school-based intervention programs was associated with reduced opiate use. However, all of the school-based intervention programs when used in conjunction with the family-based interventions were successful in significantly reducing the nonmedical use of opiates.

Illegal Drugs

Although the greatest costs to individuals and society come from the use and abuse of legal drugs, there is little doubt that the use and abuse of illicit drugs take a tremendous human and financial toll. In this section, we highlight the war on drugs, deregulation and legalization, and other federal and state initiatives in the effort to control illegal substances.

The War on Drugs. In the 1980s, the federal government declared a "war on drugs," which was based on the belief that controlling drug availability would limit drug use and, in turn, drug-related problems. In contrast to a position of **harm reduction**, which focuses on minimizing the costs of drug use for both user and society (e.g., distributing clean syringes to decrease the risk of HIV infection), this "zero-tolerance" approach advocated get-tough law enforcement policies, and it is responsible for the dramatic increase in the jail and prison population. In 1980, there were an estimated 41,000 drug offenders in jail or prison; by 2011, there were nearly half a million (The Sentencing Project 2013).

Race, Ethnicity, and the War on Drugs. The harsher penalties enacted as part of the "war on drugs" required prison sentences for almost all drug offenders—first-time or repeat—and limited judicial discretion in deciding what best served the public's interest. The "Rockefeller drug laws," as they were called, resulted in a disproportionate number of Hispanics and African Americans receiving "excessively long and unnecessary prison sentences" for even the most minor drug offenses (HRW 2007, p. 1).

In response to public outcries and accusations of institutional racism (see Chapter 9), reform of such laws began (Peters 2009). Nonetheless, discriminatory practices continue. For example, African American teens are 10 times more likely to be arrested for drug-related crimes than whites (Sánchez-Moreno 2012) despite research that indicates they are less likely to use drugs and less likely to have drug problems than their white counterparts (Wu et al. 2011). Alexander (2010) argues that the war on drugs simply replicated Jim Crow laws by labeling, disproportionately, African Americans as felons, thereby justifying discrimination in all its forms, as well as denying the right to vote and the right to serve on a jury.

WHAT do you THINK?

In a recent interview Marianne Alexander, a civil rights activist and law professor at Ohio State University, in reference to the trend toward the legalization of marijuana made the observation that "Forty years of impoverished black kids getting prison time for selling weed, and their families and futures destroyed.... Now, white men are planning to get rich precisely doing the same thing" (quoted in Jarecki 2014, p. 1). What do you think of Alexander's comments?

Gender and the War on Drugs. At year-end 2012, nearly a quarter of women in state prisons and over half of women in federal prisons were incarcerated for drug violations; over half were mothers (Drug Policy Alliance [DPA] 2015a). As with racial and ethnic minorities in general, women sentenced to prison for drug offenses were disproportionately black and Hispanic. As a result, children with drug-related incarcerated mothers are more likely to be racial and ethnic minorities. Even though there is no statistically significant difference by race/ethnicity for drug-related offenses, African American women are twice as likely and Hispanic women 25.0 percent more likely to be incarcerated than white women (DPA 2015b).

The number of women in prison has increased dramatically over the past two decades. Merolla (2008), using conflict theory, argues that the increase, to a large extent, is a function of the war on drugs rather than behavioral changes in women—that is, their increased use of drugs. Using data from government sources, Merolla concludes that the war on drugs, in and of itself, increased arrest rates of women in two ways. First, the war on drugs redefined through the media and in many other ways who a criminal is in nongendered terms (i.e., men *and* women). Second, as a consequence of this redefinition, law enforcement practices changed and more aggressively target female drug users.

Cost of the War on Drugs. Although the focus here is on the war in drugs in the United States, there remains a global initiative to control drug production, trafficking, and use. One collaborative project, *Count the Costs, is* dedicated to changing the punitive approach to drug control policy and, with it, the reduction of costs associated with such an

harm reduction A recent public health position that advocates reducing the harmful consequences of drug use for the user as well as for society as a whole.

emphasis (Count the Costs 2015). Count the Costs has identified seven categories of costs associated with the global war on drugs:

- Undermining development and creating conflict (e.g., use of military to fight drug cartels)
- Threatening public health, spreading disease and death (e.g., scarce resources are channeled toward law enforcement rather than treatment)
- Undermining human rights (e.g., execution of drug users in some countries)
- Promoting stigma and discrimination (e.g., some drug use is culturally embedded)
- Creating crime, enriching criminals (e.g., prohibiting illicit drugs may reduce supply, but with constant demand, prices increase)
- Deforestation and pollution (e.g., aerial spraying of drug crops)
- Wasting billions of dollars (e.g., spending on law enforcement)

The United States spends more than $51 billion a year on the war on drugs (DPA 2015b).

Losing the War on Drugs: Drug Policy Reform. There is little debate "that the global prohibition of certain drugs and the war on drugs have largely failed to reach their stated goals" (Chouvy 2013, p. 216). In fact, a CNN article called the war on drugs the "trillion-dollar failure" (Branson 2012), and a recent report with signatories including everyone from Nobel laureates to former presidents to renowned social scientists and human rights experts called for "ending the drug wars" (LSE Expert Group 2014). Consequently, beginning in 2010, the federal government began to reform national drug policy. In general, the U.S. policy on fighting drugs is two-pronged. First is **demand reduction**, which entails reducing the demand for drugs through treatment and prevention. The second strategy is **supply reduction**. A much more punitive strategy, supply reduction relies on international efforts, interdiction, and domestic law enforcement to reduce the supply of illegal drugs.

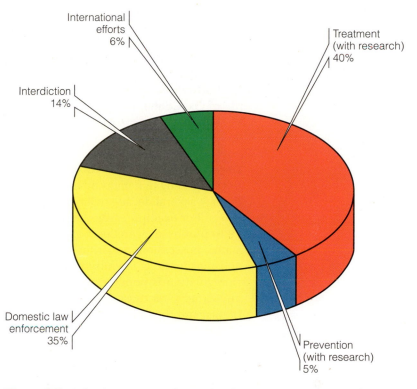

Figure 3.7 Federal Drug Control Spending by Function, Fiscal Year 2016
SOURCE: White House 2015.

The 2014 *National Drug Control Strategy* contains a blueprint for addressing illegal drugs in the United States "through a modern, evidence-based approach, encompassing prevention, early intervention, treatment, recovery support, criminal justice reform, effective law enforcement, and international cooperation" (White House 2014, p. 4). The allocation of funds for each of these strategies has changed over time as a more balanced and less punitive approach has been adopted (see Figure 3.7). For example, the budget allocation for treatment increased by 7.0 percent between funding year 2015 and funding year 2016 (White House 2015).

Numerous other examples of a shift in public policy have occurred. In 2014, Congress passed a bill ending federal raids on medical marijuana dispensaries (Halper 2014). The U.S. Sentencing Commission reduced the penalties for nonviolent drug offenders. Former U.S. Attorney General Eric Holder instituted clemency efforts for nonviolent drug offenders imprisoned in federal facilities (Associated Press 2014). Legislators in at least 30 states passed bills designed to reduce the nation's rate of incarceration including the establishment of diversion and reentry programs for drug offenders, and the District of Columbia eliminated criminal penalties for marijuana use (Porter 2015) (see Chapter 4).

demand reduction One of two strategies in the U.S. war on drugs (the other is supply reduction), demand reduction focuses on reducing the demand for drugs through treatment, prevention, and research.

supply reduction One of two strategies in the U.S. war on drugs (the other is demand reduction), supply reduction concentrates on reducing the supply of drugs available on the streets through international efforts, interdiction, and domestic law enforcement.

Reform of U.S. drug policy is in part a reflection of the failure of the war on drugs, but it is also a response to increasing international pressure. In 2013, a collaborative organization between North and South American countries issued a "game-changing report" (Doward 2013). The report (Organization of American States 2013) describes the negative impact of U.S. drug policies on Latin American countries, including increased crime and violence, economic hardship, and institutional corruption. There is little evidence to the contrary. Of the "world's eight most murderous countries, seven lie on the cocaine-trafficking route from the Andes to the United States and Europe. Only war zones are more violent than Honduras. More than 7,000 of its 8 [million] citizens are murdered each year. In the European Union, with a 500 [million] population, the figure is under 6,000" (*The Economist* 2013).

WHAT do you THINK?

Americans consume more than 80 percent of the world's supply of OxyContin, a drug so dangerous that it is responsible for 16,000 deaths—more than all deaths from heroin and cocaine combined. So, as Fang (2014) notes, "it's more than a little odd that…groups leading the fight against relaxing marijuana laws derive a significant portion of their budget from opioid manufacturers." For example, the Partnership for Drug-Free Kids' largest benefactor is Purdue Pharma, the maker of OxyContin. Why do you think companies that manufacture drugs oppose the legalization of marijuana?

Deregulation or Legalization. Given the questionable successes of the war on drugs, it is not surprising that many advocate alternative measures to the rather punitive emphasis of the last several decades. **Deregulation** is the reduction of government control over certain drugs. For example, although individuals must be 21 years old to purchase alcohol and 18 to purchase cigarettes, both substances are legal and can be purchased freely. In some states, possession of marijuana in small amounts is a misdemeanor rather than a felony, and marijuana is lawfully used for medical purposes in over 20 states (see Chapter 2). Deregulation is popular in other countries as well. For example, personal possession of any drug, even those considered the most dangerous, is legal in Spain, Italy, the Baltic States, and the Czech Republic.

Proponents for the **legalization** of drugs, who range on a continuum from no government regulation to legalization under various scenarios (e.g., age limits, required warning labels), affirm the right of adults to make informed choices. They also argue that the tremendous revenues realized from drug taxes could be used to benefit all citizens, that purity and safety controls could be implemented, and that legalization would expand the number of distributors, thereby increasing competition and reducing prices.

Those in favor of legalization also suggest that the greater availability of drugs would not increase demand, pointing to countries where some drugs have already been decriminalized. **Decriminalization**, which entails removing state penalties for certain drugs, promotes a medical rather than criminal approach to drug use that encourages users to seek treatment and adopt preventive practices. Portugal, which abolished all criminal penalties for personal possession of any drug, is a case in point (Szalavitz 2009). Despite fears of negative consequences, an evaluation of the "Portugal experiment" 12 years after decriminalization concluded that although the number of adults who report lifetime use of a controlled substance has increased, as it has throughout Europe, (1) the number of teenagers who report lifetime us of a controlled substance has decreased, (2) the number of HIV-positive drug addicts has decreased, and (3) the number of addicts who have undergone treatment has significantly increased (Hollersen 2013).

Opponents of legalization argue that it would be construed as government approval of drug use and, as a consequence, drug experimentation and abuse would increase. Furthermore, although the legalization of drugs could result in substantial revenues for the government, drug trafficking and black markets would still flourish because all drugs would not be decriminalized. Legalization would also require an extensive and costly bureaucracy to regulate the manufacture, sale, and distribution of drugs. Finally, the

deregulation The reduction of government control over, for example, certain drugs.

legalization Making prohibited behaviors legal—for example, legalizing drug use or prostitution.

decriminalization Entails removing state penalties for certain drugs; promotes a medical rather than criminal approach to drug use that encourages users to seek treatment and adopt preventive practices.

position that drug use is an individual's right cannot guarantee that others will not be harmed. It is illogical to assume that a greater availability of drugs will translate into a safer society.

Understanding Alcohol and Other Drug Use

In summary, substance abuse—that is, drugs and their use—is socially defined. As the structure of society changes, the acceptability of one drug or another changes as well. As theorists assert, the status of a drug as legal or illegal is intricately linked to those who have the power to define acceptable and unacceptable drug use. There is also little doubt that rapid social change, anomie, alienation, and inequality further drug use and abuse. Symbolic interaction also plays a role in the process: If people are labeled "drug users" and are expected to behave accordingly, then drug use is likely to continue. Thus, the theories of drug use complement rather than contradict one another.

Two issues need to be addressed in understanding drug use. The first is at the micro level: why does a given individual use alcohol or other drugs? Many individuals at high risk for drug use have been "failed by society"—they are living in poverty, unemployed, victims of abuse, dependents of addicted and neglectful parents, and the like. Despite the social origins of drug use, many treatment alternatives, emanating from a clinical model of drug use, assume that the origin of the problem lies within the individual rather than in the structure and culture of society. Although the problem may admittedly lie within the individual when treatment occurs, policies that address the social causes of drug abuse must be a priority in dealing with the drug problem in the United States.

The second question, related to the first, is why does drug use vary so dramatically across societies, often independent of a country's drug policies? The United States has historically some of the most severe penalties for drug violations in the world, but has one of the highest rates of marijuana and cocaine use. On the other hand, as mentioned earlier, Portugal decriminalized personal possession of all drugs, and youth drug use has decreased (Hollersen 2013). Most compellingly, a 2008 World Health Organization survey of 17 countries concluded that there is no link between the harshness of drug policies and the consumption rates of its citizenry (Degenhardt et al. 2008).

That said, what is needed is a more balanced approach—one that acknowledges that not all drugs have the same impact on society or on the individuals who use them. The present administration appears to be leaning this way. For example, in a break from the previous administration, President Obama supported federally funded needle exchange programs—a staple of harm reduction advocates—and has stated that it is "entirely appropriate" to use marijuana for the same purposes and under the same controls as other drugs prescribed by a physician (Dinan and Conery 2009; Heinrich 2009). In addition, the 2015 distribution of funding for harm reduction versus supply reduction is the most balanced in U.S. history (White House 2015). Only time will tell if this new approach to drug control, one that reflects both a public health and criminal justice position, will be more successful than previous policies.

> [A]...World Health Organization survey of 17 countries concluded that there is no link between the harshness of drug policies and the consumption rates of its citizenry.

Chapter Review

- **What is a drug, and what is meant by drug abuse?**
Sociologically, the term *drug* refers to any chemical substance that (1) has a direct effect on the user's physical, psychological, and/or intellectual functioning; (2) has the potential to be abused; and (3) has adverse consequences for the individual and/or society. Drug abuse occurs when acceptable social standards of drug use are violated, resulting in adverse physiological, psychological, and/or social consequences.

- **How do the three sociological theories of society explain drug use?**
Structural functionalists argue that drug abuse is a response to the weakening of norms in society, leading to

a condition known as anomie or normlessness. From a conflict perspective, drug use occurs as a response to the inequality perpetuated by a capitalist system as societal members respond to alienation from their work, family, and friends. Symbolic interactionism concentrates on the social meanings associated with drug use. If the initial drug use experience is defined as pleasurable, it is likely to recur, and over time, the individual may earn the label of "drug user."

- **What are the most frequently used legal and illegal drugs?**

Alcohol is the most commonly used and abused legal drug in the United States, with 66 percent of the adult population reporting current alcohol use. Although tobacco use in the United States has been declining, the use of tobacco products is globally very high, with 80 percent of the world's over one billion smokers living in low- or middle-income countries.

- **What are the consequences of drug use?**

The consequences of drug use are fivefold. First is the cost to the family and children, often manifesting itself in higher rates of divorce, spouse abuse, child abuse, and child neglect. Second is the relationship between drugs and crime. Those arrested have disproportionately higher rates of drug use. Although drug users commit more crimes, sociologists disagree as to whether drugs actually "cause" crime or whether, instead, criminal activity leads to drug involvement. Third are the economic costs (e.g., loss of productivity), which are in the billions. Then there are the health costs of abusing drugs, including shortened

life expectancy; higher morbidity (e.g., cirrhosis of the liver and lung cancer); exposure to HIV infection, hepatitis, and other diseases through shared needles; a weakened immune system; birth defects such as fetal alcohol syndrome; drug addiction in children; and higher death rates. Finally, illegal drug production takes its toll on the environment, which impacts all Americans.

- **What treatment alternatives are available for drug users?**

Although there are many ways to treat drug abuse, three methods stand out: The inpatient–outpatient model entails medical supervision of detoxification and may or may not include hospitalization. Twelve-step programs such as Alcoholics Anonymous (AA) and Narcotics Anonymous (NA) use a supportive community and peer leaders to help members reduce substance abuse, and drug courts divert offenders from the criminal just system through a multi-pronged treatment program.

- **What can be done about the drug problem?**

First, there are government regulations limiting the use (e.g., the law establishing the 21-year-old drinking age) and distribution (e.g., prohibitions about importing drugs) of legal and illegal drugs. The government also imposes sanctions on those who violate drug regulations and provides treatment facilities for other offenders. Economic incentives (e.g., cost) and prevention programs have also been found to impact consumption rates. Finally, legal action holding companies responsible for the consequences for their product—for example, class-action suits against tobacco producers—have been fairly successful.

Test Yourself

1. "Cannabis cafés" are commonplace throughout England.
   a. True
   b. False
2. The most used illicit drug in the world is
   a. heroin.
   b. marijuana.
   c. cocaine.
   d. methamphetamine.
3. What theory would argue that the continued legality of alcohol is a consequence of corporate greed?
   a. Structural functionalism
   b. Symbolic interactionism
   c. Reinforcement theory
   d. Conflict theory
4. Cigarette smoking is
   a. the third leading cause of preventable death in the United States.
   b. not addictive.
   c. the most common use of tobacco products.
   d. increasing in the United States.
5. In the United States, drinking is highest among young, nonwhite males.
   a. True
   b. False
6. Which European country decriminalized personal possession of all drugs, including marijuana, cocaine, heroin, and methamphetamine?
   a. Spain
   b. France
   c. Portugal
   d. The Netherlands
7. The active ingredient in marijuana, THC, can act as a sedative or a hallucinogen.
   a. True
   b. False
8. According to the proposed 2015 budget, most federal drug control dollars are allocated to
   a. international efforts.
   b. domestic law enforcement.
   c. prevention and research.
   d. treatment and research.

9. Decriminalization refers to the removal of penalties for certain drugs.
   a. True
   b. False

10. The two-pronged drug control strategy of the U.S. government entails supply reduction and harm reduction.
    a. True
    b. False

Answers: 1. B; 2. B; 3. D; 4. C; 5. B; 6. C; 7. A; 8. D; 9. A; 10. B.

Key Terms

binge drinking 71
chemical dependency 67
crack 77
decriminalization 94
demand reduction 93
deregulation 94
drug 64

drug abuse 67
drug courts 86
e-cigarettes 73
fetal alcohol syndrome 82
gateway drug 76
harm reduction 92
heavy drinking 71

legalization 94
meta-analysis 73
psychotherapeutic drugs 75
supply reduction 93
synthetic drugs 78

 Unjust social arrangements are themselves a kind of extortion, even violence."

JOHN RAWLS
A Theory of Justice

4

Crime and Social Control

Learning Objectives

After studying this chapter, you will be able to . . .

1 List patterns of crime and criminals that are consistent across global regions.

2 Argue which of the various methods of measuring crime is the most accurate.

3 Identify the crime prevention implications of each sociological theory of crime.

4 Differentiate between and give examples of each of the types of crimes discussed.

5 Using each of the demographic variables discussed, explain their relationship to the likelihood of criminal behavior and victimization.

6 Assess the relative consequences to society of crimes in the streets and crimes in the suites.

7 Analyze the various social forces that lead to and away from the punitive criminal justice policies of previous decades.

IN SEPTEMBER 2014, an 18-year-old University of Virginia student disappeared from a Charlottesville pub. Hannah Graham had gone to dinner with friends, and before heading back to her apartment, she made several stops. Video cameras and witnesses showed Hannah leaving a downtown bar at around 2:00 A.M. followed by a man who was later identified as Jesse Matthew (Martinez 2014). Five weeks later, the decomposed body of the college sophomore was found in an abandoned lot eight miles from where she had last been seen. Jesse Matthew was charged with abduction with intent to defile and first-degree murder (Alsup 2015). Upon confirmation of their daughter's death, Hannah's parents stated, "Although the waiting has ended for us, there are other families . . . who have not been as fortunate in that their loved ones are still missing. Please continue to hold these families in your thoughts and prayers" (quoted in Keneally 2014, p. 1).

Handout/Reuters

In 2015, the man charged with the death of University of Virginia student Hannah Graham, pictured, went on trial for the 2005 abduction, rape, and attempted murder of a 28-year-old woman as she returned from the grocery store (Burdine and Pettus 2015). Although a conviction in this case could lead to a life sentence, a conviction in the killing of Hannah Graham could result in the death penalty.

Three municipal police departments, four county sheriff's offices, the Texas and Virginia State Police, the Federal Bureau of Investigation (FBI), and the Virginia Chief Medical Examiner's Office worked together to find Hannah Graham. Such is the bureaucracy of the criminal justice system, which includes not only local, state, and federal law enforcement agencies, but the court system Jesse Mathew will be tried in and the correctional facility where he may be confined. In this chapter, we examine the criminal justice system as well as theories, types, and demographic patterns of criminal behavior. The economic, social, and psychological costs of crime are also examined. The chapter concludes with a discussion of social control, including policies and prevention programs designed to reduce crime in America.

The Global Context: International Crime and Social Control

Several facts about crime are true throughout the world. First, crime is ubiquitous—there is no country where crime does not exist. Second, most countries have the same components in their criminal justice systems: police, courts, and prisons. Third, adult males make up the largest category of crime suspects worldwide. Fourth, in all countries, theft is the most common crime committed, whereas violent crime is a relatively rare event.

Despite these similarities, dramatic differences exist in international crime rates, although comparisons are made difficult by variations in measurement and crime definitions. Nonetheless, looking at global statistics as a whole, we can make some statements about crime with confidence.

First, violent crime rates vary significantly by region and by country. Examining homicide rates is a case in point. Because homicide rates are correlated with other violent crimes—for example, as robbery rates increase, homicide rates increase—global homicide rates are very telling. In 2013, globally, there were over 435,000 homicides (i.e., the intentional killing of another), according to the most recent global surveys available (United Nations Office on Drug and Crimes [UNODC] 2014a). More than a third of the *number* of those deaths, 36.0 percent, took place in the Americas, followed by Africa (31.0 percent), Asia (28.0 percent), Europe (5.0 percent), and Oceania (0.3 percent).

However, crime *rates*, usually expressed per 100,000 population, take into consideration differences in population size and allow more accurate comparisons. Worldwide, the homicide rate is 6.3 homicides per 100,000 population, but several regions of the world—Southern Africa and Central America—have four times the global rate. Since the 1950s, the Americas have had homicide rates that are between five and eight times higher than in Europe and Asia, and those rates are increasing. Other regions of the world, such as Oceania and Europe, have had decreasing homicide rates since the 1990s. The gap in homicide levels between high-homicide countries and low-homicide countries has been getting wider (UNODC 2014a).

Globally, property crimes, most often entailing some kind of theft, have decreased significantly (UNODC 2014b). For example, over the last decade, motor vehicle theft and burglary have declined in the Americas, Europe, Asia, and Oceania. Furthermore, a comparison of 36 developed countries indicates that where violent crime as measured by homicide rates are high, property crimes tended to be low (Civitas 2012).

Violent crime and property crimes represent just two types of crime that take place worldwide. Although we are concerned about these types of crimes and the possibility of victimization, globalization and technological know-how have fueled the development of new categories of crimes (e.g., human trafficking) and expanded the prevalence of others (e.g., counterfeiting, cybercrime).

Transnational crime refers to organized criminal activity that crosses one or more national boundaries for the "purpose of obtaining power, influence, monetary and/or commercial gains . . . through a pattern of corruption and/or violence" (U.S. Department of Justice [USDOJ] 2012b). The significance of transnational crime should not be minimized. As Shelley (2007) states:

> Transnational crime will be a defining issue of the 21st century for policy makers—as defining as the Cold War was for the 20th century and colonialism was for the 19th. Terrorists and transnational crime groups will proliferate because these crime groups are major beneficiaries of globalization. They take advantage of increased travel, trade, rapid money movements, telecommunications and computer links, and are well positioned for growth. (p. 1)

As Table 4.1 indicates, transnational crime primarily functions to provide illicit goods and services, and to infiltrate businesses or government entities (Albanese 2012). Internet child pornography is an example of transnational crime. Prichard et al. (2013) note that

Although human trafficking is often thought of as something that exists outside the United States, it is estimated that 100,000 children annually are sexually exploited, and thousands more are forced into domestic servitude and involuntary labor in the United States.

TABLE 4.1 Types of Transnational Crime

Provision of Illicit Goods	Provision of Illicit Services	Infiltration of Business or Government
Drug trafficking	Human trafficking	Extortion and racketeering
Stolen property	Cybercrime and fraud	Money laundering
Counterfeiting	Commercial vices (e.g., illegal sex and gambling)	Corruption

SOURCE: Albanese 2012.

transnational crime Criminal activity that occurs across one or more national borders.

A young boy, a victim of human trafficking, is forced to work in a balloon factory outside of Dhaka, Bangladesh. Human trafficking is estimated to be a multibillion-dollar enterprise, approaching the international value of drug and arms trafficking according to Interpol.

Internet child pornography is increasing and that such code words as *Pthc* (preteen hard core), *Lolita*, *teen*, and *12yo* are routinely found in the top 1,000 terms used in international search engines.

Human trafficking is another example of transnational crime. Despite common misconceptions, to constitute human trafficking, a victim need not be transported from one location to another. According to federal law, "severe forms of trafficking in persons" include (U.S. Department of State 2014)

- "sex trafficking in which a commercial sex act is induced by force, fraud, or coercion, or in which the person induced to perform such an act has not attained 18 years of age; or
- the recruitment, harboring, transportation, provision, or obtaining of a person for labor or services, through the use of force, fraud, or coercion for the purpose of subjection to involuntary servitude, peonage, debt bondage, or slavery" (p. 9).

Globally, it is estimated that as many as 21 million men, women, and children are human trafficking victims, resulting in an estimated $150 billion in illegal profits annually (International Labour Office [ILO] 2014). Although human trafficking is often thought of as something that exists outside the United States, it is estimated that more than 100,000 children annually are sexually exploited, and thousands more are forced into domestic servitude and involuntary labor in the United States (Polaris Project 2013). It is also estimated that about 80 percent of the victims of human trafficking are women and girls, and that 50 percent are minors (U.S. Department of State 2014).

WHAT do you THINK?

According to the International Maritime Bureau (IMB), 245 piracy incidents were recorded in 2014—183 vessels were boarded by pirates, 13 were fired upon, and 21 were hijacked (Commercial Crime Services 2015). In 2013, three Somali pirates were convicted for killing four Americans after hijacking their yacht and were sentenced to life in prison (Daugherty 2013). In federal law, death resulting from aircraft piracy or attempted hijacking of an aircraft carries the death penalty. Do you think that deaths that result from maritime hijacking should carry the death penalty? What if the death is not of a hostage but of a fellow hijacker?

Sources of Crime Statistics

The U.S. government spends millions of dollars annually to compile and analyze crime statistics. A **crime** is a violation of a federal, state, or local criminal law. For a violation to be a crime, however, the offender must have acted voluntarily and with intent and have no legally acceptable excuse (e.g., insanity) or justification (e.g., self-defense) for his or her behavior. The three major types of statistics used to measure crime are official statistics, victimization surveys, and self-report offender surveys.

Official Statistics

Local sheriffs' departments and police departments throughout the United States collect information on the number of reported crimes and arrests and voluntarily report them to the FBI. The FBI then compiles these statistics annually and publishes them, in summary form, in the Uniform Crime Reports (UCR). The UCR lists **crime rates** or the number of crimes committed per 100,000 population, the actual number of crimes,

crime An act, or the omission of an act, that is a violation of a federal, state, or local criminal law for which the state can apply sanctions.

crime rate The number of crimes committed per 100,000 population.

the percentage of change over time, and clearance rates. **Clearance rates** measure the percentage of cases in which an arrest and official charge have been made and the case has been turned over to the courts.

These statistics have several shortcomings. For example, research indicates that just over a third of hate crimes (see Chapter 9) are reported to the police and that the number reported is decreasing rather than increasing (Yost 2013). Further, a report by the Justice Department based on data from 1985 through 2013 indicates that 80.0 percent of campus sexual assaults are not reported to the police (Hefling 2014).

Even if a crime is reported, the police may not record it. Alternatively, some rates may be exaggerated. Motivation for such distortions may come from the public (e.g., demanding that something be done), from political officials (e.g., election of a sheriff), and/or from organizational pressures (e.g., budget requests). For example, a police department may "crack down" on drug-related crimes in an election year. The result is an increase in the recorded number of these offenses for that year. Such an increase reflects a change in the behavior of law enforcement personnel, not a change in the number of drug violations. Thus, official crime statistics may be a better indicator of what police are doing rather than of what criminals are doing.

The National Incident-Based Reporting System (NIBRS), "implemented to improve the overall quality of crime data collected by law enforcement, captures details on each single crime incident—as well as on separate offenses within the same incident" (FBI 2014d, p. 1). Unlike the UCR, data on "victims, known offenders, relationships between victims and offenders, arrestees, and property involved in the crimes" are collected (p. 1). Only about one-third of the more than 18,000 law enforcement agencies participating in the UCR also participate in the NIBRS.

Victimization Surveys

Acknowledging "**the dark figure of crime**"—that is, the tendency for so many crimes to go unreported and thus undetected by the UCR—the U.S. Department of Justice conducts the National Crime Victimization Survey (NCVS). Begun in 1973 and conducted annually, the NCVS interviews 90,000 households comprising more than 160,000 people 12 and older about their experiences as victims of crime. Interviewers collect a variety of information, including the victim's background (e.g., age, race and ethnicity, sex, marital status, education, and area of residence), relationship to the offender (stranger or nonstranger), characteristics of the offender (e.g., age, race, gender), and the extent to which the victim was harmed (Bureau of Justice Statistics [BJS] 2014a).

In 2013, the latest year for which victimization data are available, there were 6.1 million violent crimes and 16.7 million property crimes, six and two times more, respectively, than estimates using official statistics (FBI 2014a; Truman and Langton 2014). However, there was no statistically significant change in the rate of firearm violence or the rate of *serious* violent crime—that is, serious (1) domestic violence, (2) stranger violence, (3) violent crime with the use of a weapon, or (4) violent crime resulting in injury.

Although adding an important dimension to the study of crime, the NCVS is not without problems. In 2008, a panel of experts offered recommendations "to improve its methodology, assure its sustainability, increase its value to national and local stakeholders, and better meet the challenges of measuring the extent, characteristics, and consequences of criminal victimization" (BJS 2013b, p. 1). The Bureau of Justice Statistics, in response to the recommendations, continues to assess the various methodologies used to collect self-report data.

Self-Report Offender Surveys

Self-report surveys ask offenders about their criminal behavior. The sample may consist of a population with known police records, such as a prison population, or it may include respondents from the general population, such as college students.

Self-report data compensate for many of the problems associated with official statistics but are still subject to exaggerations and concealment. In general, self-report surveys

> In 2013, the latest year for which victimization data are available, there were 6.1 million violent crimes and 16.7 million property crimes, six and two times more, respectively, than estimates using official statistics.

clearance rate The percentage of crimes in which an arrest and official charge have been made and the case has been turned over to the courts.

dark figure of crime The volume of crime that goes unreported.

reveal that virtually every adult has engaged in some type of criminal activity. Why, then, is only a fraction of the population labeled criminal? Like a funnel, which is large at one end and small at the other, only a small proportion of the total population of law violators are ever convicted of a crime. For individuals to be officially labeled criminals, (1) the behavior must become known to have occurred; (2) the behavior must come to the attention of the police, who then file a report, conduct an investigation, and make an arrest; and finally, (3) the arrestee must go through a preliminary hearing, an arraignment, and a trial and may or may not be convicted. At every stage of the process, offenders may be "funneled" out.

Sociological Theories of Crime

Some explanations of crime focus on psychological aspects of the offenders, such as psychopathic personalities, unhealthy relationships with parents, and mental illness. Other crime theories focus on the role of biological variables, such as central nervous system malfunctioning, stress hormones, vitamin or mineral deficiencies, chromosomal abnormalities, and a genetic predisposition toward aggression. Sociological theories of crime and violence emphasize the role of social factors in criminal behavior and societal responses to it.

Structural-Functionalist Perspective

According to Durkheim and other structural functionalists, crime is functional for society. One of the functions of crime and other deviant behavior is that it strengthens group cohesion: "The deviant individual violates rules of conduct that the rest of the community holds in high respect; and when these people come together to express their outrage over the offense, they develop a tighter bond of solidarity than existed earlier" (Erikson 1966, p. 4).

Crime can also lead to social change. For example, an episode of local violence may "achieve broad improvements in city services, be a catalyst for making public agencies more effective and responsive, for strengthening families and social institutions, and for creating public-private partnerships" (National Research Council 1994, pp. 9–10).

Although structural functionalism as a theoretical perspective deals directly with some aspects of crime, it is not a theory of crime per se. Three major theories of crime have developed from structural functionalism, however. The first, called strain theory, was developed by Robert Merton (1957) and uses Durkheim's concept of *anomie*, or normlessness. Merton argued that, when the structure of society limits legitimate means (e.g., a job) of acquiring culturally defined goals (e.g., money), the resulting strain may lead to crime. For example, rapid economic social change in Russia, and specifically high rates of unemployment, led to increases in the homicide rates (Pridemore and Kim 2007).

Individuals, then, must adapt to the inconsistency between means and goals in a society that socializes everyone into wanting the same thing but provides opportunities for only some (see Figure 4.1). *Conformity* occurs when individuals accept the culturally defined goals and the socially legitimate means of achieving them. Merton suggested that most individuals, even those who do not have easy access to the means and goals, remain conformists. *Innovation* occurs when an individual accepts the goals of society but rejects or lacks the socially legitimate means of achieving them. Innovation, the mode of adaptation most associated with criminal behavior, explains the high rate of crime committed by uneducated and poor individuals who do not have access to legitimate means of achieving the social goals of wealth and power.

Another adaptation is *ritualism*, in which, for example, individuals accept a lifestyle of hard work but reject the cultural goal of monetary rewards. *Ritualists* go through the motions of getting an education and working hard, yet they are not committed to the goal of accumulating wealth or power. *Retreatism* involves rejecting both the cultural goal of success and the socially legitimate means of achieving it. *Retreatists* withdraw or retreat

Rebel: rejects wealth and work; substitutes new goals and new means	Culturally Defined Goals:	
Structurally Defined Means:	Acceptance of (+)	Rejection of (−)
Accepts/has access to (+)	Conformist: desires wealth, works hard, saves money	Ritualist: rejects wealth, goes through motions of working hard
Rejects/lacks access to (−)	Innovator: desires wealth, teller at bank, embezzles money	Retreatist: rejects wealth and means to achieve it, turns to drugs

Figure 4.1 Merton's Modes of Adaptation Applied to Crime
SOURCE: Adapted from Merton, 1957.

from society and may become alcoholics, drug addicts, or vagrants. Finally, *rebellion* occurs when individuals reject both culturally defined goals and means and substitute new goals and means. For example, rebels may use social or political activism to replace the goal of personal wealth with the goal of social justice and equality.

Robert Agnew developed a General Strain Theory (GST) based, in part, on Merton's theory of anomie. Agnew argues that when a person experiences strain—that is, a negative condition or event—it increases the likelihood of criminal behavior (Agnew 2013). Strains vary in magnitude, duration, frequency, and so forth, as does the likelihood of a criminal response. For example, being fired (strain) may lead to anger and frustration (negative emotional responses), which in turn lead to stealing from the former employer (crime). Whether a criminal response is forthcoming, however, depends on the strength of the individual's support system, whether there are criminal incentives, the level of social control, ties to criminal or noncriminal significant others, and alternative coping mechanisms (Agnew 2013).

Whereas strain theory often explains criminal behavior as a result of blocked opportunities, subcultural theories argue that certain groups or subcultures in society have values and attitudes that are conducive to crime and violence. Members of these groups and subcultures, as well as other individuals who interact with them, may adopt the crime-promoting attitudes and values of the group. For example, Kubrin and Weitzer (2003) found that retaliatory homicide is a response to subcultural norms of violence that exist in some neighborhoods. However, other theorists argue that subcultural theorists incorrectly assume that subcultures are "coherent, consistent, and homogeneous . . . with stable sets of shared beliefs, practices and identities" (Debies-Carl 2013, p. 118).

If blocked opportunities and subcultural values are responsible for crime, why don't all members of the affected groups become criminals? Control theory may answer that question. Consistent with Durkheim's emphasis on social solidarity, Hirschi (1969) suggests that a strong **social bond** between individuals and the social order constrains some individuals from violating social norms. Hirschi identified four elements of the social bond: *attachment* to significant others, *commitment* to conventional goals, *involvement* in conventional activities, and *belief* in the moral standards of society. Several empirical tests of Hirschi's theory support the notion that the higher the attachment, commitment, involvement, and belief, the higher the social bond and the lower the probability of criminal behavior. Bell (2009) reports that a weaker attachment to parents is associated with a greater likelihood of gang membership for both males and females. Similarly, Gault-Sherman (2012), after analyzing the responses of a national sample of more than 12,500 adolescents, reports a negative association between a youth's attachment to parents and self-reported delinquency—the higher the attachment, the lower the prevalence of delinquency.

social bond The bond between individuals and the social order that constrains some individuals from violating social norms (i.e., committing crime).

Conflict Perspective

Conflict theories of crime suggest that deviance is inevitable whenever two groups have differing degrees of power; in addition, the more inequality there is in a society, the greater the crime rate in that society. Social inequality leads individuals to commit crimes such as larceny and burglary as a means of economic survival. Other individuals, who are angry and frustrated by their low position in the socioeconomic hierarchy, express their rage and frustration through crimes such as drug use, assault, and homicide. In Argentina, for example, the soaring violent crime rate is hypothesized to be "a product of the enormous imbalance in income distribution between the rich and the poor" (Pertossi 2000). Furthermore, an examination of homicide rates in 165 countries indicates that the greater the income inequality, the higher the homicide rate (Ouimet 2012).

According to the conflict perspective, those in power define what is criminal and what is not, and these definitions reflect the interests of the ruling class. Laws against vagrancy, for example, penalize individuals who do not contribute to the capitalist system of work and consumerism. Furthermore, D'Alessio and Stolzenberg (2002, p. 178) found that "in cities with high unemployment, unemployed defendants have a substantially higher probability of pretrial detention" than employed defendants. Rather than viewing law as a mechanism that protects all members of society, conflict theorists focus on how laws are created by those in power to protect the ruling class. Wealthy corporations contribute money to campaigns to influence politicians to enact tax laws that serve corporate interests (Reiman and Leighton 2013).

In addition, conflict theorists argue that law enforcement is applied differentially, penalizing those without power and benefiting those with power.

> [T]he history of racism . . . has led society to choose to fight racial economic equality using the criminal justice system (i.e., incarceration) instead of choosing to reduce racial disparities through consistent investment in social programs. . . . In other words, society chose to use incarceration as a welfare program to deal with the poor, especially since the underprivileged are disproportionately people of color. . . . Given all of the documented social and economic costs of mass incarceration, . . . it can be concluded that it has helped to maintain the economic hierarchy, predicated on race, in the United States. (Cox 2015, p. 2)

Women, too, are the victims of power differentials. Female prostitutes are more likely to be arrested than are the men who seek their services. Similarly, "rape myths" are perpetuated by the male-dominated culture to foster the belief that women are to blame for their own victimization, thereby, in the minds of many, exonerating the offenders. Such beliefs have very real consequences. Men who believe that "women secretly want to be raped," "rape is not harmful," and "some women deserve to be raped" are more likely to commit rape than men who do not endorse such beliefs (Mouilso and Calhoun 2013).

Aneok De Groot/AFP/Getty Images

Under a Dutch plan called *Coalition Project 2012*, prostitutes can no longer advertise by standing or sitting in residential windows or storefronts and, thus, use art objects as symbolic representations. Governmental officials in the Netherlands, which legalized prostitution in 2000, are limiting the number of prostitutes and cannabis cafés in the hopes of reducing crime.

Symbolic Interactionist Perspective

Two important theories of crime emanate from the symbolic interactionist perspective. The first, labeling theory, focuses on two questions: How do crime and deviance come to be defined as such, and what are the effects of being labeled criminal or deviant? According to Howard Becker (1963):

> Social groups create deviance by making rules whose infractions constitute deviance, and by applying those rules to particular people and labeling them as outsiders. From this point of view, deviance is not a quality of the act a person commits, but rather a consequence of the application by others of rules and sanctions to an "offender." The deviant is one to whom the label has successfully been applied; deviant behavior is behavior that people so label. (p. 238)

Labeling theorists make a distinction between **primary deviance**, which is deviant behavior committed before a person is caught and labeled an offender, and **secondary deviance**, which is deviance that results from being caught and labeled. After a person violates the law and is apprehended, that person is stigmatized as a criminal. This deviant label often dominates the social identity of the person to whom it is applied and becomes the person's "master status"—that is, the primary basis on which the person is defined by others.

In research conducted by Pickett et al. (2012), white respondents were asked to estimate the percentage of criminals who are black that commit various serious crimes including robbery with a firearm, residential burglary, commercial burglary, selling illegal drugs, juvenile crime, and violent crime in general. In each case, based on comparisons with official data, respondents overestimated the percent of African Americans involved in crime. For example, when asked, "If you think about crime and criminals, what percent of criminals who commit violent crime in the country are Black?" averaged a *perception* rating of 46.4 percent compared to the actual percentage of 21.0.

Being labeled as deviant often leads to further deviant behavior because (1) the person who is labeled as deviant is often denied opportunities for engaging in nondeviant behavior, and (2) the labeled person internalizes the deviant label, adopts a deviant self-concept, and acts accordingly. For example, a teenager who is caught selling drugs at school may be expelled and thus denied opportunities to participate in nondeviant school activities (e.g., sports and clubs) and to associate with nondeviant peer groups. A review of the literature on labeling and juvenile delinquency lends support for the theory. One researcher concludes, "[R]ather than discourage participation in conventional activities by labeling and isolating offenders, juvenile crime policy should be remedial and foster reintegration" (Ascani 2012, p. 83).

The assignment of meaning and definitions learned from others is also central to the second symbolic interactionist theory of crime, differential association. Edwin Sutherland (1939) proposed that, through interaction with others, individuals learn the values and attitudes associated with crime as well as the techniques and motivations for criminal behavior. Individuals who are exposed to more definitions favorable to law violation (e.g., "Crime pays") than to unfavorable ones (e.g., "Do the crime, you'll do the time") are more likely to engage in criminal behavior.

Unfavorable definitions come from a variety of sources. Of particular concern in recent years is the role of video games in promoting criminal or violent behavior, such as Hatred, in which players kill everyone they can in a variety of gruesome ways (Stuart 2015). In response to this and other violent video games, many states now require a video rating system that differentiates between cartoon violence, fantasy violence, intense violence, and sexual violence. In 2011, the U.S. Supreme Court heard arguments concerning the constitutionality of a California law that banned the sale of violent video games to minors. The court concluded that, like books, movies, or other art, "video games are a creative, intellectual, emotional form of expression and engagement, as fundamentally human as any other" (Schiesel 2011, p. 1).

Types of Crime

The FBI identifies eight index offenses as the most serious crimes in the United States. The **index offenses**, or street crimes, as they are often called, can be against a person

Individuals who are exposed to more definitions favorable to law violation (e.g., "Crime pays") than to unfavorable ones (e.g., "Do the crime, you'll do the time") are more likely to engage in criminal behavior.

primary deviance Deviant behavior committed before a person is caught and labeled an offender.

secondary deviance Deviant behavior that results from being caught and labeled as an offender.

index offenses Crimes identified by the FBI as the most serious, including personal or violent crimes (homicide, assault, rape, and robbery) and property crimes (larceny, motor vehicle theft, burglary, and arson).

TABLE 4.2 Index Crime Rates, Percentage Change, and Clearance Rates, 2013

	Rate Per 100,000 2013	Percentage Change in Rate 2004–2013	Percentage Cleared, 2013
Violent crime			
Murder	4.5	−12.1	64.1
Forcible rape	25.2	−22.1	40.6
Robbery	109.1	−20.2	29.4
Aggravated assault	229.1	−20.6	57.7
Total	367.9	−20.6	48.1
Property crime			
Burglary	610.0	−16.5	13.1
Larceny/theft	1,899.4	−19.6	22.4
Motor vehicle theft	221.3	−47.5	14.2
Arson	N/A*	N/A	20.7
Total†	2,730.7	−22.3	19.7

*Arson rates per 100,000 are calculated independently because population coverage for arson is lower than for the other index offenses.

†Property crime totals do not include arson.

SOURCE: FBI 2014b.

(called violent or personal crimes) or against property (see Table 4.2). Other types of crime include vice crime (such as drug use, gambling, and prostitution), organized crime, white-collar crime, computer crime, and juvenile delinquency. Hate crimes are discussed in Chapter 9.

Street Crime: Violent Offenses

The most recent data available from the FBI's Uniform Crime Report indicates that violent crime decreased by 4.0 percent between 2012 and 2013 (FBI 2014a). Remember, however, that crime statistics represent only those crimes *reported* to the police: 1.2 million violent crimes in 2013 (FBI 2014a). The reasons why people do not report violent victimization include feeling that (1) nothing will or can be done about it, (2) the offender(s) might retaliate, (3) the victimization isn't important enough to report, and (4) there is a better way to handle it (Greenwood 2015; Langton et al. 2012).

Violent crime includes homicide, assault, rape, and robbery. Between 2012 and 2013, all four of the violent offenses declined (FBI 2014a). *Murder* refers to the willful or nonnegligent killing of one human being by another individual or group of individuals. Although murder is the most serious of the violent crimes, it is also the least common, accounting for 1.2 percent of all violent crimes in 2013 (FBI 2014a). A typical homicide scenario includes a male killing a male with a handgun after a heated argument. The victim and offender are disproportionately young and of minority status. When a woman is murdered and the victim–offender relationship is known, she is most likely to have been killed by her husband or boyfriend (FBI 2014a).

Mass murders have more than one victim in a killing event. In 2015, nine members of a historically black church in South Carolina were killed during a prayer meeting. In 2013, Adam Lanza, after murdering his mother, entered Sandy Hook Elementary School in Newtown, Connecticut, and killed 20 first grade students and 6 staff members.

The Sandy Hook killings, in conjunction with the 2012 murder of 12 theater-goers in Aurora, Colorado, and the murder of 6 people attending a constituent meeting held by the former Arizona representative "Gabby" Giffords, led to a national debate about gun control and specifically the availability and use of high-capacity ammunition magazines and assault weapons (see "Gun Control" section).

Unlike mass murder, serial murder is the "unlawful killing of two or more victims by the same offender(s), in separate events" (U.S. Department of Justice 2008, p. 1). The most well-known serial killers, who were responsible for some of the most horrific episodes of homicide, are Ted Bundy, Kenneth Bianchi, and Jeffrey Dahmer. One of the most prolific serial killers is Gary Ridgeway, also known as the "Green River Killer," who was convicted of killing 49 women in 2003. In a 2013 interview, Ridgway claimed that the total number of his victims was somewhere between 75 and 80 (Dolak 2013).

WHAT do you THINK?

It is often assumed that serial killers are exclusively male. Although there are more male serial killers than female serial killers, 20.0 percent of serial murders in the United States are committed by women (Bonn 2014). In fact, this is a higher proportion of women than we would expect given that only 10.0 percent of the total number of homicide offenders in the United States are women. What do you think is the origin of the male serial killer stereotype?

Another form of violent crime, *aggravated assault*, involves attacking a person with the intent to cause serious bodily injury. Like homicide, aggravated assault occurs most often between members of the same race as a result of some kind of argument (FBI 2014d) and, as with violent crime in general, is more likely to occur in warm-weather months. In 2013, the assault rate was over 50 times the murder rate, with assaults making up an estimated 62.3 percent of all violent crimes (FBI 2014a).

It is estimated that nearly 80 percent of all rapes are **acquaintance rapes**—rapes committed by a family member, intimate partner, friend, or acquaintance (BJS 2013a). Although acquaintance rapes are the most likely to occur, they are the least likely to be reported and the most difficult to prosecute. Unless the rape is what Williams (1984) calls a **classic rape**—that is, the rapist was a stranger who used a weapon, and the attack resulted in serious bodily injury—women hesitate to report the crime out of fear of not being believed. The increased use of "rape drugs," such as Rohypnol, may lower reporting levels even further. This chapter's *The Human Side* poignantly describes the impact of rape on one woman's life.

Robbery, unlike simple theft, also involves force or the threat of force or putting a victim in fear and is thus considered a violent crime. Officially, in 2013, more than 345,000 robberies took place in the United States. Robberies are most often committed using "strong-arm" tactics and occur disproportionately in southern states (FBI 2014a). According to the FBI, in 2013, the average dollar value lost per robbery was $1,170, with bank robberies reporting the highest average loss per incident—$3,542 (FBI 2014a). In 2013, eight robbers dressed as police officers and carrying machine guns stole over 100 packages of diamonds valued at $50 million from the tarmac of the Brussels international airport (Michaels 2013).

Street Crime: Property Offenses

Property crimes are those in which someone's property is damaged, destroyed, or stolen; they include larceny, motor vehicle theft, burglary, and arson. The number of property crimes has declined 16.3 percent since 2004, with a 4.1 percent decrease between 2012 and 2013 (FBI 2014a). **Larceny**, or simple thefts (e.g., shoplifting, purse snatching), accounts for more than two-thirds of all property arrests (FBI 2014a) and is the most common index offense. In 2013, the average dollar value lost per larceny incident was $1,259.

Larcenies involving automobiles and auto accessories are the largest known category of thefts. However, because of the cost involved, *motor vehicle theft* is considered a separate index offense. Numbering nearly 700,000 in 2013, the motor vehicle theft rate has

acquaintance rape Rape committed by someone known to the victim.

classic rape Rape committed by a stranger, with the use of a weapon, resulting in serious bodily injury to the victim.

larceny Larceny is simple theft; it does not entail force or the use of force, or breaking and entering.

the HUMAN side

The Hidden Consequences of Rape

In many jurisdictions, victims are allowed to make statements to the court describing the impact of their victimization on their life and the lives of friends and family. Here, 23-year-old Jessica poignantly describes the often unseen consequences of violent crime—in this case, the impact of a brutal rape.

My name is Jessica. I once knew what that meant. Now all I can tell you is that I am still Jessica; however, this no longer holds any meaning to me. I have lost my identity in the cruelest of ways.

I was a 23-year-old single mother, a sister, a daughter, a girlfriend, partner, friend, and confidant to many people. I was strong, independent, and willing to make an effort. I was attempting to hold down a second job to make a better life for my child and a better person out of me. I was destroyed.

I was raped. Not once, not twice, but so many times and in so many ways that it all becomes a blur. These images haunt my days, my nights, my dreams, and my realities. I am no longer the person I was before. I was once the person that people could rely on. Now I am a shell of my former self, a speck of the brave person that was Jessica. I had my way of life, my self-esteem, my respect, and my dignity stripped from me in the most terrifying of situations.

My trust in people is all but destroyed. I even have trouble enjoying a quiet drink with my partner, family, or friends without feeling anxious and wary. I am constantly looking over my shoulder, fearing there is someone there who wants to hurt me. This is only the beginning. I fluctuate from extreme insomnia to extreme fatigue. My motivation is gone. My joy of motherhood is waning. My ability to love and care for others are disappearing. My trust in the system is all but gone. I feel like I tried so hard and was beaten down. I feel weak and vulnerable.

I suffered physical trauma to my shoulder, knees, and feet from being dragged along carpet. I suffered cuts and bruising to my genitals from continuous rapes. I suffered marks to my ankles and wrists from being bound, and I lost chunks of hair from being gagged with tape around my head. I thought I was going to die. I take medication to sleep, to relax, and to stay relatively sane. I live as a corpse with no visible future to aim for. I have cut myself several times to try and take away the pain in my heart. I don't have the strength to end my own life.

What I experienced was like looking into the eyes of Satan himself. I am not a religious person but I know that I have seen the very depths of hell.

I wish that my loved ones didn't know what happened to me. To see the horror in their eyes and feel the pain in their hearts is unbearable. My beautiful daughter doesn't know the beginning of what I suffered but she suffered too. She feels like she was abandoned by her mother and required counseling.

The trauma happened to my partner. I can see it in the way he looks at me. Our communication, love, sex, and friendship is going to take a long time to repair. He has been so deeply affected by my experience but I can't help him and he can't help me. Nobody can.

I was burgled at work on my third day of my second job. I was bound and gagged. I was abducted. I was lied to. I was confused. I was cold and alone. I WAS RAPED. This man, who does not deserve a name, hurt me in the most unimaginable of ways. Yes I survived. Yes I am alive, I just don't live. I am existing. I am empty. . . .

I am a real person who went through torture. I am not a statistic or a nameless face on the street. I am your mother, your sister, your daughter, and your friend. What happened to me was real. At least give me the satisfaction of seeing this man put in jail for the maximum term of 25 years. After all, if I don't qualify as a victim of the most violent, serious, and heinous of crimes, then who the hell does?

I am Jessica.

SOURCE: *Herald Sun* 2008.

decreased 47.5 percent since 2004 (FBI 2014a). Despite the dramatic decrease in car theft, a motor vehicle is stolen every 45 seconds in the United States. Across all model years, the most frequently stolen motor vehicles in 2013 were the Honda Accord, followed by the Honda Civic, Chevrolet Silverado, Ford F-150, and Toyota Camry (Gorzelany 2014).

Burglary, which is the second most common index offense after larceny, entails entering a structure, usually a house, with the intent to commit a crime while inside. The number of burglaries decreased by 8.6 percent between 2012 and 2013 (FBI 2014a). Official statistics indicate that, in 2013, nearly 2 million burglaries occurred. Most burglaries are residential rather than commercial and take place during the day when houses are unoccupied. According to the Department of Justice, an estimated 230,000 guns are stolen annually in home burglaries and other property crimes, and commercial burglaries (e.g., gun stores) are responsible for another 25,000 lost or stolen guns a year (Thomas et al. 2013).

Arson involves the malicious burning of the property of another. Estimating the frequency and nature of arson is difficult given the legal requirement of "maliciousness." Of the reported cases of arson, almost half involved a structure, most of which were residential, and about a quarter involved movable property (e.g., boat or car), with the remainder being miscellaneous property (e.g., crops or timber). In 2013, the average dollar amount of damage as a result of arson was $14,390 (FBI 2014a).

Vice Crime

Vice crimes, often thought of as crimes against morality, are illegal activities that have no complaining participant(s) and are often called **victimless crimes**. Examples of vice crimes include using illegal drugs, engaging in or soliciting prostitution, illegal gambling, and pornography.

Most Americans view illicit drug use, with the exception of marijuana, as socially disruptive (see Chapter 3). There is less consensus, however, nationally or internationally, that gambling and prostitution are problematic. For example, Germany legalized prostitution in 2002, and it is estimated that there are now more than 400,000 sex workers, two-thirds of whom are from overseas. That said, one of the arguments commonly voiced against prostitution is that it makes it easier for women and girls to be forced into prostitution by traffickers. In a study of 116 countries, Cho et al. (2013) conclude that countries where prostitution is legal have higher rates of human trafficking than countries in which prostitution is illegal.

Alternatively, in Canada, where prostitution laws have traditionally been liberal, a 2014 law "criminalizes the purchase of sex, as well as things like advertising or other forms of communication related to its sale" (Canadian Press 2014). In the United States, prostitution is illegal with the exception of several counties in Nevada. Despite its illegal status, it is a multimillion-dollar industry, with over 48,620 people arrested for prostitution and commercial vice in 2013 (FBI 2014a). Human trafficking for purposes of prostitution occurs both *between* the United States and other countries, as well as *within* the United States. Children are particularly vulnerable.

> Pimps and traffickers sexually exploit children through street prostitution, in adult strip clubs, brothels, sex parties, motel rooms, hotel rooms, and other locations throughout the United States. Many recovered American victims are runaways or throwaway youth who often suffer from a history of physical abuse, sexual abuse, and family abandonment issues. This population is seen as an easy target by pimps because the children are generally vulnerable, without dependable guardians, and suffer from low self-esteem. (USDOJ 2012a)

In 2003, the FBI established the Innocence Lost National Initiative to address domestic sex trafficking of children. Federal, state, and local law enforcement efforts have resulted in the recovery of 3,400 child victims (FBI 2014c).

Gambling is legal in many U.S. states, including casinos in Nevada, New Jersey, Connecticut, North Carolina, and other states, as well as state lotteries, bingo parlors, horse and dog racing, and jai alai. Of late, there are new concerns about online gambling. In 1961, Congress passed the Federal Wire Act that prohibited interstate gambling but, in 2011, it was interpreted as permitting online gambling with the exception of sports betting. In 2015, the Restoration of America's Wire Act was introduced into Congress, which, if passed, would criminalize all Internet gambling (Devaney 2015). Opponents of the law argue that the bill is nothing more than "a ploy by casinos to lockdown the competition from up-and-coming websites" (p. 1). Some have argued that there is little differences between gambling and other risky ventures such as investing in the stock market, other than societal definitions of acceptable and unacceptable behavior. Conflict theorists are quick to note that the difference is who is making the wager.

Pornography, particularly Internet pornography, is a growing international problem. Regulation is made difficult by fears of government censorship and legal wrangling as to what constitutes "obscenity." For many, the concern with pornography is not its

victimless crimes Illegal activities that have no complaining participant(s) and are often thought of as crimes against morality, such as prostitution.

consumption per se but the possible effects of viewing or reading pornography—increased sexual aggression. The literature on the relationship between pornography and sexual aggression is mixed. Malamuth et al. (2012), using a representative sample of American men, conclude that some men actively seek out violent pornographic material, which in turn leads to sexually aggressive attitudes toward women. Alternatively, Diamond et al. (2011) report that there were no significant differences in the number of rapes and other sex crimes before and after the legalization of pornography in the Czech Republic.

Organized Crime

Traditionally, **organized crime** refers to criminal activity conducted by members of a hierarchically arranged structure devoted primarily to making money through illegal means. For many people, organized crime is synonymous with the Mafia, a national band of interlocked Italian families, or the "Irish mob" made famous by such movies as *The Road to Perdition* and *Gangs of New York*. The Irish mob is one of the oldest organized crime groups in the United States. Fitting the stereotype of the "gangster" is James (Whitey) Bulger who, in 2013, was charged with 19 murders, drug trafficking, extortion, bribery, bookmaking, and loan sharking, just to name a few of the crimes listed in the 32-count indictment. He was convicted of 31 counts and sentenced to two consecutive life terms in prison (Murphy and Valencia 2013; Seelye 2013).

Organized crime also occurs in other countries. With nearly 60,000 members, the Japanese Yakuza is a network of gangs involved in drugs, prostitution, extortion, and white-collar crime (French Press Agency 2014). The largest of the Yakuza crime groups, with 40.0 percent of the total membership, is the Yamaguchi-gumi syndicate which, in the face of tougher anti-gang laws and concerns over recent acts of violence, published a public relations magazine aimed at winning back public support and recruiting new members. Whereas membership in the Yakuza is not illegal, the activities that members engage in are (McCurry 2013).

Although traditional organized crime groups still exist, Jay Albanese, a noted criminologist, contends that globalization and technology have facilitated a change from the "old" organized crime groups to *transnational* organized crime (TOC) groups (see Table 4.3). These groups continue to supply many of the same products and services (e.g., prostitution, drugs, gambling) and engage in many of the same behaviors (e.g., extortion, money laundering, theft) as their predecessors (Albanese 2012). Transnational crime organizations are a growing threat to the United States and to global security.

TABLE 4.3 The Evolution of Organized Crime

Original Activity	Modern Version
Local numbers and lottery gambling	Internet gambling at international sites outside national regulation
Heroin and cocaine tracking	Synthetic drugs (less vulnerable to supply problems)
Street prostitution	Internet prostitution and trafficking in human beings
Extortion of local businesses for protection money	Extortion of corporations, kidnappings, and piracy for profit
Loan sharking (exchanging money at interest rates above the rate permitted by law)	Laundering of money, precious stones, and commodities
Theft and fencing stolen property	Theft of intellectual property, Internet scams, and trafficking globally available goods (e.g., weapons, natural resources)

SOURCE: Albanese 2012.

organized crime Criminal activity conducted by members of a hierarchically arranged structure devoted primarily to making money through illegal means.

WHAT do you THINK?

As a result of well-financed organized transnational crime groups in Africa, the Middle East, and around the world, the United Nations Security Council passed a resolution expressing its concerns with "financing obtained by terrorist groups through illicit activities—such as the trafficking of drugs, people, arms, and artifacts" (UN 2014, p.1). Do you think members of organized transnational crime groups whose profits finance state-sponsored terrorism should be treated as criminals or as soldiers fighting a holy war (e.g., Jihadists)?

White-Collar Crime

White-collar crime includes *occupational crime*, in which individuals commit crimes in the course of their employment, and *corporate crime*, in which corporations violate the law in the interest of maximizing profit. Occupational crime is motivated by individual gain. Employee theft of merchandise, embezzlement, and insurance fraud are examples of occupational crime. Price fixing, antitrust violations, and security fraud are all examples of corporate crime, that is, crime that benefits the organization (see Table 4.4).

Recently, several bankers have been charged with criminal offenses. Officials of the French bank BNP Paribas pled guilty to criminal charges that they had violated U.S. money laundering laws. The bank was fined nearly $9 billion (Thompson and Perez 2015). Leaked files from the Swiss subsidiary of the British bank HSBC, the second-largest bank in the world, revealed that bank officials had "helped wealthy customers dodge taxes and conceal millions of dollars of assets, doling out bundles of untraceable cash and advising clients on how to circumvent domestic tax authorities" (Leigh et al. 2015, p.1). HSBC officials state that they are cooperating with authorities in investigating the misconduct.

No doubt triggered by investigations of bank fraud on the heels of the subprime mortgage crisis and a host of other financial scandals, the Obama administration has increased its attention on white-collar crime. In President Obama's 2012 State of the Union Address, he "vowed to establish a new financial crimes unit dedicated to investigating and prosecuting 'large-scale' financial fraud" (quoted in Carter and Berlin 2012). Nonetheless, according to one federal judge, "The lack of charges against high profile individuals involved in the financial crisis . . . [is] one of the most egregious failures of the criminal justice system in many years" (quoted in Cohn 2014, p. 1).

White-collar criminals go unpunished for a variety of reasons. First, many companies, not wishing the bad publicity surrounding a scandal, simply dismiss the parties involved and/or pay large fines to the government in lieu of prosecution. For example, despite strong evidence of securities fraud by JPMorgan and a ready-to-testify whistleblower, federal prosecutors entered into a $13 billion civil settlement with the bank, and neither JPMorgan nor any of its executives were criminally charged (McCoy 2014).

Second, many white-collar crimes, as with traditional crimes, go undetected. In a survey of a representative sample of 2,503 U.S. households, the National White Collar Crime Center (NWCCC) found that nearly one in four households (24 percent) had been the victim of some type of white-collar crime in the previous year. However, only 11.7 percent of the crimes were brought to the attention of a law enforcement agency (NWCCC 2010). Americans' lingering perceptions that white-collar crime is not "real crime" and, therefore, is less serious (Holtfreter et al. 2008) may have contributed to this lack of reporting. In a study on sentencing outcomes for white-collar offenders versus

TABLE 4.4 Types of White-Collar Crime

CRIMES AGAINST CONSUMERS	CRIMES AGAINST EMPLOYEES
Deceptive advertising	Health and safety violations
Antitrust violations	Wage and hour violations
Dangerous products	Discriminatory hiring practices
Manufacturer kickbacks	Illegal labor practices
Physician insurance fraud	Unlawful surveillance practices
CRIMES AGAINST THE PUBLIC	**CRIMES AGAINST EMPLOYERS**
Toxic waste disposal	Embezzlement W
Pollution violations	Pilferage
Tax fraud	Misappropriation of government funds
Security violations	Counterfeit production of goods
Police brutality	Business credit fraud

© Cengage Learning 2017.

white-collar crime Includes both *occupational crime*, in which individuals commit crimes in the course of their employment, and *corporate crime*, in which corporations violate the law in the interest of maximizing profit.

property offenders, Van Slyke and Bales (2012) conclude that, despite structured sentencing guidelines designed to prevent sentencing disparities, white-collar offenders received preferential treatment when compared to property offenders.

Third, federal prosecutions of white-collar criminals have generally decreased. Few believe the decrease is a result of a lower prevalence of white-collar crime offenses. Two forces appear to be in operation. First, white-collar crimes are becoming increasingly complex, making prosecution a time and resource-intensive endeavor particularly in times of intense budget cuts (Cohn 2014). Second, the decrease in white-collar crime prosecutions represents a shift in priorities (Marks 2006). After the events of 9/11, nearly one-third of FBI agents were moved from criminal programs to terrorism and intelligence duties, leaving "the bureau seriously exposed in investigating areas such as white-collar crime" (Lichtblau et al. 2008, p. 1).

Corporate Violence. **Corporate violence**, a form of corporate crime, refers to the production of unsafe products and the failure of corporations to provide a safe working environment for their employees. Corporate violence is the result of negligence, the pursuit of profit at any cost, and intentional violations of health, safety, and environmental regulations. The automobile industry provides several examples of corporate violence.

Air Force Lt. Stephanie Erdman's eye was injured by shrapnel when the airbag in her 2002 Honda Civic exploded. Lt. Erdman testified before the U.S. Senate Committee on Commerce, Science, and Transportation on the defective Takata airbags and the need for vehicle recalls.

Jim Watson/AFP/Getty Images

In 2014 and 2015, 11 years after ignition system problems were discovered, General Motors recalled over 2.6 million vehicles (Geier 2015; NBC News 2014b). To date, there have been 251 death claims and over 2,000 injury claims against the automobile manufacturer (NBC News 2014a). Similarly, Honda failed to account for 1,729 death and injury claims and to submit "early warning reports identifying potential or actual safety issues" to the National Highway Traffic Safety Administration (NHTSA) (Sanders 2015, p. 1).

In the largest automobile recall in history, nearly 30 million cars across 10 different car manufacturers were recalled by the NHTSA and/or car manufacturers for dangerous airbags (Consumer Reports 2015). The injuries from the airbags, which exploded during minor accidents or, in some cases, for no apparent reason, were so traumatic that investigating police officers often thought that the victims had been shot or stabbed (Isidore 2014). To date, there have been eight deaths and more than a hundred injuries as a result of the airbags deploying (Consumer Reports 2015). Workers at Takata Corporation, makers of the airbags, say officials knew of the problem as early as 2004 when after-hours and weekend tests were performed in secret. Upon discovery of the problem, instead of alerting safety officials, "Takata executives discounted the results and ordered the lab technicians to delete the testing data from their computers and dispose of the airbag inflators in the trash" (Tabuchi 2014).

Computer Crime

corporate violence The production of unsafe products and the failure of corporations to provide a safe working environment for their employees.

computer crime Any violation of the law in which a computer is the target or means of criminal activity.

hacking Unauthorized computer intrusion.

Computer crime refers to any violation of the law in which a computer is the target or means of criminal activity. Sometimes called *cybercrime*, computer crime is one of the fastest-growing crimes in the United States. **Hacking**, or unauthorized computer intrusion, is one type of computer crime. In 2013, sometimes called the "year of the retailer breach," a variety of outlets were hacked, including the *New York Times*, CNN, *The Guardian*, the *Wall Street Journal*, Microsoft, Adobe, Apple, Home Depot, Nordstrom, and Target (Verizon 2014).

Furthermore, in 2015, one of the largest hacking heists in history took place as attackers penetrated the administrative systems of more than 100 banks and other financial institutions in 30 countries. Estimates of the "take" are as high as $1 billion (Sanger and Perlroth 2015). Not surprisingly, in a 2014 survey of Americans, 69.0 percent reported that they "frequently or occasionally" worried about having their credit card information stolen and used by hackers (Riffkin 2014).

The Consumer Sentinel Network logged nearly 7 million consumer complaints between 2012 and 2014, with identity theft being the single-largest category of complaints (Federal Trade Commission 2015). **Identity theft**, the second type of computer crime, is the use of someone else's identification (e.g., Social Security number or birth date)

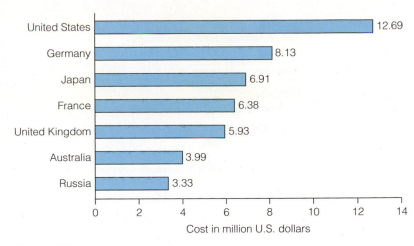

Figure 4.2 Average Costs of Cybercrime in Selected Countries as of June 2014 (in million U.S. dollars)
SOURCE: Statista 2015.

to obtain credit or other economic rewards. In 2015, in the largest government cyber-intrusion in history, Social Security numbers, birth dates, and other personnel data of an estimated 18 million former, present, and prospective federal employees were compromised (Dilanian 2015).

A third type of computer crime is Internet fraud. According to the Internet Crime Complaint Center (ICCC 2014a), common types of Internet fraud include nondelivery of merchandise or payment, debt elimination fraud, auction fraud, and investment fraud. Other Internet fraud categories include check fraud, Nigerian letter fraud (i.e., a letter offering the recipient the "opportunity" to share in millions of dollars being illegally transferred to the United States—just send us your bank account numbers!), computer fraud, and credit/debit card fraud.

A fourth type of computer crime is online child pornography and child sexual exploitation. In 2013, the National Center for Missing and Exploited Children's' (NCMEC) Cyber Tipline received more than 500,000 reports of suspected child pornography and/or child sexual exploitation (NCMEC 2014). Of that number, 97 percent of the reports entailed the "production, manufacture and/or distribution of child pornography" (NCMEC 2014, pp. 8–9). The NCMEC, in conjunction with local law enforcement services, the FBI, and the U.S. Department of Justice, also deals with child sex trafficking cases. The data indicate that in 2014, one in six runaways were likely to have been the victims of sex trafficking, and 60 percent of the trafficking victims had been in the care of social services or a foster home prior to victimization (NCMEC 2015).

Finally, Verizon's (2014) *Data Breach Investigations Report* provides an excellent summary of cyberattacks around the world. The report analyzed 1,367 confirmed data breaches and over 63,437 security incidents from organizations, governments, corporations, and banks in 95 countries. The results indicate that the United States has suffered more data beaches than any other country in the world (see Figure 4.2). The report also reveals that the most common type of computer breaches are hacking, followed by malware, social (e.g., impersonating someone to gain access to a computer), and physical (e.g., stealing the hard drive).

Juvenile Delinquency and Gangs

In general, children younger than age 18 are handled by the juvenile courts, either as status offenders or as delinquent offenders. A *status offense* is a violation that can be committed only by a juvenile, such as running away from home, truancy, and underage drinking. A *delinquent offense* is an offense that would be a crime if committed by an adult, such as the eight index offenses. The most common status offenses handled in juvenile court are underage drinking, truancy, and running away. In 2013, 9.7 percent of

identity theft The use of someone else's identification (e.g., Social Security number, birth date) to obtain credit or other economic rewards.

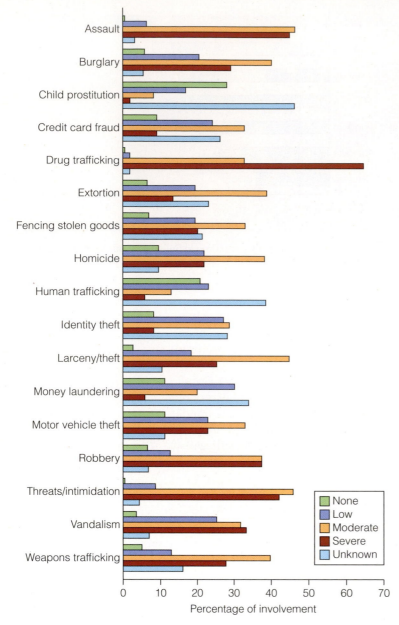

Percentage of involvement

Figure 4.3 Law Enforcement Ratings of Gang Level Involvement in Various Crimes
SOURCE: NGR 2014.

all arrests (excluding traffic violations) were of offenders younger than age 18 (FBI 2014a). As is the case with adults, juveniles commit more property crimes than violent crimes, and males are more likely to be arrested than females.

The number of juveniles *arrested* for violent crimes decreased by 8.6 percent between 2012 and 2013 (FBI 2014b). Nonetheless, the National Gang Report (NGR 2014), based on a survey of 631 U.S. law enforcement agencies, estimates that gang membership and particularly female gang membership continues to grow. The growth of gangs is, in part, a function of two interrelated social forces: the increased availability of guns in the 1980s, and the lucrative and expanding drug trade. Recently, however, gangs have expanded into nontraditional gang activities such as white-collar crime (e.g., counterfeiting, extortion), smuggling, prostitution, and human trafficking, often working with national gangs, drug trafficking organizations, and criminal syndicates (see Figure 4.3) (NGR 2014).

In 2013, of all gang members, 88 percent were in neighborhood-based gangs (NBGs), the most violent and destructive to communities; 2.5 percent were in outlaw motorcycle gangs; and 9.5 percent in prison gangs, considered jurisdictionally to be the least problematic (NGR 2014). Gang members who infiltrate military bases, correctional facilities, and government and law enforcement agencies use their positions to gain advantage over rival gangs and police, and to disrupt criminal investigations (NGR 2014). Furthermore, gangs are estimated to be in 80 percent of public schools and are a formidable obstacle for educators, law enforcement, and other youth-service professionals. Gang presence on college campuses is also a concern as more gang members are gravitating toward colleges to escape gang life, join college athletic programs, or to acquire advanced skill sets for their gang.

Demographic Patterns of Crime

Although virtually everyone violates a law at some time, individuals with certain demographic characteristics are disproportionately represented in the crime statistics (see Figure 4.4). Victims, for example, are disproportionately young, lower-class, minority males from urban areas. Similarly, the probability of being an offender varies by gender, age, race, social class, and region. This section ends with a discussion of crime and victimization.

Gender and Crime

It is a universal truth that women everywhere are less likely to commit crime than men. In the United States, both official statistics and self-report data indicate that females commit fewer violent crimes than males. In 2013, males accounted for 73.5 percent of

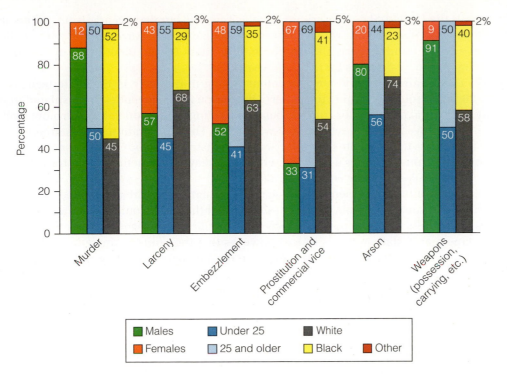

Figure 4.4 Percentage of Arrests for Selected Crimes by Sex, Age, and Race, 2013
SOURCE: FBI 2014a.

all arrests, 79.9 percent of all arrests for violent crime, and 62.2 percent of all arrests for property crimes (FBI 2014a). Not only are females less likely than males to commit serious crimes, the monetary value of female involvement in theft, property damage, and illegal drugs is typically far less than that for similar offenses committed by males.

Nonetheless, rates of female criminality have increased over the last decade. Between 2004 and 2013, arrest rates for women increased for burglary and larceny, contributing to the 15.5 percent increase of women arrested for property crimes (FBI 2014a). Over the same time period, female arrests increased over 15.0 percent for robbery, driving while intoxicated, and gambling. Heimer et al. (2005) hypothesize that such increases are a function of the *economic marginalization* of women relative to men; that is, female criminality goes up when women's economic circumstances in relation to men's goes down.

The increase in crimes committed by females has led to the growth of feminist criminology. **Feminist criminology** focuses on how the subordinate position of women in society affects their criminal behavior and victimization. For example, arrest rates for runaway juvenile females are higher than those for males not only because girls are more likely to run away as a consequence of sexual abuse in the home but also because police with paternalistic attitudes are more likely to arrest female runaways than male runaways (Chesney-Lind and Sheldon 2004; Sheldon 2013; Zahn et al. 2010). Feminist criminology, concentrating on gender inequality in society, thus adds insights into understanding crime and violence that are often neglected by traditional theories of crime. Feminist criminology has also had an impact on public policy. Mandatory arrest for domestic violence offenders, the development of rape shield laws, public support for battered women's shelters, laws against sexual harassment, and the repeal of the spousal exception in rape cases are all, according to Winslow and Zhang (2008), outcomes of feminist criminology.

Age and Crime

In general, criminal activity is more prevalent among younger people than among older people. In 2013, 37.5 percent of all arrests in the United States were of people younger than age 25 (FBI 2014a). Although those younger than age 25 made up over half of all

[T]he development of rape shield laws, public support for battered women's shelters, laws against sexual harassment, and the repeal of the spousal exception in rape cases are all . . . outcomes of feminist criminology.

feminist criminology
An approach that focuses on how the subordinate position of women in society affects their criminal behavior and victimization.

> [African Americans] have 3.7 times the arrest rate for possession of marijuana, are six times more likely to be admitted to prison, and, if admitted to prison for a violent crime, receive longer sentences than their white counterparts.

arrests in the United States for crimes such as robbery, burglary, vandalism, and arson, those younger than age 25 were significantly less likely to be arrested for white-collar crimes such as fraud, forgery, and counterfeiting. Those 65 and older made up less than 1 percent of the total arrests for the same year.

Why is criminal activity more prevalent among individuals in their late teens and early 20s and rapidly declining thereafter? One reason is that juveniles are insulated from many of the legal penalties for criminal behavior. Younger individuals are also more likely to be unemployed or employed in low-wage jobs. Thus, as strain theorists argue, they have less access to legitimate means for acquiring material goods. On the other hand, the decline in criminal offenses associated with aging may be a function of the transition to conventional roles—employee, spouse, and parent.

Other hypothesized reasons for the age–crime relationship are also linked to specific theories of criminal behaviors. For example, conflict theorists would argue that teenagers and young adults have less power in society than their middle-aged and elderly counterparts. One manifestation of this lack of power is that the police, using a mental map of who is a "typical offender," are more likely to have teenagers and young adults in their suspect pool. With increased surveillance of teenagers and young adults comes increased detection of criminal involvement—a self-fulfilling prophecy.

In the hopes of resolving the theoretical debate over the relationship between age and crime, Sweeten, Piquero, and Steinberg (2013) examined more than 40 independent variables and their association with self-reported delinquency. Using a sample of 1,300 serious offenders, the researchers interviewed youth at the age of 16 and every six months thereafter for a period of seven years. Although there was some level of support for each of the theories tested, the strongest theoretical explanation of the relationship between age and delinquency was Sutherland's learning theory—the differential association theory—confirming the positive association between criminality and antisocial peers.

WHAT do you THINK?

In a recent study of newly admitted 16- to 18-year-olds to the New York City jail system, it was revealed that half of the males and females had traumatic brain injury (TBI), defined as "one or more injuries with altered mental state" (Kapa et al. 2014, p. 616). At least in the case of assaults, which comprised over half of the circumstances surrounding the young inmates' injuries, the TBI was caused by violence. In turn, research indicates that inmates with TBI are more likely to (1) violate facility rules, (2) be drug involved, (3) repeat criminal acts, and (4) have great difficulty acclimating to life outside confinement (Kaba et al. 2014). What do you think can be done to stop the violence–TBI–crime cycle?

Race, Social Class, and Crime

Race is a factor in who gets arrested, convicted, and incarcerated. For example, African Americans represent about 14.0 percent of the population but account for 38.7 percent of all arrests for violent index offenses, and 29.0 percent of all arrests for property index offenses (FBI 2014a). They have 3.7 times the arrest rate for possession of marijuana, are six times more likely to be admitted to prison, and, if admitted to prison for a violent crime, receive longer sentences than their white counterparts (American Civil Liberties Union [ACLU] 2013; Cox 2015; Hartney and Vuong 2009).

Nevertheless, it would be inaccurate to conclude that race and crime are causally related. First, official statistics reflect the behaviors and policies of criminal justice actors. Thus, the high rate of arrests, conviction, and incarceration of minorities may be a consequence of individual and institutional bias against minorities. For example, an analysis of FBI arrest data from 3,538 police departments indicated that in less than 0.5 percent of the departments was there racial equity in arrests (Heath 2014).

Racial bias, sometimes called **racial profiling**, is the practice of targeting suspects on the basis of race. Proponents of the practice argue that because race, like gender, is a significant predictor of who commits crime, the practice should be allowed. Opponents hold that racial profiling is little more than discrimination, often based on

racial profiling The law enforcement practice of targeting suspects on the basis of race.

stereotypes, and thus should be abolished. A survey of Seattle residents lends support to such a contention (Drakulich 2013). The results indicate that "crime stereotypes about racial and ethnic minorities are associated with reduced perceptions of neighborhood safety and increased anxieties about victimization" among white respondents (p. 322) (see also this Chapter's *Social Problems Research Up Close*). Presently, more than 25 states have laws that prohibit racial profiling and/or require that state jurisdictions collect data on police stops and searches (Resource Center 2013).

In 2014, thousands of Americans—young and old, black and white, male and female—participated in "days of resistance" demonstrations after grand jury decisions failed to hold police officers criminally responsible for the deaths of two unarmed black men, Michael Brown and Eric Garner. Demonstrators made "hands-up" gestures referring to witnesses' accounts that Michael Brown was surrendering when shot, and carried signs reading "I can't breathe," Eric Garner's last words. With signs reading "Ferguson Is Everywhere," in 2015, demonstrators protested the killing of other black males, including Freddie Gray and 12-year-old Tamir Rice, both of which were ruled homicides. A 2015 study found that, adjusting for the U.S. population, black Americans are twice as likely to be shot by the police as white and Hispanic Americans (Laughland et al. 2015). African Americans are also significantly more likely to, when shot, be unarmed.

On December 12, 2014, peaceful protesters gathered in Los Angeles, protesting jury decisions not to indict police officers who were responsible for the deaths of several black males including Trayvon Martin, Michael Brown, and Eric Garner. Today, Black Lives Matter is a social movement advocating for criminal justice reform and police accountability.

A second reason for the higher rates of African American involvement in the criminal justice system is that nonwhites are overrepresented in the lower classes. Because lower-class members lack legitimate means to acquire material goods, they may turn to instrumental, or economically motivated, crimes. In addition, although the haves typically earn social respect through their socioeconomic status, educational achievement, and occupational role, the have-nots more often live in communities where respect is based on physical strength and violence, as subcultural theorists argue. For example, Kubrin (2005) examined the "street code" of inner-city black neighborhoods by analyzing rap music lyrics. Her results indicate that "lyrics instruct listeners that toughness and the willingness to use violence are central to establishing viable masculine identity, gaining respect, and building a reputation" (p. 375).

Some research indicates, however, that even when social class backgrounds of blacks and whites are comparable, blacks have higher rates of criminality. Using data from nearly 3,000 respondents aged 18 to 25, Sampson et al. (2005) found that the likelihood of self-reported violence by blacks was 85 percent higher than for whites. Interestingly, the likelihood of Latino self-reported violence was 10 percent less than that reported by whites.

Region and Crime

In general, crime rates and, in particular, violent crime rates increase as population size increases. For example, in 2013, the violent crime rate in metropolitan statistical areas (MSAs) was 397.4 per 100,000 inhabitants; in cities outside of metropolitan areas, it was 376.7 per 100,000 inhabitants; and, finally, in nonmetropolitan counties, the crime rate was 180.8 per 100,000 inhabitants (FBI 2014a). Furthermore, the National Crime Victimization Survey indicates that violent crime, serious violent crime, and property crime each decrease as one moves from urban to suburban to rural areas (Truman and Langton 2014).

Higher crime rates in urban areas result from several factors. First, social control is a function of small intimate groups that socialize their members to engage in law-abiding behavior, expressing approval for their doing so and disapproval for their noncompliance. In large urban areas, people are less likely to know one another and thus are

As discussed in Chapter 1, experiments are often used in social science research. You should remember that in experimental designs, the researcher manipulates the independent variable and measures its effect on the dependent variable, keeping all other variables constant. The present research, using an experimental design, investigates one of the most important issues in the sociological study of crime, criminals, and social control: the role of race in criminal justice policies.

Study 1 Sample and Methods

At a time when California was considering amending their three-strikes policy, Hetey and Eberhardt (2014) recruited 62 California voters to participate in a study in return for a $10 gift card. Using an iPad, subjects "watched a 40-s[econd] video in which 80 color photographs (actual mugshots) of black and white male inmates sequentially flashed across the screen" (p. 1950). In the *less black* condition, 25.0 percent of the photographs were of black inmates, the actual percent of African Americans in California prisons at that time. In the *more black* condition, 45.0 percent of the photographs were of African Americans males, the actual percent of black males in prison under California's three-strikes policy. Participants were then informed about the policy, which mandated a 25-year to life sentence for a third felony if the first two were serious or violent. Subjects were then asked how harsh they thought the current three-strikes policy was on a scale from 1 (not punitive enough) to 10 (too punitive). The researchers also asked the participants if they wanted to sign "a real petition for a statewide ballot initiative that would lessen the severity of the nation's most punitive three-strikes law" (p. 1950).

Results of Study 1

A participant's perception of the California prison population as *more black* or *less black* had no effect on perceived punitiveness of the existing three-strikes law. However, a participant's perception of the prison population as *more black* or *less black* was predictive of who signed the petition. Subjects who believed the prison population to be *more black* were less likely to be in favor of lowering the severity of the punishment than those who perceived the prison population to be *less black*.

Study 2 Sample and Methods

In Study 2, the researchers examined the relationship between fear of crime

not influenced by the approval or disapproval of strangers. Demographic factors also explain why crime rates are higher in urban areas: Large cities have large concentrations of poor, unemployed, and minority individuals. Some cities, including the 10 most dangerous cities in the United States, have crime rates as much as five times the national average.

Crime rates also vary by region of the country. In 2013, both violent and property crimes were highest in southern states, with 43.8 percent of all murders, 37.8 percent of all rapes, and 43.0 percent of all aggravated assaults recorded in the South (FBI 2014a). The high rate of southern lethal violence has been linked to high rates of poverty and minority populations in the South, a southern "subculture of violence," higher rates of gun ownership, and a warmer climate that facilitates victimization by increasing the frequency of social interaction.

Crime and Victimization

The National Crime Victimization Survey (NCVS) indicates a slight decline in violent crime, from 26.1 to 23.2 victimizations per 1,000 population, and a more dramatic decrease in property crime, from 155.8 to 131.4 victimizations per 1,000 household (Truman and Langston 2014). Males have a slightly higher rate of violent victimization than females, racial and ethnic minorities have a higher rate of violent victimization than white non-Hispanics, and people between the ages of 12 to 17 have the highest rate of violent victimization when compared to other age groups.

and acceptance of punitive criminal justice policies. The researchers recruited 164 white adult New York City residents to complete an online study. Rather than photographs, respondents were given statistics of the percentage of black and white inmates in New York prisons. In the *less black* condition, participants were told that the prison population was about 40.0 percent black, about the distribution of the inmate population in the United States, and 32.0 percent white. In the *more black* condition respondents were told that the prison population was 60.0 percent black, about the percent of black inmates in the New York City Department of Corrections facilities, and 12 percent were white. A Crime Concern Scale was created using four questions (e.g., "To maintain safety, how justified is it to use stop-and-frisk tactics?"), with possible responses ranging from "not at all" (1) to "extremely" (6). Subjects were then asked how punitive they thought the stop-and-frisk program was, and whether or not they would sign a petition supporting the end of the stop-and-frisk program.

Results of Study 2

Consistent with the results of Study 1, participants in the *more black* condition were significantly less likely to sign a petition to end the stop-and-frisk policy then participant in the *less black* condition. Furthermore, participants in the *more black* condition were significantly more concerned about crime and, in turn, being more concerned about crime was linked to a lower likelihood of signing a petition to end the stop-and -frisk policy.

Conclusion

As Hetey and Eberhardt (2014) hypothesized, the perception of a greater racial disparity in prisons, i.e., a higher percentage of blacks, was associated with (1) a lower willingness to amend California's three-strikes law and (2) a higher willingness to retain New York City's stop-and-frisk policy. The authors acknowledge that in the United States, as around the world, minorities are disproportionately treated punitively. Although the present research may shed light on some of the reasons for racial disparities, the authors fear that "exposure to extreme racial disparities may make the public less, not more, responsive to attempts to lessen the severity of policies that help maintain those disparities" (p. 1952).

SOURCE: Hetey and Eberhardt 2014.

Because minorities are disproportionately offenders and violent victimization is often directed toward family members, friends, and acquaintances, it is not surprising that African Americans, Hispanics, and American Indian/Alaska natives are disproportionately victimized. Familiarity also explains the high rate of victimization of males and young adults between the ages of 12 and 17.

According to the National Criminal Victimization Survey, in 2013, reports of victimization were highest for serious violent crime, followed by violent crimes, and property crimes. Rape and sexual assault were the least likely violent crimes to be reported (34.8 percent), followed by simple assault (38.5 percent). Theft was the least likely reported property crime (28.6 percent) and motor vehicle theft the highest (75.5percent).

Females are particularly vulnerable to certain kinds of crimes, most notably those that entail sexual violence. In 2013, Ariel Castro was charged "... with 512 counts of kidnapping, 446 cases of rape, seven counts of gross sexual imposition, six counts of felonious assault, [and] three counts of child endangerment..." (Welsh-Huggins 2013, p.1). Castro, who had held three young women as captives for nearly 10 years, was sentenced to life in prison but committed suicide just a month after sentencing (Boyle 2015). Although males are more likely than females to be victims of violent crime, females.

Children, too, are especially vulnerable, both directly and indirectly. According to the NIBRS (FBI 2014d), in 2013, there were 695 homicides of victims under the age of 21, 18.1 percent of the total number of homicides in that year. Of the 695 homicides, 28.8 percent were 10 years old or younger, 8.9 percent were between the ages of 11 and 15, and 61.3 percent between the ages of 16 and 20. Children 10 years old and younger comprised nearly one-third of sex offense victims in 2013.

The Societal Costs of Crime and Social Control

As this chapter has demonstrated, there are a variety of types of crimes and criminals. Nonetheless, often based on media stereotypes, many people think only of street crimes and criminals as predominantly "young black men living in poor urban neighborhoods committing violent and drug-related crimes" (Leverentz 2012, p. 348). The costs of crime—and many of them are incalculable—go far beyond those perpetrated by what is thought to be the "typical offender." For example, transnational organized crime, perhaps the most insidious of all crime categories, "threatens peace and human security, leads to human rights being violated, and undermines the economic, social, cultural, political, and civil development of societies around the world" (UNODC 2012). Although these costs are impossible to quantify, the following section discusses other costs associated with crime and criminals, including physical injury and loss of life, economic losses, social and psychological costs, and the costs to families and children.

Physical Injury and the Loss of Life

Crime often results in physical injury and the loss of life. For example, homicide is the third-most common cause of death among 15- to 24-year-olds, exceeded only by accidental death and suicide (Centers for Disease Control and Prevention 2014). In 2013, 14,196 people were the victims of known homicides in the United States (FBI 2014a). That number is dwarfed, however, by the deaths that take place as a result of white-collar crimes including occupational hazards (e.g., black lung disease), environmental factors (e.g., pollution), medical malpractice (e.g., unnecessary surgery), and consumer safety violations (e.g., defective products) (Mokhiber 2007; Reiman and Leighton 2013).

"Green criminologists" study one example of the impact of white-collar crime and, specifically, corporate violence. In a review of the literature, Katz (2012) identifies several important findings from the existing research on corporate pollution and health and, specifically, cancer mortality:

> First, a variety of research illustrates multinational corporate culpability in the proliferation of environmental pollution both in the USA and globally. Second, this has resulted in increasing cancer mortality rates across the globe among minority communities as well as non-western developing nation states. Third, corporate environmental pollution has been facilitated through the proliferation of international free trade agreements and international financial loans from international financial institutions.

Using a conflict theory perspective, with its emphasis on the problems created by capitalism, Katz (2012) examines cancer mortality rates and the location of the richest transnational corporations in the world. She reports that 10 of the 152 countries studied have significantly higher cancer mortality rates than the remainder and, of that number, 6 are the locations of the headquarters of the largest transnational corporations in the world. On closer examination, Katz (2012) argues, the data reveal that most of the industries in the 10 nations where the cancer mortality rate is the highest are chemical, energy, water, and oil. Furthermore, Dow Chemical Company, a U.S.-based corporation that Katz (2012) calls "the primary global polluter," has operations in 9 of the 10 nations.

Finally, the U.S. Public Health Service now defines violence as one of the top health concerns facing Americans. Health initiatives related to crime include reducing drug and alcohol use and the deaths and diseases associated with them, lowering rates of domestic violence, preventing child abuse and neglect, and reducing violence through public health interventions. It must also be noted that crime has mental as well as physical health consequences (see the section titled "Social and Psychological Costs").

The High Price of Crime

Conklin (2007, p. 50) suggests that the financial costs of crime can be classified into at least six categories. First are *direct losses* from crime, such as the destruction of buildings through arson, of private property through vandalism, and of the environment by polluters. In 2014, the average dollar loss of destroyed or damaged property as a result of arson was $14,390 (FBI 2014b). Computer crime also has direct costs for those who are victimized. According to the Internet Crime Complaint Center, the average cost of crimes perpetrated on the Internet (e.g., fraud, nondelivery of goods) was $6,245 (ICCC 2014b).

Second are costs associated with the *transferring of property*. Bank robbers, car thieves, and embezzlers have all taken property from its rightful owner at tremendous expense to the victims and society. According to the Uniform Crime Reports, the total cost of stolen goods in 2013 was over $10 billion, robberies alone accounting for $404 million in losses. Of the index offenses, the largest single loss for individuals was from motor vehicle theft, which includes, for example, the theft of "sport utility vehicles, automobiles, trucks, buses, motorcycles, motor scooters, all-terrain vehicles, and snowmobiles." The average dollar loss per vehicle was $5,972 (FBI 2014a). Moreover, the average cost of computer crime to U.S-based organizations is $12.7 million (Ponemon Institute 2014).

A third major cost of crime is that associated with *criminal violence*, including the medical cost of treating crime victims. The National Crime Prevention Council (2005) estimates that the average cost for *each* criminal incident of rape or sexual assault is $7,700, including expenses related to medical and mental health care, law enforcement, and victim and social services. More recently, Anderson (2012) estimates that crime-related deaths and injuries cost society, directly and indirectly, over $750 billion annually. Deaths from firearms alone cost the U.S. economy and the U.S. health care system $37 billion in 2005, the most recent year for which data are available. In 2013, for "a patient with a gunshot wound, a single surgery followed by two days in the intensive care unit [costs] . . . about $100,000" (Young 2013, p. 1). Finally, it is estimated that the criminal justice costs of violent crime surpass $22 billion annually, the cost of corrections exceeding the cost of the police and the courts combined (Shapiro and Hassett 2012).

Fourth are the costs associated with the production and sale of illegal goods and services—that is, *illegal expenditures*. The expenditure of money on drugs, gambling, and prostitution diverts funds away from the legitimate economy and enterprises, and lowers property values in high-crime neighborhoods.

Fifth is the cost of *prevention and protection*—the billions of dollars spent on locks and safes, surveillance cameras, self-defense products, guard dogs, insurance, counseling and rehabilitation programs, and the like. For example, more specifically, it is estimated that Americans spend $16.2 billion annually on security systems alone (U.S. Department of Commerce 2011).

Finally, there is the cost of *controlling crime*—$260 billion annually (Roeder et al. 2015). As indicated by Figure 4.5, however, criminal justice spending changes over time increasing in some functions (i.e., law enforcement) and decreasing in others (i.e., victim services) (Austin 2015). Concern over criminal justice costs has led to some states passing laws that require that a judge be informed of the costs of the sentences they impose prior to a final disposition (Eisen 2013).

Social and Psychological Costs

Crime entails social and psychological costs as well as economic costs. One such cost is fear. The Gallup Organization regularly asks a sample of Americans whether they "feel safe walking alone at night in the city or area where you live?" (Jones 2013a). Gallup then "compiled the data for residents of the 50 most populous metropolitan statistical areas [MSA]" (p. 1). Based on respondents' answers, the three safest MSAs are Minneapolis–St. Paul–Bloomington (Minnesota), Denver-Aurora (Colorado), and Raleigh-Cary (North Carolina). The three least safe MSAs of the 50 largest in the United States are Memphis, TN–MS–AR (includes parts of Tennessee, Mississippi, and Arkansas), New Orleans–Metairie–Kenner (Louisiana), and Riverside–San Bernardino–Ontario (California).

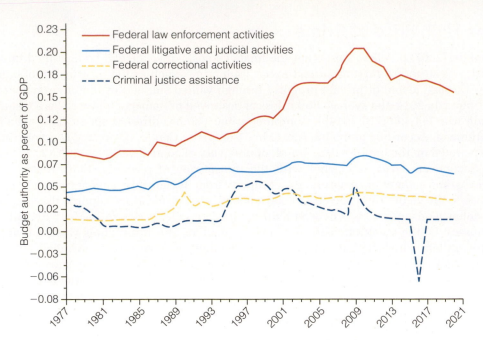

Figure 4.5 Administration of Justice Costs by Function, 1977–2020 (2016–2020 projected)*

*Discretionary Budget Authority as a percentage of GDP, FY1977–FY2020.
SOURCE: Austin 2015.

In a recent survey, 75.0 percent of Americans responded that they "feel safe walking alone at night in the city or area" where they live (Dugan 2014). Global data from 133 countries indicate that, on the average, respondents were more fearful than Americans, with 64 percent responding they felt safe. Furthermore, respondents from the wealthy nations of the Organisation for Economic Cooperation and Development also were more fearful than Americans. It should also be noted, however, that an examination of the global data indicates that the United States has one of the highest **safety gender gaps** of the countries surveyed; that is, women expressed significantly higher rates of fear than men. (See the *Self and Society* feature.)

In 2014, the U.S. Department of Justice conducted research on the socioemotional impact of violent crime (i.e., rape/sexual assault, robbery, aggravated assault, simple assault, violence including a weapon, and violence involving an injury) using data from the NCVS (Langton and Truman 2014). Socioemotional problems (SEPs) were defined as "the experience of one or more of the following: moderate to severe distress; significant problems with work or school, such as trouble with a boss, co-worker, or peers; or significant problems with family members and friends, including more arguments than before the victimization, an inability to trust, or not feeling as close after the victimization" (p. 1). Results of the investigation indicate that (1) 57.0 percent of violent crime victims experienced SEP; (2) emotional symptoms (e.g., angry, worried, depressed) of SEP were more common than physical symptoms (e.g., insomnia, upset stomach, headaches); (3) victims of intimate/partner violence compared to other victim–offender relationships had the highest rates of SEP; (4) regardless of the type of crime or victim–offender relationship, females were more likely to report SEP than males; and (5) experiencing SEP was associated with higher rates of reporting the crime to the police and receiving victim services

Although socioemotional responses vary by type of offense, it should not be concluded that white-collar crimes do not carry a social and psychological toll. In 2009, Bernard Madoff pled guilty to 11 felony counts related to a massive Ponzi scheme run through his investment firm. The scheme to defraud investors of $65 billion took place over a 20-year period and involved thousands of victims. The 71-year-old Madoff was sentenced to 150 years in prison, leaving behind him a trail of misery. The following are excerpts

safety gender gap The difference between women's expressed rate of fear and men's.

Fear of Criminal Victimization

Using the scale below, place the number that corresponds to your response in the blank to the right of the question. Answer the questions as honestly and accurately as possible. When finished, compare your responses to the results from a random sample of U.S. adults.

1 = VERY AFRAID   2 = AFRAID   3 = SOMEWHAT AFRAID   4 = NOT AFRAID AT ALL

QUESTION   **RESPONSE**

How afraid are you of being victimized in the following ways?

1. Being mugged   _____
2. Identity theft/Credit card fraud   _____
3. Being stalked   _____
4. Murder   _____
5. Police brutality   _____
6. Sexual assault   _____
7. Racial/Hate crime   _____
8. Random/Mass shooting   _____

Results of a National Survey in Percentages, 2014 | (N = 1572)

	VERY AFRAID	AFRAID	SOMEWHAT AFRAID	NOT AT ALL AFRAID
1. Mugging	5.5	14.9	37.5	39.9
2. Identity theft/Credit card fraud	19.6	30.1	34.7	13.2
3. Being stalked	5.1	12.4	24.1	44.2
4. Murder	8.0	12.5	20.1	48.5
5. Police brutality	5.0	12.3	24.5	54.8
6. Sexual assault	7.0	12.4	25.7	52.1
7. Racial/Hate crime	6.0	13.9	23.8	53.6
8. Random/Mass shooting	8.9	15.8	32.5	39.9

SOURCE: Chapman University 2014.

from just seven of the thousands of victims impacted by his crime. They attest to the social and psychological costs of white-collar crime (Victim Statements 2009):

- "He robbed us not only of our money, but of our faith in humanity, and in the systems in place that were supposed to protect us."
- "I can't tell you how scattered we feel—it goes beyond financially. It reaches the core and affects your general faith in humanity, our government and basic trust in our financial system."
- "I am constantly nervous and anxious about my future. I jump at the slightest noise. I can't sleep and all I do is worry about what will happen to us."
- "I don't know which emotion is more destructive, the fear and anxiety or the major depression that I experienced daily."
- "[W]hen you sentence Madoff, I trust that you touch on the loss of money, the loss of dignity, the loss of freedom from financial worries and possible financial ruin."

- "At this point we cannot trust anyone."
- "How do I live the rest of my life?"

Research on the impact of fraud (e.g. , credit card fraud, lottery scam, insurance fraud) in Great Britain reveals similar sorts of consequences, the most severe including feelings of anger and stress, psychological turmoil, problems in relationship, and physical health problems (Button et al. 2014).

The Cost to Children and Families

The dramatic increase in incarceration in recent years has impacted millions of families, particularly spouses/partners and children of inmates. For example, paternal incarceration is associated with disrupted relationships between parents, particularly if the partners lived together prior to the father's incarceration, family economic insecurity, and maternal neglect and physical aggression toward children (Travis et al. 2014; Turney 2014; Wildeman et al. 2012). Paternal incarceration, in part as a result of the changing nature of the parents' relationship, also impacts the quality and frequency of father–child contact, his financial contribution to the child's welfare, and involvement in co-parenting (Travis et al. 2014; Turney and Wildman 2013).

Young children with incarcerated fathers are significantly more likely to be retained sometime between kindergarten and third grade (Turney and Haskins 2014). Although research is mixed, in general, children of incarcerated fathers have higher rates of depression, irritability, and anxiety than children who do not have incarcerated fathers. The strongest and most consistent relationship is between fathers' incarceration and behavioral problems, particularly among boys, including delinquency and aggression (Travis et al. 2014).

As noted earlier, in recent years, there's been a dramatic increase in the number of incarcerated women, resulting in an increase in the number of children with incarcerated mothers, three-quarters of whom lived with their children prior to confinement. Although a less researched topic, because mothers are often the primary caregivers, the consequences of an incarcerated mother have only recently been investigated. What research exists suggests that maternal incarceration is associated with poor academic performance, school suspension, and higher rates of delinquency (Travis et al. 2014). In a review of the literature, Smyth (2012) concludes that "children with incarcerated mothers are at heightened risk for attachment disturbance, leading to depression, anxiety, and other traumatic-related stress . . . [and] . . . are often subject to frequent changing of caregivers within the foster care system, which exasperates these problems" (p. 33).

Strategies for Action: Crime and Social Control

The extent to which crime is prevented or recidivism reduced is difficult to assess. On the one hand, if a crime prevention program is initiated, and the crime rate goes down, it would be tempting to conclude that the program was a success. But what if the crime rate would have gone down anyway, with or without the prevention program? On the other hand, if a crime prevention program is initiated, and the crime rate goes up, many would conclude that the program was a failure. But what if the crime rate was lower than it would have been if the program had not been initiated?

Crime and social control are social phenomena and, as society changes, the social forces that impede or enhance the likelihood of crime also change. Consider the aging population (see Chapter 13). Crime tends to be a young person's activity. For example, of the more than 9 million people arrested in the United States in 2013, arrestees 65 and older accounted for just 1.0 percent of the total arrests for violent crime and 0.7 percent of the total arrests for property crimes (FBI 2014a). Fewer young people, less crime.

Nonetheless, as a society, we want to be able to "do" something about the crime problem. One way to combat crime is to attack the social problems that contribute to its existence. Moreover, when a random sample of Americans was asked which of two views—increasing law enforcement or resolving social problems—came closer to their own in dealing with the crime problem, the majority of respondents (64.0 percent) selected resolving social problems (Gallup Poll 2011).

In addition to policies that address social problems, numerous policies and programs have been initiated to alleviate the crime problem. These policies and programs include local initiatives, criminal justice policies, legislative action, and international efforts in the fight against crime.

Local Initiatives

Increasingly, municipalities are embracing technological innovations in their efforts to prevent crime. Additionally, local youth programs such as the Boys and Girls Clubs and community programs that involve families and schools are an effective "first line of defense" against crime and juvenile delinquency.

Technology and Crime. Increasingly, police are turning to technology in the fight against crime. Byrne and Marx (2011) make a needed distinction between hard and soft technologies. Hard technologies, in the context of crime prevention, include such things as metal detectors in schools, home security systems, and police drones. Soft technologies "involve the strategic use of information to prevent crime . . . and to improve the performance of the police" (p. 19). Crime analysis programs, Amber alerts, facial recognition software, and social media are examples of soft technologies.

In varying degrees, cities throughout the United States are using these innovations to fight crime. PredPol, for example, is a software program used in Los Angeles, that predicts crime "based on times and locations of previous crimes, combined with sociological information about criminal behavior and patterns" (Kelly 2014). The ShotSpotter system, through the placement of microphones throughout an area, can detect gunshots and calculate their location within 40 to 50 feet. The use of Facebook to post pictures of suspects, and GPS devices and cell phone tracking techniques, makes finding criminals and documenting their locations easier.

One controversial "hard technology" is the use of video cameras in public places. Chicago, for example, has more than 25,000 cameras distributed throughout the city, and all of the cameras, both public and private, are linked into a single system called Operation Virtual Shield. Recently, the Chicago Transit Authority (CTA) added thousands of cameras to its trains. On the basis of official statistics, after two consecutive years of increased serious crime, in 2014, serious crime (e.g., murder, sexual assault, robbery) on the CTA decreased by 26.0 percent (Rossi 2015). Despite the enormity of Operation Virtual Shield, the United Kingdom has nearly 6 million video cameras— 1 for every 11 citizens (Patterson 2014).

Although the video cameras in Chicago were installed initially to monitor traffic safety, three-quarters of the cameras have now been upgraded to 360-degree viewing capabilities. There are now surveillance cameras in neighborhoods, on buses and trains, and in schools, leading one journalist to conclude, "[T]here are few places you can go in Chicago without being monitored" (Tobin 2015, p. 1). The Illinois ACLU has objected to the cameras on the grounds that their ability to "pan-tilt-zoom," use face recognition software, and automatically track someone from one camera to the next is a violation of civil liberties, including the right to privacy (ACLU 2011). What do you think?

Youth Programs. Early intervention programs acknowledge that preventing crime is better than "curing" it once it has occurred. Fight Crime: Invest in Kids is a nonpartisan, nonprofit anti-crime organization made up of more than 5,000 law enforcement leaders, prosecutors, and violence survivors (Fight Crime 2014). The organization advocates a four-step plan: (1) provide all families with access to high-quality early care and education, (2) offer parent coaching to at-risk parents of young children to prevent child abuse and neglect, (3) ensure all school-age children have access to effective programs during and after school, and (4) identify at-risk and delinquent children and provide them and their parents with effective interventions.

The HighScope Perry Preschool Project is another example of an early intervention program. After a sample of 123 African American children was randomly assigned to either a control group or an experimental group, the experimental group members received academically oriented interventions for one to two years, frequent home visits, and weekly parent–teacher conferences. The control group members received no interventions. The control group members and the experimental group members were then compared at various points in time between the ages of 3 and 40. As adults, experimental group members had higher employment and homeownership rates, and significantly lower violent and property crime rates (Schweinhart 2007). In a cost–benefit analysis, Heckman et al. (2010) conclude that the financial benefit of the program outweighed the initial investment.

Finally, many youth programs are designed to engage juveniles in noncriminal activities and integrate them into the community. There are more than 11 million school-age children who are alone and unsupervised in the hours after school, primarily between 3:00 and 6:00 p.m., and in the summer months, which are peak times for delinquency. Engaging students in after-school programs is not only essential but highly successful (Afterschool Alliance 2014). For example, an evaluation of a California after-school program called LA's BEST concluded that children who attended the program were 30 percent less likely to engage in criminal behavior than those who did not attend the program (Afterschool Alliance 2013).

Community Programs. Neighborhood watch programs involve local residents in crime prevention strategies. For example, MAD DADS (Men against Destruction—Defending against Drugs and Social Disorder) patrol the streets in high-crime areas of the city on weekend nights, providing positive adult role models and fun community activities for troubled children. Members also report crime and drug sales to police, paint over gang graffiti, organize gun buyback programs, and counsel incarcerated fathers. At present, 75,000 men, women, and children are in MAD DADS in 60 chapters in 17 states throughout the United States (MAD DADS 2014).

The National Association of Town Watch (NATW) is "a network of law enforcement agencies, neighborhood watch groups, civic groups, state and regional crime prevention associations and concerned citizens" that are "dedicated to the development and promotion of crime prevention in communities across the nation" (NATW 2014, n.p.). NATW began the "National Night Out" event in 1984 and, today, over 38 million people in more than 16,000 communities in 50 states participate in the event in which citizens, businesses, neighborhood organizations, and local officials join together in outdoor activities to heighten awareness of neighborhood problems, promote anticrime messages, and strengthen community ties (NATW 2014).

Criminal Justice Policy

The criminal justice system is based on the principle of **deterrence**—the use of harm or the threat of harm to prevent unwanted behaviors. The criminal justice system assumes that people rationally choose to commit crime, weighing the rewards and consequences of their actions. Thus, "get-tough" measures hold that maximizing punishment will increase deterrence and cause crime rates to go down. Yet, most recently, incarceration rates have declined and crime has not increased, calling into question the principle of deterrence. Moreover, 30 years of get-tough policies have not significantly reduced recidivism rates but have merely led to prison overcrowding and a host of criminal justice problems. Presently, based on U.S. Bureau of Justice statistics, 17 states have over 100 percent occupancy in their correctional facilities, including North Dakota (150.5 percent), Illinois (151.7 percent), and California (142.7 percent) (Wilson 2014).

deterrence The use of harm or the threat of harm to prevent unwanted behaviors.

Given overcrowding and high recidivism rates, experts are scratching their collective heads and asking, "What works?"

Law Enforcement Agencies. In 2011, Anders Behring Breivik killed 69 people at a youth camp where more than 700 teenagers and young adults were meeting before Norwegian police shot him. A 2012 commission investigating the mass murder concluded that "the police and security services could and should have done more to avert the crisis" (Greene 2012).

The United States had nearly 1 million full-time law enforcement officers and full-time civilian employees in 2013 (e.g., clerks, meter attendants, correctional guards), yielding an estimated 3.4 law enforcement employees per 1,000 inhabitants and 2.3 sworn officers per 1,000 inhabitants (FBI 2014a). There are more than 14,000 law enforcement agencies in the United States, including municipal (e.g., city police), county (e.g., sheriff's department), state (e.g., highway patrol), and federal agencies (e.g., FBI), often with overlapping jurisdictions.

In 2011, the latest year for which national data are available, 26 percent of the 16 and older U.S. resident population had contact with the police (BJS 2013c). According to the Police-Public Contact Survey, a supplement to the National Crime Victimization Survey (NCVS), contacts were equally initiated by citizens (e.g., request for police assistance) and police (e.g., traffic stop). Results of the survey indicate that blacks and Hispanics were more likely to receive traffic tickets than whites and to be stopped by the police for a traffic violation. When a traffic stop occurred, black and Hispanic drivers were also more likely to be searched or frisked than white drivers.

For decades, accusations of racial profiling, police brutality, and discriminatory arrest practices have made police–citizen cooperation in the fight against crime difficult. Concerns have been fed by such highly publicized cases as the killing of Michael Brown, Eric Garner, and Trayvon Martin. There is some empirical support for such contentions. For example, one study concluded that arrests for the possession of marijuana, over 50 percent of all drug arrests, are not only exorbitantly expensive but that "enormous disparities exist in states and counties nationwide between arrest rates of blacks and whites for marijuana possession" (ACLU 2013). Not surprisingly, when combining survey data from 2011 through 2014, we note that 59 percent of whites reported having a "great deal/quite a lot" of confidence in the police compared to 37 percent of blacks—a statistically significant difference (Newport 2014).

The Violent Crime Control and Law Enforcement Act of 1994 established the Office of Community Oriented Policing Services (COPS), in part, out of concern for police–citizen relations. Community policing emphasizes that

> trust and mutual respect between police and the communities they serve is critical to public safety. This concept is the foundation of "community policing," and ensures that police and community stakeholders partner in solving our nation's crime challenges. Community policing is a law enforcement philosophy that focuses on community partnerships, problem-solving and organizational transformation. (COPS 2014)

Such an approach often includes "problem-oriented policing," a model that includes analyzing the underlying causes of crime, looking for solutions, and actively seeking out alternatives to standard law enforcement practices. One success story is in Miami, Florida, where, over 25 years ago, the Miami Police Department developed a program for school-age children called "Do the Right Thing" (DTRT). The initiative not only

Office of Community Oriented Policing Services

Here, police officer Jackelyn Burgos receives hugs and well wishes after G.R.E.A.T (Gang Resistance Education and Training) classes in the Cleveland Public Schools. This photo, one of the winners in a COPS photo contest, depicts the importance of COPS programs—integration of police into the community.

reinforces positive youthful behaviors but also provides the opportunity for the development of meaningful relationships between children and the police (DTRT 2015).

Rehabilitation versus Incapacitation. An important debate concerns the primary purpose of the criminal justice system: Is it to rehabilitate offenders or to incapacitate them through incarceration? Both rehabilitation and incapacitation are concerned with **recidivism** rates, or the extent to which criminals commit another crime. Advocates of **rehabilitation** believe that recidivism can be reduced by changing the criminal, whereas proponents of **incapacitation** think that recidivism can best be reduced by placing offenders in prison so that they are unable to commit further crimes against the general public.

Fear of crime, historically, has led to a public emphasis on incapacitation and a demand for tougher mandatory sentences, longer time served, a reduction in the use of probation and parole, support for truth-in-sentencing laws, and life without the possibility of parole sentencing (Travis 2014; Travis et al. 2014). As a result of these policies, the United States is one of the most punishment-oriented countries in the world. For example, when comparing the U.S. incarceration rate to the 115 nations in the Organisation for Economic Cooperation and Development (OECD), the United States' incarceration rate is more than 6 times higher than the average OECD country (Kearney et al. 2014).

According to Travis and colleagues (2014), the fear of crime is only one force in a series of social forces that lead to the draconian criminal justice policies of the last three decades. Social and political unrest in the 1960s (e.g., Vietnam War protests), rising crime rates, heightened concerns over the "race problem" following the civil rights movement and urban riots, and the perception that Democrats and liberal courts were "soft on crime" all contributed to the public embracing "law and order" candidates. Consequently, Republicans gained political inroads into the South, previously dominated by the Democrats, and solidified their political base. It is notable that "Republican Party control, especially at the state level, generally has been associated with larger expansions of the prison population" (p. 120).

Of late, however, public attitudes toward get-tough measures and the skyrocketing incarceration rates have changed. Research indicates that Americans believe that the United States imprisons too many people and at too high a cost, and that nonviolent offenders should be diverted from the prison system (Tides Center 2015). Furthermore, support for criminal justice policy reform crosses age, gender, and racial/ethnic group as well as region of the country and political party (Mellman Group 2012).

These shifts in public opinion coincide with concerns by politicians, academicians, and criminal justice officials about the efficacy of mass incarceration. Their concerns can be classified into five categories. First, research indicates that incarceration may not always deter crime. Between 1994 and 2012, the crime rates in New York and Florida decreased by 54.0 percent. During the same time period, New York had *reduced* its incarceration rate (less punitive), and Florida had *increased* its incarceration rate (more punitive). Most important, across all states, in general, the "crime drop since 1994 has been bigger in states that cut imprisonment rates" (Pew 2014, p.1).

Second is the accusation that get-tough measures are unevenly applied. For example, 3.0 percent of black males in the United States are in prison compared to 0.5 percent of white males (Carson 2014). In addition, African Americans disproportionately receive death sentences, particularly when there is a white victim. For example, although half of murder victims in the United States are white, since 1976, when capital punishment was reinstated, 77.0 percent of persons executed in the United States received the death penalty for killing a white person. Conversely, blacks constitute half of all murder victims yet, since 1976, only 13 percent of those executed have been convicted of killing a black person (The Sentencing Project 2013). States' adoption of sentencing guidelines over the last several decades was intended to reduce the disparate treatment of minorities. Similar policy reforms such as the adoption of mandatory minimums and truth in sentencing laws were intended "to constrain judicial discretion and create greater fairness, transparency, and equality in punishment" (Johnson and Lee 2013, p. 503).

recidivism A return to criminal behavior by a former inmate, most often measured by rearrest, reconviction, or reincarceration.

rehabilitation A criminal justice philosophy that argues that recidivism can be reduced by changing the criminal through such programs as substance abuse counseling, job training, education, and so on.

incapacitation A criminal justice philosophy that argues that recidivism can be reduced by placing offenders in prison so that they are unable to commit further crimes against the general public.

However, in a review of the literature on the effectiveness of sentencing guidelines in reducing racial and ethnic bias, Johnson and Lee (2013) conclude that disparities in criminal sentencing, although less obvious and direct, remain.

Third, there are concerns that putting people in prison may actually escalate their criminal behaviors once released. Inmates may learn new and better techniques of committing crime and, surrounded by other criminals, further internalize the attitudes toward and motivation for continued criminal behavior. Such a possibility is called the **breeding ground hypothesis**. Furthermore, as labeling theorists suggest, the stigma associated with being in prison makes reentry into society difficult. Unable to find employment, return to school, and meet new friends, former prisoners may feel they have no choice but to return to crime.

Fourth, in an environment of budget deficits and legislative cuts, states simply can no longer afford the policies of decades ago. The average annual cost of incarcerating an offender, more than $27,744 for federal inmates and $29,141 for state inmates, is actually much higher than often stated (Bureau of Prisons 2014; National Association of State Budget Officers 2013). Additional costs, often not part of the calculations, include the cost of pensions, health care, and other benefits for correctional employees, construction and renovation costs, administrative and legal expenses, and the cost of rehabilitation programs, hospital care, and private prisons for inmates.

Finally, some are questioning the logic of increased criminal justice spending when the crime rate has declined 45 percent or more in the last two decades (Roeder et al. 2015). At the same time crime has been decreasing, the average prison sentence has been increasing. For example, the average length of time served in prison was nine months longer in 2009 than in 1990 (Pew 2012). For some, the inverse relationship between incarceration rates and crime is an indication that get-tough policies work. Yet, researchers and criminal justice experts, armed with 20 years of crime and incarceration data concluded that since "about 1990, the effectiveness of increased incarceration on bringing down crime has been essentially zero" (Roeder et al. 2015, p. 23).

As a result of these and other concerns, the U.S. Department of Justice conducted a thorough review of the federal criminal justice system in 2013 (U.S. Department of Justice 2014). The "Smart on Crime Initiative" resulted in significant reforms in federal policies. Low-level, nonviolent drug offenders are no longer subject to excessive mandatory minimum sentences and, whenever possible, are diverted to alternative programs such as drug courts (see Chapter 3). Elderly inmates, inmates with dependents and no other possible caretaker, and nonviolent offenders who have served a significant portion of their sentences may receive early release. There is also a greater focus on reentry issues, particularly lowering recidivism rates, and a shift in resources to preventing violent crime among vulnerable populations.

Clearly, sentencing offenders for long periods of time to confinement enhances incapacitation. However, given budget cuts and concerns over the effectiveness of the get tough policies of the past, many states, as well as the federal government, are revisiting the ideals of rehabilitation. Rehabilitation assumes that criminal behavior is caused by sociological, psychological, and/or biological forces rather than being solely a product of free will. If such forces can be identified, the necessary change can be instituted. Many rehabilitation programs focus on helping inmates reenter society. In 2008, President Bush signed the Second Chance Act, which supports reentry programs in the hopes of reducing recidivism (Greenblatt 2008).

Andrew Burton/Getty Images

Anthony Alvarez, 82 years old, has been in prison for 42 years. Although Alvarez has never committed a violent offense, burglaries, possession of illegal firearms, and escaping from county jails led to a life sentence as a result of California's three strike law. The elderly are the fastest growing demographic group in prisons.

breeding ground hypothesis A hypothesis that argues that incarceration serves to increase criminal behavior through the transmission of criminal skills, techniques, and motivations.

WHAT do you THINK?

Just 7 years ago, one in 20 prisoners were over the age of 55; today, that number is one in 10. Prisons are not designed for "inmates who need wheelchairs, walkers, portable oxygen tank; who cannot get dressed without help or haul themselves to the top bunk; who can't hear prison officer's commands; who are incontinent, forgetful, or infirm" (Fellner 2015, p. 1). Do you think geriatric prisoners should receive some special dispensation—that is, is there any age when they should simply be released? Do you think there should be high-security nursing homes?

Corrections. Prison population rates vary dramatically by regions of the world but share at least one similarity: they are growing (Walmsley 2012). An examination of global rates reveals that the United States' incarceration rate at 707 per 100,000 population exceeds many times over those of other countries; for example, the rate in Russia is 467 and Iraq 133; Mexico's rate is 214 and China's 120; the rate for England and Wales is 148; for Canada, 118; and Germany, 76 (ICPS 2015).

The recent three-year decline in the U.S. prison population, which had grown by 436 percent between 1979 and 2009, ended in 2013 with a slight increase in the prison population (Carson 2014; Human Rights Watch 2014). The U.S. state and federal prison population is estimated at 1.5 million, 8 percent of which are in private facilities (see Figure 4.6). Fifty-four percent of the state prison population were sentenced for violent offenses, and over half of both male and female inmates were 39 years old or younger. Non-Hispanic blacks were the largest category of male inmates (37 percent) in 2013, compared to Hispanic males (22 percent) and non-Hispanic white males (32 percent). The incarceration rate for white females was less than half the incarceration rate for black females (Carson 2014).

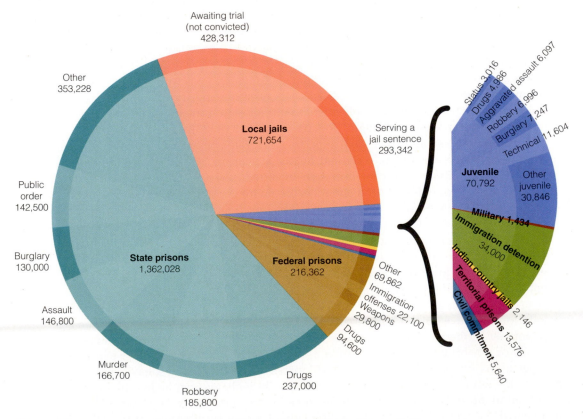

Figure 4.6 Distribution of Total Correctional Population, Offense Categories*

*Most recent data available
SOURCE: Wagner and Sakala 2014.

Probation entails the conditional release of an offender who, for a specific time period and subject to certain conditions, remains under court supervision in the community. **Parole** entails release from prison, for a specific time period and subject to certain conditions, before the inmate's sentence is finished. Varying by race, age, and gender, nearly 5 million people were on probation or parole in the United States in 2013 (BJS 2014a). The reincarceration rate of parolees is nearly twice the reincarceration rate of probationers.

Capital Punishment. With **capital punishment**, the state (the federal government or a state) takes the life of a person as punishment for a crime. In 2014, 22 countries carried out 607 executions excluding the thousands of people who are thought to have been executed in China (Amnesty International 2015). At the end of 2014, 19,094 people were under death sentences. Contrary to in the United States, globally, offenders may be sentenced to death for crimes other than murder including adultery, blasphemy, treason, and drug offenses. The global trend, consistent with U.S. public opinion since 1994, is away from capital punishment (Amnesty International 2015; Jones 2014).

In 2014, 35 executions were carried out in the United States with Texas, Missouri, and Florida responsible for carrying out 80.0 percent of the death sentences (Death Penalty Information Center [DPIC] 2014). Of the 31 states that have the death penalty, the majority and the federal government almost exclusively use lethal injection as the method of execution. Three problems, however, have been raised with the use of lethal injection, leading some states to halt executions.

First is the question of whether or not death by lethal injection violates the Eighth Amendment's prohibition against cruel and unusual punishment. In 2007, a district court judge held that Tennessee's lethal injection procedures "present a substantial risk of unnecessary pain" that could "result in a terrifying, excruciating death" (Schelzig 2007, p. 1). Subsequently, inmates in Oklahoma challenged the use of midazolam, the first drug in a three-drug protocol, arguing that it did not "render a person insensate to pain." In 2015, the U.S. Supreme Court upheld the use of the drug (*Glossip v. Gross* 2015).

The second issue concerns the role of physicians in state executions. According to the American Medical Association's Medical Code of Ethics, a "physician, as a member of a profession dedicated to preserving life when there is hope of doing so, should not be a participant in a legally authorized execution" and, therefore, should not be involved in prescribing or administering medications, monitoring vital signs, or attending or observing executions (AMA 2015, Section 2.06).

The third issue surrounding the continued use of capital punishment concerns the availability of the drugs used in lethal injection. Pharmaceutical companies throughout Europe, where capital punishment is banned, have refused to supply the drugs under threat of legal sanctions (Berman 2014). Furthermore, in 2015, Roche Pharmaceuticals, the manufacturer of midazolam, released a statement saying, "Roche did not supply midazolam for death penalty use and would not knowingly provide any of [their] medicines for this purpose. We support a worldwide ban on the death penalty."

Proponents of capital punishment argue that executions of convicted murderers are necessary to convey public disapproval and intolerance for such heinous crimes. Those against capital punishment believe that no one, including the state, has the right to take another person's life and that putting convicted murderers behind bars for life is a "social death" that conveys the necessary societal disapproval.

Proponents of capital punishment also argue that it deters individuals from committing murder. Critics of capital punishment hold, however, that because most homicides are situational and are not planned, offenders do not consider the consequences of their actions before they commit the offense. Critics also point out that the United States has a higher murder rate than most Western European nations that do not practice capital punishment, and that death sentences are racially discriminatory. For example, in 2012, a North Carolina judge ruled that an inmate's death sentence should be commuted to life in prison after it was found that prosecutors "deliberately excluded qualified black jurors from jury service . . . and [that] there was evidence this was happening in courts throughout the state" (Eng 2012, p. 1).

probation The conditional release of an offender who, for a specific time period and subject to certain conditions, remains under court supervision in the community.

parole Parole entails release from prison, for a specific time period and subject to certain conditions, before the inmate's sentence is finished.

capital punishment The state (the federal government or a state) takes the life of a person as punishment for a crime.

There are nearly 130,000 federally licensed gun stores in the United States, over 10 times the number of McDonald's. According to the Bureau of Alcohol, Tobacco, Firearms, and Explosives, 8.6 million new firearms were manufactured in the United States in 2012.

Capital punishment advocates suggest that executing a convicted murderer relieves taxpayers of the costs involved in housing, feeding, guarding, and providing medical care for inmates. Opponents of capital punishment argue that financial considerations should not determine the principles that decide life and death issues. In addition, taking care of convicted murderers for life may actually be less costly than sentencing them to death because of the lengthy and costly appeals process for capital punishment cases. As of 2015, 19 states have abolished the death penalty, including Connecticut, Illinois, New York, Michigan, and Wisconsin. Note, however, no southern state to date has repealed its death penalty statute.

Those in favor of capital punishment argue that it protects society by preventing convicted individuals from committing another crime, including the murder of another inmate or prison official. One study of the deterrent effect of capital punishment concluded that each execution is associated with at least eight fewer homicides (Rubin 2002). Opponents contend that capital punishment may result in innocent people being sentenced to death. According to the Innocence Project, there have been 330 post-conviction exonerations using DNA evidence since 1989. The most common reasons for wrongful convictions, in order, are (1) eyewitness misidentification, (2) faulty forensic science, (3) false confessions and incriminating statement, and (4) informants (e.g., paid informants lying) (The Innocence Project 2015).

Legislative Action

Legislative action is one of the most powerful methods of fighting crime. Federal and state legislatures establish criminal justice policy by the laws they pass, the funds they allocate, and the programs they embrace.

WHAT
do you
THINK?

In 2014, the pro-gun Republican-controlled legislature in Georgia passed what is being called the "gun everywhere" law. With a permit, residents can now carry guns into school classrooms, nightclubs, bars, Transportation Security Administration (TSA) checkpoints in airports, and, with the permission of the religious leaders, churches, temples, and other places of worship (Richinick 2014). What do you think? Should American citizens, with a permit, be able to carry a gun anywhere they want to?

Gun Control. According to a report by the FBI, in 2013, firearms were used in 69.0 percent of the nation's murders, 40.0 percent of robberies, and 21.6 percent of aggravated assault (FBI 2014a). Males, African Americans, and people under the age of 30 were the most likely to be victims of fatal firearm violence. Firearm homicides are highest in the South and lowest in the Northeast regions of the United States. Globally, the United States has the highest percentage of murders committed with a firearm in the world's developed nations (Masters 2013).

An examination of data collected from interviews with 3,243 U.S. adults in 2014 indicates the extent to which gun ownership ("gun, rifle, or pistol") is predictable using sociological variables. Gun ownership was higher in southern compared to northern

TABLE 4.5 Five States with Highest and Lowest Gun Death Rates by Household Gun Ownerhip, 2013

STATES WITH THE FIVE *HIGHEST* GUN DEATH RATES				STATES WITH THE FIVE *LOWEST* GUN DEATH RATES			
Rank	State	Household Gun Ownership	Gun Death Rate per 100,000	Rank	State	Household Gun Ownership	Gun Death Rate per 100,000
1	Alaska	60.6 percent	19.59	50	Hawaii	9.7 percent	2.71
2	Louisiana	45.6 percent	19.15	49	Massachusetts	12.8 percent	3.18
3	Alabama	57.2 percent	17.79	48	New York	18.1 percent	4.39
4	Mississippi	54.3 percent	17.55	47	Connecticut	16.2 percent	4.48
5	Wyoming	62.8 percent	17.51	46	Rhode Island	13.3 percent	5.33

SOURCE: Violence Policy Center 2015.

states and over twice as high among white compared to black or Hispanic respondents. A social portrait of the group most likely to own guns includes older, white, Republican males from rural areas in the South who identify themselves as conservatives. The group least likely to own guns includes white liberal Democrats from non-southern urban areas (Morin 2015).

Those against gun control argue that not only do citizens have a constitutional right to own guns but also that more guns may actually lead to less crime as would-be offenders retreat in self-defense when confronted (Lott 2003). Advocates of gun control, however, insist that the estimated 300 million privately owned firearms in the United States (Arnold 2013) significantly contribute to the violent crime rate and distinguish the United States from other industrialized nations. In fact, recent evidence would suggest they are correct. The "more guns, less crime" hypothesis put forth by Lott (2003) was addressed again in 2013 and researchers concluded just the opposite—more guns, more crime (see Table 4.5). States with right to carry laws had significantly higher rates of violent crime (Anega et al. 2014).

After a seven-year battle with the National Rifle Association (NRA), gun control advocates achieved a small victory in 1993, when Congress passed the Brady Bill. The law initially required a five-day waiting period on handgun purchases so that sellers can screen buyers for criminal records or mental instability. The law was amended in 1998, to include an instant check of buyers and their suitability for gun ownership.

Today, the law requires background checks of not just handgun users but also those who purchase rifles and shotguns. For example, if a person wants to buy a gun, their name and other personal information is entered into the National Instant Criminal Background Check System via the Internet or a toll-free number to check whether the buyer is eligible. The same information is then run through several FBI-managed databases such as the National Crime Information Center, which searches federal and state criminal records for information on the applicant (Jones 2013b).

Fueled by the escalating number of gun deaths in the United States and prompted by the killings at Sandy Hook Elementary School in Newtown, Connecticut, in 2012, President Obama established a task force to examine gun violence in the United States. Although supporters of gun control argued that the task force's recommendations didn't go far enough, recommending only the reinstatement of the assault weapons ban, limiting magazine rounds to 10, and requiring universal background checks, the U.S. Senate nonetheless failed to pass the proposed legislation (Barrett and Cohen 2013).

Fear that the legislation might pass led to record sales of firearms and, with it, record profits for gun manufacturers (O'Toole 2013). The National Rifle Association (NRA) and the firearm industries it represents have spent over $80 million on congressional and presidential elections since 2000. Political candidates, as conflict theorists would argue, are thus hesitant to vote against the interests of such a powerful group—and one that provides financial support for their reelection campaigns (Berlow and Witkin 2013).

Opinion polls on gun control are often misleading. When a representative sample of Americans was asked which was more important—gun rights or gun control—52 percent responded gun rights and 46 percent responded gun control (Doherty 2015). Yet, in a survey where policy options were presented, over half of respondents favored background checks, prohibiting gun ownership by the mentally ill, federal databases to track gun sales, armed officials at schools, and bans on semiautomatic and assault-style weapons (Pew 2013).

Other Crime and Social Control Legislation. There have been several landmark legislative initiatives, including the 1994 Violent Crime Control and Law Enforcement Act, which created community policing, and the 2006 Adam Walsh Child Protection and Safety Act, which, when enacted, created a national registry of substantiated cases of child abuse and neglect (Fact Sheet 2012). A sample of significant crime-related legislation presently before Congress includes the following bills:

- *Corporate Crime Database Act.* This bill, if passed would (1) create a database of all judicial proceedings against corporations or corporate officials; (2) develop a public website containing the improper conduct of corporations with revenues exceeding $1 billion annually; and (3) require that a report be submitted to Congress identifying the number, scope, and outcomes of all law enforcement actions brought against corporations or corporate officials.
- *End Racial Profiling Act.* If passed, this bill would prohibit racial profiling on the basis of race, ethnicity, national origin, and other minority statuses by federal, state, county, and municipal law enforcement.
- *Safety for Our Schoolchildren Act of 2015.* This bill, in part, would require state and local educational agencies to obtain FBI background checks prior to hiring school employees, prohibits hiring anyone convicted of a felony, and requires reporting anyone identified as a sexual predator to the police.
- *Domestic Violence Enhanced Penalty Act of 2015 or Candace's Law.* If passed, this bill would require that person convicted of committing or attempting to commit domestic violence in the presence of a minor child shall receive an enhanced sentence (U.S. Congress 2015–2016a).

International Efforts in the Fight against Crime

Europol is the European law enforcement organization that handles criminal intelligence. Unlike the FBI, Europol officers do not have the power to arrest; they predominantly provide support services for law enforcement agencies of countries that are members of the European Union. For example, Europol coordinates the dissemination of information, provides operational analysis and technical support, and generates strategic reports (Europol 2015). Europol, in conjunction with law enforcement agencies in member states, fights against transnational crimes such as facilitating illegal immigration, illicit drug trafficking, cigarette smuggling, child pornography, human trafficking, money laundering, and counterfeiting of the euro.

Interpol, the International Criminal Police Organization, was established in 1923 and is the world's largest international police organization, with 190 member countries (Interpol 2014). Similar to Europol, Interpol provides support services for law enforcement agencies of member nations. It has four core functions. First, Interpol operates a secure global law enforcement network that connects the member nations' information systems "allowing them to instantly access, request, and submit vital data" (Interpol 2014, p. 1). Second, Interpol provides 24/7 assistance and support to member countries including, but not limited to, crisis response, investigative expertise, security at high-profile events, and identification of disaster victims. Third, Interpol provides police training and development, and technical assistance to help member states better fight the increasingly

Interpol The largest international police organization in the world.

complex and globalized nature of crime. Finally, through extensive databases and forensic expertise (e.g., DNA profiles, fingerprints, suspected terrorists), Interpol ensures that police get the information they need to investigate existing crime and prevent new crime from occurring.

The International Centre for the Prevention of Crime (ICPC) is a consortium of policy makers, academicians, police, governmental officials, and nongovernmental agencies from all over the world. It is the only "global non-governmental organization (NGO) focused exclusively on crime prevention and community safety" (ICPC 2014, p. 1). In fulfilling such a task, the ICPC seeks to (1) raise awareness of and access to crime prevention knowledge and enhance community safety; (2) promote the exchange of information between researchers, practioners, policy makers, and other stakeholders; and (3) respond to calls for technical assistance.

Understanding Crime and Social Control

What can we conclude from the information presented in this chapter? Research on crime and violence supports the contentions of both structural functionalists and conflict theorists. Inequality in society, along with the emphasis on material well-being and corporate profit, produces societal strains and individual frustrations. Poverty, unemployment, urban decay, and substandard schools—the symptoms of social inequality—in turn lead to the development of criminal subcultures and conditions favorable to law violation. Furthermore, criminal behavior is encouraged by the continued weakening of social bonds among members of society and between individuals and society as a whole, the labeling of some acts and actors as "deviant," and the differential treatment of minority groups by the criminal justice system.

Recently, there has been a general decline in crime, making it tempting to conclude that get-tough criminal justice policies are responsible for the reductions. Other valid explanations exist and are likely to have contributed to the falling rates: changing demographics, rising incomes, community policing, stricter gun control, and a reduction in the use of crack cocaine.

Concerns over the cost of "nail 'em and jail 'em" policies, overcrowded prisons, and high recidivism rates have some policy makers looking elsewhere. Many states are already expanding the use of community-based initiatives and developing evidence-based reentry programs. The National Criminal Justice Commission Act of 2015, if passed, would establish a commission to "undertake a comprehensive review of the criminal justice system, make recommendations for Federal criminal justice reform to the President and Congress, and disseminate findings and supplemental guidance to the Federal Government, as well as to State, local and tribal governments" (U.S. Congress 2015–2016b, p. 1).

Rather than getting tough on crime after the fact, some advocate getting serious about prevention. Prevention programs are not only preferable to dealing with the wreckage crime leaves behind, but they are also cost-effective. For example, the Perry Preschool Project, as discussed earlier, cost $15,166 per participant but produced savings of $258,888 per participant. Of that savings, 88 percent was associated with a reduction in costs related to criminal justice (Schweinhart 2007).

Lastly, the movement toward **restorative justice**, a philosophy primarily concerned with repairing the victim–offender–community relation, is in direct response to the concerns of an adversarial criminal justice system that encourages offenders to deny, justify, or otherwise avoid taking responsibility for their actions. Restorative justice holds that the justice system, rather than relying on "punishment, stigma, and disgrace" (Siegel 2006, p. 275), should "repair the harm" (Sherman 2003, p. 10). Key components of restorative justice include restitution to the victim, remedying the harm to the community, and mediation. Restorative justice is increasingly used in schools to resolve conflict and mediate grievances.

restorative justice A philosophy primarily concerned with reconciling conflict among the victim, the offender, and the community.

Chapter Review

- **Are there any similarities between crime in the United States and crime in other countries?**

  All societies have crime and have a process by which they deal with crime and criminals; that is, they have police, courts, and correctional facilities. Worldwide, most offenders are young males, and the most common offense is theft; the least common offense is murder.

- **How can we measure crime?**

  There are three primary sources of crime statistics. First are official statistics—for example, the FBI's Uniform Crime Reports, which are published annually. Second are victimization surveys designed to get at the "dark figure" of crime, crime that official statistics miss. Finally, self-report studies have all the problems of any survey research. Investigators must be cautious about whom they survey and how they ask the questions.

- **What sociological theory of criminal behavior blames the schism between the culture and structure of society for crime?**

  Strain theory was developed by Robert Merton (1957) and uses Durkheim's concept of *anomie*, or normlessness. Merton argued that, when the structure of society limits legitimate means (e.g., a job) of acquiring culturally defined goals (e.g., money), the resulting strain may lead to crime. Individuals, then, must adapt to the inconsistency between means and goals in a society that socializes everyone into wanting the same thing but provides opportunities for only some.

- **What are index offenses?**

  Index offenses, as defined by the FBI, include two categories of crime: violent crime and property crime. Violent crimes include murder, robbery, assault, and rape; property crimes include larceny, car theft, burglary, and arson. Property crimes, although less serious than violent crimes, are the most numerous.

- **What is meant by white-collar crime?**

  White-collar crime includes two categories: occupational crime—that is, crime committed in the course of one's occupation; and corporate crime, in which corporations violate the law in the interest of maximizing profits. In occupational crime, the motivation is individual gain.

- **How do social class and race affect the likelihood of criminal behavior?**

  Official statistics indicate that minorities are disproportionately represented in the offender population. Nevertheless, it is inaccurate to conclude that race and crime are causally related. First, official statistics reflect the behaviors and policies of criminal justice actors. Thus, the high rate of arrests, conviction, and incarceration of minorities may be a consequence of individual and institutional bias against minorities. Second, race and social class are closely related in that nonwhites are overrepresented in the lower classes. Because lower-class members lack legitimate means to acquire material goods, they may turn to instrumental, or economically motivated, crimes. Thus, the apparent relationship between race and crime may, in part, be a consequence of the relationship between these variables and social class.

- **What are some of the economic costs of crime?**

  First are direct losses from crime, such as the destruction of buildings through arson or of the environment by polluters. Second are costs associated with the transferring of property (e.g., embezzlement). A third major cost of crime is that associated with criminal violence (e.g., the medical cost of treating crime victims). Fourth are the costs associated with the production and sale of illegal goods and services. Fifth is the cost of prevention and protection. Finally, there is the cost of the criminal justice system, law enforcement, litigation and judicial activities, corrections, and victims' assistance.

- **What is the present legal status of capital punishment in this country?**

  Of the 31 states that have the death penalty, the majority and the federal government almost exclusively use lethal injection as the method of execution. Concerns over the constitutionality of lethal injection led to inmates in Oklahoma questioning the use of midazolam, the first drug in a three drug protocol. In 2015, the U.S. Supreme Court upheld the use of the drug in executions.

Test Yourself

1. The United States has the highest incarceration rate of developed countries.
   a. True
   b. False
2. The Uniform Crime Reports is a compilation of data from
   a. the U.S. Census Bureau.
   b. law enforcement agencies.
   c. victimization surveys.
   d. the Department of Justice.
3. According to _____, crime results from the absence of legitimate opportunities as limited by the social structure of society.
   a. Hirschi
   b. Marx
   c. Merton
   d. Becker

4. Which of the following is not an index offense?
   a. Drug possession
   b. Homicide
   c. Rape
   d. Burglary
5. The economic costs of white-collar crime outweigh the costs of traditional street crime.
   a. True
   b. False
6. Women everywhere commit less crime than men.
   a. True
   b. False
7. Probation entails
   a. early release from prison.
   b. a suspended sentence.
   c. court supervision in the community in lieu of incarceration.
   d. incapacitation of the offender.

8. Europol is an advisory and support law enforcement agency for European Union members.
   a. True
   b. False
9. Lethal injection
   a. violates the Eighth Amendment of the U.S. Constitution.
   b. has been determined to be painless.
   c. by physicians is approved by the American Medical Association.
   d. is the most common method of execution in the United States.
10. The crime rate in the United States has steadily increased over the last several decades.
    a. True
    b. False

Answers: 1. A; 2. B; 3. C; 4. A; 5. A; 6. A; 7. C; 8. A; 9. D; 10. B.

Key Terms

> " We must recognize that there are healthy as well as unhealthy ways to be single or to be divorced, just as there are healthy and unhealthy ways to be married."
>
> STEPHANIE COONTZ
> family historian

5

Family Problems

Learning Objectives

After studying this chapter, you will be able to . . .

1 Give examples of how family forms and norms vary around the world.

2 Describe at least five trends, patterns, and/or variations in contemporary U.S. families, and differentiate between the marital decline and marital resiliency perspectives.

3 Explain how structural functionalism, conflict theory, and symbolic interaction help us understand the family institution, divorce, and domestic violence and abuse.

4 Discuss the social causes of divorce and its consequences for children and adults.

5 Identify the different forms of abuse in relationships, and describe the effects of violence and abuse on victims and the factors that contribute to domestic violence and abuse.

6 Discuss strategies to strengthen families, including expanding the definition of family, workplace and economic supports, relationship literacy education, covenant marriage, divorce mediation, divorce education programs, and domestic violence and abuse prevention and policies.

7 List ways in which divorce and domestic violence impact society and identify what many family scholars say is the "fundamental issue" in solving family problems.

In 2014, Baltimore Ravens running back Ray Rice was caught on a surveillance camera dragging his unconscious then-fiancée Janay Palmer from an elevator in Atlantic City. After Rice was indicted for assault charges, he and Palmer—who had a child together—got married. Rice pleaded not guilty to assault charges, enrolled in an intervention program, avoided jail time, and had his arrest record expunged. The NFL initially sanctioned Rice with a two-game suspension, but months later, another video was released that showed what happened on the elevator: Ray Rice hit Janay so hard that she collapsed, and then he appeared to spit on her body. Hours after this video went viral, the Ravens terminated Rice and NFL Commissioner Roger Goodall suspended Rice indefinitely from the league, ending his lucrative football career.

A year after the assault charge, Janay Rice said of her husband, "He's not a wife-beater, he's someone who made a mistake" (quoted in NESN Staff 2015).

AP Images/Mel Evans

The highly publicized story of Ray and Janay Rice brought national attention to the problem of domestic violence. In this chapter, we turn our attention to family problems, focusing on violence and abuse in intimate and family relationships, and problems associated with divorce and its aftermath. Many of the problems families face—such as health problems, poverty, job-related problems, drug and alcohol abuse, discrimination, and military deployment of a spouse—are dealt with in other chapters in this text. We begin by sampling the global diversity in family life and noting patterns, trends, and variations in contemporary U.S. families.

The Global Context: Family Forms and Norms Around the World

The U.S. Census Bureau defines *family* as a group of two or more people related by blood, marriage, or adoption. Sociology offers a broader definition of family: A **family** is a kinship system of all relatives living together or recognized as a social unit, including adopted people. This broader definition recognizes foster families, unmarried same-sex and opposite-sex couples with or without children, and any relationships that function and feel like a family. Next we describe global variation in family forms and norms.

Monogamy and Polygamy In many countries, including the United States, the only legal form of marriage is **monogamy**—a marriage between two partners. The term *monogamy* also refers to sexual relationships in which both partners have sex only with each other. In U.S culture and throughout much of the world, monogamy is considered morally superior to other forms of marital and sexual relationships. A common variation of monogamy is **serial monogamy**—a succession of marriages in which a person has more than one spouse over a lifetime but is legally married to only one person at a time.

In some cultures, law or custom allows the practice of **polygamy**—a form of marriage in which a person has more than one spouse. **Polygny,** the most common form of polygamy, involves one husband having more than one wife and is most common in traditional African cultures and in some Muslim populations. **Polyandry**—the concurrent marriage of one

family A kinship system of all relatives living together or recognized as a social unit, including adopted people.

monogamy Marriage between two partners; the only legal form of marriage in the United States.

serial monogamy A succession of marriages in which a person has more than one spouse over a lifetime but is legally married to only one person at a time.

polygamy A form of marriage in which one person may have two or more spouses.

polygyny A form of marriage in which one husband has more than one wife.

polyandry The concurrent marriage of one woman to two or more men.

woman to two or more men—is less common than polygyny, and is found in northern India, Nepal, and Tibet, as well as in 53 other societies. Polyandrous cultures tend to be small-scale egalitarian societies that obtain food by hunting and gathering, foraging, and horticulture and that have an imbalanced sex ratio favoring males (Starkweather and Hames 2012).

The U.S. Congress outlawed polygamy in the late 1800s. Under U.S. law, being married to more than one person at a time is a crime referred to as **bigamy**. Although the Mormon Church has officially banned polygamy, it is still practiced among some members of the fundamentalist Mormon movement, including the Church of Jesus Christ Latter-Day Saints (LDS)—a Mormon splinter group that believes that a man must marry at least three wives to go to heaven (Zeitzen 2008). In 1879, the Supreme Court in *Reynolds v. United States* ruled that, although people have religious freedom, religious belief is not justification for acting against the law. In 2013, a federal judge ruled that part of Utah's anti-polygamy law prohibiting "cohabitation" violates the Constitution, granting polygamist families the legal right to live together, as long as they do not obtain multiple marriage licenses.

Kody Brown, along with his four wives, challenged the Utah law banning cohabitation, claiming that it violated his right to privacy and religious freedom. In 2013, a federal judge ruled in Brown's favor.

Polygamy in the United States also occurs among some immigrants who come from countries where polygamy is accepted, such as West African countries. Immigrants who practice polygamy generally keep their lifestyle a secret because polygamy is grounds for deportation under U.S. immigration law.

Arranged Marriages versus Self-Choice Marriages. In some countries, young adults do not choose their spouses; rather, their parents or other third party arrange their marriage. In arranged marriages, couples are not expected to fall in love before they get married, but they are expected to develop emotional attachment following marriage. In many arranged marriages, the bride and groom do not meet until a few weeks before the wedding, or on the wedding day. In 2005, less than 5 percent of ever-married women ages 25 to 49 in India had a primary role in choosing their husband, and only 22 percent knew their husband for more than a month before they married (Allendorf 2013). Parents arrange marriages based on social class, religion, politics, aristocracy, and/or occupation.

In recent decades, young adults living in societies that have traditionally practiced arranged marriage are increasingly selecting their own spouses. Factors associated with this transition from traditional to modern marriage norms include increased education, urbanization, and exposure to Western cultural norms and values (Allendorf 2013).

Division of Power in the Family. In many societies, male dominance in the larger society is reflected in the dominance of husbands over wives in the family (see also Chapter 10, "Gender Inequality"). In many societies, for example, men make decisions about their wife's health care, when their wife may visit relatives, and/or household purchases (Population Reference Bureau 2011). In sub-Saharan Africa, male dominance is linked to widespread wife abuse. According to Nigeria's minister for women's affairs, "It is like it is a normal thing for women to be treated by their husbands as punching bags. . . . The Nigerian man thinks that a woman is his inferior" (quoted by LaFraniere 2005, p. A1). In some societies, many women and men believe that it is acceptable for a man to beat his wife if she argues with him and/or refuses to have sex with him (Population Reference Bureau 2011).

bigamy The criminal offense in the United States of marrying one person while still legally married to another.

In developed Western countries, although gender inequality persists (see Chapter 10), marriages have become more egalitarian, which means women and men view each other as equal partners who share decision making, housework, and child care. In one study of U.S. couples, nearly a third (31 percent) reported joint decision making for most decisions; 43 percent reported that the woman makes decisions in more areas than the man; and 26 percent reported that men make more of the decisions in the family (Pew Research Center 2008). As couples become more egalitarian, they share more in household and child care responsibilities, with the roles of moms and dads converging: Moms are participating more in earning income for the family and dads are doing more housework and child care (Parker and Wang 2013).

Social Norms Related to Childbearing. In less developed societies, where social expectations for women to have children are strong, women on average have four to five children in their lifetime. In developed societies, where women may view having children as optional—as a personal choice—women have, on average, fewer than two children.

Norms regarding appropriate childbearing age also vary. More than half of women in least developed countries were married before age 18—many before they turned 15 (UNICEF 2014). In countries where child marriages are common, teenage births are considered normal, even though they are associated with increased risk of health problems and poverty. In the United States and other developed countries, teenage childbearing is discouraged.

Norms about childbirth out of wedlock also vary worldwide (Child Trends 2014). The percent of children born to unmarried mothers ranges from 1 percent in India to 84 percent in Colombia. In many poor countries, nonmarital childbirth is rare because girls often marry at a young age (before they are 18). Rates of nonmarital childbearing are highest in Central/South America, followed by Northern and Western Europe where between one-third and one-half of children are born to unmarried women. In the United States, 4 in 10 births in 2014 were to unmarried women (Hamilton et al. 2015).

Interracial and interethnic couples are more common today than in previous generations.

Same-Sex Couples. Norms, policies, and attitudes concerning same-sex intimate relationships also vary around the world. In some countries, homosexuality is punishable by imprisonment or even death. First legalized in the Netherlands in 2001, same-sex marriages are, as of this writing, now legal in at least 19 countries. In 2015, the U.S. Supreme Court ruled by a 5 to 4 vote that same-sex couples have the constitutional right to be legally married in all 50 states. See Chapter 11, "Sexual Orientation and the Struggle for Equality," for in-depth information about same-sex couples and families.

Looking at families from a global perspective underscores the fact that families are shaped by the social and cultural context in which they exist. As we discuss the family problems related to divorce, violence, and abuse, we refer to social and cultural forces that shape these events and the attitudes surrounding them. Next, we look at patterns, trends, and variations in U.S. families.

Contemporary U.S. Families: Patterns, Trends, and Variations

Family forms and norms vary not only across societies; they also change across time. In the following section, we identify some of the changing patterns, trends, and variations in contemporary U.S. families and households.

- *Increased singlehood and older age at first marriage.* In 1960, about 1 in 10 U.S. adults ages 25 and older had not been married; by 2012, this number had doubled to 1 in 5. About half (53 percent) of never-married adults in 2012 said they would like to marry eventually, down from 61 percent in 2010 (Wang and Parker 2014). U.S. women and men are staying single longer and tying the knot—if they do so at all—at older ages. Between 1960 and 2014, the median age at first marriage for U.S. women rose from 20.3 years to 27.0 years; for men, it rose from 22.8 to 29.3 (U.S. Census Bureau 2014). One benefit of delayed marriage is that it improves women's financial situation: Women who marry after age 30 earn more per year than those who marry before age 20 (Hymowitz et al. 2013).

- *Increased interracial/interethnic couples.* Between 2000 and 2010, the percentage of U.S. opposite-sex marriages that involved husbands and wives of different races or Hispanic ethnicity increased from 7 percent to 10 percent—a 28 percent increase (U.S. Census Bureau 2012) (see also Chapter 9). The percentage of interracial/interethnic couples is even higher for unmarried opposite-sex couples (18 percent) and unmarried same-sex couples (21 percent).

WHAT do you THINK?

Why do you think cohabiting couples are more likely than married couples to be in an interracial or interethnic relationship?

- *Delayed childbearing or remaining child-free.* The average age for first-time mothers in the United States rose from 21.4 in 1970 to 25.8 in 2012 (Shah 2014). The first birthrate for women aged 40 to 44 more than doubled between 1990 and 2012 (Mathews and Hamilton 2014). Delayed childbearing enables women to pursue education and careers, thus establishing more financial security and independence. On the negative side, women having their first child after age 40 are at higher risk for health complications for themselves and their infant. Among U.S. adults ages 18 to 29, 74 percent say they want children, 19 percent aren't sure they want to be parents, and 7 percent say they don't want to have children (Wang and Taylor 2011).

- *Increased heterosexual and same-sex cohabitation.* The percentage of people who cohabited with their spouses before marriage more than doubled between 1980 and 2000, rising from 16 percent to 41 percent (Amato et al. 2007). Today, about one-fourth of never-married young U.S. adults ages 25 to 34 are cohabiting with a partner (Wang and Parker 2014). Couples live together to test their relationship, to reduce or share expenses, to co-parent, or to avoid losing pensions or alimony from prior spouses (although in some situations, living with a partner can be grounds for terminating or reducing alimony). For some, cohabitation is an alternative to marriage. Cohabitation among opposite-sex couples is illegal in three states (Mississippi, Florida, and Michigan) and punishable by fines and prison time, although the laws are rarely enforced.

- *Increased births to unmarried women.* The percentage of births to unmarried women has risen to historic levels. About 4 in 10 U.S. births are to unmarried women (Hamilton et al. 2015). This trend may be changing; the birthrate for unmarried women dropped for six consecutive years and in 2014 was the lowest since 2002 (Hamilton et al. 2015). The highest rates of nonmarital births are among blacks, American Indian/Alaskan natives, and Hispanics (see Figure 5.1). Six in 10 (61 percent) of U.S. adults

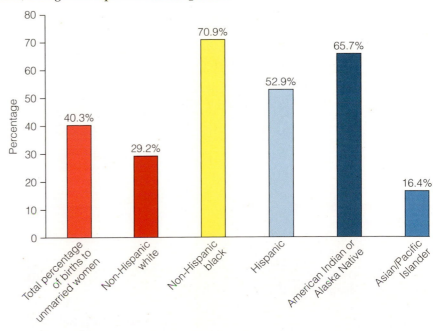

Figure 5.1 Percentage of All Births to Unmarried U.S. Women by Race and Hispanic Origin, 2014
SOURCE: Hamilton et al. 2015.

say having a baby outside marriage is "morally acceptable"—up from 45 percent in 2002 (Jones 2015). Not all unwed mothers are single parents; many unwed mothers are cohabiting with their partners when their children are born. Indeed, "shotgun cohabitations"—moving in together after a pregnancy—are now more common than "shotgun marriages" (Wang and Parker 2014). Nevertheless, children born to unwed mothers are more likely to experience poverty, poorer physical and mental health, and what Sawhill (2014) calls the "family-go-round": instability, less parenting, parents' multiple partners, or a series of cohabiting partners.

- *Stabilized divorce rate.* The **refined divorce rate**—the number of divorces per 1,000 married women—more than doubled between 1960 and 1980, but it has declined and stabilized since the early 1980s. One reason the divorce rate has stabilized is more unmarried couples are living together, and when these relationships do not work out, they break up instead of divorce. In addition, Americans are getting married at older ages, which is associated with a lower risk of divorce than marrying in one's teens or early 20s.

- *Increased gray divorces.* Although the divorce rate has declined and stabilized over the past three decades, **gray divorces**—divorces among people age 50 and older— have increased. Married people age 50 and older are twice as likely to get divorced today as they were in 1990 (Brown and Lin 2014).

- *Increased blended families.* Many previously married adults, including divorced and widowed adults, remarry. One in four marriages in the United States involves at least one previously married partner (Wang and Parker 2014).

WHAT do you THINK?

A national survey found that more than half (54 percent) of divorced women said they do not want to remarry, versus only 30 percent of divorced men (Wang and Parker 2014). Why do you think divorced women are less likely than divorced men to want to remarry?

Remarriages that involve children from a previous relationship are known as **blended families**, or stepfamilies. Six percent of U.S. children live with a step-parent (Livingston 2014). More than 4 in 10 U.S. adults have at least one step-relative in their family—a stepparent, a step- or half-sibling, or a stepchild (see Figure 5.2) (Parker 2011).

Stepparents and stepchildren do not have the same legal rights and responsibilities as biological/adopted children and biological/adoptive parents. Stepchildren do not automatically inherit from their stepparents, and courts have been reluctant to give stepparents legal access to stepchildren in the event of a divorce (Sweeney 2010).

- *Increased education, employment, and income of wives and mothers.* In earlier decades, husbands typically had more education than their wives. But among

refined divorce rate The number of divorces per 1,000 married women.

gray divorces Divorces among people ages 50 and older.

blended families Also known as *stepfamilies*, families involving children from a previous relationship.

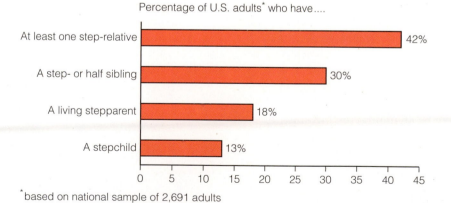

Percentage of U.S. adults* who have....

At least one step-relative	42%
A step- or half sibling	30%
A living stepparent	18%
A stepchild	13%

*based on national sample of 2,691 adults

Figure 5.2 Percentage of U.S. Adults with Step-Relatives
SOURCE: Parker 2011. From "A Portrait of Step Families," January 13, 2011, Social and Demographic Trends, a project of Pew Research Center. Reprinted with permission.

David Knox

David Knox

One in ten U.S. children lives with a grandparent.

couples married in recent years, U.S. wives have, on average, more education than their husbands (Schwartz and Han 2014). Mothers today are less likely than in the past to stay at home to raise children and take care of domestic chores. Labor force participation (either employed or looking for employment) of all U.S. mothers with children under age 18 rose from 47 percent in 1975 to 70 percent in 2014 (Bureau of Labor Statistics 2010, 2015b). The percentage of wives earning more than their husbands increased from 24.8 percent in 1990 to 38.1 percent in 2013 (Bureau of Labor Statistics 2015c). Issues related to balancing work and family are discussed in Chapter 7.

- *Grandparents living with grandchildren.* Grandparents have become more involved in their grandchildren's lives, in part due to increased life expectancy, single-parent families, and female employment. One in 10 U.S. children live with a grandparent (with one parent, two parents, or no parents living with them) (Ellis and Simmons 2014). Although grandparents may live with their adult children because the grandparent needs care and assistance with daily living (see also Chapter 12), more commonly grandparents are providing needed assistance with child care. **Grandfamilies**—families in which children reside with and are being raised by grandparents—include multigenerational families where grandparents care for grandchildren so the parent(s) can work or go to school. Grandfamilies also form in response to the parent(s) experiencing job loss, out-of-state employment, military deployment, divorce, deportation, illness, death, substance abuse, incarceration, or mental illness (Generations United 2014). Some grandparents are caring for their grandchildren to keep them out of foster care or to serve as foster parents themselves.

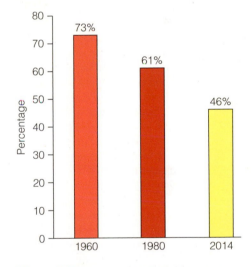

Figure 5.3 Percent of U.S. Children Living in a "Traditional"* Family
*A home with two married heterosexual parents in their first marriage
SOURCE: Based on Livingston, 2014.

- *Fewer children living in "traditional" families.* Many of the trends in families we describe here mean that fewer children are living in "traditional" families—a home with two married heterosexual parents in their first marriage. Fewer than half of U.S. children under age 18 live in a traditional family (see Figure 5.3).

- *A new family form: Living apart together.* Some couples live apart in different cities or states because of their employment situation. Known as "commuter marriages," these couples generally would prefer to live together, but their jobs require them to live apart. However, other couples (married or unmarried) live apart in separate residences out of choice. Family scholars have identified this arrangement as an

grandfamilies Families in which children reside with and are being raised by grandparents; parents may or may not also live in the household.

Elisabeth Sheff began studying polyamorists while a graduate student. Her research evolved into a 15-year study of polyamorous families with children, published in the book *The Polyamorists Next Door* (Sheff 2014). After briefly describing her sample and research methods, we present some of Sheff's findings.

Sample and Methods

Sheff's sample consisted of persons who self-identified as polyamorous or who were members of poly families. The sample included about 500 participants whom Sheff observed, and 131 persons whom she interviewed, including 22 children. Most participants (89 percent) were white; had attained a bachelor's or graduate degree; and had a professional job, such as teacher, health care professional, or information technology specialist.

Sheff's research utilized the following methods: (1) participant observation, in which Sheff observed and interacted with poly people in various settings, including poly support groups, meetings, private dinner parties, and other events; (2) content analysis, in which Sheff identified common themes in poly people's books, magazines, and blogs; (3) interaction with people on the Internet list-serve "Polyfamilies"; (4) in-depth interviews with poly people; and (5) written questionnaires asking for demographic information (e.g., age, sex, race, sexual orientation, occupation). Sheff interviewed children who were 5 years old and older. She observed and asked parents questions about children younger than 5.

Selected Findings and Conclusions

Sheff's research sheds light on many aspects of polyamory, including its many variations, who practices it and why, issues facing polyamorous relationships, and benefits and difficulties in polyamorous family life, some of which are described here.

Benefits of Poly Families

1. *Shared resources.* Poly family members in Sheff's study described sharing resources as the single most important advantage of living in a poly family. "From shared income to increased personal time for adults and more attention for children, having numerous adults in the family allows members to distribute tasks so that (ideally) no one person has to take the brunt of family care" (p. 196). One member of a poly family relayed the following:

   It has been amusing having monogamous friends who first asked questions about polyamory, like "isn't that complicated or a lot of work?" Then as we all got to child-rearing age our friends have changed their tune or seemed a little jealous and talk about how wonderful it would be to have more parents—single parents want more help and the couples want another parent as well. . . . Our friends tease us but wish they had more help too. (p. 201)

2. *Honesty and emotional intimacy among family members.* In Sheff's study, "Parents emphasized honesty with their children as a key

element of their overall relationship philosophy and parenting strategy" and "characterize honesty as the primary factor that cultivates emotional intimacy" (p. 191). Sheff interviewed children and teens who valued the open and honest relationships they had with their parents. Speaking of her relationship with her parents, one girl said:

   We have a good dialog, there is nothing I would keep from them. We are just very open people; there is no need to hide anything. . . . Some of my friends, things are bad for them at home, they can't and don't want to talk to their parents. It is kinda sad; they don't think they can trust their parents. (pp. 194–195)

   Children in poly families are also exposed to other qualities that are important in maintaining honest and intimate relationships, such a willingness to meet others' needs and skillful communication and negotiation.

3. *Multiple role models for children.* Children in poly families benefit from the multiple role models available in poly families. "Many parents say that their children's lives, experiences, and self-concepts are richer for the multiple loving adults in their families" (p. 201).

Difficulties in Poly Families

1. *Social stigma.* Due to the social stigma associated with polyamory, members of poly families may experience rejection from family members and risk discrimination in

---

living apart together (LAT) relationships An emerging family form in which couples—married or unmarried—live apart in separate residences.

polyamory Multiple intimate sexual and/or loving relationships with the knowledge and consent of all partners involved.

emerging family form known as **living apart together (LAT) relationships**. Couples may choose this family form for a number of reasons, including wanting to maintain a measure of independence, not feeling ready to cohabit, resuming a marital or cohabiting relationship after it broke up, one partner having a job or going to school a distance from the other partner, or prioritizing other responsibilities such as children. This new family form has been observed in Western Europe as well as in the United States (Lara 2005; Levin 2004).

- *Polyamorous couples and families.* **Polyamory** refers to multiple intimate sexual and/or loving relationships with the knowledge and consent of all partners involved. Polyamory can be viewed as a relationship practice, a philosophy, a lovestyle, an

the workplace, housing, and child custody matters. Sheff noted that "people in poly families were . . . aware that the stigma of being a sexual minority made them more vulnerable to accusations of poor parenting or questionable family situations from relatives, neighbors, teachers, and Child Protective Services" (p. 225). Although children ages 5 to 8 generally were not aware that their family was different from other families, older children and teens knew that their families were different and were aware of the potential for social stigma. One child reported he did not invite friends to his house because he did not want to explain the multiple adults living in his household. A 15-year-old male said, "It's kind of weird to live with a secret, something you can't tell any of your friends cause they wouldn't understand" (p. 144). But other children are comfortable with their family situation. One 17-year-old male explained:

If a friend came over and both of my dads were there I would just tell them straight out, that's my other dad. If they were a little weirded out, I didn't take it as anything. I was never ashamed about it or never thought it was weird. (p. 143)

2. *Teenagers' leverage against poly parents.* Raising teenagers can be challenging for any parent, but poly parents have the additional possibility that a disgruntled teen can blackmail

their parents, threatening to reveal their unconventional lifestyle to authorities, employers, or teachers.

3. *Lack of access to legal divorce.* One poly family in Sheff's study consisted of Alicia, Ben, Monique, and Edward. Monique and Edward had biological children together. Alicia was not employed and cared for the home and for Monique and Edward's children. When this family broke up after 11 years of being in a poly family:

Alicia had no access to the usual recourses available to women whose monogamous legal marriages end. . . . Although legal protections would not have shielded Alicia from the emotional impact of the family's dissolution, they would at least have allowed her visitation of the children she reared and financial compensation for the years she spent raising them and maintaining the household. (p. 180)

Conclusions

While Sheff does not identify as polyamorous herself, she sees polyamory as "a legitimate relationship style that can be tremendously rewarding for adults and provide excellent nurturing for children" (p. x). She also notes that "polyamory is not for everyone. Complex, time-consuming, and potentially fraught with emotional booby traps, polyamory is tremendously rewarding for some people and a complete disaster for others" (p. ix).

Sheff's investigation of poly families highlights that the nonsexual emotional

ties that bind people in poly families together are far more important than the sexual connections between the adults:

While the sexual relationships polys establish with each other get the most attention from the media, . . . they are not the . . . most important aspect of poly relationships. . . . Much like heterosexual families, poly families spend far more time hanging out together, doing homework, making dinner, carpooling, folding laundry, and having family meetings or relationship talks than they do having sex. (pp. 206–207)

Sheff found that the children in the poly families she studied "seemed remarkably articulate, intelligent, self-confident, and well adjusted" and "appeared to be thriving with the plentiful resources and adult attention their families provided" (p. 135). The well-being of the children could also be due, in part, to their privileged upper-middle-class, predominantly white background.

Does being raised in a poly family lead children to reject monogamy in their own lives? Sheff found that some teens were definitely not interested in polyamory in their own relationships, others envisioned trying polyamory in their adult lives, and still others were undecided, but "none of them reported feeling pressured to become polyamorous in the future or feeling like their choices were constrained" (p. 158).

SOURCE: Sheff 2014. Used by permission.

identity, or as a sexual orientation (Klesse 2014). Polyamory includes a wide range of multiple intimate involvements, including polyfidelity (or group "marriage"), primary relationships open to other intimate or sexual relationships, and casual sexual involvements with multiple partners. Polyamory is most common in the United States, Canada, Europe, and Australia (Sheff 2014). Poly people may be in the "closet," as openly polyamorous individuals are subject to stigmatization and discrimination. Multi-partner relationships that raise children and function as families are known as **poly families** (see this chapter's *Social Problems Research Up Close*).

- *Intentional communities.* Some individuals, couples, and families choose to live in **intentional communities**—communities where people live together on the basis

poly families Multi-partner relationships that raise children and function as families.

intentional communities Communities where people live together based on shared values and goals.

of shared values and goals. There are more than 1,000 intentional communities in North America, including income-sharing communes, ecovillages, spiritual or religious-based communities, residential land trusts, and cohousing. One of the oldest U.S. intentional communities (started in 1967) is Twin Oaks, which includes about 90 adults and 15 children who live on 350 acres in Louisa, Virginia. Twin Oaks embraces diversity and includes adults who are single, married, in committed but nonmarital relationships, in same-sex relationships and families, and in polyamorous relationships and families. Twin Oaks is based on values of equality, nonviolence, ecological sustainability, and income sharing. One of the most interesting aspects of Twin Oaks is its economic system, where each working age adult is expected to work 42 hours a week in exchange for housing, food, medical care, and a small monthly allowance. In the Twin Oaks system, domestic work (such as laundry, child care, cooking, and housework) is valued equally with other jobs (including managing Twin Oaks' tofu-making or hammock-making businesses). In other words, doing one hour of work managing a business receives the same "labor credit" as doing one hour of cleaning toilets.

> Whereas once the main purpose of marriage was to have and raise children, today women and men want marriage to provide adult intimacy and companionship.

Marital Decline? Or Marital Resiliency?

Does the trend toward diversification of family forms mean that marriage and family are disintegrating, falling apart, or even disappearing? Or has family simply undergone transformations in response to changes in socioeconomic conditions, gender roles, and cultural values?

According to the **marital decline perspective**, (1) personal happiness has become more important than marital commitment and family obligations, and (2) the decline in lifelong marriage and the increase in single-parent families have contributed to a variety of social problems, such as poverty, delinquency, substance abuse, violence, and the erosion of neighborhoods and communities (Amato 2004). According to the **marital resiliency perspective**, "poverty, unemployment, poorly funded schools, discrimination, and the lack of basic services (such as health insurance and child care) represent more serious threats to the well-being of children and adults than does the decline in married two-parent families" (Amato 2004, p. 960). According to this perspective, many marriages in the past were troubled, but because divorce was not socially acceptable, these problematic marriages remained intact. Rather than the view of divorce as a sign of the decline of marriage, divorce provides adults and children an escape from dysfunctional home environments.

Although the high rate of marital dissolution seems to suggest a weakening of marriage, divorce may also be viewed as resulting from placing a high value on marriage, such that a less than satisfactory marriage is unacceptable. People who divorce may be viewed not as incapable of commitment but as those who would not settle for a bad marriage. Indeed, the expectations that young women and men have of marriage have changed. Whereas once the main purpose of marriage was to have and raise children, today women and men want marriage to provide adult intimacy and companionship (Coontz 2000).

The high rate of childbirths out of wedlock and single parenting does not necessarily indicate a decline in the value of marriage. In interviews with low-income single women with children, most women said they would like to be married but just have not found "Mr. Right" (Edin 2000). Low-income single mothers in Edin's study were reluctant to marry the father of their children because these men had low economic status, traditional notions of male domination in household and parental decisions, and patterns of untrustworthiness and even violent behavior. Given the low level of trust these mothers have of men and given their view that husbands want more control than the women are willing to give them, women realize that a marriage that is also economically strained is likely to be conflictual and short-lived. "Interestingly, mothers say they reject entering into economically risky marital unions out of respect for the institution of marriage, rather than because of a rejection of the marriage norm" (Edin 2000, p. 130).

marital decline perspective A view of the current state of marriage that includes the beliefs that (1) personal happiness has become more important than marital commitment and family obligations, and (2) the decline in lifelong marriage and the increase in single-parent families have contributed to a variety of social problems.

marital resiliency perspective A view of the current state of marriage that includes the beliefs that (1) poverty, unemployment, poorly funded schools, discrimination, and the lack of basic services (such as health insurance and child care) represent more serious threats to the well-being of children and adults than does the decline in married two-parent families; and (2) divorce provides adults and children an escape from dysfunctional home environments.

Is the well-being of a family measured by the degree to which that family conforms to the idealized married, two-parent, stay-at-home mom model of the 1950s? Or is family well-being measured by function rather than form? As suggested by family scholars Mason et al. (2003), "the important question to ask about American families . . . is not how much they conform to a particular image of the family, but rather how well do they function—what kind of love, care, and nurturance do they provide?" (p. 2).

It is also important to have a perspective that takes into account the historical realities of families. Family historian Stephanie Coontz (2004) explained:

> [M]any things that seem new in family life are actually quite traditional. Two-provider families, for example, were the norm through most of history. Stepfamilies were more numerous in much of history than they are today. There have been several times and places when cohabitation, out-of-wedlock births, or nonmarital sex were more widespread than they are today. (p. 974)

In sum, although the institution of marriage and family has undergone significant changes in the last few generations, it is not clear whether these changes are for the better or for the worse. Coontz (2005) noted, "Marriage has become more joyful, more loving, and more satisfying for many couples than ever before in history. At the same time it has become optional and more brittle" (p. 306).

Sociological Theories of Family Problems

Three major sociological theories—structural functionalism, conflict and feminist theories, and symbolic interactionism—help to explain different aspects of the family institution and the problems in families today.

Structural-Functionalist Perspective

The structural-functionalist perspective views the family as a social institution that performs important functions for society, including producing and socializing new members, regulating sexual activity and procreation, and providing physical and emotional care for family members. This perspective views the high rate of divorce and the rising number of single-parent households as constituting a "breakdown" of the family institution. This supposed breakdown of the family is considered a primary social problem that leads to secondary social problems such as crime, poverty, and substance abuse.

According to the structural-functionalist perspective, traditional gender roles contribute to family functioning: Women perform the "expressive" role of managing household tasks and providing emotional care and nurturing to family members, and men perform the "instrumental" role of earning income and making major family decisions. According to this view, families have been disrupted and weakened by the change in gender roles, particularly women's participation in the workforce—an assertion that has been criticized and refuted, but is still supported by some "family values" scholars and supporters who advocate a return to traditional gender roles in the family.

Structural functionalism also looks at how changes in other social institutions affect families. For example, research has found that changes in the economy—specifically falling wages among unskilled and semiskilled men—contribute to both intimate partner abuse and the rise in female-headed single-parent households (Edin 2000).

Conflict and Feminist Perspectives

Conflict theory focuses on how capitalism, social class, and power influence marriages and families. Feminist theory is concerned with how gender inequalities influence and are influenced by marriages and families. Feminists are critical of the traditional male domination of families—a system known as **patriarchy**—that is reflected in the tradition

patriarchy A male-dominated family system reflected in the tradition of wives taking their husband's last name and children taking their father's name.

of wives taking their husband's last name and children taking their father's name. Patriarchy implies that wives and children are the property of husbands and fathers.

A U.S. poll found that nearly a third (31 percent) of people believe a woman should take her husband's last name after marriage (Adams 2014). What do you think about the tradition of women taking their husband's last name?

The overlap between conflict and feminist perspectives is evident in views on how industrialism and capitalism have contributed to gender inequality. With the onset of factory production during industrialization, workers—mainly men—left the home to earn incomes and women stayed home to do unpaid child care and domestic work. This arrangement resulted in families founded on what Friedrich Engels calls "domestic slavery of the wife" (quoted by Carrington 2002, p. 32). Modern society, according to Engels, rests on gender-based slavery, with women doing household labor for which they receive neither income nor status, whereas men leave the home to earn an income. Times have certainly changed since Engels made his observations, with most wives today leaving the home to earn incomes. However, wives employed full-time still do the bulk of unpaid domestic labor, and women are more likely than men to compromise their occupational achievement to take on child care and other domestic responsibilities. The continuing unequal distribution of wealth that favors men contributes to inequities in power and fosters economic dependence of wives on husbands. When wives do earn more money than their husbands, the divorce rate is higher—the women can afford to leave abusive or inequitable relationships (Jalovaara 2003).

Economic factors have also influenced norms concerning monogamy. In societies in which women and men are expected to be monogamous within marriage, there is a double standard that grants men considerably more tolerance for being nonmonogamous. Engels explained that monogamy arose from the concentration of wealth in the hands of a single individual—a man—and from the need to bequeath this wealth to children of his own, which required that his wife be monogamous. The "sole exclusive aims of monogamous marriage were to make the man supreme in the family and to propagate, as the future heirs to his wealth, children indisputably his own" (quoted by Carrington 2002, p. 32).

Feminist and conflict perspectives on domestic violence suggest that the unequal distribution of power among women and men and the historical view of women as the property of men contribute to wife battering. When wives violate or challenge the male head of household's authority, the male may react by "disciplining" his wife or using anger and violence to reassert his position of power in the family (see Figure 5.4).

Although modern gender relations within families and within society at large are more egalitarian than in the past, male domination persists, even if less obvious. Lloyd and Emery (2000) noted that "one of the primary ways that power

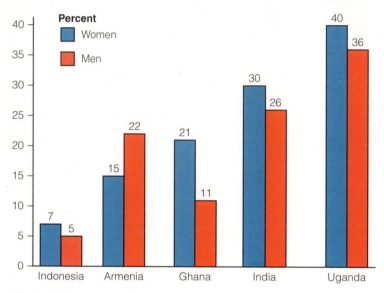

Figure 5.4a Percent of Women and Men Who Agree That Wife Beating Is Acceptable if a Wife Argues with Her Husband
SOURCE: Population Reference Bureau 2011.

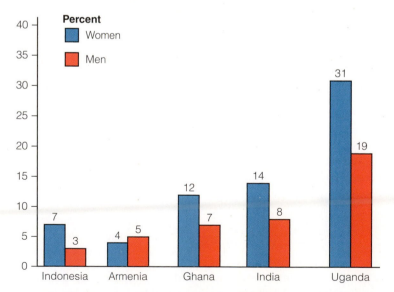

Figure 5.4b Percent of Women and Men Who Agree That Wife Beating Is Acceptable if a Wife Refuses Sex with Her Husband
SOURCE: Population Reference Bureau 2011.

disguises itself in courtship and marriage is through the 'myth of equality between the sexes.' . . . The widespread discourse on 'marriage between equals' serves as a cover for the presence of male domination in intimate relationships . . . and allows couples to create an illusion of equality that masks the inequities in their relationships" (pp. 25–26).

Conflict theorists emphasize that powerful and wealthy segments of society largely shape social programs and policies that affect families. The interests of corporations and businesses are often in conflict with the needs of families. Corporations and businesses strenuously fought the passage of the 1993 Family and Medical Leave Act, which gives people employed full-time for at least 12 months in companies with at least 50 employees up to 12 weeks of unpaid time off for parenting leave, illness or death of a family member, and elder care. Government, which corporate interests largely influence through lobbying and political financial contributions, often enacts policies and laws that serve the interests of for-profit corporations rather than families.

Symbolic Interactionist Perspective

The symbolic interactionist perspective emphasizes that interaction with family members, including parents, grandparents, siblings, and spouses, has a powerful effect on our self-concepts. For example, negative self-concepts may result from verbal abuse in the family, whereas positive self-concepts may develop in families in which interactions are supportive and loving. The importance of social interaction in children's developing self-concept suggests a compelling reason for society to accept rather than stigmatize nontraditional family forms. Imagine the effect on children who are called "illegitimate" or who are teased for having two moms or dads.

The symbolic interactionist perspective is useful in understanding the dynamics of domestic violence and abuse. For example, some abusers and their victims learn to define intimate partner violence as an expression of love (Lloyd 2000). Emotional abuse often involves using negative labels (e.g., *stupid, whore*) to define a partner or family member. Such labels negatively affect the self-concept of abuse victims, often convincing them that they deserve the abuse. Sibling violence is pervasive and widely tolerated because it is commonly viewed as "a harmless and inconsequential form of family aggression" (Khan and Rogers 2015, p. 437).

The symbolic interactionist insight that labels affect meaning and behavior can be applied to issues related to divorce. For example, when a noncustodial divorced parent (usually a father) is awarded "visitation" rights, he may view himself as a visitor in his children's lives. The meaning attached to the visitor status can be an obstacle to the father's involvement because the label *visitor* minimizes the importance of the noncustodial parent's role (Pasley and Minton 2001). Fathers' rights advocates suggest replacing the term *visitation* with terms such as *parenting plan* or *time-sharing arrangement*, because these terms do not minimize either parent's role. Many states have dropped the term *custody* in favor of more family-friendly terms such as *parenting plan* to reduce conflict among parents who might otherwise fight over who "wins" custody.

Problems Associated with Divorce

More than 40 percent of first marriages and 60 percent of second marriages today end in divorce (Hawkins and VanDenBerghe 2014). Divorce is considered problematic because of the negative effects it has on children, adults, and the larger society. In some societies, legal and social barriers to divorce are considered problematic because such barriers limit the options of spouses in unhappy and abusive marriages. Ireland did not allow divorce until 1995, and Chile did not allow divorce until 2004.

Social Causes of Divorce

When we think of why a particular couple gets divorced, we typically think of a number of individual and relationship factors that might have contributed to the marital breakup: incompatibility in values or goals; poor communication; lack of conflict resolution skills; sexual incompatibility; extramarital relationships; substance abuse; emotional or physical abuse or neglect; boredom; jealousy; and difficulty coping with change or stress related to parenting, employment, finances, in-laws, and illness. From a larger perspective, divorce is linked to the following social and cultural factors:

1. *Changing function of marriage.* Before the Industrial Revolution, marriage functioned as a unit of economic production and consumption that was largely organized around producing, socializing, and educating children. However, the institution of marriage has changed over the last few generations:

   > Marriage changed from a formal institution that meets the needs of the larger society to a companionate relationship that meets the needs of the couple and their children and then to a private pact that meets the psychological needs of individual spouses. (Amato et al. 2007, p. 70)

   When spouses do not feel that their psychological needs—for emotional support, intimacy, affection, love, or personal growth—are being met in the marriage, they may consider divorce with the hope of finding a new partner to fulfill these needs.

2. *Increased economic autonomy of women.* Before 1940, most wives were not employed outside the home and depended on their husband's income. Today, the majority of married women are in the labor force, and in 2013, 38.1 percent of wives earned more than their husbands (Bureau of Labor Statistics 2015c). A wife who is unhappy in her marriage is more likely to leave the marriage if she has the economic means to support herself (Jalovaara 2003). An unhappy husband may also be more likely to leave a marriage if his wife is self-sufficient and can contribute to the support of the children.

3. *Increased work demands and economic stress.* Another factor influencing divorce is increased work demands and the stresses of balancing work and family roles. Some workers are putting in longer hours, often working overtime or taking second jobs, while others face job loss and unemployment. As discussed in Chapters 2, 6, and 7, many families struggle to earn enough money to pay for rising housing, health care, and child care costs. Financial stress can cause marital problems. Couples with an annual income under $25,000 are 30 percent more likely to divorce than couples with incomes over $50,000 (see Table 5.1).

4. *Inequality in marital division of labor.* Many employed parents, particularly mothers, come home to work a **second shift**—the work involved in caring for children and household chores (Hochschild 1989). Women are more likely than

TABLE 5.1 **Factors that Decrease Women's Risk of Separation or Divorce During the First 10 Years of Marriage**

Factor	Percent Decrease in Risk of Divorce Or Separation
Annual income over $50,000 (versus under $25,000)	−30
Having a baby 7 months or more after marriage (versus before marriage)	−24
Marrying over 25 years of age (versus under 18)	−24
Having an intact family of origin (versus having divorced parents)	−14
Religious affiliation (versus none)	−14
Some college (versus high school dropout)	−25

SOURCE: National Marriage Project and the Institute for American Values 2012.

second shift The household work and child care that employed parents (usually women) do when they return home from their jobs.

men to spend time, on an average day, doing housework, food preparation, and childcare (Bureau of Labor Statistics 2015a). This unequal division of household chores negatively affects marital satisfaction among wives, particularly those who believe that marital roles should be more equal (Ogolsky et al. 2014).

5. *No-fault divorce laws.* Before 1970, the law required a couple who wanted a divorce to prove that one of the spouses was at fault and had committed an act defined by the state as grounds for divorce—adultery, cruelty, or desertion. In 1969, California became the first state to initiate **no-fault divorce**, which permitted a divorce based on the claim that there were "irreconcilable differences" in the marriage. Today, all 50 states recognize some form of no-fault divorce. No-fault divorce law has contributed to the U.S. divorce rate by making divorce easier to obtain.

6. *Increased individualism.* U.S. society is characterized by **individualism**—the tendency to focus on one's individual self-interests and personal happiness rather than on the interests of one's family and community. "Marital commitment lasts only as long as people are happy and feel that their own needs are being met" (Amato 2004, p. 960). Belief in the right to be happy, even if it means getting divorced, is reflected in social attitudes toward divorce: Most U.S. adults (67 percent of men and 75 percent of women) say that divorce is morally acceptable (Dugan 2015).

7. *Weak social ties.* Couples with strong social ties have a network of family and friends who provide support during difficult times and who express disapproval for behavior that threatens the stability of the marriage. Couples who live in the same community for a long time have the opportunity to develop and maintain strong social ties. But many Americans move from place to place during their adult years, more so than people in other countries, which may help explain why the U.S. divorce rate is higher than in other countries. Cherlin (2009) explains that "moving from one community to another could affect marriages because it disrupts social ties. Migration can separate people from friends and relatives who could help them through family crises" (pp. 148–149).

8. *Longer life expectancy.* Finally, more marriages today end in divorce, in part, because people live longer than they did in previous generations, and "'til death do us part" involves a longer commitment than it once did.

Hill Creek Pictures/UpperCut Images/Getty Images

Wives tend to be less happy in marriage than husbands when they perceive the division of labor to be unfair.

Consequences of Divorce

When parents have bitter and unresolved conflict or if one parent is abusing a child or the other parent, divorce may offer a solution to family problems. However, divorce often has negative effects for ex-spouses and their children and also contributes to problems that affect society as a whole.

Physical and Mental Health Consequences. Numerous studies show that divorced individuals have more health problems and a higher risk of mortality than married individuals; divorced individuals also experience lower levels of psychological well-being,

no-fault divorce A divorce that is granted based on the claim that irreconcilable differences exist within a marriage (as opposed to one spouse being legally at fault for the marital breakup).

individualism The tendency to focus on one's individual self-interests and personal happiness rather than on the interests of one's family and community.

Some men and women experience a decline in well-being after divorce; others experience an improvement.

including more unhappiness, depression, anxiety, and poorer self-concepts (Amato 2003; Kamp Dush 2013). Unmarried people are more likely than married people to have low social attachment, low emotional support, and increased economic hardship (Walker 2001). Some research suggests that divorce leads to higher levels of depressive symptoms for women but not for men (Kalmijn and Monden 2006), especially when young children are in the family (Williams and Dunne-Bryant 2006). This finding is probably due to the increased financial and parenting strains experienced by divorced mothers who have custody of young children.

Some studies have found that divorced individuals report higher levels of autonomy and personal growth than married individuals do (Amato 2003). For example, many divorced mothers report improvements in career opportunities, social lives, and happiness after divorce; some divorced women report more self-confidence; and some men report more interpersonal skills and a greater willingness to self-disclose. For people in a poor-quality marriage, divorce has a less negative or even a positive effect on well-being (Amato 2003; Kalmijn and Monden 2006). However, leaving a bad marriage does not always result in increased well-being because "divorce is a trigger for even more problems after the divorce" (Kalmijn and Monden 2006, p. 1210). In sum, some men and women experience a decline in well-being after divorce; others experience an improvement.

Economic Consequences. Following divorce, there tends to be a dramatic drop in women's income and a slight drop in men's income (Gadalla 2009). Compared with married individuals, divorced individuals have a lower standard of living, have less wealth, and experience greater economic hardship, although this difference is considerably greater for women than for men (Amato 2003). The economic costs of divorce are often greater for women and children because women tend to earn less than men (see Chapter 10) and because mothers devote substantially more time to household and child care tasks than fathers do. The time women invest in this unpaid labor restricts their educational and job opportunities as well as their income.

In earlier decades, when most women worked in the home raising children, divorced women were commonly awarded permanent alimony—payments from the ex-husband that continued until death, the woman's remarriage, or a significant change in circumstances. Today, nearly half the workforce is women, and more than a third of employed married women earn more than their spouse. Although alimony laws vary by state, the general trend is for judges to award **rehabilitative alimony** for a specified length of time to allow the recipient time to find a job or to complete education or job training (Amundsen and Kelly 2014).

After divorce, both parents are responsible for providing economic resources to their children. However, some nonresident parents fail to provide child support. In some cases, failure to pay child support is not due to parents being "deadbeats" but rather to the fact that many parents are "dead broke." Parents who are unemployed or who have low-wage jobs may be unable to make child support payments.

Effects on Children and Young Adults. Parental divorce is a stressful event for children that often involves continuing conflict between parents, a decline in the standard of living, moving and perhaps changing schools, separation from the noncustodial parent (usually the father), and/or parental remarriage. These stressors place children of divorce at higher risk for a variety of emotional and behavioral problems. Children with divorced parents score lower on measures of academic success, psychological adjustment, self-concept, social competence, and long-term health; they also have higher levels of aggressive behavior and depression (Amato 2003; Wallerstein 2003). However, children of divorced parents who live in shared parenting families—where children live with each parent at least 35 percent of the time—may have fewer negative effects of parental divorce. Compared with children who live primarily with one divorced parent, children who live in shared parenting families tend to have higher levels of emotional, behavioral,

rehabilitative alimony
Alimony that is paid to an ex-spouse for a specified length of time to allow the recipient time to find a job or to complete education or job training.

and psychological well-being, better physical health, and better relationships with their mothers and fathers (Nielsen 2014).

Many of the negative effects of divorce on children are related to the economic hardship associated with divorce. Economic hardship is associated with less effective and less supportive parenting, inconsistent and harsh discipline, and emotional distress in children (Demo et al. 2000).

Experiencing parental divorce during one's adult years can also affect one's well-being. While some adult children of divorce are not traumatized by their parents' divorce, and some are even relieved by it, others struggle emotionally, especially when they are put in the middle of their parents' conflict or are forced to takes sides between parents (Greenwood 2014).

Despite the adverse effects of divorce on children, research findings suggest that "most children from divorced families are resilient, that is, they do not suffer from serious psychological problems" (Emery et al. 2005, p. 24). Other researchers conclude that "most offspring with divorced parents develop into well-adjusted adults," despite the pain they feel associated with the divorce (Amato and Cheadle 2005, p. 191).

Divorce can also have positive consequences for children and young adults. In highly conflictual marriages, divorce may improve the emotional well-being of children relative to staying in a conflicted home environment (Jekielek 1998). In interviews with 173 grown children whose parents divorced years earlier, Ahrons (2004) found that most of the young adults reported positive outcomes for their parents and for themselves. Although many young adults who have divorced parents fear that they, too, will have an unhappy marriage (Dennison and Koerner 2008), such a fear can also lead young adults to think carefully about their choices regarding marriage.

Effects on Father–Child Relationships. Children who live with their mothers may have a damaged relationship with their nonresidential father, especially if he becomes disengaged from their lives. Some research has found that young adults whose parents divorced are less likely to report having a close relationship with their father compared with children whose parents are together (DeCuzzi et al. 2004). However, in another study of adult children of divorce, more than half felt that their relationships with their fathers improved after the divorce (Ahrons 2004). Children may benefit from having more quality time with their fathers after parental divorce. Some fathers report that they became more active in the role of father after divorce. One study found that the older the child is at the time of parental separation, the greater the amount of time children spend with their fathers (Swiss and Le Bourdais 2009).

Women who have primary custody of the children serve as gatekeepers for the relationship their children have with their fathers (Trinder 2008). As gatekeepers, custodial mothers can either keep the gate open and encourage contact between the children and their father, or make every effort to keep the gate closed, cutting off the children's contact with their father. Some noncustodial divorced fathers discontinue contact with their children as a coping strategy for managing emotional pain (Pasley and Minton 2001). Many divorced fathers are overwhelmed with feelings of failure, guilt, anger, and sadness over the separation from their children (Knox 1998). Hewlett and West (1998) explained that "visiting their children only serves to remind these men of their painful loss, and they respond to this feeling by withdrawing completely" (p. 69)

Parental Alienation. **Parental alienation** (PA) occurs when one parent makes intentional efforts to turn a child against the other parent and essentially destroy any positive relationship a child has with the other parent (see this chapter's *The Human Side* feature). Bernet and Baker (2013), two experts on parental alienation, define PA as a "mental condition in which a child, usually one whose parents are engaged in a high-conflict separation or divorce, allies himself strongly with one parent (the preferred parent) and rejects a relationship with the other parent (the alienated parent) without legitimate justification" (p. 99).

parental alienation The intentional efforts of one parent to turn a child against the other parent and essentially destroy any positive relationship a child has with the other parent.

TABLE 5.2 Parental Alienation Behaviors

Limiting the child's contact with the other parent
Interfering with communication between the child and the other parent
Destroying gifts or memorabilia from the other parent
Failing to display any positive interest in the child's activities or experiences with the other parent
Expressing disapproval or dislike of the child spending time with the other parent
Limiting mention and photographs of the other parent
Withdrawal of love or expressions of anger if the child indicates positive feelings for the other parent
Telling the child that the other parent does not love him or her
Forcing the child to choose between his or her parents
Creating the impression that the other parent is dangerous (when the parent is not)
Forcing the child to reject the other parent
Asking the child to spy on the other parent
Asking the child to keep secrets from the other parent
Referring to the other parent by his or her first name
Referring to a stepparent as "Mom" or "Dad" and encouraging the child to do the same
Withholding medical, social, or academic information from the other parent and keeping the other parent's name off of such records
Changing the child's name to remove association with the other parent
Denigrating the other parent, the other parent's extended family, and any new partner/spouse

© Cengage Learning

When a parent influences his or her child to hate or fear the other parent, both the alienated parent and child may be said to be victims of parental alienation. Parents may alienate their child from the other parent by engaging in a variety of behaviors (see Table 5.2) (Baker and Chambers 2011).

Parental alienation is a disputed concept. Critics have argued that there is insufficient research to document its existence, but Bernet and Baker (2013) present numerous examples of research documenting parental alienation in both the United States and in other countries. Nevertheless, some lawyers, judges, therapists, and family scholars challenge the validity of parental alienation, claiming that it does not really exist. Baker (2006) explains:

> Perhaps such skeptics hold the belief that a parent must have done *something* to warrant their child's rejection and/or the other parent's animosity. This is the double victimization of [parental alienation]: the shame and frustration of being misunderstood in addition to the grief and anger associated with being powerless to prevent the alienation in the first place. (p. 192)

Other critics fear abusive parents can misuse the concept of "parental alienation" by claiming that children who do not want to visit them (because of past mistreatment) are victims of parental alienation. However, it is unlikely that a judge would ignore evidence of abuse and grant abusive parents unsupervised access to their children.

As we have seen, the effects of divorce on adults and children are mixed and variable. In a review of research on the consequences of divorce for children and adults, Amato (2003) concluded that "divorce benefits some individuals, leads others to experience temporary decrements in well-being that improve over time, and forces others on a downward cycle from which they might never fully recover" (p. 206).

Amy Baker (2007) conducted interviews with 40 individuals who believed that they had been turned against one parent by the other parent. This feature presents excerpts from these interviews, providing a glimpse into the experience of parental alienation from the child's perspective.

Larissa (speaking about her mother): *She'd always made both my brother and me feel that our father was somehow to blame for everything. Every day there'd be some attempt by her, some tale she'd tell me, to turn me against my father, so many incidents, it's simply impossible to list them all. . . . I grew to detest him, with a truly visceral hate. I couldn't stand to be in the same room with him, or to even talk to him or have him talk to me.* (pp. 30–31)

Bonnie: *My dad tried several different tactics to talk to me but my mom foiled his plans, I guess you could say. At the time I sort of knew what he was doing. . . . It felt like he was trying to contact me but she would stop it. For example, when I was in middle school in 6th grade, I think, my dad went up to the school to have lunch with me and she had told the school that if he would ever go up there they were to immediately call the police and I remember it was a big production. They came and got me out of class and they put me in the principal's office and they locked the door. I found out later he was escorted off the grounds by the police.* (p. 126)

Maria: *Basically I was a chess piece between the two of them, and it was very hard. . . . I started picking up on signals from my mother that showing any kind of affection or love for my father could be a problem for me. If I showed any positiveness toward him, it was, "How could you do this to me? . . . You are betraying me." When I go to visit him and I come back, to this day, my mother still says, "How can you do this to me? You are betraying me. You are slapping me in the face any time you have anything to do with this man. How could you do this to me?"* (p. 139) *. . . . My dad and I were very close and then after a time and all of this stuff from my mother and then I think I took on a lot of her anger toward him, feeling responsible for her because she was so, "Oh poor me. I am so broken and so wounded." . . . And even if I would have desired anything to do with him, I would not have acted on it because of how my mother would have reacted.* (pp. 139, 141)

Domestic Violence and Abuse

Domestic violence, which includes acts of violence committed by intimate partners and family members, accounts for 21 percent of all violent crime in the United States (Morgan and Truman 2014). Before reading further, you may want to take the Abusive Behavior Inventory in this chapter's *Self and Society* feature.

Intimate Partner Violence and Abuse

Abuse in relationships can take many forms, including emotional and psychological abuse, physical violence, and sexual abuse. **Intimate partner violence (IPV)** refers to actual or threatened violent crimes committed against individuals by their current or former spouses, cohabiting partners, boyfriends, or girlfriends. Emotional abuse can take many forms, including yelling and screaming, withholding physical contact, isolating someone from their family and friends, belittling or insulting a person, restricting a person's activities, controlling a person's behavior (e.g., telling them what to wear, what to eat, where they can go, what time they have to be home, and so on), and exhibiting unreasonable jealousy (Bureau of Justice Statistics 2011; Follingstad and Edmundson 2010).

Intimate partner violence is widespread. A study of young U.S. adults ages 17 to 24 found that 4 in 10 experienced physical violence and 5 in 10 experienced verbal abuse in their current or most recent relationship (Halpern-Meekin et al. 2013). Globally, one woman in every three has experienced physical or sexual violence by a partner (World Health Organization 2014b). There has been ongoing debate about whether more IPV is perpetrated by women or men. U.S. government data find that about four in five victims of IPV are female, suggesting men engage in more acts of IPV than do women (Catalano

intimate partner violence (IPV) Actual or threatened violent crimes committed against individuals by their current or former spouses, cohabiting partners, boyfriends, or girlfriends.

Circle the number that best represents your closest estimate of how often each of the behaviors happens in your relationship with your partner or happened with a former partner during the previous six months (1, never; 2, rarely; 3, occasionally; 4, frequently; 5, very frequently).

1. Called you a name and/or criticized you   1 2 3 4 5
2. Tried to keep you from doing something you wanted to do (e.g., going out with friends, going to meetings)   1 2 3 4 5
3. Gave you angry stares or looks   1 2 3 4 5
4. Prevented you from having money for your own use   1 2 3 4 5
5. Ended a discussion with you and made the decision himself/herself   1 2 3 4 5
6. Threatened to hit or throw something at you   1 2 3 4 5
7. Pushed, grabbed, or shoved you   1 2 3 4 5
8. Put down your family and friends   1 2 3 4 5
9. Accused you of paying too much attention to someone or something else   1 2 3 4 5
10. Put you on an allowance   1 2 3 4 5
11. Used your children to threaten you (e.g., told you that you would lose custody, said he/she would leave town with the children)   1 2 3 4 5
12. Became very upset with you because dinner, housework, or laundry was not done when he/she wanted it done or done the way he/she thought it should be   1 2 3 4 5
13. Said things to scare you (e.g., told you something "bad" would happen, threatened to commit suicide)   1 2 3 4 5
14. Slapped, hit, or punched you   1 2 3 4 5
15. Made you do something humiliating or degrading (e.g., begging for forgiveness, having to ask permission to use the car or to do something)   1 2 3 4 5
16. Checked up on you (e.g., listened to your phone calls, checked the mileage on your car, called you repeatedly at work)   1 2 3 4 5
17. Drove recklessly when you were in the car   1 2 3 4 5
18. Pressured you to have sex in a way you didn't like or want   1 2 3 4 5
19. Refused to do housework or child care   1 2 3 4 5
20. Threatened you with a knife, gun, or other weapon   1 2 3 4 5
21. Spanked you   1 2 3 4 5
22. Told you that you were a bad parent   1 2 3 4 5
23. Stopped you or tried to stop you from going to work or school   1 2 3 4 5
24. Threw, hit, kicked, or smashed something   1 2 3 4 5
25. Kicked you   1 2 3 4 5
26. Physically forced you to have sex   1 2 3 4 5
27. Threw you around   1 2 3 4 5
28. Physically attacked the sexual parts of your body   1 2 3 4 5
29. Choked or strangled you   1 2 3 4 5
30. Used a knife, gun, or other weapon against you   1 2 3 4 5

Scoring

Add the numbers you circled and divide the total by 30 to find your score. The higher your score, the more abusive your relationship.

SOURCE: Melanie F. Shepard and James A. Campbell, 1992 (September), "The Abusive Behavior Inventory: A Measure of Psychological and Physical Abuse," *Journal of Interpersonal Violence* 7(3):291–305. Copyright © 1992 by Sage Publications. Used by permission.

2012). But other research has found that the rate of female perpetration of IPV is equivalent to or exceeds the rate of male perpetration (Cui et al. 2013; Fincham et al. 2013). Some researchers suggest that women are more likely than men to *report* their acts of IPV because there is less stigma associated with women's violence against men compared to men's violence against women. Anderson (2013) explains that men who abuse women are depicted in U.S. culture as dangerous brutes, whereas women's violence against men is often depicted as funny or as justified. Some research has found that when women assault their male partners, these assaults tend to be acts of retaliation or self-defense, or to protect others, particularly children (Johnson 2001; Simiao et al. 2015; Swan et al. 2008). Regardless of which gender engages in more acts of IPV, both male and female violence and abuse in intimate relationships are problems that warrant attention.

Four Types of Intimate Partner Violence. One of the reasons that researchers disagree about which gender perpetrates more acts of IPV is that researchers use different methods for defining and measuring IPV, and these methods may not account for the different types of IPV. Johnson and Ferraro (2003) identified the following four types of partner violence:

1. *Common couple violence* refers to occasional acts of violence arising from arguments that get "out of hand." Common couple violence usually does not escalate into serious or life-threatening violence.
2. *Intimate terrorism* is motivated by a wish to control one's partner and involves not only violence but also economic subordination, threats, isolation, verbal and emotional abuse, and other control tactics. Intimate terrorism is almost entirely perpetrated by men and is more likely to escalate over time and to involve serious injury.
3. *Violent resistance* refers to acts of violence that are committed in self-defense, usually by women against a male partner.
4. *Mutual violent control* is a rare pattern of abuse "that could be viewed as two intimate terrorists battling for control" (p. 169).

Intimate partner abuse also takes the form of sexual aggression, which refers to sexual interaction that occurs against one's will through use of physical force, threat of force, pressure, use of alcohol or drugs, or use of position of authority. In most cases of rape or sexual assault of females the offender is an intimate partner, family member, friend, or acquaintance (Planty et al. 2013).

Three Types of Male Perpetrators of Intimate Partner Violence. Because male perpetrators of IPV tend to inflict more physical harm on victims than do female perpetrators, most research on perpetrators has focused on males. Researchers have identified three types of male abusers (Fowler and Westen 2011):

1. *The psychopathic abuser* is generally violent, impulsive, and lacks empathy and remorse. Men in this subgroup often experienced high rates of childhood abuse and neglect.
2. *The hostile/controlling abuser* is suspicious, hypersensitive to perceived criticism, holds grudges, feels misunderstood, and has few friends. He is a "controlling" person who, instead of taking responsibility for his actions, blames and attacks his partner.
3. *The borderline/dependent abuser* is unhappy, depressed, and prone to emotions that spiral out of control. This type of abuser suffers from deep fears of abandonment and lashes out at the person he loves and needs the most. Men in this subgroup appear more fragile than violent, and after abusive episodes, they tend to feel self-loathing and guilt and are likely to beg for forgiveness.

Effects of Intimate Partner Violence and Abuse. IPV results in bruises, broken bones, traumatic brain injury, cuts, burns, other physical injuries, and death. When women are abused during pregnancy, the result is often miscarriage or birth defects in the child. Victims of intimate partner violence are at higher risk for depression, anxiety, suicidal thoughts and attempts, lowered self-esteem, inability to trust men, fear of intimacy, substance abuse, and harsh parenting practices (Gustafsson and Cox 2012).

Battering also interferes with women's employment. Some abusers prohibit their partners from working. Other abusers "deliberately undermine women's employment by depriving them of transportation, harassing them at work, turning off alarm clocks, beating them before job interviews, and disappearing when they promise to provide child care" (Johnson and Ferraro 2003, p. 508). Battering also undermines employment by causing repeated absences, impairing women's ability to concentrate, and lowering their self-esteem and aspirations.

Abuse is also a factor in many divorces, which often results in a loss of economic resources. Women who flee an abusive home and who have no economic resources may find themselves homeless.

Children who witness domestic violence are at risk for emotional, behavioral, and academic problems as well as future violence in their own adult relationships (Kitzmann et al. 2003; Parker et al. 2000). Children may also commit violent acts against a parent's abusing partner. Children who witness domestic violence are also at increased risk of being violent toward animals (Phillips 2014).

Why Do Some Adults Stay in Abusive Relationships? Adult victims of abuse are commonly blamed for tolerating abusive relationships and for not leaving the relationship as soon as the abuse begins. However, from the point of view of the victims, there are compelling reasons to stay, including economic dependency, emotional attachment, commitment to the relationship, guilt, fear, hope that things will get better, and the view that violence is legitimate because they "deserve" it. Having children complicates the decision to leave an abusive relationship. Although some abuse victims leave to protect the children from harm by the abuser, others may stay because they cannot support and/or raise the children alone, or may fear losing custody of the children if they leave (Lacey et al. 2011). Many victims of intimate partner abuse stay because they fear retribution from their abusive partner if they leave. Some victims also delay leaving a violent home because they fear the abuser will hurt or neglect a family pet (Fogle 2003).

WHAT do you THINK?

A major reason that abused women stay in abusive relationships is economic dependency. Yet, in a national study of abused women, researchers found that among women of color, those with higher socioeconomic status were more likely to stay in abusive relationships than those with lower socioeconomic status (Lacey et al. 2011). How might this unexpected finding be explained?

Victims also stay because abuse in relationships is usually not ongoing and constant but rather occurs in cycles. The **cycle of abuse** involves a violent or abusive episode followed by a makeup period when the abuser expresses sorrow and asks for forgiveness and "one more chance." The makeup period may last for days, weeks, or even months before the next violent outburst occurs. Nevertheless, most abused women eventually leave their abusers for good, although leaving tends to be a process rather than a one-time event, and the process often involves leaving and returning multiple times (Lacey et al. 2011).

Researchers in one study that investigated the impact of technology on the maintenance of abusive relationships found that undergraduates check their text messages frequently and feel compelled to read their text messages—even those sent by an abusive partner or ex-partner (Halligan et al. 2013). Abusive partners and ex-partners try to maintain or resume the relationship by sending texts that include apologies, recognition of hurt they caused, promises to never be abusive again, and pleas to resume the relationship.

Child Abuse

Child abuse refers to the physical or mental injury, sexual abuse, negligent treatment, or maltreatment of a child under the age of 18 by a person who is responsible for the child's welfare. Worldwide, nearly one in four adults have been physically abused as children (World Health Organization 2014a). In low- and middle-income countries, three-quarters of children between 2 and 14 experience violent discipline in the home (Levtov et al. 2015). The most common form of child maltreatment is **neglect**—the caregiver's failure to provide adequate attention and supervision, food and nutrition, hygiene, medical care, and a safe and clean living environment (see Figure 5.5).

The highest rates of victimization are for children under age 1, and for children who are African American, multiracial, or American Indian/Alaska Native children. Very low rates of child abuse are found among Asian children and among children living in female

cycle of abuse A pattern of abuse in which a violent or abusive episode is followed by a makeup period when the abuser expresses sorrow and asks for forgiveness and "one more chance," before another instance of abuse occurs.

child abuse The physical or mental injury, sexual abuse, negligent treatment, or maltreatment of a child younger than age 18 by a person who is responsible for the child's welfare.

neglect A form of abuse involving the failure to provide adequate attention, supervision, nutrition, hygiene, health care, and a safe and clean living environment for a minor child or a dependent elderly individual.

same-sex couple families (Gartrell et al. 2010; National Center for Injury Prevention and Control 2014a).

Perpetrators of child abuse are most often the parents of the victim. Abuse is more common in families where there is a lot of stress that can result from a family history of violence, drug or alcohol abuse, poverty, chronic health problems, and social isolation.

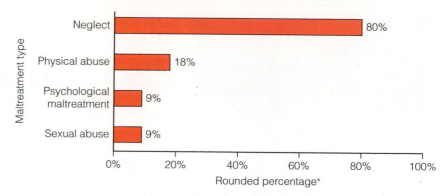

*Percentages sum to more than 100% because children may experience more than one type of maltreatment.

Figure 5.5 U.S. Distribution of Child Maltreatment, by Type
SOURCE: U.S. Department of Health and Human Services. 2015. Child Maltreatment 2013. Available at http://www.acf.hhs.gov

Effects of Child Abuse. Physical injuries sustained by child abuse cause pain, disfigurement, scarring, physical disability, and death. In 2013, an estimated 1,520 U.S. children, most of whom were under age 3, died of abuse and/or neglect U.S. Department of Health and Human Services 2015. One of the most deadly forms of child abuse is **abusive head trauma**, commonly known as **shaken baby syndrome**, which is a form of inflicted traumatic brain injury resulting from violent shaking or blunt impact. Shaking a baby causes the brain to bounce back and forth against the child's skull, causing bruising, swelling, and bleeding, which can lead to death. Abusive head trauma is a leading cause of death in children under 1 year (Reese et al. 2014). Those who survive may suffer from severe brain damage and permanent disabilities, including cerebral palsy, seizure disorders, and blindness. Babies up to 4 months old are at greatest risk, and inconsolable crying is the most common trigger for shaking a baby (James 2014).

Adults who were abused as children have an increased risk of a number of problems, including depression, smoking, alcohol and drug abuse, eating disorders, obesity, high-risk sexual behavior, and suicide (Centers for Disease Control and Prevention 2014). Sexual abuse of young girls is associated with decreased self-esteem, increased levels of depression, running away from home, alcohol and drug use, and multiple sexual partners (Jasinski et al. 2000; Whiffen et al. 2000). Sexual abuse of boys produces many of the same reactions that sexually abused girls experience, including depression, sexual dysfunction, anger, self-blame, suicidal feelings, guilt, and flashbacks (Daniel 2005). Married adults who were physically and sexually abused as children report lower marital satisfaction, higher stress, and lower family cohesion than married adults with no abuse history (Nelson and Wampler 2000).

Elder, Parent, Sibling, and Pet Abuse

Domestic violence and abuse may involve adults abusing their elderly parents or grandparents, children abusing their parents, and siblings abusing each other. Pets (or companion animals) are also victimized by domestic violence.

Elder Abuse. **Elder abuse** includes physical abuse, sexual abuse, psychological abuse, financial abuse (such as improper use of the elder person's financial resources), and neglect. The most common form of elder abuse is neglect—failure to provide basic health and hygiene needs, such as clean clothes, doctor visits, medication, and adequate nutrition. Neglect also involves unreasonable confinement, isolation of elderly family members, lack of supervision, and abandonment.

Older women are far more likely than older men to suffer from abuse or neglect. Two out of every three cases of elder abuse reported to state adult protective services involve women. Although elder abuse also occurs in nursing homes, most cases of elder abuse occur in a domestic setting. The most likely perpetrators are adult children, followed

abusive head trauma A form of inflicted brain injury resulting from violent shaking or blunt impact; a leading cause of death in children under 1 year (see also **shaken baby syndrome**).

shaken baby syndrome A form of potentially fatal brain damage resulting from violently shaking a baby (see also **abusive head trauma**).

elder abuse The physical, sexual or psychological abuse, financial exploitation, or medical abuse or neglect of the elderly.

by other family members and spouses or intimate partners (Teaster et al. 2006).

Parent Abuse. Parent abuse, committed by children and teens under age 18, includes physical violence, threats, intimidation, and damage to property. Parent abuse is most common by teen sons against single mothers (Santich 2014). In some cases of children being violent toward their parents, the parents had been violent toward the children, but many abused parents have done nothing wrong to provoke the abuse. Rosemary Pate reportedly did everything she could to help her 19-year-old son, who, after years of abusing his mother, fatally stabbed her in her own bedroom (Davis 2014). In response to Rosemary's tragic death, Florida lawmakers considered, but did not pass, a bill that would have established a separate legal category for parental abuse by a teen, required mandatory reporting of suspected parental abuse to the authorities, and allocated additional resources and police protection for abused parents (Santich 2014).

Sibling Abuse. Also referred to as "sibling bullying," sibling abuse is "a widespread and serious problem . . . and, arguably, the most frequent type of aggression in American society" (Skinner and Kowalski, 2013, p. 1727). A national survey found that nearly 30 percent of children 2 to 17 years old had been physically assaulted by a sibling (Finkelhor et al. 2005). Another study found that 48 percent of children 3 to 9 years old and 29 percent of adolescents 10 to 17 years old had experienced physical, property, or psychological aggression from a sibling in the past year (Tucker et al. 2014). The most common form of sibling abuse is emotional abuse, such as verbal teasing or excluding or ignoring a sibling (Skinner and Kowalski 2013). Sexual abuse also occurs in sibling relationships.

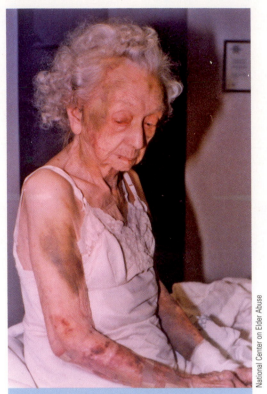

National Center on Elder Abuse

If you suspect elder abuse or are concerned about the well-being of an older person, call your state abuse hotline immediately.

WHAT do you THINK?

In recent years, the problem of bullying among peers in schools has received increased attention, and those concerned about bullying have worked to spread the message that bullying is not acceptable behavior. In contrast, sibling bullying is commonly viewed as acceptable and expected (Skinner and Kowalski 2013). Why do you think peer bullying is viewed as a serious social problem, whereas sibling bullying is generally not viewed as a serious social concern?

Pet Abuse. The majority of U.S. households (68 percent) have a pet (Phillips 2014). Because pets are often viewed as "members of the family," abused pets (or companion animals) can be considered victims of family violence (see also this chapter's *Animals and Society* feature). Abusers threaten to hurt pets to control their victims, and all too often their threats are carried out. One victim of domestic violence revealed how her abuser killed her cat as a means of inducing fear and intimidation:

> The very last thing he did to my cat hurt my heart so bad. He had me stand here and . . . she was tied to the tree [with] . . . fishing wire or . . . thread or something. And he . . . turned her around, stuffed [fireworks] in her behind and lit it. And I had to stand there and watch my cat explode in my face. And he was like, "That could happen to you." (quoted in Hardesty et al. 2013, p. 9)

Factors Contributing to Domestic Violence and Abuse

In many ways, U.S. culture tolerates and even promotes violence in sports (e.g. football), television and movies, video games, and as a means of dealing with international conflict

ANIMALS and society

Pets and Domestic Violence

In 2005, an 81-year-old woman walked nearly a mile from her house to a restaurant, pleading for help to escape the abuse she endured. She said her son repeatedly beat her and her dog. The restaurant owner called the local sheriff's office, and a deputy arrested the son. The woman relocated but could not take her dog, a 2-year-old 60-pound Catahoula, with her. Kathy Cornwell, one of the victim advocates who helped the woman, adopted the dog. Kathy and the dog were in the bedroom when her husband came home. At the sight of a strange man, the dog jumped onto the bed and laid his body across Kathy. When Kathy later spoke with the abuse victim, she learned the dog had always tried to protect the woman from her abusive son, covering her body and absorbing the blows. As a result of the beatings,

the dog had permanent nerve damage in his eyes and must wear goggles when he is outside because his pupils cannot dilate. The dog, known as Little Horatio, goes to universities and other public forums on abuse. Cornwell explains that Horatio gives abuse victims hope: "This dog has gone through so much. And look at how happy he is. He's not stuck in the past. It shows people that it's hard, but you can move on" (quoted in Sullivan 2010, n.p.).

Domestic violence and animal abuse are linked: Violence against animals is a predictor that the abuser is also violent toward people, and vice versa (Phillips 2014). Animal abuse is used to threaten human victims and to prevent family members from leaving an abusive home out of fear for the safety of their pets. Fear of their companion animals being harmed or killed creates a significant

barrier that prevents abused women from escaping their violent situations (Ascione 2007).

Because animal cruelty and family violence commonly co-occur, cross-reporting and cross-training have been instituted in many communities to teach human service personnel how to recognize and report cases of animal abuse and, conversely, to teach animal welfare personnel to recognize and report child, spousal, and elder abuse. Pets may be included in domestic violence protective orders, and seven states (Arizona, Colorado, Indiana, Maine, Nebraska, Nevada, and Tennessee) have added "cruelty to animals" to the definition of domestic violence when committed to intimidate or coerce a partner (Phillips 2014). Finally, some programs provide procedures for housing pets in domestic violence situations at animal shelters or with animal rescue groups. The American Humane Association PAWS (Pets and Women's Shelters) program provides grants to shelters to create housing where abuse victims and their pets can be together. Carol Wick, CEO of Harbor House, a domestic violence shelter in Orange County, Florida, describes the benefits of providing shelters for abuse victims and their pets:

> Every week we receive calls from domestic violence victims who want to leave their situation but are fearful of what will happen to their pets if they are left behind. . . . The new on-site kennel will allow women and children to safely leave an abusive situation and bring their pet with them—so no one gets left behind. (quoted in American Humane Association 2010)

Little Horatio was a victim of domestic violence.

Kathleen E. Cornwell

(Walby 2013). Other factors that contribute to domestic violence and abuse are discussed in the following sections.

Individual, Relationship, and Family Factors. Although we all have the potential to "lose our cool" and act abusively toward an intimate partner or family member, some individuals are at higher risk for being abusive. Risk factors include having witnessed or been a victim of abuse as a child, past violent or aggressive behavior, lack of employment and other stressful life events or circumstances, and drug and alcohol use (National Center for Injury Prevention and Control 2014b). Alcohol and other drugs increase aggression in some individuals and enable the offender to avoid responsibility by blaming the violent behavior on drugs or alcohol.

Although abuse in adult relationships occurs among all socioeconomic groups, it is more prevalent among the poor. However, most poor people do not maltreat their children, and poverty, per se, does not cause abuse and neglect; the correlates of poverty, including stress, drug abuse, and inadequate resources for food and medical care, increase the likelihood of maltreatment (Kaufman and Zigler 1992).

Nearly half of young adults experience "churning" in their romantic relationships—a pattern of instability in which the partners break up and then reconcile. A study of young adults aged 17 to 24 found that "relationship churners" were much more likely to report physical and verbal abuse in their most current relationship than were young adults who were either stably together or stably broken up (Halpern-Meekin et al. 2013).

Gender Inequality and Gender Socialization.
In the United States before the late 19th century, a married woman was considered the property of her husband. A husband had a legal right and marital obligation to discipline and control his wife through the use of physical force. This traditional view of women as property may contribute to men's doing with their "property" as they wish. In a study of men in battering intervention programs, about half of the men viewed battering as acceptable in certain situations (Jackson et al. 2003).

The view of women and children as property also explains marital rape and father–daughter incest. Historically, the penalties for rape were based on property rights laws designed to protect a man's property—his wife or daughter—from rape by other men (Russell 1990). Although a husband or father "taking" his own property in the past was not considered rape; today, marital rape and incest is considered a crime in all 50 states. Acquaintance and date rape can be explained in part by the fact that, in U.S. culture, "men receive support, even permission, from each other to sexually assault women through their direct encouragement or by ignoring problematic behavior" (Foubert et al. 2010, p. 2238).

Traditional male gender roles have taught men to be aggressive and to be dominant in male–female relationships. Traditional male gender socialization also discourages men from verbally expressing their feelings, which increases the potential for violence and abusive behavior (Umberson et al. 2003). Traditional female gender roles have also taught some women to be submissive to their male partner's control.

Acceptance of Corporal Punishment.
Corporal punishment—the intentional infliction of pain intended to change or control behavior—is widely accepted as a parenting practice. In the United States, corporal punishment is (1) almost universal—94 percent of toddlers are spanked; (2) chronic—toddlers are often spanked three or more times a week; (3) often severe, with more than one in four parents using an object such as a paddle or belt to punish their children; and (4) of long duration—13 years for a third of U.S. children, and 17 years for 14 percent of U.S. children (Straus 2010).

The acceptance of corporal punishment in Black American culture can be traced to historical mistreatment of black slaves by white slave owners and to the ongoing mistreatment of blacks by police. Dyson (2014) explains:

> Black parents beat their children to keep them from misbehaving in the eyes of whites who had the power to send black youth to their deaths for the slightest offense. Today, many black parents fear that a loose tongue or flash of temper could get their child killed by a trigger-happy cop. They would rather beat their offspring than bury them. (p. A33)

corporal punishment The intentional infliction of pain intended to change or control behavior.

Although not everyone agrees that all instances of corporal punishment constitute abuse, some episodes of parental "discipline" are undoubtedly abusive. Many mental health and child development specialists advise against physical punishment, arguing that it is ineffective and potentially harmful to the child (Straus 2000). One study of 3,870 families found that children who were spanked at age 1 were more likely to be aggressive at age 3 and depressed, anxious, and/or withdrawn at age 5 (Gromoske and Maguire-Jack 2012). Although the percentage of U.S. women and men who agree that children sometimes need a "good hard spanking" declined from 1986 to 2012 (see Figure 5.6), the majority of U.S. adults still view spanking as sometimes necessary (Child Trends 2013).

Inaccessible or Unaffordable Community Services. Many cases of child and elder abuse and neglect are linked to inaccessible or unaffordable health care, day care, elder care, and respite care facilities. Failure to provide medical care to children and elderly family members (a form of neglect) may result from lack of accessible or affordable health care services in the community. Failure to provide supervision for children and adults may result from inaccessible day care and elder care services. Without elder care and respite care facilities, socially isolated families may not have any help with the stresses of caring for elderly family members and children with special needs.

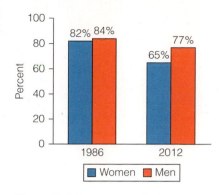

Figure 5.6 Percent of U.S. Adults Who Agree That a Child Sometimes Needs a "Good Hard Spanking," 1986–2012
SOURCE: Child Trends 2013.

Strategies for Action: Strengthening Families

Each chapter in this text discusses problems that affect families. Thus, addressing social problems such as inadequate health care (Chapter 2), substance abuse (Chapter 3), crime (Chapter 4), gender discrimination (Chapter 10), racial discrimination (Chapter 9), and so forth, also strengthens families. Our discussion here focuses on expanding the definition of family, relationship literacy education, workplace and economic supports (see also Chapters 6 and 7), reducing unplanned nonmarital pregnancies, covenant marriage, divorce education programs, divorce mediation, and domestic violence prevention and policies.

Expanding the Definition of Family

Increasingly, government and workplace definitions of "family" are expanding to include unmarried couples so that they and their children can access benefits and protections. Legal recognition of varied family forms is important for the well-being of children, as more than a third of unmarried cohabiting couples have children under age 18 living in the home, and many cohabiting same-sex couples are raising children (U.S. Census Bureau 2014). When children are denied a legal relationship to both parents because of the parents' unmarried status, they may be denied Social Security survivor benefits, health care insurance, or the ability to have either parent authorize medical treatment in an emergency, among other protections.

Some states, cities, counties, and employers allow unmarried partners (same-sex and/or heterosexual partners) to apply for a **domestic partnership** designation, which grants them some legal entitlements, such as health insurance benefits and inheritance rights that have traditionally been reserved for married couples. Some states have revised definitions of family members, partners, and intimate relationships to provide domestic violence protection to victims regardless of marital status, sexual orientation, or gender identity.

Relationship Literacy Education and the Healthy Marriage Initiative

Relationship literacy education includes voluntary, community-based programs that teach youth, young adults, cohabiting couples, and engaged and married couples the

domestic partnership A status that some states, counties, cities, and workplaces grant to unmarried couples, including gay and lesbian couples, that conveys various rights and responsibilities.

relationship literacy education Voluntary, community-based educational programs that teach youth, young adults, cohabiting couples, and engaged and married couples what healthy relationships look like and the skills and attitudes that make them work.

skills and values to form and maintain healthy relationships. Relationship literacy education programs cover topics such as these (Hawkins and VanDenBerghe 2014):

- Danger signs of violence and abuse in dating relationships
- Interpersonal communication skills
- Problem-solving skills
- Realistic marriage expectations
- How to peacefully end an unhealthy relationship
- Co-parenting skills
- How to enrich intimacy

The Healthy Marriage Initiative (HMI) provides federal funds to support research and programs to encourage healthy marriages and promote involved and responsible fatherhood, particularly among low-income populations who have the highest divorce rates and lowest marriage rates. Funds may be used for marriage and premarital education, public advertising campaigns that promote healthy marriage, high school programs on the value of marriage, marriage mentoring programs, and parenting skills programs. Compared with high-income adults, low-income adults have similar or more traditional family values and also have more money problems in their relationships, which suggests that government initiatives to strengthen marriage among low-income groups should focus on improving their economic conditions, rather than on educating them on the value of marriage (Trail and Karney 2012).

Workplace and Economic Supports

The most important strategies for forming and strengthening families may be those that maximize employment and earnings. A third of never-married U.S. adults said that financial insecurity is the main reason for not being married (Wang and Parker 2014). U.S. divorce rates are higher and marriage rates are lower among low-income populations (Trail and Karney 2012). The low rate of marriage among black women is affected by the high unemployment rate of black men: Among never-married blacks aged 25 to 34, there are only 51 employed men for every 100 women (Wang and Parker 2014). Supports such as job training, employment assistance, flexible workplace policies that decrease work–family conflict, affordable child care, and economic support, such as the earned income tax credit, are discussed in Chapters 6 and 7. In addition, policy makers should take a hard look at policies that penalize poor couples for marrying. Poor couples who marry are often penalized by losing Medicaid benefits, food stamps, and other forms of assistance.

> The most important pro-marriage and divorce prevention measures may be those that maximize employment and earnings.

Reduce Unplanned Nonmarital Childbearing

One way to reduce the negative effects of nonmarital childbearing and single parenting (discussed earlier) is to reduce the number of unplanned and nonmarital births. Sawhill (2014) argues that instead of planning when to have children, young women tend to "drift" into parenthood. The "drifters" could be turned into "planners" by providing them with long-acting reversible contraception, allowing women to delay childbearing until they are better able to provide a more stable home environment for the child.

Covenant Marriage

covenant marriage A type of marriage (offered in a few states) that requires premarital counseling and that permits divorce only under condition of fault or after a marital separation of more than two years.

In 1996, Louisiana became the first state to offer a **covenant marriage**, which permits divorce only under condition of fault (e.g., abuse, adultery, or felony conviction) or after a two-year separation. Couples who choose a covenant marriage must also get premarital counseling. Variations of the covenant marriage have also been adopted in Arizona and Arkansas.

The covenant marriage option was designed to strengthen marriages and decrease divorce. Research has found that covenant marriages are less likely to end in divorce, but marital satisfaction is only slightly increased, only for husbands, due to the requirement for premarital counseling (DeMaris et al. 2012). Critics argue that covenant marriage may increase family problems by making it more difficult to terminate a problematic marriage and by prolonging the exposure of children to parental conflict (Applewhite 2003).

Strategies to Strengthen Families during and after Divorce

Children are more traumatized by postdivorce conflict between parents than by the divorce itself (Ahrons 2004). Children benefit when divorced (or divorcing) parents minimize their conflict and develop a cooperative co-parenting relationship, which may be achieved through divorce education programs and divorce mediation.

Divorce Education Programs. Divorce education programs are designed to help parents who are divorced or planning to divorce reduce parental conflict, foster cooperative parenting, and understand and respond to their children's reactions to the divorce. Many counties and states require divorcing spouses to attend a divorce education program, whereas it is optional in other jurisdictions. Despite the expectation that cooperative postdivorce parenting benefits children, one study found that parents' perceptions of their children's psychological, behavioral, and social well-being was not associated with the parents' perceptions of their coparenting relationship (Beckmeyer et al. 2014). The researchers concluded that "divorcing parents can effectively rear children even when coparenting is limited or conflictual" (p. 533), and that what is more important for children's functioning is the parents' individual parenting skills and quality or quality of parent–child relationship. This is reassuring news for divorced parents who are unable to establish peaceful, positive, and cooperative coparenting with their ex-spouse.

Divorce Mediation. In **divorce mediation**, divorcing couples meet with a neutral third party, a mediator, who helps them resolve issues of property division, child and spousal support, and assist in developing a parenting plan in a way that minimizes conflict and encourages cooperation. It is generally best for parents and their children if the parents can agree on a parenting plan. Otherwise, "judges routinely decide where the children of divorced parents will attend school, worship and receive medical care; judges may even decide whether they plan soccer or take piano lessons" (Emery 2014, n.p.). Mediation can not only speed settlement, save money on attorney and court fees, and increase compliance, it can also result in improved relationships between nonresidential parents and children, as well as between divorced parents 12 years after the dispute settlement (Emery et al. 2005).

Many jurisdictions and states have mandatory child custody mediation programs, whereby parents in a custody or visitation dispute must attempt to resolve their dispute through mediation before a court will hear the case. Mediation is *not* recommended for couples in which one or both partners are unwilling to negotiate in good faith, have a mental illness, or have been violent or abusive in the relationship.

divorce mediation A process in which divorcing couples meet with a neutral third party (mediator) who assists the individuals in resolving issues such as property division, child custody, child support, and spousal support in a way that minimizes conflict and encourages cooperation.

Domestic Violence and Abuse Prevention Strategies

Domestic violence in the United States declined 67 percent from 1994 to 2013 (Planty 2014). Abuse prevention strategies include reducing violence-provoking stress by reducing poverty and unemployment and providing adequate housing, child care programs

The movement to ban corporal punishment has resulted in 46 countries banning it.

and facilities, nutrition, medical care, and educational opportunities. Strengthening the supports for poor families with children reduces violence-provoking stress and minimizes neglect that results from inaccessible or unaffordable community services. Public education and media campaigns may help to reduce domestic violence by conveying the criminal nature of domestic assault and offering ways to prevent abuse. And parent education teaches parents realistic expectations about infant and child behavior and methods of child discipline that do not involve corporal punishment.

At the global level, an international campaign called "MenCare" promotes men and boys' involvement in equitable, nonviolent caregiving. With initiatives in 30 countries, MenCare promotes positive parenting, violence prevention, and sharing domestic and caregiving work with women (Levtov et al. 2015).

Banning Corporal Punishment. Corporal punishment can be physically and emotionally abusive and reinforces the cultural acceptance of violence in family relationships. In 1979, Sweden became the first country in the world to ban corporal punishment in all settings, including the home. By 2015, 46 countries banned corporal punishment in all settings (Global Initiative to End All Corporal Punishment of Children 2015). It is legal in all 50 U.S. states for a parent to spank, hit, belt, paddle, whip, or otherwise inflict punitive pain on a child, so long as the corporal punishment does not meet the individual state's definition of child abuse. Corporal punishment is banned in public schools in 31 U.S. states and the District of Columbia, and it is banned in public and private schools in New Jersey and Iowa.

WHAT do you THINK?

A U.S. federal bill called Ending Corporal Punishment in Schools Bill 2015, originally introduced in 2010, would ban corporal punishment in public schools. Are you in favor of this bill passing? Why or why not?

The Period of PURPLE Crying® Program. A program to help prevent abusive head trauma to infants, the **Period of PURPLE Crying** program, is a parent education program that teaches parents and caregivers about normal infant crying and the dangers of shaking an infant. The program teaches methods for coping with infant crying, and emphasizes that if the baby is inconsolable, it is OK for the caregiver to leave the baby in a safe place and walk away. The acronym PURPLE is used to describe characteristics of an infant's crying during this phase (see Figure 5.7). The Period of PURPLE Crying program has been implemented in over 800 hospitals and organizations in 49 states, eight Canadian provinces and 1 territory (see www.purplecrying.info). In one study evaluating the effectiveness of the program, most mothers (76 percent) who received the Period of PURPLE Crying education rated it as very useful (Reese et al. 2014).

Help for Adult Abuse Victims

Domestic violence and abuse victims may seek help from local domestic violence programs. The National Domestic Violence Hotline (1-800-799-SAFE) is a 24-hour, toll-free service that provides information, support, and referrals to local domestic violence programs which offer emergency shelter; transitional housing; legal services; counseling;

period of PURPLE crying The phase at 2 weeks to 3 to 4 months of age when infants' crying is characterized by the acronym PURPLE: **P**eaks in crying that are **U**nexplained, **R**esists soothing, are accompanied by a **P**ain-like face, are **L**ong-lasting, and occur in the **E**vening and late afternoon.

and/or assistance with employment, transportation, food, medical care, and child care. Some programs offer a safe shelter for pets of victims of domestic violence. Due to inadequate funding and staff, local domestic violence programs receive thousands of requests for housing and other services that cannot be provided because the programs do not have the resources (National Network to End Domestic Violence 2014). When services are not available, abuse victims often return to the abuser, become homeless, or live in their cars. Some states have laws that allow victims to (1) take a leave of absence from their job without fear of losing their employment, (2) change their phone number free of charge, and (3) terminate their housing lease early without penalty (National Conference of State Legislators 2014a).

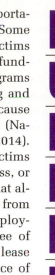

Peak of Crying:
Babies tend to cry the most from age 2 weeks to 4 months of age.

Unexpected:
Babies can cry suddenly and without any apparent reason.

Resists Soothing:
Babies can continue to cry no matter what the caregiver does to try to soothe the baby.

Pain-Like Face:
Crying babies may look like they are in pain, even when they are not.

Long-Lasting:
Crying can last as much as 5 hours a day or longer.

Evening:
Babies may cry more in the later afternoon or evening.

Figure 5.7 Period of PURPLE Crying
SOURCE: Adapted from http://purplecrying.info/what-is-the-period-of-purple-crying.php

Help for Child Abuse/Neglect Victims

In 48 states, the law requires members of certain professions (such as teachers, doctors, counselors) to report suspected child abuse/neglect, and in 18 states, any person who suspects child abuse or neglect must report it to a law enforcement authority, a hotline (e.g., 1-800-422-4453), or to Child Protective Services (National Conference of State Legislators 2014b). Child Protective Services, sometimes referred to as the Department of Children and Family Services, is the state agency that investigates reports of child abuse and neglect. Unfortunately, the child protective services system is understaffed and suffers from an overload of cases. An investigation by the Associated Press revealed that over a six-year period, at least 768 U.S. children died from child abuse or neglect while their families were being investigated or while the children were under some form of protection services because of abuse or neglect in the home (Mohr and Burke 2014).

Children who are abused or neglected in the family, or who live in a home where abuse is occurring, may be removed from their homes and placed in government-supervised foster care. Foster care placements include grandparents and other family members, certified foster parents, group homes, and other institutional facilities. Hundreds of thousands of children are in foster care, waiting to be reunited with their families or adopted. There is a shortage of people willing to adopt foster children because foster children tend to be older and are more likely to have emotional or physical problems (Koch 2009). Every year, thousands of children who have not been adopted or reunited with their families must leave foster care because they turn 18. In some states, youth who age out of the foster care system can receive government aid such as housing assistance and Medicaid. However, many youth who age out of foster care struggle to fend for themselves, and many become homeless.

Another problem that plagues the foster care system is that, although it is intended to protect children from abuse, foster parents or caregivers sometimes abuse or neglect the children. Placing a child in foster care can also result in severing ties not only with the child's family but also with the child's culture, language, and identity, as often happens when Native American children are placed in foster care (Cross 2014).

Legal Action against Abusers

Domestic violence and abuse are crimes for which individuals can be arrested, jailed, and/or ordered to leave the home or enter a treatment program. Many states and Washington, DC, have laws that require police to arrest abusers, and prosecutors to prosecute those arrested for abuse, even if the victim does not want to press charges. This removes responsibility for arrest and prosecution from the victim, who may fear that taking legal action against her abuser may result in retaliation. Abuse victims may also obtain a restraining order that prohibits the perpetrator from going near the abused partner.

WHAT do you THINK?

Just over half (55 percent) of domestic violence incidents involving victims age 12 and older are reported to police (Morgan and Truman 2014). Why do you think many domestic violence victims do not call the police?

Abusers may also be required by courts to receive treatment, although some abusers may enter treatment voluntarily (Carter 2010). Treatment for abusers typically involves group and/or individual counseling, substance abuse counseling, and/or training in communication, conflict resolution, and anger management. The success rate of these intervention programs is relatively low, and some experts say that treatment should not replace punishment (Cohn 2014).

Understanding Family Problems

Families are influenced by the larger society and culture of which they are a part. Just as societies vary around the world and change across time, so do family norms and structures vary from region to region and from generation to generation.

What we consider to be family problems also varies over time, across cultures, and even within cultures. For example, many Americans view physical punishment of children as necessary for effective child discipline; others view physical punishment as harmful to child development. For some, the solution to family problems implies encouraging marriage and discouraging other family forms, such as single parenting, cohabitation, polyamorous families, and same-sex unions. However, many family scholars argue that the fundamental issue is making sure that children are well cared for, regardless of their parents' marital status, sexual orientation, or lifestyle. "Refocusing laws, regulations, and policies on children, rather than the adults to whom they are related, would far better serve the diverse families that actually live in the United States today" (Sheff 2014, p. 285). Some even suggest that marriage is part of the problem, not part of the solution. Martha Fineman of Cornell Law School said, "This obsession with marriage prevents us from looking at our social problems and addressing them. . . . Marriage is nothing more than a piece of paper, and yet we rely on marriage to do a lot of work in this society: It becomes our family policy, our policy in regard to welfare and children, the cure for poverty" (quoted by Lewin 2000, p. 2).

Strengthening marriage is a worthy goal because strong marriages offer many benefits to individuals and their children. But strengthening marriage does not mean that other family forms should not also be supported. The reality is that contemporary families come in many forms, each with its strengths, needs, and challenges. Given the diversity of families today, social historian Stephanie Coontz (2004) suggested that "the appropriate question . . . is not what single family form or marriage arrangement we would prefer in the abstract, but how we can help people in a wide array of different committed relationships minimize their shortcomings and maximize their solidarities" (p. 979). She further argued that

> [i]f we withdrew our social acceptance of alternatives to marriage, marriage itself might suffer. . . . The same personal freedoms that allow people to expect more from their married lives also allow them to get more out of staying single and give them more choice than ever before in history about whether or not to remain together. (Coontz 2005, p. 310)

The family problems emphasized in this chapter—problems of divorce, and domestic violence and abuse—have something in common: Economic hardship and poverty can be a contributing factor and a consequence of each of these problems. In the next chapter, we turn our attention to poverty and economic inequality, problems that are at the heart of many other social ills.

Chapter Review

- **What are some examples of diversity in families around the world?**
  Some societies recognize monogamy as the only legal form of marriage, whereas other societies permit polygamy. Societies also vary in their norms and policies regarding arranged marriage versus self-chosen marriage, same-sex couples, childbearing, and the division of power in the family.

- **What are some patterns, trends, and variations in contemporary U.S. families?**
  Patterns, trends, and variations in U.S. families include increased singlehood and older age at first marriage; more interracial and interethnic relationships; delayed childbearing or remaining child-free; increased heterosexual and same-sex cohabitation; increased births to unmarried women; stabilized divorce rate; increased "gray" divorces; more blended families; greater education, employment, and income of wives and mothers; increase in grandfamilies; fewer children living in "traditional" families; the emergence of living apart together (LAT) relationships; polyamorous couples and families; and couples and families living in intentional communities.

- **What is the marital decline perspective? What is the marital resiliency perspective?**
  According to the marital decline perspective, the recent transformations in American families signify a collapse of marriage and family in the United States. According to the marital resiliency perspective, poverty, unemployment, poorly funded schools, discrimination, and the lack of basic services (such as health insurance and child care) are more harmful to the well-being of children and adults than is the decline in married two-parent families.

- **How do the three main sociological theories (structural functionalism, conflict theory, or symbolic interactionism) view family problems?**
  Structural functionalism views the family as a social institution that performs important functions for society, and views traditional gender roles as contributing to family functioning. It also looks at how changes in other social institutions affect families. Conflict theory focuses on how capitalism, social class and power influence families and emphasizes that powerful and wealthy groups shape social programs and policies that affect families. Both feminist and conflict theories are concerned with how gender inequality influences and results from marriage and family patterns. Symbolic interaction looks at how family interactions affect self-concept and draws attention to the effects of labels and meanings on self-concept and behavior related to family forms, domestic violence, and issues related to divorce.

- **What are the social causes of divorce discussed in your text?**
  Social causes of divorce include (1) the changing function of marriage, (2) increased economic autonomy of women, (3) greater work demands and economic stress, (4) inequality in marital division of labor, (5) no-fault divorce laws, (6) increased individualism, (7) weak social ties, and (8) longer life expectancy.

- **What are some of the effects of divorce on children?**
  Reviews of research on the consequences of divorce for children find that children with divorced parents score lower on measures of academic success, psychological adjustment, self-concept, social competence, and long-term health and that they have higher levels of aggressive behavior and depression. Such effects are related to the economic hardship associated with divorce, the reduced parental supervision resulting from divorce, and parental conflict during and after divorce. In highly conflictual marriages, divorce may actually improve children's emotional well-being relative to staying in a conflicted home environment.

- **What are the four types of partner violence that Johnson and Ferraro (2003) identified?**
  The four patterns of partner violence are (1) common couple violence (occasional acts of violence arising from arguments that get "out of hand"), (2) intimate terrorism (violence that is motivated by a wish to control one's partner), (3) violent resistance (acts of violence that are committed in self-defense), and (4) mutual violent control (both partners battling for control).

- **Why do some abused adults stay in abusive relationships?**
  Adult victims of abuse are commonly blamed for choosing to stay in their abusive relationships. From the point of view of the victim, reasons to stay in the relationship include economic dependency, emotional attachment, commitment to the relationship, guilt, fear, hope that things will get better, and the view that violence is legitimate because they "deserve" it. Some victims with children leave to protect the children, but others stay because they need help raising the children or fear losing custody of them if they leave. Some victims stay because they fear the abuser will abuse or neglect a pet.

- **What is the most common form of child abuse? What is one of the most deadly forms of child abuse and is the leading cause of death in children under age 1 year?**
  The most common form of child abuse is neglect. One of the most deadly forms of child abuse, particularly in children under age 1 year, is abusive head trauma, also known as shaken baby syndrome.

- **What factors contribute to domestic violence and abuse?**
  Factors that contribute to domestic violence and abuse include (1) individual, relationship, and family factors (drug/alcohol use, stressful life events, having witnessed or experienced child abuse as a child, relationship churning); (2) gender inequality and gender socialization; (3) acceptance of corporal punishment; and (4) inaccessible or unaffordable community services.

- **What strategies to strengthen families are discussed in your text?**
  Strategies to strengthen families include (1) expanding the definition of family; (2) workplace and economic supports; (3) relationship literacy education; (4) covenant marriage, divorce mediation, and divorce education programs; and (5) domestic violence and abuse prevention and policies.

Test Yourself

1. The United States has the highest nonmarital birthrate of any country in the world.
   a. True
   b. False
2. Multipartner families that raise children and function as families are called
   a. communal families.
   b. extended families.
   c. functional families.
   d. poly families.
3. Two perspectives on the state of marriage in the United States are the marital decline perspective and the marital ___ perspective.
   a. health
   b. resiliency
   c. incline
   d. stability
4. How many states recognize no-fault divorce?
   a. None
   b. 5
   c. 25 and the District of Columbia
   d. All 50
5. Most U.S. adults believe that divorce is morally acceptable.
   a. True
   b. False

6. In the majority of married-couple families with children under 18, both parents are employed.
   a. True
   b. False
7. The purpose of divorce mediation is to help couples who are considering divorce repair their relationship and stay together.
   a. True
   b. False
8. Babies go through a period of PURPLE
   a. Crying.
   b. Poopies.
   c. Language.
   d. Parents.
9. The majority of U.S. adults view spanking as
   a. appropriate only for children under the age of 10.
   b. harmful.
   c. sometimes necessary.
   d. the responsibility of the father, not the mother.
10. Treatment for abusers—such as group and/or individual counseling; substance abuse counseling; and/or training in communication, conflict resolution, and anger management is regarded as relatively unsuccessful.
   a. True
   b. False

Answers: 1. B; 2. D; 3. B; 4. D; 5. A; 6. A; 7. B; 8. A; 9. C; 10. A.

Key Terms

"Overcoming poverty is not a gesture of charity. It is an act of justice."

NELSON MANDELA

Economic Inequality, Wealth, and Poverty

Learning Objectives

After studying this chapter, you will be able to . . .

1 Describe the extent of economic inequality, wealth, and poverty around the world.

2 Summarize different ways of defining and measuring poverty, and discuss criticisms of the U.S. poverty line.

3 Analyze how the structural-functionalist, conflict, and symbolic interactionist perspectives view the nature, causes, and consequences of economic inequality, wealth, and/or poverty.

4 Describe the extent of economic inequality and poverty in the United States, and discuss patterns that characterize the wealthy "1 percent" as well as patterns of poverty.

5 Explain how economic inequality and poverty affect health and hunger, housing and homelessness, legal and political inequality, war and social conflict, vulnerability to natural disasters, education, marriage opportunity and family problems, and intergenerational poverty.

6 Discuss international and U.S. strategies aimed at reducing poverty and economic inequality.

7 Identify and describe various U.S. public assistance and welfare programs, and discuss at least four myths about welfare in the United States.

8 Understand the implications of blaming poverty on the poor, and explain why alleviating poverty and reducing economic inequality are worthy goals.

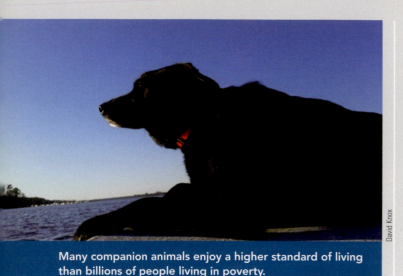

Many companion animals enjoy a higher standard of living than billions of people living in poverty.

MY BLACK LAB Cosmo enjoys two good meals a day in addition to his special doggy biscuit "treats," and there is always clean water in his bowl. Cosmo lives in a temperature-controlled house and has a comfortable place to sleep—actually, two: downstairs he sleeps in my bedroom on a down-filled ottoman converted to a dog bed, and upstairs (in a room that I call his "apartment") Cosmo sleeps on a small Pier 1 sofa. Cosmo has access to health care, with regularly scheduled visits to the veterinarian. He has received "education" both in a group dog-training "classroom" setting and in private "tutoring" from a skilled dog trainer who taught Cosmo how to catch a Frisbee in midair. Cosmo has a job that is very important to him and that gives him great satisfaction: He barks when someone approaches the house and so announces the arrival of a guest or the delivery of a package. His loud bark also serves as our home alarm system, offering protection against unwanted intruders. Cosmo also enjoys recreation: the daily walk around the neighborhood, and the occasional ride in the car or on a boat.

Like many companion animals, Cosmo—a dog—enjoys a higher standard of living than billions of people who live in poverty. Adequate food and clean water; a comfortable shelter; access to health care, education, and recreation; and meaningful work—these are basic needs that go unfulfilled for too many people who live in or near poverty. In this chapter, we examine economic inequality, wealth, and poverty globally and in the United States.

The Global Context: Economic Inequality, Wealth, and Poverty Around the World

One billion people—14.5 percent of the world's population—are extremely poor, living on less than $1.25 a day (World Bank 2015). Most of the world's poor—80 percent—live in sub-Saharan Africa or South Asia (World Bank Group and International Monetary Fund 2015). The very poor suffer from a myriad of problems including malnutrition, lack of access to clean water and sanitation, inadequate housing, lack of education, and poor health— problems we discuss later in this chapter.

In contrast to the extreme poverty that plagues the approximately one in seven people living on this planet, in 2014, there were 1,645 billionaires in the world, nearly a third (30 percent) of whom are U.S. citizens (Dolan and Kroll 2014). Living opulent, lavish lifestyles that include luxuries most of us can only imagine, billionaires have access to anything money can buy: yachts and private jets, multiple homes around the world, the best education and health care, and, as we discuss later, influence in political affairs.

Economic inequality—the wide gap that divides the rich and the poor—includes inequality in both income and wealth. **Wealth** refers to the total assets of an individual or household minus liabilities (mortgages, loans, and debts). Wealth includes the value of a home, investment real estate, cars, unincorporated business, life insurance (cash value), stocks, bonds, mutual funds, trusts, checking and savings accounts, individual retirement accounts (IRAs), and valuable collectibles.

wealth The total assets of an individual or household minus liabilities.

The inequality in the world distribution of wealth is reflected in the following data (Credit Suisse Research Institute 2014; Hardoon 2015):

- A person needs just $3,650 (in U.S. dollars) of wealth to be among the wealthiest half of the world's population; $77,000 to be in the top 10 percent; and $798,000 to be among the richest 1 percent worldwide.
- In 2014, the richest 1 percent of adults (ages 20 and older) in the world owned nearly half (48 percent) of global wealth; the richest 10 percent owned 87 percent of global wealth.
- In 2014, the richest 80 people in the world owned as much wealth as the 3.5 billion people at the bottom half of the world's population.

People in this tented village in New Delhi, India, live in extreme poverty.

- By 2016, the top 1 percent is expected to own more than half of all global wealth, which means that the wealthiest 1 percent will own more than the rest of us combined.
- The United States has the highest number of millionaires of any country. In 2014, 41 percent of the world's millionaires lived in the United States. Japan had the second highest number of millionaires (8 percent), followed by France (7 percent), Germany, and the United Kingdom (6 percent).
- People with wealth totaling more than $1 million represent just 0.7 percent of the global population, but they own 41 percent of the world's wealth. The more than two-thirds of the world's population (69 percent) who have a net worth of less than $10,000 collectively own only 3 percent of global wealth.

Defining and Measuring Poverty

Absolute poverty refers to the lack of resources necessary for well-being—most important, food and water, but also housing, sanitation, education, and health care. In contrast, **relative poverty** refers to the lack of material and economic resources compared with some other population. If you are a struggling college student living on a limited budget, you may feel as though you are "poor" compared with the middle- or upper-middle-class lifestyle to which you may aspire. However, if you have a roof over your head; access to clean water, toilets, and medical care; and enough to eat, you are not absolutely poor; indeed, you have a level of well-being that millions of people living in absolute poverty may never achieve.

Global Measures of Poverty

The most widely used standard to measure **extreme poverty** in the developing world is $1.25 or less in income per day. Based on this measure, about a billion people, or one in seven people on this planet, live in extreme poverty (World Bank 2015).

A measure of relative poverty is based on comparing the income of a household to the median household income in a specific country. According to this relative poverty measure, members of a household are considered poor if their household income is less than 50 percent of the median household income in that country.

absolute poverty The lack of resources necessary for material well-being—most important, food and water, but also housing, sanitation, education, and health care.

relative poverty The lack of material and economic resources compared with some other population.

extreme poverty Living on less than $1.25 a day.

TABLE 6.1 Three Dimensions of Multidimensional Poverty

1. Health (nutrition, child mortality)

2. Education (years of schooling, school attendance)

3. Standard of living (cooking fuel, toilet, water, electricity, floor, assets)

© Cengage Learning

Low income is only one indicator of impoverishment. The **Multidimensional Poverty Index** is a measure of serious deprivation in the dimensions of health, education, and living standards that combines the number of deprived and the intensity of their deprivation (see Table 6.1). About 1.5 billion people in the world are multidimensionally poor (United Nations Development Programme [UNDP] 2014).

WHAT do you THINK?

The next section describes how poverty is measured in the United States by comparing the annual pretax income of a household to the official U.S. poverty line—a dollar amount that determines who is considered poor. Before reading further, answer these questions: How much annual income do you think a household with one adult needs to earn to avoid living in poverty? What about a household with two adults? One adult and one child? Two adults and one child? Compare your answers with the official poverty thresholds in Table 6.2.

U.S. Measures of Poverty

In 1964, the Social Security Administration devised a poverty index based on data that indicated that families spent about one-third of their income on food. The official poverty level was set by multiplying food costs by three. Since then, the poverty level has been updated annually for inflation; it differs by the number of adults and children in a household and by the age of the head of household, but it is the same across the continental United States (see Table 6.2). Anyone living in a household with pretax income below the official poverty line is considered "poor." Individuals living in households with incomes that are above the poverty line, but not very much above it, are classified as "near-poor," and those living in households with income below 50 percent of the poverty line live in "deep poverty," also referred to as "severe poverty." A common working definition of "low-income" households is households with incomes that are between 100 percent and 200 percent of the federal poverty line or up to twice the poverty level.

The U.S. poverty line has been criticized on several grounds. First, the official poverty line is based on pretax income so tax burdens, as well as tax credits, are disregarded. Family wealth, including savings and property, are also excluded in official poverty calculations, and noncash government benefits that assist low-income families—food assistance, Medicaid, and housing and child care assistance—are not taken into account. In addition, the federal poverty line is a national standard that does not reflect the significant variation in the cost of living from state to state and between urban and rural areas. Finally, the poverty line underestimates the extent of material hardship in the United States because it is based on the assumption that low-income families spend one-third of their household income on food. That was true in the 1950s, but because housing, medical care, child care, and transportation costs have risen more rapidly than food costs, low-income families today spend far less than one-third of their income on food.

The Economic Policy Institute's Family Budget Calculator measures the income families of different sizes need in order to obtain "a secure yet modest living standard" based on estimated costs of housing, food, child care, transportation, health care, other necessities, and taxes in specific geographic locations (Gould et al. 2013). For a two-parent, two-child family, the basic family budget ranges from $48,144 in Marshall County, Mississippi, to $93,502 in New York City. The Family Budget Calculator finds that families need, at *minimum*, twice the poverty-level income to meet their basic needs.

TABLE 6.2 Poverty Thresholds, 2014 (Householder Younger Than 65 Years)

Household Makeup	Poverty Threshold
One adult	$12,316
Two adults	$15,853
One adult, one child	$16,317
Two adults, one child	$19,055
Two adults, two children	$24,088

SOURCE: U.S. Census Bureau 2015b.

Multidimensional Poverty Index A measure of serious deprivation in the dimensions of health, education, and living standards that combines the number of deprived and the intensity of their deprivation.

Sociological Theories of Economic Inequality, Wealth, and Poverty

Americans are taught that we live in a **meritocracy**—a social system in which individuals get ahead and earn rewards based on their individual efforts and abilities (McNamee and Miller 2009). In a meritocracy, everyone has an equal chance to succeed; those who are "$ucce$$ful" are smart and talented and have worked hard and deserve their success, while those who fail to "make it" have only themselves to blame. This individualistic perspective views economic inequality as the result of some people developing their potential and working hard and earning their success, while others don't measure up, make bad choices, don't work hard enough, and have only themselves to blame for their predicament. In contrast to the individualistic perspective, structural functionalism, conflict theory, and symbolic interactionism offer sociological insights into the nature, causes, and consequences of poverty and economic inequality.

Structural-Functionalist Perspective

According to the structural-functionalist perspective, poverty results from institutional breakdown: economic institutions that fail to provide sufficient jobs and pay, educational institutions that fail to equip members of society with the skills they need for employment, family institutions that do not provide two parents, and government institutions that do not provide sufficient public support. Sociologist William Julius Wilson explains:

> Where jobs are scarce . . . and where there is a disruptive or degraded school life purporting to prepare youngsters for eventual participation in the workforce, many people eventually lose their feeling of connectedness to work in the formal economy; they no longer expect work to be a regular, and regulating, force in their lives. (Wilson 1996, pp. 52–53)

More than 60 years ago, Davis and Moore (1945) presented a structural-functionalist explanation for economic inequality, arguing that a system of unequal pay motivates people to achieve higher levels of training and education and to take on jobs that are more important and difficult by offering higher rewards for higher achievements. However, this argument is criticized on the grounds that many important occupational roles, such as child care workers and nurse assistants, have low salaries, whereas many individuals in nonessential roles (e.g., professional sports stars and entertainers) earn outrageous sums of money. The structural-functionalist argument that CEO pay is high to reward strong performance is shattered by the fact that CEOs are paid huge salaries and bonuses even when they contribute to the economic failure of their corporation and/or to the problem of unemployment.

In his classic article "The Positive Functions of Poverty," sociologist Herbert Gans (1972) draws on the structural-functionalist perspective to identify ways in which poverty can be viewed as functional for the nonpoor segments of society. For example, having a poor population ensures that society has a pool of low-cost laborers willing to do unpleasant jobs, providing a labor pool for jobs ranging from the military to prostitution. Poor populations also provide labor for the affluent in the form of domestic work, such as maids and gardeners. The poor help keep others employed in jobs such as policing, prison work, and social work. And the poor provide a pool of consumers for used goods and second-rate service providers. However, the structural-functionalist view of poverty also highlights the ways in which poverty and economic inequality are *dysfunctional* for society. For example, as we discuss later in this chapter, poverty and economic inequality are linked to crime and violence, family instability, and social conflict and war (see also Chapters 4, 5, and 16).

Conflict Perspective

Karl Marx (1818–1883) proposed that economic inequality results from the domination of the *bourgeoisie* (owners of the factories, or "means of production") over the *proletariat* (workers). In a capitalist economy, the bourgeoisie accumulate wealth as they profit from

meritocracy A social system in which individuals get ahead and earn rewards based on their individual efforts and abilities.

the labor of the proletariat, who earn wages far below the earnings of the bourgeoisie. Marx predicted that inequality resulting from capitalism would lead to the collapse of society. Based on an ambitious analysis of data on income and wealth in 20 countries over a period of three centuries, French economist Thomas Piketty (2014) concluded that inequality is intrinsic to capitalism and is likely to increase to levels that threaten democracy.

Modern conflict theorists recognize that the power to influence economic outcomes arises not only from ownership of the means of production but also from management position, interlocking board memberships, control of media, financial contributions to politicians, and lobbying. The conflict perspective views money as a tool that can be used to achieve political interests. Wealthy corporations and individuals use financial political contributions to influence political elections and policies in ways that benefit the wealthy. The interests of the wealthy include things such as keeping taxes low on capital gains, and the wealthy are more likely than the general public to oppose increasing the minimum wage and other policies that would create upward mobility among low-income Americans (Callahan and Cha 2013).

The power of the wealthy to influence political outcomes was reinforced by the 2010 Supreme Court ruling (5 to 4) in *Citizens United v. Federal Election Commission* that corporations have a First Amendment right to spend unlimited amounts of money to support or oppose candidates for elected office. Senator Bernie Sanders, who wants to overturn the Supreme Court's *Citizens* ruling, explained:

> What the Supreme Court did in Citizens United is to tell billionaires like the Koch brothers and Sheldon Adelson, "You own and control Wall Street. You own and control coal companies. You own and control oil companies. Now, for a very small percentage of your wealth, we're going to give you the opportunity to own and control the United States government." That is the essence of what Citizens United is all about. (quoted in Easley 2015, n. p.)

Super PACs also give the wealthy increased leverage in the political process. A Super PAC is a political action committee that is allowed to raise and spend unlimited amounts of money for the purpose of supporting or defeating a political candidate, as long as the monies are not given to any political candidate's campaign. During the 2012 election cycle, two wealthy Americans (Sheldon and Miriam Adelson) gave a combined $91.8 million to Super PACs. "The Adelsons gave more to shape the 2012 federal elections than all the combined contributions from residents in 12 states: Alaska, Delaware, Idaho, Maine, Mississippi, Montana, New Hampshire, North Dakota, Rhode Island, South Dakota, Vermont, and West Virginia" (Callahan and Cha 2013, pp. 18–19).

Laws and policies that favor the rich, such as tax breaks that benefit the wealthy, are sometimes referred to as **wealthfare**. For example, the richest fifth of the U.S. population receives housing subsidies through the mortgage interest tax deduction that amounts to nearly four times the housing assistance provided to the poorest fifth (Garfinkel 2013). **Corporate welfare** refers to government subsidies to corporations, including direct payments and tax breaks. The profitable oil and gas industries receive large federal subsidies, and many states give corporations tax breaks as part of their economic development efforts to entice businesses to locate operations in their state.

Although the official federal corporate tax rate is 35 percent, legal tax loopholes enable corporations to pay significantly less than the 35 percent rate. Many corporations pay a tax rate of less than 20 percent and some have paid no taxes in a given year (Americans for Tax Justice 2014). One example of a corporate tax loophole is the practice known as **corporate tax inversion**, in which a company lowers its taxes by merging with a foreign company and changing to an offshore address. Inversions largely occur on paper and typically do not involve moving operations overseas.

Conflict theorists also note that "free-market" trade and investment economic policies, which some claim to be a solution to poverty, primarily benefit wealthy corporations. Trade and investment agreements enable corporations to (1) expand production and increase economic development in poor countries, and (2) sell their products and services to consumers around the world, thus increasing poor populations' access to goods and services. Yet, such policies also enable corporations to relocate production

wealthfare Laws and policies that benefit the rich.

corporate welfare Laws and policies that benefit corporations.

corporate tax inversion The practice in which a company lowers its taxes by merging with a foreign company and changing to an offshore address.

to countries with abundant supplies of cheap labor, which leads to a lowering of wages, and a resultant decrease in consumer spending, which leads to more industries closing plants, going bankrupt, and/or laying off workers.

Furthermore, most trade and investment agreements include a key provision that allows corporations to take legal action against governments with policies that protect the public, but at a cost to corporate profits. In 2012, in 70 percent of cases where corporations took legal action against governments for violating trade agreements, the World Bank's trade court ruled in favor of the corporation, and governments had to pay tens or even hundreds of millions of dollars—money that could otherwise go toward education, health care, and other public investments to improve the lives of the public, and especially the poor (McDonagh 2013).

When corporations claim that their products or services are essential in the fight against poverty, a conflict perspective might reveal a different story. For example, powerful food and biotech corporations such as Monsanto, Cargill, and Archer Daniels Midland have used their economic and political power to impose a system of agriculture based on intensive chemical use and patented and genetically modified seeds (McDonagh 2013). These corporations assert that their model of agriculture, which requires farmers to purchase their chemicals and seeds, yields more and better food, and thus is important in the global fight against hunger and poverty. Yet, this corporate control of agriculture has resulted in farmers' dependence and debt (and an epidemic of suicides among poor farmers), environmental degradation (through the increased use of chemicals), and health risks associated with chemicals and genetically modified foods.

Symbolic Interactionist Perspective

Symbolic interactionism focuses on how meanings, labels, and definitions affect and are affected by social life. This view calls attention to ways in which wealth and poverty are defined and the consequences of being labeled "poor." Individuals who are poor are often viewed as undeserving of help or sympathy; their poverty is viewed as due to laziness, immorality, irresponsibility, lack of motivation, or personal deficiency (Katz 2013). Wealthy individuals, on the other hand, tend to be viewed as capable, motivated, hardworking, and deserving of their wealth.

The language we use to label things can have a profound influence on how we view things. A Canadian study found that the public expressed higher support for government "spending on the poor" than for spending on "welfare" (Harell et al. 2008). If *welfare* is a "dirty word," as this study suggests, then it makes a difference whether one uses the term *welfare* versus other terms such as *assistance to the poor, public assistance,* or *safety net.*

Meanings and definitions of wealth and poverty vary across societies and across time. Although many Americans think of poverty in terms of income level, for millions of people, poverty is not primarily a function of income. For indigenous women living in the least developed areas of the world, poverty and wealth are determined primarily by access to and control of their natural resources (such as land and water), which are the sources of their livelihoods (Susskind 2005).

Finally, the symbolic interactionist perspective is concerned with how social conditions come to be viewed as social problems. Economic inequality rose to the forefront of world citizens' awareness when **Occupy Wall Street (OWS)**, a decentralized protest movement concerned with economic inequality, greed, corruption, and the influence of corporations on government, gained media attention in 2011. OWS started as a wave of park occupations and spread around the world, branching out into a number of areas, all focused on the concerns of the "99 percent"—that is, the concerns of regular, hardworking folks versus the "1 percent"—the wealthy. The Occupy Wall Street movement helped to frame economic inequality as a national and global social problem.

The publication of French economist Thomas Piketty's (2014) book *Capital in the Twenty-first Century* has further increased public awareness of economic inequality as a social problem. Piketty's book essentially argues that economic inequality will continue to increase to levels that threaten social stability unless governments take such actions as enforcing a global wealth tax.

Occupy Wall Street (OWS) A protest movement that began in 2011 and is concerned with economic inequality, greed, corruption, and the influence of corporations on government.

Economic Inequality, Wealth, and Poverty in the United States

The United States has the greatest degree of income inequality and the highest rate of poverty of any industrialized nation. After looking at U.S. income and wealth inequality and the "1 percent," we look at patterns of U.S. poverty.

U.S. Income Inequality

In 2012, the top 1 percent of U.S. taxpayers earned 22.5 percent of all U.S. income (Sommeiller and Price 2015). Wages used to be tied to worker productivity—the amount of goods and services produced per hour worked. From 1948 to 1973, worker productivity increased 97 percent and wages increased nearly as much (91 percent). But from 1973 to 2013, although worker productivity increased by 74 percent, wages rose by only 9 percent (Mishel 2015). The wage stagnation of middle- and low-income earners is in stark contrast to the huge increase in wages of the top earners. From 1973 to 2013, wages of the top 1 percent grew 138 percent, while wages for the bottom 99 percent rose by only 15 percent (Mishel et al. 2015).

One reason why workers' wages have not increased in sync with their productivity is that CEOs are taking a bigger piece of the pie. The most extreme wage inequality is found between the compensation (salaries, bonuses, stock options, and so on) of chief executive officers (CEOs) and the average employee. In 2014, CEOs at the top 350 U.S. corporations received, in salaries and other compensation (such as bonuses and stocks), 303 times the average compensation of U.S. workers (Mishel and Davis 2015). That means that a typical worker would have to work 303 years to earn what a CEO makes in 1 year! Table 6.3 shows the dramatic increase in the ratio of CEO pay to average worker pay since 1965.

TABLE 6.3 Ratio of CEO Compensation* to Average Worker Pay, U.S., 1965–2014

1965	20:1
1978	30:1
1995	123:1
2000	376:1
2014	303:1

*Rounded; includes pay and stock options; based on the top 350 U.S. firms, excluding Facebook due to its high-outlier CEO compensation. SOURCE: Mishel and Davis 2015.

WHAT do you THINK?

Most Americans agree that CEOs *should* earn more than typical workers, but how much more? What do you think the ideal ratio of CEO pay to worker pay should be, and why? Most Americans view the ideal ratio of CEO to worker pay to be much lower than it is: 7 to 1 (Norton 2014).

U.S. Wealth Inequality

Wealth in the United States, like in the rest of the world, is unevenly distributed and concentrated at the top (see Figure 6.1). The wealthiest 1 percent has received much attention as a result of the Occupy Wall Street's "99 percent versus 1 percent" slogan. But inequality exists even among the rich. More than 40 percent of U.S. wealth in 2012 was owned by the top 1 percent, but more than half of that wealth was owned by the top 0.1 percent (Saez and Zucman 2014).

There is a saying: The best way to make a million dollars is to start out with $900,000! Wealth tends to snowball, and the bigger the snowball you start off with, the bigger it grows. Consider that between 1963 and 2013 (Urban Institute 2015),

- families at the 99th percentile saw their wealth increase six-fold.
- families at the 90th percentile (those wealthier than 90 percent of families) quadrupled their wealth.
- families in the *middle* of the wealth distribution roughly doubled their wealth.
- families at the *bottom* 10 percent of the wealth distribution went from having no wealth, on average, to being about $2,000 in debt.

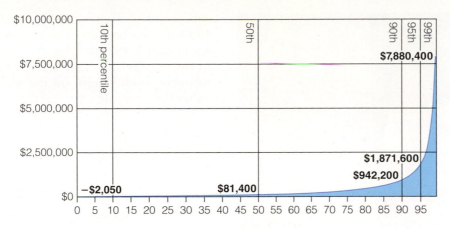

Figure 6.1 Average Wealth of U.S. Families, 2013
SOURCE: Based on Urban Institute 2015.

Disparities in income and wealth show an economic advantage of whites over racial/ethnic minorities. In 2014, median household income for non-Hispanic whites was $60,256 compared with $42,491 for Hispanic households and $35,398 for black households (DeNavas-Walt and Proctor 2015). The median wealth of white households is 13 times the median wealth of black households and more than 10 times the wealth of Hispanic households (Kochhar and Fry 2015) (see Figure 6.2). Whites are more likely than blacks or Hispanics to own homes, which for many Americans, is their most valuable asset. And white families are five times more likely than Hispanics or blacks to receive large gifts or inheritances, which can be used to pay for college, a down payment on a home, and other wealth-building assets (Urban Institute 2014).

The Wealthiest Americans

In 2014, Bill Gates was ranked as the wealthiest person not only in the United States but in the world (Dolan and Kroll 2014). If Gates cashed in all his wealth and spent $1 million each day, it would take him 218 years to spend it all (Oxfam 2014).

In the 2011 Forbes 400 annual list of the wealthiest Americans, more than 70 percent of the 282 billionaires on the list were described as "self-made," suggesting that these individuals achieved financial success on their own, without assistance from family or society. But the notion that wealthy individuals have created their own financial success ignores the importance of gender, race, and family background as well as the role that tax policies play in creating wealthy individuals. United for a Fair Economy (2012) examined the 2011 Forbes 400 list of wealthiest Americans and found that

- 17 percent of the Forbes 400 have family members who are also on the list;
- about 40 percent of the 2011 Forbes 400 list inherited a "sizeable asset" from a spouse or family member;
- more than one in five of the Forbes 400 inherited enough wealth to make the list;
- just one African American is on the list, and of the women on the list (who comprise just 10 percent of the list), 88 percent inherited their fortune; and
- 60 percent of the income owned by those on the Forbes 400 list comes from capital gains (investments) that are taxed at a lower rate than other income.

There are, indeed, true "rags to riches" success stories in the United States that exemplify the idea that anyone can achieve the American dream. Approximately one-third of the individuals on the 2011 Forbes 400 list came from a lower- or middle-class background. Oprah Winfrey, for example—the only black person

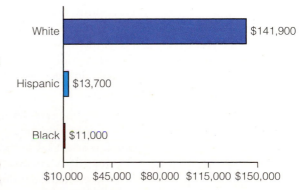

Figure 6.2 Median Household Wealth by Race/Hispanic Origin, 2013
SOURCE: Kochhar and Fry, 2014.

and one of 40 women on the 2011 Forbes 400 list—was born to a poor unwed teenage mother, yet she developed a successful career in television, film, and publishing. However, such stories are the exception rather than the rule.

David Knox

Children are more likely than adults to live in poverty.

Patterns of Poverty in the United States

Although poverty is not as widespread or severe in the United States as it is in many other parts of the world, the United States has the highest rate of poverty among wealthy countries belonging to the Organisation for Economic Cooperation and Development (OECD). In 2014, 46.7 million Americans—14.8 percent of the U.S. population—lived below the poverty line (DeNavas-Walt and Proctor 2015). More than half (58 percent) of Americans between the ages of 20 and 75 will spend at least one year in poverty, and one in three Americans will experience a full year of extreme poverty at some point in adult life (Pugh 2007).

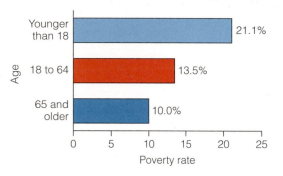

Figure 6.3 U.S. Poverty Rates by Age, 2014
SOURCE: DeNavas-Walt and Proctor 2015.

Age and Poverty. If the poverty statistics for adults are troubling, the statistics for children are even worse (see Figure 6.3). About a third of U.S. children experience poverty for at least part of their childhood, and 10 percent of children are persistently poor, spending at least half their childhood living in poverty (Ratcliffe and McKernan 2010). About one-third of the U.S. poor population are children (DeNavas-Walt and Proctor 2015). Compared with other industrialized countries, the United States has the highest child poverty rate. Childhood poverty is particularly problematic because "[g]rowing up in poverty can cast a shadow over the rest of a person's life" (Golden 2013, n.p.).

In our sociology classes, we introduce the topic of U.S. poverty by asking students to think of an image of a person who represents poverty in the United States and to draw that imaginary person. Students are asked to give the person a name (to indicate the person's sex) and to write down the age of the person. Most students draw a picture of a middle-aged man. Yet U.S. poverty statistics reveal that the higher poverty rates are among women, not men, and among youth, not middle-aged adults. Why do you think the most common image of a U.S. poor person is a middle-aged man?

feminization of poverty
The disproportionate distribution of poverty among women.

Sex and Poverty. Women are slightly more likely than men to live below the poverty line—a phenomenon referred to as the **feminization of poverty**. In 2014, 16.1 percent of women and 13.4 percent of men were living below the poverty line (DeNavas-Walt and Proctor 2015). As discussed in Chapter 10, women are less likely than men to pursue advanced educational degrees and tend to have low-paying jobs, such as service and clerical jobs. However, even with the same level of education and the same occupational role, women still earn significantly less than men. Women who are racial or ethnic minorities and/or who are single mothers are at increased risk of being poor.

Education and Poverty. Education is one of the best insurance policies for protecting an individual against living in poverty. In general, the higher a person's level of educational attainment, the less likely that person is to be poor (see Figure 6.4). The relationship between educational attainment and poverty points to the importance of fixing our educational system so that students from all socioeconomic backgrounds have access to quality education (see also Chapter 8). But we also need to consider the fact that many jobs do not require advanced education. Indeed, the majority of job growth through 2018 will involve jobs that do not require a four-year college degree (Wider Opportunities for Women 2010). Wright and Rogers (2011) suggest that "poverty in a rich society does not simply reflect a failure of equal opportunity to acquire a good education; it reflects a social failure in the creation of sufficient jobs to provide an adequate standard of living for all people regardless of their education or levels of skills" (p. 224).

Family Structure and Poverty. Poverty is much more prevalent among female-headed single-parent households than among other types of family structures (see Figure 6.5). In other industrialized countries, poverty rates of female-headed families are lower than those in the United States. Unlike the United States, other developed countries offer a variety of supports for single mothers, such as income supplements, tax breaks, universal child care, national health care, and higher wages for female-dominated occupations.

In general, same-sex couples are more likely than heterosexual couples to be poor. Children in same-sex couple families are nearly twice as likely to be poor as children of married different-sex couples (Badgett et al. 2013).

Race or Ethnicity and Poverty. As displayed in Figure 6.6, poverty rates are higher among racial and ethnic minority groups than among non-Hispanic whites. As discussed in Chapter 9, past and present discrimination has contributed to the persistence of poverty among minorities. Other contributing factors include the loss of manufacturing jobs from the inner city, the movement of whites and middle-class blacks out of the inner city, and the resulting concentration of poverty in predominantly minority inner-city neighborhoods (Massey 1991; Wilson 1987, 1996). Finally, blacks and Hispanics are more likely to live in female-headed households with no spouse present—a family structure that is associated with high rates of poverty.

Labor Force Participation and Poverty. A common image of the poor is that they are jobless and unable or unwilling to work. Although the poor in the United States are primarily children and adults who are not in the labor force, many U.S. poor are classified as **working poor**—individuals who spend at least 27 weeks per year in the labor force (working or looking for work), but whose income falls below the official poverty level.

> Education is one of the best insurance policies for protecting an individual against living in poverty.

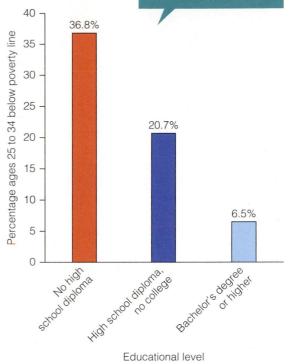

Figure 6.4 Relationship between Education and Poverty, 2013
SOURCE: Gabe 2015.

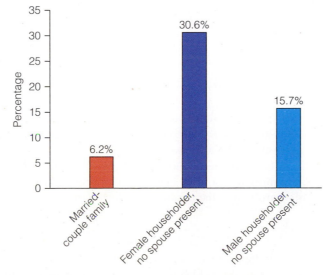

Figure 6.5 U.S. Poverty Rates by Family Structure, 2014
SOURCE: DeNavas-Walt and Proctor 2015.

working poor Individuals who spend at least 27 weeks per year in the labor force (working or looking for work) but whose income falls below the official poverty level.

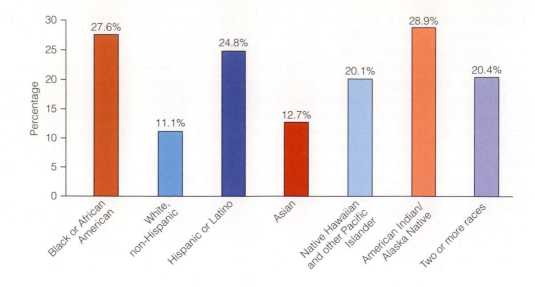

Figure 6.6 U.S. Poverty Rates by Race and Hispanic Origin, 2013
SOURCE: U.S. Census Bureau 2014.

TABLE 6.4 States with the Highest and Lowest Poverty Rates: 2013

Highest Poverty Rates (%)	Lowest Poverty Rates (%)
Mississippi: 24.0	New Hampshire: 8.7
New Mexico: 21.9	Alaska: 9.3
Louisiana: 19.8	Maryland: 10.1
Arkansas: 19.7	Connecticut: 10.7
Georgia: 19.0	Hawaii: 10.8
District of Columbia: 18.9.	Wyoming: 10.9
Kentucky: 18.8	Minnesota: 11.2
Alabama: 18.7	New Jersey: 11.4
South Carolina: 18.6	Virginia: 11.7
Arizona: 18.6	Massachusetts: 11.9
North Carolina: 17.9	Vermont: 12.3
Tennessee: 17.8	Delaware: 12.4

SOURCE: Bishaw and Fontenot (2014).

Region and Poverty. Poverty rates vary considerably by region of the United States, with the highest rates being in the South and West, and the lowest rates being in the Northeast. In 2013 the rates of poverty ranged from a low of 8.7 percent in New Hampshire to a high of 24.0 percent in Mississippi (see Table 6.4).

Poverty is increasingly found in the suburbs. The number of suburban poor surpassed the number of urban poor in the 2000s (Semuels 2015). Compared with poor urban dwellers, the suburban poor have less access to public transit and social safety-net programs.

Consequences of Economic Inequality and Poverty

From one point of view, economic inequality and poverty are problematic because they contradict the values of fairness, justice, and equality of opportunity, and they constitute a moral violation of basic human rights. Economic inequality and poverty are also viewed as problems because they have economic and social consequences that affect the whole society. For example, when income is concentrated toward the top, less money circulates in the local economy because while money earned by low- and middle-income households is likely to be spent on goods and services that benefit the local economy, money among the rich is often invested in other regions and spent on luxuries (Talberth et al. 2013). The larger the segment of the population that is in the lowest income brackets, the more our society is affected by problems that plague the poor, but that also affect us all—problems discussed in the following sections.

Health Problems, Hunger, and Poverty

In developing countries, absolute poverty is associated with high rates of maternal and infant deaths, indoor air pollution from heating and cooking fumes, and unsafe water and sanitation (see also Chapter 2). In wealthy countries, such as the United States, we take for granted the availability of bathrooms and toilets, and safe drinking water that is piped into our homes. But more than 1 in 10 people in the world (700 million people) do not have access to safe drinking water, and more than one-third of the world's population—2.5 billion people—do not use improved sanitation facilities—those that ensure hygienic separation of human excreta from human contact (World Health Organization and UNICEF 2014). Lack of access to clean water and sanitation facilities is a major cause of disease and death. Inadequate sanitation and hygiene contributes to the spread of diseases such as Ebola, and poor sanitation causes diarrheal diseases, which are the second leading cause of death among young children in developing countries (Agazzi 2012).

In sub-Saharan Africa, one in four people are undernourished.

Living in poverty is also linked to hunger and malnourishment. In 2014, more than 1 in 10 people globally were chronically undernourished, with the highest rate of hunger in sub-Saharan Africa, where one in four people are undernourished (Food and Agriculture Organization 2014). Inadequate nutrition hampers the ability to work and generate income, and it can produce irreversible health problems such as blindness (from vitamin A deficiency) and physical stunting (from protein deficiency).

Hunger in the United States is measured by the percentage of households that are "food insecure," which means that the household had difficulty providing enough food for all its members due to a lack of resources. In 2013, about 14 percent of U.S. households were food insecure at some time during the year (Coleman-Jensen et al. 2014). Assess your own degree of food security in this chapter's *Self and Society* feature.

In the United States, low-wage earners have higher rates of obesity, hypertension, diabetes, arthritis, and premature death (Leigh 2013). Compared with middle- and upper-income adults, U.S. adults living in poverty are more likely to experience extreme or chronic pain, worry and mental distress, sadness, anger, and stress (Graham 2015). Poor U.S. children and adults tend to receive inadequate and inferior health care, which exacerbates their health problems. Minimal income means that people may not have funds to purchase medicine to control their cholesterol, high blood pressure, and other health problems. As discussed in Chapter 2, people with limited incomes may not have access to or be able to afford healthier foods such as fresh produce, which tend to be more expensive than processed convenience foods that are higher in calories, sugar, salt, and fats. Finally, many people partially assess their self-worth based on their income, and long-term feelings of low self-worth also have negative consequences for health (Leigh 2013).

Economic inequality is also linked to health problems. In a comparison of 30 wealthy countries, researchers found an association between greater economic inequality and a higher overall death rate (Kondo et al. 2009). Another study by the U.S. National Research Council and Institute of Medicine (2013) compared health outcomes in the United States with those of 16 other high-income, industrialized countries and found that Americans die sooner and have higher rates of disease or injury. One explanation for this finding is that although the income of Americans is higher on average than in other countries, the United States has higher rates of poverty (especially child poverty), more income inequality, and less social mobility.

Substandard Housing and Homelessness

Having a roof over one's head is considered a basic necessity. However, for the poor, that roof may be literally caving in. In addition to having leaky roofs, housing units of the poor often have holes in the floor and open cracks in the walls or ceiling. Low-income

The U.S. Department of Agriculture conducts national surveys to assess the degree to which U.S. households experience food security, food insecurity, and food insecurity with hunger. To assess your own level of food security, respond to the following items and use the scoring key to interpret your results:

1. In the last 12 months, the food that (I/we) bought just didn't last, and (I/we) didn't have money to get more.
   a. **Often true**
   b. **Sometimes true**
   c. Never true
2. In the last 12 months, (I/we) couldn't afford to eat balanced meals.
   a. **Often true**
   b. **Sometimes true**
   c. Never true
3. In the last 12 months, did you ever cut the size of your meals or skip meals because there wasn't enough money for food?
   a. **Yes**
   b. No (skip Question 4)

4. If you answered yes to Question 3, how often did this happen in the last 12 months?
   a. **Almost every month**
   b. **Some months but not every month**
   c. Only 1 or 2 months
5. In the last 12 months, did you ever eat less than you felt you should because there wasn't enough money to buy food?
   a. **Yes**
   b. No
6. In the last 12 months, were you ever hungry but didn't eat because you couldn't afford enough food?
   a. **Yes**
   b. No

Scoring and Interpretation

The answer responses in boldface type indicate affirmative responses. Count the number of affirmative responses you gave to the items, and use the following scoring key to interpret your results.

Number of Affirmative Responses and Interpretation

0 or 1 item: *Food secure* (In the last year, you have had access to enough food for an active, healthy life.)

2, 3, or 4 items: *Food insecure* (In the last year, you have had limited or uncertain availability of food and have been worried or unsure you would get enough to eat.)

5 or 6 items: *Food insecure with hunger evident* (In the last year, you have experienced more than isolated occasions of involuntary hunger as a result of not being able to afford enough food.)

If you scored as food insecure (with or without hunger), you might consider exploring whether you are eligible for public food assistance (e.g., food stamps) or whether there is a local food assistance program (e.g., food pantry or soup kitchen) that you could use.

SOURCE: Based on the short form of the 12-month Food Security Scale found in Bickel et al. 2000.

Many Americans would be shocked to see the conditions under which many poor people in this country live.

AP Photo/Ric Feld

housing units often lack central heating and air conditioning, sewer or septic systems, and electric outlets in one or more rooms. Housing for the poor is also often located in areas with high crime rates and high levels of pollution.

Concentrated areas of poverty and poor housing in urban areas are called **slums.** One-third of urban populations in developing regions are living in slums. In sub-Saharan Africa, two-thirds of urban populations are living in slums (UN-Habitat 2010).

Over the course of a lifetime, an estimated 9 percent to 15 percent of the U.S. population becomes homeless (Hoback and Anderson 2007). In January 2014, there were 578,424 homeless people in the United States: Of those, 37 percent were families with children, 9 percent were veterans, and almost 8 percent were unaccompanied youth (National Alliance to End Homelessness 2015). Although the majority of the homeless population stays in shelters or transitional housing, about a third lives on the street, in a car, in an abandoned building, or in places not meant for human habitation, such as storage units or makeshift dwellings

slums Concentrated areas of poverty and poor housing in urban areas.

Thousands of people in the United States are homeless on any given day.

made of a variety of discarded materials such as pieces of wood and boards, cardboard, mattresses, fabric, and plastic tarps.

WHAT
do you
THINK?

Many cities have passed laws that prohibit homeless people from begging as well as sleeping, sitting, and/or "loitering" in public. Do you think that such laws unfairly criminalize the homeless? Or are these laws necessary to protect the public?

Causes of homelessness include poverty, unemployment, eviction, domestic violence, mental illness and substance abuse, and lack of affordable housing. Housing is considered affordable when a household pays no more than 30 percent of its income on housing expenses. In 2014, 6.4 million renters paid more than half of their income on housing (National Alliance to End Homelessness 2015).

For people living on the street, every day can be a struggle for survival. In recent years, there has been a surge in unprovoked violent, and sometimes fatal, attacks against homeless individuals (National Coalition for the Homeless 2014b). In most cases, the attacks are by teenage and young adult males. Many acts of violence toward the homeless are not reported to the police, so documented cases may be just the tip of the iceberg. During the years he lived homeless on the street, David Pirtle was attacked five times, and he did not report the attacks to police. "I was struck on the back, kicked, urinated on, spray-painted. . . . A lot of people who are homeless go through it, and it's just the way it is" (quoted in Dvorak 2009, p. DZ01).

There is also a new fascination with "bum bashing" or "bum fight" videos on YouTube—videos shot by young men and boys who are seen beating the homeless or who pay homeless people a few dollars to fight each other or to do dangerous stunts like banging their heads through glass windows and going down stairs in a shopping cart. Individuals who find the idea of bum bashing entertaining can also purchase bum fight DVDs and play web-based bum-bashing games. The National Coalition for the Homeless (2014b) reports that there is a correlation between watching videos that show violence against homeless individuals and committing "copycat" acts of violence against the homeless.

WHAT do you THINK?

Under hate crime laws, violators are subject to harsher legal penalties if their crime is motivated by the victim's race, religion, national origin, or sexual orientation. The number of fatal attacks on the homeless is nearly triple the number of hate crime deaths based on race, religion, ethnicity, and sexual orientation *combined* (National Coalition for the Homeless 2014b). A handful of states have added homeless status to their hate crime law, and several cities and counties have also taken measures to recognize homeless status in their laws or procedures. Proposed legislation to add homelessness to the federal hate crime law has not, as of this date, passed. Do you think that violent acts toward homeless individuals should be categorized as hate crimes and be subject to harsher penalties? Why or why not?

Legal Inequality

The American ideal of "justice for all" may be more accurately described as "justice for those who can afford to pay for it." Many poor defendants are held in pretrial detention because they cannot afford to post bail (Human Rights Watch 2015). Although the Supreme Court ruled in 1963 (*Gideon v. Wainwright*) that criminal defendants who cannot afford to hire an attorney have the constitutional right to a public defense, public defender offices are overworked and underfunded, and often spend only minutes per case due to their unrealistic caseloads (Giovanni and Patel 2013). Without the resources for effective legal representation, poor defendants often accept unfair plea bargains, and "the systemic result is harsher outcomes for defendants and more people tangled in our costly criminal justice system" (Giovanni and Patel 2013, p. 1). The economic inequality embedded in the U.S. legal system is problematic not only for the poor, but for the entire society, as it contributes to the social and economic costs of mass incarceration in the United States (see also Chapter 4).

Political Inequality and Alienation

Economic inequality also contributes to political inequality, as expressed in a version of "the golden rule": "He who has the gold makes the rules." Although the United States represents itself as a democracy whose government represents all citizens, research shows that a small group of economic elite has more influence over political outcomes than do ordinary citizens (Gilens and Page 2014). Thus, instead of being a true democracy, the United States can be described as a **plutocracy**: a country governed by the wealthy. The poor and even middle classes feel that their interests are not represented by their elected politicians. In countries around the world, people at the bottom of the inequality spectrum often feel as though they do not have a say in the policies that govern them. Hence, those in the lower socioeconomic classes are vulnerable to experiencing **political alienation**—a rejection of or estrangement from the political system accompanied by a sense of powerlessness in influencing government. The poor face obstacles in running for political office, as money and connections are needed to run for office. The poor are less likely than the affluent to vote: In 2012, less than half of eligible voters with family incomes under $20,000 voted, compared with about 80 percent of those with annual incomes of $100,000 or more (Weeks 2014). The poor have a lower voting turnout than the wealthier segments of the population, in part due to political alienation, but also because of obstacles such as difficulty taking time off of work, transportation problems getting to the polls, and lack of a required form of identification.

plutocracy A country governed by the wealthy.

political alienation A rejection of or estrangement from the political system accompanied by a sense of powerlessness in influencing government.

Crime, Social Conflict, and War

Poverty and economic inequality are linked to crime and violence (see also Chapters 4 and 16). For example, homicide rates are nearly four times higher in countries with extreme economic inequality than in more equal countries (Oxfam 2014). The denial of a political voice or influence to masses of people at the bottom of the wealth distribution can cause social tensions, political instability, and violent conflict (United Nations 2013).

Economic inequality and poverty are often root causes of conflict and war within and between nations. Poorer countries are more likely than wealthier countries to be involved in civil war, and countries that experience civil war tend to become and/or remain poor. Armed conflict and civil war are generally more likely to occur in countries with extreme and growing inequalities between ethnic groups (United Nations 2005). Conversely, countries with higher levels of equality are more likely to be peaceful (Institute for Economics and Peace 2013). A United Nations (2005) report suggested that "the most effective conflict prevention strategies . . . are those aimed at achieving reductions in poverty and inequality, full and decent employment for all, and complete social integration" (p. 94).

Not only does poverty breed conflict and war, but war also contributes to poverty. War devastates infrastructures, homes, businesses, and transportation systems. In the wake of war, populations often experience hunger and homelessness.

Natural Disasters, Economic Inequality, and Poverty

Although natural disasters such as hurricanes, tsunamis, floods, and earthquakes strike indiscriminately—rich and poor alike—poverty increases vulnerability to devastation from such disasters. In 2010, both Chile and Haiti experienced major earthquakes, but the damage in Haiti was much more severe, with the death toll magnitudes higher than that in Chile. The reason Haiti suffered more was, in part, due to the fact that Haiti is much poorer than Chile. Chileans had the advantage of having homes and offices with steel skeletons designed to withstand earthquakes—even low-income housing was built to be earthquake resistant. In contrast, there is no building code in Haiti and homes crumbled and collapsed in the earthquake (Bajak 2010). Wealthy countries also have more resources than poor countries for natural disaster relief efforts, such as rebuilding infrastructure, providing medical care for the injured, and offering food and shelter for people who have lost their homes.

But even in wealthy countries, the poor are more vulnerable to natural disasters, while the more affluent have resources that enable them to cope with natural disasters. Columnist David Rohde (2012) wrote about how economic inequality affected people dealing with Hurricane Sandy, which devastated the northeastern United States in 2012:

> Divides between the rich and the poor are nothing new in New York, but the storm brought them vividly to the surface. There were residents like me who could invest all of their time and energy into protecting their families. And there were New Yorkers who could not. Those with a car could flee. Those with wealth could move into a hotel. Those with steady jobs could decline to come into work. But the city's cooks, doormen, maintenance men, taxi drivers, and maids left their loved ones at home. . . . In the Union Square area, New York's privileged—including myself—could have dinner, order a food delivery, and pick up supplies an hour or two before Sandy made landfall. The cooks, cashiers, and hotel workers who stayed at work instead of rushing home made that possible. (n.p.)

Educational Problems and Poverty

In many countries, children from the poorest households have little or no schooling, and enter their adult lives without basic literacy skills (see also Chapter 8). In the United States, children from disadvantaged

Leonard Zhukovsky/Shutterstock.com

The effects of natural disasters, such as Hurricane Sandy of 2012, are more devastating for the poor.

homes perform less well in school on average than children from more advantaged households (Ladd 2012). Children who grow up in poverty tend to receive lower grades, receive lower scores on standardized tests, are less likely to finish high school, and are less likely to attend or graduate from college than their nonpoor peers.

The poor often attend schools that are characterized by lower-quality facilities, overcrowded classrooms, and a higher teacher turnover rate (see also Chapter 8). Although other rich countries invest more money in education for disadvantaged children, the United States spends more on schools in wealthy districts because public schools are funded largely by local property tax money (Straus 2013).

Children who grow up in poverty suffer more health problems that contribute to their lower academic achievement. Because poor parents have less schooling on average than do nonpoor parents, they may be less able to encourage and help their children succeed in school. Children from poor households have limited access to high-quality preschools, books and computers, and enriching after-school and summer activities including tutoring, travel, lessons (music, dance, sports, etc.), and camps (Ladd 2012; Sobolewski and Amato 2005). With the skyrocketing costs of tuition and other fees, many poor parents cannot afford to send their children to college. Although some students have wealthy parents who write out tuition checks, other students are graduating from college with substantial college debt.

Marriage Opportunity Gap and Family Problems Associated with Poverty and Economic Inequality

The erosion of legal prohibitions against same-sex marriage in recent years reflect, in part, the cultural value and belief that everyone should have the opportunity to marry and to have children within a stable, socially recognized family household. But poverty and economic inequality create a marriage opportunity gap. The top one-third of U.S. households are more likely to enjoy the benefits of being a "two-two-two-one household"— having two parents, two college degrees, two incomes, and one stable marriage (Blankenhorn et al. 2015). But among lower- and even middle-income households, marriage rates are low, divorce rates are high, and nonmarital childbearing rates are high (see also Chapter 5).

The stresses associated with low income contribute to substance abuse, domestic violence, and child abuse and neglect (see also Chapters 3 and 5). Children in poor households are more likely to experience harsh or neglectful parenting (Lanier et al. 2014). Without access to affordable child care and medical care, poor parents may leave their children at home without adult supervision or fail to provide needed medical care.

Poverty is also linked to teenage pregnancy and unintended childbearing. Poor women are more than five times as likely as affluent women to have an unintended birth, because they are less likely to use contraception and are less likely to have an abortion once pregnant (Reeves and Venator 2015). Poor adolescent teenagers are at higher risk of having babies than their nonpoor peers. Early childbearing is associated with increased risk of premature babies or babies with low birth weight, dropping out of school, and lower future earning potential as a result of lack of academic achievement. Luker (1996) noted that "the high rate of early childbearing is a measure of how bleak life is for young people who are living in poor communities and who have no obvious arenas for success" (p. 189).

> Having a baby is a lottery ticket for many teenagers: It brings with it at least the dream of something better, and if the dream fails, not much is lost. . . . In a few cases it leads to marriage or a stable relationship; in many others it motivates a woman to push herself for her baby's sake; and in still other cases it enhances the woman's self-esteem, since it enables her to do something productive, something nurturing and socially responsible. (Luker 1996, p. 182)

Intergenerational Poverty

Problems associated with poverty, such as health and educational problems, create a cycle of poverty from one generation to the next. Nearly half of U.S. children born to low-income parents will become low-income adults (Oxfam 2014). Poverty that is transmitted from one generation to the next is called **intergenerational poverty**.

The top one-third of U.S. households are more likely to enjoy the benefits of being a "two-two-two-one household"— having two parents, two college degrees, two incomes, and one stable marriage.

intergenerational poverty
Poverty that is transmitted from one generation to the next.

Intergenerational poverty creates a persistently poor and socially disadvantaged population, referred to as the underclass. Although the underclass is stereotyped as being composed of minorities living in inner-city areas, the underclass is a heterogeneous population that includes poor whites living in urban and nonurban communities (Alex-Assensoh 1995). Intergenerational poverty and the underclass are linked to a variety of social factors, including the decrease in well-paying jobs and their movement out of urban areas, the resultant decline in the availability of marriageable males able to support a family, dropping marriage rates and an increase in out-of-wedlock births, the migration of the middle class to the suburbs, and the effect of deteriorating neighborhoods on children and youth (Wilson 1987, 1996).

Strategies for Action: Reducing Poverty and Economic Inequality

Because poverty and economic inequality are primary social problems that cause many other social problems, strategies to reduce or alleviate problems related to poverty and economic inequality include those discussed in other chapters of this text that deal with such issues as health (Chapter 2), work (Chapter 7), education (Chapter 8), racial discrimination (Chapter 9), and gender discrimination (Chapter 10). Here we briefly outline a number of strategies aimed at reducing poverty and economic inequality in the United States, and internationally.

International Responses to Poverty and Economic Inequality

In 2000, leaders from 191 United Nations member countries pledged to achieve eight Millennium Development Goals—an international agenda for reducing poverty and improving lives. One of the Millennium Development Goals (MDGs) was to halve, between 1990 and 2015, the proportion of people who live in severe poverty and who suffer from hunger. The MDG poverty reduction goal was met in 2010—five years ahead of schedule. The MDGs expired at the end of 2015 and were replaced with a new, expanded set of 17 goals, collectively called the **Sustainable Development Goals** (SDGs) (see Table 6.5).

In 2014, Oxfam International launched a global campaign called Even It Up, calling for governments, corporations, and institutions to reduce economic inequality. Next we discuss some approaches for reducing poverty and economic inequality throughout the world.

Taxes on the Wealthy. Oxfam (2014) calculated that adding a 1.5 percent tax on the world's billionaires could raise $74 billion in tax revenue—enough to fill the annual gaps in funding needed to provide education to every child and deliver health care services in the poorest 49 countries. Another way to gain tax revenue that could both alleviate poverty and reduce economic inequality is to close tax loopholes that enable the wealthy to avoid paying taxes on much of their income. In 2013, the world lost an estimated $156 billion in tax revenue due to wealthy individuals hiding their money in offshore tax havens (Oxfam 2014).

Economic Development. Increasing the economic output or the gross domestic product of a country can bring needed economic resources into poor countries. However, economic development does not always reduce poverty; in some cases, it increases it. Policies that involve cutting government spending, privatizing basic services, liberalizing trade, and producing goods primarily for export may increase economic growth at the national level, but the wealth ends up in the hands of the political and corporate elite at the expense of the poor. Economic growth does not help poverty reduction when public spending is diverted away from meeting the needs of the poor and instead is used to pay international debt, finance military operations, and support corporations that do not pay workers fair wages.

Sustainable Development Goals A set of 17 goals that comprise an international agenda for reducing poverty and economic inequality and improving lives.

TABLE 6.5 Sustainable Development Goals

Goal 1. End poverty in all its forms everywhere.

Goal 2. End hunger, achieve food security and improved nutrition and promote sustainable agriculture.

Goal 3. Ensure healthy lives and promote well-being for all at all ages.

Goal 4. Ensure inclusive and equitable quality education and promote lifelong learning opportunities for all.

Goal 5. Achieve gender equality and empower all women and girls.

Goal 6. Ensure availability and sustainable management of water and sanitation for all.

Goal 7. Ensure access to affordable, reliable, sustainable and modern energy for all.

Goal 8. Promote sustained, inclusive and sustainable economic growth, full and productive employment and decent work for all.

Goal 9. Build resilient infrastructure, promote inclusive and sustainable industrialization and foster innovation.

Goal 10. Reduce inequality within and among countries.

Goal 11. Make cities and human settlements inclusive, safe, resilient and sustainable.

Goal 12. Ensure sustainable consumption and production patterns.

Goal 13. Take urgent action to combat climate change and its impacts.

Goal 14. Conserve and sustainably use the oceans, seas and marine resources for sustainable development.

Goal 15. Protect, restore and promote sustainable use of terrestrial ecosystems, sustainably manage forests, combat desertification, and halt and reverse land degradation and halt biodiversity loss.

Goal 16. Promote peaceful and inclusive societies for sustainable development, provide access to justice for all and build effective, accountable and inclusive institutions at all levels.

Goal 17. Strengthen the means of implementation and revitalize the global partnership for sustainable development.

SOURCE: United Nations 2014.

> The human development approach views people—not money—as the real wealth of a nation.

green growth Economic growth that is environmentally sustainable.

Another problem with economic development is that the environment and natural resources are often destroyed and depleted in the process of economic growth. The environmental problems caused by economic growth can be minimized if governments and corporations embrace "**green growth**," which is economic growth that is environmentally sustainable (World Bank Group and International Monetary Fund 2015) (see also Chapter 13).

Economic development also threatens the lives and cultures of the 370 million indigenous people who live in 70 countries around the world. Indigenous people who live on land that is rich in natural resources are displaced by corporations that want access to the land and its natural resources, and by government forces that help the corporations expand their activities (Ramos et al. 2009). As remote areas are "developed," many indigenous people are forced to either leave their land or give up their traditional ways of life and become assimilated into the dominant culture.

Human Development. Unlike the economic development approach to poverty alleviation, the human development approach views people—not money—as the real wealth of a nation:

> The central contention of the human development approach . . . is that well-being is about much more than money. . . . Income is critical but so are having access to education and being able to lead a long and healthy life, to influence the decisions of society, and to live in a society that respects and values everyone. (UNDP 2010, p. 114)

In many poor countries, large segments of the population are illiterate and without job skills and/or are malnourished and in poor health. Investments in human development involve programs and policies that provide adequate nutrition, sanitation, housing, health care (including reproductive health care and family planning), and educational and job training.

Microcredit Programs. The old saying "It takes money to make money" explains why many poor people are stuck in poverty: They have no access to financial resources and services. **Microcredit programs** refer to the provision of loans to people who are generally excluded from traditional credit services because of their low socioeconomic status. Microcredit programs give poor people the financial resources they need to become self-sufficient and to contribute to their local economies.

The Grameen Bank in Bangladesh, started in 1976, has become a model for the more than 3,000 microcredit programs that have served millions of poor clients (Roseland and Soots 2007). To get a loan from the Grameen Bank, borrowers must form small groups of five people "to provide mutual, morally binding group guarantees in lieu of the collateral required by conventional banks" (Roseland and Soots 2007, p. 160). Initially, only two of the five group members are allowed to apply for a loan. When the initial loans are repaid, the other group members may apply for loans.

Reducing U.S. Poverty and Economic Inequality

Strategies to reduce poverty and economic inequality include those discussed in other chapters, such as improving the quality and *equality* of health care and education to ensure that these services and resources are not unfairly skewed toward children and young adults from more affluent families (see Chapters 2 and 8). Chapter 7 discusses strategies that also impact poverty and economic inequality, such as job creation and training, and strengthening labor unions. Here, we address issues concerning minimum wage and living wages, tax reform efforts, political reforms, and efforts to reduce wage theft.

Tax Reforms. The wealthy investor Warren Buffet pointed to the unfairness of the U.S. tax system when he famously remarked that he paid a lower tax rate than his secretary because his capital gains earnings are taxed at a lower rate than ordinary income. While most politicians agree that the tax system needs fixing—that it is currently too complicated and/or unfair—there is ongoing partisan disagreement about how to reform the tax system.

One way to reduce the gap between the top and the bottom of the economic system is to make the tax system more progressive. **Progressive taxes** are those in which the tax rate increases as income increases, so that those who have higher incomes are taxed at higher rates. A more progressive tax system would increase taxes on the wealthy. Other tax reforms that could help reduce economic inequality include increasing estate taxes (labeled by opponents as "death taxes") and gift taxes, as well as capital gains taxes. Other tax reform proposals include limiting itemized deductions for the wealthy (such as the mortgage interest deduction), increasing the cap on Social Security taxes (in 2015, Social Security taxes applied only to the first $118,500 of earned income), and closing corporate tax loopholes that enable many corporations to pay less than their "fair share" of taxes. Increasing taxes on corporations and the rich does not necessarily result in simple redistribution of income or wealth from the rich to the poor. Rather, revenue from increasing taxes on the rich could be directed to public projects that would provide more equal access to education, health care, public transportation, and other services that would give low-income Americans the resources they need to improve their economic situation (McNamee and Miller 2009).

Political Reform. Given the unfair advantage the wealthy have in influencing the political process, another key strategy in reducing economic inequality is to reduce the influence of money in politics. Some lawmakers are calling for an amendment to the Constitution to overturn the 2010 Citizens United ruling that gave corporations unlimited

microcredit programs The provision of loans to people who are generally excluded from traditional credit services because of their low socioeconomic status.

progressive taxes Taxes in which the tax rate increases as income increases, so that those who have higher incomes are taxed at higher rates.

political spending power. Reducing the political inequality that perpetuates economic inequality also necessitates enacting limits on the amount of money that wealthy individuals can spend on politics.

Minimum Wage Increase and "Living Wage" Laws. As of this writing, the federal minimum wage is $7.25. At this hourly rate, a person working full-time with two children earns $14,500 (before taxes)—an income well below the poverty line. The Fair Minimum Wage Act, proposed by Senator Tom Harkin (D-Iowa) and Representative George Miller (D-Calif.) would raise the federal minimum wage to $10.10 and increase the annual income of a full-time worker to $20,200, which is above the poverty line for one adult with two children. If passed, the Fair Minimum Wage Act would lift about 2.4 million people out of poverty (Gould 2014). As of January 1, 2015, 29 states and the District of Columbia have mandated a minimum wage that is higher than the federal $7.25.

Many cities and counties throughout the United States have **living wage laws** that require state or municipal contractors, recipients of public subsidies or tax breaks, or, in some cases, all businesses to pay employee wages that are significantly above the federal minimum, enabling families to live above the poverty line. Research findings show that businesses that pay their employees a living wage have lower worker turnover and absenteeism, reduced training costs, higher morale and productivity, and a stronger consumer market (Kraut et al. 2000).

Reduce Wage Theft. **Wage theft** is the failure to pay what workers are legally entitled to. Wage theft occurs when employers require workers to work off the clock or refuse to pay them for overtime when their weekly work hours exceed 40. Wage theft is widespread: In nearly 9,000 investigations of the restaurant industry, the U.S. Department of Labor found that more than 80 percent of the restaurants investigated had wage and hour violations (Gould 2014).

WHAT do you THINK?

All the robberies, burglaries, larcenies, and motor vehicle theft that occurred in the United States in 2012 cost their victims less than $14 billion. In contrast, wage theft costs U.S. workers more than $50 billion a year (Meixell and Eisenbrey 2014). Why do you think wage theft gets so little attention in the media compared with other types of theft?

Because wage theft affects primarily low-wage workers, reducing this illegal practice would help lift the incomes of those at the bottom. Combating wage theft calls for increasing penalties to deter companies and employers from engaging in this practice; denying federal contracts to companies found guilty of wage and hour violations; and increasing the number of investigators working for the Department of Labor's Wage and Hour Division (Meixell and Eisenbrey 2014).

The Safety Net: Public Assistance and Welfare Programs in the United States

Public assistance, or "welfare" programs in the United States are aimed at providing a safety net for adults and children who are economically disadvantaged. Many assistance programs are **means-tested programs** that have eligibility requirements based on income and/or assets. Public assistance programs designed to help the poor include the earned income tax credit, Supplemental Security Income, Temporary Assistance for Needy Families, food programs, housing assistance, medical care, educational assistance, and child care assistance.

Earned Income Tax Credit. The federal **earned income tax credit (EITC)**, created in 1975, is a refundable tax credit based on a person's income, marital status, and number of children. The EITC is designed to offset Social Security and Medicare payroll taxes

living wage laws Laws that require state or municipal contractors, recipients of public subsidies or tax breaks, or, in some cases, all businesses to pay employees wages that are significantly above the federal minimum, enabling families to live above the poverty line.

wage theft Occurs when employers "steal" workers' wages by requiring them to work off the clock or refusing to pay them for overtime.

means-tested programs Assistance programs that have eligibility requirements based on income.

earned income tax credit (EITC) A refundable tax credit based on a working family's income and number of children.

on working poor families and to encourage and reward work, and lifts millions of U.S. children and adults out of poverty each year. About half the states have their own EITCs to supplement the federal EITC.

WHAT
do you
THINK?

Childless workers—adults without children and noncustodial parents—receive little or no benefit from the EITC and, as a result, "are the sole group that the federal tax system taxes into (or deeper into) poverty" (Center on Budget and Policy Priorities 2015, n.p.). Some lawmakers have proposed expanding and increasing EITC benefits to childless workers. Would you support or oppose such proposals? Why?

Supplemental Security Income. Supplemental Security Income (SSI), administered by the Social Security Administration, provides a minimum income to poor people who are age 65 or older, blind, or disabled. SSI is not the same as Social Security: A millionaire can collect Social Security, but a person must be either elderly or disabled *and* must have limited income and assets to collect SSI.

Temporary Assistance for Needy Families. In 1996, the U.S. social welfare system was dramatically changed with the passage of the Personal Responsibility and Work Opportunity Reconciliation Act (PRWORA), which replaced the cash assistance program Aid to Families with Dependent Children (AFDC) with a new program, **Temporary Assistance for Needy Families (TANF)**. Although the previous AFCD program provided a more reliable safety net to the poorest of Americans, the current TANF program is a cash assistance program for the poor that offers more limited assistance, with time limits and work requirements. Within two years of receiving benefits, adult TANF recipients must be either employed or involved in work-related activities, such as on-the-job training, job search, and vocational education. A federal lifetime limit of five years is set for families receiving benefits, and able-bodied recipients age 18 to 50 without dependents have a two-year lifetime limit. Some exceptions to these rules are made for individuals with disabilities, victims of domestic violence, residents in high unemployment areas, and those caring for young children. The success of the TANF program is measured not by how many low-income families move into careers that provide a living wage, but rather by the number of people leaving the TANF program, regardless of their reason for doing so or their well-being thereafter (Green 2013).

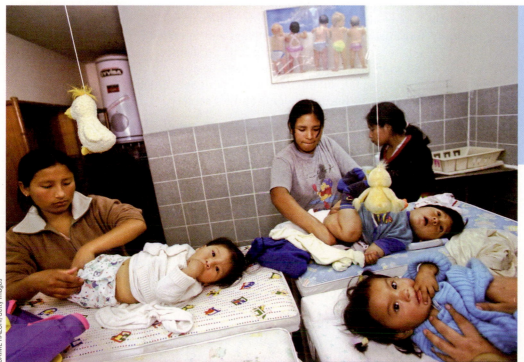

Some parents who cannot afford an adequate supply of diapers leave their babies in a soiled diaper for much too long. There is no federal assistance program that provides diapers to low income parents.

Temporary Assistance for Needy Families (TANF)
A federal cash welfare program that involves work requirements and a five-year lifetime limit.

Food Assistance. The largest food assistance program in the United States is the **Supplemental Nutrition Assistance Program (SNAP)** (formerly known as the Food Stamp Program), followed by school meals and the Special Supplemental Nutrition Program for Women, Infants, and Children (WIC). SNAP issues monthly benefits through coupons or an electronic debit card. The U.S. Census Bureau (2015a) reported that one in five U.S. children receives SNAP benefits.

In 2014, the average benefit for an individual receiving SNAP was equal to $125.35 per month, or about $4.00 per day (USDA Food and Nutrition Service 2015). Unemployed adults aged 18 to 49 who are not physically or mentally disabled or caring for a minor child are allowed to receive only three months of SNAP benefits in a three-year period, although states can allow a temporary waiver of the three-month limit in areas with persistent high unemployment (Bolen 2015). To supplement SNAP, school meals, and WIC, many communities have food pantries (which distribute food to poor households), "soup kitchens" (which provide cooked meals on site), and food assistance programs for the elderly population (such as Meals on Wheels).

WHAT do you THINK?

In the United States, one out of three low-income families struggles to buy diapers, which cost up to $100 a month per baby (National Diaper Bank Network n.d.). Some babies are left in a soiled diaper for an entire day or longer because their parents cannot afford to buy an adequate supply of diapers. Babies who stay in a dirty diaper for too long are exposed to potential health risks, and mothers who lack an adequate supply of diapers are more likely to report depression and anxiety, and are more likely to abuse their child (Mencimer 2013). SNAP benefits cannot be used to purchase diapers—like alcohol and cigarettes, diapers are considered disallowed purchases. Do you think SNAP recipients should be allowed to use their benefits to purchase diapers? Why or why not?

Housing Assistance. The biggest expense for most families is housing. Lack of affordable housing is not just a problem for the poor living in urban areas. "The problem has climbed the income ladder and moved to the suburbs, where service workers cram their families into overcrowded apartments, college graduates have to crash with their parents, and firefighters, police officers, and teachers can't afford to live in the communities they serve" (Grunwald 2006).

Federal housing assistance programs include public housing, Section 8 housing, and other private project-based housing. The **public housing** program, initiated in 1937, provides federally subsidized housing that is owned and operated by local public housing authorities (PHAs). To save costs and avoid public opposition, high-rise public housing units were built in inner-city projects. These have been plagued by poor construction, managerial neglect, inadequate maintenance, and rampant vandalism. Poor-quality public housing has serious costs for its residents and for society:

> Distressed public housing subjects families and children to dangerous and damaging living environments that raise the risks of ill health, school failure, teen parenting, delinquency, and crime—all of which generate long-term costs that taxpayers ultimately bear. . . . These severely distressed developments are not just old, outmoded, or run down. Rather, many have become virtually uninhabitable for all but the most vulnerable and desperate families. (Turner et al. 2005, pp. 1–2)

Section 8 housing involves federal rent subsidies provided either to tenants (in the form of certificates and vouchers) or to private landlords. Unlike public housing that confines low-income families to high-poverty neighborhoods, the aim with Section 8 housing is to disperse low-income families throughout the community. However, because of opposition by residents in middle-class neighborhoods, most Section 8 housing units remain in low-income areas.

A major barrier to building affordable housing is zoning regulations that set minimum lot size requirements, density restrictions, and other controls. Such zoning regulations serve the interests of upper-middle-class suburbanites who want to maintain

Supplemental Nutrition Assistance Program (SNAP) The largest U.S. food assistance program.

public housing Federally subsidized housing that is owned and operated by local public housing authorities (PHAs).

Section 8 housing A housing assistance program in which federal rent subsidies are provided either to tenants (in the form of certificates and vouchers) or to private landlords.

their property values and keep out the "riffraff"—the lower-income segment of society who would presumably hurt the character of the community. Thus, one answer to the housing problem is to change zoning regulations that exclude affordable housing and to require developers to reserve a percentage of units for affordable housing (Grunwald 2006).

Alleviating and Preventing Homelessness. Programs to temporarily alleviate homelessness include "homeless shelters" that provide emergency shelter beds, and transitional housing programs, which provide time-limited housing and services designed to help individuals gain employment, increase their income, and resolve substance abuse and other health problems. However, the number of homeless individuals exceeds the number of beds available (National Alliance to End Homelessness 2015). Resolving homelessness also requires strategies to *prevent* homelessness from occurring in the first place, such as increasing employment and living wages, providing tax benefits to renters (not just to homeowners), providing more affordable housing, protecting homeowners and renters against foreclosures, providing treatment for mental illness and substance abuse, and expanding programs to house victims of domestic violence.

Some clinics offer free veterinary care to companion animals of homeless individuals. One could argue that homeless individuals, who cannot take care of their own needs, should not take on the responsibility of caring for an animal. But as this chapter's *Animals and Society* feature reveals, companion animals can play an important role in the lives of homeless individuals.

Massachusetts has a law guaranteeing emergency housing for homeless families that qualify. New York City, which by law must provide emergency shelter to anyone who needs it, has a program called Homebase that tries to keep people in their homes by offering renters on the verge of eviction services such as financial counseling, job training, and help with contesting eviction in court ("Nipped in the Bud" 2015).

Earlier we mentioned that some cities have laws that prohibit the homeless from sleeping, asking for money ("panhandling"), or sitting, or "loitering" in public places. In addition, at least 31 cities have rules that restrict or prohibit individuals or groups (e.g., faith-based and other nonprofit organizations) from providing food to homeless individuals in public places (National Coalition for the Homeless 2014a). This chapter's *The Human Side* features one college student's protest and call for action against policies that, according to advocates for the homeless, unfairly criminalize and punish homeless individuals.

Medicaid. Medicaid is a government program that provides medical services and hospital care for the poor through reimbursements to physicians and hospitals (see also Chapter 2). States vary in rules about who is eligible for Medicaid; many low-income individuals and families do not qualify for Medicaid.

Educational Assistance. Educational assistance includes Head Start and Early Head Start programs and college assistance programs (see also Chapter 8). Head Start and Early Head Start programs provide educational services for disadvantaged infants, toddlers, and preschool-age children and their parents, and are designed to improve children's cognitive, language, and social-emotional development and strengthen parenting skills (Administration for Children and Families 2002).

To help low-income individuals wanting to attend college, the federal government offers grants, loans, and work opportunities. The Pell Grant program aids students from low-income families. The federal college work–study program provides jobs for students with "demonstrated need." The guaranteed student loan program enables college students and their families to obtain low-interest loans with deferred interest payments. However, mounting student debt has reached disturbing levels. The average undergraduate who borrows money to pay for college graduates with nearly $30,000 in debt (Carey 2015).

Child Care Assistance. In the United States, lack of affordable, good child care is a major obstacle to employment for single parents and a tremendous burden on dual-income

Leslie Irvine, professor of sociology at University of Colorado in Boulder, conducted qualitative interviews with homeless individuals who had pets. Irvine found her sample by approaching homeless people in a downtown park and at street clinics that provide veterinarian services for companion animals of homeless individuals. Using an analytic approach called "personal narrative analysis," Irvine uncovered themes in the stories homeless people told about their lives with animals. In the following excerpts, homeless individuals describe how their companion animals improved the quality of their lives in ways that were life changing, and in some cases, lifesaving (Irvine 2013).

Donna's Story
Donna's life consisted of homelessness, prostitution, drug addiction, and abusive relationships. . . . Eventually, she "got the virus," referring to HIV, and she still does not know whether it came from "the sex or the needles.". . . "I would never, ever shove a needle in my arm anymore," she told me. I asked her how she quit. She paused while tears sprang up in her eyes. She said softly, "Athena." She paused again, then looked at me and said, "She was the love of my life." Athena, a German shepherd/Labrador

retriever mix, was Donna's companion for ten years. . . .

Donna became Athena's guardian through a woman named Sita, long a common denominator between homeless people and homeless animals in San Francisco. About a decade ago, Donna lived with her abusive boyfriend in a garbage-strewn encampment under a freeway. Worn out from addiction and hard living, they began camping in Donna's mother's backyard. Sita and Donna knew each other from the streets, and, as Donna explained, "Sita said, 'You need a dog in your life.'" Sita had rescued three-year-old Athena from a shelter. Although it might not seem that a homeless drug addict in an abusive relationship would make the best guardian for a dog, the match saved two lives. As Donna recalled, "Athena did everything for me. She got me out of an abusive relationship. And it was either the dog or him, and I chose the dog. . . . Everything. The guy used to beat me up, and Sita told me it was either the man or the dog, so I chose Athena. I got the dog. Got rid of the man." With the boyfriend out of the picture, Donna moved into her mother's house, to a space she described as "the upstairs." But Sita had also said, "You

have to be clean to have the dog." Her mother agreed, so Donna faced a decision. "I realized Athena meant everything to me," she told me. "I said to myself, 'My dog comes first in my life. Would I rather use drugs, or feed my dog?' And I fell in love with Athena, so I gave up the needle. Gave up the pipe. I gave up liquor. Everything." . . . Donna also credits Athena with improving her HIV status. After becoming clean and sober, she felt better and began taking care of herself.

Trish's Story
I met Trish on a cold December day in Boulder. She stood on the median at the exit of a busy shopping center with her Jack Russell terrier bundled up in a dog bed beside her. She was "flying a sign," or panhandling, with a piece of cardboard neatly lettered in black marker to read, "Sober. Doing the best I can. Please help." . . . Trish's dog Pixel came from a pet store where Trish had worked eight years ago. Then a puppy, Pixel had contracted parvovirus and survived through Trish's diligent care. But the storeowner no longer considered Pixel sellable, and offered him to Trish. Although she could hardly afford to feed herself, she had always loved animals, and the

---

families and employed single parents. Child Care Aware of America (2014) reported the following:

- The cost of full-time center-based care for two children is the highest single household expense for families living in the Northeast, Midwest, and South (Child Care Aware of America 2014). In the West, child care costs for two children are second only to average housing costs.
- The annual cost of child care in a day care center for a 4-year-old child ranges from $4,515 in Tennessee to $12,320 in Massachusetts.
- Across the country, the annual cost of center-based infant care cost an average of more than 40 percent of the state median income for a single parent, ranging from 23.8 percent in South Dakota to 60.6 percent in Massachusetts.
- In 31 states and the District of Columbia, infant care in a day care center costs more than a year's tuition and fees at a four-year public college.

Some public policies provide limited assistance with child care, such as tax relief related to child care expenses and public funding for child care services for the poor (in conjunction with mandatory work requirements). However, child care assistance is inadequate;

two have remained inseparable ever since. . . . Trish told me that she had been homeless "off and on" for over 10 years. For her, not being home-less meant sleeping in a car or in the back of a store where she briefly held a job. In her younger days, she had followed the Grateful Dead around the country, and eventually landed in Boulder. By then, she had become a heroin addict. . . . She and Pixel slept under a bridge well known in Boulder as a homeless camp. When I met her, she had been clean and sober for two years... She supported herself by working various jobs that paid under the table, mostly cleaning houses. . . . "It took about six months, and I got us off the streets. We lived out of my car."

. . . Trish said that Pixel kept her going, even during her darkest times. She even said that Pixel kept her alive. . . . " I was [on the streets]. I hated it. I was totally at rock bottom. I just wanted to die. . . . But I couldn't give up because I had something else to take care of besides myself. So he kept me alive.... I didn't care about myself, but I had to care about him, you know? He got me through a really tough spot. If I would've had to be without him

out there before, I don't think I would have made it, at all." In the two years since that "tough spot," Trish credits Pixel with helping her stay sober. "He definitely helps keep me on the straight and narrow," she said. She claims that Pixel "hates the smell of alcohol," and he keeps her away from "bad elements, or groups of people, because of the alcohol, the drugs, and all that." She claims that the dog will nip the heels of people who approach smelling of alcohol. "He's an awesome judge of character," she said. "He just knows." She looked down at Pixel and added, "Right, buddy?" The little dog never took his eyes off her.

Denise's Story

White, middle-aged, slender, and nicely dressed, Denise looked nothing like the stereotypical homeless person. Ivy, a tiny, black-and-white cat, sat sphinx-like in a carrier, her white paws tucked neatly underneath her. . . .

Denise had worked as a self-employed graphic artist, but severe depression caused her to miss dead-lines. Major clients lost faith in her, and the accounts gradually dwindled. She fell behind on rent, and her landlord evicted her from her apartment. . . . She put her belongings in storage and

moved into her car, where she and Ivy had lived for over eight months when we met. "Half of the car is taken up by her stuff," Denise said of Ivy. "There's a big carrier, which she sleeps in, and then her litter box, and her bowls, so she essentially has the whole back seat, and then I'm in the front seat." . . . The eviction has made it difficult for Denise to find another apartment. Many land-lords simply will not rent to prospec-tive tenants who have an eviction on their records. . . . She never imagined she would live in her car for over eight months.

. . . People often said she had no right to keep a cat in her circumstances. . . . When I asked how she responded to this, she explained: "I have a history with depression up to suicide ideation, and Ivy, I refer to her as my suicide barrier. And I don't say that in any light way. I would say most days, she's the reason why I keep going, because I made a commitment to take care of her when I adopted her. So she needs me, and I need her. She is the only source of daily, steady affection and compan-ionship that I have. The only one. I can't imagine being without her, wanting to go on at all, without her."

SOURCE: Irvine 2013.

many states have waiting lists for child care assistance. Only about one in every six eligi-ble children receive federally funded child care assistance (Child Care Aware of America 2014). Finally, many families earn more than the eligibility limit, but not enough to afford child care expenses.

Welfare in the United States: Myths and Realities

Mitt Romney, candidate in the 2012 presidential election, famously referred to the 47 percent of Americans who, according to Romney, would vote for President Obama because they rely on public assistance. Romney remarked, "[T]here are 47 percent . . . who are dependent upon government, who believe that they are victims, who believe the government has a responsibility to care for them, who believe that they are entitled to health care, to food, to housing, to you-name-it" (quoted in Plumer 2012, n.p.). Nega-tive attitudes toward welfare assistance and welfare recipients, such as those conveyed in Romney's comments, are not uncommon (Epstein 2004). But these negative images are grounded in myths and misconceptions about welfare. For example, Romney was right that about half of Americans live in a household that receives some kind of federal

the HUMAN side

A Student Activist Speaks Up against Criminalizing the Homeless

To my classmates,

Spring Break is right around the corner and it is on all of our minds. Will you join your peers and flock to warmer climates and sandy beaches? We all deserve some time off to relax and turn off our brains. But while I'm still studying away, I've been thinking about how my decisions as a consumer impact the town I'm visiting and the people who live there. I'm willing to drop a cool $100 for a view of the beach, but at what human cost?

Are drinks by the pool and tickets to concerts the only cost of spring break?

Many vacation and spring break destinations compete for our business. For them, reputation is everything, and unfortunately being nice to homeless people isn't exactly a trait we tourists are often looking for. For that reason, cities introduce ordinances to keep the streets clean of all visible reminders that poverty exists. They don't want us to share food with the homeless people

near our hotels, beaches and restaurants. They don't want us to see a man sleeping on the park bench or a mother and her child asking for money on the public transportation system. Public lands near our vacation hot spots are no longer a place of rest for the homeless. New laws passed by cities throughout the country ban sleeping outside, asking for money and prohibit private citizens from sharing food with the homeless. These ordinances make criminals out of people who are homeless.

The most outrageous thing is that they are even punishing the people who are trying to help the homeless! Fort Lauderdale is one mean city that continuously threatens people with massive fines and jail-time for feeding others in public spaces. The city wants them to move indoors and out of sight, not considering how difficult it is for the homeless people to get around the city to all the many places they must go to try to get help.

Do you really want to go to a beach that only allows people who can afford a $14 daiquiri to enjoy the view? I don't—and I know what I am going to do about it. I won't give them my business! I will not condone a city starving its most vulnerable residents for my sake. I will not visit Fort Lauderdale until they repeal the cruel food-sharing ban. Instead, I pledge to support cities that work to end homelessness by creating affordable housing, job training programs, access to affordable health care and an increase in the availability in public assistance.

Be a Student Promoting Fairness! . . . Don't let cities profit from discrimination and criminalization of the homeless!

Deirdre Walsh
Student Activist

SOURCE: National Coalition for the Homeless, 2015 (February 11), "Students Promoting Fairness—#SPF15." Available at www.nationalhomeless.org

benefit, but a big chunk of these benefits go to support the elderly and disabled. Nearly a third of U.S. households in 2011 received Medicare and Social Security—benefits for the elderly and the disabled (Plumer 2012). Medicare and Social Security are provided to older Americans across the economic spectrum; they are not programs designed to target the poor. Indeed, although government assistance programs are often referred to as "entitlement programs," labeling Social Security an "entitlement program" may be misleading because retirees collect the money that they and their employers have contributed over their work history. As we discuss in Chapter 12, Social Security is a form of retirement insurance administered by the government, which is substantially different from public assistance to the poor that is funded by tax dollars. The following discusses other myths about welfare:

Myth 1. People who receive welfare are lazy, have no work ethic, and prefer to have a "free ride" on welfare rather than work.

Reality. Three-quarters of recipients of TANF are children, and nearly half of TANF cases are child-only cases where no adult is involved in the benefit calculation and only children are aided (Office of Family Assistance 2014). Because we do not expect children to work, we can hardly think of children in need of assistance as "lazy."

Adults receiving public assistance are lazy? More than 1 in 10 TANF families have earned income from employment, but with average monthly earnings of only $838 (in 2011), they could not survive on the income from their jobs. Other adult welfare recipients are participating in work activities, including job training or education and job searches. Unemployed adult welfare recipients experience a number of barriers that prevent them from working, including disability and poor health, job scarcity, lack of transportation, and lack of education. Parents with infants or young children may be unable to work because they cannot afford child care. Finally, most adult welfare recipients would rather be able to support themselves and their families than rely on public assistance. The image of a welfare "freeloader" lounging around enjoying life is far from the reality of the day-to-day struggles and challenges of supporting a household on a monthly TANF check of $387, which was the average monthly cash assistance to families receiving TANF assistance in 2011 (Office of Family Assistance 2014). This chapter's *Social Problems Research Up Close* feature presents research that looks at the challenges that low-income mothers who rely on public assistance face for survival.

Myth 2. Most welfare mothers have large families with many children.

Reality. In 2011, the average number of children in families that receive TANF was only 1.8; half of families receiving TANF had only one child, and less than 8 percent of families had more than three children (Office of Family Assistance 2014).

Myth 3. Welfare benefits are granted to many people who are not really poor or eligible to receive them.

Reality. Although some people obtain welfare benefits through fraudulent means, it is much more common for people who are eligible to receive welfare not to receive benefits. Only about a third of families who are eligible to receive TANF are receiving TANF benefits (Office of Family Assistance 2014) (see Table 6.6).

One reason for not receiving benefits is lack of information; some people do not know about various public assistance programs, or even if they know about a program, they do not know they are eligible. Another reason that many people who are eligible for public assistance do not apply for it is because they desire personal independence and do not want to be stigmatized as lazy people who just want a "free ride" at the taxpayers' expense. Others have difficulty navigating the complex administrative processes involved in applying for assistance. Assistance programs are administered through separate offices at different locations, have various application procedures and renewal deadlines, and require different sets of documentation.

Green (2013) explains that:

> [a]s low-income mothers struggle to meet the intense demands of balancing work and family, they also have to continue the time-intensive task of piecing together in-kind and cash benefits to pad their low wages. Doing so involves traveling from one office to another; repeatedly disclosing intimate and personal information; and documenting, in a detailed paper trail, the legitimacy of one's story. . . . Although there is little time to spare in this world, each office treats clients as if they have endless time to waste. Furthermore, poor families are often at the mercy of buses that are late, babysitters who do not show up, over-worked caseworkers who misplace documents, and other similar barriers to the successful performance of the role of "good client." This situation can lead to extreme levels of personal frustration, which add to the hardship and defeatism experienced while engaging this system. (p. 55)

Navigating through the "system"—getting time off work to meet with caseworkers and finding child care and transportation—can produce so much frustration that some people who are eligible for assistance just give up on the system.

Finally, some individuals who are eligible for public assistance do not receive it because it is not available. In cities across the United States, thousands of eligible low-income households are on waiting lists for public housing assistance because there are not enough public housing units available, and some cities have even stopped accepting housing applications. Even when people receive

TABLE 6.6 Percentage of Individuals Living below Poverty Level in Households That Receive Means-Tested Assistance, 2013

Type of Assistance	Percentage
Any type of assistance	73.8
Medicaid	61.3
SNAP (formerly called "food stamps")	49.5
Cash assistance	17.4
Housing assistance	14.8

SOURCE: Gabe 2015.

social problems RESEARCH UP CLOSE

Patchwork: Poor Women's Stories of Resewing the Shredded Safety Net

Courtesy of Autumn Green

Autumn Green, a sociologist whose research and advocacy focuses on low-income families, conducted research that explores how poor women who receive public assistance manage to navigate the system—how they piece together, in a patchwork fashion, resources so that they and their children can survive. Patchwork quilting is traditionally women's work, and so the metaphor of "patchwork" reflects the gendered nature of poverty—the highest rates of poverty are among women and single female-headed households with children. The patchwork metaphor also conveys that "the daily work of surviving in poverty can be likened to the creation of a complex tapestry drawing from many resources, performing a multiplicity of tasks and duties, and piecing together one's own subsistence with the work of caregiving and nurturing for others in meaningful and artistic ways" (Green 2013, p. 53).

Green's research and advocacy interests stem from her own life experiences as a former teenage parent, low-income mother, and public assistance recipient. Green describes how she once mapped out the various patches of the "quilt" that represented the responsibilities she faced as a low-income, single mother in graduate school. With the help of a friend, Green composed a 12-page list that she taped together like a quilt spread across the dining room table, on which she wrote out and categorized the different patches that made up the quilt:

One square of paper read "Benefits" at the top. "Assemble packet for Section 8 recertification was written on the first line" with subcategories: "(1) get paperwork from landlord, (2) collect proof-of-income letter from work, (3) get letter from daughter's therapist for a three-bedroom disability accommodation," each line representing several hours of work to complete. Then the list continued with similarly mapped steps for upcoming deadlines for fuel assistance, SNAP, and MassHealth recertification. And this was only one category at one moment in time listing only those items with immediate deadlines. Work also had a category, then school. Caring for my disabled daughter entailed a long list of tasks, from scheduling school meetings, to therapy appointments, to finding summer camp programs (and, of course, the money to pay for them). And the list went on for 12 pages and was still far from complete. Few low-income women have the training, resources, and connections even to map out their obligations, let alone meet them without feeling frustrated or defeated. (Green 2013, pp. 60–61)

Method and Sample

From 2004 through 2011, Green conducted in-depth interviews with 10 low-income Boston-area single mothers who were current or former college students. These women were recruited through (1) flyers that were distributed to programs serving low-income mothers, and (2) network sampling through Green's personal connections as a local welfare rights advocate. Six participants identified as African American, two as white, one as Latina, and one—an immigrant from Haiti—considered her race to be Haitian. The ages of the participants ranged from 19 to 49 (two were 19, three were in their 20s, three were in their 30s, and two were in their 40s). Five of the participants had only one child; one had two children. The children ranged from 5 months to 22 years old; however, most of the children were younger than age 8. All the participants had at least some college or vocational training beyond high school, but none had completed a four-year degree. Six participants were currently enrolled in college courses, one participant had graduated with a vocational degree, two had previously attended vocational

programs and were trying to reenroll in college, and one had recently dropped out of college.

Selected Findings

Green's interviews yielded qualitative data about the lived experiences of poor mothers juggling their roles as parent, worker, and recipient of various public assistance benefits. After briefly describing one of Green's participants, we discuss the policy implications of Green's research.

Research Participant: Jessica's Story
Jessica is a 24-year-old Caucasian mother who recently graduated from a cosmetology program, ended a relationship with her daughter's father, and moved back to Boston from New York. She found a three-bedroom apartment, sight unseen, for $1,000 per month, which is considered very low rent in Boston. Jessica got a job at a salon near her home, but she was paid as an independent contractor, so her income varied depending on the number of clients she served in a given week and the cost of services. Jessica struggled to find child care for her 2-year-old daughter Beth, and first hired a neighbor's daughter to babysit while she was at work, but Jessica's income was low and inconsistent, and it was hard to pay the babysitter. So Jessica began asking friends, family members, and neighbors to watch Beth as a favor on a day-to-day basis. Although Jessica tried to access the state's child care assistance voucher program, she was told that it would be a two-year wait before she could receive a voucher. Green, who combined her advocacy work with her research, informed Jessica that if she received TAFDC benefits (Massachusetts' TANF cash assistance program), she could get a voucher immediately. Jessica applied for TAFDC benefits, and although she worked full time, she was deemed eligible because her wages were low and inconsistent. Through the TAFDC program, she was able to receive cash assistance, SNAP, and MassHealth insurance (i.e., Medicaid) benefits, as well as an immediate child care voucher. Jessica used the voucher to send her daughter to a child care provider in the neighborhood who Jessica said, "isn't perfect, but I just try not to think about it because I don't have another choice" (p. 59).

Jessica had difficulty meeting her monthly expenses, so she found a lower-cost apartment but still struggled to break even, let alone have anything extra. Jessica applied for public housing and Section 8 in the Greater Boston Area, but she was told that the waiting list was three to five years for public housing and even longer for Section 8. During the workday, on breaks and between clients, Jessica made phone calls to local charities to inquire about emergency assistance programs to prevent the shutoff of utilities or eviction, but these calls provided nothing but dead ends. With the help of a friend, Jessica figured out how to get an application for fuel assistance. This application required her to produce documentation of her income, so she typed up a letter for her boss to sign, photocopied her paychecks, and took extra time during her lunch break to deliver the paperwork.

Green explains that TAFDC participants and those transitioning from "welfare to work" have the most integrated system of case management, with one caseworker who administers their cash, SNAP, MassHealth, and child care subsidy. But as Jessica's income went up, she transitioned off the TAFDC program and then had to maintain each of her state benefits separately (SNAP, MassHealth, and subsidized child care). For working-poor families not receiving cash assistance, these other social services are administered through separate offices, have various renewal deadlines, and require different sets of documentation. "This situation requires the working poor, whose time is already extremely stretched with the balance of work and family, to engage separate case managers, visit separate locations, individually apply for services, and provide duplicate documentation" (Green 2013, p. 60).

The social service benefits Jessica received, plus her wages, still fell short of supporting her basic needs. So on her days off, Jessica visited family members in the suburbs to do her laundry, traveled to local food pantries, searched for emergency assistance programs, and attended mandatory nutrition classes and recertification appointments for her food benefits. These tasks were in addition to the other generalized tasks of motherhood: cleaning, grocery shopping, cooking, playing with her toddler, scheduling doctor's appointments, and so on. To help with the cost of rent, Jessica decided to bring in a roommate to rent the extra bedroom in her apartment. She had concerns about bringing a stranger into her home, but she had few options available.

Discussion

Although each of Green's participants had a different story, their strategies for survival were similar: They each used a patchwork strategy to piece together resources for survival. Jessica, the research participant profiled here, managed to stay afloat by using a patchwork of strategies, but she did so at great personal cost. She was continuously exhausted; felt torn away from spending time with her daughter; and she felt isolated and lonely with few opportunities to socialize and little time to take care of herself.

Green compared the patchwork strategy she observed her research participants engage in to a literal type of quilt called a *crazy quilt*—a quilt made from random strips of fabric of various sizes and shapes that are pieced together without any cohesive pattern or organization. Green explains, "I know from my first experience trying to make a crazy quilt that sometimes the pieces do not quite fit together, and you get little holes between the patches. . . . [A] patchwork system leaves holes that allow families to fall through the cracks too easily" (p. 60).

Green's interviews with her research participants highlights how the patchwork social service system, with each service having different requirements and guidelines, adds another layer of difficulty for women already struggling against the odds of economic disadvantage.

If, for example, the utility assistance office is open only from 1 p.m. until 4 p.m. and [a client's] workfare hours are from 1 p.m. to 5 p.m., what are her options for preventing sanction from the Department of Transitional Assistance while still getting the electricity turned back on? If [a client] is reported for bringing [her child] to work with her, she could be sanctioned and lose her cash assistance, but she has been offered no help to find a caregiver who will take her voucher and provide good-quality care for [her child]. Where is he [or she] to go while she completes the work hours required to keep the funds that pay her rent and support her children? (p. 57)

Green concludes that, "[u]ltimately, the patchwork safety net system is ineffective and unreasonably demanding. We eat up women's time with the work of survival and thus make it impossible for women to work on self-improvement, personal empowerment, or education, thus trapping them in perpetual poverty. The system also takes away from caregiving and thus is a detriment for children" (p. 61).

Green's research concludes with a number of broad policy recommendations, a few of which will be mentioned here. First, Green suggests that public assistance should be provided through a more centralized system that would consider benefit levels in relation to one another, and would implement benefit reductions due to wage increases more gradually, allowing a greater opportunity for economic advancement. "Within this system, families would spend less energy obtaining social services and could use their time better toward meeting immediate family needs, nurturing their children, improving job performance (through increased availability and decreased absenteeism), investing in self-care (e.g., receiving counseling or addressing health issues), or meeting long-term goals through education and training" (p. 62).

Another suggestion for welfare reform involves integrating the various forms of public assistance so that clients can simultaneously apply for multiple social services in a single location and/or through a single case manager. Streamlining social services into a one-stop approach not only would be helpful for clients using public assistance, but would also cut costs and reduce errors and inconsistencies in program administration.

Given the difficulty Green's participants had in finding landlords who would accept housing vouchers, Green says we should investigate why providers are not willing to accept state assistance vouchers: Low reimbursement rates? Excessive bureaucratic requirements? Negative stereotypes of voucher holders as bad tenants?

Green argues that, given the recent recession and the large number of families struggling for basic survival, we should expand social services, as well as opportunities for living wage work and higher education, and increase our value for women's work in the home. At a minimum, says Green, current funding must be maintained, rather than allow further cuts in programs that are already in crisis.

Finally, Green says that mending the tattered safety net of public assistance "requires embracing a universal perspective on meeting human services needs as a societal duty to all citizens" (p. 62):

Understanding that these issues address universal human needs, rather than make up for the faults of the "morally deficient," we can change the terms of human services and thus the way in which we provide them. The last of these options ultimately requires a transformative shift in societal values that, given recent political contention, will most likely be an uphill battle. (p. 62)

benefits, using them may be difficult. For example, individuals with Section 8 vouchers for housing may have a hard time finding a landlord who will accept them, even though it is against the law to refuse a Section 8 renter. Individuals who have Medicaid may have difficulty finding a doctor who will take Medicaid patients. Low-income parents who receive child care assistance vouchers are often unable to find an available "voucher slot" and may be on a waiting list at a child care center for more than a year.

Myth 4. There is widespread abuse and fraud in SNAP by beneficiaries, who use their food stamp benefits to purchase beer, wine, liquor, cigarettes, and/or tobacco, or who sell their food stamp benefits for cash.

Reality. First, the SNAP program strictly prohibits beneficiaries from purchasing alcoholic beverages and tobacco products, as well as any nonfood items such as pet food, cosmetics, and paper products, with their benefits card. And due to increased government oversight and the introduction of the Electronic Benefit Transfer (EBT) card system, fraud in the SNAP program has decreased considerably (Blumenthal 2012).

Myth 5. Immigrants place a huge burden on our welfare system.

Reality. Low-income noncitizen immigrants, including adults and children, are less likely to receive public benefits than those who are native born. Moreover, when noncitizen immigrants receive benefits, the value of benefits they receive is lower than the value of benefits received by those born in the United States (Ku and Bruen 2013). Federal rules restrict immigrants' eligibility for public benefit programs, and undocumented immigrants are generally ineligible to receive benefits from Medicaid, SNAP, and TANF, although some benefit programs, such as the National School Lunch Program, the Women, Infants, and Children Nutrition Program (WIC), and Head Start, do not include immigration status as an eligibility factor. And although children born in the United States are considered citizens and are therefore eligible for public assistance, undocumented parents often do not apply for assistance for their children because they either do not know their children can receive benefits, or they fear that applying for benefits for their children will result in their deportation (see also Chapter 9).

Understanding Economic Inequality, Wealth, and Poverty

As we have seen in this chapter, the quality of our lives is intricately related to the economic resources we have—resources that buy access to goods and services such as housing, food, education, health care, resources that influence virtually every aspect of our lives. On a positive note, significant gains have been made in improving the standard of living for populations living in absolute poverty. But at the same time, economic inequality has reached unprecedented levels in the world, and in the United States, as the "rich get richer."

A common belief among U.S. adults is that the rich are deserving and the poor are failures. Blaming poverty on the individual rather than on structural and cultural factors implies not only that poor individuals are responsible for their plight but also that they are responsible for improving their condition. If we hold individuals accountable for their poverty, we fail to make society accountable for making investments in human development that are necessary to alleviate poverty, such as providing health care, adequate food and housing, education, child care, job training and job opportunities, and living wages. Lastly, blaming the poor for their condition diverts attention away from the recognition that the wealthy—individuals and corporations—receive far more benefits in the form of wealthfare or corporate welfare, without the stigma of welfare.

Efforts to alleviate poverty and reduce economic inequality are often motivated by a sense of moral responsibility. But alleviating poverty and reducing economic inequality also makes sense from an economic standpoint. According to one study, if economic inequality in Maryland was lowered to the level it was in 1968—the year that economic inequality was at its lowest level in modern U.S. history—the economic benefits would be equivalent to adding 22 percent to Maryland's annual gross state product in the form of personal consumption expenditures, decreased social and environmental costs, increased access to higher education, and additional spending by the poor (Talberth et al. 2013). The cost of poverty in the United States is more than $500 billion (4 percent of the GDP) per year due to increased health care costs, increased crime-related costs, and lowered productivity (vanden Heuvel 2011). Unused labor potential caused by extreme poverty worldwide costs an estimated $3.5 trillion, or 5 percent of global gross domestic product (Davis 2015). According to Oxfam International (2013), the $240 billion net income of the richest 100 billionaires in 2012 is *four times* the amount of money needed to eradicate extreme poverty worldwide.

Ending or reducing poverty begins with the recognition that doing so is a worthy ideal and an attainable goal. Imagine a world where everyone had comfortable shelter, plentiful food, clean water and sanitation, adequate medical care, and education. If this imaginary world were achieved and if absolute poverty were effectively eliminated, what would be the effects on social problems such as crime, drug abuse, family problems (e.g., domestic violence, child abuse, and divorce), health problems, prejudice and racism, and international conflict? In the current global climate of conflict and terrorism, we might consider that "reducing poverty and the hopelessness that comes with human deprivation is perhaps the most effective way of promoting long-term peace and security" (World Bank 2005). Instead of asking if we can afford to eradicate poverty, we might consider: Can we afford not to?

Chapter Review

• **What is the extent of poverty, wealth, and economic inequality in the world?**
One billion people—about one in seven—are extremely poor, living on less than $1.25 a day. In stark contrast to this extreme poverty, in 2014, there were 1,645 billionaires in the world, and the richest 1 percent of adults (ages 20+) owned nearly half of global wealth.

• **How is poverty measured?**
The most widely used standard to measure extreme poverty in the developing world is $1.25 per day. According to measures of relative poverty, members of a household are considered poor if their household income is less than 50 percent of the median household income in that country. The Multidimensional Poverty Index considers health, education, and living standards in measuring who is poor. Each year, the U.S. federal government establishes "poverty thresholds" that differ by the number of adults and children in a family and by the age of the family head of household. Anyone living in a household with pretax income below the official poverty line is considered "poor." The Economic Policy Institute's Family Budget Calculator finds that U.S. families need at minimum twice the poverty level income to meet a family's basic economic needs.

• **How do the structural-functionalist, conflict, and symbolic interactionist perspectives view wealth, poverty, and economic inequality?**
According to the structural-functionalist perspective, poverty results from institutional breakdown: economic institutions that fail to provide sufficient jobs and pay, educational institutions that fail to equip members of society with the skills they need for employment, family institutions that do not provide two parents, and government institutions that do not provide sufficient public support. The structural-functionalist perspective also focuses on both the functions and dysfunctions of poverty.

The conflict perspective is concerned with how capitalism has contributed to poverty and economic inequality and how of wealthy corporations and individuals use their wealth to influence elections and policies in ways that benefit the wealthy. The conflict perspective is also critical of how trade and investment policies benefit wealthy corporations and negatively affect the poor.

The symbolic interactionist perspective calls attention to ways in which wealth and poverty are defined and the effects of being labeled "poor." This perspective is also concerned with how economic inequality has come to be more widely recognized as a social problem as a result of media attention stemming from the Occupy Wall Street movement and the recent publication of Thomas Piketty's book *Capital in the Twenty-first Century*.

• **What is the extent of economic inequality and poverty in the United States?**
The United States has the highest level of income inequality and the highest rate of poverty of any industrialized nation. In 2012, the top 1 percent of U.S. taxpayers earned 22.5 percent of all U.S. income, and the wealthiest 1 percent owned 40 percent of U.S. wealth. In the United States, groups that are at increased risk of poverty include children, women (especially those who are single parents), those with low levels of educational attainment, and racial and ethnic minorities.

- **What are some of the consequences of poverty and economic inequality for individuals, families, and societies?**

  Poverty is associated with health problems and hunger; substandard housing and homelessness; unequal treatment in the legal system; political inequality and alienation; crime, social conflict, and war; increased vulnerability to natural disasters; problems in education; and unequal marriage opportunity and other family problems including higher rates of divorce, child abuse/neglect, and teenage and unintended pregnancy. These various problems are interrelated and contribute to the perpetuation of poverty across generations, feeding a cycle of intergenerational poverty.

- **What are some global strategies for reducing poverty and economic inequality?**

  Global approaches for reducing poverty and economic inequality include increasing taxes on the wealthy, promoting economic development, investing in human development, and providing microcredit programs that provide loans to poor people.

- **What are some strategies to reduce U.S. poverty and economic inequality?**

  Strategies to reduce U.S. poverty and economic inequality include the earned income tax credit, tax reform efforts that make taxes more progressive and that close corporate tax loopholes, political reforms to reduce influence of money in politics, increasing the minimum wage and establishing living wage laws, and reducing wage theft so that workers receive the income they are legally entitled to.

- **What are some public assistance and welfare programs in the United States?**

  U.S. public assistance and welfare programs include Supplemental Security Income, Temporary Assistance for Needy Families (TANF), food programs (such as school meal programs and SNAP), housing assistance, Medicaid, educational assistance (such as Pell Grants), child care assistance, and the earned income tax credit (EITC).

- **What are five common myths about welfare and welfare recipients?**

  Common myths about welfare and welfare recipients are (1) that welfare recipients are lazy, have no work ethic, and prefer to have a "free ride" on welfare rather than work; (2) that most welfare mothers have large families with many children; (3) that welfare benefits are granted to many people who are not really poor or eligible to receive them; (4) that there is widespread abuse and fraud in the SNAP program; and (5) that immigrants place an enormous burden on our welfare system.

Test Yourself

1. The _____ Poverty Index is a measure of serious deprivation in the dimensions of health, education, and living standards that combines the number of deprived and the intensity of their deprivation.
   a. Relative
   b. Human
   c. International
   d. Multidimensional

2. According to the 2014 official U.S. poverty threshold guidelines, a single adult earning $13,000 a year is considered "poor."
   a. True
   b. False

3. In a meritocracy, individuals get ahead and earn rewards based on their
   a. political power.
   b. inherited wealth.
   c. luck.
   d. efforts and abilities.

4. In 2013, a typical U.S. worker would have to work _____ years to make what a typical CEO of a large U.S. corporation earns in one year.
   a. 15
   b. 99
   c. 211
   d. 296

5. Corporate welfare refers to which of the following?
   a. Taxes corporations pay that provide most of the funding for federal welfare programs for the poor
   b. Tax-deductible contributions that corporations make to charitable organizations
   c. Laws and policies that benefit corporations
   d. Employee assistance programs offered by corporations to help employees who are struggling with debt

6. What age group in the United States has the highest rate of poverty?
   a. Younger than 18
   b. 30 to 44
   c. 45 to 64
   d. Older than 65

7. According to the text, the wealthy are hardest hit by natural disasters because they have more to lose than do the poor.
   a. True
   b. False

8. A plutocracy is a country ruled by
   a. the wealthy.
   b. the people.
   c. women.
   d. religion.

9. SNAP benefits can be used to purchase
   a. alcohol.
   b. tobacco.
   c. diapers.
   d. none of the above

10. Women receiving TANF typically have at least three children.
    a. True
    b. False

Answers: 1. D; 2. B; 3. D; 4. D; 5. C; 6. A; 7. B; 8. A; 9. D; 10. B.

210 | **CHAPTER 6** Economic Inequality, Wealth, and Poverty

Key Terms

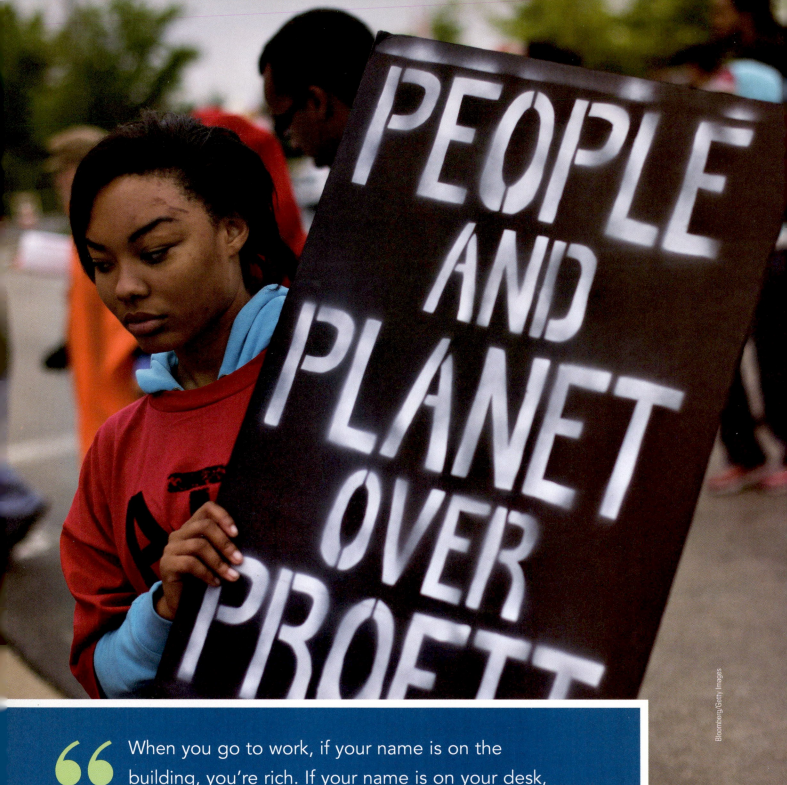

PEOPLE AND PLANET OVER PROFIT

" When you go to work, if your name is on the building, you're rich. If your name is on your desk, you're middle class. If your name is on your shirt, you're poor."

RICH HALL
writer and performer

7

Work and Unemployment

Learning Objectives

After studying this chapter, you will be able to . . .

1 Explain how the 2007 U.S. economic crisis affected the global economy, describe the differences between capitalism and socialism, and discuss the pros and cons of free trade agreements and transnational corporations.

2 Give examples of how structural functionalism, conflict theory, and symbolic interactionism view work and the economic institution.

3 Understand problems associated with work in the United States and around the world, including unemployment, slavery, sweatshop labor, child labor, health and safety in the workplace, work/life conflict, alienation, job stress, and issues concerning labor unions and workers' rights.

4 Describe strategies for reducing unemployment; creating alternatives to capitalism through worker cooperatives; ending slavery, child labor, and sweatshop labor; making the workplace safer; helping workers achieve work/life balance; and strengthening labor unions.

5 Identify employment-related human rights issues and present examples of how and why these human rights are being violated in workplaces around the world.

When an 8-story garment industry building collapsed, 1,100 garment workers lost their lives.

the $20 billion garment export industry in Bangladesh, which is second only to China. When the building began to rumble, some workers wanted to leave, but managers ordered them to stay at their sewing machines because they had important clothing orders to fill. As the building began to shake, one worker jumped from a fifth-floor window, breaking her back and causing her baby to miscarry (International Federation of Social Workers 2015). When the eight-story building collapsed, 1,100 workers lost their lives and 2,500 were injured. Unsafe working conditions in Bangladesh's garment industry are not uncommon. Independent factory inspections in 2013 found the following: "Dangerously heavy storage loads sent cracks down walls and stressed sagging support beams. In some cases, basic fire equipment was missing, and exit routes didn't lead outside" (Motlagh 2014). In 2012, a fire at Tazreen Fashions claimed the lives of 112 workers. The tragedies at Rana Plaza and Tazreen Fashions were well publicized in the media, drawing attention to the problem of unsafe working conditions in the overseas garment industry.

IN APRIL 2013, 5,000 workers in Bangladesh went to work in a building known as Rana Plaza, where they spent 16 hours a day making name-brand clothing for

Health and safety hazards in the workplace, often due to employers' willful violations of health and safety regulations, are among the work-related problems discussed in this chapter. Other problems we examine include unemployment, forced labor, child labor, sweatshop labor, alienation, work/life conflict, and declining labor strength and representation. We set the stage with a brief look at the global economy.

The Global Context: The New Global Economy

In recent decades, innovations in communication and information technology have spawned the emergence of a **global economy**—an interconnected network of economic activity that transcends national borders and spans the world. The globalization of economic activity means that our jobs, the products and services we buy, and our nation's economic policies and agendas influence and are influenced by economic activities occurring around the world.

In 2007, the economic situation around the world took a downward turn, as banks faltered, credit froze, businesses closed, unemployment rates soared, and investments plummeted. This global financial crisis, which originated in the United States and spread throughout the world, illustrates the globalization of the **economic institution**. Some say that the cause of the global economic crisis was the lack of U.S. financial regulatory oversight that enabled financial institutions to engage in predatory and subprime lending during the housing boom in the early 2000s. As the housing boom turned to bust, and adjustable rate mortgages were reset to higher rates, millions of homeowners were unable to keep up with mortgage payments, and foreclosures skyrocketed. Homeowners lost their homes, renters lost their leases, and banks suffered because the foreclosed homes they now owned were often worth less than the mortgages owed on them. As banks lost revenue, they had less money to lend, so credit froze, consumer spending

global economy An interconnected network of economic activity that transcends national borders.

economic institution The structure and means by which a society produces, distributes, and consumes goods and services.

plummeted, businesses went bust, and stockholders watched their investments and retirement accounts take a nosedive. The whole banking system was faltering—some got bailed out; others (e.g., Bear Stearns) went bust. All this happened in the United States, but in this new global economy, what happens in Vegas does not stay in Vegas—the crisis spread around the world. This is because those risky subprime and adjustable rate mortgages were packaged and resold as "mortgage-backed securities" to financial institutions around the world.

The U.S. economic crisis that began in 2007 also triggered a huge drop in world trade. When the United States went into recession and consumer spending declined, all the countries that depended on U.S. consumers to buy their goods and services lost a major source of revenue.

The global economic crisis reignited debate between those who view U.S. capitalism as the cause of economic problems in the world, and those who hail capitalism as "the greatest engine of economic progress and prosperity known to mankind" (Ebeling 2009). After summarizing capitalism and socialism—the two main economic systems in the world—we look at the emergence of free trade agreements and transnational corporations.

> In this new global economy, what happens in Vegas does not stay in Vegas.

Capitalism and Socialism

Under **capitalism**, private individuals or groups invest capital (money, technology, machines) to produce goods and services to sell for a profit in a competitive market. Capitalism is characterized by economic motivation through profit, the determination of prices and wages primarily through supply and demand, and the absence of government intervention in the economy. **Socialism** is an economic system in which the means of production (factories, machinery, land, stores, offices, etc.) are socially owned by the public, worker collectives, or the state. In a socialist economy, theoretically, goods and services are distributed according to the needs of the citizens, with some goods and services—such as water, education, health care, and/or child care—provided to all citizens through a socially funded system of taxation. Whereas capitalism emphasizes individualistic pursuit of profit and individual freedom, socialism emphasizes collective well-being and social equality.

Socialism is often confused with communism, and some people use the two terms interchangeably. But socialism is more accurately an economic system that most often exists with some variation of a democratic political system, whereas **communism** is a political system that usually takes on the form of totalitarianism—a one-party system where the state controls all aspects of public and private life, permitting no individual freedom. In communism, there is no private property (all property is commonly owned), there are no class distinctions among people (goods and services are distributed equally so that no one has more or less than anyone else), and goods and services are distributed directly to people without the use of money. Some people view communism as an extreme form of socialism.

In reality, there are no pure socialist or capitalistic economies. Rather, most countries have mixed economies, incorporating elements of both capitalism and socialism. Most developed countries, for example, have both privately owned and publicly owned enterprises, as well as a social welfare system. The U.S. economy is dominated by capitalism, but it also includes elements of socialism, such as the provision of fire and police services, public roads, education, a postal service, public assistance programs for the poor, government subsidies and low-interest loans to industry, fiscal stimulus money, and bailout money to the auto industry and banks.

Critics of socialism argue that socialism creates excessive government control, reduces work incentives and technological development, and lowers the standard of living. A national survey of U.S. adults found that 39 percent say they have a positive image of socialism, compared with 54 percent who have a negative image of socialism (Newport 2012).

Although a majority (61 percent) of Americans say they have a positive image of capitalism, nearly a third (31 percent) have a negative image of capitalism (Newport 2012). And while 70 percent of Americans agree people are better off in a free-market (i.e., capitalistic) economy, even though some people are rich and some are poor, one in four

capitalism An economic system characterized by private ownership of the means of production and distribution of goods and services for profit in a competitive market.

socialism An economic system characterized by state ownership of the means of production and distribution of goods and services.

communism A political system that usually takes on the form of totalitarianism—a one party system where the state controls all aspects of public and private life, permitting no individual freedom.

Americans *disagreed* (Pew Research Center 2014). Critics of capitalism point to a number of social ills linked to this economic system, including high levels of inequality; economic instability; job insecurity; pollution and depletion of natural resources; and corporate dominance of media, culture, and politics. Capitalism is also criticized as violating the principles of democracy by allowing (1) private wealth to affect access to political power (see also Chapter 6); (2) private owners of property to make decisions that affect the public (such as when the owner of a factory decides to move the factory to another country); and (3) workplace dictatorships where workers have little say in their working conditions, thus violating the democratic principle that people should participate in collective decisions that significantly affect their lives (Wright 2013). Wolff (2013a, 2013b) explains that major shareholders and boards of directors within corporations make key decisions about what products the corporation will produce, what technologies will be used, where production will occur, and how the revenues will be distributed—decisions that profoundly affect workers who have no say in these decisions:

> The most important activity of an adult's life in this country is work. It's what we do five days out of every seven. If democracy belongs anywhere, it belongs in the workplace. Yet we accept, as if it were a given, that once we cross the threshold of our store, factory, or office, we give up all democratic rights. (Wolff, quoted in Barsamian 2012, p. 12)

WHAT do you THINK?

Economist Richard Wolff points out that capitalism is an institution, like education and health care, but although it is considered appropriate to debate whether our schools and health care system are working properly and meeting our needs, it is taboo to ask whether the way we organize the production and distribution of goods and services is meeting our needs (Barsamian 2012). Do you view Americans who critically examine capitalism as being "un-American" or disloyal to the United States? Why or why not?

The Globalization of Trade and Free Trade Agreements

The globalization of trade refers to raw materials, manufactured goods, and agricultural products being bought and sold across national and hemispheric borders. The first set of global trade rules were adopted through the General Agreement on Tariffs and Trade (GATT) in 1947. In 1995, the World Trade Organization (WTO) replaced GATT as the organization overseeing the multilateral trading system.

In the 1980s and early 1990s, U.S. officials began negotiating regional free trade agreements that would open doors to U.S. goods in neighboring countries and reduce the growing U.S. trade deficit. A **free trade agreement (FTA)** is a pact between two countries or among a group of countries that makes it easier to trade goods across national boundaries. Free trade agreements reduce or eliminate foreign restrictions on exports, reduce or eliminate tariffs (or taxes) on imported goods, and prevent technology from being copied and used by competitors through protection of "intellectual property rights." Treaties such as the Canada–U.S. Free Trade Agreement, the North American Free Trade Agreement (NAFTA), the Free Trade Area of the Americas (FTAA), and the Central America Free Trade Agreement (CAFTA) were designed to accomplish these trade goals. As this book goes to press, Congress is considering a new trade agreement called the Trans-Pacific Partnership (TPP).

Although free trade agreements have expanded trading opportunities, benefiting large export manufacturing and service industries in the global north, they have also undermined the ability of national, state, and local governments to implement laws designed to protect workers, consumers, and the environment. Trade agreements include provisions that supersede national and local laws, including the U.S. Constitution, and give foreign corporations the right to sue governments for lost profits due to environmental, worker safety, food or product safety, and other laws that hurt corporate profits (see also Chapters 6 and 13; Korten 2015; Scott and Ratner 2005).

free trade agreement (FTA) A pact between two or more countries that makes it easier to trade goods across national boundaries by reducing or eliminating restrictions on exports and tariffs (or taxes) on imported goods and protecting intellectual property rights.

Free trade agreements have also hurt both U.S. and foreign workers. Before the U.S.–Korea free trade agreement took effect in 2012, the United States had a trade deficit with Korea, meaning that the United States imported more goods from Korea than it exported to Korea. The U.S.–Korea FTA was implemented with the promise that it would reduce the trade deficit, and that it would lead to increased U.S. exports to Korea, which means more U.S. jobs. But the U.S.–Korea FTA also resulted in increased Korean imports, which reduced the demand for domestic goods and services and resulted in more than 75,000 U.S. jobs lost in the first three years after the agreement took effect (Scott 2015). The North American Free Trade Agreement (NAFTA) allowed U.S. corn growers to sell their corn in Mexico, but Mexican corn farmers could not compete with the cheap price of U.S. corn, which put many Mexican corn growers out of business (Scott and Ratner 2005). Free trade agreements have also made it easier for U.S. companies to move jobs "offshore," usually to countries where wages are low and there are few environmental, health, or safety regulations with which to comply. Although offshoring jobs increases profits to corporations, it also takes jobs away from U.S. workers.

Transnational Corporations

Although free trade agreements have increased business competition around the world, resulting in lower prices for consumers for some goods, they have also opened markets to monopolies (and higher prices) because they have facilitated the development of large-scale transnational corporations. **Transnational corporations**, also known as *multinational corporations*, are corporations that have their home base in one country and branches, or affiliates, in other countries.

Transnational corporations provide jobs for U.S. managers, secure profits for U.S. investors, and help the United States compete in the global economy. Transnational corporations benefit from increased access to raw materials, cheap foreign labor, and the avoidance of government regulations. They can also avoid or reduce tax liabilities by moving their headquarters to a "tax haven." But the savings that transnational companies reap from cheap labor and reduced taxes are not passed on to consumers. "Corporations do not outsource to far-off regions so that U.S. consumers can save money. They outsource in order to increase their margin of profit" (Parenti 2007). For example, shoes made by Indonesian children working 12-hour days for 13 cents an hour cost only $2.60 but are still sold for $100 or more in the United States.

General Electric (GE) is one of the biggest transnational corporations in the world. More than half of GE's workforce is based outside the United States.

Danny Lehman/Terra/Corbis

Transnational corporations contribute to the trade deficit in that more goods are produced and exported from outside the United States than from within. Transnational corporations also contribute to the budget deficit, because the United States does not get tax income from U.S. corporations abroad, yet transnational corporations pressure the government to protect their foreign interests; as a result, military spending increases. Transnational corporations contribute to U.S. unemployment by letting workers in other countries perform labor that U.S. employees could perform. Finally, transnational corporations are implicated in an array of other social problems, such as poverty resulting from fewer jobs, urban decline resulting from factories moving away, and racial and ethnic tensions resulting from competition for jobs.

transnational corporations
Also known as *multinational corporations*, corporations that have their home base in one country and branches, or affiliates, in other countries.

Sociological Theories of Work and the Economy

In sociology, structural functionalism, conflict theory, and symbolic interactionism serve as theoretical lenses through which we may better understand work and economic issues and activities.

Structural-Functionalist Perspective

According to the structural-functionalist perspective, the economic institution is one of the most important of all social institutions, as it functions to provide the basic necessities common to all human societies, including food, clothing, and shelter and thus contributes to social stability. After the basic survival needs of a society are met, surplus materials and wealth may be allocated to other social uses, such as maintaining military protection from enemies, supporting political and religious leaders, providing formal education, supporting an expanding population, and providing entertainment and recreational activities. Societal development is dependent on an economic surplus in a society (Lenski and Lenski 1987).

As noted in Chapter 6, structural-functionalist theorists Davis and Moore (1945) argued that jobs differ in their pay so that workers will be motivated to achieve higher levels of education and training, and that jobs that offer higher rewards are those that are more important and difficult. This argument falls apart when one considers that many very important jobs have low pay. As discussed in Chapter 6, the inequality associated with salaries and wages is often considered dysfunctional, rather than functional, for society. The economic institution can also be dysfunctional when it fails to provide members with jobs and the goods and services they need; when the production, distribution, and consumption of goods and services depletes and pollutes the environment; when participation in the labor force leads to alienation and work/life conflict (discussed later in this chapter); and when it includes practices that violate basic human rights, such as slavery and unsafe working conditions (also discussed later in this chapter).

The structural-functionalist perspective also focuses on how the economy affects and is affected by changes in society. For example, as societies become more economically developed, infant and child death rates drop, women have fewer children, and life expectancy increases, which leads to the aging of the world's population. Population aging affects the economy because as the proportion of older people increases, there are fewer working-age adults to fill jobs and to support the elderly population (see also Chapter 12).

Conflict Perspective

According to the conflict perspective, the ruling class controls the economic system for its own benefit and exploits and oppresses the working masses. The conflict perspective is critical of ways that the government caters to the interests of big business at the expense of workers, consumers, and the public interest. This system of government that serves the interests of corporations—known as **corporatocracy**—involves ties between government and business. For example, in Chapter 2, we discussed how the pharmaceutical and health insurance industries influence politicians on matters related to health care.

Corporations influence government through lobbying, and donations to politicians, campaigns, and Super PACs. As we discussed in Chapter 6, in 2010, corporations gained power to influence political outcomes when the Supreme Court ruled (5 to 4) in *Citizens United v. Federal Election Commission* that corporations, as well as unions, have a First Amendment right to spend unlimited amounts of money to support or oppose candidates for elected office. That decision also freed corporations and unions to "educate" workers about elections through voter guides, get-out-the-vote activities, and other tools. So a company can legally inform its workers that if they vote for candidate A, the company will suffer and jobs will be lost. Is this education? Or is it the company intimidating or coercing the workers to vote a certain way? The line is murky (Carney 2012).

corporatocracy A system of government that serves the interests of corporations and that involves ties between government and business.

The pervasive influence of corporate power in government exists worldwide. The policies of the International Monetary Fund (IMF) and the World Bank pressure developing countries to open their economies to foreign corporations, promoting export production at the expense of local consumption, encouraging the exploitation of labor as a means of attracting foreign investment, and hastening the degradation of natural resources as countries sell their forests and minerals to earn money to pay back loans. In his book *Confessions of an Economic Hit Man*, John Perkins (2004) described his prior job as an "economic hit man"—a highly paid professional who would convince leaders of poor countries to accept huge loans (primarily from the World Bank) that were much bigger than the country could possibly repay. The loans would be used to help develop the country by paying for needed infrastructure, such as roads, electrical plants, airports, shipping ports, and industrial plants. One of the conditions of the loan was that the borrowing country had to give 90 percent of the loan back to U.S. companies (such as Halliburton or Bechtel) to build the infrastructure. The result: The wealthiest families in the country benefit from additional infrastructure, and the poor masses are stuck with a debt they cannot repay. The United States uses the debt as leverage to ask for "favors," such as land for a military base or access to natural resources such as oil. According to Perkins, large corporations want "control over the entire world and its resources, along with a military that enforces that control" (quoted by MacEnulty 2005, p. 10).

Symbolic Interactionist Perspective

According to symbolic interactionism, the work role is a central part of a person's self-concept and social identity. When making a new social acquaintance, one of the first questions we usually ask is "What do you do?" The answer largely defines for us who that person is. An individual's occupation is one of the person's most important statuses; for many, it represents a "master status," that is, the most significant status in a person's social identity. This chapter's *Social Problems Research Up Close* feature describes a study that looks at how job loss of white-collar professionals in midlife affects their self-concepts and attitudes about work and unemployment.

Symbolic interactionism emphasizes the fact that attitudes and behavior are influenced by interaction with others. The applications of symbolic interactionism in the workplace are numerous: Employers and managers use interpersonal interaction techniques to elicit the attitudes and behaviors they want from their employees; union organizers use interpersonal interaction techniques to persuade workers to unionize. And, as noted in the *Social Problems Research Up Close* feature, parents teach their young adult children important lessons about work and unemployment through interaction with them.

Symbolic interaction also focuses on the importance of labels. Although the concepts of capitalism and free enterprise are theoretically similar, Americans tend to view the term *capitalism* more negatively than *free enterprise*. Thus, "politicians seeking the most positive overall reaction from voters should choose to use the term 'free enterprise' rather than 'capitalism' in describing America's prevailing economic system" (Newport 2012, n.p.).

Problems of Work and Unemployment

In this section, we examine unemployment and other problems associated with work. Poverty, minimum wage and living wage issues, workplace discrimination, and retirement and employment concerns of older adults are discussed in other chapters. We discuss problems concerning unemployment, child labor, forced labor, sweatshop labor, health and safety hazards in the workplace, alienation, job stress, work/life conflict, and labor unions and the struggle for workers' rights.

Unemployment

In 2014, more than 200 million people worldwide—nearly 6 percent of the global labor force—were unemployed. In some countries in Africa and the Middle East, up to 30 percent of the workforce was unemployed. The unemployment rate for youths

social problems
RESEARCH
UP CLOSE | Job Loss at Midlife

Research presented here investigates how job loss affected the self-concepts and views about employment among a sample of unemployed professionals in midlife (Mendenhall et al. 2008).

Sample and Methods
The sample consisted of 77 men and women who were recruited through two Chicago-area networking groups that help unemployed managers and executives find employment, and also through announcements at local churches and posted flyers at cafés and public libraries. Participants had to meet four eligibility criteria: They had to (1) have been unemployed for at least three months during the past year, (2) have been married at the time of their job loss, (3) have children between the ages of 12 and 18 living at home, and (4) live in the Chicago area. Most participants were male (83 percent) and white (80 percent). Respondents had been unemployed for an average of 15 months at the time of their first interview. More than half had earned annual salaries of $100,000 or more; no one earned less than $50,000, and most had jobs with generous benefits.

The researchers interviewed participants at libraries, cafés, or on the University of Chicago campus. Interview topics included the job loss event, how the job loss affected family relationships, and educational plans for their children. Participants also completed a survey about their job loss experience, family economic circumstances, family relationships, and their own health and well-being. About a year later, participants were interviewed again, and completed another questionnaire.

Selected Findings
This study revealed some interesting patterns in how job loss among professionals in midlife affected (1) their self-concept, (2) their job-seeking strategies, and (3) the messages about employment they conveyed to their children. The participants viewed their job termination as evidence of a lack of employer loyalty and a change from a lifetime employment contract to one in which even high-level employees can be terminated without warning—a shift in which participants came to view themselves as free agents who effectively "rent" their services to employers. As free agents, there is no expectation of permanent employment or loyalty in the employee–employer relationship.

More than half of the participants 50 years and older said they perceived age discrimination in the job market, in that employers signaled that they were looking for entry-level employees, which is code for "young." In response, participants often deemphasized their age and experience by omitting graduation dates and some of their work history on their resumes—a phenomenon known as "deprofessionalization." A former CEO explained that, in the first several months of job searching, he listed the date of his college graduation on his résumé and got no responses. Then he deleted the date and got a half dozen responses.

Many participants viewed their job loss experience as providing an opportunity to teach real-life lessons to their children about the world of work. A 50-year-old project manager said:

> I think [that my son getting a good glimpse into the reality of life is] a positive because . . . if he goes out there with rose-tinted glasses, he's going to get smacked upside the head real hard some day. (p. 203)

One 47-year-old information technology consultant used his job loss experience to teach his children to prepare for a job market in which neither employee nor employer expects a lifetime employment contract. He told his children:

> For most jobs now, you need to view [it] like the movie industry where . . . you're rented[,] . . . you're contracted to do that movie, whether you're the lights or the cameraman or whatever, and then you're out of work again. And that's more the way most jobs are now where you should view a job as just your job until you're out of work again. (p. 204)

Respondents advised their children to take specific steps to prevent being overly dependent on employers. Some advised their children to own their own businesses; others urged their children to develop skills that could be "transferable" to a new employer. Others encouraged their children to choose careers where their potential client base was diversified:

> I've started talking to them about how . . . they should start thinking about [whether] they want to work for big companies or . . . go into a field where their client base is very much diversified, which is lawyers, doctors, CPAs, therapists. . . . Because then if someone fires you, you don't care because you have another 100 [clients] whereas when you work for a company, and . . . your boss . . . fires you . . . you're out of a job. (p. 204)

SOURCE: Mendenhall et al. 2008.

---

unemployment To be currently without employment, actively seeking employment, and available for employment, according to U.S. measures of unemployment.

(aged 15 to 24) was nearly three times higher than for adults (International Labour Organization [ILO] 2015b).

In the United States, the official **unemployment** rate includes individuals who are currently unemployed, are available for employment, and have looked for work in the past 4 weeks. The U.S. unemployment rate dipped to a 31-year low of 4 percent in 2000, but it rose to 10 percent in 2009 as a result of the economic recession that began in 2007 as companies went out of business and plants closed. A **recession** refers to a significant decline in economic activity spread across the economy and lasting for at least six months. In communities hardest hit by the recession, unemployment rates rose to over 20 percent. By 2015 (March), U.S. unemployment had dropped to 5.5 percent (Bureau of Labor Statistics 2015b). Rates of unemployment are higher among certain groups, including racial and ethnic minorities (see also Chapter 9) and among those with lower levels of education.

The official unemployment rate does not include people who are "marginally attached" to the labor force—individuals who want a job, have searched for work in the past 12 months, but have not looked for a job in the past 4 weeks. These "marginally attached" persons have not looked for a job in the past 4 works because they are discouraged over job prospects or because they have family responsibilities, school or job training commitments, or ill health or disability. The official unemployment rate also does not include involuntary part-time workers—those who want to work full-time but who settle for a part-time job, nor does it include highly skilled and highly educated individuals who are working in low skill, low-paying jobs. Because the official unemployment rate does not include these marginally attached and involuntary part-time workers, it grossly undercounts the percentage of people whose employment needs are not being met.

Causes of Unemployment. A primary cause of unemployment is lack of available jobs. In December 2000, jobs were plentiful: For every job opening, the ratio of job seekers to job openings was 1 to 1, meaning there was one available job for every person seeking employment. But between 2008 and 2013, the ratio of job seekers to job openings fluctuated from 3 to 1 to more than 6 to 1 (Shierholz 2011, 2013). In early 2015, the ratio of job seekers to jobs fell below 2 to 1 (Bureau of Labor Statistics 2015c). Even when jobs are plentiful, some amount of unemployment is inevitable because it takes time for job seekers to find the right job and for employers to find the right workers. Jobs and workers may also be in different locations and may need time to find each other, during which time jobs are not filled, and workers are unemployed. This type of inevitable unemployment is known as *frictional unemployment* (Palley 2015).

Another cause of U.S. unemployment is **job exportation**, also referred to as **offshoring**—the relocation of jobs to other countries. Jobs most commonly offshored have been in manufacturing, but offshoring also occurs with service jobs, including information technology, human resources, finance, purchasing, and legal services. Exporting jobs enables corporations to maximize their profits by reducing their costs of raw materials and labor. By offshoring service jobs, employers cut labor costs by about 75 percent (Davidson 2012). **Outsourcing**, which involves a business subcontracting with a third party to provide business services, saves companies money as they pay lower salaries and no benefits to those who provide outsourced services. Many commonly outsourced jobs—including accounting, web development, information technology, telemarketing, and customer support—are outsourced to non-U.S. workers.

Automation, or the replacement of human labor with machinery and equipment, also contributes to unemployment, although automation creates new jobs too. ATMs have reduced the need for bank tellers, but at the same time, they have created jobs for workers who produce and service ATMs. A technology called **3-D printing** is likely to have a profound effect on manufacturing jobs. The revolutionary 3-D printing involves downloading a digital file containing a design for a product. The printer reads the file and then shoots out the product (made of specialized plastic or other raw materials) through a heated nozzle. For example, General Electric uses 3-D printing to manufacture parts for jet engines; Nike uses it to make soccer cleats. A *Bloomberg View* report concluded that

> 3-D printing seems likely to throw a lot of people out of work in the medium term, especially in industries that depend on assembly line labor. Eventually, as with most technological breakthroughs, it will probably create new jobs in new industries. But that transition period will be hazardous, and displaced workers will need help to navigate it. (The Editors 2013, n.p.)

recession A significant decline in economic activity spread across the economy and lasting for at least six months.

job exportation The relocation of jobs to other countries where products can be produced more cheaply.

offshoring The relocation of jobs to other countries.

outsourcing A practice in which a business subcontracts with a third party to provide business services.

automation The replacement of human labor with machinery and equipment.

3-D printing A revolutionary manufacturing technology that involves downloading a digital file containing a design for a product. A printer reads the file and then shoots out the product (made of specialized plastic or other raw materials) through a heated nozzle.

3-D printers, such as the one depicted here, are expected to have a major impact on manufacturing jobs.

Another cause of unemployment is increased global and domestic competition. Mass layoffs in the U.S. automobile industry have occurred, in part, due to competition from makers of foreign cars who, unlike U.S. automakers, do not have the burden of providing health insurance for their employees. Unemployment also results from mass layoffs that occur when plants close and companies downsize or go out of business.

Effects of Unemployment on Individuals, Families, and Societies. One study found that poor-quality jobs—those with high demands, low control over decision making, high job insecurity, and low pay—had more negative effects on mental health than having no job at all (Butterworth et al. 2011). Nevertheless, unemployment has been linked to depression, low-self esteem, and increased mortality rates (Turner and Irons 2009). One study found that, among workers with no preexisting health conditions, losing a job because of a business closure increased the risk for a new health problem by 85 percent, with the most common health problems including hypertension, heart disease, and arthritis (Strully 2009). Unemployment is also a risk factor for homelessness, substance abuse, and crime, as some unemployed individuals turn to illegitimate, criminal sources of income, such as theft, drug dealing, and prostitution.

Long-term unemployment—which is considered as being unemployed for 27 weeks or longer—can have lasting effects, such as increased debt, diminished retirement and savings accounts (which are depleted to meet living expenses), home foreclosure, and/or relocation from secure housing and communities to unfamiliar places to find a job. But, even when individuals who are laid off find another job in one or two weeks, they still suffer damage to their self-esteem from having been told they are no longer wanted or needed at their workplace. And being fired affects worker trust and loyalty in future jobs. Employees who are not fired during a mass layoff are also affected, as they worry that "their job could be next" (Uchitelle 2006).

In families, unemployment is also a risk factor for child and spousal abuse and marital instability. When an adult is unemployed, other family members are often compelled to work more hours to keep the family afloat. And unemployed noncustodial parents, usually fathers, fall behind on their child support payments.

Plant closings and large-scale layoffs affect communities by lowering property values and depressing community living standards. High numbers of unemployed adults create a drain on societies that provide support to those without jobs. Job displacement of the nonelderly is linked to lower rates of social participation in church groups, charitable organizations, youth and community groups, and civic and neighborhood groups (Brand and Burgard 2008). Globally, the high numbers of young adults without jobs create a risk for crime, violence, and political conflict (United Nations 2005).

Unemployment creates a vicious cycle: The unemployed (as well as those who fear job loss) cut back on spending, which hurts businesses that then must cut jobs to stay afloat. During times of economic stress or uncertainty, have you cut back on everyday spending? See this chapter's *Self and Society* feature.

Employment Concerns of Recent College Grads

Most college students seek a college degree to improve their job opportunities. But having a college degree is no guarantee of employment in a struggling economy. The unemployment rate for young college graduates (without an advanced degree and not currently

> One study found that poor-quality jobs—those with high demands, low control over decision making, high job insecurity, and low pay—had more negative effects on mental health than having no job at all.

long-term unemployment Unemployment that lasts for 27 weeks or more.

How Do Your Spending Habits Change in Hard Economic Times?

Difficult economic times make people reconsider what is important for them to have in their daily lives, and what they can live without or change to save money. How do your spending habits change in hard economic times? For each of the following spending categories, place a check mark next to those items that describe changes in your spending habits in response to tough economic times.

1. Buy more generic brands. _____
2. Take lunch from home instead of buying lunch out. _____
3. Go to hairdresser/barber/stylist less often. _____
4. Stopped purchasing bottled water; switched to refillable water bottle. _____
5. Cancelled one or more magazine subscriptions. _____
6. Cancelled or cut back on TV/cable service. _____
7. Stopped buying coffee in the morning. _____
8. Cut down on dry cleaning. _____
9. Changed or cancelled cell phone service. _____
10. Began carpooling or using mass transit. _____

The list below presents the percentages of responses to each of the 10 previous items from a 2014 online Harris Poll of U.S. adults. The most common changes in spending behavior include purchasing more generic brands and "brown-bagging" lunch (taking lunch from home instead of buying lunch out).

TABLE 1 Cutting Back on Spending over Past Six Months, in Percentages

1. 54%	6. 21%
2. 37%	7. 16%
3. 33%	8. 15%
4. 31%	9. 14%
5. 22%	10. 11%

SOURCE: Birth 2014.

enrolled in further education) tends to be more than twice the overall unemployment rate (Shierholz et al. 2014). Some college grads question whether going to college was worth the expense and the effort. In 2014, 46 percent of employed college grads under age 27 were working in jobs that did not require a college degree (Davis et al. 2015).

College grads who find employment are not receiving the same wages or benefits as they used to. Between 2000 and 2014, the inflation-adjusted wages of young college grads

> Most college students seek a college degree to improve their job opportunities. But having a college degree is no guarantee of employment in a struggling economy.

College grads hope their education will pay off. But many recent college grads remain unemployed or work in a low-paying job that does not require a college degree.

Forced prison labor is a type of forced labor that is controlled by the state. Forced prison labor is particularly widespread in China.

Mark Peterson/Corbis News/Corbis

declined 7.7 percent, and the share of young college grads who received pension benefits from their employer dropped from 41.5 percent in 2000 to 27 percent in 2012 (Shierholz et al. 2014). Due to their high unemployment rate, the low wages of many jobs, the high cost of living, and student loans and other college debts, many recent college graduates end up living with their parents or other family, instead of on their own. In 2013, 28 percent of young adults ages 24 to 34 who were living with their parents were college graduates (Jones 2014b).

Slavery

Slavery, also known as **forced labor**, refers to any work that is performed under the threat of punishment and is undertaken involuntarily. An estimated 35.8 million people throughout the world live in slavery, more than half of whom live in just five countries: India, China, Pakistan, Uzbekistan, and Russia (*Global Slavery Index* 2014). Slaves are found working in agriculture, domestic work, food service, construction, garment manufacturing, the sex industry, factories, mining, and other sectors. In South Asia, where sex slavery is most common, either girls are forced into prostitution by their own husbands, fathers, and brothers to earn money to pay family debts, or they are lured by offers of good jobs and then are forced to work in brothels under the threat of violence.

The form of slavery most people are familiar with is **chattel slavery**, in which slaves are considered property that can be bought and sold. Although chattel slavery still exists in some areas, most forced laborers today are not "owned" but are rather controlled by violence, the threat of violence, and/or debt. The most common form of forced labor today is called *bonded labor*. Bonded laborers are poor individuals who take out a loan simply to survive or to pay for a wedding, funeral, medicines, fertilizer, or other necessities. Debtors must work for the creditor to pay back the loan, but they are often unable to repay it. Creditors can keep debtors in bondage indefinitely by charging the debtors illegal fines (for workplace "violations" or for poorly performed work) or charge laborers for food, tools, and transportation to the work site while keeping wages too low for the debt to ever be repaid.

There are more than 60,000 slaves in the United States, most of whom were trafficked into the country and forced into slavery, most commonly in domestic work, farm labor, and the sex industry (*Global Slavery Index* 2014; Skinner 2008). Migrant workers are tricked into working for little or no pay as a means of repaying debts from their transport across the U.S. border. Migrant workers are particularly vulnerable because if they try to escape and report their abuse, they risk deportation. Traffickers posing as employment agents lure women into the United States with the promise of good jobs and education but then place them in "jobs" where they are forced to do domestic or sex work.

slavery Any work that is performed under the threat of punishment and is undertaken involuntarily; also known as *forced labor*.

forced labor See **slavery**.

chattel slavery A form of slavery in which slaves are considered property that can be bought and sold.

WHAT do you THINK?

In U.S. state and federal prisons, inmates work for little to no pay, often under conditions that expose the inmates to toxic chemicals without proper safety equipment (Flounders 2013). In some states, prisoners who refuse to work are sent to disciplinary housing and lose canteen privileges as well as "good time" credit that could lower their prison sentence. Some prisoners claim that they were beaten for refusing to work. In U.S. federal prisons, inmates earn 23 cents an hour producing night-vision goggles, camouflage uniforms, and high-tech components for military aircraft and helicopters, guns, and land mine sweepers. Large corporations purchase these fruits of prison labor for a low price, and then make a huge profit from contracts with the U.S. military. Should large corporations be allowed to profit from prison labor? Should prisoners have the same workplace protections as other workers? Why or why not?

Sweatshop Labor

Millions of people worldwide work in **sweatshops**—work environments that are characterized by less than minimum wage pay, excessively long hours of work (often without overtime pay), unsafe or inhumane working conditions, abusive and discriminatory treatment of workers by employers, and/or the lack of worker organizations aimed at negotiating better working conditions. Sweatshop labor conditions occur in a wide variety of industries, including garment production, manufacturing, mining, and agriculture.

More than 97 percent of all clothing purchased in the United States is imported, often made under sweatshop conditions (Institute for Global Labour and Human Rights 2011). Garment workers at a factory in Ha-Meem, Bangladesh, work 12- to 14-hour shifts, 7 days a week, with 1 day off a month. Young women making garments for Gap, J.C. Penney, Phillips-Van Heusen (PVH), Target, and Abercrombie & Fitch make 20 to 26 cents an hour (Institute for Global Labour and Human Rights 2011).

Sweatshop labor commonly occurs in the garment industry.

Sweatshop Labor in the United States. Sweatshop conditions in overseas industries have been widely publicized. However, many Americans do not realize the extent to which sweatshops exist in the United States. For example, in 2012, the Department of Labor found widespread "sweatshop-like" labor violations in the Los Angeles garment industry (Miles 2012). Ten garment contractors, producing clothing for more than 20 retailers, were found to be paying less than minimum wage, not paying overtime, and/or falsifying time cards or failing to maintain records of employees' hours. Most garment workers in the United States are immigrant women who typically work 60 to 80 hours a week, often earning less than minimum wage with no overtime, and many face verbal and physical abuse.

Immigrant farmworkers, who process most of the fruits and vegetables grown in the United States, also work under sweatshop conditions. Many live in substandard and crowded housing provided by their employer and lack access to safe drinking water as well as bathing and sanitary toilet facilities. Farmworkers commonly suffer from heat exhaustion, back and muscle strains, injuries resulting from the use of sharp and heavy farm equipment, and illness resulting from pesticide exposure (Austin 2002). Working 12-hour days under hazardous conditions, farmworkers have the lowest annual family incomes of any U.S. wage and salary workers, and more than 60 percent of them live in poverty (Thompson 2002).

Child Labor

Child labor involves a child performing work that is hazardous; that interferes with a child's education; or that harms a child's health or physical, mental, social, or moral development. Even though virtually every country in the world has laws that prohibit or limit the extent to which children can be employed, child labor persists throughout the world. More than half of the 168 million children who are engaged in child labor are involved in hazardous work (ILO 2015a). Some of the worst forms of child labor are forced labor, drug trafficking, sexual exploitation, and armed conflict.

Child labor is involved in many of the products we buy, wear, use, and eat. Child laborers work in factories, workshops, construction sites, mines, quarries, fields, and on fishing boats. Child laborers make bricks, shoes, soccer balls, fireworks and matches, furniture,

sweatshops Work environments that are characterized by less than minimum wage pay, excessively long hours of work (often without overtime pay), unsafe or inhumane working conditions, abusive treatment of workers by employers, and/or the lack of worker organizations aimed to negotiate better working conditions.

child labor Involves a child performing work that is hazardous, that interferes with a child's education, or that harms a child's health or physical, mental, social, or moral development.

Hundreds of thousands of U.S. child workers labor on commercial farms (Human Rights Watch 2014). Child farmworkers in the United States are typically of Hispanic origin; many are U.S. citizens. Girls who work on U.S. farms are sometimes victims of sexual harassment.

toys, rugs, and clothing. They work in the manufacturing of brass, leather goods, and glass. They tend livestock and pick crops. Child laborers work long hours with few (or no) breaks or days off, often in unsafe conditions where they are exposed to toxic chemicals and/or excessive heat, and they endure beatings and other forms of mistreatment from their employers, all for as little as a dollar a day. A former textile worker in Bangladesh described conditions at a company called Harvest Rich, where clothing is sewn for U.S. firms including Walmart and J.C. Penney. She testified that hundreds of children, some as young as 11 years old, were illegally working at Harvest Rich, sometimes for up to 20 hours a day: "Before clothing shipments had to leave for the United States, there are often mandatory 19 to 20-hour shifts from 8:00 a.m. to 3:00 or 4:00 a.m. . . . The workers would sleep on the factory floor for a few hours before getting up for their next shift in the morning. If they did anything wrong, they were beaten every day." Workers had two days off a month and were paid $3.20 a week (Tate 2007). At another garment factory that made fleece jackets for Walmart, 14- or 15-year-old kids worked 18- or 20-hour shifts, from 8:00 a.m. to midnight or 4:00 a.m., seven days a week. When they passed out, rulers struck them to wake them up. Some of the girls were raped by management (Tate 2007).

Child labor also exists in the United States in restaurants, grocery stores, meatpacking plants, garment factories, and agriculture. Several hundred thousand children work on U.S. commercial farms (Human Rights Watch 2014). Child farmworkers are typically children of immigrants; many are U.S. citizens. Farmworkers—both adults and children—are exposed to chemicals, sun, and temperature extremes; they work with sharp tools and heavy machinery, climb tall ladders, and carry heavy buckets and sacks. They frequently work 12-hour days without adequate access to drinking water, toilets, or hand-washing facilities. Although children are not legally permitted to purchase tobacco products, they are allowed to work on tobacco farms where they are vulnerable to "green tobacco sickness," a poisoning that occurs when workers absorb nicotine through the skin while handling tobacco plants.

In most jobs, U.S. law requires children to be at least 16 years old to work, with a few exceptions, such as that 14- and 15-year-olds can work as cashiers, grocery baggers, and car washers. But in agricultural jobs, U.S. law allows children of any age to work on small farms as long as they have a parent's permission. Children ages 12 and older may be hired on any farm with their parent's consent or if they work on the same farm with their parents. U.S. law allows children at age 14 to work on any farm, even without parental permission.

Health and Safety in the U.S. Workplace

Although many workplaces are safer today than in generations past, fatal and disabling occupational injuries and illnesses still occur in troubling numbers. In 2013, 4,585 U.S. workers—most of whom were men—died of fatal work-related injuries (Bureau of Labor Statistics 2015a). The most common type of job-related fatality involves transportation accidents (see Figure 7.1).

Nonfatal occupational injuries and illnesses in U.S. workplaces include sprains, strains, and cuts. Workers who do repeated motions such as typing or assembly line work are prone to repetitive strain injuries (see this chapter's *The Human Side* feature).

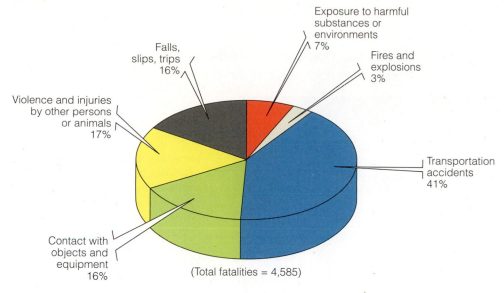

Falls,
slips, trips
16%

Exposure to harmful
substances or
environments
7%

Fires and
explosions
3%

Violence and injuries
by other persons
or animals
17%

Transportation
accidents
41%

Contact with
objects and
equipment
16%

(Total fatalities = 4,585)

Figure 7.1 Causes of Workplace Fatalities, 2014
SOURCE: Bureau of Labor Statistics, 2015a.

The incidence of occupational injury and illnesses is probably much higher than the reported statistics show. Government data on workplace illness and injury exclude many categories of workers including self-employed workers, federal government agency employees, farms with fewer than 11 workers, and private household workers (AFL-CIO 2015). Long-term illnesses caused by, for example, exposure to carcinogens often are difficult to relate to the workplace and are not adequately recognized and reported. And employees don't always report workplace injuries or illnesses for fear of being disciplined, losing their jobs, or in the case of undocumented immigrants, fear of being deported. Employers discourage reporting of injuries by offering financial rewards to workers and departments when no injuries are reported for a specified period of time. Even when workers report their injuries, employers don't always keep accurate records because they fear increased workers' compensation costs, increased scrutiny from government workplace inspectors, and fines for possible violations of health and safety regulations.

Workplace inspections reveal that many workplace injuries, illnesses, and deaths are due to employers' willful violations of health and safety regulations. The Occupational Safety and Health Administration (OSHA), created in 1970, develops, monitors, and enforces health and safety regulations in the workplace. But inadequate funding leaves OSHA unable to hire enough workplace inspectors to do the job effectively. The International Labour Organization recommends one inspector per 10,000 workers; in 2015, there was one federal or state OSHA inspector for every 71,695 workers. At current staffing levels, it would take federal OSHA inspectors, on average, 140 years to inspect each workplace just once (AFL-CIO 2015).

Suppose that a corporation is guilty of a serious violation of health and safety laws, in which "serious violation" is defined as one that poses a substantial probability of death or serious physical harm to workers. What penalty do you think that corporations should pay for such a violation? Serious violations of workplace health and safety laws in 2014 carried an average federal penalty of $1,972 or a state penalty of $1,043 (AFL-CIO 2015). Even when violations result in worker fatalities, the median federal penalty in 2014 was only $5,050; median state penalty was $4,438. Under criminal law, a willful violation that results in a worker's death is considered a misdemeanor. Since 1970, only 88 cases of willful workplace health and safety violations have been prosecuted, with defendants serving a total of 100 months in jail (AFL-CIO 2015). Why do you think penalties for violating workplace health and safety laws are so weak?

WHAT
do you
THINK?

Poultry industry workers often suffer from work-related injuries.

Poultry production is big business in the United States. Americans consume more chicken than anyone else in the world—nearly 84 pounds per year per capita, and more than any other type of meat (National Chicken Council 2012).

The Southern Poverty Law Center and Alabama Appleseed conducted interviews with 302 workers currently or previously employed in the poultry industry in Alabama—one of the top three poultry-producing states. Participants were asked about safety procedures and equipment, safety rights and enforcement, their experience with injuries and employer response to injuries, workplace discrimination and harassment, and working conditions in general. The following excerpts describe the experiences of five poultry industry workers.*

Oscar's Story: When Oscar heard that a poultry processing plant in Alabama was looking for workers, he thought he could apply the skills he learned from studying mechanical engineering in Cuba. "I thought maybe . . . that I could work with the machinery, given my abilities and my hands," he said. . . . He was denied two positions where he could apply his mechanical skills and instead was asked to fold chicken wings on the production line. As bird carcasses sped by him on the line, Oscar had to grab the wings and twist them into the position the company wanted, folding fast enough to meet a quota of approximately 40 chicken wings per minute—or roughly 18,000 wings per day. . . . As he repeated this motion thousands of times, it put pressure on his hands and wrists. After about a month, he developed serious hand and wrist pain. . . . Oscar was diagnosed with tendinitis and carpal tunnel syndrome. When his injuries made him no longer useful to the company, he was fired.

Natashia's Story: Natashia Ford had been a healthy person all her life. But after spending six years deboning chickens at a poultry processing plant in North Alabama, she's a different person. She's been diagnosed with histoplasmosis, a lung disease similar to tuberculosis that's caused by breathing airborne spores at the plant. Eight nodules are growing within her lungs, and they cannot be removed. The company she worked for resisted paying for any of her medical expenses, such as her inhaler and medications. . . . She said the company didn't provide her or her co-workers with face masks as they worked on the processing line. Chicken juices would get into Natashia's ears, nose, and mouth. . . . Workers would process 30,000 to 60,000 birds per shift as they raced to keep pace with the mechanized line. If a chicken became lodged in the machinery, the line would stop so it could be dislodged. Hurt workers couldn't count on the same mercy. The processing line never slowed or stopped for them, she said. It didn't matter if they were cut, hurt, or sick. . . . The machinery kept churning—even when Natashia was so sick that she had to be picked up and carried off. Natashia eventually sued her employer, a rare occurrence in this industry. The company ultimately paid for some of her medical bills, but not all of them. Today, Natashia can't stand for more than 15 minutes. She wears knee and back braces and walks with a cane.

Job Stress

A Gallup Poll found that 28 percent of U.S. workers are completely or somewhat dissatisfied with the amount of stress in their jobs, making job stress workers' most common complaint (dissatisfaction with the amount of money they make is workers' second most common complaint) (Jones and Saad 2014). Prolonged job stress, also known as

"No line shut down for a human, but it'd shut down for a bird," she said.

Horacio's Story: Horacio was only 18 when he began working as a chicken catcher in Alabama. . . . Chicken catchers—the workers who catch birds in chicken houses and load them onto trucks bound for processing plants—encounter many of the same problems as plant workers. These problems include repetitive motion injuries, respiratory ailments and supervisors who have little concern for their safety. . . . His crew typically filled 14 or 15 trailers with chickens during each shift. Each trailer held about 4,400 chickens. Horacio would carry about seven chickens at a time—roughly 63 pounds total. It's a feat he would perform more than 100 times for each trailer. For Horacio to carry seven chickens at a time, he had to pick the birds up by their feet and place the feet between the fingers of his hand until he held four live, squawking, scratching, pecking chickens. He then had to grab three more birds and secure their feet between the fingers of his other hand without dropping the first four chickens. Given these conditions, it's no surprise that chicken catchers often develop the same types of back, arm, wrist, and hand injuries other poultry workers suffer, though the damage is often more severe. Chicken catchers have reported their hands have swollen to twice their normal size. . . . Chicken catchers may also develop respiratory ailments, due to the poor air quality in chicken houses. . . . Horacio, like other chicken catchers, said the dust and fecal matter in the air made his eyes burn and skin itch. . . . He had a protective mask to wear, but it was so heavy he didn't use it. None of the workers wore the masks because they inhibited their breathing—preventing them from working fast enough to meet their boss's demands. Today, after 19 years as a chicken catcher, 37-year-old Horacio exhibits a telltale sign of the profession. Both of his arms are in constant pain. He also walks with a limp—a painful reminder of the time a truck ran over his foot as it backed into a chicken house. His boss insisted that he keep working through the pain.

Juan's Story: Juan . . .worked for six years in a poultry plant. He worked primarily in stacking jobs that required him to lift, carry, and stack two 80-pound boxes of chicken a minute. While lifting a box of chicken, he became dizzy, slipped, and fell to the floor. He was told to go right back to working despite being in great pain. Juan's back pain worsened and the swelling became constant. He was unable to sleep. When he was finally able to get X-rays, they revealed that he had two lumbar vertebrae fractures from the fall. He was eventually fired. Juan has yet to recover. His employer never paid for any medical treatments.

Marta's Story: Marta couldn't take it anymore. She picked up the phone and called her company's human resources hotline. She had endured several years of sexual harassment from her supervisor at the processing plant in southeast Alabama where she was a sanitation worker. He had repeatedly pressured the 48-year-old Latina to have sex with him, telling her that she could have any job she wanted—if she gave in to his advances. She was finally reporting him. But Marta's phone call didn't end her ordeal. In fact, it made matters worse. She was accused of inventing the story and was transferred to a lower-paying job. Her two sons, who also worked at the plant, received job transfers that cut their pay as well. A year later, Marta was fired. She was told she was fired over her immigration status—after seven years at the company. Her harasser, who kept his job, made it clear that immigration wasn't the real issue: He told her that if she had agreed to sleep with him, she'd still have her job.

Conclusion

The problems experienced by the workers described here are not uncommon in the poultry industry. One in five workers in the survey cited here said they or someone they knew was subjected to unwelcome touching of a sexual nature, and about a third of the workers said they or someone they knew had been subjected to unwelcome sexual comments. Nearly three-quarters of the poultry workers interviewed for this report described suffering from some type of significant work-related injury or illness, such as debilitating pain in their hands and gnarled fingers from carpal tunnel syndrome, chemical burns, and respiratory problems.

*Some names have been changed to protect the identity of the participants.

SOURCE: Based on Southern Poverty Law Center and Alabama Appleseed, 2013, *Unsafe at These Speeds: Alabama's Poultry Industry and Its Disposable Workers.* Available at www.splc .org. Used by permission.

job burnout, can cause or contribute to physical and mental health problems, such as high blood pressure, ulcers, headaches, anxiety, and depression. A U.S. study of 10 common job stressors (see Table 7.1) found that more than 120,000 people die each year from physical and mental health problems associated with workplace-related stress (Goh et al. 2015). Taking time off to heal and "recharge one's batteries" is not an option for many workers who do not have paid sick leave or paid vacation. The United States

job burnout Prolonged job stress that can cause or contribute to high blood pressure, ulcers, headaches, anxiety, depression, and other health problems.

is the only advanced nation that does not mandate a minimum number of vacation days; in the European Union countries, employers are required to give workers at least four weeks of vacation each year (some countries mandate five or six weeks of vacation) (Greenhouse 2008). And, even when workers are on vacation, they are often still connected to their jobs by their smartphones and laptops.

Work/Life Conflict

A major source of stress for U.S. workers is **work/life conflict**—the day-to-day struggle to simultaneously meet the demands of work and other life responsibilities and goals, including family, education, exercise, and recreation. Employed spouses strategize ways to coordinate their work schedules to have vacation time together, or simply to have meals together. Among dual-income spouses with children, 56 percent of moms and 50 percent of dads report that it is very or somewhat difficult to juggle work and family life (Parker and Wang 2013). Employed parents with young children must find and manage arrangements for child care and negotiate with their employers about taking time off to care for a sick child or to attend a child's school event or extracurricular activity. Some employers post their workers' weekly schedules only a few days in advance, making it difficult to arrange child care. In some two-parent households, spouses or partners work different shifts so that one adult can be home with the children. However, working different shifts strains marriage relationships, because the partners rarely have time off together. Some employed parents who cannot find or afford child care leave their children with no adult supervision.

Employees with elderly and/or ill parents worry about how they will provide care for their parents, or arrange for and monitor their care, while putting in a 40-hour (or more) workweek. Full-time workers who are caregivers to an older adult typically spend 16 hours a week providing hands-on elder care, including housework, meal preparation, physical care, and transportation (Matos 2014).

Balancing the responsibilities of a job with the demands of school is also a challenge for many adults. Millions of college students, both traditional age (18 to 24) and over age 25, work either part-time or full-time jobs while pursuing their college degree.

TABLE 7.1 Ten Common Work-Related Stressors

1. Unemployment
2. Lack of health insurance
3. Shift work
4. Long working hours
5. Job insecurity
6. Work/family conflict
7. Low job control
8. High job demands
9. Low social support at work
10. Low organizational justice (perception of unfairness at work)

SOURCE: Goh et al. 2015.

Alienation

Work in industrialized societies is characterized by a high degree of division of labor and specialization of work roles. As a result, workers' tasks are repetitive and monotonous and often involve little or no creativity. Limited to specific tasks by their work roles, workers are unable to express and utilize their full potential—intellectual, emotional, and physical. According to Marx, when workers are merely cogs in a machine, they become estranged from their work, the product they create, other human beings, and themselves. Marx called this estrangement "alienation."

Alienation in the workplace has four components: (1) *Powerlessness* results from working in an environment in which one has little or no control over the decisions that affect one's work; (2) *meaninglessness* results when workers do not find fulfillment in their work; (3) workers may experience *normlessness* if workplace norms are unclear or conflicting, such as when companies have nondiscrimination policies yet they practice discrimination; and (4) *self-estrangement* may stem from the workers' inability to realize their full human potential in their work roles.

work/life conflict The day-to-day struggle to simultaneously meet the demands of work and other life responsibilities and goals, including family, education, exercise, and recreation.

The fast-food industry epitomizes working conditions that lead to alienation. But the principles that characterize the fast-food industry also characterize many other workplaces, a phenomenon known as **McDonaldization** (Ritzer 1995). McDonaldization involves four principles:

Efficiency. Tasks are completed in the most efficient way possible by following prescribed steps in a process overseen by managers.

Calculability. Quantitative aspects of products and services (e.g., portion size, cost, and the time it takes to serve the product) are emphasized over quality.

Predictability. Products and services are uniform and standardized. A Big Mac in Albany is the same as a Big Mac in Tucson. Workers behave in predictable ways. For example, servers at McDonald's learn to follow a script when interacting with customers.

Control through technology. Automation and mechanization are used in the workplace to replace human labor.

What are the effects of McDonaldization on workers? In a McDonaldized workplace, employees are not permitted to use their full capabilities, be creative, or engage in genuine human interaction. Workers are not paid to think, just to follow a predetermined set of procedures. Because human interactions are unpredictable and inefficient (they waste time), "we're left with either no interaction at all, such as at ATMs, or a 'false fraternization.' Rule number 17 for Burger King workers is to smile at all times" (Ritzer, quoted by Jensen 2002, p. 41). Workers may also feel that they are merely extensions of the machines they operate. The alienation that workers feel—the powerlessness and meaninglessness that characterize a "McJob"—may lead to dissatisfaction with one's job and, more generally, with one's life.

McFlexible

We have shift patterns to suit the lifestyles of all our Crew. **Not bad for a McJob**

We know it is important to have a life outside of work, so many of our Crew work shift patterns that allow them to pursue their hobbies and interests. They also enjoy some great discounts as well as thorough training, in addition to access to our free private healthcare scheme after they've been with us for 3 years.
www.mcdonalds.co.uk

This poster is part of McDonald's advertising campaign to discredit the negative image implied by the term McJob.

The slang term *McJob* is found in several dictionaries. For example, the *Oxford English Dictionary* defines a McJob as "an unstimulating, low-paid job with few prospects, especially one created by the expansion of the service sector." *Merriam-Webster* defines a McJob as "a low-paying job that requires little skill and provides little opportunity for advancement." Not surprisingly, McDonald's is opposed to such characterizations of its employment conditions and has fought back with an advertising campaign in which posters at more than 1,200 restaurants play up "the positive aspects of working for McDonald's" and include the phrase, "Not bad for a McJob." Do you think that the *Oxford* and *Merriam-Webster* dictionary definitions of McJob are accurate and fair?

Labor Unions and the Struggle for Workers' Rights

Having a job is no guarantee of having favorable working conditions and receiving decent pay and benefits. **Labor unions** are worker advocacy organizations that developed to protect workers and represent them at negotiations between management and labor.

Benefits and Disadvantages of Labor Unions to Workers. Labor unions have played an important role in fighting for fair wages and benefits, healthy and safe work environments,

WHAT
do you
THINK?

McDonaldization The process by which principles of the fast-food industry (efficiency, calculability, predictability, and control through technology) are being applied to more sectors of society, particularly the workplace.

labor unions Worker advocacy organizations that developed to protect workers and represent them at negotiations between management and labor.

and other forms of worker advocacy. Compared with nonunion workers, unionized workers tend to have higher earnings, better insurance and pension benefits, and more paid time off. Union members also have higher life satisfaction than nonunion members, which is not due to higher income, but rather to the fact that union members (1) experience greater satisfaction with their work experiences, (2) feel greater job security, (3) have more opportunities for social interaction and integration, and (4) benefit from democratic participation and citizenship within their union (Flavin and Shufeldt 2014).

One of the disadvantages of unions is that members must pay dues and other fees, and these dues have been rising in recent years. Union members resent the high salaries that many union leaders make. Another disadvantage for unionized workers is the loss of individuality. Unionized workers are members of an overall bargaining unit in which the majority rules. Decisions made by the majority may conflict with individual employees' specific employment needs.

Declining Union Density. The strength and membership of unions in the United States have declined over the last several decades. **Union density**—the percentage of workers who belong to unions—grew in the 1930s and peaked in the 1940s and 1950s, when 35 percent of U.S. workers were unionized. In 2014, the percentage of U.S. workers belonging to unions had fallen to just under 11.1 percent, down from 20.1 percent in 1983 (Bureau of Labor Statistics 2015d). Union membership among public sector workers is more than five times the rate of union membership among private sector employees.

One reason for the decline in union representation is the loss of manufacturing jobs, which tend to have higher rates of unionization than other industries. In addition, globalization has led to layoffs and plant closings at many unionized work sites as a result of companies moving to other countries to find cheaper labor. Corporations also take active measures to discourage workers from unionizing.

Weak U.S. Labor Laws and Anti-union Legislation. The 1935 National Labor Relations Act (NLRA) guarantees the right to unionize, bargain collectively, and to strike against private-sector employees. However, in addition to excluding public-sector workers, the law excludes agricultural and domestic workers, supervisors, railroad and airline employees, and independent contractors. As a result, millions of workers do not have the right under U.S. law to negotiate their wages, hours, or employment terms.

Anti-union legislation refers to any law that undermines or weakens unions. Republicans are more likely than Democrats to support anti-union legislation. In 2011, Republican governor Scott Walker (Wisconsin) signed legislation to weaken unions representing state and local government employees. The legislation prohibited collective bargaining by most public workers for issues beyond wages, required unions to hold annual votes on whether they should remain in existence, and increased workers' contributions for pensions and health care. In 2015, Walker signed legislation making Wisconsin the 25th state to pass a law that allows workers to have jobs in unionized workplaces without having to join the union or pay union dues (Jones 2014a). Supporters of right-to-work laws argue that workers should not be required to join any private organization, such as a labor union, against their will, and that such laws attract business. Opponents of these laws warn that "right-to-work" laws weaken unions, leading to lower wages and benefits for workers, and also argue that nonunion workers at unionized workplaces are

Union members in Michigan protest against "right-to-work" legislation.

union density The percentage of workers who belong to unions.

"free riders" who don't pay union dues but nevertheless benefit from the better wages and benefits negotiated by the union.

Labor Union Struggles around the World. In 1949, the International Labour Office established the Convention on the Right to Organise and Collective Bargaining. About half of the world's workforce lives in countries that have not ratified this convention, including China, India, Mexico, Canada, and the United States. According to the International Trade Union Confederation (2014), 87 countries exclude certain workers from the right to strike and at least 37 countries impose fines or prison sentences for workers who go on strike. In at least 9 countries, murder or disappearance of workers is a common practice used to intimidate workers.

Strategies for Action: Responses to Problems of Work and Unemployment

Government, private business, human rights organizations, labor organizations, college student activists, and consumers play important roles in responding to problems of work and unemployment.

Strategies for Reducing Unemployment

Some economists argue that full employment should be a top priority of U.S. economic policy. **Full employment** is achieved when unemployment rates are below 5 percent, taking into account that some amount of unemployment is unavoidable (Palley 2015). Full employment not only ensures that jobs are available for all workers but also creates labor scarcity, which gives workers more bargaining power for decent wages.

One strategy for reducing unemployment is to make sure that current and future workers have the education and skills needed for employment. Efforts to prepare high school students for work include the establishment of technical and vocational high schools and high school programs and school-to-work programs. School-to-work programs involve partnerships between business, labor, government, education, and community organizations that allow high school students to explore different careers, and provide job skill training and work-based learning experiences (Bassi and Ludwig 2000; Leonard 1996).

Although educational attainment is often touted as the path to employment and economic security, more than a third of college-educated workers under age 25 are working at jobs that don't require a college degree (Luhby 2013). More than half of all U.S. jobs require only a high school diploma or less (Lockard and Wolf 2012); thus, it is important to ensure that all jobs pay a minimum "living wage" (see also Chapter 6). As long as our economy allows people who work full-time to earn poverty-level wages, having a job is not necessarily the answer to economic self-sufficiency. As David Shipler (2005) explained in his book *The Working Poor:*

> A job alone is not enough. Medical insurance alone is not enough. Good housing alone is not enough. Reliable transportation, careful family budgeting, effective parenting, effective schooling are not enough when each is achieved in isolation from the rest. There is no single variable that can be altered to help working people move away from the edge of poverty. Only where the full array of factors is attacked can America fulfill its promise. (p. 11)

Workforce Development. The federally funded **Workforce Investment Act (WIA)** awards stipends to some unemployed persons to help pay for further education and job training. However, the cost of further education and training often exceeds the stipend, and many unemployed who go through an education or job training program end up with massive debt and no job offer in their chosen field (Williams 2014). Some workforce development

full employment Exists when there are jobs available for all workers; usually when unemployment rates are below 5 percent.

Workforce Investment Act (WIA) Legislation passed in 1998 that provides a wide array of programs and services designed to assist individuals to prepare for and find employment.

programs focus on strategies to improve the employability of hard-to-employ individuals through providing targeted interventions such as substance abuse treatment, domestic violence services, prison release reintegration assistance, mental health services, and homelessness services, in combination with employment services (Martinson and Holcomb 2007).

The Workforce Innovation and Opportunity Act, implemented in 2015, provides English instruction and vocational training to immigrants. Programs that focus on integrating immigrants into the workforce are important, given that immigrants are projected to account for nearly all of the nation's labor force growth through 2050, when more than a third of the U.S. population is projected to be foreign born or the children of immigrants (Wilson 2015).

Job Creation and Preservation. In a national poll, one in four Americans said that the best way to create more U.S. jobs is to keep manufacturing jobs in the United States and stop sending work overseas (Newport 2011). Although both Democrat and Republican respondents cited keeping manufacturing jobs in the United States as the best way of creating jobs, Republicans also favored job creation strategies that involve reducing taxes and limiting government involvement in regulating business, whereas Democrats were more likely to favor using government to create jobs through infrastructure projects.

The 2010 Small Business Jobs Act provides tax breaks and better access to credit for small businesses so that small businesses can create new jobs. The 2010 Hiring Incentives to Restore Employment (HIRE) Act provided tax incentives to companies that hired employees who had been looking for work for 60 days or more.

The Global Jobs Pact. In 2009, the International Labour Organization adopted a Global Jobs Pact designed to guide national and international policies aimed at stimulating the economy, creating jobs, and providing protection to workers and their families. The Global Jobs Pact calls for a wide range of measures to retain workers, sustain businesses, create jobs, and provide social protections to workers and the unemployed. The pact also calls for more stringent supervision and regulation of the financial industry so that it better serves the economy and protects individuals' savings and pensions. The pact urges a number of measures, including (1) a shift to a low-carbon, environmentally friendly economy that will help create new jobs; (2) investments in public infrastructure; and (3) increases in social protection and minimum wages (to reduce poverty, increase demand for goods, and stimulate the economy). The Global Jobs Pact provides a vision for a healthier economy that meets the needs of workers and consumers. The hard work of translating the Global Jobs Pact into reality falls to employers, trade unions, and especially governments.

Worker Cooperatives: An Alternative to Capitalism

The global financial crisis that began in 2007 stimulated public discussions about the problems that stem from capitalism and renewed efforts to consider alternative business models. Economist Richard Wolff explains:

> We believe the capitalist organization of production has now finished its period of usefulness in human history. It is now no longer able to deliver the goods. It's bringing profits and prosperity to a tiny portion of the population, and delivering not the goods but the "bads" to most people. Jobs are steadily more insecure, unemployment is high[,] . . . benefits are increasingly being reduced, the prospects for our children are even worse, as more of them go deeper and deeper in debt to get the degrees that do not provide them with the jobs and incomes to get out of that debt. . . . It's longer overdue that we face honestly that the crisis we endure is the product of an economic system whose organization is something we should question, debate, and change. (quoted in Rampell 2013, n.p.)

One alternative to the capitalistic economic model is the cooperative workplace model that uses democratic methods in its business organization. **Worker cooperatives or workers' self-directed enterprises**—also known as "co-ops"—are businesses that are owned and democratically governed by their employees; workers participate in deciding what, how, and where to produce, and they would decide how to distribute the surpluses (or profits) generated in and by their enterprise. One of the most successful cooperative businesses is Mondragon, a corporation in Spain that started out as a small 7-member cooperative and, over 60 years, has grown into a corporation with over 10,000 workers. In 2013, the Mondragon Corporation won the Drivers of Change category at the annual Boldness in Business awards, sponsored by the London newspaper the *Financial Times*. At Mondragon, the workers hire the managers, the exact opposite of a capitalistic corporation. The workers decided to adopt a rule specifying that the highest-paid worker cannot be paid more than six and a half times what the lowest-paid worker earns, unlike many U.S. CEOs, who make up to a few hundred times more than the *average* worker (not even the lowest paid). Wolff is not surprised by this rule:

> If all the workers in any office, store, or factory, got together and had the power— which in a co-op they would—to decide what the wages and salaries of everybody are, do you think they'd give a handful of people at the top tens of millions of dollars, while everybody else is scrambling and unable to pay for their kids' college education, etc.? That's not going to happen. Even if you decide to pay some people more you're not going to live in the world of extreme inequality the way that's normal and typical for capitalism. . . . (quoted in Rampell 2013, n.p.)

Efforts to End Slavery, Child Labor, and Sweatshop Labor

In at least 25 countries, slave trafficking is actively prosecuted and treated as a serious crime. However, slave traffickers often avoid punishment because, as a former official of the U.S. Agency for International Development explained, "government officials in dozens of countries assist, overlook, or actively collude with traffickers" (quoted in Cockburn 2003, p. 16). In many countries, the justice system is more likely to jail or expel sex slaves than to punish traffickers ("Sex Trade Enslaves Millions of Women, Youth" 2003). In the United States, the Victims of Trafficking and Violence Protection Act, passed by Congress in 2000, protects slaves against deportation if they testify against their former owners. Convicted slave traffickers in the United States are subject to prison sentences.

In 1989, the General Assembly of the United Nations adopted the Convention on the Rights of the Child, which asserts the right that children should not be engaged in work deemed to be "hazardous or to interfere with the child's education, or to be harmful to the child's health." The International Labour Office has taken a leading role in enforcing these rights, leading efforts to prevent and eliminate child labor. Although almost every country has laws prohibiting the employment of children below a certain age, some countries exempt certain sectors—often the very sectors where the highest numbers of child laborers are found. Efforts to prevent child labor also focus on increasing penalties for violating child labor laws, which are often weak and poorly enforced.

Promoting Education for All—a global movement advocating that every child has access to free, basic education (primary school plus two or three years of secondary school)—is critical in the effort to reduce child labor because children with no access to education have little choice but to enter the labor market. In most countries, primary education is not free and parents must pay for costs such as uniforms and books. Parents who cannot afford these fees are also likely to view their children's labor as a necessary source of income to the household.

In the United States, the Children's Act for Responsible Employment (CARE Act) is a bill that proposes to raise the minimum age of farmworkers from 12 to 14 (except for children working on their parents' farm). The bill also contains provisions to protect young farmworkers from hazardous work and exposure to pesticides. As of this writing, the CARE Act has not passed.

worker cooperatives Democratic business organizations controlled by their members, who actively participate in setting their policies and making decisions; also known as *workers' self-directed enterprises*.

workers' self-directed enterprises See **worker cooperatives**.

Student protesters from United Students Against Sweatshops protest against Hanes' use of sweatshop labor.

Janet Mayer/Splash News/Newscom

Combating sweatshop labor conditions requires inspections and monitoring of factories and other workplaces to ensure compliance with labor standards. Favorable inspections, however, are not always accurate, given that employers sometimes know in advance that an inspector is coming. Corruption is also a problem, with some factory managers giving money to inspectors in exchange for a favorable report (Human Rights Watch 2015b). But inspecting and monitoring factories has resulted in exposing sweatshop conditions and the major companies that have contracts with these factories, leading to some improvements in working conditions. The Fair Labor Association (FLA) works to promote adherence to international labor standards and improve working conditions worldwide. More than 40 leading companies that sell brand-name apparel, footwear, and other goods voluntarily participate in FLA's factory monitoring system, which requires factories to meet minimum labor standards, such as not requiring workers to work more than 60 hours a week. In addition, nearly 200 colleges and universities require their collegiate licensees (companies that manufacture logo-carrying goods for colleges and universities) to participate in FLA's monitoring system.

Student Activism. United Students Against Sweatshops (USAS), formed in 1997, is a grassroots organization of youth and students who fight against labor abuses and for the rights of workers around the world, including campus workers, fast-food workers, and garment workers who make collegiate licensed apparel. USAS student activists have influenced more than 180 colleges and universities to affiliate with the Worker Rights Consortium (WRC), which investigates factories that produce clothing and other goods with school logos to make sure that the factory meets the code of conduct developed by each school. A typical code of conduct includes fair wages, a safe working environment, a ban on child labor, and the right to be represented by a union or other form of employee representation. If the WRC investigation finds that a factory fails to meet the code of conduct, the companies—often well-known international brands—who purchase items from that factory are warned that their contract with the school will be terminated if working conditions at the factory do not improve.

WHAT
do you
THINK?

> Do you know where the clothing with your college or university logo is made? Do you think most students care if the college or university logo clothing or products they buy are made under sweatshop conditions?

Sodexo—a profitable multinational food service company—serves food to more U.S. college students than any other company. In 2009, United Students Against Sweatshops began a Kick Out Sodexo campaign after discovering that Sodexo paid its campus food service and janitorial workers poverty-level wages and interfered with employees' attempts to form a union. Students who are active in the Kick Out Sodexo campaign have written letters to Sodexo, expressing their concerns about the company's labor practices, and have campaigned to convince college and university administrators to cancel their food service contracts with Sodexo and to do business instead with a food service company that meets certain labor standards.

Legislation. Perhaps the most effective strategy against sweatshop work conditions is legislation. Some U.S. states, cities, counties, and school districts have passed "sweat-free" procurement laws that prohibit public entities (e.g., schools, police, and fire departments) from purchasing uniforms and apparel made under sweatshop conditions (SweatFree Communities n.d.).

Establishing and enforcing labor laws to protect workers from sweatshop labor conditions is difficult in a political climate that offers more protections to corporations than it does to workers. Companies have demanded and won all sorts of intellectual property and copyright laws to defend their corporate trademarks, labels, and products. Yet, the corporations have long said that extending similar laws to protect the human rights of a 16-year-old girl in Bangladesh who sews the garment would be "an impediment to free trade." Under this distorted sense of values, the label is protected, but not the human being, the worker who makes the product (National Labor Committee 2007).

Responses to Workplace Health and Safety Concerns

In 2014, the Mine Safety and Health Administration (MSHA) issued new standards to reduce coal miners' exposure to coal dust that causes "black lung disease." But many other efforts to improve worker health and safety have stalled. The Protecting America's Workers Act (PAWA), introduced in Congress in 2009 and 2013, proposed to strengthen OSHA by extending coverage to uncovered workers, including state and local public employees; enhance whistle-blower protections so that workers who report workplace safety violations would have job protection; and increase penalties for serious and willful violations and in cases of worker death. The bill did not pass through Congress. Other proposed legislation would lower the acceptable level of worker exposure to silica dust; strengthen OSHA's authority to shut down operations that pose an imminent danger to workers; prohibit employer policies and practices that discourage injury reporting or retaliate against workers who report injuries; and improve reporting and tracking of workplace injuries and illnesses (AFL-CIO 2015). These and other initiatives to strengthen workplace health and safety are opposed by industry groups, and politicians who argue that excessive regulation is burdensome to business, hampers investment, and hurts job creation.

In developing countries, governments fear that strict enforcement of workplace regulations will discourage foreign investment. Investment in workplace safety in developing countries, whether by domestic firms or foreign multinationals, is far below that in the rich countries. Unless global standards of worker safety are implemented and enforced in all countries, millions of workers throughout the world will continue to suffer under hazardous work conditions. Low unionization rates and workers' fears of losing their jobs—or their lives—if they demand health and safety protections leave most workers powerless to improve their working conditions.

something else/Alamy

Behavior-Based Safety Programs. Instead of examining how work processes and conditions compromise health and safety on the job, **behavior-based safety programs** attribute workplace illnesses and injuries to workers' own carelessness and unsafe acts (Frederick and Lessin 2000). These programs focus on teaching employees and managers to identify, "discipline," and change unsafe worker behaviors that cause accidents and encourage a work culture that recognizes and rewards safe behaviors.

Critics contend that behavior-based safety programs divert attention away from the employers' failures to provide safe working conditions. They also say that the real goal of behavior-based safety programs is to discourage workers from reporting illness and injuries. Workers whose employers have implemented behavior-based safety programs describe an atmosphere of fear in the workplace, such that workers are reluctant to report injuries and illnesses for fear of being labeled "unsafe workers."

Work-Life Policies and Programs

Policies that help women and men balance their work and family responsibilities are referred to by a number of terms, including *work-family, work-life,* and *family-friendly* policies. As shown in Table 7.2, the United States lags far behind many other countries in national work-family provisions.

Federal and State Family and Medical Leave Initiatives. The **Family and Medical Leave Act (FMLA)**, signed into law in 1993, requires employers (with 50 or more employees who worked at least 1,250 hours in the preceding year) to provide up to 12 weeks of job-protected, *unpaid* leave so that they can care for a seriously ill child, spouse, or parent; stay home to care for their newborn, newly adopted, or newly placed foster child; or take time off when they are seriously ill. A 2008 amendment to the FMLA requires employers to provide up to 26 weeks of unpaid leave to employees to care for a seriously ill or injured family member who is in the armed forces, including the National Guard or Reserves. However, only about half of workers are eligible for FMLA benefits (Miller 2015). Some eligible workers do not use their FMLA benefit because they

behavior-based safety programs A strategy used by business management that attributes health and safety problems in the workplace to workers' behavior, rather than to work processes and conditions.

Family and Medical Leave Act (FMLA) A federal law that requires public agencies and companies with 50 or more employees to provide eligible workers with up to 12 weeks of job-protected, unpaid leave so that they can care for an ill child, spouse, or parent; stay home to care for their newborn, newly adopted, or newly placed foster child; or take time off when they are seriously ill, and up to 26 weeks of unpaid leave to care for a seriously ill or injured family member who is in the armed forces, including the National Guard or Reserves.

TABLE 7.2 How Do U.S. Work Policies Compare with Other Countries?

Policy	United States	Other Countries
Paid childbirth leave	No federal policy	168 countries offer paid leave to women; 98 countries offer 14 or more weeks of paid leave
Right to breast-feed at work	No federal policy	107 countries protect working women's right to breast-feed
Paid sick leave	No federal policy	145 countries provide paid sick leave
Paid annual leave	No federal policy	137 countries require employers to provide paid annual leave; 121 countries guarantee 2 weeks or more
Guaranteed leave for major family events (e.g., weddings, funerals)	No federal policy	49 countries guarantee leave for major family events (leave is paid in 40 countries)

SOURCE: Based on Heymann et al. 2007.

cannot afford to take leave without pay, and/or they fear they will lose their job if they take time off. Some employers do not comply with the FMLA either because they are unaware of their responsibilities under FMLA, or because they are deliberately violating the law. A national survey of employers found that one in five employers who are supposed to comply with FMLA fails to do so (Matos and Galinsky 2014).

Out of 185 countries, only two lack guaranteed paid maternity leave: the United States and Papua New Guinea (Walsh 2015). Several international human rights and labor treaties that call for paid maternity leave have not been ratified by the United Sates (Human Rights Watch 2015a). California, New Jersey, and Rhode Island offer workers paid family leave benefits that are financed through payroll taxes deductions (no employer contribution) and offer eligible employees a portion of their salary for four to six weeks. Opponents of paid family leave say it is an economic burden on businesses, but in California, most employers say paid leave has had no effect or a positive effect on productivity, turnover, and morale. (Miller 2015). A 2015 bill introduced in Congress—the Family and Medical Insurance Leave Act—would create a national paid family and medical leave program enabling workers to receive partial pay (66 percent of wages, up to a cap) for up to 12 weeks to care for new children, ill family members, or their own health problems.

As shown in Table 7.2, there is no U.S. federal policy requiring employers to provide workers with any paid sick leave; however, Connecticut, California, Massachusetts, as well as several cities have passed laws requiring some employers (e.g., with 50 employees or more) to provide paid sick leave to their workers. The Healthy Families Act proposes to establish a federally mandated paid sick day policy. Employees could take the benefit if they or a family member is sick. Supporters of the Healthy Families Act point out that, when sick employees go to work because they cannot afford to take unpaid sick days or fear losing their job by taking a sick day, they risk spreading infectious diseases at the workplace. And sending sick children to school because their parents cannot afford to take unpaid sick days to stay home with them risks spreading infectious diseases at school.

Employer-Based Work/Life Policies. Aside from government-mandated work–family policies, some corporations and employers have "family-friendly" work policies and programs, including unpaid or paid family and medical leave, child care assistance, assistance with elderly parent care, and flexible work options such as allowing workers to control when they start and end their workday (see Table 7.3).

Efforts to Strengthen Labor

Just over half of Americans (53 percent) approve of labor unions, down from 75 percent in the 1950s (Jones 2014a). The lack of widespread support for labor unions is a challenge for those who want to strengthen labor unions in order to remedy many of the problems facing workers.

In an effort to strengthen their power, some labor unions have merged with one another. Labor union mergers result in higher membership numbers, thereby increasing the unions' financial resources, which are needed to recruit new members and to withstand long strikes. Because workers must fight for labor protections within a globalized economic system, their unions must cross national boundaries to build international cooperation and solidarity. Otherwise, employers can play working and poor people in different countries against each other.

Strengthening labor unions requires combating the threats and violence against workers who attempt to organize or who join unions. One way to do this is to pressure

TABLE 7.3 Employer-Based Work/Life Benefits and Policies, U.S., 2014

Benefit or Policy	Percentage of Employers* That Provide Benefit/Policy
Full pay for maternity leave	5%
Private space for breast-feeding	74%
Some paid time off for spouses/ partners of women who give birth	14%
Health insurance for unmarried partners	43%
Allow some employees to periodically change their arrival and departure time	81%
Provide information about elder care services	43%
Child care at or near worksite	7%
Dependent Care Assistance Programs that help employees pay for child care with pretax dollars	61%

*Employers with 50 or more employees.
SOURCE: Matos and Galinsky 2014.

governments to apprehend and punish the perpetrators of such violence. Another tactic is to stop doing business with countries where government-sponsored violations of free trade union rights occur.

Understanding Work and Unemployment

On December 10, 1948, the General Assembly of the United Nations adopted and proclaimed the Universal Declaration of Human Rights. Among the articles of that declaration are the following:

> Article 23. Everyone has the right to work, to free choice of employment, to just and favourable conditions of work and to protection against unemployment.
>
> Everyone, without any discrimination, has the right to equal pay for equal work.
>
> Everyone who works has the right to just and favourable remuneration ensuring for himself and his family an existence worthy of human dignity, and supplemented, if necessary, by other means of social protection.
>
> Everyone has the right to form and to join trade unions for the protection of his interests.
>
> Article 24. Everyone has the right to rest and leisure, including reasonable limitation of working hours and periodic holidays with pay.

More than a half century later, workers around the world are still fighting for these basic rights as proclaimed in the Universal Declaration of Human Rights.

To understand the social problems associated with work and unemployment, we must first recognize that corporatocracy—the ties between government and corporations—serves the interests of corporations over the needs of workers. Although some people argue that the growth of multinational corporations brings economic growth, jobs, lower prices, and quality products to consumers throughout the world, others view global corporations as exploiting workers, harming the environment, dominating public policy, and degrading cultural values.

Decisions made by U.S. corporations about what and where to invest influence the quantity and quality of jobs available in the United States. As conflict theorists argue, such investment decisions are motivated by profit, which is part of a capitalist system. Profit is also a driving factor in deciding how and when technological devices will be used to replace workers and increase productivity. If goods and services are produced too efficiently, however, workers are laid off and high unemployment results. When people have no money to buy products, sales slump, recession ensues, and social welfare programs are needed to support the unemployed. When the government increases spending to pay for its social programs, it expands the deficit and increases the national debt. Deficit spending and a large national debt make it difficult to recover from the recession, and the cycle continues.

What can be done to break the cycle? Those adhering to the classic view of capitalism argue for limited government intervention on the premise that business will regulate itself by means of an "invisible hand" or "market forces." But Americans are growing increasingly skeptical of the notion that big business, which has caused many of the economic problems in our country and throughout the world, can solve the very problems it creates. New models of business, such as the workplace cooperative or workers' self-directed enterprises, offer visions of what the workplace could look like. The Occupy Wall Street social movement (see also Chapter 6) is part of a growing consciousness that job insecurity, worker exploitation, and extreme levels of wage inequality are not consistent with the American dream; there must be a better way. Participants in the Occupy Wall Street protests include students, the unemployed, union members, professionals, and others who are fed up with corporate greed and the widening gap between the rich and the poor. They are frustrated by high unemployment and the erosion of workers' salaries, benefits, and rights. Although this growing movement has not formulated any specific demands, what is certain is that Americans want change in the economy and in the workplace. Many of us spend much of our time at our jobs, working. And how we spend our time is how we spend our lives.

Chapter Review

- **The economy has become globalized. What does that mean?**
  In recent decades, innovations in communication and information technology have led to the globalization of the economy. The global economy refers to an interconnected network of economic activity that transcends national borders and spans the world. The globalized economy means that our jobs, the products and services we buy, and our nation's economic policies and agendas influence and are influenced by economic activities occurring around the world.

- **What are some of the criticisms of capitalism?**
  Capitalism is criticized for creating high levels of inequality; economic instability; job insecurity; pollution and depletion of natural resources; and corporate dominance of media, culture, and politics. Capitalism is also criticized as violating the principles of democracy by allowing (1) private wealth to affect access to political power; (2) private owners of property to make decisions that affect the public (such as when the owner of a factory decides to move the factory to another country); and (3) workplace dictatorships where workers have little say in their working conditions, thus violating the democratic principle that people should participate in collective decisions that significantly affect their lives.

- **What are transnational corporations?**
  Transnational corporations are corporations that have their home base in one country and branches, or affiliates, in other countries.

- **How do structural functionalism, conflict theory, and symbolic interactionism view work and the economy?**
  The structural-functionalist perspective focuses on the functions and dysfunctions of the economic institution and ways in which the economy affects and is affected by changes in society. According to the conflict perspective, the ruling class controls the economic system for its own benefit and exploits and oppresses the working masses. The conflict perspective is critical of ways that the government caters to the interests of big business at the expense of workers, consumers, and the public interest. According to symbolic interactionism, the work role is an important part of a person's self-concept and social identity. Symbolic interactionism emphasizes that job-related attitudes and behaviors are influenced by social interaction.

- **What are some of the causes of unemployment?**
  Causes of unemployment include not enough jobs, job exportation or "offshoring" (the relocation of jobs to other countries), automation (the replacement of human labor with machinery and equipment), increased global competition, and mass layoffs as plants close and companies downsize or go out of business.

- **Does slavery still exist today? If so, where?**
  Slavery exists today all over the world, including the United States. Most forced laborers work in agriculture, mining, prostitution, and factories.

- **Why are reported statistics on injuries and illnesses in U.S. workplaces inaccurately low?**
  Government data on workplace illness and injury exclude many categories of workers. Long-term illnesses caused by, for example, exposure to carcinogens often are difficult to relate to the workplace and are not adequately recognized and reported. Employees don't always report workplace injuries or illnesses for fear of being disciplined, losing their jobs, or fear of being deported. Employers discourage reporting of injuries by offering financial rewards to workers and departments when no injuries are reported for a specified period of time. And employers don't always keep accurate records because they fear increased workers' compensation costs, increased scrutiny from government workplace inspectors, and fines for possible violations of health and safety regulations.

- **According to a Gallup Poll, what is the most common complaint among U.S. workers?**
  A Gallup Poll found that job stress is the most common complaint of workers (followed by the amount of money they earn).

- **What is work/life conflict?**
  Work/life conflict is the day-to-day struggle to simultaneously meet the demands of work and other life responsibilities, including family, education, exercise, and recreation.

- **From the point of view of workers, what are the advantages and disadvantages of labor unions?**
  Labor unions have benefitted workers by fighting for fair wages and benefits, healthy and safe work environments, and other forms of worker advocacy. Compared with non-union workers, unionized workers tend to have higher earnings, better insurance and pension benefits, more paid time off, and higher life satisfaction.

  One disadvantage of unions is that members must pay dues and other fees. Another disadvantage for unionized workers is the loss of individuality. Unionized workers are members of an overall bargaining unit in which the majority rules. Decisions made by the majority may conflict with individual employees' specific employment needs.

- **What are worker cooperatives?**
  Worker cooperatives are businesses that are owned and democratically governed by their employees.

- **What is the federal Family and Medical Leave Act?**
  The 1993 Family and Medical Leave Act (FMLA) requires all companies with 50 or more employees to provide eligible workers with up to 12 weeks of job-protected, unpaid leave so that they can care for a seriously ill child, spouse, or parent; stay home to care for their newborn, newly adopted, or newly placed foster child; or take time off when they are seriously ill. The United States is the only advanced nation that does not mandate paid leave for family and medical reasons.

Test Yourself

1. According to this text, corporations outsource labor to other countries so that U.S. consumers can save money on cheaper products.
   a. True
   b. False

2. Which of the following statements is/are true?
   a. The U.S. economy is a mix of capitalism and socialism.
   b. Communism is more a political system than an economic system.
   c. Both a and b
   d. Neither a nor b

3. The unemployment rate for young college graduates tends to be _____ the overall unemployment rate.
   a. half that of
   b. twice that of
   c. about the same as
   d. slightly less than

4. Under federal law, willful violation that results in a worker's death is considered which of the following?
   a. A misdemeanor with a maximum prison sentence of 1 year for a first conviction
   b. A misdemeanor with a minimum prison sentence of 6 years for a first conviction
   c. A felony with a maximum prison sentence of 26 years for a first conviction
   d. A felony with a minimum prison sentence of 6 years for a first conviction

5. The United States is the only advanced nation that does not mandate a minimum number of vacation days.
   a. True
   b. False

6. Which of the following is *not* one of the four principles of McDonaldization?
   a. Affordability
   b. Predictability
   c. Calculability
   d. Efficiency

7. In states with "right-to-work" laws, workers are allowed to work at unionized workplaces
   a. only if they pay union dues.
   b. only if they agree not to go on strike.
   c. without having to join the union and pay union dues.
   d. both a and b

8. Worker cooperatives are businesses that are owned by
   a. investors.
   b. managers.
   c. workers.
   d. banks and other financial institutions.

9. The Children's Act for Responsible Employment (CARE Act) proposes to raise the minimum age of children working _____ from 12 to 14.
   a. on farms
   b. as babysitters
   c. delivering newspapers
   d. in restaurants

10. As of 2015, the United States did not guarantee workers paid maternity leave.
   a. True
   b. False

Answers: 1. B; 2. C; 3. B; 4. A; 5. A; 6. A; 7. C; 8. C; 9. A; 10. A.

Key Terms

3-D printing   221
automation   221
behavior-based safety programs   238
capitalism   215
chattel slavery   224
child labor   225
communism   215
corporatocracy   218
economic institution   214
Family and Medical Leave Act
   (FMLA)   238

forced labor   224
free trade agreement (FTA)   216
full employment   233
global economy   214
job burnout   229
job exportation   221
labor unions   231
long-term unemployment   222
McDonaldization   231
offshoring   221
outsourcing   221

recession   221
slavery   224
socialism   215
sweatshops   225
transnational corporations   217
unemployment   220
union density   232
worker cooperatives   235
workers' self-directed enterprises   235
Workforce Investment Act (WIA)   233
work/life conflict   230

Bloomberg/Getty Images

> 66 Schools need stability, adequate resources, well-prepared and experienced educators, community support, and a clear vision of what good education is."
>
> **DIANE RAVITCH**
> education historian

8

Problems in Education

Learning Objectives

After studying this chapter, you will be able to . . .

1 Identify educational trends in countries that participate in the Organisation for Economic Cooperation and Development (OECD).

2 Use structural functionalism and conflict theory to analyze the presence of corporate sponsors in public schools.

3 Discuss each of the independent variables associated with school achievement.

4 Describe the social problems (e.g., the dropout rate) associated with schools, teachers, and students.

5 Summarize educational policies and practices in the United States.

MARC MAGELLAN, a music professor, is considered by many to be one of the best teachers at Miami-Dade College with high marks on ratemyprofessor. com to prove it. One evening, while walking through a campus parking garage, Professor Magellan heard someone call his name. As he turned around, he was immediately punched in the face and, after falling to the ground, had his head repeatedly pounded against the concrete floor (Collman 2014). Nothing was stolen, but the professor suffered from a broken nose, broken hand, various bruises, and other injuries to his skull, face, arms, knees, and feet. Although no arrests have been made, a detective for the Miami-Dade Police Department concluded that the assault was the "result of a student being ejected from the professor's class just days earlier" (Herrera 2014, p. 1).

The assault of Professor Magellan represents just one of the many problems in education today—school violence. Too often, graduates are ill prepared for the 21st-century workplace. Frustrated, teachers leave the profession for better salaries and the fulfillment they once thought teaching would provide. The public questions whether their children are safe at school, and dropout rates for minority and poor students remain high. Solutions to these problems seem out of reach as budgets are slashed, political wrangling over the best way to "fix" schools heats up, and an ever-growing student population puts additional strain on school resources.

Yet, education is often claimed as a panacea—the cure-all for poverty and prejudice, drugs and violence, war and hatred, and the like. Can one institution, riddled with problems, be a solution for other social problems? In this chapter, we focus on this question and on what is being called the educational crisis. We begin with a look at education around the world.

> Nearly 17 percent of adults cannot read or write—775 million people, two-thirds of them women.

The Global Context: Cross-Cultural Variations in Education

Looking only at the American educational system might lead one to conclude that most societies have developed some method of formal instruction for their members. After all, the United States has more than 140,000 schools, 4.6 million primary and secondary schoolteachers and college faculty, 3.1 million public school administrators and support staff, and 77.8 million students (National Center for Education Statistics [NCES] 2015a, 2014c). In reality, many societies have no formal mechanism for educating the masses. Nearly 17 percent of adults in the world cannot read or write—775 million people, two-thirds of them women. An estimated 122 million children in the world are illiterate and, according to the most recent data available, 57.8 million out-of-school children are of elementary school age (UNESCO 2014a, 2014b) (see Figure 8.1).

Education at a Glance, a publication of the Organisation for Economic Cooperation and Development (OECD), reports education statistics on more than 34 countries (OECD 2014). Some interesting findings are revealed. First, in general, educational levels are rising. For example, many countries saw an increase in the proportion of young people attending colleges and universities. However, large numbers of people do not graduate from tertiary institutions. On the average, just 33.3 percent of all adults attain a college degree.

WHAT do you THINK?

Education for children under the age of 5 is not a priority in many low- and middle-income countries. However, *Sesame Street* is broadcast in over 130 countries around the world with millions of children watching daily. In a meta-analysis of over 20 studies encompassing 10,000 children in 15 countries, Mares and Pan (2013) report that children with higher exposure to *Sesame Street* had higher learning outcomes (e.g., number recognition). Did you watch *Sesame Street,* and if so, did it have an impact on your life?

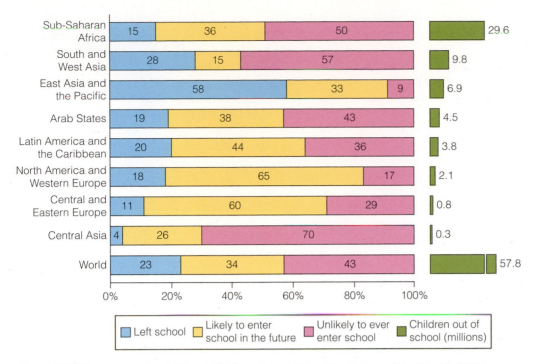

Figure 8.1 Elementary-Aged School Children Out-of-School by Region of the World, 2012*
*Or most recent data available.
SOURCE: UNESCO 2014b.

Second, there is a clear link between education and income, and between education and employment. In general, across the OECD-participating countries, the more education people have, the higher their income and the greater the likelihood they will be employed. For example, the unemployment rates of college-educated people is nearly half of those with less than a high school degree. Furthermore, investment in a college degree has an **"earnings premium"**; that is, the benefits of having a college degree far outweigh the cost of getting one. Today, the average U.S. college graduate recoups their investment in about 10 years and has a lifetime earnings premium of approximately $300,000 (Abel and Deitz 2014).

Third, across all OECD-participating countries, an average of $9,487 is spent per student each year they are in school, from elementary school to college. Spending, although increasing in general, varies widely by country. In Brazil, Indonesia, Mexico, and Turkey, less than $4,000 is spent per student per year. However, Australia, France, Japan, Sweden, Ireland, and the United Kingdom spend over $10,000, and Switzerland and the United States spend over $15,000, on the average, per student per year (OECD 2014).

Fourth, in reference to teachers, the average student–teacher ratio in elementary schools in OECD-participating countries is less than 16:1, but it ranges from a high of 28:1 in Mexico to 11:1 or less in Poland, Norway, and Saudi Arabia. Average elementary school class sizes range from a high of 52 students in China to a low of 17 students in Luxembourg (OECD 2014).

Globally, few countries have the quality of schools or accessibility to education as enjoyed in the United States. Here, students in Lahtora, India, have outdoor lessons because their classroom is too crowded and too cold to have classes inside.

earnings premium The benefits of having a college degree far outweigh the cost of getting one.

Teachers' salaries represent the single largest category in educational spending. Between 2008 and 2012, teachers' salaries rose in most countries with the notable exceptions of Hungary and Italy. In general, the higher the level of class taught, the higher the teacher's salary. Despite the fact that public school teachers in the United States, on the average, earned $56,382 in 2013 (NCES 2015b), near the OECD average, teachers in the United States work more hours than in any of the other OECD-participating countries. For example, elementary school teachers in the United States, on the average, teach 1,131 hours annually compared to the OECD average of 782 annual work hours (Klein 2014).

Fifth, educational attainment in OECD-participating countries has increased over the last decade. The percentage of people without a high school education has declined, and the percentage of people with a college education has grown. However, generational and regional differences exist. For example, more than 40 percent of 25- to 34-year-olds in OECD countries have college degrees, but fewer than 25 percent of 55- to 64-year-olds have completed college degrees. In the United States, the difference between the two age groups is slightly more than 1 percent; in Korea, over 50 percent (OECD 2014).

Finally, the Program for International Student Assessment (PISA) is designed "to evaluate education systems worldwide by testing the skills and knowledge of 15-year-old students" (OECD 2013, p. 1). A random sample of more than 510,000 15-year-old students from 65 countries (34 OECD countries and 31 partner countries) participate in the PISA. The results of the most recent study indicate that of the OECD-participating countries, students in Shanghai–China scored the highest in science and mathematics, and students in Hong Kong–China, Japan, and Singapore the highest in reading (PISA 2014). Furthermore, in the Trends in International Mathematics and Science Study (TIMSS), which uses data from more than 50 countries, the United States scored higher than the international TIMSS average for both 4th and 8th graders in mathematics and science (Kastberg et al. 2013).

Sociological Theories of Education

The three major sociological perspectives—structural functionalism, conflict theory, and symbolic interactionism—are important in explaining different aspects of American education.

Structural-Functionalist Perspective

According to structural functionalism, the educational institution serves important tasks for society, including instruction, socialization, the sorting of individuals into various statuses, and the provision of custodial care (Sadovnik 2004). Many social problems, such as unemployment, crime and delinquency, and poverty, can be linked to the failure of the educational institution to fulfill these basic functions (see Chapters 4, 6, and 7). Structural functionalists also examine the reciprocal influences of the educational institution and other social institutions, including the family, political institutions, and economic institutions.

Instruction. A major function of education is to teach students the knowledge and skills that are necessary for future occupational roles, self-development, and social functioning. Although some parents teach their children basic knowledge and skills at home, most parents rely on schools to teach their children to read, spell, write, tell time, count money, and use computers. As discussed later, many U.S. students display low levels of academic achievement. The failure of many schools to instruct students in basic knowledge and skills both causes and results from many other social problems.

Socialization. The socialization function of education involves teaching students to respect authority—behavior that is essential for social organization (Merton 1968). Students learn to respond to authority by asking permission to leave the classroom, sitting quietly at their desks, and raising their hands before asking a question. Students

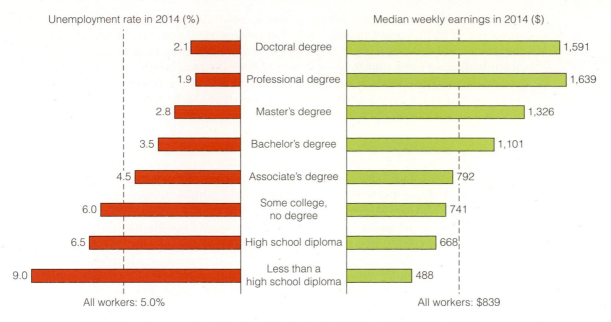

Unemployment rate in 2014 (%)

Median weekly earnings in 2014 ($)

	Unemployment rate	Median weekly earnings
Doctoral degree	2.1	1,591
Professional degree	1.9	1,639
Master's degree	2.8	1,326
Bachelor's degree	3.5	1,101
Associate's degree	4.5	792
Some college, no degree	6.0	741
High school diploma	6.5	668
Less than a high school diploma	9.0	488

All workers: 5.0%   All workers: $839

Note: Data are for persons age 25 and over. Earnings are for full-time wage and salary workers.

Figure 8.2 Unemployment Rate and Median Weekly Earnings for Individuals 25 and Over by Highest Level of Education, 2014
SOURCE: Bureau of Labor Statistics 2015.

who do not learn to respect and obey teachers may later disrespect and disobey employers, police officers, and judges.

The educational institution also socializes youth into the dominant culture. Schools attempt to instill and maintain the norms, values, traditions, and symbols of the culture in a variety of ways, such as celebrating holidays (e.g., Martin Luther King Jr. Day and Thanksgiving); requiring students to speak and write in standard English; displaying the American flag; and discouraging violence, drug use, and cheating.

Some have described schools as "spaces of cultural conflict where different values, beliefs, and norms create discord" (Fraise and Brooks 2015, p. 6). This is particularly true as the size of racial and ethnic minority groups continues to increase. American schools are thus faced with a dilemma: Should they promote only one common culture, or should they emphasize the cultural diversity reflected in the U.S. population? Some evidence suggests that most Americans believe that schools should do both—they should promote one common culture and emphasize diverse cultural traditions. **Multicultural education**—that is, education that includes all racial and ethnic groups in the school curriculum—promotes awareness and appreciation for cultural diversity and, ideally, leads to **cultural competence** among teachers and students alike (also see Chapter 9).

Sorting Individuals into Statuses. Schools sort individuals into statuses by providing credentials for individuals who achieve various levels of education at various schools within the system. These credentials sort people into different statuses—for example, "high school graduate," "Harvard alumna," and "English major." In addition, schools sort individuals into professional statuses by awarding degrees in fields such as medicine, engineering, and law. The significance of such statuses lies in their association with occupational prestige and income—the higher one's education, the higher one's income. Furthermore, unemployment rates and earnings are tied to educational status, as seen in Figure 8.2.

Custodial Care. The educational system also serves the function of providing custodial care by providing supervision and care for children and adolescents until they are 18 years old (Merton 1968). Despite 12 years and almost 13,000 hours of instruction, some school

multicultural education
Education that includes all racial and ethnic groups in the school curriculum, thereby promoting awareness and appreciation for cultural diversity.

cultural competence
In reference to education, refers to having an awareness of and appreciation for different cultural groups in order to inform student–teacher and student–student interaction.

districts are increasing the number of school hours and/or days beyond the "traditional" schedule. In 2013, Arne Duncan, the U.S. secretary of education and an advocate for a longer school year, argued that "whether educators have more time to enrich instruction or students have more time to learn how to play an instrument and write computer code, adding meaningful in-school hours is a critical investment that better prepares children to be successful in the 21st century" (quoted in Smyth 2013). To date, more than 1,000 U.S. schools have extended their academic year.

Conflict Perspective

Conflict theorists emphasize that the educational institution solidifies the class positions of groups and allows the elite to control the masses. Although the official goal of education in society is to provide a universal mechanism for achievement, in reality, educational opportunities and the quality of education are not equally distributed.

Conflict theorists point out that the socialization function of education is really indoctrination into a capitalist ideology (Sadovnik 2004). In essence, students are socialized to value the interests of the state and to function to sustain it. Such indoctrination begins in early childhood education. Rosabeth Moss Kanter (1972) coined the term the *organization child* to refer to the child in nursery school who is most comfortable with supervision, guidance, and adult control. Teachers cultivate the organization child by providing daily routines and rewarding those who conform. In essence, teachers train future bureaucrats to be obedient to authority.

In addition, to conflict theorists, education serves as a mechanism for **cultural imperialism**, or the indoctrination into the dominant culture of a society. When cultural imperialism exists, the norms, values, traditions, histories, and languages of minorities have been historically ignored. A European-Mexican American recalls the richness of European history and that the little time spent on ancient Mexico and Central American societies was

> largely spent establishing their many evils: Blood sacrifices to hungry gods, an insatiable lust for power and conquest, and a morbid, crude fascination with death. . . . The text compared these primitive societies with the more advanced Europeans, making list of things they didn't have. They didn't have beasts of burden. They didn't have guns. They didn't have steel. And, of course, they didn't have wheels. (Brammer 2014, p. 1)

The conflict perspective also focuses on what Kozol (1991) called the "savage inequalities" in education that perpetuate racial disparities. Kozol documented gross inequities in the quality of education in poorer districts, largely composed of minorities, compared with districts that serve predominantly white middle-class and upper-middle-class families. For example, in high-concentration poverty schools, California teachers report that such chronic conditions as insufficiently qualified substitutes; noisy, dirty, and disrupted classrooms; inadequate access to the school library; nonteaching duties (e.g., janitorial tasks); and a lack of computers reduce the average instructional time for poor students by two weeks when compared to students in low-concentration poverty schools (Rogers and Mirra 2014).

Kozol also revealed that schools in poor districts tend to receive less funding and to have inadequate facilities, books, materials, equipment, and personnel. It should be noted, however, that such inequities exist in other countries as well as evidenced by, for example, per-pupil expenditures in developed versus developing countries (OECD 2014).

Symbolic Interactionist Perspective

Whereas structural functionalism and conflict theory focus on macro-level issues, such as institutional influences and power relations, symbolic interactionism examines education from a micro-level perspective. This perspective is concerned with individual and small-group issues, such as teacher–student interactions and the self-fulfilling prophecy.

cultural imperialism The indoctrination into the dominant culture of a society.

To cover financial costs associated with the increasing needs of educational programs, school systems often find it necessary to contract with major corporations such as Coca-Cola. However, a recent analysis of the impact of commercialization in schools concluded that "the potential threat to children posed by marketing in schools is great enough that . . . the default assumption must be that marketing in schools is harmful unless explicitly proven otherwise" (National Education Policy Center 2013).

Teacher–Student Interactions. Symbolic interactionists examine the consequences of the social meaning conveyed in teacher–student interactions. From the teachers' point of view, middle-class students may be defined as easy and fun to teach. They grasp the material quickly, do their homework, and are more likely to "value" the educational process. Students from economically disadvantaged homes often bring fewer social and verbal skills to those same middle-class teachers, who may, inadvertently, hold up social mirrors of disapproval. Teacher disapproval contributes to lower self-esteem among disadvantaged youth.

The differences in interactions may be subtle and are often unconscious. Consider the impact of two only slightly different statements: "You only have a B?" (message: you can do better) versus "I'm so proud of you; you got a B!" (message: you can't do much better) (Anderson 2014; Avery 2014). Because students from economically advantaged homes are more likely to bring to the classroom social and verbal skills that elicit approval from teachers, they may be more likely to invoke the former, whereas students from economically disadvantaged homes may invoke the latter.

Self-Fulfilling Prophecy. The example above may lead to a **self-fulfilling prophecy,** a tendency for people to act in a manner consistent with the expectations of others. If I'm already working to my potential, there's no need for me to try harder. Teachers who define a student as a slow learner may be less likely to call on that student or to encourage the student to pursue difficult subjects. The teacher may also be more likely to assign the student to lower-ability groups or curriculum tracks (Riehl 2004). As a consequence of the teacher's behavior, the student is more likely to perform at a lower level.

self-fulfilling prophecy
A concept referring to the tendency for people to act in a manner consistent with the expectations of others.

A classic study by Rosenthal and Jacobson (1968) provides empirical evidence of the self-fulfilling prophecy in the public school system. Five elementary school students in a San Francisco school were selected at random and identified for their teachers as "spurters." Such a label implied that they had superior intelligence and academic ability. In reality, they were no different from the other students in their classes. At the end of the school year, however, these five students scored higher on their intelligence quotient (IQ) tests and made higher grades than their classmates who were not labeled as spurters. In addition, the teachers rated the spurters as more curious, interesting, and happy, and more likely to succeed than the nonspurters. Because the teachers expected the spurters to do well, they treated the students in a way that encouraged better school performance.

For years, ability grouping—which takes place within classes, in contrast to tracking, which takes place between classes—has been criticized for its potential negative effect on students and, consequently, went out of favor several decades ago (Loveless 2013). The fear, still held by some today, was that ability grouping not only reflects differences in race and class but also contributes to those differences; that is, it perpetuates inequality. Nonetheless, a recent study by the Brown Center on Education Policy reports a resurgence of grouping students based on ability. For example, using data from the United States, 61 percent of fourth grade teachers and 76 percent of eighth grade teachers report creating math groups based on student achievement (Loveless 2013).

Who Succeeds? The Inequality of Educational Attainment

Figure 8.3 shows the extent of the variation in highest level of education attained by individuals 25 years of age and older in the United States. As noted earlier, conflict theory focuses on such variations in discussions of educational inequalities. Educational inequality is based on social class and family background, race and ethnicity, and gender. Each of these factors influences who succeeds in school.

Social Class and Family Background

One of the best predictors of educational success and attainment is socioeconomic status. Children whose families are in middle and upper socioeconomic brackets are more likely to perform better in school and to complete more years of education than children from families of lower socioeconomic classes. For example, 15-year-old students from disadvantaged backgrounds in OECD countries are one and half times more likely to repeat a grade than their advantaged counterpart (Avvisati 2014).

Globally, in general, children of professionals outperform children of manual workers. That said, there are exceptions. In reference to mathematics performance, the children of cleaners in Shanghai–China outperform the children of professionals in the United States. Although there is a strong relationship between parents' occupations and students' academic performance, "the fact that students in some education systems, regardless of what their parents do for a living, outperform children of professionals in other countries [indicates] . . . that it is possible to provide children of factory workers the same high-quality education opportunities that children of lawyers and doctors enjoy" (Biecek and Borgonovi 2014, p. 4).

On standardized tests such as the SAT and the ACT, "children from the lowest-income families have the lowest average test scores, with an incremental rise in family income associated with a rise in test scores" (Corbett et al. 2008, p. 3). Muller and Schiller (2000) report that students from higher socioeconomic backgrounds are more likely to enroll in advanced courses for mathematics credit and to graduate from high school—two indicators of future educational and occupational success. In addition, a report to Congress on educational inequality acknowledges that students from high-income families compared to low- and middle-income families are not only more likely to

One of the best predictors of educational success and attainment is socioeconomic status. Children whose families are in middle and upper socioeconomic brackets are more likely to perform better in school and to complete more years of education than children from families of lower socioeconomic classes.

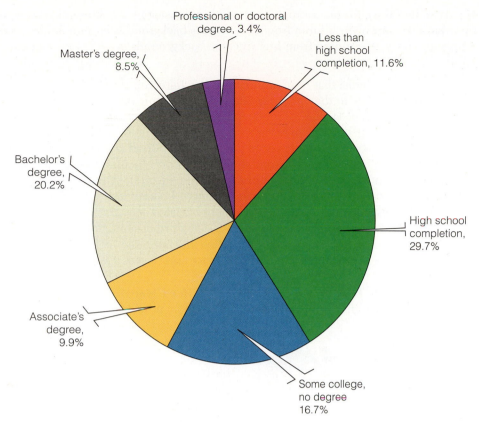

Professional or doctoral degree, 3.4%

Master's degree, 8.5%

Less than high school completion, 11.6%

Bachelor's degree, 20.2%

High school completion, 29.7%

Associate's degree, 9.9%

Some college, no degree 16.7%

Note: Percentages do not sum to 100 percent because of rounding.

Figure 8.3 Highest Level of Education Attained by Individuals Age 25 Years Old and Older, 2014
SOURCE: U.S. Census Bureau 2015.

attend college, they are also more likely to complete it (Advisory Committee on Student Financial Assistance 2010).

WHAT do you THINK?

> Reardon (2013) argues that the growing achievement gap is a consequence of "rich students . . . increasingly entering kindergarten much better prepared to succeed in school" (p. 1), thus creating a "rug rat race" (Ramey and Ramey 2010). The president's 2016 budget includes a plan to provide "all four-year-olds from low- and moderate-income families with access to high-quality preschool" (Office of Management and Budget 2015). Do you think the federal government should pay for universal preschool for all 4-year-olds?

In one of the most significant studies to date on the relationship between family background and educational success, sociologist Carl Alexander and his colleagues followed nearly 800 Baltimore students from 1982 when they entered first grade until young adulthood (Alexander et al. 2014). Over the course of 25 years, the researchers obtained school records and interviewed and surveyed the children, their parents, and teachers. Research findings include the following:

- Only 4 percent of children from low-income backgrounds completed college compared to 45 percent of children from high-income backgrounds.
- Of those who did not attend college, white men from low-income backgrounds secured higher-paying blue-collar jobs than black men from low-income backgrounds.
- White women from low-income backgrounds had higher incomes than black women from low-income backgrounds.

- Nearly half of the children sampled remained in the same socioeconomic status as their parents—only 33 children from low-income backgrounds became high-income adults and only 19 children from high-income backgrounds became low-income adults.

The lead author concludes that the "implication is where you start in life is where you end up in life," contrary to the "popular ethos that we are makers of our own fortune" (quoted in Rosen 2014, p. 1).

Families with low incomes have fewer resources to commit to educational purposes—less money to buy books or computers or to pay for tutors. Disadvantaged parents are less involved in learning activities. As parental education and income levels increase, the likelihood of a parent taking a child to a library, play, concert or other live show, art gallery, museum, or historical site also increases. Parents who have less education and lower income levels are less likely to be involved in school-related activities such as attending a school event or volunteering at a school (NCES 2011). Furthermore, low-income students are less likely to be engaged in the kind of activities that build the "soft skills" (e.g., teamwork) necessary to be successful as an adult (Venator and Reeves 2015).

Head Start and Early Head Start. In 1965, Project **Head Start** began to help preschool children from the most disadvantaged homes. Head Start provides an integrated program of health care, parental involvement, education, and social services for qualifying children. In 2014, nearly 1 million children, birth to age five, and pregnant women in 50 states participated in the program. Pregnant women comprise only 1 percent of the serviced group, whereas 81 percent were 3- and 4-year-old preschoolers (Head Start 2015).

Assessments of Head Start and Early Head Start, a program for infants and toddlers from low-income families, generally conclude that children and families benefit from inclusion in the programs. There are, however, some concerns as to whether those benefits are sustainable. Results from a "large-scale, randomized, controlled study of nearly 5,000 children from low-income families . . . found that the positive effects on literacy and language development demonstrated by children who entered Head Start at age 4 had dissipated by the end of the third grade" (Maxwell 2013, p. 1).

Students get in line for breakfast at the start of class rather than before class so that students who are late won't go hungry. Research indicates that hungry students are more likely to perform poorly, lack concentration, and often act out in class.

AP Images/Trent Nelson

School District Funding. Local dollars make up 44 percent of school funding but varies dramatically by district based on socioeconomic variables (Federal Education Budget Project [FEBP] 2014). Local expenditures on schools come from taxes, usually property taxes, and as housing prices decline, property taxes decline. The U.S. system of decentralized funding for schools has several additional consequences.

First, school districts with low socioeconomic status are more likely to be in urban areas and in inner cities where the value of older and dilapidated houses has depreciated; less desirable neighborhoods are hurt by "white flight," with the result that the tax base for local schools is lower in deprived areas. Second, school districts with low socioeconomic status are less likely to have businesses or retail outlets where revenues are generated; such businesses have closed or moved away. Third, because of their proximity to the downtown area, school districts with low socioeconomic status are more likely to include hospitals, museums, and art galleries, all of which are tax-free facilities.

Head Start Begun in 1965 to help preschool children from the most disadvantaged homes, Head Start provides an integrated program of health care, parental involvement, education, and social services for qualifying children.

These properties do not generate revenues. Fourth, neighborhoods with low socioeconomic status often need the greatest share of city services; fire and police protection, sanitation, and public housing consume the bulk of the available revenues, often leaving little for education. Finally, in school districts with low socioeconomic status, a disproportionate amount of the money has to be spent on maintaining the school facilities, which are old and in need of repair.

Although states provide additional funding to supplement local taxes, this funding is not always enough to lift schools in poorer districts to a level that even approximates the funding available to schools in wealthier districts. For example, in *Leandro v. State* (1997), the North Carolina Supreme Court held that the state constitution required that all schools must provide adequate resources to fully educate disadvantaged students (i.e., those who are poor, in special education, and have limited English proficiency). Yet, over two decades later, problems continue to exist. In 2014, a county in North Carolina filed a motion alleging that the state had failed to fulfill its obligation "to provide all children in North Carolina the opportunity for a sound basic education and evidence regarding their ability to fulfill such a plan" (Education Justice 2015). One county in North Carolina was doing such a poor job of educating its disadvantaged students that a judge called it "academic genocide" (Waggoner 2009).

Race and Ethnicity

It is projected that by 2024, racial and ethnic minorities will comprise 54 percent of the pre-kindergarten through 12th grade student population (NCES 2015a). To a large extent, this demographic shift in the racial and ethnic makeup of public school students is a result of the growth in the number of Hispanic students. The white student population is expected to decrease from 60 percent in 2001 to 46 percent in 2024. Alternatively, the number of Hispanic students in public schools is projected to increase from 17 percent in 2001 to 29 percent in 2024.

In comparison to whites, blacks, Hispanics, and Native Americans are less likely to succeed at every level. As early as the start of kindergarten, black children have lower reading, math, and working memory scores when compared to their white counterparts, much of the difference explained by student socioeconomic status and school quality (Quinn 2015). By fourth grade, approximately 80 percent of black, Hispanic, and Native American students are reading below grade level compared to 55 percent of white students. Similarly, by eighth grade, 86 percent of black, and 79 percent of Hispanic and Native American students compared to 56 percent of white students are below grade level in mathematics (Kids Count 2014). As Table 8.1 indicates, although educational attainment has increased over time for all groups, racial and ethnic disparities remain.

It is important to note that socioeconomic status interacts with race and ethnicity. Because race and ethnicity are so closely tied to socioeconomic status—that is, a disproportionate number of racial and ethnic minorities are poor—it *appears* that race or ethnicity determines school success. For example, cities have the highest proportion of minority students as well as the highest proportion of students receiving free or reduced-price lunch, a proxy for socioeconomic status (NCES 2015a). Although race and ethnicity may have an independent effect on educational achievement, their relationship is largely a result of the association between race and ethnicity, and socioeconomic status.

In addition to the socioeconomic variables, there are several reasons that minority students have academic difficulty. First, minority children may be English language learners (ELLs). The National Center for Education Statistics estimates that there are more than 4.4 million English language learners in the United States. In six states, including Alaska, California, Colorado, Nevada, New Mexico, and Texas, and the District of Columbia, 10 percent or more of the public school population are English language learners (NCES 2015a). ELLs come from over 400 different language backgrounds, the overwhelming majority being native Spanish speakers.

> Although race and ethnicity may have an independent effect on educational achievement, their relationship is largely a result of the association between race and ethnicity and socioeconomic status.

Now imagine, as Goldenberg (2008) suggests,

you are in second grade and don't speak English very well and are expected to learn in one year . . . irregular spelling patterns, diphthongs, syllabication rules, regular and irregular plurals, common prefixes and suffixes, antonyms and synonyms; how to follow written instructions, interpret words with multiple meanings, locate information in expository texts . . . read fluently and correctly at least 80 words per minute, add approximately 3,000 words to your vocabulary . . . and write narratives and friendly letters using appropriate forms, organization, critical elements, capitalization, and punctuation, revising as needed. (p. 8)

Not surprisingly, ELLs score significantly below non-ELLs on standardized tests in both reading and mathematics (NCES 2014a). To help ELLs, some educators advocate **bilingual education**, teaching children in both English and their non-English native language.

Advocates claim that bilingual education results in better academic performance of minority students, enriches all students by exposing them to different languages and cultures, and enhances the self-esteem of minority students. Critics argue that bilingual education limits minority students and places them at a disadvantage when they compete outside the classroom, reduces the English skills of minorities, costs money, and leads to hostility with other minorities who are also competing for scarce resources.

The second reason that racial and ethnic minorities don't perform well in school, and compounding the difficulty ELLs have, is that many of the tests used to assess academic achievement and ability are biased against minorities. Questions on standardized tests often require students to have knowledge that is specific to the white middle-class culture—knowledge that racial and ethnic minorities may not have.

The disadvantages that minority students face on standardized tests have not gone unnoticed. In 2012, a complaint was filed with the U.S. Department of Education alleging civil rights violations in New York City's elite public high schools that base admissions decisions solely on standardized tests scores (Rooks 2012). In the fall of 2013, although 70 percent of the city's public high school students were black or Hispanic, they received fewer than 12 percent of the invitations to attend New York City's elite public high schools (Hewitt et al. 2013).

A third factor that hinders minority students' academic achievement is overt racism and discrimination. Discrimination against minority students may take the form of unequal funding, as discussed earlier, as well as racial profiling, school segregation, and teacher and peer bias. In a qualitative study of racial discrimination in the classroom, black students recounted incidents of being disciplined more severely for the same behavior than white students, the expectation of failure by their teachers, and a lack of institutional support for racial inclusiveness (Hope et al. 2015).

Racial Integration. In 1954, the U.S. Supreme Court ruled in *Brown v. Board of Education* that segregated education was unconstitutional because it was inherently unequal. In 1966, a landmark study titled *Equality of Educational Opportunity* (Coleman et al. 1966) revealed that almost 80 percent of all U.S. schools attended by whites contained 10 percent or fewer blacks and that, with the exception of Asian Americans, whites outperformed minorities on standardized tests. Coleman and colleagues emphasized that the only way to achieve quality education for all racial groups was to desegregate the schools. This recommendation, known as the **integration hypothesis**, advocated busing to achieve racial balance.

Despite the Coleman report, court-ordered busing, and a societal emphasis on the equality of education, U.S. public schools remain largely segregated. The majority of black and Hispanic students attend schools that are predominantly minority in enrollment. This is particularly true in large urban areas such as New York City, Chicago, Philadelphia, and Los Angeles where minorities comprise over 85 percent of the student population.

A study by the Civil Rights Project at UCLA indicates that not only does school segregation exist, it's increasing (Orfield et al. 2014). In the United States, the typical white student in a classroom of 30 could expect to have 22 white, 2 black, 4 Hispanic, 1 Asian,

bilingual education In the United States, teaching children in both English and their non-English native language.

integration hypothesis A theory that the only way to achieve quality education for all racial and ethnic groups is to desegregate the schools.

and 1 "other" classmates. Alternatively, a typical black or Hispanic student in a classroom of 30 could expect to have 8 white classmates and 20 or more black and/or Hispanic classmates. Attending a desegregated school, although having no negative achievement effect on white students, has been found to have a positive effect on black and Hispanic students in terms of learning and graduation rates (Orfield et al. 2014).

Socioeconomic Integration. In 2007, the U.S. Supreme Court held that public school systems "cannot seek to achieve or maintain integration through measures that take explicit account of a student's race" (Greenhouse 2007, p. 1). At the time, the Court's decision reflected a general trend toward using socioeconomic or income-based integration rather than race-based integration variables.

Increasingly, school classrooms will be characterized by racial and ethnic diversity. By 2040, less than half of all school-age children will be non-Hispanic whites.

Kahlenberg (2006, 2013) has long advocated this approach for several reasons. First, "socioeconomic integration more directly and effectively achieves the first aim of racial integration: raising the achievement of students" (Kahlenberg 2006, p. 10). Second, socioeconomic integration, because of the relationship between race and income, achieves racial integration, and racial integration, in turn, fosters racial tolerance and social cohesion. Third, unlike race-based integration that is subject to "strict scrutiny" by the government, school assignments based on socioeconomic status are perfectly legal. Fourth, the problem of low-income students in schools, regardless of race and ethnicity, is growing as poverty spreads beyond urban areas and into suburban neighborhoods. Finally, evidence suggests socioeconomic integration is a more cost-effective means of raising student achievement than spending additional dollars in high-poverty schools (Kahlenberg 2013).

WHAT
do you
THINK?

The arguments for single-sex classrooms (SSCs) are fairly simple: Girls would participate more in the learning process, particularly in **STEM** courses (i.e., those in science, technology, engineering, and math), without being intimidated by confident, overbearing boys (Gross-Loh 2014). Boys would be less distracted and less likely to feel that they can't compete with their higher-performing female classmates. What do you think about single-sex education?

Gender

Not only do women comprise two-thirds of the world's illiterate adults, but girls comprise more than 62 percent of the 125 million children who don't attend school (UNESCO 2014b). Although progress in reducing the education gender gap has been made, gender parity in primary and secondary schools has not been achieved. Sixty percent of countries for which there is data have achieved gender parity at the elementary school level, but just 20 percent in low-income countries. Just over a third of countries have achieved gender parity at the high school level (UNESCO 2014c).

Historically, U.S. schools have discriminated against women. Before the 1830s, U.S. colleges accepted only male students. In 1833, Oberlin College in Ohio became the first college to admit women. Even so, in 1833, female students at Oberlin were required to wash male students' clothes, clean their rooms, and serve their meals and were forbidden to speak at public assemblies (Fletcher 1943; Flexner 1972).

STEM An acronym for science, technology, engineering, and mathematics.

TABLE 8.1 Educational Attainments among Persons Aged 25 and Over, by Race, Ethnicity, and Sex, 1970 and 2013

	1970		2013	
	Males	Females	Males	Females
High school completion or higher				
White	54.0	55.0	92.7	93.2
Black	30.1	32.5	84.9	86.6
Hispanic	37.9	34.2	64.6	67.9
Asian and Pacific Islander	61.3	63.1	91.6	89.0
Total	51.9	52.8	87.6	88.6
Bachelor's or higher degree				
White	14.4	8.4	36.0	34.4
Black	4.2	4.6	20.2	23.4
Hispanic	7.8	4.3	13.9	16.2
Asian and Pacific Islander	23.5	17.3	55.1	50.2
Total	13.5	8.1	32.0	31.4

SOURCE: NCES 2015b

> Boys are more likely to lag behind girls in the classroom, be diagnosed with attention-deficit/hyperactivity disorder (ADHD), have learning disabilities, feel alienated from the learning process, and drop out or be expelled from school.

In the 1960s, the women's movement sought to end sexism in education. Title IX of the Education Amendments of 1972 states that no person shall be discriminated against on the basis of sex in any educational program receiving federal funds (see Chapter 10, "Gender Inequality"). These guidelines were designed to end sexism in the hiring and promoting of teachers and administrators. Title IX also sought to end sex discrimination in granting admission to college and awarding financial aid. Finally, the guidelines called for an increase in opportunities for female athletes by making more funds available for their programs.

Gender inequality in education continues to be a problem worldwide. High-performing girls underachieve compared to boys when "they are asked to think like scientists, such as when they are asked to formulate situations mathematically or interpret phenomena scientifically" (Borgonovi and Achiron 2015, p. 1). A self-fulfilling prophecy is likely at hand: parents are less likely to expect their daughters compared to their sons to go into STEM occupations. However, in general, international data suggest that 15-year-old boys are less likely to achieve academically than girls of the same age, particularly in reading.

In the United States, the push toward equality has had considerable effect. For example, in 1970, nearly twice as many men than women had four or more years of college; by 2013, those differences significantly declined (see Table 8.1). Furthermore, scores on the National Assessment of Educational Progress (NAEP) exam indicate that the gender gap in both mathematics and reading scores has decreased over the last few decades. However, significant differences remain and, in general, follow educational stereotypes—boys outscore girls in mathematics, and girls outscore boys in reading (Perry 2014).

Traditionally, the gender gap in achievement has been explained by differences in gender role socialization (see Chapter 10). More recently, however, popular portrayals of male and female differences in achievement have been attributed to "hardwiring," as though there are "pink and blue brains." As neuroscientist Lise Eliot (2010) notes:

> A closer look reveals that the gaps vary considerably by age, ethnicity, and nationality, for example, among the countries participating in PISA, the reading gap is more than twice as large in some countries (Iceland, Norway, and Austria) as in others (Japan, Mexico, and Korea); for math, the gap ranges from a large male advantage in certain countries (Korea and Greece) to essentially no gap in other countries—or even reversed in girls' favor (Iceland and Thailand). What's more, a recent analysis of PISA data found that higher female performance in math correlates with higher levels of gender equity in individual nations. (p. 32)

Much of the research on gender inequality in the schools focuses on how female students are disadvantaged in the educational system. But what about male students? Boys are more likely to lag behind girls in the classroom, be diagnosed with attention-deficit/hyperactivity disorder (ADHD), have learning disabilities, feel alienated from the learning process, and drop out or be expelled from school (Dobbs 2005; Mead 2006; Tyre 2013). Furthermore, black and Hispanic males compared to white males score lower on the NAEP, are less likely to be in gifted and talented programs or to be in advanced placement (AP) classes, and are less likely to graduate from high school or college (Schott Report 2015). Thus, the problems boys have in school may indeed require schools to devote more resources and attention to them (e.g., recruit male teachers).

For example, the District of Columbia has proposed an all-boys public high school in an effort to increase achievement levels and graduation rates of African American boys (Mitchell 2015).

Problems in the American Educational System

For years, the PDK/Gallup annual survey on education asks a random sample of U.S. adults to "grade" our nation's schools (Bushaw and Calderon 2014). In 2014, 29 percent of respondents assigned a letter grade of D or F to the nation's schools, and another 51 percent assigned a C. Given budget cuts, low academic achievement, high dropout rates, questionable teacher training, school violence, and the challenges of higher education, such concern may be warranted.

Lack of Financial Support

When a national sample of public school parents was asked the biggest problems facing the public schools in their communities, the most common response was a "lack of financial support" (Bushaw and Calderon 2014). Despite the importance the American public places on education, efforts to increase state funding are often rejected. In Utah, a proposed increase in the state income tax that would have raised $400 million a year for education was defeated (Fox News 2015).

Per-pupil spending on public school students varies dramatically by state. For example, New York and Washington, DC, spend more than $19,000 per pupil per year. At the other extreme, Idaho and Utah spend $6,500 per pupil per year (Klein 2015). Despite a healthier economy and a growing need for a highly educated workforce, in 2014 over a third of states' per-pupil expenditures were lower than before the 2007 recession. Furthermore, in states where education spending has increased, it has not increased enough to make up for the budget cuts in previous years. A decrease in state education spending means that local school districts must reduce their educational services, raise their local tax base, or both. Between 2008 and 2013, local school districts eliminated more than 300,000 jobs (Leachman and Mai 2014).

Not supporting education financially is shortsighted and costly, and it perpetuates further deficits. For example, narrowing the achievement gap between children of different socioeconomic statuses by as little as one-third would yield $50 billion in fiscal savings and $200 billion (e.g., reduced cost of unemployment) in savings for taxpayers. But rather than invest in low-income schools, "we heap demands on those schools, deprive them of the resources they urgently need, and then declare them to be 'failing schools' when they don't perform miracles" (Welner 2014, p. 2).

Low Levels of Academic Achievement

The Educational Research Center uses several indicators to measure K–12 achievement in public schools, including current levels of performance, improvement over time, and the achievement gap between poor and nonpoor learners called the *poverty gap* (Hightower 2013). Based on a 100-point scale, the achievement average for the nation in 2015 was 70.2 (C–) and ranged from a high of 86.2 (B) for Massachusetts and a low of 64.2 (F) for Mississippi (Quality Counts 2015).

One way to measure performance is to look at the results of the National Assessment of Educational Progress (NAEP) of public and private school students, over time. National trends indicate that reading and mathematics scores between 1998 and 2013 have improved but not for all students. In general, black and Hispanic students made larger gains during the same time period than white students, narrowing the reading and math gap between whites and minorities. Nonetheless, blacks and Hispanics remain significantly behind their white counterparts in both reading and math. The gender gap has also narrowed

The following research was conducted by the Center for Promise (CFP 2014) at Tufts University and examines the social context in which students decide to leave school. As the authors note, the "decision" to separate from school is a process rather than a decision, and the result of a combination of individual and contextual variables.

Sample and Methods

The researchers used a mixed-design study. The first part of the research included an exploratory design, appropriate when "not enough is known about a given phenomenon to develop theories or hypotheses with confidence" (p. 19). Between June and September 2013, researchers interviewed groups of youth in 16 communities throughout the United States. The interview sample was made up of 212 youth, 55 percent male with an average age of 19, recruited from America Promise Alliance organizations, which serve youth who are not in school or who are otherwise unproductive.

A survey was also used, composed of 58 questions based on input from interviewers, contextual elements, existing literature, and prior questionnaires used with similar populations. Respondents were solicited by email and phone, the final sample comprised of 1,936 respondents who had left high school for at least one semester. As the researchers note, the students who were interviewed are not representative of all students who have left high school for at least one semester although the survey respondents are a more diverse and representative sample.

Findings and Conclusions

The results of the research yielded four major findings. First, leaving school and returning to it are the result of a "cluster of factors" such as homelessness, becoming a caregiver for a family member, living in a dangerous neighborhood, or being the victim of a violent crime. Both survey data and interviews support this conclusion. As one 16-year-old respondent recounts:

> Well, my mom's laid up in bed, can't pay the bills. And I turned sixteen and I started roofing [working construction]. Well, eventually weed and everything played into it so much that I seen so much money in my hands that when I went to school it seemed to be a waste of time. . . .

I eventually dropped out just 'cause the bills weren't getting paid and I knew I could pay the bills, step up. I never took on responsibility like that before in my life. (p. 13)

The second finding is that people who leave school are often living in a "toxic environment"—emotional, sexual, and physical abuse, bullying, and parental absence or neglect (p. 20). Using both the qualitative and quantitative data, the researchers identified the three most pervasive sources of toxicity: (1) exposure to or victimization of violence, (2) personal and family health trauma, and (3) unsupported or unresponsive school policies (p. 21).

> My mom had a hernia and needed an operation to get rid of it. . . . I went to go ask if I could get a month off school to help out with my mom and I was told that if I left to help my mom that I would have to stay for two more years in school and I was already on my last year so I just dropped out. (p. 20)

Third, within these challenging environments, respondents who left school sought to establish connections

over time although, in 2013, females still outperformed males in reading and males still outperformed females in mathematics (NAEP 2014).

It is important to note that these gains, even if statistically significant, may mask poor performances. Nine-year-olds are typically in the 4th grade, 13-year-olds in the 8th grade, and 17-year-olds in the 11th grade. Because long-term assessments are based on age, not grade level, NAEP is able to assess the percentage of students at each age who are performing below their typical grade level. In 2013, for example, 58 percent or more of fourth graders and eighth graders performed below grade level in both reading and mathematics. Just 38 percent of high school seniors were proficient in reading, and about a quarter were proficient in math (Hefling 2014).

U.S. students are also outperformed by many of their foreign counterparts—something particularly troubling in a knowledge-based global economy that emphasizes STEM proficiency. The most recent results available from the *Trends in International Mathematics and Science Study* (TIMSS) documents that U.S. students, although scoring above the global average, are consistently outperformed by students in several other countries (Kastberg et al. 2013). For example, nearly half of eighth graders in South Korea, Singapore, and Taiwan scored in the "advanced" level in math in 2011, compared to only 7 percent of American students (Mullis et al. 2012).

with others (e.g., family members, teachers, peers) and those connections either lead a young person toward school or further away from it. Nongraduates often reported feeling alienated from school, bored, and disinterested, unable to connect with teachers or the curriculum.

> The basic reason why I dropped out was because—a traditional high school setting like, I just couldn't learn very much. I guess I learn really hands-on and if it's shown to me in a really creative way then I get it right away. But, in traditional high school you sit down and read a book and hopefully you learn this. I just couldn't do that. I used to really like to read things but once I got into high school and that's all I was doing, I started hating reading. (pp. 28–29)

Forty-one percent of the survey respondents mentioned encouragement by others as a reason they returned to school, and 27 percent credited family support as an important factor in the decision to reenroll. Unable to feel "connected" as a result of family abandonment (e.g., death), instability (e.g., homelessness), or a lack of school or community support, caring relationships can also contribute to leaving school. As

one respondent related, "The gangs showed me love, showed me the ropes, showed me how to get money. After that I was like, what do I need school for?" (p. 30).

Finally, the authors describe nongraduates as "persistently resilient in their day-to-day lives; they are bouncing back, but need additional support to 'reach up' toward positive youth development" (p. 32). Interviewees did not want to be considered failures, and consistently said that they did not "drop out" but rather simply stopped going to school.

> The barriers, the barriers was myself, cause . . . when I get behind on something I doubt myself, and that was my tendency when I was younger. I just wasn't believin' in myself to do my own school work, so I prevented myself from going to school basically . . . . my skin is a lot thicker now and I'm thinking straight, that's what maturity did to me. I'm a grown man.

The authors conclude that both the interview responses and survey data support the conclusion that, in general, youth left school as a consequence of difficult social circumstances that required an immediate response to an immediate problem whether it

be earning money for family bills, taking care of a sick mother, or physical, sexual, or emotional abuse. They also acknowledge that many nongraduates demonstrated enormous strength in overcoming difficult social circumstances. Of the nearly 2,000 survey respondents who had left school, nearly two-thirds had returned to and graduated from high school, and 18 percent had completed at least some college. About half were employed, and of those who were not employed 23 percent were attending school. Lastly, the authors issue a challenge:

> Researchers, practitioners, and policymakers still have much more to learn about the most effective ways to intervene in the multi-faceted challenges that many young people in urban communities face. Nevertheless, we know enough to take positive steps forward now by giving extra attention to students who are navigating challenging circumstances, leveraging the strengths young people bring to their struggles, and building on the promising and evidence-based practices that are already helping young people stay in or return to school. (p. 37)

SOURCE: CFP 2014.

Lastly, the results of *Education at a Glance* indicate that the United States ranks below the OECD global average, ranking 22nd out of the 29 countries for which graduation rates were available. With an 80 percent high school graduation rate, the United States ranks below Germany, the United Kingdom, Japan, Canada, Israel, Poland, and Chile (OECD 2014). The only countries with lower high school graduation rates are Sweden, China, Greece, Luxembourg, Austria, Turkey, and Mexico (OECD 2014).

School Dropouts

The *status dropout rate* is the percentage of 16- to 24-year-olds that is not in school and has not earned a high school degree or its equivalent. In the last several decades, the male and female status dropout rates have declined significantly to 7 percent and 6 percent, respectively (NCES 2015a). Nonetheless, more than 1 million students drop out of school every year (NCES 2015c). Figure 8.4 graphically portrays the percentage of ninth graders in 2009 who dropped out of high school, by sex and parents' socioeconomic status.

Dropping out of school is associated with increased costs of public assistance, poorer health, higher medical costs, and dying at a younger age. Because 45 percent of graduates with a GED (general educational development) smoke compared to 10 percent of college

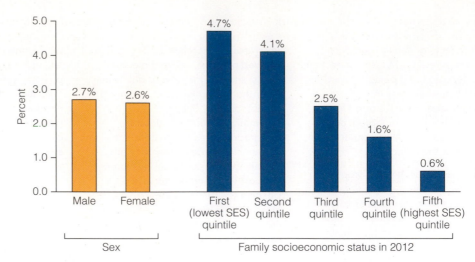

Figure 8.4 Percentage of U.S. Ninth Graders who Dropped Out of High School before Graduating, by Sex and Socioeconomic Status (SES)*
*Socioeconomic status was based on parents' education, occupational prestige, and family income.
SOURCE: NCES 2015c.

> In the last several decades, the male and female status dropout rates have declined significantly to 7 percent and 6 percent, respectively... Nonetheless, more than 1 million students drop out of school every year.

graduates, high school dropouts cost Medicare $20 billion a year (Segal 2013). The dropout rate is also associated with lower incomes (Alliance for Excellent Education [AEE] 2014a). In general, dropouts, earn $8,000 a year less than their high school graduate counterparts, and $26,500 a year less than college graduates. Furthermore, it is estimated if the male dropout rate was reduced by just 5 percent,

> the nation could save as much as $18.5 billion in annual crime costs. . . . [T]he nation would also see a decrease in annual incidences of assault by nearly 60,000; larceny by more than 37,000; motor vehicle theft by more than 31,000; and burglaries more than 17,000. It would also prevent nearly 1,300 murders, more than 3,800 occurrences of rape, and more than 1,500 robberies. (AEE 2014a, p. 1)

Second-chance initiatives such as GED certification allow students to complete their high school requirements. Other dropout interventions include early or middle school college programs that allow dropouts to enroll in community colleges or, in some cases, four-year degree programs. There, they receive a secondary school education, earn a high school diploma, and often accrue college credits. There are also efforts, at the state and federal levels, to increase the age of compulsory school attendance in the hopes that it will deter students from dropping out of school.

Crime, Violence, and School Discipline

Learning is impacted by any number of social variables—the quantity and quality of teachers, home environment, availability of resources such as computers and textbooks, parental involvement, and the like. One important variable is the extent to which a student feels safe and supported. Research indicates that "high stress environments in which students feel chronically unsafe and uncared for make it physically and emotionally harder for them to learn and more likely for them to act out or drop out" (Sparks 2013, p. 16).

Crime and Violence against Students. Despite the horrors of high-publicity school killings such as those at Columbine, Virginia Tech, and Sandy Hook Elementary School, the chance of a student dying at school is quite rare. Less than 2 percent of the total youth homicide rate, and proportionately even fewer suicides, take place on school property (Robers et al. 2015). The unlikelihood of such an event is reflected in students' perceptions of safety. According to the most recent data available, 3 percent of students report fear of attack or harm on their way to or at school. Fear is higher among minority students in urban public schools compared to their counterparts (Robers et al. 2015).

In 2013, 4 percent of the student population between 12 and 18 years of age reported being the victim of a nonviolent crime, resulting in 1.4 million in-school incidents (NCES 2015a). The most common in-school offense was theft. Violent (e.g., robbery) victimization of 12- to 18-year-olds was higher at school than away from school, higher for males than females, and higher for younger students (aged 12 to 14) than older students (aged 15 to 18). Eight percent of high school students reported being in a physical fight on school grounds within the last year, and nearly a quarter reported that they had been offered, sold, or given drugs while on school property (Robers et al. 2015).

Sexual Assault on Campus. According to the White House Task Force to Protect Students from Sexual Assault (2014b), an estimated 20 percent of women on campus, most often in their freshman or sophomore year, will be sexually assaulted while in college. Many will be victims of an **"incapacitated assault**; they are sexually abused while drugged, drunk, passed out, or otherwise incapacitated" (p. 6). The victim will probably know her attacker but will not report what happened.

President Obama has also launched federal investigations into sexual assault complaints at more than 100 college campuses across the country, both public and private, including the University of California, Ohio State, Knox College in Illinois, Harvard, and Catholic University of America. The investigations are being conducted under Title IX, which prohibits gender discrimination at colleges or universities receiving federal funds (Dockterman 2015; Williams and Hefling 2014). According to a national survey of 440 colleges and universities, 40 percent of schools had not conducted a single investigation of sexual assault in the five years prior to the survey, and even fewer provided sexual assault training for faculty and staff (Majority Staff 2014). According to two documentary filmmakers who traveled the country interviewing both sexual assault victims and assailants for their film *The Hunting Ground,* the assaults are usually premeditated; the "survivor was picked out, plied with alcohol and set up to be assaulted" (quoted in Dockterman 2015).

Crime and Violence against Teachers. Teachers may also be victimized in schools. In 2013, the most recent year for which data are available, teachers were more likely to be threatened or injured in urban schools than in suburban or rural schools, and in public schools than in private schools (Robers et al. 2015). Female teachers reported being physically attacked by a student at higher rates than male teachers; and, somewhat surprisingly, a greater percentage of elementary school teachers reported being physically attacked than high school teachers. There were also differences in the likelihood of being threatened or physically attacked by race and ethnicity of the teacher. African American teachers were more likely to be threatened by a student and/or physically attacked by a student than white or Hispanic teachers.

School Discipline. Across grade levels, black, Hispanic, and Native American boys and girls are disproportionately suspended and expelled when compared to their white counterparts. For example, on average, 16.4 percent of African American students are suspended compared to 4.6 percent of white students (U.S. Department of Education 2014a). In addition, white students represent 51.0 percent of the public school population yet 39.0 percent of those arrested for a school-related incident. Black students, on the other hand, represent 16.0 percent of the public school population and 31.0 percent of those subjected to a school-related arrest (U.S. Department of Education 2014a). The debate, of course, is whether these disparities can be attributed to overt discrimination or a reflection of differences in behavior.

Disciplinary practices differ between schools and school districts, with some expelling or suspending 90 percent of the student body at least once. Such rates are the result of *zero-tolerance disciplinary policies.* Although the origin of the phrase is from the 1980s and fears over drug use and students bringing guns to school, today's zero-tolerance policies may include suspensions or expulsions for "bringing a cell phone to school, public displays of affection, truancy, or repeated tardiness" (National Public Radio 2013, p. 1).

Increasingly, educators, counselors, teachers, and psychologists "denounce such practices [suspensions and expulsions] as harmful to students academically and socially, useless

incapacitated assault.
An assault that takes place when the victim is unable to make an informed decision because of alcohol or other drugs, being unconscious, and the like.

as prevention tools, and unevenly applied" (Shah 2013, p. 1). Further motivation for school reform comes from the **school-to-prison pipeline**—the established relationship among severe disciplinary practices, increased rates of dropping out of school, lowered academic achievement, and court or juvenile detention involvement. The school-to-prison pipeline disproportionately hurts minority students who are more likely to be suspended or expelled, and when "forced out of school become stigmatized and fall behind in their studies, . . . drop out of school altogether, and . . . may commit crimes in the community" (Amurao 2013, p. 1).

Bullying. Bullying "involves targeted intimidation or humiliation . . . a physically stronger or socially more prominent person (ab)uses her/his power to threaten, demean, or belittle another" (Juvonen and Graham 2014, p. 161). Bullying may be direct (e.g., hitting someone) or indirect (e.g., spreading rumors) and is considered a form of aggression. Whether examining bullying by age, race/ethnicity, culture, country, or gender, boys are more likely than girls to participate in direct bullying such as kicking, hitting, and pushing.

Just over 20 percent of 12- to 18-year-olds report being bullied at school (Robers et al. 2015). Female students have higher rates of being the subject of rumors, being "made fun of, called names, or insulted," and being "excluded from activities on purpose" than male students (p. 46). Males students were more likely to be "pushed, shoved, tripped or spit on." 21 percent of whom reported being injured as a result of the incident (p. 46). Bullying is most likely to occur in hallways or stairwells and, somewhat surprisingly, in classrooms.

Three white students were expelled and one suspended from San Jose State University for allegedly racially bullying their black roommate. The students are accused of using racial slurs, forcing the victim to wear a U-shaped bicycle lock around his neck, displaying a Confederate flag as well as swastikas and pictures of Adolf Hitler, and preventing him from leaving his room (Cordell 2013; Kaplan and Murphy 2014). Some are calling the incident a bullying hate crime. What do you think—as alleged, is it a case of bullying or a hate crime or both?

Inadequate School Facilities

Almost half of all schools in the United States were built in the 1940s and 1950s, and many need costly repairs and renovations. School enrollment is projected to increase through 2019, state and local building funds remain low, and federal spending is about half what it was prior to the 2007 recession, further exacerbating the problem. The American Society of Civil Engineers, assigning a letter grade of D (poor) to the infrastructures of American schools, estimates that at least $270 billion or more is needed to "modernize and maintain" our nation's schools (American Society of Civil Engineers 2014).

Nationally representative data collected by the U.S. Department of Education's National Center for Education Statistics reports similar concerns (Alexander et al. 2014). Results indicate that in 25 percent or more of public schools, the windows, plumbing/lavatories, heating and air-conditioning systems, ventilation/filtration systems, security systems and exterior lighting, roofs, and communication systems were rated as fair or poor. Furthermore, 27 percent or more of outdoor features such as parking lots, fencing, athletic facilities, sidewalks, and playgrounds were also rated as fair or poor. In general, the lower the socioeconomic status of the student population as measured by the percentage of students eligible for free or reduced-priced lunch, the worse the overall condition of the facilities.

Because of development patterns, many of the most disadvantaged schools are in large urban areas. A study of the condition of New York City public schools found that it is directly associated with poverty—the more impoverished the neighborhood, the worse the school (Service Employees International Union 2013). In 2013, the city of Philadelphia closed 23 public schools, some because of underuse and others because they needed millions of dollars in repair (Hurdle 2013). Because courts have consistently held that the quality of building facilities is part and parcel of equal educational opportunities, state governments monitor spending on infrastructure needs to ensure equitable distributions of funds (American Federation of Teachers 2009).

school-to-prison pipeline The established relationship between severe disciplinary practices, increased rates of dropping out of school, lowered academic achievement, and court or juvenile detention involvement.

bullying Bullying occurs when "a physically stronger or socially more prominent person (ab)uses her/his power to threaten, demean, or belittle another" (Juvonen and Graham 2014, p. 161).

the HUMAN side | I Used to Be a Teacher

To The Editor:

I used to be a teacher. My dream was to reach the hard-to-reach learners. I was going to change the world. You can do that, change the world, when you are a teacher. . . .

The day I realized that dream would be lost was a dark one. I had a good cry.

Teachers work 9 a.m. to 3 p.m. and get summers off. This is the argument I hear most often around teaching and why individuals that do not work in the school and have no idea what they are talking about feel teachers are overpaid. Teachers do not work 9 to 3. A typical day in a teacher's world is 7 a.m. to 9 or 10 p.m. Why so late? Lesson plans are not given to a teacher like many feel it is. Lessons do not typically come out of a book. Lessons are created by teachers. . . . Weekends and holidays are spent grading papers, creating exams, creating help sheets, worksheets, projects and lesson planning as well. Teachers never put in 40 hour weeks—the weeks are closer to 70 or more hours, depending.

Being a teacher used to be something respectable. Now if you indicate you are a teacher it comes with eye rolls and judgment. Teachers no longer have the same freedoms they used to have creating lessons. They are not supposed to teach to the standardized tests, but if they do not teach to these tests the children will not pass them and now it is a reflection of how the teacher teaches. You are graded by how well your students do regardless of whether you are in a well off area versus a struggling area. You are told you get paid too much and work too little. You are told to teach this way, teach this content and you are not teaching the way you should be by individuals that never went to school for teaching and have no background in education at all. You are constantly watched. You are constantly stressed. You work harder than the majority of people work and it is not seen or acknowledged. . . .

I used to be a teacher. I used to be someone to look up to. Someone that molded minds and helped create futures. I used to be a teacher that loved her job, loved her students and even loved the hours I put in because I knew what my students would get out of my hard work was worth it. . . .

I used to be a teacher. I used to be following my dream. The day I realized that dream would be lost was a dark one. I had a good cry. I had several good cries. I let go of the dream because I had to. I am not a teacher. I work in insurance. I answer phones and assist others in their claim status and benefit information. I do enjoy my work but my heart is heavy. I want to be out there changing the world. I try to look at my job as if it is a teaching experience. I am often complimented at how easy I make their insurance sound when they called in without a clue. "You should have been a teacher" I have heard several times. Each time I hear it I cringe. I want to scream that I should be a teacher. I should be teaching! But right now, that is not possible. This is not the time for educators. This is the time for individuals that have nothing to do with education telling educators what they should be doing and who they should be.

I'm not the only one that had to give up my dream. There are thousands more of me out there—amazing teachers that never had a chance to reach the stars. I think of my third grade teacher and wish I could have told her my dream was realized. Instead, you can call the number on the back of your health insurance card and I would be happy to assist you with your insurance needs.

Danielle Caryl

SOURCE: Caryl 2013.

A considerable body of evidence documents the relationship between the school physical environment and academic achievement. Milkie and Warner (2011), after interviewing more than 10,000 parents and teachers of first grade students, conclude that students' stress levels are negatively impacted by deteriorated school facilities. Tanner (2008) found a significant relationship between school environment (e.g., space, movement patterns, light, etc.) and academic achievement of third graders even when controlling for the socioeconomic status of the school. Air quality, noise, overcrowding, inadequate space, environmental contaminants, and lighting all affect a child's ability to learn, a teacher's ability to teach, and a staff member's ability to be effective.

Recruitment and Retention of Quality Teachers

School districts with inadequate funding and facilities, low salaries, lack of community support, and minimal professional development have difficulty attracting and retaining qualified school personnel. A study by the Alliance for Excellent Education in conjunction with the New Teacher Center (NTC) found that 13 percent of teachers either move schools or leave the profession every year (AEE 2014b). The percent is even higher in high-poverty schools—about 50 percent higher than low-poverty schools—with an annual

Many school buildings and facilities are in need of repair. Mold, defective ventilation systems, faulty plumbing, and the like are not uncommon. Still, quality education is expected to continue in the classrooms despite such deplorable conditions.

Mark Richards/PhotoEdit

turnover rate of 20 percent. Teacher movement tends to be from urban to suburban districts, and from schools with a high percentage of poor and/or minority students to schools with a low percentage of poor and/or minority students.

High teacher turnover is a problem in several ways. First, newer teachers are less experienced and often less effective. Effective teachers, those who are ranked in the top 20 percent of performance, add an average of 5.5 more months of learning each year compared to low performing teachers. (U.S. Department of Education 2014b). Second, teacher turnover contributes to a lack of continuity in programs and educational reforms. Finally, recruiting and training expenses, in addition to the time and effort devoted to replacing teachers who have left the profession, is considerable. The estimated cost to states of teacher turnover is upward of $2.2 million a year (Witt 2014).

Because teacher salaries are the largest component of a school district's costs, poor school districts have less money to offer qualified teachers (FEBP 2014). Thus, students in poorer school districts are more likely to be taught by (1) teachers with less than three years of experience, (2) teachers assigned to areas outside their specialty, (3) substitute teachers, and (4) less effective teachers, when compared to students in more affluent districts (Barton 2004; FEBP 2014; Goldhaber et al. 2015; Sawchuk 2011). Furthermore, teachers in poorer school districts are more likely to be impacted by conditions that make teaching difficult (e.g., noisy classrooms, too few computers) and to be tasked with noninstructional duties (e.g., janitorial duties, lunch/yard supervision) (Rogers and Mirra 2014).

Recruiting and retaining quality teachers in poverty-level schools is critical to the success of its students. Because minority students disproportionately populate poor school districts, it is also important to recruit and retain teachers who meet the needs of children from diverse backgrounds and of varying abilities. The number of minority teachers who can serve as role models, have similar life experiences, and have similar language and cultural backgrounds is far too few for the number of minority students. Called the "**diversity gap**," less than 20 percent of teachers are non-Hispanic, nonwhite, despite the fact that nearly half of public school students are minorities (Boser 2014). Not surprisingly, a survey of a representative sample of high school teachers in California found that teachers in schools with concentrated poverty reported higher rates of students with economic and social stressors (Rogers and Mirra 2014) (see Figure 8.5).

Teacher Effectiveness. Recruiting and retaining quality teachers may be made more difficult with the recent emphasis on accountability and implementation of **value-added measurement (VAM)**. VAM is the use of student achievement data to assess a teacher's effectiveness. Although the majority of Americans believe that using teacher evaluations to improve teaching skills and document ineffectiveness is very important, 61 percent of Americans are opposed to using test results in evaluating teachers (Bushaw and Calderon 2014), and not without reason. The American Statistical Association (ASA), in a 2014 open letter on educational assessment, note that the majority of differences in standardized test scores between students are "attributable to factors outside of the teacher's control, such as student and family background, poverty, curriculum, and [other] unmeasured influences" (ASA 2014, p. 7).

Assessing teachers based on student performance assumes all else is constant and ignores the reality of student differences in such nonschool factors as family life, poverty, emotional and physical obstacles, and the like (see Figure 8.5). In addition, there are concerns that teachers, fearing for their jobs and concerned about merit-based pay, may

> [I]f a child from a poor family has a good teacher for five consecutive years, the achievement gap between that child and a child from a higher-income family would be closed.

diversity gap The difference between the racial, ethnic, and socioeconomic statuses of public school teachers versus public school students.

value-added measurement (VAM) VAM is the use of student achievement data to assess teacher effectiveness.

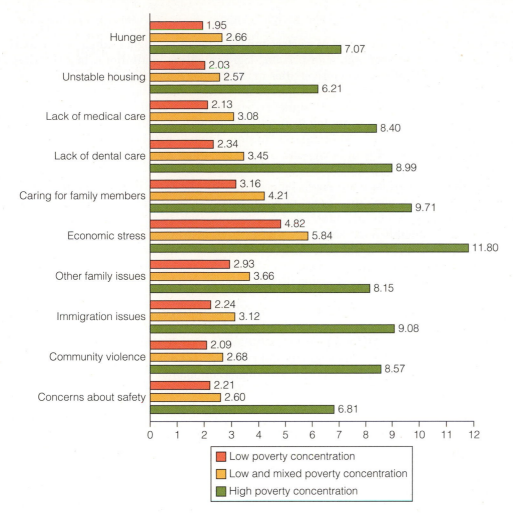

Figure 8.5 Teachers' Estimates of the Number of High School Students Affected by Stressors in a Typical California Classroom by Poverty Concentration (N = 783), 2013
SOURCE: Rogers and Mirra 2014.

begin "teaching to the test" and worse. In 2015, 10 former Atlanta public school educators were convicted for providing students with answers to standardized test questions and changing incorrect answers to correct ones.

WHAT do you THINK?

In 2014, a California judge held that tenure, the school policy that protects teachers from dismissal without cause, was ruled in violation of the state's constitution (Martinez 2014). At the college level, tenure assures that professors cannot be fired for, as examples, teaching unpopular ideas or holding political views deemed by others as "radical." Tenure provides university faculty academic freedom. Do you think the system of tenure in public school, including colleges and universities, should be eliminated?

To place quality teachers in the classroom, many states have implemented mandatory competency testing (e.g., the Praxis Series). The need for teachers who are officially classified as "highly qualified" is tied to federal mandates that emphasize the importance of having licensed teachers in the classroom. Additionally, teachers who have a bachelor's degree and have been in the classroom for three or more years are also eligible for national board certification. Eighty-one percent of Americans believe that board certification should be required (Bushaw and Calderon 2014). Some studies indicate that students of "highly qualified" teachers and/or board-certified teachers perform better on standardized tests and have shown greater testing gains than students of teachers who are not "highly qualified" and/or board certified (National Board for Professional Teaching Standards 2015).

The Challenges of Higher Education

Although there are many types of postsecondary education, higher education usually refers to two- or four-year, public or private, degree-granting institutions. In 2014, there were nearly 5,352 colleges and universities in the United States (Ginder et al. 2014).

Of the 24.5 million undergraduate and 3.8 million graduate students enrolled in U.S. colleges and universities, full-time students, women, and racial and ethnic minorities have disproportionately contributed to enrollment growth over the years. However, projected enrollment through 2022 is expected to slow down in part because of demographic changes—the number of people who are of typical college age, 18 to 24, is expected to decline (Lederman 2014).

Over the last decade, there has been a significant decrease in full-time tenured or tenure-track faculty, once the "core" of academia, and significant increases in noninstructional staff (e.g., administrators) and non-tenure-track, part- or full-time instructors. In general, as academic rank increases—that is, from instructor to assistant professor, to associate professor, to full professor—salaries increase and the proportion of women and minorities decrease. In academic year 2013–2014, the average salary for female faculty was $70,400; for male faculty, $85,500 (NCES 2015a).

Cost of Higher Education. In the academic year 2013–2014, the total cost for a first time, full-time student, who lived on campus and paid in-state tuition was $22,190 at a public institution and $44,370 for a private institution (NCES 2015a). Out-of-state students' costs are even higher. For many students and families, without financial aid, the expense of a four-year degree would make a college degree unobtainable. Nearly three-quarters of college students graduate with some kind of debt averaging $29,400 (White House 2014a), 40 million graduates are still paying on their student loans, and another 7 million have defaulted (Dynarski 2014).

As student debt surpassed the $1 trillion mark, a total larger than the sum of all mortgage debt or all credit card debt (Dynarski 2014), President Obama signed an executive order designed to ease the burden of borrowers. Among its many provisions, it will cap loan payments at 10 percent of the borrower's income and, by working with the private sector, help students with payment options by expanding the financial counseling students receive.

"Degrees of Inequality." In a recent book, *Degrees of Inequality: How the Politics of Higher Education Sabotaged the American Dream* (Mettler 2014), the author argues that higher education perpetuates inequality rather than expanding opportunities. Although the book outlines many ways in which this is the case, central arguments include (1) the growth and expense of for-profit colleges and universities (e.g., the University of Phoenix, Kaplan Higher Education) that predominantly serve students from low-income families; (2) the dismal graduation rates—25 percent on the average—at for-profit colleges and universities; (3) the increased cost of going to college at a time when student aid programs cover less and less of the expense; and (4) the underfunding of community colleges at a time when federal loans to students at for-profit institutions increases student debt (i.e., they are more expensive that nonprofit institutions), often without a degree, and costs taxpayers millions of dollars in defaulted loans.

For example, evidence indicates that the number of students receiving merit-based aid rather than need-based aid is growing (Marcus and Hacker 2014). From a college or university's perspective, spreading what aid there is to more people in smaller amounts both increases enrollment rates and the institutions' reputation as grade point averages and the proportion of incoming freshmen who graduate increases. By definition, those who receive merit-based aid are high achievers and more likely to have graduated from better-funded high schools where enrollment for students from low-income backgrounds is small. Thus, not only is need-based aid for low-income high school graduates less available, but these graduates are less likely to be eligible for merit-based aid.

Because socioeconomic status and race/ethnicity are related, the same social forces behind "degrees of inequality" that impact low-income students disproportionately affects racial and ethnic minorities. Of the total fall enrollment at four-year degree-granting institutions in 2013, only 14.3 percent of students were black and 16.5 percent Hispanic

SELF and society | Transitions

Using the scale below, answer each of the questions by placing the number that corresponds to your answer in the blank provided. When you are finished, compare your answers to those from a national sample of U.S. adults (*N* = 1,001) .

1 = very important; 2 = somewhat important; 3 = not very important; 4 = not at all important

How important are each of the factors in helping a high school student get a good job one day?	Your Response (1–4)
1. Performing well on standardized tests, such as the ACT or SAT.	
2. Earning a B or higher GPA on completed coursework.	
3. Having a mentor or advisor.	
4. Working on a real-world project that takes at least six months to complete.	
5. Learning skills like dependability, persistence, and teamwork.	
6. How important is a college education today?	

Results of a National Sample

How important are each of the factors in helping a high school student get a good job one day?	Percentage in Each Response Category			
	Very Important	Somewhat Important	Not Very Important	Not At All Important
1. Performing well on standardized tests, such as ACT or SAT.	24%	44%	43%	9%
2. Earning a B or higher GPA on completed coursework.	44%	45%	9%	1%
3. Having a mentor advisor.	59%	37%	4%	0%
4. Working on a real-world project that takes at least six months to complete.	42%	42%	11%	4%
5. Learning skills like dependability, persistence, and teamwork.	86%	13%	1%	0%
6. How important is a college education today?	43%	48%	9%	1%

SOURCE: Bushaw and Calderon 2014.

(NCES 2015a). Note that minority enrollment at four-year for-profit institutions and two-year for-profit institutions was 52 percent and 60 percent, respectively (NCES 2015a).

Additionally, although college enrollments of racial and ethnic minorities have increased over the last several decades, the increases have been accompanied by "separate and unequal" access to "selective and well-funded four-year colleges" (Carnevale and Strohl 2013, p. 6). Between 1995 and 2009, 82 percent of incoming white college students were enrolled at the top 468 colleges and universities in the United States, compared to 9 percent of African American college students and 13 percent of Hispanic college students.

Lastly, African American and Hispanic students with A averages in high school are more likely to be enrolled in community colleges compared to their similarly qualified white counterparts. The 468 most selective four-year colleges where white students disproportionately attend have (1) more financial resources, (2) higher graduation rates, (3) higher enrollment in and completion of graduate degrees, and (4) graduates with greater future earnings (Carnevale and Strohl 2013). (See Chapter 9 for a discussion of race, ethnicity, and affirmative action.)

Community Colleges. Despite being called the nation's "unsung heroes of the American education system" (Obama quoted in Adams 2011, p. 1), just over half of the respondents in a survey of U.S. adults agreed with the statement that "community colleges offer high-quality education" (Lumina 2013). In contrast to such perceptions, enrollment in 2-year colleges continues to increase, with 7 million students in 2013, expected to increase to over 8 million by 2024. The number of associate degrees awarded in the same year was nearly twice that awarded a decade ago (NCES 2015a).

Community colleges play a vital role in U.S. educational policy. They are starting points for many Americans who hope to eventually transfer to baccalaureate institutions. Minority and low-income students and women are more likely to enroll in community colleges than their white, male middle-class counterparts for a variety of reasons. Community colleges are often closer to where students live, and they offer a more flexible schedule, allowing students to work full- or part-time while attending school and avoiding the cost of room and board by living at home.

Because of the importance of community colleges and the role they play in higher education, in 2015, President Obama proposed that, for some students, community college be tuition-free (White House 2015). Called America's College Promise, this federal government initiative would work with states to waive tuition for responsible students who maintain a minimum grade point average and are in high-performing programs. Additionally, as proposed, there would be institutional reforms that would help ensure that community college students complete an associate's degree. As many as 9 million students could benefit from America's College Promise, with a full-time community college student, on the average, saving $3,800 in tuition per year.

Strategies for Action: Trends and Innovations in American Education

Americans consistently rank improving education as one of their top priorities (Pew Research Center 2015). Recent attempts to improve schools include raising graduation requirements, barring students from participating in extracurricular activities if they are failing academic subjects, providing three-year bachelor's degree programs, lengthening the school day, prohibiting dropouts from obtaining driver's licenses, implementing year-round schooling, and extending the number of years permitted to complete a high school degree.

However, educational reformers on both sides of the political aisle continue to call for changes that go beyond these get-tough policies. On the one hand, many Republicans believe that increasing competition and accountability lead to better schools, teachers, and students. Democrats, on the other hand, argue that increased funding and a commitment to ending educational inequality is what is needed. Differences aside, all agree that something needs to be done.

Educational Policy across the States

The challenges facing national educational policies are considerable. In 2011, President Obama, speaking at Kenmore Middle School in Arlington, Virginia, stated that

> unfortunately too many students aren't getting a world-class education today. As many as a quarter of American students aren't finishing high school. The quality of our math and science education lags behind many other nations. And America has fallen to 9th in the proportion of young people with a college degree. Understand, we used to be first, and we now rank 9th. That's not acceptable. (quoted in Adams 2011, p. 1)

Although it may be difficult to predict the totality of this administration's educational policy and its impact, there is evidence of both significant changes and "things as they were" politics.

Reforming No Child Left Behind. No Child Left Behind (NCLB) was signed into law in 2002. It is the most current version of the Elementary and Secondary Education Act (ESEA) originally passed in 1965. As of midyear 2015, the act had not been reauthorized, predominantly because of political wrangling. Points of contention, in various degrees, include the adoption of Common Core State Standards, standardized testing and the federal role in developing accountability systems, and privatization of schools.

Common Core State Standards. In 2008, the National Governors Association and the Council of Chief State School Officers convened to adopt a common set of academic standards for mathematic and language arts to be used across the states as a means of standardizing educational requirements. Known as the Common Core State Standards (CCSS), the initiative was motivated by concerns that students in different states were not being prepared equally for postsecondary education and/or jobs in the global workforce.

CCSS have been implemented or are in some stage of implementation in 44 states and the District of Columbia. Nearly 80 percent of teachers in schools where they have been implemented report feeling prepared to teach CCSS, and 84 percent of teachers who have one year or more of full implementation report being enthusiastic about the new standards. Most important, the majority of teachers report that CCSS are having a positive impact on their students (Primary Sources 2014) (see Figure 8.6).

Public support for CCSS has declined over the years, with only 40 percent of Americans favoring the new standards. In a 2014 telephone survey of 1,001 Americans, of those in favor of CCSS, the most important reason for its support was that it would help students moving between schools and/or school districts. Of those who were opposed, respondents feared its adoption would limit a teacher's ability to teach what they think is best for their students (Bushaw and Calderon 2014).

Testing and Accountability. The emphasis on teacher and school accountability requires that students take multiple standardized tests to allow comparisons within and between school districts, and between states. Teachers, students, and even schools are then "graded" based on the results.

Criticisms of "high-stakes tests" and their uses are numerous. First, and perhaps most important, standardized tests encourage rote learning, superficial thinking, and memorization rather than critical thinking skills (Harris et al. 2011; National Center for Fair and Open Testing [NCFOT] 2012). That said, second, there is concern that using standardized tests scores, either in whole or in part, to determine student placements, teachers' salaries, or school closures is highly unreliable (ASA 2014; Rothstein et al. 2010).

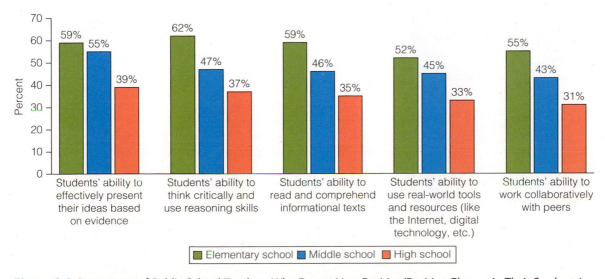

Figure 8.6 Percentage of Public School Teachers Who Report Very Positive/Positive Change in Their Students' Abilities in Each Area as a Result of CCSS, by Grade(s) Taught (N = 167), 2014
SOURCE: Primary Sources 2014.

Third, there are concerns about the content of standardized tests. The NCFOT argues that "standardized tests are not objective. . . . [D]ecisions on what to include, how questions are worded, [and] which answers are 'correct' . . . are made by subjective human beings" (NCFOT 2012). Often, questions reflect the cultural biases of the test maker. Finally, standardized testing is not cost-effective—a particularly salient concern in these times of budget cuts and sequestering. The cost of tests, testing services, and test preparatory materials, according to one estimate, is more than $2.3 billion a year and is rapidly increasing (Gardner 2013).

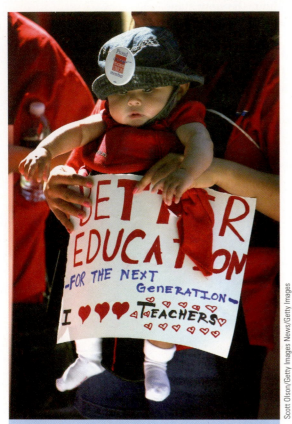

In 2012, over 25,000 Chicago Public School teachers went on strike over salaries, benefits, and employment security. The strike lasted eight days, and a compromise was reached after hours of negotiations. The teachers were heavily criticized for striking, and many Americans do not believe that teachers, like the police, should be able to unionize.

Scott Olson/Getty Images News/Getty Images

parent trigger laws State legislation that allows parents to intervene in their children's education and schooling.

Advocacy and Grassroots Movements

Concerns about failing schools, standardized testing, and value-added measurement (VAM, or the practice of teachers' salaries being tied to students' performances) have led to a growing movement of parents and educators "opting out" of high-stakes testing. There have been writing protests, new campaigns by teacher unions, boycotts of standardized testing by teacher and student groups, and even Facebook pages with names like "Parents and Kids against Standardized Testing" and "Scrap the MAP"—the Measurement of Academic Progress (Toppo 2013).

Parent Revolution, whose mission is to "transform underperforming public schools by empowering parents to advocate what is good for children," began in 2009 in Southern California and was instrumental in passing the first **parent trigger law** (Parent Revolution 2014, p. 1). The principle behind parent trigger laws is that, with a sufficient number of signatures, parents can intervene on behalf of their children by taking one or more actions as statutorily permitted. In general, these actions include replacing administrators, principals, and/or teachers, converting the school to a charter school (see the section titled "The Debate over School Choice" in this chapter) or closing the school altogether.

To date, seven states have enacted parent trigger laws, sometimes called empowerment laws, with more than 24 states considering their adoption (Lindstrom 2015). For example, Ohio schools that rank "in the lowest five percent in performance statewide for three or more consecutive years" are eligible for intervention based on the state's trigger laws (National Conference of State Legislatures [NCSL] 2013a, p. 1). Options for action include converting the school to a charter school, replacing "at least 70 percent of the school's personnel related to its poor performance," turning control of the school over to the state, and privatization (NCSL 2013a, p. 1). Based on Ohio's parent trigger law, 20 percent of Columbus, Ohio, schools now qualify for key leadership changes (Prothero 2014).

As a structural functionalist would argue, trigger laws empower parents who heretofore had little to say about the quality of their children's education. Compatible with conflict theory, opponents hold that trigger laws are part of a political agenda funded by wealthy donors to privatize schools (Lu 2013, p. 1). Ironically, because privatization "outsources school governance to educational management organizations (EMOs) who have no obligation to (and often no physical presence in) the community, the parent trigger ultimately thwarts continued, sustained community and parental involvement" (Lubienski et al. 2012, p. 2).

The Network for Public Education (NPE) is an "advocacy group whose goal is to fight to protect, preserve and strengthen our public school system, an essential institution in a democratic society" (NPE 2015, p. 1). The NPE endorses political candidates, provides support and information for similarly minded "friends and allies," and issues grassroots reports on essential issues in public education. Recent NPE actions include protesting the role of business leaders (e.g., the Gates Foundation) in education policy, the continued increase in class size, the closing of schools in Chicago, high-stakes testing and their use in teacher evaluations, and charter schools that divert needed funds from public schools.

Character Education

Character education entails teaching students to act morally and ethically, including the ability to "develop just and caring relationships, contribute to community, and assume the responsibilities of democratic citizenship" (Lickona and Davidson 2005). Despite most schools' emphasis on academic achievement, knowledge without character is potentially devastating. Sanford McDonnell (2009), the former CEO of McDonnell Aircraft Corporation and chairman emeritus of the Character Education Partnership, recounts a letter written by a principal and former concentration camp survivor to his teachers at the start of a new school year:

> My eyes saw what no person should witness: gas chambers built by learned engineers, children poisoned by educated physicians, infants killed by trained nurses, women and babies shot and burned by high school and college graduates. Your efforts must never produce learned monsters and skilled psychopaths. Reading, writing, and arithmetic are important only if they serve to make our children more humane. (p. 1)

It is difficult, however, to create "a culture of integrity; [it] is not quick or easy, particularly when a long-standing and well-entrenched culture of cheating is already in place" (Stephens and Wangaard 2013).

In 2010, the National Collegiate Athletic Association (NCAA) began an investigation into the University of North Carolina football program focusing on "impermissible benefits from sports agents," and in 2012, they sanctioned the football program. In the same year, an investigation began into one academic department's role in fraudulently assigning grades. The results indicated that over the course of two decades, there were over 200 "make believe" courses and more than 500 unexplained grade changes involving over 1,500 student-athletes, and several administrators, faculty members, academic advisors, and coaches (Beard 2014; Friedlander 2014). In 2015, the NCAA acknowledged that academic misconduct is rising in college athletics and that it is currently handling more than 20 investigations (Russo 2015).

Unfortunately, the scandal at UNC, although highly publicized, is just one of many examples of academic integrity violations that occur at all education levels. Seventy-five percent of students report cheating at least once in their college career (Buchmann 2014), and cheating among high school students is at an all-time high (Dorff 2014). Furthermore, technological advances have made academic dishonesty easier. Khan and Balasubramanian (2012) report that over a third of undergraduate students surveyed reported cheating in the traditional way (e.g., copying someone's homework), while over three-quarters admitted to cheating using various technological devises (i.e., cell phones, iPods).

Given the lack of academic integrity plaguing schools today, school violence, bullying, and crime and violence, character education is a much needed and essential ingredient of a positive school environment. This is particularly true given that a lack of academic integrity is associated with a higher likelihood of dishonesty after graduation (King and Case 2014). On the other hand, character education in schools is associated with "higher academic performance, improved attendance, reduced violence, fewer disciplinary issues, reduction in substance abuse, and less vandalism" (Lahey 2013, p. 1).

> Despite most schools' emphasis on academic achievement, knowledge without character is potentially devastating.

character education
Education that emphasizes the moral and ethical aspects of an individual.

Distance Education

The number of students enrolled in online classes and degree programs continues to rise (NCES 2014a). In 2012, the latest year for which data are available, 14.2 percent of undergraduates were enrolled in at least one online course, and 11 percent were enrolled exclusively in online courses. In general, the larger the college or university, the higher the percentage of online students. Over half of students enrolled exclusively in online classes are located in the same state or jurisdiction as the institution they attend (NCES 2014b).

Online education serves a number of functions. First, it often serves a segment of the population that would not otherwise be able to attend school—older, married, full-time employees, and people from remote areas. Thus, for example, 30 states have complete school curriculums online allowing students anywhere in the state to attend classes (Watson et al. 2014). Second, some research suggests that online learning benefits those who have historically been disadvantaged in the classroom. A study by DeNeui and Dodge (2006) indicates that females in a blended course were more likely to use a learning management system (e.g., Blackboard) than males and to significantly outperform them as measured by final grades in an introductory psychology class. Third, a survey of high school administrators indicates that offering online courses is important because they allow schools to offer courses that would otherwise not be available, permit flexibility in schedules, extend the school day and year, are financially beneficial, build linkages with colleges, and prepare students for 21st-century jobs (Picciano and Seaman 2010).

Despite the many functions online learning serves, a survey of 656 employers and 215 community college students indicates that there are concerns about the quality of online education. Over half of the employers reported that they would prefer to hire a graduate from an *average* school with all face-to-face courses rather than a student from a *top* school who completed their degree entirely online (Public Agenda 2013). Furthermore, 42 percent of community college students indicate that students learn less in online-only classes compared to face-to-face classes, and nearly half expressed a desire to take fewer online courses. Finally, in a survey of more than 2,800 colleges and university officials, two-thirds of the respondents indicated that "concerns about the relative quality of online courses" will continue over the next five years (Allen and Seaman 2014).

The Debate over School Choice

Historically, children have gone to public schools in the district where they live. However, a 2014 survey of more than 5,000 U.S. adults indicates that over a quarter of American families have a school age child who is or has been educated in a nontraditional school setting (Henderson et al. 2015). School vouchers, charter schools, and private schools provide parents with alternative school choices for their children (see Figure 8.7).

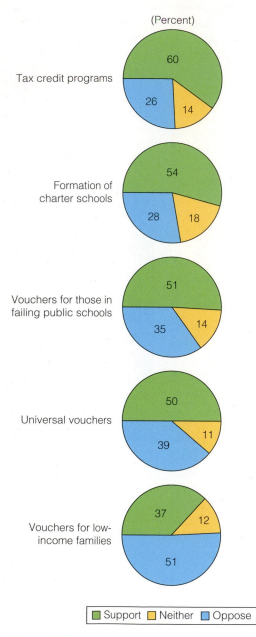

Figure 8.7 Public Support for the Expansion of School Choice (N = 2,269), 2014
SOURCE: Henderson et al. 2015.

school vouchers Tax credits that are transferred to the public or private school that parents select for their child.

School Vouchers. School vouchers and state tax credits allow families to send their children to private schools. **School vouchers** are state-funded "scholarships" that allow public school students to attend private schools. Eligibility for vouchers is usually confined to special populations, for example, low-income students, special education students, and students in chronically low-performing schools (NCSL 2013b). Similarly, "tax credit programs . . . allow businesses or individuals to contribute to organizations that distribute private-school scholarships to low income families" (Henderson et al. 2015, p. 15). Presently, about half the states have some kind of school voucher or tax credit program.

Proponents of the voucher system argue that it increases the quality of schools by creating competition for students. Those who oppose the voucher system argue that it will drain needed funds and the best students away from public schools. Opponents also argue that vouchers increase segregation because white parents use the vouchers to send their children to private schools with few minorities.

The Milwaukee Parental Choice Program is the "oldest and largest publicly funded voucher program in the United States," beginning in 1991 and continuing to the present (Cowen et al. 2013). Wisconsin state law requires an evaluation of the voucher program and its impact on academic achievement. To assess attainment, students attending voucher schools were matched (e.g., percentage black, percentage female, etc.) with those attending public schools so that an accurate comparison could be made. The results indicate that students who were exposed to a voucher environment in the eighth and ninth grades were more likely to graduate from high school, enroll at a four-year institution, and stay in school beyond their first year of college (Cowen et al. 2013).

Charter Schools. In some states, vouchers can be used for charter schools. **Charter schools** originate in contracts, or charters, which articulate a plan of instruction that local or state authorities must approve. Although foundations, universities, private benefactors, and entrepreneurs can fund charter schools, many are supported by tax dollars. There are about 6,400 charter schools in the United States with a student population of 2.5 million (Center for Research on Educational Outcomes [CREDO] 2015).

Charter schools, like school vouchers, were designed to expand schooling options and to increase the quality of education through competition. Like vouchers, charter schools have come under heavy criticism for increasing school segregation, reducing public school resources, and "stealing away" top students. Proponents argue that charter schools encourage innovation and reform, and increase student learning outcomes. Fifty-four percent of Americans are in favor of the formation of charter schools (Henderson et al. 2015) (see Figure 8.7).

A CREDO study of 41 major U.S. urban areas indicates that, overall, charter school students outperform students in traditional public schools. In 2014, students in urban charter schools had higher levels of annual growth in both mathematics and reading than their traditional public school counterparts. Such growth is equivalent to students in the urban charter schools receiving 40 days of additional learning in math and 28 days of additional learning in reading a year. Furthermore, the learning gains measured were the highest for black, Hispanic, low-income, and special education students (CREDO 2015).

Privatization of Schools. Another school choice parents can make is to send their children to a private school. The number of students enrolled in private schools in 2012, the most recent year for which data are available, was 5.3 million, a decrease from previous years (NCES 2015a). The decline in private school enrollment is associated with an increase in students attending charter schools (Ewert 2013).

Parents send their children to private schools for a variety of reasons, including the availability of academic programs and extracurricular activities, smaller class size and a lower student–teacher ratio, religious instruction, and dissatisfaction with public schools (Ewert 2013). Many people believe that private schools are superior to public schools in terms of academic achievement, and some evidence suggests that is the case. For example, private school students outperform public school students on the National Assessment of Educational Progress (NAEP), their average score being higher on the 4th grade reading test and the 4th, 8th, and 12th grade mathematics and science tests (Chen 2015).

charter schools Schools that originate in contracts, or charters, which articulate a plan of instruction that local or state authorities must approve.

In addition to traditional private schools, there's a growing movement for local or state governments to "contract out" schools and other educational services to private for-profit corporations. Some experts are "alarmed at what they see as increasingly aggressive moves by companies to make money from the K–12 system; others say the expanding role of for-profit ventures is just a natural evolution of the interplay between the private and public sectors in efforts to improve schools" (Davis 2013, p. 52). Research comparing Michigan public schools, nonprofit managed charter schools, and for-profit charter schools operated by education management organizations (EMOs) indicates that (1) charter schools, in general, have a higher proportion of black students than traditional public schools; (2) charter schools have a lower proportion of Hispanic students than traditional public schools; and (3) for-profit charter schools are less likely to enroll poor students than nonprofit charter schools. The authors suggest that the different enrollment patterns of poor students between nonprofit and for-profit charter schools may be a result of their need to maximize revenues and minimize costs (Ertas and Roch 2015).

WHAT do you THINK?

An alternative to sending your children to school is parent-based education in the home (Ray 2015). Public funds are not used for the roughly 2.2 million children who are homeschooled, saving taxpayers $24 billion a year. Do you think families who homeschool their children should be financially compensated in some way by their state or local school district?

Understanding Problems in Education

Educational reform continues to be the focus of legislators and governments across the country and the world. Although there is disagreement as to what needs to be done and how, all can agree that significant reform is needed to meet the needs of a global economy in the 21st century and, perhaps more importantly, to fulfill Horace Mann's dream of education as the "balanced wheel of social machinery," equalizing social differences among members of an immigrant nation.

First, we must invest in teacher education and in teaching practices that have been empirically documented to work in raising student outcomes. Teachers' salaries also need to better reflect the priority Americans place on children, education, and the education of children and should not be tied to student performance. As a society, we need to help teachers overcome the obstacles that have led to a growing teacher dropout crisis (see Figure 8.8).

Second, the "savage inequalities" in education, primarily based on race, ethnicity, and socioeconomic status, must be addressed. Segregation, rather than decreasing, is increasing—a reflection of housing patterns, local school districts' heavy reliance on property taxes, and immigration patterns. Public schools should provide all U.S. children with the academic and social foundations necessary to participate in society in a productive and meaningful way; however, for many children, schools perpetuate an endless downward cycle of failure, alienation, and hopelessness.

Third, the general public needs to become involved, not just in their children's education but also in the *institution* of education. An uneducated and unthinking populace hurts all of society, particularly in terms of global competitiveness. As Kohn (2011) observes, children from low-income families continue to be taught "the pedagogy of poverty" (Haberman 1991) or what has been called the "McEducation of the Negro" (Hopkinson 2011). Like Big Macs, children are being packaged in one-size-fits-all wrappers—with learning how to think, explore, question, and debate being replaced by worksheets and standardized tests. Sadly, as education historian Diane Ravitch notes, the present reform movement that "once was an effort to improve the quality of education [has] turned into an accounting strategy" (Ravitch 2010, p. 16).

Fourth, as conflict theorists would note, we must be wary of market principles in schools and of those who advocate them. Educational policy in the United States is

> Like Big Macs, children are being packaged in one-size-fits-all wrappers—with learning to think, explore, question, and debate being replaced by worksheets and standardized tests.

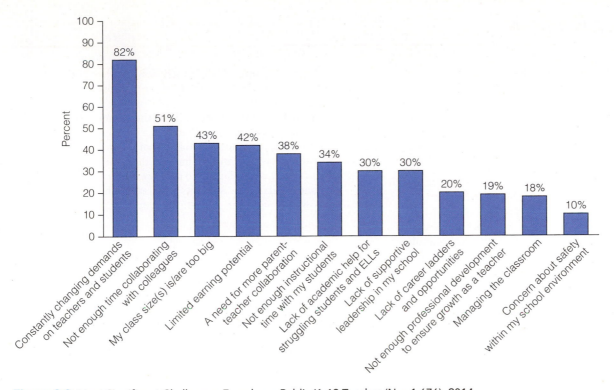

Figure 8.8 Most Significant Challenges Faced as a Public K–12 Teacher (N = 1,676), 2014
SOURCE: Primary Sources 2014.

increasingly influenced by K–12 corporate providers, many of which have political ties and vested interests. The number of schools operated by for-profit education management companies has increased from 6 in 1996 to 758 in 2012 (Davis 2013).

Finally, as a society, we must attend to and be cognizant of the importance of early childhood development (UNESCO 2014c). As Poliakoff (2006) notes:

> Children's physical, emotional, and cognitive development are profoundly shaped by the circumstances of their preschool years. Before some children are even born, birth weight, lead poisoning, and nutrition have taken a toll on their capacity for academic achievement. Other factors—excessive television watching, little exposure to conversation or books, parents who are absent or distracted, inadequate nutrition—further compromise their early development. (p. 10)

We must provide support to families so that children grow up in healthy, safe, and nurturing environments. Children are the future of our nation and of the world. Whatever resources we provide to improve the lives and education of children are sure to be wise investments in our collective future.

Students work on their laptops at the Philadelphia School of the Future. Sponsored as part of a public/private partnership between the city of Philadelphia and Microsoft, this model for future schools opened on September 6, 2006. The school is 95 percent minority, 47 percent male and 53 percent female, and 41 percent of its students are enrolled in AP courses.

Chapter Review

- **Do all countries educate their citizens?**
  No. Many societies have no formal mechanism for educating the masses. As a result, millions of adults around the world are illiterate. The problem of illiteracy is greater in developing countries than in developed nations and, worldwide, disproportionately affects women more than men.

- **According to the structural-functionalist perspective, what are the functions of education?**
  Education has four major functions. The first is instruction—that is, teaching students knowledge and skills. The second is socialization that, for example, teaches students to respect authority. The third is sorting individuals into statuses by providing them with credentials. The fourth function is custodial care—a babysitting agency, of sorts.

- **What is a self-fulfilling prophecy?**
  A self-fulfilling prophecy occurs when people act in a manner consistent with the expectations of others.

- **What variables predict school success?**
  Three variables tend to predict school success. Socioeconomic status predicts school success: The higher the socioeconomic status, the higher the likelihood of school success. Race and ethnicity predict school success, with nonwhites and Hispanics having more academic difficulty than whites and non-Hispanics. Gender also predicts success, although it varies by grade level.

- **What are the four reasons given as to why black and Hispanic Americans, in general, do not perform as well in school as their white and Asian counterparts?**
  First, because race and ethnicity are so closely tied to socioeconomic status, it appears that race or ethnicity alone can determine school success when, in fact, it may be socioeconomic status. Second, many minorities are not native English speakers, making academic achievement significantly more difficult. Third, standardized tests have been demonstrated to be culturally biased favoring those in the upper and middle classes. Finally, racial and ethnic minorities may be the victims of racism and discrimination.

- **What are some of the conclusions of the study summarized in the *Social Problems Research Up Close* feature?**
  The results of the study indicate that when a student leaves school, it is a process rather than a decision, often based on social circumstances that are out of the student's control—criminal victimization, illness in the family, housing instability, and a failure to connect with caring and concerned others. Students who leave school are resilient, however, and nearly two-thirds return to graduate.

- **What are some of the problems associated with the American school system?**
  One of the main problems is the lack of funding and the resulting reduction in programs and personnel. Low levels of academic achievement in our schools are also of some concern—particularly when U.S. data are compared with data from other industrialized countries. Minority dropout rates are high, and school violence, crime, and discipline problems continue to be a threat. School facilities are in need of repair and renovations, and personnel, including teachers, have been found to be deficient. Higher education must also address several challenges.

- **What is meant by value-added measurement (VAM), and why are there concerns about its use?**
  VAM is the use of student achievement data to assess teachers' effectiveness. Critics of VAM argue that assessing teachers based on student performance assumes all else constant, and ignores the reality of student differences in such nonschool factors as family life, poverty, emotional and physical obstacles, and the like.

- **What are the arguments for and against school choice?**
  Proponents of school choice programs argue that they reduce segregation and that schools that have to compete with one another will be of a higher quality. Opponents argue that school choice programs increase segregation and treat disadvantaged students unfairly. Low-income students cannot afford to go to private schools, even with vouchers. Furthermore, those opposed to school choice are quick to note that using government vouchers to help pay for religious schools is unconstitutional.

Test Yourself

1. All societies have some formal mechanism to educate their citizenry.
   a. True
   b. False
2. According to structural functionalists, which of the following is not a major function of education?
   a. Teach students knowledge and skills
   b. Socialize students into the dominant culture
   c. Indoctrinate students into the capitalist ideology
   d. Provide custodial care for children
3. Common Core State Standards have been embraced by and implemented in all states.
   a. True
   b. False
4. Which of the following statements is true about dropouts in the United States?
   a. Dropout rates in the United States are increasing.
   b. Students who drop out of school, on the average, make more money because they've been working longer than their high school graduate counterparts.

c. Dropout rates in the United States are similar between racial and ethnic subgroups.

d. Dropouts have poorer health and shorter life expectancy.

5. Which of the following statements about bullying is not true?

a. Student bullies tend to be marginal students.

b. Bullying may be direct or indirect.

c. There are serious consequences for the student victims of bullying.

d. Over 50 percent of 12- to 18-year-olds report being bullied at school.

6. Character education is associated with increased student achievement.

a. True

b. False

7. Over the next decade, the need for teachers is likely to

a. decrease because of the baby boom retirements.

b. increase because of immigration patterns.

c. remain stable.

d. decrease because of the aging population.

8. Online students compared to face-to-face students are more likely to be

a. single.

b. part-time employees.

c. inner-city dwellers.

d. older.

9. Nonprofit schools are increasing at a faster rate than for-profit schools.

a. True

b. False

10. School vouchers are state-funded "scholarships" that are transferred to the private school that parents select for their child.

a. True

b. False

Answers: 1. B; 2. C; 3. B; 4. D; 5. D; 6. A; 7. B; 8. D; 9. B; 10. A.

Key Terms

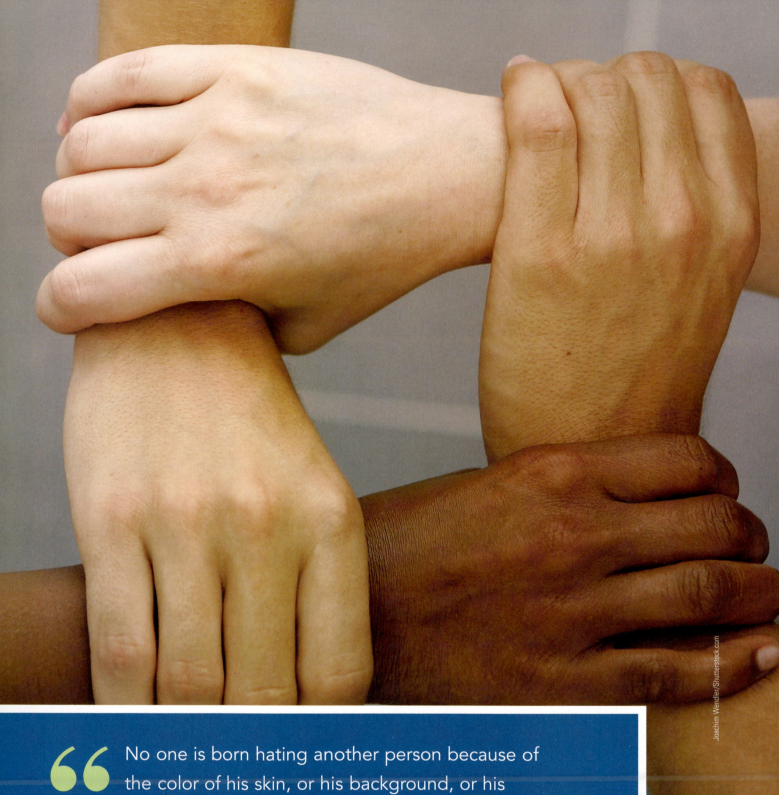

Joachim Wendler/Shutterstock.com

"No one is born hating another person because of the color of his skin, or his background, or his religion. People must learn to hate, and if they can learn to hate, they can learn to love..."

NELSON MANDELA

Race, Ethnicity, and Immigration

Learning Objectives

After studying this chapter, you will be able to . . .

1 Explain the idea that race and ethnicity are socially constructed, and identify and give examples of six patterns of race and ethnic group interaction.

2 Describe racial and ethnic diversity in the United States and the degree to which U.S. adults view race relations as problematic.

3 Give examples of key historical and current immigration policies, and discuss myths versus facts about immigration and immigrants.

4 Compare and contrast how structural-functionalism, conflict theory, and symbolic interactionism view issues related to race, ethnicity, and/or immigration.

5 Define and give examples of different forms of racism and prejudice, and explain how prejudice can be learned through socialization and the media.

6 Differentiate between individual and institutional discrimination, and describe discrimination in employment, housing, and education; racial microaggressions, and hate crimes.

7 Distinguish the Equal Employment Opportunity Commission's role in responding to employment discrimination, examine the issue of affirmative action, and discuss educational strategies to promote diversity and multicultural awareness, and apologies and reparations as a means of achieving racial reconciliation.

8 Identify at least two things that are important for achieving racial and ethnic equality.

Each year Native Americans organize protests against the annual holiday that celebrates Christopher Columbus.

AP Images/Elaine Thompson

COLUMBUS DAY is a federal holiday to commemorate the arrival of Christopher Columbus to the "New World" in 1492. But a growing number of schools, cities, and states are abandoning the traditional celebration of Christopher Columbus and are, instead, celebrating Indigenous People's Day, or Native Americans' Day (Grinberg 2014). In 2014, Minneapolis and Seattle joined the anti–Columbus Day movement and celebrated their first-ever Indigenous Peoples' Day. Seattle city councilmember Kshama Sawant explained why she supported the city's rejection of Columbus Day, saying, "We're making sure that we acknowledge the absolute horrors of colonization and conquering that happened in the Americas at the hands of the European so-called 'explorers,' . . . Columbus did not discover America. . . . He plundered it and he brutalized its people" (quoted in Futon 2014).

In his diary, Columbus describes the Arawak people he encountered when he landed on the Bahama Islands:

They were well-built, with good bodies and handsome features. . . . They do not bear arms, and do not know them, for I showed them a sword, they took it by the edge and cut themselves out of ignorance. They have no iron. Their spears are made of cane. . . . They would make fine servants. . . . With fifty men we could subjugate them all and make them do whatever we want. (quoted in Halper 2014)

Bartolome de las Casas, a priest who journeyed with Columbus to Cuba, described the brutal murder of the island's native people:

Endless testimonies . . . prove the mild and pacific temperament of the natives. . . . But our work was to exasperate, ravage, kill, mangle and destroy. . . . And the Christians . . . began to carry out massacres and strange cruelties against them. They attacked the towns and spared neither the children nor the aged nor pregnant women nor women in childbed, not only stabbing them and dismembering them but cutting them to pieces as if dealing with sheep in the slaughter house. They laid bets as to who, with one stroke of the sword, could split a man in two or could cut off his head or spill out his entrails with a single stroke of the pike. They took infants from their mothers' breasts, snatching them by the legs and pitching them head first against the crags or snatched them by the arms and threw them into the rivers, roaring with laughter and saying as the babies fell into the water, "Boil there, you offspring of the devil!" Other infants they put to the sword along with their mothers and anyone else who happened to be nearby. They made some low wide gallows on which the hanged victim's feet almost touched the ground, stringing up their victims in lots of thirteen, in memory of Our Redeemer and His twelve Apostles, then set burning wood at their feet and thus burned them alive. (quoted in Halper 2014)

Throughout history, humans have encountered other humans who are different from them—different in language, beliefs, norms, and physical appearance. And we don't have to travel to another continent or country to find these differences: we find religious, racial, and ethnic diversity in our own communities, and even in our own families. This chapter is concerned with racial and ethnic differences globally and within the United States. We focus on the nature and consequences of racism, prejudice, and discrimination toward minority groups, and strategies to reduce these problems. A **minority group** is a category of people who have unequal access to positions of power, prestige, and wealth and are targets of prejudice and discrimination. Minority status is not based on numerical representation in society but rather on social status. For example, although Hispanic individuals outnumber non-Hispanic whites in California, Texas, and New Mexico, they are considered a "minority" because they are underrepresented in positions of power, prestige, and wealth, and because they are targets of racism, prejudice, and discrimination. In this chapter, we also examine issues related to U.S. immigration, because immigrants often bear the double burden of being minorities *and* foreigners who are not welcomed by many native-born Americans. Note that some very important issues related to race are not covered in this chapter because they are discussed in other chapters. Chapter 4, for example, discusses race and crime, treatment by police, and the criminal justice system, and Chapter 6 covers race and poverty.

The Global Context: Diversity Worldwide

All human beings belong to the same species. Yet, because of differences in physical appearance and culture, humans are often classified into categories based on race and ethnicity. After examining the social construction of race and ethnicity, we review patterns of interaction among racial and ethnic groups and examine racial and ethnic diversity in the United States.

The Social Construction of Race and Ethnicity

The concept of "race" has been described as "one of the most misunderstood, misused, and often dangerous concepts of the modern world" (Marger 2012, p. 12). The term *race* has been used to describe people of a particular nationality (the Mexican "race"), religion (the Jewish "race"), skin color (the white "race"), and even the entire human species (the human "race"). Confusion around the term *race* stems from the fact that it has both biological and social meanings.

Race as a Biological Concept. As a biological concept, *race* refers to a classification of people based on hereditary physical characteristics such as skin color, hair texture, and the size and shape of the eyes, lips, and nose. But there are no clear guidelines for distinguishing racial categories on the basis of visible traits. Skin color is not black or white but rather ranges from dark to light with many gradations of shades. Noses are not either broad or narrow but come in a range of shapes. Physical traits come in an infinite number of combinations. For example, a person with dark skin can have a broad nose (a common combination in West Africa), a narrow nose (a common combination in East Africa), or even blond hair (a combination found in Australia and New Guinea).

Another problem with race as a biological concept is that the physical traits used to mark a person's race are arbitrary. What if we classified people into racial categories based on eye color instead of skin color? Or hair color? Or blood type? What if all dark-haired individuals were considered to belong to one race, and all light-haired people to another race? Is there any scientific reason for selecting certain traits over others in determining racial categories? The answer is no. As a biological concept, "races are not scientifically valid because there are no objective, reliable, meaningful criteria scientists can use to construct or identify racial groupings" (Mukhopadhyay et al. 2007, p. 5).

The science of genetics also challenges the biological notion of race. Geneticists have discovered that the genes of any two unrelated people, chosen at random from around the globe, are 99.9 percent alike (Ossorio and Duster 2005). Furthermore, "most human genetic variation—approximately 85 percent—can be found between any two individuals from the same group (racial, ethnic, religious, etc.). Thus, the vast majority of variation is within-group variation" (Ossorio and Duster 2005, p. 117). Finally, classifying people into different races fails to recognize that, over the course of human

> What if we classified people into racial categories based on eye color instead of skin color?

minority group A category of people who have unequal access to positions of power, prestige, and wealth in a society and who tend to be targets of prejudice and discrimination.

history, migration and intermarriage have resulted in the blending of genetically transmitted traits:

> To summarize, races are unstable, unreliable, arbitrary, culturally created divisions of humanity. This is why scientists...have concluded that race, as scientifically valid biological divisions of the human species, is fiction not fact. (Mukhopadhyay et al. 2007, p. 14)

Race as a Social Concept. The idea that race is socially created is one of the most important lessons in understanding race from a sociological perspective. The social construction of race means that "the actual meaning of race lies not in people's physical characteristics, but in the historical treatment of different groups and the significance that society gives to what is believed to differentiate so-called racial groups" (Higginbotham and Andersen 2012, p. 3). The concept of race grew out of social institutions and practices in which groups defined as "races" have been enslaved or otherwise exploited.

People learn to perceive others according to whatever racial classification system exists in their culture. Systems of racial classification vary across societies and change over time. For example, as late as the 1920s, U.S. Italians, Greeks, Jews, Irish, and other "white" ethnic groups were not considered to be white. Over time, the category of "white" changed so that it included these groups. As an example of cross-cultural variation in racial categories, Brazilians use dozens of terms to racially categorize people based on various combinations of physical characteristics, although officially, the major racial categories in Brazil are *brancos* (white), *pardos* (brown or mulatto), *pretos* (black), and *amarelos* (yellow).

Rachel Dolezal "passed" as African-American before she was outed by her parents as being white.

Incorporating both biological and social meanings of race, we define **race** as a category of people who are perceived to share distinct physical characteristics that are deemed socially significant. The significance of race is not biological but social and political, because race is used to separate "us" from "them" and becomes a basis for unequal treatment of one group by another. Despite increasing acceptance that "there is no biological justification for the concept of 'race'" (Brace 2005, p. 4), its social significance continues throughout the world.

WHAT do you THINK?

Rachel Dolezal identified as black (or biracial) for years. She attended a historically black university, married a black man, and was a civil rights activist, serving as president of her local chapter of the NAACP. In 2015, Dolezal's parents revealed that Rachel is not black. "She's clearly our birth daughter, and we're clearly Caucasian—that's just a fact," said Rachel's father (quoted in Pérez-Peña 2015). Reactions to the public "outing" of Ms. Dolezal ranged from condemnation to applause. Some criticized Dolezal for dishonesty and disrespect of black culture, while others admired her dedication to civil rights activism and argued that just as Caitlyn Jenner is transgender and can say she is a woman (see Chapter 10), so can Dolezal be "transracial" and say she is black. How do you view Rachel Dolezal's choice to identify as black?

race A category of people who are perceived to share distinct physical characteristics that are deemed socially significant.

ethnicity A shared cultural heritage or nationality.

Ethnicity as a Social Construction. **Ethnicity**, which refers to a shared cultural heritage, nationality, or lineage, is also partly socially constructed. Ethnicity can be distinguished on the basis of language, forms of family structures and roles of family members, religious beliefs and practices, dietary customs, forms of artistic expression such as music and dance, and national origin or origin of one's parents.

Although the Census Bureau defines Hispanic or Latino as "a person of Cuban, Mexican, Puerto Rican, South or Central American, or other Spanish culture or origin regardless of race" (Ennis et al. 2011, p. 2), when it comes down to collecting census data on the U.S. population, a person is Hispanic if they say they are Hispanic (see Table 9.1). Hence, ethnicity is socially constructed.

Patterns of Racial and Ethnic Group Interaction

When two or more racial or ethnic groups come into contact, one of several patterns of interaction occurs; these include genocide, expulsion, segregation, acculturation, pluralism, and assimilation.

Genocide refers to the deliberate, systematic annihilation of an entire nation or people. The European invasion of the Americas, beginning in the 16th century, resulted in the decimation of most of the original inhabitants of North and South America. Some native groups were intentionally killed; others fell victim to diseases brought by the Europeans. In the 20th century, Hitler led the Nazi extermination of 12 million people, including 6 million Jews, in what is known as the Holocaust. In the early 1990s, ethnic Serbs attempted to eliminate Muslims from parts of Bosnia—a process they called "ethnic cleansing." In 1994, genocide took place in Rwanda when Hutus slaughtered hundreds of thousands of Tutsis. Genocide is continuing in the Darfur region of Sudan, where the Sudanese government, using Arab *Janjaweed* militias, its air force, and organized starvation, is systematically killing the African Muslim communities because some among them have challenged the authoritarian rule of the Sudanese government (see also Chapter 15).

Expulsion occurs when a dominant group forces a subordinate group to leave the country or to live only in designated areas of the country. The 1830 Indian Removal Act called for the relocation of eastern tribes to land west of the Mississippi River. The movement, lasting more than a decade, has been called the Trail of Tears because tribes were forced to leave their ancestral lands and endure harsh conditions of inadequate supplies and epidemics that caused illness and death. After Japan's attack on Pearl Harbor in 1941, 120,000 Japanese Americans, who became viewed as threats to national security, were forced from their homes and into evacuation camps surrounded by barbed wire.

Segregation refers to the physical separation of two groups in residence, workplace, and social functions. Segregation can be *de jure* (Latin meaning "by law") or *de facto* ("in fact"). Between 1890 and 1910, a series of U.S. laws, which came to be known as *Jim Crow laws,* were enacted to separate blacks from whites by prohibiting blacks from using "white" buses, hotels, restaurants, and drinking fountains. In 1896, the U.S. Supreme Court (in *Plessy v. Ferguson*) supported de jure segregation of blacks and whites by declaring that "separate but equal" facilities were constitutional. Blacks were forced to live in separate neighborhoods and attend separate schools. Beginning in the 1950s, various rulings overturned these Jim Crow laws, making it illegal to enforce racial segregation. Although de jure segregation is illegal in the United States, de facto segregation still exists in the tendency for racial and ethnic groups to live and go to school in segregated neighborhoods.

Acculturation refers to adopting the culture of a group different from the one in which a person was originally raised. Acculturation may involve learning the dominant language, adopting new values and behaviors, and changing the spelling of the

TABLE 9.1 Who's Hispanic? The U.S. Census Bureau Approach to Defining Who Is Hispanic

Q.	I immigrated to Phoenix from Mexico. Am I Hispanic?
A.	You are if you say so.
Q.	My parents moved to New York from Puerto Rico. Am I Hispanic?
A.	You are if you say so.
Q.	My grandparents were born in Spain but I grew up in California. Am I Hispanic?
A.	You are if you say so.
Q.	I was born in Maryland and married an immigrant from El Salvador. Am I Hispanic?
A.	You are if you say so.
Q.	My mom is from Chile and my dad is from Iowa. I was born in Des Moines. Am I Hispanic?
A.	You are if you say so.
Q.	I was born in Argentina but grew up in Texas. I don't consider myself Hispanic. Does the Census count me as Hispanic?
A.	Not if you say you aren't.

SOURCE: Passel and Taylor 2009.

genocide The deliberate, systematic annihilation of an entire nation or people.

expulsion Occurs when a dominant group forces a subordinate group to leave the country or to live only in designated areas of the country.

segregation The physical separation of two groups in residence, workplace, and social functions.

acculturation The process of adopting the culture of a group different from the one in which a person was originally raised.

These Native American children were forced to adopt white U.S. culture as part of their "education" in white-run schools.

family name. In some instances, acculturation may be forced. In the 1800s, the U.S. federal government began allowing religious institutions and other private organizations to forcibly remove American Indian children from their homes and transport them to boarding schools hundreds or even thousands of miles away. By 1887, over 14,000 Indian children were living in the more than 200 boarding schools where the children were forced to abandon their Indian language, culture, and religion and adopt White ways. "The premise of these schools was that Indian children needed a 'proper' education and that Indian religion and culture were inferior to white religion and culture" (American Civil Liberties Union 2014, p. 56).

Pluralism refers to a state in which racial and ethnic groups maintain their distinctness but respect each other and have equal access to social resources. In Switzerland, for example, four ethnic groups—French, Italians, Swiss Germans, and Romansch—maintain their distinct cultural heritage and group identity in an atmosphere of mutual respect and social equality.

Assimilation is the process by which formerly distinct and separate groups merge and become integrated as one. *Primary assimilation* occurs when members of different racial or ethnic groups are integrated in personal, intimate associations, as with friends, family, and spouses. *Secondary assimilation* occurs when different groups become integrated in public areas and in social institutions, such as neighborhoods, schools, the workplace, and in government.

Assimilation is sometimes referred to as the "melting pot," whereby different groups come together and contribute equally to a new, common culture. Although the United States has been referred to as a melting pot, in reality, many minorities have been excluded or limited in their cultural contributions to the predominant white Anglo-Saxon Protestant culture.

Racial and Ethnic Group Diversity in the United States

The first census in 1790 divided the U.S. population into four groups: free white males, free white females, slaves, and other people (including free blacks and Indians). To increase the size of the slave population, the *one-drop rule* specified that even one drop of "negroid" blood defined a person as black and therefore eligible for slavery. The "one-drop rule" is still operative today: Biracial individuals are typically seen as a member of whichever group has the lowest status (Wise 2009).

In 1960, the census recognized only two categories: white and nonwhite. In 1970, the census categories consisted of white, black, and "other" (Hodgkinson 1995). In 1990, the U.S. Census Bureau recognized four racial classifications: (1) white; (2) black; (3) American Indian, Aleut, or Eskimo; and (4) Asian or Pacific Islander. The 1990 census also included the category of "other." Beginning with the 2000 census, racial categories expanded to include Native Hawaiian or other Pacific Islander, and also allowed individuals the option of identifying themselves as being more than one race rather than checking only one racial category (see Figure 9.1).

Although U.S. citizens come from a variety of ethnic backgrounds, the largest ethnic population in the United States is of Hispanic origin. The Census Bureau began

pluralism A state in which racial and ethnic groups maintain their distinctness but respect each other and have equal access to social resources.

assimilation The process by which formerly distinct and separate groups merge and become integrated as one.

collecting data on the U.S. Hispanic population in 1970.

As you read the following section on U.S. census data on race and Hispanic origin, keep in mind that the use of racial and ethnic labels is often misleading and imprecise. The ethnic classification of "Hispanic/Latino," for example, lumps together disparate groups such as Puerto Ricans, Mexicans, Cubans, Venezuelans, Colombians, and others from Latin American countries. The racial term *American Indian* includes more than 300 separate tribal groups that differ enormously in language, tradition, and social structure. The racial label *Asian* includes individuals from China, Japan, Korea, India, the Philippines, or one of the countries of Southeast Asia. The term *Asian American* is used to describe people with Asian racial features who are born in the United States, as well as those who immigrate to the United States.

U.S. Census Data on Race and Hispanic Origin

Census data show that the U.S. population is becoming increasingly diverse: By 2044, more than half of all Americans are projected to belong to a racial/ethnic minority group and in 2020, more than half of U.S. children are expected to be a racial or ethnic minority (Colby and Ortman 2015). In 19 of the largest (in population) U.S. counties, whites make up less than half of the population (Krogstad 2015). Currently, racial and ethnic minority populations outnumber non-Hispanic whites in four states: California, New Mexico, Hawaii, and Texas. In these states, Hispanics are the largest minority group, except for Hawaii, where the largest minority group is Asian American (Humes et al. 2011). And in 10 states, racial and ethnic minority youth outnumber non-Hispanic white youth (Tavernise 2011). Figure 9.2 shows the U.S. population by race and Hispanic origin in 2014 and projected in 2060.

More than half of the U.S. Hispanic population is Mexican, with other Hispanics having ties to Puerto Rico, Central America, South America, Cuba, and other countries. More than half of the U.S. Hispanic population lives in just three states: California, Texas, and Florida.

One of the most common confusions about Hispanic origin is the question of whether "Hispanic" is a race or an ethnicity. According to the federal government, Hispanic origin is considered an ethnicity, not a race. But many people—Hispanic and non-Hispanic—think otherwise. Although the 2010 census included instructions that stated, "for this census, Hispanic origins are not races," many Hispanics identified their race as "Latino," "Mexican," "Puerto Rican," "Salvadoran," or other ethnicity or national origin, which the census classified in the category of "Some Other Race" (see Table 9.2). Because of the widespread view of "Hispanic" as a race, the Census Bureau is considering whether to make "Hispanic" a racial instead of an ethnic category for the 2020 census (Ayala and Huet 2013). The current Census Bureau classification system does not allow people of mixed Hispanic or Latino ethnicity to identify themselves as such. Individuals with one Hispanic and one non-Hispanic parent still must say that they are either Hispanic or not Hispanic.

> ➔ **NOTE: Please answer BOTH Question 5 about Hispanic origin and Question 6 about race. For this census, Hispanic origins are not races.**
>
> 5. **Is this person of Hispanic, Latino, or Spanish origin?**
>   - ☐ **No,** not of Hispanic, Latino, or Spanish origin
>   - ☐ Yes, Mexican, Mexican Am., Chicano
>   - ☐ Yes, Puerto Rican
>   - ☐ Yes, Cuban
>   - ☐ Yes, another Hispanic, Latino, or Spanish origin — *Print origin, for example, Argentinean, Colombian, Dominican, Nicaraguan, Salvadoran, Spaniard, and so on.*↳
>
> 6. **What is this person's race?** Mark ☒ one or more boxes.
>   - ☐ White
>   - ☐ Black, African Am., or Negro
>   - ☐ American Indian or Alaska Native — *Print name of enrolled or principal tribe.*↳
>
>   - ☐ Asian Indian   ☐ Japanese   ☐ Native Hawaiian
>   - ☐ Chinese   ☐ Korean   ☐ Guamanian or Chamorro
>   - ☐ Filipino   ☐ Vietnamese   ☐ Samoan
>   - ☐ Other Asian — *Print race, for example, Hmong, Laotian, Thai, Pakistani, Cambodian, and so on.*↳   ☐ Other Pacific Islander — *Print race, for example, Fijian, Tongan, and so on.*↳
>
>   - ☐ Some other race — Print race.↳

Figure 9.1 2010 Census Questions
SOURCE: Humes et al. 2011.

> One of the most common confusions about Hispanic origin is the question of whether "Hispanic" is a race or an ethnicity.

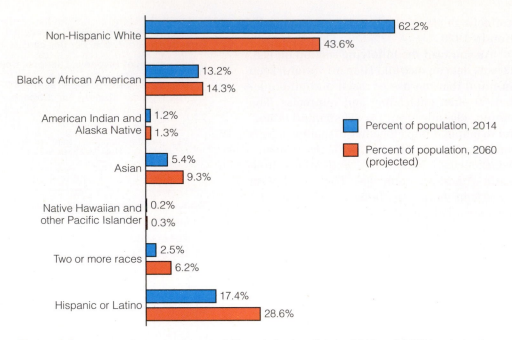

Figure 9.2 U.S. Population by Race and Hispanic/Latino Origin, 2014 and 2060 (projected)
SOURCE: Colby and Ortman 2015.

TABLE 9.2 Racial Identification of Hispanics/ Latinos in the United States, 2013

White	65.9%
Black or African American	2.1%
American Indian/Alaskan Native	0.9%
Asian	0.3%
Native Hawaiian/Pacific Islander	0.1%
Some Other Race	26.2%
Two or More Races	4.5%

SOURCE: Stepler and Brown 2015.

Mixed-Race Identity

Individuals who identify as "two or more races" comprise the fastest-growing racial group in the United States (Frey 2015). The number of persons identifying as both black and white more than doubled between 2000 and 2010 (Frey 2015).

The multiracial population has grown as mixed-race marriages have increased over recent years. Until 1967, 16 states had **antimiscegenation laws** banning interracial marriage of whites and nonwhites—primarily blacks, but in some cases, also Native Americans and Asians. In 1967, the Supreme Court (in *Loving v. Virginia*) declared these laws unconstitutional. In 2010, 15 percent of new marriages in the United States were between spouses with different racial or ethnic identities—more than double the percentage in 1980 (6.7 percent) (Wang 2012; see Figure 9.3).

Attitudes toward interracial marriages have changed dramatically over the last few generations. The percentage of U.S. adults who disapproved of marriage between blacks and whites decreased from 94 percent in 1958, to 29 percent in 2002, to 11 percent in 2013 (Gallup Organization 2013). More than 4 in 10 Americans view the increase in intermarriages as a change for the better in society; 1 in 10 say it has been a change for the worse (the remaining share say it doesn't make a difference). When asked, "How would you react if a member of your family were going to marry someone of a different race or ethnicity?" 63 percent of Americans say they would be fine with it. Individuals most likely to have positive attitudes about intermarriage are minorities, younger, more educated, liberal, and living in the eastern or western states (Wang 2012).

Race and Ethnic Group Relations in the United States

Despite significant improvements over the last two centuries, race and ethnic group relations continue to be problematic. In response to a question asking whether relations between blacks and whites will always be a problem for the United States or whether a solution will eventually be worked out, 4 in 10 U.S. adults said that race relations will

antimiscegenation laws
Laws banning interracial marriage until 1967, when the Supreme Court (in *Loving v. Virginia*) declared these laws unconstitutional.

always be a problem (Gallup Organization 2013). A Gallup Poll asked a national sample of U.S. adults to rate relations between various groups in the United States. Table 9.3 presents results of this survey. Relations between various racial and ethnic groups are influenced by prejudice and discrimination (discussed later in this chapter). Race and ethnic relations are also complicated by issues concerning immigration—the topic we turn to next.

> A Gallup Organization poll asked U.S. adults, "If blacks and whites honestly expressed their true feelings about race relations, do you think this would do more to bring races together or cause greater racial division?" More than half (56 percent) replied "bring races together," and about a third (37 percent) said "cause greater division" (7 percent had no opinion) (Gallup Organization 2013). What would your answer be? Why?

Immigrants in the United States

The growing racial and ethnic diversity of the United States is largely the result of immigration as well as the higher average birthrates among many minority groups. Immigration generally results from a combination of "push" and "pull" factors. Adverse social, economic, and/or political conditions in a given country "push" some individuals to leave that country, whereas favorable social, economic, and/or political conditions in other countries "pull" some individuals to those countries.

U.S. Immigration: A Historical Perspective

For the first 100 years of U.S. history, all immigrants were allowed to enter and become permanent residents. The continuing influx of immigrants, especially those coming from nonwhite, non-European countries, created fear and resentment among native-born Americans, who competed with immigrants for jobs and who held racist views toward some racial and ethnic immigrant populations. America's open-door policy on immigration ended in 1882 with the Chinese Exclusion Act, which suspended the entrance of the Chinese to the United States for 10 years and declared Chinese ineligible for U.S. citizenship (this act was repealed in 1943). The Immigration Act of 1917 required all immigrants to pass a literacy test before entering the United States. Immigration legislation passed in the 1920s established a quota system, limiting the numbers of immigrants from specific countries. Under the quota system, 70 percent of immigrant slots were allotted to people from just three countries: United Kingdom, Ireland, and Germany. In 1965, the passage of the Hart-Celler Act abolished the national origins quota system that had been in place since the 1920s, and it instituted a system that gave preference to immigrants who had family in the United States or who had job skills. The Hart-Celler Act was an extension of the civil rights movement, as it was designed to end discrimination based on race and ethnicity, and it continues to shape immigration law today.

The percentage of the U.S. population that is foreign-born is expected to increase from 13.3 percent in 2014 to 18.8 percent in 2060 (Colby and Ortman 2015). More than three out of four foreign-born residents of the United States are either naturalized citizens or in the United States

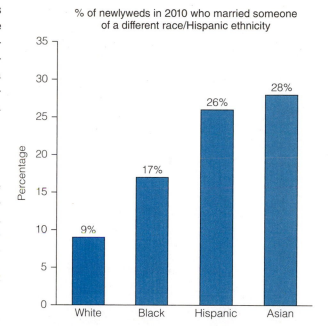

Figure 9.3 U.S. Intermarriage Rates, by Race and Hispanic Origin
SOURCE: Wang 2012.

TABLE 9.3 Perceptions of Race and Ethnic Relations in the United States, 2013

A national sample of U.S. adults was asked to rate relations between various groups in the United States. The results are depicted in this table.

	Very or Somewhat Good	Very or Somewhat Bad
Whites and blacks	70%	30%
Whites and Hispanics	70%	29%
Blacks and Hispanics	60%	32%
Whites and Asians	87%	10%

Note: Percentages do not add up to 100 because of "no opinion" responses.
SOURCE: Gallup Organization 2013.

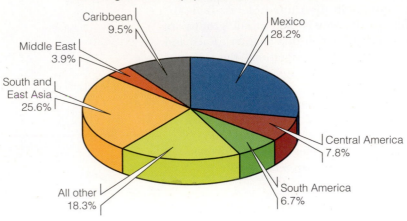

Total foreign-born U.S. population in 2012: 40.7 million

- Mexico 28.2%
- Central America 7.8%
- South America 6.7%
- All other 18.3%
- South and East Asia 25.6%
- Middle East 3.9%
- Caribbean 9.5%

Figure 9.4 U.S. Foreign-Born Residents by Region of Birth, 2012
SOURCE: Brown and Patten 2014.

legally (Pew Hispanic Center 2013). More than half of the U.S. foreign-born population came from either Mexico or South or East Asia (see Figure 9.4).

Guest Worker Program

The United States has two guest worker programs that allow employers to import unskilled labor for temporary or seasonal work: the H-2A program for agricultural work and the H-2B program for nonagricultural work. H-2 visas generally do not permit guest workers to bring their families to the United States.

Immigrant "guest workers" are hardly treated like "guests"; these workers are systematically exploited and abused. Guest workers are bound to the employers who "import" them so they are not able to change jobs if they are mistreated. Guest workers are often cheated out of pay and forced to live in squalid conditions; and although they perform some of the most difficult and dangerous jobs in the United States, many who are injured on the job are unable to obtain medical treatment and workers' compensation benefits. Immigrant women working at low-wage jobs are often targets of sexual violence. If guest workers complain about mistreatment, they are threatened with deportation or being "blacklisted" and unable to find another job. Rampant with labor and human rights abuses, the guest worker program was described as a modern-day system of indentured servitude (Southern Poverty Law Center [SPLC] 2013). In 2015, new rules were added to the H-2B program to protect workers from some of these abuses.

Illegal Immigration

Illegal immigration occurs when immigrants enter the United States without going through legal channels such as the H-2 visa program, and when immigrants who were admitted legally stay past the date they were required to leave. There were 11.2 million unauthorized immigrants living in the United States in 2012 (Pew Research Center 2015). More than half of the total unauthorized immigrant population lives in just four states: California, Texas, Florida, and New York. The majority of unauthorized immigrants (59 percent) are from Mexico, followed by El Salvador, Guatemala, Honduras, and China.

Border Crossing. U.S. Customs and Border Protection, an agency within the Department of Homeland Security, has more than 21,000 border patrol agents who are responsible for patrolling 6,000 miles of Canadian and Mexican borders and more than 2,000 miles of coastal waters around Florida and Puerto Rico (U.S. Customs and Border Protection 2013). Despite the border patrol, along with a "fence" at the U.S.–Mexico border, people continue to find ways to illegally cross the U.S. border.

This photo depicts the kind of substandard housing in which many immigrant guest workers are forced to live.

Southern Poverty Law Center

Some people cross (or attempt to cross) the U.S.–Mexican border with the help of a *coyote*—a hired guide who typically charges $3,000 to $5,000 to lead people across the border (Maril 2011). Crossing the border illegally involves a number of risks, including death from drowning (e.g., while trying to cross the Rio Grande) or dehydration. Border crossers also risk encounters with members of **nativist extremist groups**—organizations that not only advocate restrictive immigration policy but also encourage their members to use vigilante tactics to confront or harass suspected unauthorized immigrants.

Unauthorized Immigrants in the Workforce. In 2012, there were 8.1 million undocumented immigrants in the U.S. labor force, comprising about 5 percent of the U.S. workforce (Pew Research Center 2015). Sociologist Robert Maril (2004) noted that "[t]he vast majority of illegal immigrants leave their home countries to work hard, save their money, then return to their homeland.... These individuals do not travel their difficult and dangerous journeys searching for a welfare handout; they immigrate to work" (pp. 11–12).

Chris Simoux is an example of an American who patrols the U.S.–Mexico border looking for unauthorized immigrants. Such self-appointed border patrol guards sometimes use violence when they encounter suspected undocumented immigrants.

Undocumented workers often work in sweatshop conditions (see Chapter 7), where they are exploited and mistreated by employers. Americans who feel that undocumented workers do not deserve protection from labor rights and human rights abuses might consider the argument made by economist David Cooper (2015):

> Failure to protect any group of workers—even those without lawful immigration status—is damaging to all workers. When businesses can exploit immigrant workers who cannot speak out for fear of deportation, it lowers the wages of other workers in the same or similar fields. (n.p.)

Policies Regarding Illegal Immigration. The 1986 Immigration Reform and Control Act made hiring unauthorized immigrants an illegal act punishable by fines and even prison sentences. Enforcement of this act occurs primarily through workplace raids, and the E-Verify system, which is a free online service created in 1997 by the Department of Homeland Security (DHS) that allows employers to check the legal statuses of their workers. Employers enter basic information—name, Social Security number, date of birth, alien registration number, and so forth—which is then cross-checked against Social Security Administration and DHS databases. If an employee is not in any database as a legally authorized worker, the employee has eight days to prove legal status, or the employer is mandated to fire them.

The Secure Fence Act of 2006 authorized the construction of a 700-mile "fence" along the U.S.–Mexico border to prevent unauthorized immigrant workers, as well as drug dealers and terrorists, from entering the United States. Critics of the fence argue that this barrier—miles of double chain-link and barbed wire fences with light and infrared camera poles—is expensive and ineffective in stopping illegal immigration, disrupts the environment and harms wildlife along the border, increases the risk and danger to immigrants trying to cross the border, and damages diplomatic relations with Mexico.

The Development, Relief, and Education for Alien Minors Act (DREAM Act), introduced in Congress in 2009, would permit certain immigrant students who have grown up in the United States to apply for temporary legal status and to eventually obtain permanent status and become eligible for U.S. citizenship if they go to college or serve in the U.S. military. Although the DREAM Act has not passed, at least 19 states have passed

nativist extremist groups
Organizations that not only advocate restrictive immigration policy, but also encourage their members to use vigilante tactics to confront or harass suspected undocumented immigrants.

This immigration reform rally is focused on rights for immigrant workers.

laws providing qualified undocumented students access to in-state tuition, scholarships, and/or state financial aid (National Immigration Law Center 2015).

In 2010, Arizona passed the Support Our Law Enforcement and Safe Neighborhoods Act, known as SB 1070—one of the toughest illegal immigration bills in the country. Federal law requires that legal immigrants carry registration papers with them at all times. Arizona's SB 1070 made failure to carry registration documents a state crime, and it required police to verify the legal status of a person during traffic stops, detentions, or arrests if the police suspect that person is in the country illegally. SB 1070 also criminalized anyone who transported unauthorized immigrants, which, according to the law, constituted "smuggling of human beings." Employers who transported unauthorized day laborers to the worksite, as well as church volunteers who transported unauthorized immigrants to church services, were in violation of the law. Following the passage of Arizona's SB 1070, five other states—Alabama, Georgia, South Carolina, Utah, and Indiana—enacted legislation modeled after Arizona's law, and many more states considered similar legislation. But after a series of federal and state court rulings that struck down key provisions of restrictive state immigration policies, and after Latino voters helped reelect President Obama in 2012, states shifted from pursuing restrictive immigration policies to enacting immigrant-friendly legislation, such as laws helping unauthorized immigrant students pursue higher education, and laws allowing unauthorized immigrants to apply for a driver's license (Chisti and Hipsman 2013).

In 2012, President Obama passed the Deferred Action for Childhood Arrivals (DACA) program, which grants deferred deportation (for two years) to people under age 31 who came to the United States before age 16. And in 2014, Obama created the Deferred Action for Parents of Americans and Lawful Permanent Residents (DAPA) program that allows some undocumented people to apply for work authorization and avoid deportation if they have a child who is a U.S. citizen or legal permanent resident. In 2015, a federal district court in Texas temporarily blocked DAPA, as well as the expansion of DACA's two year deportation deferment to three years. Whether an appeal to this ruling will reinstate DAPA and the DACA expansion is unknown.

Becoming a U.S. Citizen

naturalized citizens
Immigrants who apply for and meet the requirements for U.S. citizenship.

Almost half of the nearly 40 million foreign-born U.S. residents in 2010 were **naturalized citizens** (immigrants who applied and met the requirements for U.S. citizenship) (Grieco et al. 2012). Requirements to become a U.S. citizen for most immigrants include (1) having resided continuously as a lawful permanent U.S. resident for at least five years (three years for a spouse of a U.S. resident); (2) being able to read, write, speak, and understand basic English (certain exemptions apply); (3) being "a person of good moral character" (cannot have a record of criminal offenses such as prostitution, illegal gambling, failure to pay child support, drug violations, and violent crime); (4) demonstrating willingness to support and defend the U.S. Constitution by taking the Oath of Allegiance; and (5) passing an examination on English (speaking, reading, and writing) and U.S. government and history (U.S. Citizenship and Immigration Services 2011).

Myths about Immigration and Immigrants

Many foreign-born U.S. residents work hard to succeed educationally and occupationally. The percentage of foreign-born adults (age 25 and older) with at least a bachelor's degree (27 percent) matches that of native-born U.S. adults (28 percent) (Grieco et al. 2012). Despite the achievements and contributions of immigrants, many myths about immigration and immigrants persist, largely perpetuated by anti-immigrant groups and campaigns:

Myth 1. Immigrants increase unemployment and lower wages among native workers.
Reality: Most academic economists agree that immigration has a small but positive impact on the wages of native-born workers because although new immigrant workers add to the labor supply, they also consume goods and services, which creates more jobs (Shierholz 2010). Immigrants also start their own businesses at a higher rate than native U.S. residents, which increases demand for business-related supplies (such as computers and office furniture) and service providers (such as accountants and lawyers) (Pollin 2011).

Myth 2. Immigrants drain the public welfare system and our public schools.
Reality: Unauthorized and temporary immigrants are ineligible for major federal benefit programs, and even legal immigrants may face eligibility restrictions. Two benefit programs that do not have restrictions against unauthorized immigrants are the Special Supplemental Nutrition Program for Women, Infants, and Children (WIC) and the National School Lunch Program.

Regarding public education, a 1982 Supreme Court case (*Plyler v. Doe*) held that states cannot deny students access to public education, even if they are not legal U.S. residents. The Court ruled that denying public education could impose a lifetime of hardship "on a discrete class of children not accountable for their disabling status" (Armario 2011). Children of unauthorized immigrants, 73 percent of whom are U.S. citizens, comprise only 6.8 percent of students in elementary and secondary schools (Passel and Cohn 2009), although the percentage is much higher in communities with large immigrant populations.

Although the states bear the cost of education, social services, and medical services for the immigrant population, research suggests that the economic benefits that immigrants provide for the states outweigh the costs associated with supporting them. For example, a study of immigrants in North Carolina found that, over the prior 10 years, Latino immigrants had cost the state $61 million in a variety of benefits—but they also were responsible for more than $9 billion in state economic growth (Beirich 2007). Half to three-fourths of undocumented immigrants pay federal, state, and local taxes, including Social Security taxes for benefits they will never receive (*Teaching Tolerance* 2011).

Myth 3. Immigrants do not want to learn English.
Reality: In 2012, 29 percent of the U.S foreign-born population (age 5 and older) spoke English "not well" or "not at all" (Gambino et al. 2014). While children benefit from public school programs that teach English as a Second Language (see also Chapter 8), adults who want to learn English have limited educational opportunities, and there are not enough adult English education programs to meet the demand. The majority of programs designed to teach English to adults with limited English proficiency have waiting lists, with wait times ranging from a few weeks to more than three years (Wilson 2014).

Myth 4. Undocumented immigrants have children in the United States as a means of gaining legal status.

Reality: Under the Fourteenth Amendment of the U.S. Constitution, any child born in the United States is automatically granted U.S. citizenship. But having children who are U.S. citizens does not provide immigrants with a means of gaining legal status in the United States. Children under 21 are not allowed to petition for their parents' U.S. citizenship. Nevertheless, some legislators have called for ending birthright citizenship by amending the Constitution or enacting state law to limit citizenship to children who have at least one authorized parent (Dwyer 2011).

Myth 5. Immigrants have high rates of criminal behavior.

Reality: Immigrants are less likely than natives to commit crimes. Because they risk deportation, undocumented immigrants have a strong motivation to avoid involvement with the law. In 2000, the U.S. incarceration rate for native-born men aged between 18 and 39 was 3.5 percent—five times greater than that of their foreign-born counterparts (Bauer and Reynolds 2009). El Paso, Texas, a city with a high immigrant population, is among the safest big cities in the United States. Criminologist Jack Levin said, "If you want to find a safe city, first determine the size of the immigrant population. If the immigrant community represents a large proportion of the population, you're likely in one of the country's safer cities" (quoted in Balko 2009).

> Immigrants are less likely than natives to commit crimes. Because they risk deportation, undocumented immigrants have a strong motivation to avoid involvement with the law.

Sociological Theories of Race and Ethnic Relations

Some theories of race and ethnic relations suggest that individuals with certain personality types are more likely to be prejudiced or to direct hostility toward minority group members. Sociologists, however, focus on how the structure and culture of society affect race and ethnic relations.

Structural-Functionalist Perspective

The structural-functionalist perspective focuses on how parts of the whole are interconnected. This perspective reminds us that we cannot fully understand the history of U.S. civil rights in a vacuum; we need to consider how forces outside the United States affected U.S. policies and culture regarding race relations. In *Cold War Civil Rights: Race and the Image of American Democracy*, Mary Dudziak (2000) links the 1965 passage of U.S. civil rights legislation to the United States' efforts to win the Cold War. Following World War II, international public opinion was critical of the extreme racial inequality in the United States. Racial discrimination was damaging to the United States' credibility as a democracy and to U.S. foreign relations. Although some legislators who supported the passage of civil rights legislation wanted to end the injustice of discrimination, they were also responding to international pressure and seeking to bolster an image of the United States as a democracy and world leader.

The structural-functionalist perspective considers how aspects of social life are functional or dysfunctional—that is, how they contribute to or interfere with social stability. Racial and ethnic inequality are functional in that keeping minority groups in a disadvantaged position ensures that there are workers who will do menial jobs for low pay. Most sociologists emphasize the ways in which racial and ethnic inequality are dysfunctional— a society that practices discrimination fails to develop and utilize the resources of minority members (Williams and Morris 1993). Prejudice and discrimination aggravate social problems, such as crime and violence, war, unemployment and poverty, health problems, family problems, urban decay, and drug use—problems that cause human suffering as well as impose financial burdens on individuals and society. Picca and Feagin (2007) explain:

> [T]he system of racial oppression in the United States affects not only Americans of color but white Americans and society as a whole.... Whites lose when they have to pay huge taxes to keep people of color in prisons because

they are not willing to remedy patterns of unjust enrichment and...to pay to expand education, jobs, or drug-treatment programs that would be less costly. They lose by driving long commutes so they do not have to live next to people of color in cities....They lose when white politicians use racist ideas and arguments to keep from passing legislation that would improve the social welfare of all Americans. Most of all, whites lose...by not having in practice the democracy that they often celebrate to the world in their personal and public rhetoric. (p. 271)

The structural-functionalist analysis of manifest and latent functions also sheds light on issues of race and ethnic relations. For example, the manifest function of the civil rights legislation in the 1960s was, in part, to improve conditions for racial minorities. However, civil rights legislation produced an unexpected negative consequence, or *latent dysfunction.* Because civil rights legislation supposedly ended racial discrimination, whites were more likely to blame blacks for their social disadvantages and thus perpetuate negative stereotypes such as "blacks lack motivation" and "blacks have less ability" (Schuman and Krysan 1999).

Conflict Perspective

The conflict perspective examines how competition over wealth, power, and prestige contributes to racial and ethnic group tensions. Consistent with this perspective, the "racial threat" hypothesis views white racism as a response to perceived or actual threats to whites' economic well-being or cultural dominance by minorities.

For example, between 1840 and 1870, large numbers of Chinese immigrants came to the United States to work in mining (the California Gold Rush of 1848), railroads (the transcontinental railroad, completed in 1869), and construction. As Chinese workers displaced whites, anti-Chinese sentiment rose, resulting in increased prejudice and discrimination and the eventual passage of the Chinese Exclusion Act of 1882, which restricted Chinese immigration until 1924. More recently, white support for Proposition 209—a 1996 resolution passed in California that ended state affirmative action programs—was higher in areas with larger Latino, African American, or Asian American populations, even after controlling for other factors (Tolbert and Grummel 2003). In other words, opposition to affirmative action programs that help minorities was higher in areas with greater racial and ethnic diversity, suggesting that whites living in diverse areas felt more threatened by the minorities.

In another study, researchers interviewed individuals in white racist Internet chat rooms to examine the extent to which people would advocate interracial violence in response to alleged economic and cultural threats (Glaser et al. 2002). The researchers posed three scenarios that might be perceived as threatening: interracial marriage, minority in-migration (i.e., blacks moving into one's neighborhood), and job competition (i.e., competing with a black person for a job). Respondents' reactions to interracial marriage were the most volatile, followed by in-migration. The researchers concluded that violent ideation among white racists stems from perceived threats to white cultural dominance and separateness rather than from perceived economic threats.

Furthermore, conflict theorists suggest that capitalists profit by maintaining a surplus labor force—that is, by having more workers than are needed. A surplus labor force ensures that wages will remain low because someone is always available to take a disgruntled worker's place. Minorities who are disproportionately unemployed serve the interests of the business owners by providing surplus labor, keeping wages low, and, consequently, enabling them to maximize profits.

Conflict theorists also argue that the wealthy and powerful elite foster negative attitudes toward minorities to maintain racial and ethnic tensions among workers. So long as workers are divided along racial and ethnic lines, they are less likely to join forces to advance their own interests at the expense of the capitalists. In addition, the "haves" perpetuate racial and ethnic tensions among the "have-nots" to deflect attention away from their own greed and exploitation of workers.

Symbolic Interactionist Perspective

The symbolic interactionist perspective focuses on the social construction of race and ethnicity—how we learn conceptions and meanings of racial and ethnic distinctions through interaction with others—and how meanings, labels, and definitions affect racial and ethnic groups and intergroup interaction. We have already explained that contemporary race scholars agree that there is no scientific, biological basis for racial categorizations. However, people have learned to think of racial categories as real, and, as the *Thomas theorem* suggests, if things are defined as real, they are real in their consequences. Ossorio and Duster (2005) explain:

> People often interact with each other on the basis of their beliefs that race reflects physical, intellectual, moral, or spiritual superiority or inferiority....By acting on their beliefs about race, people create a society in which individuals of one group have greater access to the goods of society—such as high-status jobs, good schooling, good housing, and good medical care—than do individuals of another group. (p. 119)

The labeling perspective directs us to consider how negative stereotypes affect minorities. **Stereotypes** are exaggerations or generalizations about the characteristics and behavior of a particular group. Negative stereotyping of minorities can lead to a self-fulfilling prophecy—a process in which a false definition of a situation leads to behavior that, in turn, makes the originally falsely defined situation come true (see also Chapter 8). Marger (2012) provides the following example of how the negative stereotype of blacks as less intelligent than whites can lead to a self-fulfilling prophecy:

> If blacks are considered inherently less intelligent, fewer community resources will be used to support schools attended primarily by blacks on the assumption that such support would only be wasted. Poorer-quality schools, then, will inevitably turn out less capable students, who will score lower on intelligence tests. The poorer performance on these tests will "confirm" the original belief about black inferiority. Hence, the self-fulfilling prophecy. (p. 15)

Even stereotypes that appear to be positive can have negative effects. The view of Asian Americans as a "model minority" involves the stereotypes of Asian Americans as excelling in academics and occupational success. These stereotypes mask the struggles and discrimination that many Asian Americans experience and also put enormous pressure on Asian American youth to live up to the social expectation of being a high academic achiever (Tanneeru 2007).

The symbolic interactionist perspective also draws attention to the importance of symbols in race relations. One such symbol is the Confederate flag, which many Americans find offensive because it symbolizes slavery, racism, and oppression, while others defend the Confederate flag as a symbol of Southern pride. Following the 2015 fatal shootings of nine black churchgoers in Charleston, South Carolina, lawmakers in that state passed a bill to remove the Confederate flag from the state capitol.

As this book goes to press, a campaign is waging to remove Andrew Jackson from the $20 bill, because Jackson symbolizes race-based oppression and the mass genocide of Native Americans. Jackson was involved in brutal military campaigns against Native Americans and passed the Indian Removal Act of 1830, which forced thousands of indigenous people from their homelands to the West, an event known as the "Trail of Tears." Removing Jackson from the $20 bill is important because "by confronting and correcting the symbols of our violent and racist histories, we prompt conversation about how that legacy continues to affect marginalized communities today" (Keenan 2014). (See this chapter's *The Human Side* feature.)

The symbolic interactionist perspective is concerned with how individuals learn negative stereotypes and prejudicial attitudes through language. Different connotations of the colors white and black, for example, may contribute to negative attitudes toward people of color. The white knight is good, and the black knight is evil; angel food cake is

stereotypes Exaggerations or generalizations about the characteristics and behavior of a particular group.

white, and devil's food cake is black. Other negative terms associated with black include *black sheep, black plague, black magic, black mass, blackballed,* and *blacklisted.* The continued use of derogatory terms for racial and ethnic groups confirms the power of language in perpetuating negative attitudes toward minority group members. One of the most insulting, demeaning words used to refer to African Americans—the word *nigger*—became a topic of public discussion in 2013, after celebrity chef Paula Deen admitted in a legal deposition that she had used the "N-word."

When whites use the "N-word," it is demeaning and insulting to blacks. Yet black comedians and rappers commonly use the N-word. And some blacks, as well as people of other races, use the word *nigger* or *nigga* to refer to a buddy or friend, conveying a neutral or affectionate tone. The late rapper Tupac Shakur defined "nigger" as a black man with a slavery chain around his neck, whereas a "nigga" is a black man with a gold chain on his neck (urbandictionary.com). Tupac is also known for coming up with the acronym "N.I.G.G.A." meaning "Never Ignorant Getting Goals Accomplished." Oprah Winfrey is one of many blacks who oppose the use of the N-word (and any variation of it) by anyone—including by blacks. She explained, "When I hear the N-word, I still think about every black man who was lynched—and the N-word was the last thing he heard" (Winfrey 2009, n.p.). In an interview, Oprah asked rapper Jay-Z if using the N-word is necessary. Jay-Z explained:

> Nothing is necessary. It's just become part of the way we communicate. My generation hasn't had the same experience with that word that generations of people before us had. We weren't so close to the pain. So in our way, we disarmed the word. We took the fire pin out of the grenade.

Do you think that the attempt to redefine the N-word by Tupac and Jay-Z will effectively disarm it? Why or why not?

WHAT
do you
THINK?

Advocates for immigrant rights suggest that the terms *illegal aliens* and *illegal immigrants* are derogatory and stigmatize and criminalize people rather than their actions. In 2013, the Associated Press (AP) news agency announced it would not use the term *illegal immigrant.* AP journalists are now instructed to "use *illegal* only to refer to an action, not a person: *illegal immigration,* but not *illegal immigrant* (unless the term is used in direct quotations)" (Colford 2013). Other news agencies, including *USA Today,* have also implemented guidelines that prohibit the use of the terms *illegal alien* and *illegal immigrant.*

Bonilla-Silva (2012) gives examples of what he calls the "racial grammar" that shapes how we see or don't see race, and how we frame matters as racial or not race related. Racial grammar involves conveying and perpetuating racial meanings not only through what we say, but also by what we *don't* say. For example, "in the USA one can talk about HBCUs (historically black colleges and universities), but not about HWCUs (historically white colleges and universities) or one can refer to black movies and black TV shows but not label movies and TV shows white when in fact most are" (Bonilla-Silva 2012, p. 173).

In the next section, we explore the concepts of racism and prejudice in more depth and discuss ways in which socialization and the media perpetuate negative stereotypes.

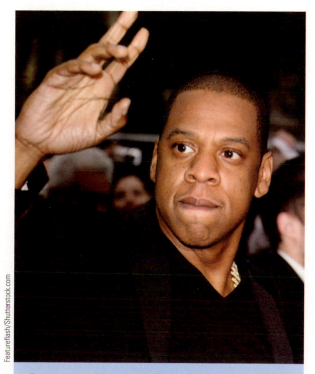

Featureflash/Shutterstock.com

Rapper Jay-Z is among those blacks who use the N-word in a way that conveys a neutral or even friendly meaning.

the HUMAN side

A Cherokee Citizen's View of Andrew Jackson

When Mary Kathryn Nagle, citizen of the Cherokee Nation of Oklahoma, was 11 years old, she and her family drove eight hours to visit the grave of President Andrew Jackson. When they arrived at the gravesite, Nagle watched her grandmother walk up to the president's grave. Nagle describes what she saw her grandmother do at Jackson's gravesite:

> She stood for half a second, then her neck arched back, her body heaved forward, and for the first time in my life, I saw my grandma spit.... It wasn't a casual spit. It was a once in a lifetime spit. I watched as the saliva of our ancestors flowed through her mouth and hit that grave.... What made my grandma, who had never so much as uttered a swear word in her life, spit on the grave of Andrew Jackson?

Later, Nagle asked her grandmother why she spat on Andrew Jackson's grave. Her grandmother pointed to photos of Nagle's great-grandfathers—members of the Cherokee Nation—and explained:

> They fought to save an entire Nation not with a gun in a battlefield, but with a petition in a court of law. In 1832, when the State of Georgia and the...federal government threatened the Cherokee Nation's very existence, my grandfathers—along with Principal Chief John Ross—took the Cherokee Nation's case to the Supreme Court of the United States. In an unprecedented decision (*Worcester v. Georgia*), Justice Marshall issued a ruling declaring the Cherokee Nation to be a sovereign, "distinct community, occupying its own territory" with "the preexisting power of the Nation to govern itself."

After this victory, one of Nagle's great-grandfathers visited the White House to ask President Jackson how he planned to enforce the Supreme Court's decision. President Jackson responded:

> "John Marshall has issued his decision. Let him enforce it." And with the turn of his hand, Andrew Jackson became the only president in the history of the United States to...defy an order from the Supreme Court....Recognizing that neither the federal government nor the State of Georgia would abide our constitutional and inherent right to exist in our homes, my grandfathers signed the Treaty of New Echota, agreeing to relinquish our cherished homeland in exchange for land in what is now Oklahoma. Following the signing of the treaty, President Jackson forcibly placed 16,000 Cherokee in concentration camps for holding until they were "sent" to what is today Oklahoma. More than 4,000 died on the journey now known as The Trail of Tears....

In 1830, Jackson signed the Indian Removal Act...which required a removal treaty with an Indian Nation before its citizens could be moved. However,...Jackson never negotiated or signed a removal treaty with the Muscogee Creek Nations; instead...removal was carried out at bayonet point....Tens of millions of acres were taken illegally.

Nagle quotes Andrew Jackson's comments to Congress in 1833—insulting words that convey utter disregard for the inherent value of Native Americans:

> "[American Indians] have neither the intelligence, the industry, the moral habits, nor the desire of improvement which are essential to any favorable change in their condition. Established in the midst of...a superior race...they must necessarily yield to the force of circumstances and ere long disappear."

. . . The passage of the Fourteenth Amendment in 1868 renders Jackson's assertions that American Indians must "ere long disappear" because of "their inferiority" not only un-democratic, but unconstitutional. Nothing could be more undemocratic in 21st century America than continuing to celebrate a president who contradicts the democratic values we cherish—and continue to fight to preserve—in our Constitution.

Nagle laments that "we still live in a country that teaches its third-graders that Native Americans are nothing more than costume pieces to wear at Halloween and Thanksgiving." The failure to recognize Andrew Jackson's extermination of Native Americans as "genocide" and remove his image from the $20 bill "impedes our ability to heal collectively and restore the true democratic ideals embodied in our Constitution."

SOURCE: Nagle 2015.

The $20 bill, with its depiction of Andrew Jackson, is a symbol of historical brutality toward Native Americans.

Sergiy Palamarchuk/Shutterstock.com

Prejudice and Racism

Prejudice refers to a preconceived opinion or bias that could be negative or positive. Prejudice can be directed toward individuals of a particular religion, sexual orientation, political affiliation, age, physical appearance, social class, sex, race, or ethnicity. Most references to prejudice refer to negative opinions and biases, as these tend to be most problematic.

Racism is generally defined as the belief that race accounts for differences in human character and ability and that a particular race is superior to others. According to this general definition, anyone of any race can be racist. However, many scholars argue that an essential element of racism is power, and so only the dominant social group—in the United States, that would be whites—can be racist. According to the "racism = prejudice + power" definition, blacks who view whites as inferior are prejudiced, but not racist because blacks as a group do not control the institutions that affect the opportunities and resources of whites. The term **institutional racism** refers to the systematic distribution of power, resources, and opportunity in ways that benefit whites and disadvantage minorities (see also "institutional discrimination" discussed later in this chapter). This chapter's *Social Problems Research Up Close* feature suggests that racism in the United States is more common than many people realize.

WHAT do you THINK?

> Which definition of racism makes more sense to you, the general definition or the "prejudice + power" definition? Why? Is black racism toward whites equivalent to white racism toward blacks? Why or why not?

Forms of Racism and Prejudice

Compared with traditional, "old-fashioned" prejudice and racism, which are blatant, direct, and conscious, contemporary forms are often subtle, indirect, and unconscious.

Implicit Prejudice. **Implicit prejudice** is a prevalent form of racial bias over which a person has little or no conscious awareness or control (Sinclair et al. 2014). You can go online to www.implicit.harvard.edu and take the Implicit Association Test (IAT), which measures implicit racial prejudices. After journalist Chris Mooney (2014) took the IAT, he wrote:

> Taking the IAT made me realize that we can't just draw some arbitrary line between prejudiced people and unprejudiced people, and declare ourselves to be on the side of the angels. Biases have slipped into all of our brains. And that means we all have a responsibility to recognize those biases—and work to change them.

Aversive Racism. **Aversive racism** represents a subtle, often unintentional, form of prejudice exhibited by many well-intentioned white Americans who possess strong egalitarian values and who view themselves as unprejudiced. The negative feelings that aversive racists have toward blacks and other minority groups are not feelings of hostility or hate but rather feelings of discomfort, uneasiness, disgust, and sometimes fear (Gaertner and Dovidio 2000). Aversive racists may not be fully aware that they harbor these negative racial feelings; indeed, they disapprove of individuals who are prejudiced and would feel falsely accused if they were labeled prejudiced. "Aversive racists find blacks 'aversive,' while at the same time find any suggestion that they might be prejudiced 'aversive' as well" (Gaertner and Dovidio 2000, p. 14).

Another aspect of aversive racism is the presence of pro-white attitudes, as opposed to anti-black attitudes. In several studies, respondents did not indicate that blacks were worse than whites, only that whites were better than blacks (Gaertner and Dovidio 2000). For example, blacks were not rated as being lazier than whites, but whites were rated as being more ambitious than blacks. Gaertner and Dovidio (2000) explain that "aversive racists would not characterize blacks more negatively than whites because that response could

prejudice Preconceived opinion or bias that could be positive or negative.

racism The belief that race accounts for differences in human character and ability and that a particular race is superior to others.

institutional racism The systematic distribution of power, resources, and opportunity in ways that benefit whites and disadvantage minorities.

implicit prejudice A prevalent form of racial bias over which a person has little or no conscious awareness or control.

aversive racism A subtle form of prejudice that involves feelings of discomfort, uneasiness, disgust, fear, and pro-white attitudes.

In a book titled *Two-Faced Racism*, sociologists Leslie Picca and Joe Feagin present research on white college students' encounters with racism in their everyday lives.

Sample and Methods
Picca and Feagin (2007) recruited 934 college students to participate in research that required students to keep a journal of "everyday events and conversations that deal with racial issues, images, and understandings" (p. 31). Students did not write their name on the journals, but they did indicate their gender, race, and age. The majority of students were from colleges and universities in the South, and were between 18 and 25 years of age. The results presented here are based on roughly 9,000 journal entries written by 626 white students (68 percent women and 32 percent men).

Selected Findings
About three-quarters of the students' journal entries described racist events. In a few of these accounts, students described their own racist actions or thoughts:

> **Kristi:** When I went to pick up the laundry, I saw a young black man

sitting in the driver's side of a minivan with the engine running. My first thought was that he was waiting for a friend to rob the store and he was the getaway driver....I am so embarrassed and saddened by my thinking. (p. 1)

More frequently, students described the racist comments or actions of friends, family members, acquaintances, and strangers. Students wrote about racist accounts targeted at African Americans, Latinos, Asian Americans, Native Americans, Jewish Americans, and Middle Easterners. The most frequently targeted minority group was black Americans. A few journal accounts described antiracist actions by whites.

Picca and Feagin used sociologist Erving Goffman's theatrical metaphors to describe and understand the racist behavior described in the journals. Goffman suggested that when people are "backstage" in private situations, they act differently than they do when they are in "frontstage" public situations. The expression of racist attitudes has gone backstage to private settings where whites are among other whites,

especially family and friends because most whites today understand that it is not socially acceptable to be openly racist. Consider the following diary entry, in which a college student describes being with some of her friends when one of the friends began telling racist jokes:

> **Hannah:** My one friend, Dylan, started telling jokes....Dylan said: "What's the most confusing day of the year in Harlem?" "Father's Day...who's your daddy?" Dylan also referred to black people as "porch monkeys." Everyone laughed a little, but it was obvious that we all felt a little less comfortable when he was telling jokes like that. ... Dylan is not a racist person. He has more black friends than I do, that's why I was surprised he so freely said something like that. Dylan would never have said something like that around anyone who was a minority....It is this sort of "joking" that helps to keep racism alive today. People know the places they have to be politically correct and most people will be. However, until this sort of "behind-the-scenes"

readily be interpreted by others or oneself to reflect racial prejudice" (p. 27). Compared with anti-black attitudes, pro-white attitudes reflect a more subtle prejudice that, although less overtly negative, is still racial bias.

Color-Blind Racism. Many whites claim that they don't see color, just people, and that minorities (especially blacks) are responsible for whatever racial tension exists in the United States (Bonilla-Silva 2013). **Color-blind racism** is based on the belief that paying attention to race is, itself, racism, and therefore, people should ignore race. Color-blindness assumes we are living in a postracial world where race no longer matters, when in reality, race continues to be a significant issue. Color-blindness is a form of racism because it prevents acknowledgment of privilege and disadvantage associated with race, and therefore it allows the continuation of cultural and structural forms of racial bias. A college student at Washington University explains how she learned the color-blind mindset (Frieden 2013):

> Growing up as a white girl in New Hampshire, I was raised with this kind of a "color-blind" mindset: a mindset that says that race should be ignored, and that racism exists because some people simply refuse to ignore race like they should. (p. 1)

color-blind racism A form of racism that is based on the idea that overcoming racism means ignoring race, but color-blindness is, in itself, a form of racism because it prevents acknowledgment of privilege and disadvantage associated with race, and therefore allows the continuation of institutional forms of racial bias.

racism comes to an end, people will always harbor those stereotypical views that are so prevalent in our country. This kind of joking really does bother me, but I don't know what to do about it. I know that I should probably stand up and say I feel uncomfortable when my friends tell jokes like that, but I know my friends would just get annoyed with me and say that they obviously don't mean anything by it. (pp. 17–18)

Hannah's journal entry illustrates the social norm that defines expressions of racism as acceptable in private backstage settings and inappropriate in public frontstage settings. Hannah's journal entry also exemplifies a theme of "white innocence," whereby whites often view racist comments not as "real racism," but as harmless remarks that are not to be taken seriously.

Although white women also display racist behavior, the journal entries in this research suggest that "white men disproportionately make racist jokes and similar racially barbed comments" (p. 133). One female student described in her journal how her white male friends referred to black people as "jigaboos" and other derogatory terms, while her female friends did not make such comments.

Some students made comments that reflected a general lack of awareness of racism. One student, for example, wrote, "This assignment made me realize how many racial remarks are said every single day and I usually never catch any of them or pay close attention" (p. 39).

Many students were offended by the racist remarks they recorded in their journals, but did not express their disapproval and, so, participated in racism as a "bystander." Some students indicated that, after participating in the journal writing assignment, they are more aware of the need to intervene in social situations where racism is being expressed:

Kyle: As my last entry in this journal, I would like to express what I have gained out of this assignment. I watched my friends and companions with open eyes. I was seeing things that I didn't realize were actually there. By having a reason to pick out the racial comments and actions, I was made aware of what is really out there. Although I noticed that I wasn't partaking in any of the racist actions or comments, I did notice that I wasn't stopping them either. I am now in a position to where I can take a stand and try to intervene in many of the situations. (p. 275)

Conclusion

The research findings presented here suggest that "the majority of whites still participate in openly racist performances in the backstage arena...and do not define such performances as problematic and deserving of action aimed at eradication" (p. 22). Picca and Feagin further note that "most of our college student diarists did not, or could not, see the connection between the everyday racial performances they recorded and the unjust discrimination and widespread suffering endured by people of color in society generally" (p. 28).

SOURCE: Picca and Feagin 2007.

Interestingly, research has found that "people who claim to be 'color-blind' and go to great pains to avoid talking about race during social interactions, are in fact perceived as more prejudiced by black observers than people who openly acknowledge race" (Apfelbaum 2011).

Learning to Be Prejudiced: The Role of Socialization and the Media

Psychological theories of prejudice focus on forces within individuals that give rise to prejudice. For example, the frustration-aggression theory of prejudice (also known as the scapegoating theory) suggests that prejudice is a form of hostility that results from frustration. According to this theory, minority groups serve as convenient targets of displaced aggression. The authoritarian-personality theory of prejudice suggests that prejudice arises in people with a certain personality type. According to this theory, people with an authoritarian personality—who are highly conformist, intolerant, cynical, and preoccupied with power—are prone to being prejudiced.

Rather than focus on individuals, sociologists focus on social forces that contribute to prejudice. Earlier, we explained how intergroup conflict over wealth, power, and prestige gives rise to negative feelings and attitudes that serve to protect and enhance dominant group interests. Prejudice is also learned through socialization and the media.

Learning Prejudice through Socialization. In the socialization process, individuals adopt the values, beliefs, and perceptions of their family, peers, culture, and social groups. Prejudice is taught and learned through socialization, although it need not be taught directly and intentionally. White parents who teach their children to not be prejudiced yet live in an all-white neighborhood, attend an all-white church, and have only white friends may be indirectly teaching negative racial attitudes to their children. Socialization can also be direct, as in the case of parents who use racial slurs in the presence of their children or who forbid their children from playing with children from a certain racial or ethnic background. Children can also learn prejudicial attitudes from their peers. Telling racial and ethnic jokes among friends, for example, perpetuates stereotypes that foster negative racial and ethnic attitudes.

Prejudice and the Media. The media—including television, radio, and the Internet—play a role in perpetuating prejudice and hate. In television and movies, racial minorities are underrepresented and appear mostly in stereotypical and secondary roles, such as thugs, buffoons, and angry people. In news media, stories of missing white women and children receive more coverage than missing minority women and children (Bonilla-Silva 2013).

Another form of media that is used to promote hatred toward minorities and to recruit young people to the white power movement is "white power music," which contains anti-Semitic, racist, and homophobic lyrics. The song "Fire Up the Ovens" by the Bully Boys has lyrics that say, "Leave no stone unturned/We're going to burn until the last Jew burns/Fire up the ovens—let's do it again," and another Bully Boys song titled "Jigrun" advocates violence toward blacks: "We're going on the town tonight/Hit and run/Let's have some fun/We've got jigaboos on the run" (Hankes 2014). Racist music, which grew into a multimillion-dollar industry in the 1990s, "has been one of the most effective tools for skinheads and other racist extremists to raise money and recruit new, particularly young, members" (Hankes, 2014 p. 36). After the Southern Poverty Law Center published a report in 2014 identifying 54 racist bands that sold music through iTunes, Apple removed racist hate music from iTunes, but as of spring 2015, racist hate music was still being sold through Amazon and Spotify (Southern Poverty Law Center 2015).

Discrimination against Racial and Ethnic Minorities

Whereas prejudice refers to attitudes, **discrimination** refers to actions or practices that result in differential treatment of categories of individuals. In the United States, discrimination is a common experience among people of color. President Obama said:

> There are very few African American men in this country who haven't had the experience of being followed when they were shopping in a department store. That includes me. There are probably very few African American men who haven't had the experience of walking across the street and hearing the locks click on the doors of cars. That happens to me—at least before I was a senator. There are very few African Americans who haven't had the experience of getting on an elevator and a woman clutching her purse nervously and holding her breath until she had a chance to get off. (quoted in *Washington Post* Staff 2013)

Individual versus Institutional Discrimination

Individual discrimination occurs when individuals treat other individuals unfairly or unequally because of their group membership. Individual discrimination can be overt or adaptive. In **overt discrimination**, individuals discriminate because of their own

discrimination Actions or practices that result in differential treatment of categories of individuals.

individual discrimination The unfair or unequal treatment of individuals because of their group membership.

overt discrimination Discrimination that occurs because of an individual's own prejudicial attitudes.

prejudicial attitudes. For example, a white landlord may refuse to rent to a Mexican American family because of a prejudice against Mexican Americans. Or a Taiwanese American college student who shares a dorm room with an African American student may request a roommate reassignment from the student housing office because of prejudice against blacks.

Suppose that a Cuban American family wants to rent an apartment in a predominantly non-Hispanic neighborhood. If the landlord is prejudiced against Cubans and does not allow the family to rent the apartment, that landlord has engaged in overt discrimination. But what if the landlord is not prejudiced against Cubans but still refuses to rent to a Cuban family? Perhaps that landlord is engaging in **adaptive discrimination**, or discrimination that is based on the prejudice of others. In this example, the landlord may fear that other renters who are prejudiced against Cubans may move out of the building or neighborhood and leave the landlord with unrented apartments. Overt and adaptive individual discrimination can coexist.

Institutional discrimination refers to institutional policies and procedures that result in unequal treatment of and opportunities for minorities. Institutional discrimination is covert and insidious and maintains the subordinate position of minorities in society. For example, when schools use standard intelligence tests to decide which children will be placed in college preparatory tracks, they are limiting the educational advancement of minorities whose intelligence is not fairly measured by culturally biased tests developed from white middle-class experiences. And the funding of public schools through local tax dollars results in less funding for schools in poor and largely minority school districts (see also Chapter 8).

Institutional discrimination is also found in the criminal justice system, which disproportionately incarcerates people of color (see also Chapter 4). Because the "war on drugs" has been waged predominately in poor communities of color, most people arrested for drug offenses are black or Latino, even though people of color are no more likely to use or sell drugs than whites (Alexander 2010). Millions of people of color are labeled as felons for relatively minor, nonviolent drug offenses. Due to felon disenfranchisement laws, those labeled felons are denied the right to vote and are automatically excluded from juries. Felons may also be legally discriminated against in employment, housing, access to education, and public benefits, much like their grandparents or great grandparents may have been discriminated against during the Jim Crow era (Alexander 2010).

Another example of institutional discrimination concerns voting rights. In June 2013, the U.S Supreme Court in *Shelby County v. Holder* ruled that Section 4 of the Voting Rights Act—which determines which states and jurisdictions are covered by Section 5—is invalid. Section 5 requires states with a historical record of discriminatory practices to receive preclearance from the federal government before these states could make changes to voting laws or procedures. Within months following the ruling, six southern states passed legislation imposing new voting restrictions that will disproportionately affect minorities. North Carolina—one of the states affected by Section 5—was the first state to approve legislation that effectively suppresses voting among minorities by curbing early and same-day registration and requiring photo IDs. Lawmakers later eased the photo ID requirement, allowing voters to claim a "reasonable impediment declaration" explaining why they lacked a photo ID.

Colorism

Colorism refers to prejudice or discrimination based on skin tone. Discussions of colorism tend to focus on the preference for lighter skin among blacks, framing colorism as intraracial "black-on-black" prejudice and discrimination. But whites also engage in colorism: One study found that whites are more likely to view African Americans and Hispanics with lighter skin as intelligent (Hannon 2015). It is important to recognize white colorism as a variation of white racism. Hannon (2015) explains:

> Consider, for example, a hypothetical case where a white employer discriminates against darker skinned African Americans for customer relations positions.

adaptive discrimination
Discrimination that is based on the prejudice of others.

institutional discrimination
Discrimination in which institutional policies and procedures result in unequal treatment of and opportunities for minorities.

colorism Prejudice or discrimination based on skin tone.

Claiming racism would be insufficient; such a claim could be countered with evidence of past (lighter skinned) African American hires. (p. 19)

Although Barack Obama is biracial, he identifies as an African American and is often referred to as the first black U.S. president. Do you think that Obama's light skin tone played a role in his getting elected? If Obama's skin tone was several shades darker, do you think he would have been elected president?

Although Barack Obama's background is biracial—he is the son of a black father and a white mother—he identifies as black. Would he have been elected president if he had darker skin?

Employment Discrimination

In a months-long search for a job, José Zamora sent out between 50 and 100 résumés a day, applying on-line to every job he found that he felt qualified for. Day after day he got no responses. So he decided to drop one letter in his name: *José* became *Joe*. A week after he began applying for jobs as "Joe," the responses started pouring in with requests for interviews. In a Buzzfeed video, Zamora explained, "Sometimes I don't even think people know or are conscious or aware that they're judging—even if it's by a name—but I think we do it all the time" (quoted in Matthews 2014).

Despite laws against it, discrimination against minorities occurs today in all phases of the employment process, from recruitment to interview, job offer, salary, promotion, and firing decisions. A sociologist at Northwestern University studied employers' treatment of job applicants in Milwaukee, Wisconsin, by dividing job applicant "testers" into four groups: blacks with a criminal record, blacks without a criminal record, whites with a criminal record, and whites without a criminal record (Pager 2003). Applicant testers, none of whom actually had a criminal record, were trained to behave similarly in the application process and were sent with comparable résumés to the same set of employers. The study found that white applicants with no criminal record were the most likely to be called back for an interview (34 percent) and that black applicants with a criminal record were the least likely to be called back (5 percent). But surprisingly, white applicants with a criminal record (17 percent) were more likely to be called back for a job interview than were black applicants *without* a criminal record (14 percent). The researcher concluded that "the powerful effects of race thus continue to direct employment decisions in ways that contribute to persisting racial inequality" (Pager 2003, p. 960).

Discrimination in hiring may be unintended. For example, many businesses rely on their existing employees to refer new recruits when a position opens up. Word-of-mouth recruitment is inexpensive and efficient; some companies offer bonuses to employees who bring in new recruits. But this traditional recruitment practice tends to exclude minority workers, because they often do not have a network of friends and family members in higher positions of employment who can recruit them (Schiller 2004).

Employment discrimination contributes to the higher rates of unemployment and lower incomes of blacks and Hispanics compared with those of whites (see Chapters 6 and 7). Lower levels of educational attainment among minority groups account for some, but not all, of the disadvantages they experience in employment and income. As shown in

AP Images/Alan Diaz

Figure 9.5, mean earnings of whites are higher than those for blacks, Hispanics, and Asians at most levels of educational attainment. In a report on the job prospects of young college graduates, the authors explained that "having an equivalent amount of higher education and a virtual blank slate of prior work experience still does not generate parity in unemployment across races and ethnicities. This suggests other factors may be at play, such as minorities not having equal access to the informal professional networks that often lead to job opportunities, and/or discrimination against racial and ethnic minorities" (Shierholz et al. 2013, p. 11).

Workplace discrimination also includes unfair treatment and harassment. African American employees working for A. C. Widenhouse—a North Carolina–based freight trucking company—were repeatedly subjected to derogatory racial comments and slurs by employees and managers. These insulting comments and slurs included "n----r," "monkey," and "boy" (Equal Employment Opportunity Commission 2011). One employee was approached by a coworker with a noose who said, "This is for you. Do you want to hang from the family tree?" A company manager allegedly told an employee, "We are going coon hunting. Are you going to be the coon?"

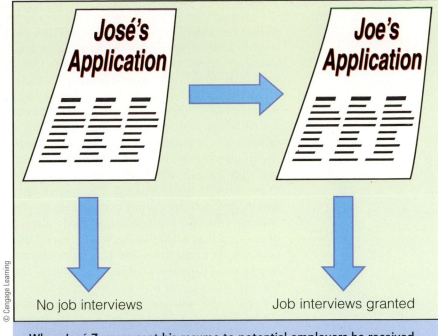

When José Zamora sent his resume to potential employers he received no responses. When he dropped the "s" in his name, changing his name from "José" to "Joe," he received many responses to his job applications.

© Cengage Learning

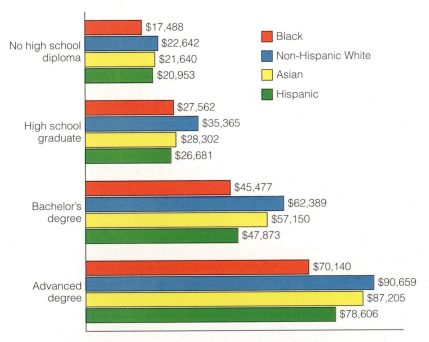

Figure 9.5 Mean Earnings of Workers 18 Years and Over, by Educational Attainment, Race, and Hispanic Origin, 2011
SOURCE: U.S. Census Bureau 2013.

Housing Discrimination and Segregation

Before the 1968 Fair Housing Act and the 1974 Equal Credit Opportunity Act, discrimination against minorities in housing and mortgage lending was common. Banks and mortgage companies commonly engaged in "redlining"—the practice of denying mortgage loans in minority neighborhoods on the premise that the financial risk was too great, and the ethical standards of the National Association of Real Estate Boards prohibited its members from introducing minorities into white neighborhoods.

Although housing discrimination is illegal today, it is not uncommon. To assess discrimination in housing, researchers use a method called "paired testing" in which two individuals—one minority and the other nonminority—are trained to pose as home seekers, and they interact with real estate agents, landlords, rental agents, and mortgage lenders to see how they are treated. The testers are assigned comparable or identical income, assets, and debt as well as comparable or identical housing preferences, family circumstances, education, and job characteristics. A paired testing study of housing discrimination found that whites in the rental market were more likely to receive information about available housing units and had more opportunities to inspect available units than did blacks and Hispanics (Turner et al. 2002). The same study found that, in the home sales market, white home buyers were more likely to be shown homes in more predominantly non-Hispanic white neighborhoods than were comparable black and Hispanic buyers. Whites were also more likely to receive information and assistance with financing.

Despite continued housing discrimination, homeownership rates among minorities and low-income groups increased substantially in the 1990s, reaching record rates in many central cities. However, minority and low-income homeowner rates still lag behind the overall homeownership rate. Also, many of the gains in minority and low-income homeownership rates are due to increases in *subprime lending*—loans with higher fees and higher interest rates that are offered to borrowers who have poor (or nonexistent) credit records (Williams et al. 2005).

Racial segregation has a long history in the United States. From 1890 to 1968, thousands of communities across the country used various means, including the law, harassment, intimidation, and even violence, to keep racial and ethnic minorities out. Communities that were purposely "all-white" are known as **sundown towns** because some of them posted signs warning blacks "Don't let the sun go down on you in this town" (Loewen 2006). The historical pattern of residential segregation has continued, increasing dramatically over the 20th century (Logan and Parman 2015).

Educational Discrimination and Segregation

Both institutional discrimination and individual discrimination in education negatively affect racial and ethnic minorities and help to explain why minorities (with the exception of Asian Americans) tend to achieve lower levels of academic attainment and success (see also Chapter 8). Institutional discrimination is evidenced by inequalities in school funding—a practice that disproportionately hurts minority students (Kozol 1991). Because minorities are more likely than whites to live in economically disadvantaged areas, they are more likely to go to schools that receive inadequate funding. Inner-city schools, which serve primarily minority students, receive less funding per student than do schools in more affluent, primarily white areas.

Another institutional education policy that is advantageous to whites is the policy that gives preference to college applicants whose parents or grandparents are alumni. The overwhelming majority of alumni at the highest-ranked universities and colleges are white. Thus, white college applicants are the primary beneficiaries of these so-called legacy admissions policies. About 10 to 15 percent of students in most Ivy League colleges and universities are children of alumni, and legacy students have a much higher rate of admission than nonlegacy applicants ("Ivy League Legacy Admissions 2015). In 2004, Texas A&M University became the first public college to abandon its legacy admittance policy.

Minorities also experience individual discrimination in the schools as a result of continuing prejudice among teachers. One survey of 1,100 educators found that more

sundown towns Communities that are purposely "all-white" and that have used various means to deliberately keep racial and ethnic minorities out.

than a quarter of survey respondents reported that they had heard racist comments from their colleagues in the past year ("Hear and Now" 2000). It is likely that teachers who are prejudiced against minorities discriminate against them, giving them less teaching attention and less encouragement.

Racial and ethnic minorities are also treated unfairly in educational materials, such as textbooks, which often distort the history and heritages of people of color (King 2000). For example, Columbus and his successors are portrayed in educational materials as navigators and discoverers, rather than as participants in Native American genocide.

Finally, racial and ethnic minorities are largely isolated from whites in a largely segregated school system. In 2014, the majority of public school students were children of color. Yet in every state except New Mexico and Hawaii, the average white student attends a school that is mostly white. And black and Latino students tend to attend schools where the majority of students are black and/or Latino (Jordan 2014). School segregation is largely due to the persistence of housing segregation and the termination of court-ordered desegregation plans. Court-mandated busing became a means to achieve equality of education and school integration in the early 1970s, after the Supreme Court (in *Swann v. Charlotte Mecklenburg*) endorsed busing to desegregate schools. But in the 1990s, lower courts lifted desegregation orders in dozens of school districts (Winter 2003). And in 2007, the United States Supreme Court issued a landmark ruling, that race cannot be a factor in the assignment of children to public schools.

Racial Microaggressions

When First Lady Michelle Obama addressed a mostly black graduating class at Tuskegee University in Alabama, she warned graduates of the "daily slights" they would experience throughout their lives (Gayle 2015). In academic literature, these daily slights are referred to as **racial microaggressions**, defined as "brief and commonplace daily verbal, behavioral, or environmental indignities, whether intentional or unintentional, that communicate hostile, derogatory, or negative racial slights and insults toward the target person or group" (Sue et al. 2007, p. 271). Although whites may experience racial microaggressions, they do so much less frequently than racial and ethnic minorities (Nadal et al. 2014). In an online survey of students of color at the University of Illinois at Urbana-Champaign, the majority of respondents reported experiencing racial microaggressions on campus (see Table 9.4) (Harwood et al. 2015). Research has found a relationship between experiencing racial microaggressions and low self-esteem (Nadal et al. 2014).

TABLE 9.4 Examples of Racial Microaggressions on Campus

- Hearing racial stereotypes in class lectures and discussions
- Being dismissed or ignored by the instructor before or after class
- Hearing inappropriate racial comments made by instructors before or after class
- Being called on in class to offer the "student of color perspective"
- Being excluded from participation in group projects
- Experiencing racial jokes and teasing
- Overhearing racist conversations between students in the classroom
- Being discouraged during meetings with one's academic advisor

SOURCE: Based on Harwood et al. 2015.

Hate Crimes

In June 2011, 49-year-old James Craig Anderson, an African American auto plant worker, was robbed, beaten, and run over by a truck near a hotel in Jackson, Mississippi. Deryl Dedmon, the 18-year-old who drove the truck that ran over Anderson, and a group of other white teenagers were reportedly out looking for a black person to assault when they found Anderson. During the brutal attack, which was captured on hotel security video, racial slurs were used, and later Dedmon bragged that he "just ran that n----- over" (Associated Press 2011).

The murder of James Craig Anderson is an example of a **hate crime**—an unlawful act of violence motivated by prejudice or bias. Examples of hate crimes, also known as "bias-motivated crimes," include intimidation (e.g., threats), destruction of or damage to property, physical assault, and murder.

According to the Federal Bureau of Investigation (FBI), between 6,000 to 10,000 hate crimes occur in the United States each year. However, FBI hate crime data undercount

racial microaggressions Brief and commonplace daily verbal, behavioral, or environmental indignities, whether intentional or unintentional, that communicate hostile, derogatory, or negative racial slights and insults toward the target person or group.

hate crime An unlawful act of violence motivated by prejudice or bias.

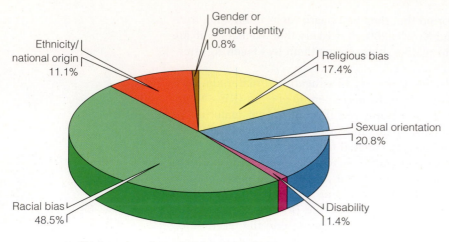

Figure 9.6 Hate Crime Incidence by Category of Bias, 2013 (rounded to nearest percent)
SOURCE: FBI 2014.

Gender or gender identity 0.8%
Ethnicity/ national origin 11.1%
Religious bias 17.4%
Sexual orientation 20.8%
Racial bias 48.5%
Disability 1.4%

Total number of reported hate crime incidents: 5,928

> Islamophobia is largely based on ignorance and misunderstanding of Islam.

the actual number of hate crimes because (1) not all U.S. jurisdictions report hate crimes to the FBI (reporting is voluntary), (2) it is difficult to prove that crimes are motivated by hate or prejudice, (3) law enforcement agencies shy away from classifying crimes as hate crimes because it makes their community "look bad," and (4) victims are often reluctant to report hate crimes to the authorities. The National Crime Victimization Survey reveals that the incidence of hate crimes is 25 to 40 times higher than the FBI numbers ("DOJ Study" 2013).

FBI hate crime data reveal that most hate crimes are based on racial bias, primarily against blacks, followed by whites (see Figure 9.6). Most hate crimes motivated by religious bias target the Jewish religion, and the majority of hate crimes based on ethnic bias are anti-Hispanic, and are largely motivated by hate toward immigrants. After the terrorist attacks of September 11, 2001, hate crimes against individuals perceived to be Muslim or Middle Eastern increased significantly. Anti-Muslim and anti-Islam bias—commonly known as **Islamophobia**—was rekindled after the proposal to build an Islamic cultural center at 9/11 Ground Zero and again after the 2013 Boston Marathon bombing, which was executed by two brothers who reportedly were motivated by an anti-American radical version of Islam. Islamophobia is largely based on ignorance and misunderstanding of Islam. Georgetown professor John L. Esposito (2011) explains:

> Mainstream American Muslims have too often been equated inaccurately with terrorists and people who reject democracy. . . . [M]ajorities of Muslims globally desire democracy and freedom and fear and reject religious extremism and terrorism.

Islamaphobia is reflected in the passage of laws banning Shariah religious law in at least seven states, including Alabama, Arizona, Kansas, Louisiana, North Carolina, South Dakota, and Tennessee. Given the U.S. Constitution's separation of church and state, such laws are unnecessary and serve only to make a political statement that promotes fear of the Muslim population (Gebreyes 2015).

Motivations for Hate Crimes. Levin and McDevitt (1995) found that the motivations for hate crimes were of three distinct types: thrill, defensive, and mission. Thrill hate crimes are committed by offenders who are looking for excitement and attack victims for the "fun of it." Defensive hate crimes involve offenders who view their attacks as necessary to protect their community, workplace, or college campus from "outsiders" or to protect their racial and cultural purity from being "contaminated" by interracial marriage and childbearing. Mission hate crimes are perpetrated by white supremacist group members or other offenders who have dedicated their lives to bigotry. Hate groups known to engage in violent crimes include the Ku Klux Klan, the Identity Church Movement, the neo-Nazis, and the skinheads. The Southern Poverty Law Center identified 784 active hate group chapters in 2014—down from a peak of 1,018 in 2011 (Potok 2015). Part of this decrease is due to many chapters of the Ku Klux Klan going underground to avoid detection. Although some members of hate groups have visible features, such as tattoos and armbands, that identify their hate group membership, criminologist Jack Levin, who studies hate crimes, noted that "many white supremacist groups are going more mainstream…eliminating the sheets and armbands….The groups realize if they want to be attractive to middle-class types, they need to look middle class" (quoted in Leadership Council on Civil Rights Education Fund 2009, p. 19). Some anti-immigrant hate groups, such as the Federation for American Immigration Reform (FAIR), the

Islamophobia Anti-Muslim and anti-Islam bias.

Protests against Islamophobia followed the shooting deaths of three Muslim students in Chapel Hill, NC, although it is disputed whether the shootings were motivated by hate or by a disagreement over parking spaces.

Center for Immigration Studies (CIS), and NumbersUSA, portray themselves as legitimate mainstream advocates against illegal immigration, but "some of these organizations have disturbing links to or relationships with extremists in the anti-immigrant movement" (p. 14).

Hate on Campus. In 2013, Oberlin College in Ohio canceled classes to address a string of reported racial incidents on campus, including several posters and fliers placed around campus containing hateful language and a reported sighting of someone dressed as a member of the Ku Klux Klan on campus (Ly 2013). In the same year, a black teddy bear was hung by a noose in the office of a black professor at Northwestern University. According to the FBI, nearly 1 in 10 hate crimes occur at schools or colleges (FBI 2012). In focus groups with students of color at a large, predominantly white university, students reported commonly hearing racial jokes and slurs and encountering racial slurs written on residence hall room doors, study rooms, and elevators (Harwood et al. 2012). Students also reported that residence halls with higher numbers of minority students are inferior (e.g., are older and lack air conditioning) and are labeled with names like "minority central" and "the projects."

Strategies for Action: Responding to Prejudice, Racism, and Discrimination

In recent years, highly publicized accounts of police violence against racial minorities fueled the growth of the Black Lives Matter movement, which began in 2012 after the murder of a young black male, Trayvon Martin (see also Chapter 4). This social movement has brought increased attention to anti-Black racism and racial inequality, and has provided impetus for other minority groups (e.g. Muslim Lives Matter) to demand just and equal treatment.

Mike Huckabee, contender for the 2016 Republican presidential nomination and former Arkansas governor, said that Dr. Martin Luther King Jr. would be "appalled" by the Black Lives Matter movement because of "the notion that we're elevating some lives above others" (quoted in Bradner 2015). Huckabee explained, "When I hear people scream, 'black lives matter,' I think, 'Of course they do.' But all lives matter. It's not that any life matters more than another." Do you think that the name "Black Lives Matter" implies that other lives don't matter? Do you think that Martin Luther King, Jr. would be "appalled " by the movement's name?

WHAT do you THINK?

Efforts to achieve racial equality in the criminal justice system are discussed in Chapter 4. In the following sections, we discuss the Equal Employment Opportunity Commission's role in responding to employment discrimination and examine the issue of affirmative action in the United States. We also discuss educational strategies to promote diversity and multicultural awareness and appreciation in schools. Finally, we look at apologies and reparations as a means of achieving racial reconciliation.

The Equal Employment Opportunity Commission

The **Equal Employment Opportunity Commission (EEOC)**, a U.S. federal agency charged with ending employment discrimination in the United States, is responsible for enforcing laws against discrimination, including Title VII of the 1964 Civil Rights Act that prohibits employment discrimination on the basis of race, color, religion, sex, or national origin. The EEOC investigates, mediates, and may file lawsuits against private employers on behalf of alleged victims of discrimination. The most frequently filed claims with the EEOC are allegations of race discrimination, racial harassment, or retaliation from opposition to racial discrimination.

In 2007, the EEOC launched a national initiative to combat racial discrimination in the workplace. The goals of this initiative, called E-RACE (Eradicating Racism And Colorism from Employment), are to (1) identify factors that contribute to race and color discrimination, (2) explore strategies to improve the administrative processing and litigation of race and color discrimination cases, and (3) increase public awareness of race and color discrimination in employment.

Affirmative Action

Affirmative action refers to a broad range of policies and practices in the workplace and educational institutions to promote equal opportunity and diversity. Affirmative action is an attempt to compensate for the effects of past discrimination and prevent current discrimination against women and racial and ethnic minorities. Veterans and people with disabilities may also qualify under affirmative action policies. Federal affirmative action policies developed in the 1960s required any employer (universities as well as businesses) who received contracts from the federal government to make "good faith efforts" to increase the pool of qualified minorities and women by expanding recruitment and training programs (U.S. Department of Labor 2002).

In higher education, affirmative action policies seek to improve access to education for those groups that have been historically excluded or underrepresented, such as women and racial/ethnic minorities. Supporters of affirmative action in higher education view the underrepresentation of minorities in higher education as a problem. Consider that, in 1965, only 5 percent of undergraduate students, 1 percent of law students, and 2 percent of medical students in the country were African American (National Conference of State Legislatures 2013). Over the last several decades, colleges and universities have implemented affirmative action policies designed to recruit and admit more minority students, and these policies have been successful in increasing the enrollment of minority students, although racial/ethnic gaps in college enrollment remain (see also Chapter 8).

In 1978, the Supreme Court's ruling in *Regents of the University of California v. Bakke* marked the beginning of a series of several court challenges to affirmative action. Although the Court ruled that affirmative action programs could not use fixed quotas in admission, hiring, or promotion policies, it affirmed the right for universities and employers to consider race as a factor in admission, hiring, and promotion to achieve diversity. Since the *Bakke* case, numerous legal battles have challenged affirmative action. In 2013, the Supreme Court ruled in *Fisher v. University of Texas* that universities may seek racial diversity but they must demonstrate that race-neutral alternatives to achieving diversity are not sufficient (Kahlenberg 2013). A number of colleges and universities have already developed creative alternatives to race-based admissions policies to increase their minority student populations. Some universities are using a "holistic admissions" approach that considers the unique circumstances of each student, prioritizing academics and treating

Equal Employment Opportunity Commission (EEOC) A U.S. federal agency charged with ending employment discrimination in the United States and responsible for enforcing laws against discrimination, including Title VII of the 1964 Civil Rights Act that prohibits employment discrimination on the basis of race, color, religion, sex, or national origin.

affirmative action A broad range of policies and practices in the workplace and educational institutions to promote equal opportunity as well as diversity.

all other factors, including race, equally. Other universities use the "10 percent plan" as a way to maintain minority enrollment. Graduates in the top 10 percent of their high school classes are admitted automatically to the public college or university of their choice; standardized test scores and other factors are not considered. Yet another approach is to increase financial aid packages to low-income students, which benefits all students from economically disadvantaged backgrounds, regardless of race/ethnicity.

Legal battles over affirmative action are likely to continue as affirmative action remains a controversial and divisive issue among Americans. A Gallup Poll found that 53 percent of whites, 77 percent of blacks, and 61 percent of Hispanics generally favor affirmative action programs for racial minorities, with Democrats being twice as likely as Republicans to favor affirmative action (Riffkin 2015). Another poll found that when questioned about college admission policies, only 28 percent of U.S. adults agree that "an applicant's racial and ethnic background should be considered to help promote diversity on college campuses." Blacks were more likely to agree (48 percent) than Hispanics (31 percent) or whites (22 percent) (Jones 2013).

Educational Strategies

Schools and universities play an important role in whether minorities succeed in school and in the job market. One way to improve minorities' chances of academic success is to reduce or eliminate disparities in school funding. As noted earlier, schools in poor districts—which predominantly serve minority students—have traditionally received less funding per pupil than do schools in middle- and upper-class districts (which predominantly serve white students). Other educational strategies focus on reducing prejudice, racism, and discrimination, and fostering awareness and appreciation of racial and ethnic diversity. These strategies include multicultural education, "whiteness studies," and efforts to increase diversity among student populations.

Multicultural Education in Schools and Communities. In schools across the nation, multicultural education, which encompasses a broad range of programs and strategies, works to dispel myths, stereotypes, and ignorance about minorities; to promote tolerance and appreciation of diversity; and to include minority groups in the school curriculum (see also Chapter 8). With multicultural education, the school curriculum reflects the diversity of U.S. society and fosters an awareness and appreciation of the contributions of different racial and ethnic groups to U.S. culture. The Southern Poverty Law Center's program Teaching Tolerance publishes and distributes materials and videos designed to promote better human relations among diverse groups. These materials are sent to schools, colleges, religious organizations, and a variety of community groups across the nation.

Many colleges and universities promote awareness and appreciation of diversity by offering courses and degree programs in racial and ethnic studies and by sponsoring

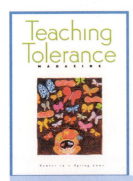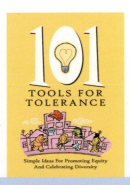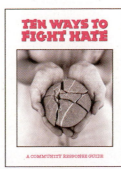

Southern Poverty Law Center

Teaching Tolerance magazine is a free resource available to teachers for tolerance education in the classroom. Other free materials available from the Southern Poverty Law Center (www.splcenter.org) include *101 Tools for Tolerance, Responding to Hate at School,* and *Ten Ways to Fight Hate.*

multicultural events and student organizations. Some colleges mandate that students take a certain number of required diversity courses to graduate.

Whiteness Studies. Traditionally, studies of and courses on race have focused on the social disadvantages that racial minorities experience, while ignoring or minimizing the other side of the racial inequality equation—the social advantages conferred upon whites. Courses in whiteness studies, which are being offered in many colleges and universities, focus on understanding "whiteness" as a social construction and fostering critical awareness of white privilege—an awareness that is limited among white students (Yeung et al. 2013). "White privilege is so fundamental as to be largely invisible, expected, and normalized" (Picca and Feagin 2007, p. 243). This chapter's *Self and Society* feature asks you to think about how you personally explain the white racial advantage.

In an often-cited paper called "White Privilege: Unpacking the Invisible Knapsack," Peggy McIntosh (1990) likened white privilege to an "invisible weightless knapsack" that whites carry with them without awareness of the many benefits inside the knapsack. Here are just a few of the many benefits McIntosh associates with being white:

- Go shopping without being followed or harassed.
- Be assured that my skin color will not convey that I am financially unreliable when I use checks or credit cards.
- Swear, or dress in secondhand clothes, or not answer letters without having people attribute these choices to the bad morals, poverty, or the illiteracy of my race.
- Avoid ever being asked to speak for all the people of my racial group.
- Be sure that, if a traffic cop pulls me over or if the IRS audits my tax return, it is not because I have been singled out because of my race.
- Take a job with an affirmative action employer without having coworkers on the job suspect that I got it because of race.

Diversification of College Student Populations. Recruiting and admitting racial and ethnic minorities in institutions of higher education can foster positive relationships among diverse groups and enrich the educational experience of all students—minority and nonminority alike (American Council on Education and American Association of University Professors 2000). Psychologist Gordon Allport's (1954) "contact hypothesis" suggested that contact between groups is necessary for the reduction of prejudice between group members. Having a diverse student population can provide students with opportunities for contact with different groups and can thereby reduce prejudice. In a study of 2,000 UCLA students, researchers found that students who were randomly assigned to roommates of a different race or ethnicity developed more favorable attitudes toward students of different backgrounds (Sidanius et al. 2010).

WHAT do you THINK?

As students progress through college, do you think that their friends become less racially diverse or more? One study found that as students advance through college, their friendship networks become *less* racially diverse (Martin et al. 2014). Between the first and second years of college, the percentage of students reporting mostly same-race friends increased from 52 to 78 percent among whites and from 36 to 64 percent among blacks. How might you explain this finding?

Retrospective Justice Initiatives: Apologies and Reparations

Various governments around the world have issued official apologies for racial and ethnic oppression. After World War II, West Germany signed a reparations agreement with Israel in which West Germany agreed to pay Israel for the enslavement and persecution of Jews during the Holocaust and to compensate for Jewish property that the Nazis stole. In 2008, the Australian government issued a formal apology for the treatment of the country's aboriginal people—specifically, for the decades during which the government removed aboriginal children from their families in a forced acculturation program.

How Do You Explain White Racial Advantage?

Directions: On average, white Americans have better jobs, income, and housing than others. How important are each of the following factors in explaining this racial advantage associated with being white? Indicate your answer using the following key, putting your answer in the space provided.

Very important = 3
Somewhat important = 2
Not very important = 1
Not important at all = 0

1. Prejudice and discrimi- _____
   nation in favor of whites

2. Laws and institutions _____
   in favor of whites more
   than other groups

3. Access to better schools _____
   and social connections

4. Effort and hard work _____

Answer

Interpretation and Comparison Data
Results from a nationally representative sample of U.S. adults is provided as follows (Croll 2013). The first three explanations for white racial advantage are structural explanations that measure a belief in white privilege and indicate an awareness of how the same factors that disadvantage minorities, benefit whites. The last explanation ("Effort and hard work") is an individualistic explanation, which explains the white racial advantage as being the result of whites working harder than other races.

	Whites	Blacks	Hispanics	Other
1. Prejudice and discrimination in favor of whites				
Very important	18%	41%	25%	29%
Somewhat important	45%	39%	47%	40%
Not very important	21%	6%	17%	16%
Not important at all	16%	14%	11%	15%
2. Laws and institutions in favor of whites more than other groups				
Very important	12%	39%	31%	29%
Somewhat important	35%	43%	42%	41%
Not very important	27%	9%	16%	15%
Not important at all	26%	9%	11%	15%
3. Access to better schools and social connections				
Very important	50%	69%	60%	58%
Somewhat important	33%	23%	26%	31%
Not very important	11%	5%	13%	4%
Not important at all	6%	3%	1%	7%
4. Effort and hard work				
Very important	57%	45%	71%	55%
Somewhat important	32%	30%	15%	26%
Not very important	7%	16%	11%	4%
Not important at all	11%	9%	3%	15%

SOURCE: Adapted from Croll 2013.

In the United States, President Gerald Ford and Congress apologized to Japanese Americans in 1976 for their internment during World War II, and reparations of $20,000 were granted to each surviving internee who was a U.S. citizen or legal resident alien at time of internment. In 1993, President Bill Clinton apologized to native Hawaiians for overthrowing the government of their nation. In 1994, the state of Florida offered monetary compensation to the survivors and descendants of the 1923 "Rosewood massacre," in which a white mob attacked and murdered black residents in Rosewood, Florida, and set fire to the town. And in 1997, the U.S. government offered monetary reparations to surviving victims of the Tuskegee syphilis study, in which blacks suffering from syphilis were denied medical treatment.

Although various forms of reparations were offered to Native American tribes to compensate for land that had been taken by force or deception, it wasn't until 2008 that the Senate Committee on Indian Affairs passed a resolution apologizing to all Indian tribes for the mistreatment and violence committed against them. In 2008, the U.S. House of Representatives issued a formal apology to African Americans for slavery; and in 2009, the U.S. Senate passed a resolution apologizing for slavery, adding the stipulation that the official apology cannot be used to support claims for restitution. The resolution "acknowledges the fundamental injustice, cruelty, brutality, and inhumanity of slavery and Jim Crow laws," and "apologizes to African Americans...for the wrongs committed against them and their ancestors who suffered under slavery and Jim Crow laws" (CNN 2009).

Supporters of the reparative justice movement believe that the granting of apologies and reparations to groups that have been mistreated promotes dialogue and healing, increases awareness of present inequalities, and stimulates political action to remedy current injustices. "Reparative justice is not an invitation to 'wallow in the past' but a way for societies to come to terms with painful histories and move forward" (Brown University Steering Committee on Slavery and Justice 2007, p. 39). Those opposed to this movement claim that "preoccupation with past injustice is a distraction from the challenge of present injustice" (Brown University Steering Committee on Slavery and Justice 2007, p. 39).

Understanding Race, Ethnicity, and Immigration

After considering the material presented in this chapter, what understanding about race and ethnic relations are we left with? First, we have seen that racial and ethnic categories are socially constructed; they are largely arbitrary, imprecise, and misleading. Although some scholars suggest that we abandon racial and ethnic labels, others advocate adding new categories—multiethnic and multiracial—to reflect the identities of a growing segment of the U.S. and world population.

Conflict theorists and structural functionalists agree that prejudice, discrimination, and racism have benefited certain groups in society. But racial and ethnic disharmony has created tensions that disrupt social equilibrium. Symbolic interactionists note that negative labeling of minority group members, which is learned through interaction with others, contributes to the subordinate position of minorities.

Prejudice, racism, and discrimination are debilitating forces in the lives of minorities and immigrants. Despite these negative forces, many minority group members succeed in living productive, meaningful, and prosperous lives. But many others cannot overcome the social disadvantages associated with their minority status and become victims of a cycle of poverty (see Chapter 6). Minorities are disproportionately poor, receive inferior education and health care, and, with continued discrimination in the workplace, have difficulty improving their standard of living.

Intentional and hateful racism still exist, but much racism is more subtle and unintentional. Attempts to be "color-blind" may be well intentioned, but if we don't "see" race, then we also do not see racial injustices or the need to remedy such injustices.

Achieving racial and ethnic equality requires first *seeing* how discrimination is institutionalized in the structure of society, and then making changes in society to eliminate institutionalized discrimination and provide opportunities for minorities—in education, employment and income, and political participation. In addition, policy makers concerned with racial and ethnic equality must find ways to reduce the racial and ethnic wealth gap and to foster wealth accumulation among minorities (Conley 1999). Social class is a central issue in race and ethnic relations. Professor and activist bell hooks

(2000) (who spells her name in all lowercase) warned that focusing on issues of race and gender can deflect attention away from the larger issue of class division that increasingly separates the haves from the have-nots. Addressing class inequality must, suggests hooks, be part of any meaningful strategy to reduce inequalities that minority groups suffer.

Making change requires members of society to recognize that change is necessary, that there is a problem that needs rectifying:

> One has to perceive the problem to embrace the solutions. If you think racism isn't harmful unless it wears sheets or burns crosses or bars blacks from motels and restaurants, you will support only the crudest antidiscrimination laws and not the more refined methods of affirmative action and diversity training. (Shipler 1998, p. 2)

As this book goes to press, Americans are experiencing a heightened awareness of racism resulting from a series of police-related incidents of fatal violence toward racial minorities (see Chapter 4) and the fatal shootings of nine black church members in South Carolina. The growing awareness of racial and ethnic prejudice, discrimination, and hate is leading citizens all across the country to raise their voices, calling for racial and ethnic justice in all aspects of U.S. culture.

Chapter Review

- **What is a minority group?**
  A minority group is a category of people who have unequal access to positions of power, prestige, and wealth in a society and who tend to be targets of prejudice and discrimination. Minority status is not based on numerical representation in society but rather on social status.

- **What is meant by the idea that race is socially constructed?**
  The concept of race refers to a category of people who are perceived to share distinct physical characteristics that are deemed socially significant. The significance of race is not biological but social and political, because race is used to separate "us" from "them" and becomes a basis for unequal treatment of one group by another. Races are cultural and social inventions; they are not scientifically valid because there are no objective, reliable, meaningful criteria scientists can use to identify racial groupings. Different societies construct different systems of racial classification, and these systems change over time.

- **What are the various patterns of interaction that may occur when two or more racial or ethnic groups come into contact?**
  When two or more racial or ethnic groups come into contact, one of several patterns of interaction occurs, including genocide, expulsion, segregation, acculturation, pluralism, and assimilation.

- **Beginning with the 2000 census, what are the five race categories used to identify the race composition of the United States?**
  Beginning with the 2000 census, the five race categories are (1) white, (2) black or African American, (3) American Indian or Alaska Native, (4) Asian, and (5) Native Hawaiian or other Pacific Islander. In addition, respondents to federal surveys and the census have the option of officially identifying themselves as being of more than one race, rather than checking only one racial category.

- **What is an ethnic group?**
  An ethnic group is a population that has a shared cultural heritage, nationality, or lineage. Ethnic groups can be distinguished on the basis of language, forms of family structures and roles of family members, religious beliefs and practices, dietary customs, forms of artistic expression such as music and dance, and national origin. The largest ethnic population in the United States is Hispanics or Latinos.

- **What percentage of the U.S. population (in 2014) was born outside the United States?**
  More than 1 in 10 U.S. residents (13.3 percent) were born in a foreign country.

- **What were the manifest function and latent dysfunction of the civil rights movement?**
  The manifest function of the civil rights legislation in the 1960s was to improve conditions for racial minorities. However, civil rights legislation produced an unexpected consequence, or latent dysfunction. Because civil rights legislation supposedly ended racial discrimination, whites were more likely to blame blacks for their social disadvantages and thus perpetuate negative stereotypes such as "blacks lack motivation" and "blacks have less ability."

- **How does contemporary prejudice differ from more traditional, "old-fashioned" prejudice?**
  Traditional, old-fashioned prejudice is easy to recognize, because it is blatant, direct, and conscious. More contemporary forms of prejudice are often subtle, indirect, and unconscious. In addition, racist expressions have gone "backstage" to private social settings.

- **Is it possible for an individual to discriminate without being prejudiced?**

Yes. In overt discrimination, individuals discriminate because of their own prejudicial attitudes. But sometimes individuals who are not prejudiced discriminate because of someone else's prejudice. For example, a store clerk may watch black customers more closely because the store manager is prejudiced against blacks and has instructed the employee to follow black customers in the store closely. Discrimination based on someone else's prejudice is called adaptive discrimination.

- **How is color-blindness a form of racism?**

Color-blindness is a form of racism because it prevents acknowledgment of privilege and disadvantage associated with race, and therefore allows the continuation of institutional racism.

- **Are U.S. schools segregated?**

Most white public school students attend schools that are largely white, and most minority public students attend schools where the majority of students are minorities. Racial and ethnic minorities are largely isolated from whites in an increasingly segregated school system. The upward trend in school segregation is due to large increases in minority student enrollment, continuing white flight from urban areas, the persistence of housing segregation, and the termination of court-ordered desegregation plans.

- **According to FBI data, the majority of hate crimes are motivated by what kind of bias?**

Since the FBI began publishing hate crime data in 1992, the majority of hate crimes have been based on racial bias.

- **What is the role of the Equal Employment Opportunity Commission (EEOC) in combating employment discrimination?**

The Equal Employment Opportunity Commission (EEOC) is responsible for enforcing laws against discrimination, including Title VII of the 1964 Civil Rights Act that prohibits employment discrimination on the basis of race, color, religion, sex, or national origin. The EEOC investigates, mediates, and may file lawsuits against private employers on behalf of alleged victims of discrimination.

- **What group constitutes the largest beneficiary of affirmative action policies?**

Affirmative action policies are designed to benefit racial and ethnic minorities, women, and, in some cases, Vietnam veterans and people with disabilities. The largest category of affirmative action beneficiaries is women.

- **What are whiteness studies?**

Courses in whiteness studies, which are being offered in many colleges and universities, focus on increasing awareness of white privilege—an awareness that is limited among white students.

Test Yourself

1. When it comes down to collecting census data on the U.S. population, a person is Hispanic if they say they are Hispanic.
   a. True
   b. False
2. Which of the following occurs when a person adopts the culture of a group different from the one in which that person was originally raised?
   a. Secondary assimilation
   b. Acculturation
   c. Genocide
   d. Pluralism
3. Individuals identifying as _____ comprise the fastest-growing racial group in the United States?
   a. non-Hispanic white
   b. black
   c. Asian
   d. two or more races
4. Compared with the U.S.-born population, immigrants have high rates of criminal behavior.
   a. True
   b. False
5. Which minority group in the United States is considered a "model minority"?
   a. Women
   b. Black women
   c. Scandinavian immigrants
   d. Asian Americans
6. Which of the following has/have moved "backstage"?
   a. Racist behavior
   b. Undocumented immigrants
   c. Multicultural education
   d. Affirmative action
7. In 2014, how many active hate group chapters were there in the United States?
   a. 8
   b. 38
   c. 211
   d. 784
8. Which of the following is responsible for enforcing laws against discrimination?
   a. Local police departments
   b. U.S. Department of Labor
   c. Equal Employment Opportunity Commission
   d. The Supreme Court
9. Colleges and universities are prohibited from requiring students to take diversity courses as a graduation requirement.
   a. True
   b. False
10. The U.S. Congress has issued an official apology to African Americans for slavery and Jim Crow laws.
    a. True
    b. False

Answers: 1. A; 2. B; 3. D; 4. B; 5. D; 6. A; 7. D; 8. C; 9. B; 10. A.

Key Terms

Sexism is a
social disease

❝ There's no excuse for sexual assault or domestic violence, there's no reason that young girls should suffer genital mutilation, there's no place in a civilized society for the early or forced marriage of children. These traditions may go back centuries; they have no place in the 21st century."

BARACK OBAMA
44th President of the United States, speaking in Nairobi, Kenya, 2015

Gender Inequality

Learning Objectives

After studying this chapter, you will be able to . . .

1 Describe some of the ways gender inequality exists around the world.

2 Identify what leads to gender inequality according to each of the three sociological and feminist theories.

3 Analyze how the structure of society contributes to gender inequality.

4 Analyze how the culture of society contributes to gender inequality.

5 Describe how traditional gender roles lead to each of the social problems discussed in this chapter.

6 Evaluate the effectiveness of U.S. laws and policies directed toward reducing gender inequality.

BORN PREMATURE and living in foster care, M.C. had a tough start in life but, as luck would have it, was adopted by a loving couple who already had two children of their own (Southern Poverty Law Center [SPLC] 2013). Three months before Pam and Mark Crawford were able to take their baby home, 16-month-old M.C., who had "ambiguous genitals" and both male and female reproductive organs, underwent sex assignment surgery. Today, at age 8, M.C. continues to live with the sexual anatomy of a girl and the gender identity of a boy. In 2013, his parents filed a lawsuit claiming that the child welfare agencies and doctors involved violated M.C.'s constitutional rights by subjecting him to unnecessary surgery without due process. Although a lower court held for the parents, in 2015 an appellate court dismissed the complaint, finding that in 2006 there was no legal precedent that performing sex assignment surgery was a violation of M.C.'s constitutional right (Largent 2015).

> Researchers, educators, and parents alike are challenging the binary concepts of sex and gender and, just as our evolving notion of sexual orientation, it has led to feelings of uneasiness for some.

This groundbreaking lawsuit, *M.C. v. Medical University of South Carolina et al.,* brings to the forefront of Americans' consciousness the significance of and difference between sex and gender. **Sex** refers to one's biological classification, whereas **gender** refers to the social definitions and expectations associated with being female or male. Today, however, researchers, educators, and parents alike are challenging the binary concepts of sex and gender and, just as our evolving notion of sexual orientation (see Chapter 11), it has led to feelings of uneasiness for some. For example, words like *transgender* and *transsexual, gender variant, androgyny, intersexed* (the condition M.C. in our opening vignette was determined to have), *boi* and *birl, metrosexual, two-spirited, gender neutrality, third sex,* and *gender bender* were not part of the American lexicon just decades ago. In most Western cultures, we take for granted that there are two categories of gender. However, in many other societies, three and four genders have been recognized:

> On nearly every continent, and for all of recorded history, thriving cultures have recognized, revered, and integrated more than two genders. Terms such as transgender and gay are strictly new constructs that assume three things: that there are only two sexes (male/female), as many as two sexualities (gay/straight), and only two genders (man/woman). (National Public Radio 2011, p. 1)

Gender identity refers to a person's "innate, deeply felt psychological identification as a man, woman or some other gender, which may or may not correspond to the sex assigned to them at birth (e.g., the sex listed on their birth certificate)" (Human Rights Campaign 2015, p. 1). A **transgender individual** (sometimes called "trans" or "gender nonconforming") is a person whose sense of gender identity is inconsistent with his or her birth (sometimes called chromosomal) sex. Transgender is not a sexual orientation, and transgender individuals may have any sexual orientation—heterosexual, homosexual, or bisexual. Lastly, **gender expression** includes the sum of an individuals' external presentation of self (e.g., dress, grooming, behaviors) as masculine, feminine, or somewhere on the continuum between the two.

There is documented movement away from a binary emphasis on sex and gender. For example, in a telephone survey of 1,000 U.S. adults between the ages of 18 and 34, 50 percent agreed with the statement "Gender is a spectrum, and some people fall outside of conventional categories" (Benenson Strategy Group 2015, p. 3). Nonetheless, most Americans still think in terms of females and males, and women and men, much the same way they continue to use such false dichotomies as black and white, gay and straight, and young and old. Although empirically inaccurate, this *social shorthand* makes conversation easier and fulfills our need to "know" and respond "appropriately." Thus, upon meeting someone, we quickly attach the label of gay, white male, or elderly African American female, though knowing little about their social biographies. Similarly, this chapter emphasizes **sexism** and gender inequality as traditionally defined—that is, as the inequality between women and men—but, wherever possible, information on transgender individuals will

sex A person's biological classification as male or female.

gender The social definitions and expectations associated with being female or male.

transgender individual A transgender individual is a person whose sense of gender identity is inconsistent with his or her birth (sometimes called chromosomal) sex (male or female).

gender expression The way in which a person presents her-or himself as a gendered individual (i.e., masculine, feminine, or androgynous) in society. A person could, for example, have a gender identity as male but nonetheless present their gender as female.

be included in the discussion. Note that civil rights issues for transgender individuals are discussed in Chapter 11.

In varying social circumstances, determining gender has been based on biological criteria such as genitalia or hormonal levels, or on who an individual identifies with (Westbrook and Schilt 2014). Which do you think should be used to determine gender? Does your position change between sex-segregated situations (e.g., bathrooms, dormitories, sports) and non-sex-segregated situations (e.g., workplace, classrooms)?

The Global Context: The Status of Women and Men

There is no country in the world in which women and men have equal status. Although much progress has been made in closing the gender gap in areas such as education, health care, employment, and government, gender inequality is still prevalent throughout the world. It is so prevalent that a survey in Great Britain of girls and young women between the ages of 11 and 21 indicates that 75 percent believe that sexism affects every aspect of their lives (Girls' Attitude Survey 2014).

The World Economic Forum assessed the gender gap in 142 countries by measuring the extent to which women have achieved equality with men in four areas: economic participation and opportunity, educational attainment, health and survival, and political empowerment (Hausmann et al. 2014). Table 10.1 presents the overall scores and rankings of (1) the 10 countries with the smallest gender gap (i.e., the least gender inequality); (2) the 10 countries with the largest gender gap (i.e., the most gender inequality); and (3) the United States, which did not rank in the top 10 or the bottom 10 but ranked number 20 of the 142 countries studied. Ties were possible. For example, several countries were tied for first place on the composite measure of educational attainment. Furthermore, note that the overall score approximates the proportion of the gender gap a country has *closed*—in the United States, 0.7463 or 74.63 percent.

Gender inequality varies across cultures, not only in its extent or degree but also in its forms. In the United States, gender inequality in family roles commonly takes the form of an unequal division of household labor and child care, with women bearing the heavier responsibility for these tasks. In other countries, forms of gender inequality in the family include the expectation

Benoit Tessier/Reuters

Andrej Pejic, a transgender model, states that "Sometimes I feel like more of a woman, other times I feel male" (quoted in Daily Mail 2011, p. 1). The 21-year-old Serbian who relocated to Australia is one of the top-ranked models in the world. He's been described as the "poster boy for fashion androgyny" (Models 2011, p. 1).

that wives ask their husbands for permission to use birth control (see Chapter 12), unequal penalties for spouses who commit adultery, with wives receiving harsher punishment, and the practice of aborting female fetuses in cultures that value male children over female children.

In a book entitled *Unnatural Selection*, author Mara Hvistendahl (2011) documented how medical technology, and specifically the increased availability of ultrasounds, has made sex-selection abortions commonplace around the world. According to the United Nations (UN), there are between 113 and 200 million "missing" girls, more than the entire female population of the United States. One researcher, upon

sexism The belief that innate psychological, behavioral, and/or intellectual differences exist between women and men and that these differences connote the superiority of one group and the inferiority of the other.

TABLE 10.1 Gender Gap Rankings: Top and Bottom 10 Countries and the United States, 2014

Country	Overall Score*	Overall Ranking	Economic Participation and Opportunity	Educational Attainment	Health and Survival	Political Empowerment
Top 10 Countries						
Iceland	0.8594	1	7	1	128	1
Finland	0.8453	2	21	1	52	2
Norway	0.8374	3	1	1	98	3
Sweden	0.8176	4	15	43	100	5
Denmark	0.8035	5	12	1	65	7
Nicaragua	0.7894	6	95	33	1	4
Rwanda	0.7854	7	25	114	118	6
Ireland	0.7850	8	28	40	67	8
Philippines	0.7814	9	24	1	1	17
Belgium	0.7819	10	27	73	52	13
United States	**0.7463**	**20**	**4**	**39**	**62**	**54**
Bottom 10 Countries						
Morocco	0.5998	133	135	116	122	98
Jordan	0.5978	134	140	74	127	119
Lebanon	0.5923	135	133	106	62	141
Cote d'Ivoire	0.5874	136	112	137	104	115
Iran	0.5811	137	139	104	89	135
Mali	0.5789	138	118	136	135	118
Syria	0.5775	139	142	101	37	126
Chad	0.5764	140	70	142	103	106
Pakistan	0.5522	141	141	132	119	85
Yemen	0.5145	142	138	140	117	138

*Overall scores are reported on a scale of 0 to 1, with 1 representing maximum gender equality.
SOURCE: Hausmann et al. 2014.

visiting a peasant family in rural China, sits with the mother-in law as her son's wife is about to give birth.

There was a low sob, and then a man's gruff voice said accusingly: "useless thing."

. . . I thought I heard a slight movement in the slop pail behind me, and automatically glanced toward it. ...To my absolute horror, I saw a tiny foot poking out of the pail. . . . Then the tiny foot twitched! The midwife must have dropped that tiny baby alive into the slop pail!

We sat in silence while I stared, sickened, at the pail. . . . The little foot was still now. . . .

"Doing a baby girl is not a big thing around here," . . . the older woman said. . . .

"That's a living child!" I said in a shaking voice. . . .

"It's not a child," she corrected me.

"What do you mean, it's not a child? I saw it."

"It's not a child. . . . If it was, we would be looking after it, wouldn't we? . . . It's a girl baby." (pp. 27–28)

A global perspective on gender inequality must also take into account the different ways in which such inequality is viewed. For example, many non-Muslims view the practice of Muslim women wearing a headscarf in public as a symbol of female subordination and oppression. To Muslims who embrace this practice (and not all Muslims do), wearing a headscarf reflects the high status of women and represents the view that women should be respected and not treated as sexual objects.

Similarly, cultures differ in how they view the practice of female genital mutilation, also known as female genital cutting or female circumcision (FGM/C). There are several forms of FGM/C, ranging from a symbolic nicking of the clitoris to removal of the clitoris and labia and partial closure of the vaginal opening by stitching the two sides of the vulva together, leaving only a small opening for the passage of urine and menstrual blood. After marriage, the sealed opening is reopened to permit intercourse and childbearing.

Nonmedical personnel perform most FGM/C procedures using unsterilized blades or string. Health risks associated with FGM/C include pain, hemorrhage, infection, shock, scarring, and infertility. Worldwide, over 125 million girls and women are estimated to have experienced FGM/C in 29 African and Middle Eastern countries (WHO 2014a).

People from countries in which FGM/C is not the norm generally view this practice as a barbaric form of violence against women. In countries where it commonly occurs, FGM/C is viewed as an important and useful practice. Although having no health benefit, in some countries it is considered a rite of passage that enhances a woman's status. In other countries, it is aesthetically pleasing. For others, FGM/C is a moral imperative based on religious beliefs (WHO 2008; Yoder et al. 2004). However, in all cases, it "reflects deep-rooted inequality between the sexes, and constitutes an extreme form of discrimination against women" (WHO 2014a, p. 1).

Inequality in the United States

Although attitudes toward gender equality are becoming increasingly liberal, the United States has a long history of gender inequality. Women have had to fight for equality: the right to vote, equal pay for comparable work, quality education, entrance into male-dominated occupations, and legal equality. As shown in Table 10.1, the United States is ranked 20th in the world in gender inequality based on the World Economic Forum's assessment of women's economic participation and opportunities, political empowerment, educational attainment, and health and survival (Hausmann et al. 2014). Most U.S. citizens agree that American society does not treat women and men equally: Women have lower incomes, hold fewer prestigious jobs, earn fewer graduate degrees, and are more likely than men to live in poverty.

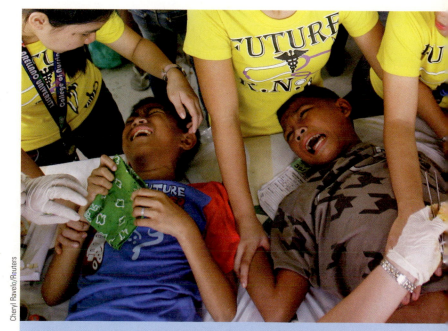

Cheryl Ravelo/Reuters

In 2011, in Manila, 1,500 boys were forced to take place in a mass circumcision as the organizers of the event tried to get into the Guinness World Records for the most circumcisions performed on boys 9 and older. Of late, the medical procedure has come under fire as questions arise about its safety, necessity and, similar to FGM/C, abuse of a child's anatomical integrity.

social problems
RESEARCH
UP CLOSE | Overdoing Gender

Willer et al. (2013) test what is called the **masculine overcompensation thesis**—the assertion "that men react to masculinity threats with extreme demonstrations of masculinity" (p. 980). The researchers evaluate this thesis using four research designs, three of which are discussed here. Two are laboratory experiments, and the third looks at the relationship between masculinity threat and attitudes associated with dominance using a national sample of American adults.

Sample and Methods
Experiments are unique methods of research in that the researchers manipulate the independent variable and measure its effect on the dependent variable. In studies 1 and 2, female and male undergraduates participated in a laboratory experiment in which, on

the basis of a "gender identity survey" administered, they were randomly assigned to one of two conditions of the independent variable. The independent variable is *gender identity threat.* In the first condition, males and females were given feedback that they were in the average female range (males' masculinity threatened). In the second condition, males and females were told they were in the average male range (females' femininity threatened). In the first study, dependent variables were assessed on a "political views survey." In the second study, which was similarly designed to the first laboratory experiment, social dominance variables were measured.

In the third study, to assess the reliability of the results of the first two studies with a larger and more diverse sample, Willer et al. (2013) used data

from the American Values Survey. Masculinity threat was measured by respondents' feelings that social change is threatening the status of men. The dependent variables were a variety of attitudinal questions related to masculinity and dominance.

Findings and Conclusions
Study 1 examined the relationship between masculinity threat and femininity threat and support for the war in Iraq, attitudes toward homosexuality, desirability of owning a SUV, and how much a subject was willing to pay for a SUV. Males who had received feedback that they had scored in the average femininity range (i.e., had their masculinity threatened) were significantly more likely when compared to male subjects who did not have their masculinity threatened to voice support for the war

Men are also victims of gender inequality. In 1963, sociologist Erving Goffman wrote that in the United States, there is only

> one complete unblushing male . . . a young, married, white, urban, northern heterosexual, Protestant father of college education, fully employed, of good complexion, weight and height, and a recent record in sports. . . . Any male who fails to qualify in one of these ways is likely to view himself . . . as unworthy, incomplete, and inferior. (p. 128)

Although standards of masculinity have relaxed, Williams (2000) argues that masculinity is still based on "success"—at work, on the athletic field, on the streets, and at home. Similarly, Vandello et al. (2008) concludes "[t]he view that manhood is tenuous, and therefore requires public proof, is consistent with research across multiple areas" (p. 1326). (See this chapter's *Social Problems Research Up Close.*)

When U.S. college students were asked to list the best and worst things about being the opposite sex, the same qualities, although in opposite categories, emerged (Cohen 2001). For example, what males listed as the best thing about being female (e.g., free to be emotional), females list as the worst thing about being male (e.g., not free to be emotional). Similarly, what females listed as the best thing about being male (e.g., higher pay), males listed as the worst thing about being female (e.g., lower pay). As Cohen (2001) noted, although "some differences are exaggerated or oversimplified, . . . we identif[ied] a host of ways in which we 'win' or 'lose' simply because we are male or female" (p. 3).

masculine overcompensation thesis The thesis that men have a tendency to act out in an exaggerated male role when believing their masculinity to be threatened.

Sociological Theories of Gender Inequality

Both structural functionalism and conflict theory concentrate on how the structure of society and, specifically, its institutions contribute to gender inequality. However, these two theoretical perspectives offer opposing views of the development and maintenance in Iraq and express negative attitudes toward homosexuality.

Men whose masculinity was threatened also reported that a SUV was more desirable than men who did not have their masculinity threatened, and they were willing to pay, on the average, $7,320 more for the SUV than non-threatened men. Femininity threat was unrelated to any of the dependent variable indicators—support for the war in Iraq, negative views of homosexuality, SUV desirability, or SUV purchase price.

The second study looked at the relationship between gender threat and dominance attitudes, including such items as "Superior groups should dominate inferior groups" and "In getting what your group wants, it is sometimes necessary to use force against other groups." Participants were also administered scales designed to measure political conservatism (e.g., support for the U.S. military, affirmative action, etc.), system justification (e.g.,

"Everyone has a fair shot at wealth and happiness"), and traditionalism (e.g., "It's better to stick with what you have than to keep trying new uncertain things").

The results of study 2 indicate that men—although not women—whose gender identity was threatened scored higher on the dominance scale than men whose gender identity was not threatened, but they did not score significantly higher on political conservatism, system justification, or traditionalism. Study 3 yielded similar results. Using a national sample of 2,210 American adults, the findings indicate that "[t]he more men felt that the status of their gender was threatened by social changes, the more they tended to support the Iraq war, hold negative views of homosexuality, believe in male superiority, and hold strong dominance attitudes" (p. 1001).

The results of these three studies lend consistent support for the

masculine overcompensation thesis. Alternatively, they do not support a feminine overcompensation thesis. The authors note, however, that this does not mean that feminine overcompensation does not exist. It may simply mean that the dependent variables used in these studies—for example, support for the war on Iraq—were not sufficient to detect feminine overcompensation. Willer et al. (2013) conclude

> that extreme masculine behaviors may in fact serve as telltale signs of threats and insecurity. Perhaps those men who appear most assuredly masculine, who in their actions communicate strength, power, and dominance at great levels, may actually be acting to conceal underlying concerns that they lack exactly those qualities they strive to project. (p. 1016)

SOURCE: Willer et al. 2013.

of gender inequality. Symbolic interactionism, on the other hand, focuses on the culture of society and how gender roles are learned through the socialization process.

Structural-Functionalist Perspective

Structural functionalists argue that preindustrial society required a division of labor based on gender. Women, out of biological necessity, remained in the home performing functions such as bearing, nursing, and caring for children. Men, who were physically stronger and could be away from home for long periods of time, were responsible for providing food, clothing, and shelter for their families. This division of labor was functional for society and became defined as both normal and natural over time.

Industrialization rendered the traditional division of labor less functional, although remnants of the supporting belief system still persist. With increased control over reproduction (e.g., contraception), declining birthrates, and fewer jobs dependent on physical size and strength, women's opportunities for education and workforce participation increased (Wood and Eagly 2002). Thus, modern conceptions of the family have, to some extent, replaced traditional ones—families have evolved from extended to nuclear, authority is more egalitarian, more women work outside the home, and greater role variation exists in the division of labor. Structural functionalists argue, therefore, that as the needs of society change, the associated institutional arrangements also change.

Conflict Perspective

Many conflict theorists hold that male dominance and female subordination are shaped by the relationships men and women have to the production process. During the hunting-and-gathering stage of development, males and females were economic equals, both controlling their own labor and producing needed subsistence. As society evolved to

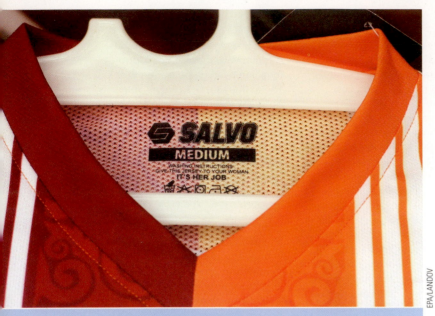

In 2015, a clothing company in Indonesia manufactured a t-shirt with the words, "Give this jersey to your woman. It's her job," under the words, "washing instructions." The company, which came under heavy criticism for the remarks, tweeted a series of apologies stating that, "The message is simple. Instead of doing it the wrong way, you might as well give it to a lady because they are more capable" (quoted in Dunn, 2015, p.1).

agricultural and industrial modes of production, private property developed and men gained control of the modes of production, whereas women remained in the home to bear and care for children. Inheritance laws that ensured that ownership would remain in their hands furthered male domination. Laws that regarded women as property ensured that women would remain confined to the home.

As industrialization continued and the production of goods and services moved away from the home, the gaps between females and males continued to grow—women had less education, lower incomes, fewer occupational skills, and were rarely owners. World War II necessitated the entry of a large number of women into the labor force, but in contrast with previous periods, many of them did not return to the home at the end of the war. They had established their own place in the workforce and, facilitated by the changing nature of work and technological advances, now competed directly with men for jobs and wages.

Conflict theorists also argue that continued domination by males requires a belief system that supports gender inequality. Two such beliefs are (1) that women are inferior outside the home (e.g., they are less intelligent, less reliable, and less rational) and (2) that women are more valuable in the home (e.g., they have maternal instincts and are naturally nurturing). Even when initiatives appear to be subverting these beliefs, patriarchy remains intact and often extended. Jitha (2013), in an investigation of micro-credit loans to women in India, concludes that women who step outside traditional gender roles are excluded from the program, the loans emphasize financial reward thereby demeaning unpaid household labor, and "it is in fact men who decide if the women can join the group, if the loan is required at all, and for what purpose it should be used" (p. 268). Thus, unlike structural functionalists, conflict theorists hold that the subordinate position of women in society is a consequence of social inducement rather than biological differences that led to the traditional division of labor.

Symbolic Interactionist Perspective

Although some scientists argue that gender differences are innate, symbolic interactionists emphasize that, through the socialization process, both females and males are taught the meanings associated with being feminine and masculine. Gender assignment begins at birth as a child is classified as either female or male. However, the learning of gender roles is a lifelong process whereby individuals acquire society's definitions of appropriate and inappropriate gender behavior.

Gender roles are taught by the family, in the school, in peer groups, and by media presentations of girls and boys and women and men (see the discussion on the social construction of gender roles later in this chapter). Most important, however, gender roles are learned through symbolic interaction as the messages that others send us reaffirm or challenge our gender performances. In an examination of parent–child interactions, Tenenbaum (2009) found that discussions regarding course selections for high school followed gender-stereotyped patterns. Here, a father talks to his fifth grade daughter (how the conversation was coded by the researchers is in brackets):

But you know, spelling is English, right? That's what English is. For the most part generally speaking, girls do better with those kind of skills, they have a

> [U]nlike structural functionalists, conflict theorists hold that the subordinate position of women in society is a consequence of social inducement rather than biological differences that led to the traditional division of labor.

harder time with math, generally, you know? [Code: Lack of ability.] Now, of course, you know it's nice to do the things you're really good at too, and you like to do. But, sometimes when you're trying to be, when you grow up to be someone in life, you also gotta take classes that are, not really, how do I say? You're not really good at, you have to put more practice in, right? [Code: Lack of ability.] So, I picked, I selected algebra, cause that's kinda, high school, mathematics. (pp. 458–459)

Although the father encouraged his daughter to take mathematics, he twice conveyed to his daughter that she is (and girls in general are) not very good in math.

Feminist theory, although also consistent with a conflict perspective, incorporates many aspects of symbolic interactionism. Feminists argue that conceptions of gender are socially constructed as societal expectations dictate what it means to be female or what it means to be male. Thus, women are generally socialized into **expressive roles** (i.e., nurturing and emotionally supportive roles), and males are more often socialized into **instrumental roles** (i.e., task-oriented roles). These roles are then acted out in countless daily interactions as boss and secretary, doctor and nurse, football player and cheerleader "do gender."

Ridgeway (2011), in her book *Framed by Gender*, begins by asking the research question, "How, in the modern world, does gender manage to persist as a basis or principal for inequality?" (p. 3). Her answer lies in the socially constructed meanings associated with gender. Despite the tremendous advances that have been made, whenever people encounter gender in a social relationship, circumstance, or situation that they are unsure of, they rely on traditional definitions as an organizing principle whereby "[g]ender inequality is rewritten into new economic and social arrangements as they emerge, preserving that inequality in modified form" (Ridgeway 2011, p. 7).

Noting that the impact of the structure and culture of society is not the same for all women and all men, feminists encourage research on gender that takes into consideration the intersection of class, race, ethnicity, and sexual orientation. In other words, to fully understand the experiences of others, we must acknowledge that it cannot be accomplished by concentrating on just one subordinate status.

In Afghanistan, little girls in families with no male children often assume the persona of little boys—short hair, traditional male clothing, and an appropriate male name (Nordberg 2010). In addition to the social pressure to have a boy child, the decision to raise a female child as a "bacha posh" is pragmatic—such a child can help "his" father at work, get a better education, and does not have to be chaperoned. The problem is that when it's time to return to being female, usually at puberty, some have difficulty transitioning. Said one young woman, "Nothing in me feels like a girl . . . for always I want to be a boy" (Nordberg 2010, p. 6). Do you think gender, not sex, is a consequence of nature or nurture?

WHAT do you THINK?

expressive roles Roles into which women are traditionally socialized (i.e., nurturing and emotionally supportive roles).

instrumental roles Roles into which men are traditionally socialized (i.e., task-oriented roles).

structural sexism The ways in which the organization of society, and specifically its institutions, subordinate individuals and groups based on their sex classification.

education dividend The associated benefits of educating women including, for example, reducing the mortality rate of children under 5 years old.

Gender Stratification: Structural Sexism

As structural functionalists and conflict theorists agree, the social structure underlies and perpetuates much of the sexism in society. **Structural sexism**, also known as *institutional sexism,* refers to the ways the organization of society and specifically its institutions subordinate individuals and groups based on their sex classification. Structural sexism has resulted in significant differences in the education and income levels, occupational and political involvement, and civil rights of women and men.

Education and Structural Sexism

As discussed in Chapter 8, not only do women comprise two-thirds of the world's illiterate adults, girls comprise more than 62 percent of the 125 million children who don't attend school (UNESCO 2014a).

Because children born to educated mothers are less likely to die at a young age, there is an **education dividend** associated with educating women. Globally, between 1990 and

2009, improved education for women of reproductive years saved over 2 million children under the age of 5 (UNESCO 2014b). Educated women are more likely to have their children immunized, have more accurate knowledge about the transmission of HIV, and are better informed about dietary requirements. If all women completed elementary school, the under-5-years-old death rate would be reduced by nearly a million children a year.

In 2013, few differences existed between men and women in their completion rates of high school and college degrees (National Center for Educational Statistics [NCES] 2014). In fact, in recent years, most U.S. colleges and universities have had a higher percentage of women than men enrolling directly from high school (see Chapter 8). Similarly, in two-parent families, 23 percent of mothers are better educated than their husbands, while only 16 percent of fathers are better educated than their wives, the remainder of spouses having similar educational background (Wang et al. 2013). This trend is causing some concern that many young American men may not have the education they need to compete in today's global economy.

Concern over the continued lack of women in STEM (science, technology, engineering, and mathematics) is also justified. As Figure 10.1 indicates, women earn 81 percent of master's degrees in library science but only 23 percent of master's degrees in engineering and engineering technologies. Reasons for the STEM gender disparity, although decreasing, include reliance on gender stereotyping ("Boys are better in math and science than girls!"), a lack of female STEM role models, little encouragement to follow STEM pursuits, and a lack of awareness about women in STEM fields.

Women earn 51 percent of doctorate degrees (e.g., JDs, MDs, PhDs)(NCES 2014). Nonetheless, unlike many of their male counterparts, many women "opt out" of labor force participation. Hersch (2013) reports that female graduates of elite institutions, those most likely to find satisfying employment in their field of study, are more likely to "opt out" of the workforce than peers from less selective institutions. Furthermore, female graduates from elite institutions who have children under the age of 18 have significantly lower levels of labor force involvement than their academic equals without children under the age of 18.

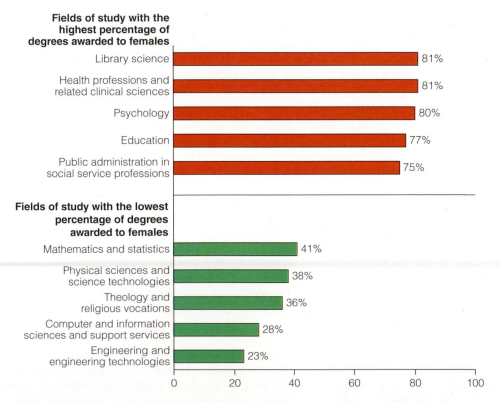

Figure 10.1 Percentage of Master's Degrees Awarded to Females in Selected Fields of Study, Academic Year, 2011–2012
SOURCE: NCES 2014.

There are also structural limitations in higher education that discourage female faculty from advancing. Women seeking academic careers may find that securing a tenure-track position is more difficult for them than it is for men, and that having children pre-tenure negatively impacts the likelihood of getting tenure; that is, there is a "pregnancy penalty" (Ceci et al. 2009). Similarly, in an analysis of data from the *National Study of Postsecondary Faculty*, Leslie (2007) reports that, as the number of children increases, the number of hours a female faculty member works decreases, and the number of hours a male faculty member works increases (see the discussion on household division of labor later in this chapter).

Research also indicates gender stereotyping in letters of recommendation. Women are more often described as socioemotive (e.g., helpful) rather than active/assertive (e.g., ambitious) and, thus, are evaluated less positively (Madera et al. 2009). However, when the letter of recommendation is for an occupation disproportionately populated by women, females are rated more highly than males (Morgan et al. 2013). Female faculty's academic products are also less likely to be cited even after controlling for tenure status, year of publication, and quality of publication (Maliniak et al. 2013); and female faculty are disproportionately assigned service duties (e.g., advising) when compared to their male counterparts (Misra et al. 2011), both of which place females at a disadvantage when being considered for tenure and/or promotion.

Work and Structural Sexism

According to an International Labour Organization (ILO) report, about half of women are in the labor force compared to nearly 80 percent of men (ILO 2015c). Women also have higher unemployment rates because of the differences in educational attainment between men and women, occupational segregation, and higher rates of exiting and reentering the labor force as result of family obligations.

Women are disproportionately employed in what is called **vulnerable employment**. Such is the case in developing countries where women are often "contributing family workers," and in developed countries where women are disproportionately working in low-wage service occupations and/or self-employed (ILO 2012, 2015c). Vulnerable employment is characterized by informal working arrangements, little job security, few benefits, and little recourse in the face of an unreasonable demand.

Finally, women are more likely to hold positions of little or no authority within the work environment and to receive lower wages than men. Over 95 percent of the chief executive officers (CEOs) of the world's largest corporations are men and, of the 500 largest corporations in the United States, just 23 have a female CEO (ILO 2015a; Van der Gaag 2014). No matter what the job, if a woman does it, it is likely to be valued less than if a man does it. For example, in the early 1800s, 90 percent of all clerks were men and being a clerk was a prestigious profession. As the job became more routine, in part because of the advent of the typewriter, the pay and prestige of the job declined and the number of female clerks increased. Today, 91.2 percent of clerks are female (Bureau of Labor Statistics [BLS] 2015), and the position is one of relatively low pay and prestige.

The concentration of women in certain occupations and men in other occupations is referred to as **occupational sex segregation**. Although occupational sex segregation remains high, as indicated by the first column in Table 10.2, it has decreased in recent years for some occupations. Between 1983 and 2014, the percentage of female physicians and surgeons more than doubled from 16 percent to 37 percent, female dentists increased from 7 percent to 29 percent, and female clergy increased from 6 percent to 18 percent (BLS 2015; U.S. Census Bureau 2009).

Although the pace is slower, men are increasingly applying for jobs that women have traditionally held. Spurred by the loss of jobs in the manufacturing sector and the economic crisis of 2007, many jobs traditionally defined as male (e.g., auto worker, construction worker) have been lost. Thus, over the last 20 years, there has been a significant increase in the number of men in traditionally held female jobs. Between 2000 and 2010, one-third of the job growth for men was in occupations that were 70 percent or more female (Dewan and Gebeloff 2012).

> Women are disproportionately employed in . . . vulnerable employment . . . characterized by informal working arrangements, little job security, few benefits, and little recourse in the face of an unreasonable demand.

vulnerable employment Employment that is characterized by informal working arrangements, little job security, few benefits, and little recourse in the face of an unreasonable demand.

occupational sex segregation The concentration of women in certain occupations and men in other occupations.

TABLE 10.2 The Wage Gap in the 10 Most Common Occupations for Women and Men (Full-Time Workers Only), 2014

	Percentage of Workers in Occupation That Are Female	Median Weekly Earnings for Women	Median Weekly Earnings for Men	Women's Earnings as Percentage of Men's
All full-time workers	44.2%	$719	$871	82.5%
10 Most Common Occupations for Women				
Secretaries and administrative assistants	94.6%	$685	$811	84.5%
Elementary and middle school teachers	80.4%	$956	$1,096	87.2%
Registered nurses	89.4%	$1,076	$1,190	90.4%
Nursing, psychiatric, and home health aides	88.0%	$466	$528	88.3%
First-line supervisors of retail sales workers	44.1%	$595	$793	75.0%
Customer service representatives	65.4%	$606	$698	86.8%
Managers, all others	37.7%	$1,153	$1,412	81.7%
Cashiers	70.6%	$387	$412	93.9%
Accountants and auditors	63.3%	$999	$1,236	80.8%
Receptionists and information clerks	90.7%	$532	$616	86.4%
10 Most Common Occupations for Men				
Driver/sales workers and truck drivers	3.90%	$545	$739	73.7%
Managers, all other	37.7%	$1,153	$1,412	81.7%
First-line supervisors of retail sales workers	44.1%	$595	$793	75.0%
Construction laborers	1.8%	—	$605	—
Janitors and building cleaners	27.4%	$415	$540	76.9%
Laborers and freight, stock, and material movers, hand	16.3%	$476	$546	87.2%
Retail salesperson	39.2%	$491	$698	70.3%
Software developers, applications and systems software	19.6%	$1,457	$1,736	83.9%
Sales representatives, wholesale and manufacturing	28.0%	$873	$1,120	77.9%
Grounds maintenance workers	5.5%	—	$454	—

SOURCE: Hegewisch and Ellis 2015.

Some evidence suggests that men in traditionally female-held jobs have an advantage in hiring, promotion, and salaries called the **glass escalator effect** (Dewan and Gebeloff 2012; Williams 2007). Table 10.2, for example, indicates that secretaries and administrative assistants are overwhelmingly female—94.6 percent. Yet the average median weekly salary for male secretaries is $811 compared to $685 for female secretaries.

Persistence of the Occupational Sex Segregation. Sex segregation in occupations continues for several reasons. First, cultural beliefs about what is an "appropriate" job for a man or a woman still exist. Snyder and Green's (2008) analysis of nurses in the United States is a case in point. Using survey data and in-depth interviews, the researchers identified patterns of sex segregation. Over 88 percent of all patient-care nurses were in sex-specific specialties (e.g., intensive care and psychiatry for male nurses, and labor or delivery and outpatient services for female nurses). Interestingly, although women rarely mentioned gender as a reason for their choice of specialty, male nurses frequently did so, acknowledging the "process of gender affirmation that led them to seek out 'masculine' positions within what was otherwise construed to be a women's profession" (p. 291).

Second, opportunity structures for men and women differ. Women and men, upon career entry, are often channeled by employers into gender-specific jobs that carry different wage and promotion opportunities. Female medical students report that male attending physicians discourage women from entering into a career as a surgeon (Bruce et al. 2015). Even women in higher-paying jobs may be victimized by a **glass ceiling**—an invisible barrier that prevents women and other minorities from moving into top corporate positions.

Working mothers are particularly vulnerable. Female lawyers returning from maternity leave found their career mobility stalled after being reassigned to less prestigious cases (Williams 2000). Using an experimental design, Correll et al. (2007) report that, even when qualifications, background, and work experience were held constant, "evaluators rated mothers as less competent and committed to paid work than non-mothers" (p. 1332). Other examples of the "**motherhood penalty**" include women who feel pressured to choose professions that permit flexible hours and career paths, sometimes known as *mommy tracks* (Moen and Yu 2000). Thus, women dominate the field of elementary education, which permits them to be home when their children are not in school. Nursing, also dominated by women, often offers flexible hours. Although the type of career pursued may be the woman's choice, it is a **structured choice**—a choice among limited options as a result of the structure of society.

Finally, Blau and Kahn (2013) argue that the comparatively low rates of female labor force participation and female labor force participation growth are the result of a lack of U.S. worker-friendly and perhaps, more important, female-friendly, employment policies. An analysis of policy data indicates that "most other countries have enacted parental leave, part-time work, and child care policies that are more extensive than in the United States, and the gap has grown over time" (p. 4). The authors conclude that such policies make part-time work more attractive and make it easier for women to "have it all"—that is, combine work and family life.

Income and Structural Sexism

In 2014, full-time working women in the United States earned, on the average, 82.5 percent of the median annual earnings of full-time working men (National Women's Law Center [NWLC] 2015a). Furthermore, cashiers, maids and household cleaners, waitresses, and personal care aids, four of the most common occupations for women, have 40-hour-a-week median incomes below the federally established poverty level for a family of four (Hegewisch and Ellis 2015).

The gender pay gap varies over time. By decade, in 1980, 1990, 2000, and 2010, women's annual earnings as a percentage of men's increased from 60 percent to 72 percent, 74 percent, and 77 percent, respectively. As indicated, closing the gender gap has slowed down since the 1980s and early 1990s (Hegewisch et al. 2015). At the present rate of progress, it is predicted that the gender gap will not be closed until 2058 (Institute for Women's Policy Research 2014).

glass escalator effect The tendency for men seeking or working in traditionally female occupations to benefit from their minority status.

glass ceiling An invisible barrier that prevents women and other minorities from moving into top corporate positions.

motherhood penalty The tendency for women with children, particularly young children, to be disadvantaged in hiring, wages, and the like, compared to women without children.

structured choice Choices that are limited by the structure of society.

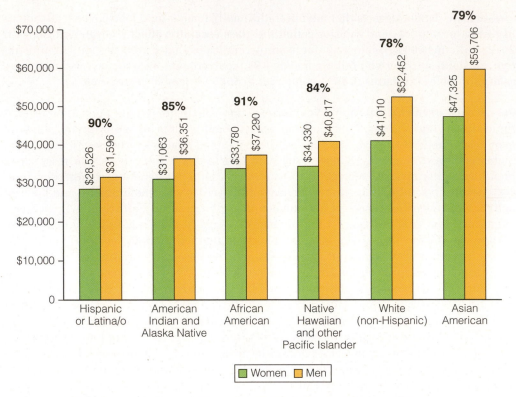

Note: Percentages in bold are women's earnings as a percentage of men's earnings.

Figure 10.2 Median Annual Earnings, by Race/Ethnicity and Gender, 2013
SOURCE: AAUW 2015.

Racial differences also exist. Although women, in general, earn 82.5 percent as much as men, black American women earn just 64 percent of white men's salaries, and Hispanic American women earn just 56 percent of white men's salaries (NWLC 2015c). Even among celebrities, a significant income gap exists. The average season salary for a National Basketball Association (NBA) player is $3 million; for a player in the Women's National Basketball Association (WNBA), $69,690 (Callaway 2015).

The gender gap not only impacts women but their families as well. Over 7 million families are headed by single working mothers with children, and over 25 percent of them are poor (see Chapter 6). Furthermore, at some point in 2013, nearly 1.5 million married couples with children and 4.2 million married couples with no children were dependent on the women's earnings alone. For the average women in 2013, closing the gender wage gap would mean an additional $10,876, which could be used for

- five months of groceries, $3,161.50;
- three months' rent and utilities, $1,950;
- three months of child care, $2,550;
- four months' health insurance premiums, $1,472;
- four months' student loan payments, $1,308;
- seven tanks of gas, $434.50 (NWLC 2015a).

The motherhood penalty not only impacts occupational sex segregation but, and in association with it, the pay gap between mothers and nonmothers. Globally, the motherhood pay gap is larger in developing countries than in developed ones, and it increases with the number of children. There is also some evidence that the motherhood pay gap is smaller when there is a female child who may take on some of the household labor and child care (ILO 2015b). Although the number of theoretical explanations of the motherhood pay gap are considerable, many social scientists hold that the gap is a consequence of child care needs that necessitate child-friendly jobs such as part-time employment, leaving and reentering the labor force after pregnancy, stereotypical decision making by employers, and occupational ghettos where women, in general, are paid less.

Why Does the Gender Pay Gap Exist? There are several arguments as to why the gender pay gap exists. One, the **human capital hypothesis**, holds that pay differences between females and males are a function of differences in women's and men's levels of education, skills, training, work experience, and the like. Bertrand et al. (2009) report that the "presence of children is associated with less accumulation of job experience, more career interruptions, and shorter work hours for female MBAs but not for male MBAs" (p. 24). Based on their analysis, the authors conclude that a decade after graduation, female MBAs earn an average annual salary of $243,481, and male MBAs earn an average annual salary of $442,353.

Lower incomes over time create a significant deficit later in life. This is particularly true given a woman's higher life expectancy and the exhaustion of household savings when her husband becomes ill. As age increases, the differences between male and female earnings increases so that by the age of 65 and over, women earn just 74 percent of their male counterparts (American Association of University Women [AAUW] 2015).

Human capital theorists also argue that women make educational choices (e.g., school attended, major, etc.) that limit their occupational opportunities and future earnings. Women, for example, are more likely to major in the humanities, education, or the social sciences rather than science and engineering, which results in reduced incomes (NCES 2014). Research also indicates, however, that after controlling for "college major, occupation, economic sector, hours worked, months unemployed since graduation, GPA, type of undergraduate institution, institution selectivity, age, geographical region, and marital status . . . a 7 percent difference in the earnings of male and female college graduates one year after graduation was still unexplained" (Corbett and Hill 2012, p. 8).

Further support for the human capital thesis comes from a survey of unemployed women and men between the ages of 25 and 54. Results indicate that 43 percent of "homemakers" who were able to work wanted a job outside the home. However, the majority indicated that they would prefer part- rather than full-time employment, and 61 percent compared to 37 percentage of men responded that they were not presently employed due to family responsibilities (Hamel et al. 2014). Although the percentage of working women in many other countries has dramatically increased between 2000 and 2013, the percentage of working women in the United States has declined from 74 percent in 2000 to 69 percent in 2013 ("The Upshot" 2014).

The second explanation for the gender gap is called the **devaluation hypothesis**. It argues that women are paid less because the work they perform is socially defined as less valuable than the work men perform. Guy and Newman (2004) argue that these jobs are undervalued in part because they include a significant amount of **emotion work**—that is, work that involves caring, negotiating, and empathizing with people, which is rarely specified in job descriptions or performance evaluations.

Finally, there is evidence that, even when women and men have equal education and experience (and, therefore, not a matter of human capital differences) and are in the same occupations (and, therefore, not a matter of women's work being devalued), pay differences remain, suggesting discriminatory occupational practices. Ten female professors at the University of Medicine and Dentistry of New Jersey, after discovering a nearly $20,000 deficit in their annual salaries compared to their male counterparts, sued the university and were awarded $4.65 million (Women in Higher Education 2013).

Politics and Structural Sexism

Women received the right to vote in the United States in 1920, with the passage of the Nineteenth Amendment. Even though this amendment went into effect more than 90 years ago, women still play a rather minor role in the political arena. In general, the more important the political office is, the lower the probability that a woman will hold it. Although women constitute over half of the population, the United States has never had a woman president or vice president and, until 2010, when Justice Elena Kagan was appointed, had only three female Supreme Court Justices in the history of the Court. The highest-ranking

human capital hypothesis
The hypothesis that pay differences between females and males are a function of differences in women's and men's levels of education, skills, training, and work experience.

devaluation hypothesis
The hypothesis that women are paid less because the work they perform is socially defined as less valuable than the work men perform.

emotion work Work that involves caring for, negotiating, and empathizing with people.

women ever to serve in the U.S. government have been former secretaries of state Madeleine Albright, Condoleezza Rice, and Hillary Clinton. In 2015, there were six female governors, and women held just 19.4 percent of all U.S. congressional seats, leading to a rank of 33 out of 49 high-income nations (Center for American Women and Politics [CAWP] 2015b; DeSilver 2015).

Women held an average of 24.2 percent of all state legislative seats, with Colorado having the highest female representation and Louisiana the lowest (CAWP 2015b, 2015c). Worldwide, women represent just 22 percent of representatives to legislative bodies (UN 2015b). The percentage, however, varies significantly by region of the world. As Figure 10.3 indicates, with the exception of Oceania, the proportion of women occupying legislative seats grew dramatically between 2000 and 2014 in developed and developing countries (UN 2014). In response to the underrepresentation of women in the political arena, today half of the world's countries have some type of legislative electoral quota in the hopes of equalizing gender representation.

The Underrepresentation of Women in Politics. The relative absence of women in politics, as in higher education and in high-paying, high-prestige jobs in general, is a consequence of structural limitations and cultural definitions of traditional gender roles (see this chapter's *Self and Society*). Running for office requires large sums of money, the political backing of powerful individuals and interest groups, and a willingness of the voting public to elect women. In fact, in a survey of present and former female politicians, money was identified as the single greatest barrier to running for office (Political Parity 2014). Disproportionately lacking these resources, minority women have even greater structural barriers to election and, not surprisingly, represent an even smaller percentage of elected officials. Of the 535 congressional representatives in 2015, women of color represent just 6.2 percent of the total (CAWP 2015a).

There is also evidence of gender discrimination against female candidates. In an experiment where two congressional candidates' credentials were presented to a sample of respondents—in one case as Ann Clark and in the other as Andrew Clark—Republican respondents were significantly more likely to say they would vote for a father with young children rather than a mother with young children. They were also more likely to vote for women without small children than women with small children—additional evidence of the "motherhood penalty." The opposite pattern was detected for Democrat respondents (Morin and Taylor 2008).

Not only do voters discriminate on the basis of gender, but political parties do as well. When women were asked whether they believed that their political party encouraged women more, encouraged women less, or equally encouraged women and men, 45 percent of the sample responded that their party was more encouraging to men than women. Only 3 percent responded that their party encouraged women more than men (Political Parity 2014).

> Running for office requires large sums of money, the political backing of powerful individuals and interest groups, and a willingness of the voting public to elect women. … Disproportionately lacking these resources, minority women have even greater structural barriers to election and, not surprisingly, represent an even smaller percentage of elected officials.

WHAT do you THINK?

In a survey by Pew Research Center (2015), a sample of Americans were asked whether "men will continue to hold more top business positions than women in the future" (53 percent) or it's "only a matter of time before women hold as many top positions as men" (44 percent) (p. 1). What do you think?

Civil Rights, the Law, and Structural Sexism

In many countries, victims of gender discrimination cannot bring their cases to court. This is not true in the United States. The 1963 Equal Pay Act and Title VII of the 1964 Civil Rights Act make it illegal for employers to discriminate in wages or employment (e.g. hiring, firing, promotions, layoffs, etc.) on the basis of sex. Despite being illegal, sex discrimination continues to occur as evidenced by the thousands of grievances filed each year with the Equal Employment Opportunity Commission (EEOC). In 2014, 26,027 grievances were filed and resolved (EEOC 2015). Known as *gender identity discrimination,* differential treatment of individuals because of their transgender status is also a violation of federal law.

Women, Men, and Leadership

Answer the following questions about women and men in politics and business. For each of the following characteristics select whether men are better at it, women are better at it, or there is no difference between the two. When you have finished, compare your responses to a sample of U.S. adults polled by the Pew Research Center.

	Men	Women	No Difference
In politics _____ are better at:			
1. Working out compromise	_____	_____	_____
2. Being honest and ethical	_____	_____	_____
3. Working to improve U.S. quality of life	_____	_____	_____
4. Standing up for beliefs	_____	_____	_____
In business, _____ in top positions are better at:			
5. Being honest and ethical	_____	_____	_____
6. Providing fair pay/benefits	_____	_____	_____
7. Mentoring employees	_____	_____	_____
8. Negotiating profitable deals	_____	_____	_____
9. Being willing to take risks	_____	_____	_____
Results:			

In **politics**, the % saying women/men in top positions are better at...

	Women are better	Men are better	No difference
Working out compromises	34%	9%	55%
Being honest and ethical	34	3	62
Working to improve U.S. quality of life	26	5	68
Standing up for beliefs	25	10	63

In **business**, the % saying women/men in top positions are better at ...

	Women are better	Men are better	No difference
Being honest and ethical	31	3	64
Providing fair pay/benefits	30	5	64
Mentoring employees	25	7	66
Negotiating profitable deals	7	18	73
Being willing to take risks	5	34	58

Results of Pew Research Center Poll, N = 1,835 U.S. Adults
SOURCE: Pew 2015.

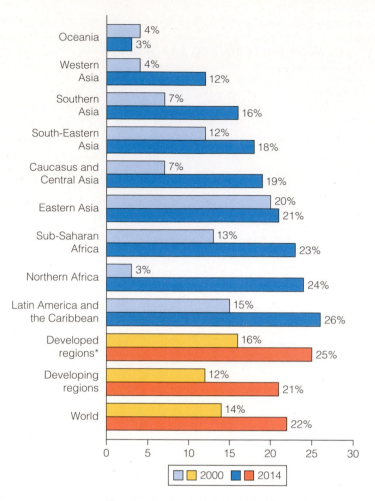

Figure 10.3 Proportion of Seats Held by Women in National Legislatures, 2000 and 2014
SOURCE: United Nations 2014.

Chart data:

Region	2000	2014
Oceania	4%	3%
Western Asia	4%	12%
Southern Asia	7%	16%
South-Eastern Asia	12%	18%
Caucasus and Central Asia	7%	19%
Eastern Asia	20%	21%
Sub-Saharan Africa	13%	23%
Northern Africa	3%	24%
Latin America and the Caribbean	15%	26%
Developed regions*	16%	25%
Developing regions	12%	21%
World	14%	22%

*Developed regions include the United States, Canada, Europe, Russian Federation, and Australia.

One technique employers use to justify differences in pay is the use of different job titles for the same type of work—for example, janitor and housekeeper. The courts have repeatedly ruled, however, that jobs that are "substantially equal," regardless of title, must result in equal pay. In addition to individual discrimination, discrimination may take place at the institutional level (see Chapter 9).

Institutional discrimination includes screening devices designed for men, hiring preferences for veterans, the practice of promoting from within an organization based on seniority, male-dominated recruiting networks, and, in many countries, laws that prohibit women from owning or inheriting assets (Head et al. 2014; Reskin and McBrier 2000). In the United States, women's lower incomes, shorter work histories, and less collateral make it difficult for some women to obtain home mortgages, rental property, or loans.

Some child brides are as young as 5 years old and rarely know the ramifications of what's happening to them. Here, a 16-year-old screams in protest as she is carried in a cart to her new husband's village.

Stephanie Sinclair/VII

There are more than 70 million child brides in the world, and more than 15 million child brides—those who married under the age of 18—are added each year. At least 90 percent had no choice in who they married (Edmeades et al. 2014). Ironically, education is one of the best deterrents of early marriage, yet early marriage most often signals the end of educational opportunities as the "skills and knowledge for a healthy transition to adulthood give way to the responsibilities of being a wife, a mother, a daughter-in-law" (Warner et al. 2014, p. 3). What do you think would be the most effective way to end this cycle?

WHAT do you THINK?

The Social Construction of Gender Roles: Cultural Sexism

As social constructionists note, structural sexism is supported by a system of cultural sexism that perpetuates beliefs about the differences between women and men. **Cultural sexism** refers to the ways the culture of society— its norms, values, beliefs, and symbols—perpetuate the subordination of an individual or group because of the sex classification of that individual or group. Cultural sexism takes place in a variety of settings, including the family, the school, and the media, as well as in everyday interactions.

Family Relations and Cultural Sexism

From birth, males and females are treated differently. **Gender roles** are patterns of socially defined behaviors and expectations, associated with being female or male.

Toys are one way that what is considered appropriate gender role behavior is often conveyed to children. A study by Professor Becky Francis of Roehampton University in England examined the impact of educational toys on 3- to 5-year-olds' learning. When parents were asked their child's favorite toys, boys' toys "involved action, construction, and machinery," whereas girls' toys were more often "dolls and perceived 'feminine' interests, such as hairdressing" (Lepkowska 2008, p. 1). After purchasing and analyzing the toys parents selected, Francis concluded that girls' toys had limited "learning potential," whereas boys' toys "were far more diverse" and propelled boys "into a world of action as well as technology" designed to "be exciting and stimulating" (quoted in Lepkowska 2008, p. 1).

Jim Pickerell/Stock Connection Blue/Alamy

Societal definitions of the appropriateness of gender roles have traditionally restricted women and men in terms of educational, occupational, and leisure-time pursuits. Fifty years ago, little boys playing with dolls and displaying nurturing behaviors would have been unheard of.

Household Division of Labor. Globally, women and girls continue to be responsible for household maintenance including cooking, gathering firewood and fetching water, and taking care of younger siblings. A survey of European women between 20 and 64 who live with their heterosexual male partners report that, on the average, they are responsible for 66 percent or more of the housework even when working full-time. The household division of labor was most equal in the Scandinavian countries (e.g., Sweden, Denmark, Finland) and the least equal in Southern European countries (e.g., Greece, Portugal, Spain) (European Social Survey [ESS] 2013).

Research in the United States documents that gendered household work patterns begin in early childhood. Wikle (2014), using longitudinal data from time-use diaries

cultural sexism The ways in which the culture of society perpetuates the subordination of an individual or group based on the sex classification of that individual or group.

gender roles Patterns of socially defined behaviors and expectations associated with being female or male.

from girls and boys who were 7 or 8 years old in 1997, examined gender and age differences in household duties (e.g., cleaning, preparing food, child care). Results indicate that a higher proportion of girls than boys participate in household labor and that the differences narrow over time. For example, in childhood, 58 percent of girls compared to 37 percent of boys engage in household activities, a difference of 21 percentage points. By adolescence, 38 percent of girls compared to 31 percent of boys participate in household labor.

At the other end of the age continuum, Leopold and Skopek (2014) investigated child care duties of more than 5,000 grandparent couples from 10 countries. The results indicate that even when both grandparents are working, grandmothers contribute more hours to caring for their grandchildren than grandfathers, even if they are the only working spouse. Although men's share of the household labor has increased over the years and women's decreased (Cohen 2004), the fact that women, even when working full-time, contribute significantly more hours to home care than men is known as the "**second shift**" (Hochschild 1989).

The traditional division of labor within the home is a consequence not only of individual choice (e.g., "opting out") but of structural constraints. Using a survey experimental design, Pedulla and Thebaud (2015) asked a sample of 329 unmarried 18- to 32-year-olds what their preferred relationship structure was—male breadwinner, female breadwinner, or egalitarian, under various levels of work constraints. Regardless of educational level, when told to imagine that their work environment was family-friendly (e.g., had subsidized child care, flexible work options, working from home), both men and women preferred a relationship structure wherein both partners were equally responsible financially and for household duties and child care. Similarly, Rehel (2013) concludes that in the absence of work constraints, fathers equally experience the responsibility of child care and become co-parents rather than simply "helpers." Discussing paternity leave, one father notes:

> Those five weeks went by so fast—we were constantly taking care of [our son]. Really made me realize to what extent taking care of a child is more than a full-time job. You don't get your 15-minute breaks, your half-hour break when you want. You don't get time off. You don't have a switch off like you do at work. Really, your attention is always—especially with a newborn—100 percent on him. (p. 122)

Explanations for the Traditional Division of Household Labor. Although some changes have occurred, the traditional division of labor remains to a large extent what one "expects" after marriage (Askari et al. 2010). Three explanations emerge from the literature. The first explanation is the *time-availability approach.* Consistent with the structural-functionalist perspective, this position claims that role performance is a function of who has the time to accomplish certain tasks. Because women are more likely to be at home, they are more likely to perform domestic chores.

A second explanation is the *relative resources approach.* This explanation, consistent with a conflict perspective, suggests that the spouse with the least power is relegated the most unrewarding tasks. Because men, on average, often have more education, higher incomes, and more prestigious occupations, they are less responsible for domestic labor. Thus, for example, because women earn less money than men do, on average, they turn down overtime and other work opportunities to take care of children and household responsibilities, which subsequently reduces their earnings potential even further (Williams 2000).

Gender role ideology, the final explanation, is consistent with a symbolic interactionist perspective. It argues that the division of labor is a consequence of traditional socialization and the accompanying attitudes and beliefs. Women and men have been socialized to perform various roles and to expect their partners to perform other complementary roles. Women typically take care of the house, and men take care of the yard. These patterns begin with household chores assigned to girls and boys and are learned through the media, schools, books, and in the family. For example, boys living in homes with a traditional division of labor are more likely to pursue male-dominated occupations than boys living in homes with greater domestic equality (Polavieja and Platt 2014).

second shift The household work and child care that employed parents (usually women) do when they return home from their jobs.

Simister's (2013) study of the division of household labor in seven countries—Cameroon, Chad, Egypt, India, Kenya, Nigeria, and the United Kingdom—is illustrative. When traditional gender roles are reversed—that is, wives earn more money than their husbands—men are more resistant to contributing to household labor than when a husband's income is greater. Similarly, Schneider (2012) reports that men in traditional female occupations spend more time on "men's work" at home; women who are in traditionally male occupations spend more time on "women's work" at home.

The findings of both of these studies are consistent with the **gender deviance hypothesis**. When there is "gender deviance" (income or occupation inconsistent with traditional gender roles), techniques are employed to neutralize the deviance (engage in traditional household division of labor), to bring it back into alignment, thereby reclaiming what is perceived to be what it means to be male and what it means to be female in American society. Note that the findings of both of these studies contradict the time availability approach and the relative resources approach.

The School Experience and Cultural Sexism

Around the world, cultural norms that further gender inequality persist. Norms favoring child marriage, quick conception, and large families, coupled with a lack of birth control, makes schooling difficult for many girls and young women, particularly in middle- and low-income countries (UN 2015a). Even when education is equally available for females and males, continued gender bias in instructional materials, student–teacher interactions, and school programs and policies may contribute to a self-fulfilling prophecy.

Instructional Material. The bulk of research on gender images in books and other instructional materials documents the way males and females are portrayed stereotypically. In a study of 200 "top-selling" children's picture books, women and girls were significantly underrepresented, with twice as many male title and main characters. Males were also more likely to be in the illustrations, to be pictured in the outdoors and, if an adult, to be visibly portrayed as employed outside the home. Both men and women were more than nine times more likely to be pictured in traditional rather than in nontraditional occupations, and "female main characters . . . were more than three times more likely than were male main characters . . . to perform nurturing or caring behaviors" (Hamilton et al. 2006, p. 761).

Bush and Furnham (2013) examined advertisements for educational and noneducational toys and games. Both educational and noneducational advertisements pictured males as central characters more often than females. The largest gender gap, however, was in *educational* toys and games wherein boys outnumbered girls as central characters by nearly 5 to one.

Student–Teacher Interactions. Gender bias is also reflected in the way that teachers treat their students. Millions of young girls are subject to sexual harassment and abuse by male students and by male teachers, who then fail them when they refuse the teachers' sexual advances (Quist-Areton 2003; UNESCO 2014b). Boys are more likely to suffer physical violence from their teachers or classmates, although sexual abuse of boys is not unheard of. In a study of schools in Vietnam, 21 percent of girls and 17 percent of boys reported being the victim of sexual violence (Nandita et al. 2014).

There is also convincing evidence that elementary and secondary school teachers are more responsive to boys than to girls—talking to them more, asking them more questions, listening to them more, counseling them more, giving them more extended directions, and criticizing and rewarding them more frequently. In *Still Failing at Fairness,* Sadker and Zittleman (2009) recount a fifth grade teacher's instructions to her students. "There are too many of us here to all shout out at once. I want you to raise your hands, and then I'll call on you. If you shout out, I'll pick somebody else." The discussion on presidents continues, with Stephen calling out:

Stephen: I think Lincoln was the best president. He held the country together during a war.

Teacher: A lot of historians would agree with you.

> There is also convincing evidence that elementary and secondary school teachers are more responsive to boys than to girls—talking to them more, asking them more questions, listening to them more, counseling them more, giving them more extended directions, and criticizing and rewarding them more frequently.

gender deviance hypothesis
The tendency to overconform to gender norms after an act(s) of gender deviance; a method of neutralization.

Kelvin (seeing that nothing happened to Stephen, calls out): I don't. Lincoln was OK but my Dad liked Reagan. He always said Reagan was a great president.

David (calling out): Reagan? Are you kidding?

Teacher: Who do you think our best president was, David?

David: FDR. He saved us from the Depression.

Max (calling out): I don't think it's right to pick one best president. There were a lot of good ones.

Teacher: That's interesting.

Rebecca (calling out): I don't think the presidents today are as good as the ones we used to have.

Teacher: Ok, Rebecca. But you forgot the rule. You're supposed to raise your hand. (pp. 65–66)

Espinoza and colleagues (2013) examined **attributional gender bias**—the practice of understanding the same behavior(s) of females and males using different explanations. Middle and high school math teachers were asked to explain the success and failure of boys and girls in math. Success in math for boys was more often attributed to their ability and for girls, to their effort. However, when boys did not perform well in math, teacher participants were more likely to attribute their failure to a lack of effort, and girls' failure to a lack of ability.

School Programs and Policies. As discussed in Chapter 8, Title IX of the 1972 Educational Amendments Act prohibits sex discrimination in educational programs and activities that receive federal financial assistance (Office of Civil Rights 2014). An evaluation of Title IX suggests that sex discrimination continues at many levels. Women's and girls' athletic programs lag behind boys' and men's programs in terms of participation, resources, and coaching. Career and technical education remain sex segregated, in part because of stereotyped career counseling in high school, needlessly limiting occupational aspirations of both males and females (National Coalition for Women and Girls in Education 2012).

Sex Segregation in the Classroom. In 2006, the U.S. Department of Education granted school districts permission to expand single-sex education if it can be shown to accomplish a specific educational objective. The federal law also requires that a "substantial relationship" exist between the educational objective and the means—in this case, single-sex classrooms. Disproportionately populated by African American and Hispanic students, as of 2014, there were 805 single-sex public schools in the United States, and the number continues to increase (Klein et al. 2014).

Evidence suggests, however, that single-sex classrooms don't enhance learning objectives and are often justified by gender stereotypes—for example, boys learn better in a competitive atmosphere and girls learn better in a helping environment. Strain (2013), studying the effects of single-sex classes on reading and math test scores, concludes that single-sex classrooms provide no advantage to students in reading outcomes, and they are associated with significantly *lower* end-of-grade math scores. As measured by the suspension rate, the results also indicate that single-sex classrooms do not improved student behavior, often a justification used by those in favor of single-sex education. Goodkind et al. (2013)—using in-depth interviews and focus groups of students, teachers, volunteers, social workers, interns, and parents at a predominantly African American low-income academy—report that single-sex classrooms in large part do not reduce student distractions and that gender stereotypes are often reinforced by curricula differences. As one teacher recounts:

And then I found out that the girls had been doing something different than the boys, you know, they had been taken to a play, to see, uh, and this is interesting, *The Scarlet Letter.* You know, well why wouldn't they take boys to see *The Scarlet Letter,* too? That's a little gender-biased there, you know, they felt as though they should warn the girls not to be sluts? You know? What is that? (p. 1178)

attributional gender bias
The practice of explaining the same behavior(s) of females and males using different explanations.

The assumption behind single-sex education is that boys and girls learn differently—have "pink and blue brains," if you will—a belief for which there is very little scientific support (see Chapter 8). Single-sex education also furthers sex and gender binaries and ignores that, "[i]n a world of ever-increasing visibility of gender diversity . . . single-sex schooling is an anachronism—one that has the potential to take us back to a time when females and males who behaved outside gender norms were perceived as 'problems' instead of as people" (Jackson 2010, p. 237).

When parents were told by officials at an elementary school that a transgender child who identifies as a girl would be using the girls' restroom, many parents objected, leading the local school board to overturn the school's decision. Said the child's father, "I . . . implore . . . all of us that as we move forward we don't trade understanding for fear and that we don't trade misconceptions for hate" (Portnoy 2015, p. 1). Only a dozen or so states have laws protecting transgender youth in schools. Do you think all states should have such protections?

WHAT do you THINK?

Self-Fulfilling Prophecy. Symbolic interactionists remind us that the expectation of an outcome increases the likelihood of that outcome occurring—that is, it is a self-fulfilling prophecy. The differing expectations and/or encouragement that females and males receive contribute to their varying abilities, as measured by standardized tests, in disciplines such as reading, math, and science. Are such differences a matter of aptitude? Social science research would indicate otherwise.

In an experiment at the University of Waterloo, male and female college students, all of whom were good in math, were shown either gender-stereotyped or gender-neutral advertisements. When, subsequently, female students who had seen the female-stereotyped advertisements took a math test, they performed significantly lower than women who had seen the gender-neutral advertisements (Begley 2000). Heyder and Kessels (2013) report that the more boys associate school as a feminine rather than masculine construct, and attribute such masculine characteristics as hostility, aggressiveness, and boastfulness to themselves, the lower their performance in language skills, a subject area typically defined as feminine.

Media, Language, and Cultural Sexism

Another concern social scientists voice is the extent to which the media portray females and males in a limited and stereotypical fashion and the impact of such portrayals. Levin and Kilbourne (2009), in *So Sexy So Soon,* document the sexualizing of young girls and boys. Advertising, books, cartoons, songs, toys, and television shows create

> [a] narrow definition of femininity and sexuality [that] encourages girls to focus heavily on appearance and sex appeal. They learn at a very young age that their value is determined by how beautiful, thin, "hot," and sexy they are. And boys, who get a very narrow definition of masculinity that promotes insensitivity and macho behavior, are taught to judge girls based on how close they come to an artificial, impossible, and shallow ideal. (p. 2)

Men are also victimized by media images. A study of 1,000 adults found that two-thirds of the respondents

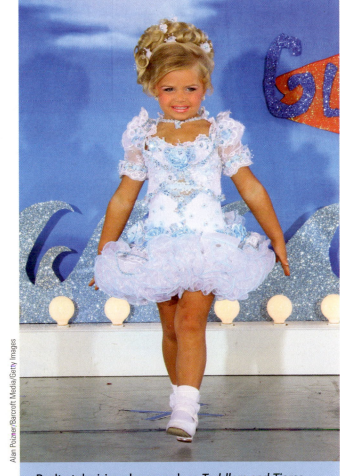

Alan Poizner/Barcroft Media/Getty Images

Reality television shows such as *Toddlers and Tiaras* and the beauty pageants they are based on hypersexualize the contestants, some as young as 6 months old. These distorted images of beauty, and the Barbie dolls and airbrushed models that follow, set an impossibly high standard for female beauty.

thought that women in television advertisements were pictured as "intelligent, assertive, and caring," whereas men were portrayed as "pathetic and silly" (Abernathy 2003). In a study of beer and liquor ads, Messner and Montez de Oca (2005) conclude that, although beer ads of the 1950s and 1960s focused on men in their work roles and only depicted women as a reflection of men's home lives, present-day ads portray young men as "bumblers" and "losers" and women as "hotties" and "bitches."

Almed and Wahab's (2014) content analysis of the 10 most popular programs on cartoon network reveals that there are significantly more male- (e.g., Batman) than female-oriented (e.g., Powder Puff Girls) cartoons, that males appear in cartoons more often than females, and that each are portrayed stereotypically—males as brave, strong, and intelligent, and females as emotional, caring, weak, dependent, and provocatively dressed. Smith et al. (2015) conducted a content analysis of female and male named characters in 120 films rated as appropriate for children between the ages of 12 and 16 from 11 countries. Results indicate the following:

- Just 30.9 percent of named characters were female.
- Thirty percent or fewer films portrayed a gender-balanced cast.
- Filmmakers, directors, producers, and writers, were 80 percent male.
- Women over 40 were less likely to be portrayed than men over 40.
- Females were twice as likely as males to be portrayed in sexually revealing clothes or partially or fully naked.
- Comments about physical attractiveness were five times more likely to be addressed toward women than men.

Figure 10.4 graphically represents work sectors represented in the films by named characters' gender. Note that there is no work sector where the number of women exceeds the number of men. It should also be noted that the two occupational categories with the highest representation of women, healthcare and education, contain fields heavily dominated by women—nurses, nurses aids/ assistants, social workers, teachers, and librarians. Alternatively, within the same work sectors, men were significantly more likely to be portrayed as doctors or pharmaceutical/healthcare managers (84.3 percent compared to 15.7 percent) and as professors (94.1 percent compared to 29.4 percent).

The Geena Davis Institute on Gender in Media, concerned about the relative absence of girls in children's programming, has sponsored public service announcements (PSAs) about "Jane"—a fictional character with a little girl's voice used as part of an awareness campaign. In one PSA, a little girl says:

Meet Jane. See Jane. See her? She makes up half the world's population. But you wouldn't know it by watching kids' media. On screen, Jane is outnumbered by a ratio of 3 to 1. When she is there, a lot of time it's purely as eye candy. And girls everywhere are watching. On average, over seven hours a day. If they see Jane, it's with little to say, few career options, and even fewer aspirations. But we can change. Meet Jane. See Jane? She is half

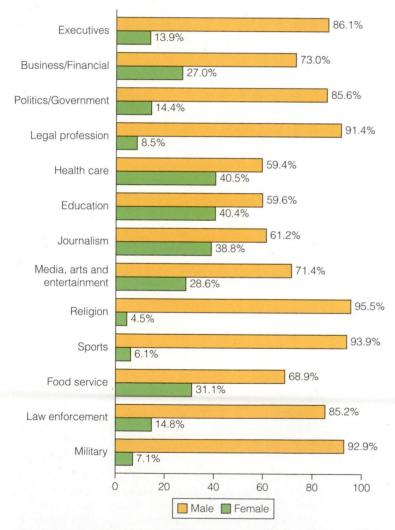

Figure 10.4 Percent of Male and Female Speaking Characters (N = 5,799) in Popular Films (N = 120) Across 11 Countries, by Work Sector, 2010–2013
SOURCE: Smith et al. 2015.

the world's population. She has important things to say. And she can be anything she wants to be. But to empower girls, we need to see Jane. See Jane.org. If she can see it, she can be it. ("See Jane" 2013)

Although media often portray men and women in traditional roles, at least one cartoon, *My Little Pony*, is "bucking the stereotypes" (Valiente and Rausmusson 2015). Nearly all of the episodes analyzed portrayed females in positions of authority, as central characters, and as the motivating force behind plot development. Given the popularity of *My Little Pony*, why do you think other media and, specifically, cartoons remain so stereotypical?

Religion and Cultural Sexism

In general, religious teachings have tended to promote traditional conceptions of gender. In 2012, the Vatican appointed an American bishop to "rein in" a group of Catholic nuns who, in the church's view, have "radical feminist themes [which are] incompatible with the Catholic faith" (Goodstein 2012, p. 1). Members of the group were cited for advocating ordination of female priests, failure to wear the traditional habit, living outside the convent, joining advocacy groups, working in academia, and focusing too much on poverty and economic injustice while remaining silent on such issues as abortion and marriage equality (Goodstein 2009, 2012). However, in 2015, Pope Francis stated that he believed that "more weight and more authority must be given to women" (quoted in Bajekal 2015, p. 1). Furthermore, in 2015, the first female bishop of the Anglican Church of England was consecrated (British Broadcasting Corporation 2015).

Gallagher (2004) found that most evangelical Christians believe that marriage should be considered an equal partnership but also believe that the male is the head of the household. Of late, however, young evangelicals have turned to "egalitarian gender theology," which views men and women as equals at work, home, and in the church (Boorstein 2014). Nonetheless, as the debate over gender roles becomes increasingly polarized in the church, the majority of U.S. evangelical churches are patriarchal, i.e., male dominated.

Although in 2013 three Orthodox Jewish women were ordained as spiritual leaders (Ungar-Sargon 2013), in general, Orthodox Judaism prohibits females from being counted as part of the *minyan* (i.e., a quorum required at prayer services), and women are not allowed to read from the Torah and are required to sit separately from men at religious services. Roman Catholic Church doctrine forbids the use of artificial forms of contraception, and Muslim women are required to be veiled in public at all times. In addition to the Catholic Church, several other religious groups prohibit women from becoming ordained religious leaders, including Mormons, Muslims, Orthodox Jews, Missouri Synod Lutherans, and members of the Orthodox Church in America and the Southern Baptists Convention (Pew 2014a).

Religious teachings are not all traditional in their beliefs about women and men. Quaker women have been referred to as the "mothers of feminism" because of their active role in the early feminist movement. Reform Judaism has allowed ordination of women as rabbis for more than 40 years and gays and lesbians for more than 15 years. In addition, the women-church movement—a coalition of feminist faith-sharing groups composed primarily of Roman Catholic women—offers feminist interpretations of Christian teachings. Within many other religious denominations, individual congregations choose to interpret their religious teachings from an inclusive perspective by replacing masculine pronouns in hymns, the Bible, and other religious readings.

Social Problems and Traditional Gender Role Socialization

One of the fundamental questions in the sociology of gender is the extent to which observed differences are a consequence of nature (i.e., innate) or nurture (i.e., environmental). It is likely, as author and neuroscientist Lise Eliot states, that "[s]ex differences are real and some are probably present at birth, but then social factors magnify them." So, she continues, "if we, as a society, feel that gender divisions do more harm than good, it would be valuable to break them down" (quoted in Weeks 2011, p. 1).

The recent acknowledgment that gender is not binary but is best conceived of as a continuum suggests that traditional gender roles are changing. However, to a large extent, the social definitions of what it means to be a woman and what it means to be a man have varied little over the decades. These definitions, in turn, are associated with several social problems, including the feminization of poverty, social-psychological costs, death and illness, conflict in relationships, and gendered violence.

The recent acknowledgment that gender is not binary but is best conceived of as a continuum suggests that traditional gender roles are changing.

The Feminization of Poverty

A large majority of the global poor are women. Women comprise 40 percent of the global nonagricultural workforce yet earn between 10 and 30 percent less than their male counterparts, have higher unemployment rates (particularly in Middle Eastern countries), and often find it more difficult to reenter the labor force. Women are also more likely than men to work in the agricultural sector in unpaid capacities, and women and girls increasingly belong to such vulnerable groups as migrants and domestic workers (UN 2014).

Women make up two-thirds of minimum wage workers in the United States and are significantly more likely to live in poverty than are men. Twenty-two percent of minimum wage workers are women of color; and, as established by federal law, at $7.25 an hour, a woman working 40 hours a week earns just $14,500 a year—$4,000 below the federally established poverty level for a mother and two children (NWLC 2014). Household incomes of transgender individuals are also low and unemployment high. Nearly four times as many transgender households live on less than $10,000 a year when compared to the general population (Grant et al. 2011), and unemployment rates are double the national average; transgender persons of color rates are four times the national average (Machitt 2014).

Two groups of women are the most likely to be poor in the United States: women over the age of 65 and women with dependent children. As Figure 10.5 indicates, U.S. women 65 years old and older are nearly twice as likely to be poor as U.S. men over the age of 65 (DeNavas-Walt and Proctor 2014). Having lower average salaries over a working life, fewer years working, and/or working in vulnerable occupations leads to smaller pensions or no retirement income at all. Although Social Security is the federally funded safety net for the elderly, the average Social Security payment for women 65 and older is just $13,500 a year compared to $17,600 a year for men 65 years and older (NWLC 2015c). Moreover, the largest source of income for women between the ages of 65 and 74 is Social Security, while for men in the same age group it is earned income (Fischer and Hayes 2013).

Women with children are far more likely to work part-time than their male counterparts (International Trade Union Commission 2011) as a result of cultural norms that dictate the primacy of a woman's role as mother and caregiver. Thirty percent of female-headed households compared to 16 percent of male-headed households are below the poverty level (DeNavas-Walt and Proctor 2014). Hispanic (38.8 percent) and black female-headed (36.7 percent) households are the poorest of all families headed by a single woman (U.S. Census Bureau 2013).

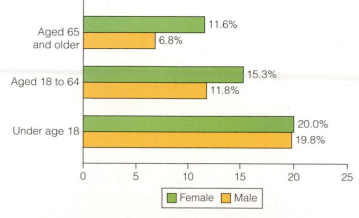

Figure 10.5 Poverty Rates by Age and Gender, 2013
SOURCE: DeNavas-Walt and Proctor 2014.

The Social-Psychological Costs of Gender Socialization

How we feel about ourselves begins in early childhood. Significant others, through the socialization process, expect certain behaviors and prohibit others based on our birth sex. Both girls and boys, often as a consequence of these expectations and the ability to perform them, feel varying degrees of self-esteem, autonomy, depression, and life dissatisfaction. Jose and Brown (2009) studied depression, stress, and rumination (i.e., worrying) in a co-ed sample of 10- to 17-year-olds. Each of the three dependent variables was significantly higher in females when compared to males.

Although it may be tempting to attribute such differences to biology, Freeman and Freeman (2013b), in their book *Stressed Sex: Uncovering the Truth about Men, Women, and Mental Health,* suggest that the higher rates of mental health problems in women is a result of "life events."

> Being judged on one's appearance and the degree to which one conforms to a largely unattainable physical "ideal," shouldering the burden of responsibility for family, home and career, growing up in a society that routinely valorizes masculinity while belittling femininity, and having to run the gauntlet of everyday sexism—all of these factors are likely to help lower women's self-esteem, increase their level of stress and leave them vulnerable to mental health problems. (2013b, p. 1)

Transgender individuals also suffer from self-esteem issues and depression (see this chapter's *The Human Side*). Caitlyn Jenner, a transgender woman and activist, discussed her own "very dark moments" in a 2015 episode of her reality show. To that point, in a survey of 6,450 transgender adults, 41 percent reported attempting suicide. The comparable number for the general population is 1.6 percent (Grant et al. 2011). For transgender individuals, family acceptance is an important buffer against societal prejudice and discrimination. Not surprisingly, transgender family members who are accepted and supported by their families have lower rates of risky behavior and negative experiences.

Pressure to conform to traditional gender roles exists not only at the individual level but at the societal level as well. After administering a questionnaire to a sample of more than 6,500 undergraduates in 13 countries, Arrindell et al. (2013) conclude that in societies where there is a strong emphasis on masculinity (i.e., in "tough countries"), masculine gender role stress is significantly higher than in countries where a masculine identity is less emphasized (i.e., in "soft countries").

The traditional male gender role places enormous cultural pressure on men to be successful in their work and to earn high incomes. Sanchez and Crocker (2005) found that, among college-aged women *and* men, the more participants were invested in traditional ideals of gender, the *lower* their self-concept and psychological well-being. Traditional male socialization also discourages males from expressing emotion and asking for help—part of what William Pollack (2000) calls the **boy code**.

It must not go unsaid, however, that although women have higher rates of some types of stress-related mental illnesses, men are more likely to abuse alcohol and drugs (see Chapter 3) and to commit suicide (see the section on "Gender-Based Violence"). Men in wealthy countries are three times more likely to commit suicide than women, 1.5 times more likely in middle- and low-income countries. Worldwide, self-harm accounts for half of all violent deaths in men (WHO 2014c).

The "Cult of Thinness." Adolescent girls are more likely to be dissatisfied with their looks, including physical attractiveness, appearance, and body weight, than adolescent boys. The Canadian nonprofit group MediaSmarts (2012) summarizes the present state of research on the "cult of thinness": Research indicates that even before being exposed to advertisements for fashion and beauty products, girls as young as 3 years old prefer game pieces that portray people who are thin rather than heavy; by age 7, they are already concerned about their appearance and can identify something they would like to change; and by 16 to 21 years old, half of young women say they would like to have cosmetic surgery to improve their bodies. Furthermore, after collecting survey data from 881 college women at Midwest universities, Eckler and colleagues (2014) conclude that time spent on Facebook is positively associated with a negative body image—the higher one, the higher the other. Respondents report comparing their bodies to friends on Facebook through posts and photos, a process known as *mass-mediated objectification.*

boy code A set of societal expectations that discourages males from expressing emotion, weakness, or vulnerability, or asking for help.

In 2014, right after Christmas, a 17-year-old committed suicide by walking in front of a truck on an Ohio freeway. Why? Because she was transgender—a girl trapped in a boy's body. Some research suggests that over 40 percent of transgender people attempt suicide sometime during their life (Haas et al. 2014). It may not be too late, however, as one commentator noted, "It may still be possible to fulfill at least one of Leelah's wishes" (Boylan 2015, p. 1).

If you are reading this, it means that I have committed suicide and obviously failed to delete this post from my queue.

Please don't be sad, it's for the better. The life I would've lived isn't worth living in . . . because I'm transgender. I could go into detail explaining why I feel that way, but this note is probably going to be lengthy enough as it is. To put it simply, I feel like a girl trapped in a boy's body, and I've felt that way ever since I was 4. I never knew there was a word for that feeling, nor was it possible for a boy to become a girl, so I never told anyone and I just continued to do traditionally "boyish" things to try to fit in.

When I was 14, I learned what transgender meant and cried of happiness. After 10 years of confusion I finally understood who I was. I immediately told my mom, and she reacted extremely negatively, telling me that it was a phase, that I would never truly be a girl, that God doesn't make mistakes, that I am wrong. If you are reading this, parents, please don't tell this to your kids. Even if you are Christian or are against transgender people don't ever say that to someone, especially your kid. That won't do anything but make them hate them self. That's exactly what it did to me.

My mom started taking me to a therapist, but would only take me to Christian therapists (who were all very biased), so I never actually got the therapy I needed to cure me of my depression. I only got more

Christians telling me that I was selfish and wrong and that I should look to God for help.

When I was 16 I realized that my parents would never come around, and that I would have to wait until I was 18 to start any sort of transitioning treatment, which absolutely broke my heart. The longer you wait, the harder it is to transition. I felt hopeless, that I was just going to look like a man in drag for the rest of my life. On my 16th birthday, when I didn't receive consent from my parents to start transitioning, I cried myself to sleep.

I formed a sort of a "fuck you" attitude towards my parents and came out as gay at school, thinking that maybe if I eased into coming out as trans it would be less of a shock. Although the reaction from my friends was positive, my parents were pissed. They felt like I was attacking their image, and that I was an embarrassment to them. They wanted me to be their perfect little straight Christian boy, and that's obviously not what I wanted.

So they took me out of public school, took away my laptop and phone, and forbid me of getting on any sort of social media, completely isolating me from my friends. This was probably the part of my life when I was the most depressed, and I'm surprised I didn't kill myself. I was completely alone for 5 months. No friends, no support, no love. Just my parents' disappointment and the cruelty of loneliness.

At the end of the school year, my parents finally came around and gave me my phone and let me back on social media. I was excited, I finally had my friends back. They were extremely excited to see me and talk to me, but only at first. Eventually they realized they didn't actually give a shit about me, and I felt even lonelier than I did before. The only friends I thought I had only liked me because they saw me five times a week.

After a summer of having almost no friends plus the weight of having

to think about college, save money for moving out, keep my grades up, go to church each week and feel like shit because everyone there is against everything I live for, I have decided I've had enough. I'm never going to transition successfully, even when I move out. I'm never going to be happy with the way I look or sound. I'm never going to have enough friends to satisfy me. I'm never going to have enough love to satisfy me. I'm never going to find a man who loves me. I'm never going to be happy. Either I live the rest of my life as a lonely man who wishes he were a woman or I live my life as a lonelier woman who hates herself. There's no winning. There's no way out. I'm sad enough already, I don't need my life to get any worse. People say "it gets better" but that isn't true in my case. It gets worse. Each day I get worse.

That's the gist of it, that's why I feel like killing myself. Sorry if that's not a good enough reason for you, it's good enough for me. As for my will, I want 100% of the things that I legally own to be sold and the money (plus my money in the bank) to be given to trans civil rights movements and support groups, I don't give a shit which one. The only way I will rest in peace is if one day transgender people aren't treated the way I was, they're treated like humans, with valid feelings and human rights. Gender needs to be taught about in schools, the earlier the better. My death needs to mean something. My death needs to be counted in the number of transgender people who commit suicide this year. I want someone to look at that number and say "that's fucked up" and fix it. Fix society. Please.

Goodbye,

(Leelah) Josh Alcorn

SOURCE: https://web.archive.org/web/20150102024359/http://lazerprincess.tumblr.com/post/106447705738/suicide-note

Boys, too, from an early age are concerned about body image, and as adults, their self-esteem is also linked to body shape and weight (Grogan 2008). Using focus groups and interview techniques, Norman (2011) describes how young men between the ages of 11 and 15, on the one hand, have concerns about their body images but, on the other, are anxious about discussing their concerns. As with females, the media often portrays an unrealistic ideal body image for males.

Dallesasse and Kluck (2013) examined muscle mass and percentage of body fat of primary male cast members in three types of reality television shows—reality drama, endurance contest, and dating/romance—on MTV, VH1, Spike TV, and the Discovery Channel. After capturing images and taking measurements, the researchers concluded that between 70 to 88 percent of the primary male characters were "somewhat muscular to very muscular," and over 90 percent had "medium to low body fat." This is in stark contrast to the 35.6 percent of 18- to 44-year-old males who strength train sufficiently enough to be considered muscular. The researchers conclude that, although "promotion of the masculine ideal *may* encourage male viewers to adopt more health-related behaviors (e.g., exercise, healthy eating habits), the ideal may also contribute to body dissatisfaction and engagement in a number of unhealthy body investment strategies among some men" (p. 314).

Research indicates that obese women and *very* obese men have lower incomes than their nonobese peers; that is, "weight-based income penalties begin at [proportionately] lower weights for women than for men" (Mason 2012, p. 424). Furthermore, the impact of body mass on men's income disadvantage disappears as years on the job increase, but does not do so for women. What do you think is responsible for these two gender differences?

Gender Role Socialization and Health Outcomes

Men are less likely to go to a doctor than women for a variety of structural and cultural reasons (WHO 2010). Men, for example, work longer hours than women and are more likely to be working full-time, making it difficult to see a doctor or attend preventive medicine programs that are often only available during the day. Men are also less likely to have a regular physician or to go to the hospital, even if time permits (White and Witty 2009).

At every stage of life, "American males have poorer health and a higher risk of mortality than females" (Gupta 2003, p. 84). Globally, as in the United States, men die about five years earlier than women, although gender differences in life expectancy have shrunk over the years (WHO 2014d). Traditionally defined gender roles for men are linked to high rates of cirrhosis of the liver (e.g., alcohol consumption), many cancers (e.g., tobacco use), and cardiovascular diseases (e.g., stress). Men also engage in self-destructive behaviors—poor diets, lack of exercise, higher drug use, refusal to ask for help or wear a seat belt, and stress-related activities—more often than women. Being married improves men's health more than it does women's (Williams and Umberson 2004), in large part because wives encourage their husbands to take better care of themselves.

The World Health Organization (WHO; 2009, 2011) has identified several ways traditional definitions of gender impact the health and well-being of women and girls. Primarily responsible for household duties, women are exposed to hundreds of pollutants as they cook that contribute to their disproportionately high rates of death from chronic obstructive pulmonary disease (COPD). Deaths from lung cancer and other tobacco-related illnesses are expected to rise as the tobacco industry targets women in developing countries. Finally, many women and girls throughout the world have a higher probability of suffering or dying from a variety of diseases because of their gender: They are more likely to be poor, are less likely to be seen as worthy of care when resources are short, and, in many countries, are forbidden to travel unaccompanied by a male, making access to a hospital difficult.

Ninety-nine percent of maternal deaths take place in developing countries. According to the World Health Organization (2014d), approximately 800 women die every day

WHAT
do you
THINK?

Men . . . work longer hours than women and are more likely to be working full-time, making it difficult to see a doctor or attend preventive medicine programs. . . . Men are also less likely to have a regular physician or to go to the hospital, even if time permits.

while giving birth, and many others "suffer serious complications from pregnancy, labor, and delivery, which can result in long-term disabilities" (U.S. Department of State 2012, p. 1). Maternal morbidity is linked to child marriage and the resulting adolescent pregnancy. In India, Ethiopia, Uganda, and Nepal, for example, nearly half of all women marry before the age of 18 (Head et al. 2014).

Gender-Based Violence

Men are more likely than women to be involved in violence—to kill and be killed; to wage war and die both as combatants and noncombatants; to take their own lives, usually with the use of a firearm; to engage in violent crimes of all types; to bully, harass, and abuse. As sociologist Michael Kimmel (2012) notes in his discussion of the mass killings at Sandy Hook Elementary School, unlike girls,

> boys learn that violence is not only an acceptable form of conflict resolution, but the one that is admired. . . . They learn it from their fathers, ...from a media that glorifies it, from sports heroes who commit felonies and get big contracts, from a culture saturated in images of heroic and redemptive violence. They learn it from each other.

Kimmel (2011) also argues that violence against transgender and gay youth is rooted in notions of masculinity and the threat that gender nonconformity poses to "real men." Consistent with this assertion, in a study of male aggression toward feminine gay, feminine heterosexual, masculine gay, and masculine heterosexual men, only feminine heterosexual men elicited a significantly higher level of aggression, suggesting a greater violation of gender conformity than that attributed to gay men (Sloan et al. 2015). Not surprisingly, a higher proportion of transgender adults compared to gender-conforming adults recount being harassed (78 percent) and physically (35 percent) and sexually (12 percent) assaulted while in school. In some cases, the victimization was so severe that 15 percent of the respondents report having to leave school (Grant et al. 2011).

Women and girls are also the victims of male violence. Worldwide, the United Nations Development Fund estimates that "at least one in every three women globally has been beaten, coerced into sex, or otherwise abused in her lifetime, with rates reaching 70 percent in some countries" (USAID 2014, p. 1; see Chapter 5). Attacks on women's and girls' bodies are thus fairly routine, often taking place in the name of religion, war, or honor. More than 5,000 women and girls are killed each year in **honor killings**—murders, often public, as a result of a female dishonoring, or being perceived to have dishonored, her family or community although the number is likely to be much higher. Although illegal throughout the world, laws prohibiting such behaviors are often not enforced or contain loopholes that permit the killer to go unpunished (Recknagel 2015).

Sonali Mukherjee, an acid attack victim, has had 27 surgeries. Three college classmates threw acid on her after she ignored their advances. The men were released from jail after serving two years.

Sajjad Hussain/AFP/Getty Images

Honor killings, although occurring throughout history and the world, are most likely to take place today in the Middle East (e.g., Syria, Jordan) and Western and Central Asia (e.g., Iraq, India). To assess contemporary attitudes toward honor killing, Eisner and Ghuneim (2013) administered a survey to 856 ninth graders in Amman, a city of 2.5 million and the capital of Jordan. The results indicate that nearly half of the boys and 20 percent of the girls surveyed condone honor killings—that is, believe that killing a woman who

honor killings Murders, often public, as a result of a female dishonoring, or being perceived to have dishonored, her family or community.

has dishonored her family is justifiable. The authors conclude that "three . . . factors are important for understanding attitudes toward honor killings, namely traditionalism, the belief in female chastity as a precious good, and a general tendency to morally neutralize aggressive behavior" (p. 413).

Globally, one of the most "severe forms of violence against women and girls" is human trafficking (ILO 2014, p. 1)—predominantly for domestic servitude and sexual slavery (see Chapter 4). Human trafficking of women and girls, as well as other forms of violence against females, is rooted in gender inequality and the lingering notion that women and girls are property—a belief tied to ancient law and many of the world's religions. Yet, even in the United States, being female is not part of the federal hate crime statutes despite the fact that, as one advocate testified before Congress, "women and girls . . . are exposed to terror, brutality, serious injury, and even death because of their sex" (Leadership Conference on Civil Rights Education Fund 2009, p. 33).

WHAT do you THINK?

Worldwide, much of the violence against women is steeped in "harmful traditional practices," including honor killings, female genital mutilation, forced marriage, and dowry killings. Dowry killings involve "a woman being killed by her husband or in-laws because her family is unable to meet their demands for her dowry—a payment made to a woman's in-laws upon her engagement or marriage as a gift to her new family" (United Nations Development Fund for Women 2007, p. 1). Although illegal in India since 1961, the number of women killed in India because of inadequate dowries has recently increased (Bedi 2013). How would you stop dowry killings?

Strategies for Action: Toward Gender Equality

In recent decades, awareness of the need to increase gender equality has grown throughout the world. Strategies to achieve this end have focused on empowering women in social, educational, economic, and political spheres and improving women's access to education, nutrition, health care, and basic human rights. But as we will see in the following section on grassroots movements, a men's movement is also concerned with gender inequities and the issues facing men.

Grassroots Movements

Efforts to achieve gender equality in the United States have been largely fueled by the feminist movement. Despite a conservative backlash, and to a lesser extent men's activist groups, feminists have made some gains in reducing structural and cultural sexism in the workplace and in the political arena.

Feminism and the Women's Movement. **Feminism** is the belief that women and men should have equal rights and responsibilities. The U.S. feminist movement began in Seneca Falls, New York, in 1848, when a group of women wrote and adopted a women's rights manifesto modeled after the Declaration of Independence. Although many of the early feminists were primarily concerned with suffrage, feminism has its "political origins . . . in the abolitionist movement of the 1830s," when women learned to question the assumption of "natural superiority" (Andersen 1997, p. 305). Early feminists were also involved in the temperance movement, which advocated restricting the sale and consumption of alcohol, although their greatest success was the passing of the Nineteenth Amendment in 1920, which recognized women's right to vote.

The rebirth of feminism almost 50 years later was facilitated by a number of interacting forces: an increase in the number of women in the labor force, the publication of Betty Friedan's book *The Feminine Mystique,* an escalating divorce rate, the socially and politically liberal climate of the 1960s, student activism, and the establishment of the Commission on the Status of Women by John F. Kennedy. The National Organization

feminism The belief that men and women should have equal rights and responsibilities.

for Women (NOW) was established in 1966, and it remains one of the largest feminist organizations in the United States, with hundreds of thousands of members in over 500 chapters across the country (NOW 2015).

One of NOW's hardest-fought battles is the struggle to win ratification of the **equal rights amendment (ERA)**, which states that "equality of rights under the law shall not be denied or abridged by the United States, or by any state, on account of sex." The proposed twenty-eighth amendment to the Constitution passed both the House of Representatives and the Senate in 1972, but failed to be ratified by the required 38 states by the 1979 deadline, which was later extended to 1982. With the exception of 2008 when presidential politics took precedence, the bill has been reintroduced into Congress every year since 1982. To date, 15 states have not ratified the ERA (only three are required for the amendment to become law), some of which have partial or inclusive guarantees of equal rights on the basis of sex within their state constitutions (United Equality 2015).

Proponents of the ERA argue that its opponents used scare tactics—saying that the ERA would lead to unisex bathrooms, mothers losing custody of their children, and mandatory military service for women—to create a conservative backlash. However, Susan Faludi, in *Backlash: The Undeclared War against American Women* (1991), contends that contemporary arguments against feminism are the same as those levied against the movement 100 years ago and that the negative consequences predicted by opponents of feminism (e.g., women unfulfilled and children suffering) have no empirical support. Proponents also argue that "without the explicit wording and intention of women's rights documented in the principles of our government, women remain second-class citizens" (Cook 2009, p. 1).

Today, a new wave of feminism is being led by young women and men who grew up with the benefits their mothers or grandmothers won. These young feminists are more inclusive than their predecessors, welcoming all who champion the cause of global equality. In a survey of U.S. adults, when a feminist was defined as "someone who believes in the social, political, and economic equality of the sexes," 60 percent of respondents considered themselves feminist (YouGov 2014).

The Men's Right's Movement. As a consequence of the women's rights movement, men began to reevaluate their own gender status. As with any grassroots movement, the men's movement has a variety of factions. One of the early branches of the men's movement is known as the mythopoetic men's movement, which began after the publication of Robert Bly's (1990) *Iron John*—a fairy tale about men's wounded masculinity that was on the *New York Times* best-seller list for more than 60 weeks (Zakrzewski 2005). Participants in the men's mythopoetic movement met in men-only workshops and retreats to explore their internal masculine nature, male identity, and emotional experiences through the use of stories, drumming, dance, music, and discussion.

Today, the men's movement includes men's organizations that advocate for gender equality and work to make men more accountable for sexism, violence, and homophobia. The National Organization for Men against Sexism (NOMAS), founded in 1975, "advocates a perspective that is pro-feminist, gay affirmative, antiracist, dedicated to enhancing men's lives, and committed to justice on a broad range of social issues including class, age, religion, and physical abilities" (NOMAS 2015, p. 1).

Some men's groups focus on issues concerning children's and fathers' rights. Groups such as the American Coalition of Fathers and Children, Dads Against Discrimination, and Fathers4Justice are attempting to change the social and legal bias against men in divorce and child custody decisions, which tend to favor women. Members of the National Coalition for Men (NCFM) are also concerned with issues surrounding fatherhood and the way "men have few or no effective choices in many critical areas of life" (NCFM 2015, p. 1).

Other concerns on the agenda of men's rights groups include the domestic violence committed against men by women, false allegations of child sexual abuse, wrongful paternity suits, and the oppressive nature of restrictive masculine gender norms. Central to the men's rights group A Voice for Men is the gynocentric nature of society—that is, the historical norm of "male sacrifice for the benefit of women," whether under

> Other concerns on the agenda of men's rights groups include the domestic violence committed against men by women, false allegations of child sexual abuse, wrongful paternity suits, and the oppressive nature of restrictive masculine gender norms.

equal rights amendment (ERA) The proposed Twenty-eighth Amendment to the Constitution, which states that "equality of rights under the law shall not be denied or abridged by the United States, or by any state, on account of sex."

350 CHAPTER 10 Gender Inequality

the banner of "honor, nobility, [or] chivalry" (Kostakis 2014, p. 1). A Voice for Men's mission statement includes, but is not limited to, (1) the elimination of male genital mutilation (i.e., circumcision); (2) ratification of the ERA; (3) abolition of selective service or inclusion of women in the draft; (4) mandatory paternity tests upon request; (5) legal punishment for false paternity allegations; and (6) the end of affirmative action based on sex, the Violence against Women Act, rape shield laws, and alimony except in special circumstances.

As A Voice for Men notes in its mission statement, just as women have fought against being oppressed by expectations to conform to traditional gender stereotypes, men, too, want the same freedom from traditional gender roles. Unfortunately, men who enter nontraditional work roles, such as nurse and primary school teacher, are often stigmatized for participating in "feminine" work. A study of men in nontraditional work roles found that these men commonly experience embarrassment, discomfort, shame, and disapproval from friends and peers (Sayman 2007; Simpson 2005). Similarly, Dunn et al. (2013), after interviewing working women with stay-at-home husbands, report that the wives' co-workers and relatives often made negative comments about their husbands.

U.S. State and National Policies

The gender pay gap varies dramatically by state. In New York, a full-time working woman earns 91 percent of what a full-time working man earns; in Colorado, 80 percent; and in Louisiana, 66 percent (AAUW 2015). Some of the variation in the gender pay gap can be explained by such things as the type of industries located in a state, the number of minimum age workers, and the strength of labor unions (Pew 2014b). In 2014, equal pay legislation was passed in three states—Illinois, Louisiana, New Hampshire—and in Puerto Rico. For example, New Hampshire S.B. 207 requires "employers to permit employees to disclose the amount of their wages and provides civil and criminal penalties for an employer who violates equal pay or non-retaliation provisions" (NCSL 2015).

A number of important federal statutes have been passed to help reduce gender inequality. They include the Equal Pay Act of 1963, Title VII of the Civil Rights Act of 1964, Title IX of the Education Amendments of 1972, and the Victims of Trafficking and Violence Protection Act of 2000. In 2009, President Obama signed the Ledbetter Fair Pay Act, which reversed the 2007 U.S. Supreme Court decision that held that victims of pay discrimination had 180 days to file a grievance after the act of discrimination. The act now defines each paycheck as a separate act of discrimination (Mehmood 2009).

One of the most contentious pieces of legislation before Congress is the proposed Paycheck Fairness Act (Congress.gov 2015), which, in part, would

- remove sex as a permissible reason for wage differences;
- prohibit employer retaliation against an employee for filing a complaint or discussing employee wages;
- make employers who engage in sex discrimination liable in civil actions;
- allow others to benefit from actions that are the result of a plaintiff's sex discrimination suit; and
- amend the Civil Rights Act of 1964 to require data on the sex, race, and national origin of employees be collected and used in determining whether pay discrimination has taken place.

Although the majority of women in both parties support passage of the act (American Women 2015), blocked by Republicans, those opposing the bill argue that any residual gender pay gap is the result of choices women and men make (e.g., one choosing a career in library science, the other in engineering) and that the law would result in the loss of jobs (Bassett 2015). In 2015, several Republicans introduced an alternative bill, the Workplace Advancement Act, which Democrats argue lacks any "teeth" for enforcement. Both bills are presently in committee.

The president's 2016 budget also contains several gender equality initiatives. The Obama administration's budget includes increased funding for Title IX and for improved criminal justice processing of sexual assault offenders. It more than doubles funding for the Campus Violence Initiative, which protects students from sexual assault on campus, and triples funding for the National Domestic Violence Hotline (NWLC 2015).

WHAT
do you
THINK?

Two college students, now the heroes of a campus rape documentary called *The Hunting Ground* (Glock 2015), filed a federal lawsuit alleging that the University of North Carolina–Chapel Hill not only failed to handle their sexual assault cases properly but also systematically underreported sexual assault cases in violation of their own campus crime reporting law (Stancill 2013). If the students' allegations are proved correct, what penalties, if any, should be imposed on the university?

Two additional gender-related areas of concern are sexual harassment and affirmative action.

Sexual Harassment. Sexual harassment is a form of sex discrimination that violates Title VII of the 1964 Civil Rights Act. The U.S. Equal Employment Opportunity Commission (EEOC) (2012) defines **sexual harassment** in the workplace as "unwelcome sexual advances, requests for sexual favors, and other verbal or physical conduct of a sexual nature . . . when this conduct explicitly or implicitly affects an individual's employment, unreasonably interferes with an individual's work performance, or creates an intimidating, hostile, or offensive work environment" (p. 1).

Any individual—female, male, or transgender—can be a victim of sexual harassment; a victim might also be someone who was not harassed but was affected by offensive conduct. There are two types of sexual harassment: (1*) quid pro quo,* in which an employer requires sexual favors in exchange for a promotion, salary increase, or any other employee benefit, and (2) the existence of a hostile environment that unreasonably interferes with job performance, as in the case of sexually explicit comments or insults being made to an employee.

Common examples of sexual harassment include unwanted touching, the invasion of personal space, making sexual comments about a person's body or attire, and telling sexual jokes (Uggen and Blackstone 2004). Sexual harassment occurs in a variety of settings, including the workplace, schools, military academies, and college campuses. Seventy percent of a sample of girls and young women between the ages of 13 and 21 report experiencing some form of sexual harassment while at school or in college. Of that number, 51 percent report experiencing "sexual jokes or taunts," 40 percent report seeing "images of girls or women that made them feel uncomfortable," 33 percent report seeing "rude or obscene graffiti about girls or women," and 28 percent report experiencing "unwanted touching" (Girls' Attitude Survey 2013 p. 9).

sexual harassment In reference to workplace harassment, when an employer requires sexual favors in exchange for a promotion, salary increase, or any other employee benefit and/or the existence of a hostile environment that unreasonably interferes with job performance.

affirmative action A broad range of policies and practices in the workplace and educational institutions to promote equal opportunity as well as diversity.

Affirmative Action. As discussed in Chapter 9, **affirmative action** refers to a broad range of policies and practices to promote equal opportunity as well as diversity in the workplace and on campuses. Affirmative action policies, developed from federal legislation in the 1960s, require that any employer, universities as well as businesses, that receives contracts from the federal government must make "good faith efforts" to increase the number of female and other minority applicants. Such efforts can be made through expanding recruitment and training programs and by making hiring decisions on a nondiscriminatory basis.

However, a 1996 California ballot initiative, the first of its kind, abolished race and sex preferences in government programs, which included state colleges and universities. Over the next three years, several other states followed suit. In 2003, the U.S. Supreme Court held that universities have a "compelling interest" in a diverse student population and therefore may take minority status into consideration when making admissions decisions. Since that time, several states have addressed the issue of affirmative action in government programs with varying results. For example, in 2006, Michigan voters passed a referendum banning the use of race or sex in state college or university admissions decisions which, in 2014, was upheld by the U.S. Supreme Court (Mears 2014).

International Efforts

International efforts to address problems of gender inequality date back to the 1979 Convention to Eliminate All Forms of Discrimination against Women (CEDAW), often referred to as the International Women's Bill of Rights, adopted by the United Nations in 1979. To date, 188 out of 194 countries have ratified this bill of rights, including every country in Europe and South and Central America. Only six countries have not ratified CEDAW: Iran, Sudan, Somalia, Palau, Tonga, and the United States.

Another significant international effort occurred in 1995, when representatives from 189 countries adopted the Beijing Declaration and Platform for Action at the Fourth World Conference on Women sponsored by the United Nations. The platform reflects an international commitment to the goals of equality, development, and peace for women everywhere. The platform identifies strategies to address critical areas of concern related to women and girls, including poverty, education, health, violence, armed conflict, and human rights.

In addition to the CEDAW and the Beijing Platform, in 2000, all of the members of the United Nations adopted the Millennium Declaration. One of the eight Millennium Development Goals, as stated in the Millennium Declaration, is the promotion of gender equality and women's empowerment by 2015, which has not been met. The secretary-general of the United Nations, on the occasion of the International Women's day, stated that "to be truly transformative, the post-2015 development agenda must prioritize gender equality and women's empowerment. The world will never realize 100 per cent of its goals if 50 per cent of its people cannot realize their full potential" (UN 2015c, p. 1).

Finally, directed toward preventing the millions of acts of violence against women and girls annually is the International Violence against Women Act (I-VAWA 2013). The act, which is largely based on the *U.S. Strategy to Prevent and Respond to Gender-Based Violence Globally* that was released in 2012, was written by Amnesty International and a consortium of other concerns. The act, which also protects men and boys, would ensure that "the U.S. government has a strategy to efficiently and effectively coordinate existing cross-governmental efforts to prevent and respond to . . . [gender-based violence] globally" (Amnesty International 2015, p. 1). With bipartisan support, the act was reintroduced into the U.S. Senate in 2015.

Understanding Gender Inequality

The traditional gender roles of men and women, and the binary concepts of sex and gender, are weakening. Women who have traditionally been expected to give domestic life first priority are now finding it more acceptable to seek a career outside the home; the gender pay gap is narrowing, albeit slowly; and significant improvements in educational disparities between men and women, boys and girls, have been made. Most of these improvements, however, are in developed countries. Globally, millions of women and girls continue to be victimized by poverty, gendered violence, illiteracy, and limited legal rights and political representation.

Gender equality is not just good for individuals; it's good for families and society as a whole. Increasing women's level of education, labor force participation, and incomes contributes to the well-being of a society, reducing infant mortality rates, poverty, the transmission of HIV, and many other social problems. In a poll of unemployed U.S. adults between the ages of 25 and 54, 61 percent of women compared to 37 percent of men reported not working because of family responsibilities. However, many reported that they would go back to work if they could work from home or had flexible hours (Hamel et al. 2014). Higher rates of female employment in developed countries—for example, Germany, Canada, France, and Great Britain—are linked to family-friendly policies such as subsidized child care, paid parental leave, and taxation of individuals rather than families (Blau and Kahn 2013). Adopting such policies in the United States would go a long way toward gender equality.

Men are also victimized by discrimination and gender stereotypes that define what they "should" do rather than what they are capable, interested, and willing to do. The National Coalition for Men (NCFM 2015) has incorporated this view into its philosophy:

> We have heard in some detail from the women's movement how such sex-stereotyping has limited the potential of women. More recently, men have become increasingly aware that they too are assigned limiting roles which they are expected to fulfill regardless of their individual abilities, interests, physical/emotional constitutions or needs. Men have few or no effective choices in many critical areas of life. They face injustices under the law. And typically they have been handicapped by socially defined "shoulds" in expressing themselves in other than stereotypical ways. (p. 1)

Increasingly, people are embracing **androgyny**—the blending of traditionally defined masculine and feminine characteristics. The concept of androgyny implies that both masculine and feminine characteristics and roles are *equally valued*. However, "achieving gender equality . . . is a grindingly slow process, since it challenges one of the most deeply entrenched of all human attitudes" (Lopez-Claros and Zahidi 2005, p. 1).

An international strategy for achieving gender equality, *gender mainstreaming* is "the process of assessing the implications for women and men of any planned action, including legislation, policies or programmes, in all areas and at all levels" (Inter-Agency Standing Committee 2009, p. 7). Difficult? Yes. But, regardless of whether traditional gender roles emerged out of biological necessity, as the structural functionalists argue, or out of economic oppression, as the conflict theorists hold, or both, it is clear that today, gender inequality carries a high price: poverty, loss of human capital, feelings of worthlessness, violence, physical and mental illness, and death. Perhaps we have reached a time when we need to ask ourselves, are the costs of traditional gender roles too high to continue to pay?

androgyny Having both traditionally defined feminine and masculine characteristics.

Chapter Review

- **Does gender inequality exist worldwide?**
There is no country in the world in which men and women are treated equally. Although women suffer in terms of income, education, and occupational prestige, men are more likely to suffer in terms of mental and physical health, mortality, and the quality of their relationships.

- **How do the three major sociological theories view gender inequality?**
Structural functionalists argue that the traditional division of labor was functional for preindustrial society and has become defined as both normal and natural over time. Today, however, modern conceptions of the family have replaced traditional ones to some extent. Conflict theorists hold that male dominance and female subordination evolved in relation to the means of production—from hunting-and-gathering societies in which females and males were economic equals, to industrial societies in which females were subordinate to males. Symbolic interactionists emphasize that, through the socialization process, both females and males are taught the meanings associated with being feminine and masculine.

- **What is meant by the terms *structural sexism* and *cultural sexism*?**
Structural sexism refers to the ways in which the organization of society, and specifically its institutions, subordinate individuals and groups based on their sex classification.

Structural sexism has resulted in significant differences between education and income levels, occupational and political involvement, and civil rights of women and men. Structural sexism is supported by a system of cultural sexism that perpetuates beliefs about the differences between women and men. *Cultural sexism* refers to the ways the culture of society—its norms, values, beliefs, and symbols—perpetuate the subordination of an individual or group because of the sex classification of that individual or group.

- **What is the difference between the glass ceiling and the glass escalator?**
The glass ceiling is an invisible barrier that prevents women and other minorities from moving into top corporate positions. The glass escalator refers to the tendency for men seeking traditionally female jobs to have an edge in hiring and promotion practices.

- **What are some of the problems caused by traditional gender roles?**
First is the feminization of poverty. Women are socialized to put family ahead of education and careers, a belief that is reflected in their less prestigious occupations and lower incomes. Second are social-psychological costs. Women and transgender individuals tend to have lower self-esteem and higher rates of depression than men. Men and boys are often subject to the emotional restrictions of the "boy code."

Third, traditional gender roles carry health costs in terms of death and illness. Finally, gendered violence is responsible for the deaths of men, women, and transgender individuals.

- **What strategies can be used to end gender inequality?** Grassroots movements, such as feminism, the women's rights movement, and the men's rights movement, have made significant inroads in the fight against gender inequality. Their accomplishments, in part, have been the result of successful lobbying for passage of laws concerning sex discrimination, sexual harassment, and affirmative action. Besides these national efforts, international efforts continue as well. One of the most important is the Convention to Eliminate All Forms of Discrimination against Women (CEDAW), also known as the International Women's Bill of Rights, which the United Nations adopted in 1979.

Test Yourself

1. The United States is the most gender-equal nation in the world.
   a. True
   b. False
2. Symbolic interactionists argue that
   a. male domination is a consequence of men's relationship to the production process.
   b. gender inequality is functional for society
   c. gender roles are learned in the family, in the school, in peer groups, and in the media.
   d. women are more valuable in the home than in the workplace.
3. Recent trends indicate that more males enter college from high school than females.
   a. True
   b. False
4. The devaluation hypothesis argues that female and male pay differences are a function of women's and men's different levels of education, skills, training, and work experience.
   a. True
   b. False
5. The glass ceiling is an invisible barrier that prevents women and other minorities from moving into top corporate positions.
   a. True
   b. False
6. Which of the following statements is true about women and U.S. politics?
   a. There are more female than male governors.
   b. In the history of the U.S. Supreme Court, there have only been four female justices.
   c. The highest-ranking woman in the U.S. government is Vice President Hillary Clinton.
   d. Women received the right to vote with the passage of the Twenty-first Amendment.
7. Which of the following statements is *not* true?
   a. Females are underrepresented in prestigious occupations in popular films.
   b. Based on a study of realty TV shows, primary male characters are most often portrayed as overweight and "paunchy."
   c. Little boys when compared to little girls are overrepresented in children's media.
   d. An analysis of prime-time television indicates that males are most often pictured in comedies.
8. The feminization of poverty refers to
   a. the tendency for women to be caretakers of the poor.
   b. feminists' criticism of public policy on poverty.
   c. the disproportionate number of women who are poor.
   d. gender role socialization of poor women.
9. *Quid pro quo* sexual harassment refers to the existence of a hostile working environment.
   a. True
   b. False
10. Feminists would argue which of the following?
    a. The ERA should be part of the Constitution.
    b. The passage of the Nineteenth Amendment was a mistake.
    c. Occupational sex segregation benefits everyone.
    d. NOMAS is a radical, sexist organization.

Answers: 1. B; 2. C; 3. B; 4. B; 5. A; 6. B; 7. B; 8. C; 9. B; 10. A.

Key Terms

affirmative action   352
androgyny   354
attributional gender bias   340
boy code   345
cultural sexism   337
devaluation hypothesis   333
education dividend   327
emotion work   333
equal rights amendment (ERA)   350
expressive roles   327
feminism   349

gender   320
gender deviance hypothesis   339
gender expression   320
gender roles   337
glass ceiling   331
glass escalator effect   331
honor killings   348
human capital hypothesis   333
instrumental roles   327
masculine overcompensation thesis   324
motherhood penalty   331

occupational sex segregation   329
second shift   338
sex   320
sexism   321
sexual harassment   352
structural sexism   327
structured choice   331
transgender individual   320
vulnerable employment 329

> " Homophobia is like racism and anti-Semitism and other forms of bigotry in that it seeks to dehumanize a large group of people, to deny their humanity, their dignity and personhood. This sets the stage for further repression and violence that spread all too easily to victimize the next minority group."
>
> **CORETTA SCOTT KING**
> Civil rights activist

Sexual Orientation and the Struggle for Equality

Learning Objectives

After studying this chapter, you will be able to . . .

1 Hypothesize reasons for the differences in acceptance of LGB people by regions of the world.

2 Describe the problems associated with doing research on gay men and women.

3 Support the position that homosexuality is a consequence of nature not nurture.

4 Using each of the sociological theories, analyze the historically negative attitudes toward gay men and women.

5 Identify the origins of anti-LGB bias.

6 Describe the contact hypothesis.

7 Summarize the arguments in favor of marriage equality.

8 List the consequences of anti-LGBT bias.

9 Evaluate the various programs and policies set in place to achieve LGBT equality.

10 Discuss the impact of *Obergefell v. Hodges* on society.

THEY MET IN 1942, the year the United States entered World War II, while growing up in Yale, Iowa. In 1947, Vivian, a 24-year-old school teacher, and "Nonie," a payroll clerk who was just a year older, moved to Davenport, Iowa, where they have lived in a committed relationship for 72 years. The two traveled extensively throughout the United States and Canada, and visited England several times. "We've had a good time," said Vivian, adding that it takes a lot of hard work and loving devotion to sustain a relationship over seven decades. Surrounded by family and friends, on September 6, 2014, Vivian and Nonie were married at First Christian Church in Davenport. Said Reverend Hunsaker, "This is a celebration of something that should have happened a very long time ago" (quoted in Geyer 2014, p. 1).

They met in 1942, during the Second World War and, after 70 years together, finally married in a small church service in Iowa. The two could never have imagined that in their lifetime same-sex marriage would be legalized in the United States.

AP Images/Thomas Geyer/The Quad City Times

sexual orientation A person's emotional and sexual attractions, relationships, self-identity, and behavior.

heterosexuality The predominance of emotional, cognitive, and sexual attraction to individuals of the opposite sex.

homosexuality The predominance of emotional, cognitive, and sexual attraction to individuals of the same sex.

bisexuality The emotional, cognitive, and sexual attraction to members of both sexes.

lesbian A woman who is attracted to same-sex partners.

gay A term that can refer to either women or men who are attracted to same-sex partners.

cisgender Refers to a person whose gender identity is consistent with his or her birth sex.

Despite the progress that has been made with respect to equal rights for same-sex-attracted individuals in recent years, the fight for equality on the basis of sexual orientation and gender identity continues to be met with opposition. The term **sexual orientation** refers to a person's emotional and sexual attractions, relationships, self-identity, and behavior. **Heterosexuality** refers to the predominance of emotional, cognitive, and sexual attraction to individuals of the opposite sex. **Homosexuality** refers to the predominance of emotional, cognitive, and sexual attraction to individuals of the same sex; and **bisexuality** is the emotional, cognitive, and sexual attraction to members of both sexes. The term **lesbian** refers to women who are attracted to same-sex partners, and the term **gay** can refer to either women or men who are attracted to same-sex partners.

Much of the current literature on the treatment and political and social agendas of individuals who are gay, lesbian, and bisexual includes transgender individuals (LGBT) while others do not (LGB). As discussed in Chapter 10, *transgender individuals* (sometimes called "*trans*") are people whose sense of gender identity—as male or female—is inconsistent with their assigned birth (sometimes called *chromosomal*) sex. Alternatively, **cisgender**—*cis* the Latin for "on the side of" as opposed to *trans*, Latin for "on the other side of"—refers to a person whose gender identity is consistent with their birth sex.

Gender nonconforming is often used synonymously with *transgender*. **Gender nonconforming** refers to displays of gender that are inconsistent with society's expectations.

The acronyms **LGBT, LGBTQ, and LGBTQI**, as well as an assortment of other variations, are used to refer collectively to individuals who are lesbian, gay, bisexual, transgender, questioning or "queer," and *intersexed*. Sometimes, one will encounter the acronym accompanied by the letter *A*, which often stands for *allies*.

This chapter focuses on sexual orientation. However, it is impossible to discuss the struggle for same-sex equality without acknowledging transgender individuals who have been part of the larger LGBT civil rights movement. Thus, some of the research reported in this chapter is on lesbians, gays, and bisexuals, whereas other studies include transgender people as participants in addition to lesbians, gays, and bisexuals. Furthermore, many laws and policies discussed in this chapter do not specifically refer to bisexuals but would, by default, apply if they were in a same-sex relationship.

In this chapter, we focus primarily on Western conceptions of diversity in sexual orientation. It is beyond the scope of this chapter to explore in depth how sexual orientation and its cultural meanings vary throughout the world. The global legal status of lesbians and gay men, however, will be summarized. We also discuss the prevalence of homosexuality and bisexuality in the United States, the beliefs about the origins of sexual orientation, and then apply sociological theories to better understand societal reactions to nonheterosexuals. After discussing the cultural origins of anti-LGB bias and the ways in which nonheterosexuals are victimized by prejudice and discrimination, we end the chapter with a discussion of strategies to reduce antigay prejudice and discrimination.

The Global Context: A Worldview of the Status of Homosexuality

Homosexuality has existed throughout human history and in most, perhaps all, human societies. A global perspective on laws and social attitudes regarding homosexuality reveals that countries vary tremendously in their treatment of same-sex sexual behavior—from intolerance, criminalization, and even death, to acceptance and legal protection. Seventy-six countries—nearly one-third of the world's nations—continue to criminalize same-sex relations (Carroll and Itaborahy 2015).

Legal penalties vary for violating laws that prohibit homosexuality. In some countries, homosexuality is punishable by prison sentences (e.g., 14 years in Gambia, a life sentence in Barbados) and/or corporal punishment, such as whipping or lashing (e.g., Malaysia). In five nations—Iran, Mauritania, Saudi Arabia, Yemen, and Sudan—people found guilty of engaging in same-sex sexual behavior may receive the death penalty. It should also be noted, however, that in some countries, regardless of official penalties, homosexuals are routinely executed, as in the case of Iraq. There are also several countries in which male homosexuality is prohibited although female homosexuality is not (e.g., India, Lebanon) and, where both are illegal, men receive more severe punishment than women (e.g., Oman) (Carroll and Itaborahy 2015).

In general, countries throughout the world are moving toward increased legal protection of nonheterosexuals, as discrimination on the basis of sexual orientation has become part of a broad international human rights agenda. In 1996, South Africa became the first country in the world to include in its constitution a clause banning discrimination based on sexual orientation. Following South Africa's lead, today six additional countries—Portugal, Sweden, Switzerland, Ecuador, Bolivia, and Mexico—have national constitutions banning discrimination based on sexual orientation (Carroll and Itaborahy 2015). Finally, in 2014, the United Nations Human Rights Council passed a resolution introduced by Brazil, Germany, Spain, and Great Britain, and 42 other co-sponsors, expressing "grave concern at acts of violence and discrimination in all regions of the world committed against individuals because of their sexual orientation and gender identity" (Valenza 2014, p. 1).

Despite these general international trends toward LGBT human rights, there have been recent setbacks in many regions of the world. In 2013, Russia, which decriminalized homosexuality in 1993, established laws prohibiting "propaganda of nontraditional sexual relations to minors"—a prohibition so broadly defined as to include public displays of affection between same-sex partners (Guillory 2013). Russian law also bans same-sex couples, or an unmarried individual from a country where same-sex marriages are legal, from adopting a Russian child (Human Rights Campaign [HRC] 2015d).

Of late, several countries in regions of the world that have traditionally been hostile to LGBT people—Africa, Asia, and the Caribbean—have renewed their anti-gay efforts. Uganda's Anti-Homosexuality Act, struck down because of a procedural irregularity, was replaced with legislation that, if passed, would impose the death penalty for consensual same-sex sexual activities. Similarly, caning (i.e., beating someone with a cane, usually made of rattan or bamboo) as penalty for same-sex sexual activity is now codified into

A global perspective on laws and social attitudes regarding homosexuality reveals that countries vary tremendously in their treatment of same-sex sexual behavior—from intolerance and criminalization to acceptance and legal protection.

gender nonconforming
Often used synonymously with transgender, gender nonconforming (sometimes called gender variant) refers to displays of gender that are inconsistent with society's expectations.

LGBT, LGBTQ, and LGBTQI
Terms used to refer collectively to lesbian, gay, bisexual, transgender, questioning or "queer," and/or intersexed individuals.

law in parts of Indonesia (Carroll and Itaborahy 2015). Harsh penalties are not always imposed by the state. A report on homophobia at a global level concludes that in Africa, "the violence against LGBT persons continues unabated and concerted efforts would have to be made to work in countries to address the hostile attitudes of communities and the lack of repercussions for persons who committed hate crimes" (Carroll and Itaborahy 2015, p. 104).

Some have argued that the legalization of gay marriage in the United States is an attack on religious freedom and vilifies those who are morally opposed to homosexuality. What do you think about this argument? What should take precedence when human rights clash with a segment of the population's religious beliefs?

Lastly, the legalization of same-sex marriages in, for example, in the United States, provides evidence of the changing status of homosexuality throughout the world, yet it remains a complicated issue that has divided jurisdictions in some countries (see this chapter's section titled "Marriage Inequality"). Some countries recognize same-sex **registered partnerships** or "civil unions," which are federally recognized relationships that convey most but not all the rights of marriage. Some countries also offer registered partnerships or civil unions to opposite-sex couples. Registered partnerships are also referred to as "domestic partnerships." Legally recognized registered partnerships for same-sex couples are available in 16 countries and in specific jurisdictions of Mexico and Australia (Carroll and Itaborahy 2015).

Homosexuality and Bisexuality in the United States: A Demographic Overview

Before looking at demographic data concerning homosexuality and bisexuality in the United States, it is important to understand the ways in which identifying or classifying individuals as homo-, hetero-, and bisexual is problematic.

Sexual Orientation: Problems Associated with Identification and Classification

The classification of individuals into sexual orientation categories (e.g., gay, bisexual, lesbian, or heterosexual) is problematic for a number of reasons. First, distinctions among sexual orientation categories are simply not as clear-cut as many people would believe. Consider the early research on sexual behavior by Alfred Kinsey and his colleagues (1948, 1953). Although 37 percent of men and 13 percent of women reported at least one homosexual encounter since adolescence, few of the individuals reported exclusive same-sex sexual behavior. These data led Kinsey to conclude that heterosexuality and homosexuality represent two ends of the continuum and that most individuals fall somewhere along this theoretical line.

More recent research has also indicated that many individuals are not exclusively heterosexual or homosexual. In a study of 243 undergraduates, Vrangalova and Savin-Williams (2010) found that 84 percent of self-identified heterosexual women and 51 percent of self-identified heterosexual men reported at least one homosexual quality—same-sex sexual attraction, fantasy, and/or behavior. Newer research (Savin-Williams and Vrangalova 2013; Vrangalova and Savin-Williams 2012) also supports a sexual orientation continuum and suggests a five-category classification of sexual identity: heterosexual, mostly heterosexual, bisexual, mostly gay/lesbian, and gay/lesbian.

The use of, as traditionally is the case, a three-category classification scheme—heterosexual, bisexual, or gay/lesbian—rather than a nuanced identity system is theoretically and methodologically problematic (*The Economist* 2015; Morgan and Thompson 2011). For example, a woman who selects "lesbian" as her sexual orientation label when

Research . . . supports a sexual orientation continuum and suggests a five-category classification of sexual identity: heterosexual, mostly heterosexual, bisexual, mostly gay/lesbian, and gay/lesbian.

registered partnerships
Federally recognized relationships that convey most but not all the rights of marriage.

only presented with three options is *presumed* to identify as lesbian, and therefore to only feel attraction to and fantasize about women, and only has or desires sexual and romantic relationships with women throughout her lifetime. Alternatively, research also indicates that not only is someone's present sexual orientation status not necessarily indicative of their past or future sexual orientation status, but that many people with same-sex attractions or who have engaged in same-sex sexual behavior do not identify or think of themselves as gay, lesbian, or bisexual (Herek and Garnets 2007; Rothblum 2000; Savin-Williams 2006).

The second factor that makes classification difficult is that research with nonheterosexual populations has tended to define sexual orientation based on *one* of three components: sexual/romantic attraction or arousal, sexual behavior, and sexual identity (Savin-Williams 2006). However, sexual orientation involves *multiple dimensions*—sexual attraction, sexual behavior, sexual fantasies, emotional attraction, self-identification, and the interrelations between these dimensions (Herek and Garnets 2007).

Lastly, a third factor complicating the identification and classification of same-sex populations is the social stigma associated with nonheterosexual identities. As a result of the stigma associated with being gay, lesbian, or bisexual, many individuals conceal or lie about their sexual orientation to protect themselves from prejudice and discrimination (Meyer 2003; Röndahl and Innala 2008; Varjas et al. 2008). This makes the recruitment of participants for research studies on sexual orientation difficult because some people are uncomfortable "outing" themselves as gay, lesbian, or bisexual. This is particularly problematic in data sets collected years ago when it was less likely that estimates were corrected for nonreporting or misreporting of sexual orientation status (*The Economist* 2015).

LGBT Individuals and Same-Sex-Couple Households in the United States

Reliable estimates of the percentage of the U.S. population that is gay, lesbian, or bisexual (LGB) are scarce. One study estimates that there are more than 8 million LGB adults in the United States, comprising 3.5 percent of the adult population. In addition, there are approximately 700,000 transgender individuals in the United States. This means that an estimated 9 million Americans identify as LGBT, a number nearly comparable to the total population of New Jersey (Gates 2011). Consistent with this estimate, one of the largest studies of the LGBT population to date reports that about 3.4 percent of the U.S. adult population are LGB, ranging from 10.0 percent in Washington, DC, to 1.7 percent in North Dakota (Gates and Newport 2013).

Self-reported LGBT identity has been found to vary across racial lines, with nonwhites more likely to identify as LGBT—for example, 4.6 percent of African Americans, 4.0 percent of Hispanics, and 4.3 percent of Asian Americans identify as LGBT compared to 3.2 percent of white Americans (Gates and Newport 2012). Furthermore, women when compared to men are significantly more likely to identify as bisexual, and estimates of individuals who identify as LGB are significantly lower than estimates of those who have experienced same-sex attraction or have engaged in homosexual behavior (Gates 2011). Table 11.1 displays the sexual orientation distribution among a sample of college students.

TABLE 11.1 Self-Described Sexual Orientation and Gender Identity of U.S. College Students, 2014*

Heterosexual	90.0 percent
Gay/lesbian	2.9 percent
Bisexual	4.6 percent
Unsure	2.5 percent
Transgender	0.3 percent

*Based on a sample of 79,266 students at 140 campuses.
SOURCE: American College Health Association 2014.

The Origins of Sexual Orientation

One of the most common questions regarding sexual orientation centers on its origin or "cause." Questions about the causes of sexual orientation are typically concerned with the origins of homosexuality and bisexuality because heterosexuality is considered normative and "natural." A wealth of biomedical and social science research has investigated the possible genetic, hormonal, developmental, social, and cultural influences on sexual

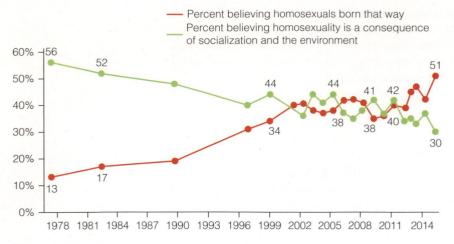

Figure 11.1 Public Opinion on the Causes of Homosexuality, 1978–2015
SOURCE: Jones 2015.

orientation, yet no conclusive findings have suggested that sexual orientation is determined by any particular factor or interplay of factors. In the absence of compelling findings, many practitioners and professionals believe that sexual orientation might be determined by the interplay of environmental and biological variables (Servick 2014).

Aside from what "causes" homosexuality, sociologists are interested in what people *believe* about the "causes" of homosexuality. People rely heavily on ideology, religion, and life experiences to form their beliefs about the causes of homosexuality (Haider-Markel and Joslyn 2008). As Figure 11.1 indicates, 51 percent of Americans believe that one's sexual orientation is a consequence of birth (i.e., nature), whereas 30 percent believe it is a matter of external variables such as socialization and the environment (i.e., nurture) (Jones 2015). Independent variables predictive of beliefs about the origins of sexual orientation include race, sex, political party affiliation, and income with whites, women, liberal Democrats, and high earners being more likely to embrace a nature argument (McCarthy 2014).

When a third belief about the "cause" of homosexuality is added as an option—that is, one *chooses* to be gay—the nature argument and the free will position ("just the way some people choose to live") are equally embraced, with fewer than 10 percent of respondents saying they believe being gay is a "result of a person's upbringing." Those who believe that being gay is a choice are significantly more likely to believe that "homosexuality should be discouraged by society" (Pew 2013a, p. 10) and to oppose marriage equality, military service, and adoption rights (Reyna et al. 2014).

Can Gays, Lesbians, and Bisexuals Change Their Sexual Orientation?

Imagine being heterosexual and being asked, in therapy, what do you think is responsible for your heterosexuality, when did you first realize you were a heterosexual, to whom have you revealed your heterosexual tendencies, and what are you doing to change your sexual orientation? For a majority of Americans such questions might be met with "What do you mean, what *caused* my heterosexuality?" followed by "I was born heterosexual!" Such a sense of indignation reflects **heteronormativity**—the cultural presumption that heterosexuality is the norm whereby other orientations, by default, become abnormal and something to change.

Various **sexual orientation change efforts (SOCE)**—whether *reparative therapy*, *conversion therapy*, or *reorientation therapy*—are dedicated to changing the sexual orientation of individuals who are nonheterosexual. Some SOCE are rooted in religious beliefs that homosexuality is inherently immoral and/or pathological. For example, "conversion" to heterosexuality, according to some SOCE, may be achieved through embracing evangelical Christianity (Cianciotto and Cahill 2007) or, if Jewish, through following a "Torah-sanctioned path" (Goldberg 2015). Some "treatments" have gone to such unethical extremes as shaming and physical punishment. In other cases, adolescents are "kidnapped" by their parents and sent to "conversion camps" or group homes.

In 2015, in a major blow to the SOCE industry, the Superior Court of New Jersey in a consumer fraud case against JONAH, a conversion therapy organization, found the defendants guilty, ruling that "[i]t is a misrepresentation in violation of the CFA [Consumer Fraud Act], in advertising or selling conversion therapy services, to describe homosexuality, not as being a normal variation of human sexuality, but as a mental illness, disease,

heteronormativity The cultural presumption that heterosexuality is the norm whereby other orientations, by default, become abnormal and something to change.

sexual orientation change efforts (SOCE) Collectively refers to reparative, conversion, and reorientation therapies.

disorder, or equivalent thereof" (Bariso 2015, p. 1). Presently, New Jersey, California, and Washington, DC, have laws banning conversion therapy for minors.

Critics of SOCE include the American Medical Association, American Academy of Pediatrics, American Academy of Child Adolescent Psychiatry, American College of Physicians, American Counseling Association, American Psychological Association, American Psychoanalytic Association, American Psychiatric Association, American School Counselor Association, American School Health Association, National Association of Social Workers, and Pan American Health Organization. All agree that homosexuality is *not* a mental disorder, that sexual orientation *cannot* be changed, and that efforts to change sexual orientation may be harmful (HRC 2015e).

Leelah Alcom, a 17-year-old transgender youth, walked in front of a semi-truck on December 27, 2014, after posting a suicide note on Tumblr (see The Human Side in Chapter 10). Forced by her parents to attend religious conversion therapy sessions, Leelah, defined as male at birth, also left a simple handwritten note: "I've had enough" (quoted in Kutner 2015, p. 1). In response to her death, a petition signed by more than 120,000 signatories, requesting a federal ban on SOCE, was sent to President Obama. In support of the ban, President Obama stated:

> Tonight, somewhere in America, a young person, let's say a young man, will struggle to fall to sleep, wrestling alone with a secret he's held as long as he can remember. Soon, perhaps, he will decide it's time to let that secret out. What happens next depends on him, his family, as well as his friends and his teachers and his community. But it also depends on us—on the kind of society we engender, the kind of future we build. (Obama 2015, p. 1)

Sociological Theories of Sexual Orientation Inequality

Sociological theories do not explain the origin or "cause" of sexual orientation diversity; rather, they help to explain societal reactions to homosexuality and bisexuality and ways in which sexual identities are socially constructed.

Structural-Functionalist Perspective

Structural functionalists, consistent with their emphasis on institutions and the functions they fulfill, emphasize the importance of monogamous heterosexual relationships for the reproduction, nurturance, and socialization of children. From a structural-functionalist perspective, homosexual relations, as well as heterosexual nonmarital relations, are "deviant" because they do not fulfill the main function of the family institution—producing and rearing children. Clearly, however, this argument is less salient in a society in which (1) other institutions, most notably schools, have supplemented the traditional functions of the family, (2) reducing (rather than increasing) population is a societal goal, and (3) same-sex couples can and do raise children.

Some structural functionalists argue that antagonisms between individuals who are heterosexual and homosexual disrupt the natural state, or equilibrium, of society. Durkheim

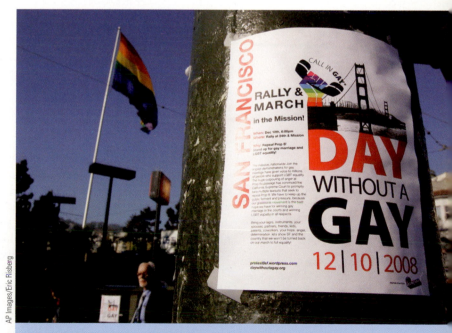

AP Images/Eric Risberg

To protest antigay initiatives in several states, activists declared a "Day without a Gay." Gay men and women were encouraged to "call in gay" and stay home from work to emphasize the economic impact of the gay community.

(1993 [1938]), however, recognized that deviation from society's norms can also be functional. Specifically, the gay rights movement has motivated many people to reexamine their treatment of sexual orientation minorities and has produced a sense of cohesion and solidarity in the gay population. Gay activism has also been instrumental in advocating HIV/AIDS prevention strategies and health services that benefit the society as a whole.

The structural-functionalist perspective is concerned with how changes in one part of society affect other aspects. For example, the worldwide increase in legal and social support for gay, lesbian, bisexual persons is likely the result of other cultural and structural changes in society. First, concern with the equality of the sexes has led to a relaxing of traditional gender roles, thereby supporting the varied expressions of female and male sexuality (see Chapter 10). Second, as gender roles have changed, so has the institution of marriage. Once an inherently unequal contractual relationship, heterosexual marriages today are most often a partnership whereby two people share, in varying degrees, household chores, financial responsibilities, and childrearing.

Third, LGB persons are more visible today than in any other time in history, leading to larger numbers of people who can "put a face" to the cause. Nearly 90 percent of Americans know someone who is gay, and 23 percent report that they know "a lot" of gay people (Drake 2013). The rejection of stereotypes (e.g., gay people don't believe in monogamy) becomes easier in the face of contradictory personal experience and is predictive of support for LBG equality (Reyna et al. 2014).

Lastly, globalization permits the international community to influence individual nations. Those advocating LGBT equality, whether in the United States or England or South Africa, are seen and heard through traditional and social media—strategies shared, petitions signed, and rage felt. Governments, corporations, or organizations that support LGBT rights may also bring pressure to bear on those that don't. In 2014, in response to Uganda's antigay laws, the United States imposed sanctions on Uganda, banning officials who were involved in "human right abuses" from entering the United States, canceling a military exercise, and reducing or discontinuing aid for some government programs (Sink 2014).

The structural-functionalist perspective is also concerned with latent functions, or unintended consequences. The movement toward LGBT equality has led to a backlash whereby, for example, over 30 states amended their state constitutions to define marriage as exclusively between a man and a women. A second unintended consequence is the economic boom enjoyed by states where marriage equality has been achieved. It is estimated that nationwide marriage equality would generate $2.6 billion in the first three years—a kind of "gay stimulus package," as some have observed—and would lead to an additional $184.7 million in tax revenues and 13,058 new jobs (Kastanis et al. 2015).

Conflict Perspective

The conflict perspective frames the gay rights movement and the opposition to it as a struggle over power, prestige, and economic resources. Homosexuals want to be recognized as full citizens deserving of all the legal rights and protections entitled to individuals who are heterosexual. A major achievement in gaining acceptance of gay and lesbian individuals occurred in 1973, when the American Psychiatric Association (APA) removed homosexuality from its official list of mental disorders (Bayer 1987).

More recently, gay and lesbian individuals have been waging a political battle to win civil rights protections against discrimination and, successfully in the United States, have fought to be allowed to marry a same-sex partner. Conflict theory helps to explain why many business owners and corporate leaders support nondiscrimination policies. First, gay-friendly work policies help employers maintain a competitive edge in recruiting and retain a talented and productive workforce. Second, many LGBT as well as heterosexual consumers prefer to purchase products and services from businesses that provide workplace protections for LGBT employees. Recent trends toward increased social acceptance of homosexuality may, in part, reflect the corporate world's competition for the gay and lesbian consumer dollar.

In Australia, business leaders have banned together and written a letter to the members of Parliament urging that the government legalize same-sex marriage. Signed by many "corporate giants," the letter, in part, reads, "Not only is marriage equality the only truly fair option, but it's also a sound economic option given that a happy workforce is a productive one. To remain competitive, and to attract top talent from around the world, organizations—and nations—must create a fair and respectful environment for all" (Mercer 2015, p. 1).

Business interests in the United States were galvanized in response to states that passed **religious freedom laws**. Opponents to religious freedom laws argue that, passed by conservative lawmakers, such laws legalize discrimination against LGBT consumers based on religious grounds. Indiana's passage of such a measure was met with vocal opposition, including threats of boycotts from organizations as diverse as the National Collegiate Athletic Association (NCAA), Angie's List, and Marriot (Socarides 2015). Over 75 executives in the technology industry representing such companies as eBay, Yelp, Apple, Twitter, Evernote, Cisco Systems, Netflix, and PayPal signed an open letter that, in part, states (Kaufman 2015):

> To ensure no one faces discrimination and ensure everyone preserves their right to live out their faith, we call on all legislatures to add sexual orientation and gender identity as protected classes to their civil rights laws and to explicitly forbid discrimination or denial of services to anyone. (p. 1)

After intense pressure from business, civic, and sports leaders, Indiana's Religious Freedom Restoration Act was amended to prohibit discrimination against homosexuals and transgender consumers (Cook et al. 2015). To date, 21 states have religious freedom laws (National Conference of State Legislatures 2015). Furthermore, in 2015, the **First Amendment Defense Act** was introduced in Congress that, if passed, would prohibit the government from penalizing individuals or organizations that discriminate against married same-sex couples if the individual or group has a "religious belief or moral conviction" that only heterosexual marriages should be recognized in law (H.R. 2802).

In 2015, several states passed "religious freedom" bills, ostensibly to protect the religious beliefs of Americans who are opposed to homosexuality and, more specifically, same-sex marriage. In general, the bills would permit, for example, a baker to refuse to provide the wedding cake at a non-heterosexual wedding.

Symbolic-Interactionist Perspective

Symbolic interactionism focuses on the meanings of heterosexuality, homosexuality, and bisexuality; how these meanings are socially constructed; and how they influence the social status, self-concepts, and well-being of nonheterosexual individuals. Examining the meaning of sexual orientations is part of a larger research area, the sociolinguistics of gay and lesbian lexicon. Just as "queer language" is learned, evolves, and is shared with others through social interaction, the meanings we associate with same-sex relations are learned—through society and its institutions.

A study of children raised by gay or lesbian parents is a case in point. Sasnett (2015) interviewed 20 adults raised in same-sex households. Although the construction of identities as children of lesbian and gay parents was sometimes difficult, the problem was not the sexual orientation of their parents. It was *when* they learned of their biological parent's sexual orientation that proved to be the "critical component for how they were able to create meanings in their lives" (p. 196).

religious freedom laws
Laws that protect business owners who discriminate against customers (e.g., gay men and women) based on religious grounds.

First Amendment Defense Act An act, if passed, that would prohibit the government from imposing penalties for discriminating against married same-sex couples if it is demonstrated that the individual or organization held a "religious belief or moral conviction" against such marriages.

Children born into same-sex households who had never known any other type of family structure experienced less stress than those who, after a divorce, learned of their parent's sexual orientation. However, the stress was often externally imposed as others offered their own definitions of the situation. Ann, 43 years old, remarks:

> My mom and dad were married for 20+ years, and I was fourteen when my mom left my dad for another woman. . . . My dad supported my mom's decision to leave, but my school counselor freaked out. When she found out about the divorce she said, "You cannot live with your mom." Her negative response sent a clear message that I could not discuss this with people. So it wasn't my parents' response, but instead the response of the counselor at my school that scared me and made me think that my family wasn't normal. (p. 213)

The negative meaning associated with homosexuality is also reflected in the current use of the phrase "That's so gay" or "You're so gay," which is meant to convey that something or someone is stupid or worthless. When a sample of LGB college students between the ages of 18 and 25 were surveyed, just over half female and 73 percent white, the results indicate that the more frequently LGB students hear the expression "That's so gay," the higher the probability they reported feeling "left out" and the more likely they were to suffer from physical illness (Woodford et al. 2012).

Sociological research has also shown that the use of such words as *fag* and *queer* by heterosexuals to insult one another facilitates social acceptance by peers, particularly among heterosexual males. The results of a study by Slaatten and Gabrys (2014) on gay-related name-calling among ninth grade students suggest that "gay-related name-calling among boys is more frequently used as a way of regulating unwanted expressions of masculinity than as a way of actively hurting or teasing someone" (p. 31). Nonetheless, gay-related name-calling comes at the expense of nonheterosexuals who are witness to or targets of such interactions.

The symbolic-interactionist perspective also points to the effects of labeling on individuals. Language is power. Once individuals become identified or labeled as lesbian, gay, or bisexual, that label tends to become their **master status**. In other words, the dominant heterosexual community tends to view "gay," "lesbian," and "bisexual" as the most socially significant statuses of individuals who are identified as such. These labels, in turn, often imbued with negative social meaning, impact an individual's self-definition and may lead to increased risk for a variety of, for example, mental and physical health problems, including substance abuse (Goldbach et al. 2014) and domestic violence (Hines 2014) (see this chapter's "The Consequences of Anti-LGB Bias").

> Sociological research has shown that the use of such words as *fag* and *queer* by heterosexuals to insult one another facilitates social acceptance by peers, particularly among heterosexual males.

Cultural Origins of Anti-LGB Bias

Anti-LGB bias, although declining, has existed for hundreds of years, as countries around the world defined LGB behavior as a symptom of mental illness, a sin, or both. In the United States, anti-LGB bias has its roots in various aspects of American culture, including religion, rigid gender roles, and the myths and negative stereotypes about nonheterosexuals.

Religion

Organized religion has been both a source of comfort and distress for many lesbian, gay, and bisexual Americans. Countless LGB individuals have been forced to leave their faith communities due to the condemnation embedded in doctrine and practice. Not surprisingly, LGB as well as transgender Americans are more likely to leave their childhood religion than their heterosexual and cisgender counterparts (Jones et al. 2014). Research has also found that higher levels of religiosity (Brown and Henriquez 2008; Lipka 2014; Pew 2013b; Shackelford and Besser 2007), religious fundamentalism (Pew 2013c; Summers 2010), and

master status The status that is considered the most significant in a person's social identity.

more conservative political beliefs (Jones et al. 2014; McCarthy 2014) are consistently associated with negative attitudes toward homosexuality.

As recently as 2003, every major religious group in the United States was opposed to marriage equality. Today, there is variability within Judeo-Christian traditions regarding attitudes toward LGB individuals. Figure 11.2 displays the major religious groups' stance on same-sex marriages, an indication of the movement toward greater LGB acceptance (Masci 2015).

The official position of a religion on same-sex marriage or on LGBT people, in general, does not necessarily represent the beliefs of the followers, in part because of the multiple statuses (e.g., white married father, college graduate) that people hold. Despite the Vatican's prohibition against same-sex marriage, the majority of Catholics— 70 percent—are accepting of homosexuality. The extent of acceptance, however, varies by age. Eight-five percent of Catholics between the ages of 18 and 29 believe that homosexuals should be accepted in society, compared to 57 percent of Catholics over the age of 65 (Lipka 2014). Similarly, 46 percent of Protestants believe that homosexuality should be accepted in society, but there are significant differences by denomination and race: Thirty percent of white evangelicals, 39 percent of black Protestants, and 65 percent of white mainline (e.g., Presbyterian, Lutheran) Protestants believe that homosexuality should be accepted in society (Drake 2013).

In response to anti-LGB bias and the resulting rejection from existing religious institutions, in 1968 a Pentecostal minister established the first of today's 201 Metropolitan Community Churches (MCCs) located in 37 countries (MCC 2015). MCCs are LGBT affirming, congregationally led denominations that "offer a picture of Christianity and religion which celebrates God's diverse creativity" (MCC 2014, p. 1). Kane (2013) argues that the formation of MCCs was an act of protest challenging the assumption that nonheterosexuality and Christianity are inherently incompatible. Confirming this belief, when the United Federation of Metropolitan Churches petitioned to join the National Council of Churches, "essentially requesting full inclusion in U.S. religious life," they were denied (Kane 2013, p. 137).

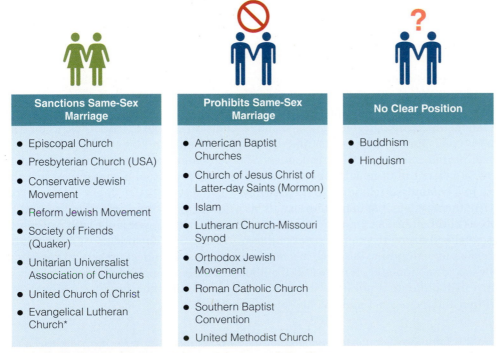

Sanctions Same-Sex Marriage	Prohibits Same-Sex Marriage	No Clear Position
• Episcopal Church	• American Baptist Churches	• Buddhism
• Presbyterian Church (USA)	• Church of Jesus Christ of Latter-day Saints (Mormon)	• Hinduism
• Conservative Jewish Movement	• Islam	
• Reform Jewish Movement	• Lutheran Church-Missouri Synod	
• Society of Friends (Quaker)	• Orthodox Jewish Movement	
• Unitarian Universalist Association of Churches	• Roman Catholic Church	
• United Church of Christ	• Southern Baptist Convention	
• Evangelical Lutheran Church*	• United Methodist Church	

Note: *The Evangelical Lutheran Church allows each congregation's minister to determine whether or not to marry same-sex couples.

Figure 11.2 Major Religious Groups' Position on Same-Sex Marriage, 2015
SOURCE: Pew Research 2015c.

Rigid Gender Roles

Disapproval of homosexuality also stems from rigid gender roles. Kimmel (2011), a sociologist who specializes in research on masculinity, writes about the relationship between perceived masculinity and bullying:

> Why are some students targeted? Because they're gay or even "seem" gay—which may be just as disastrous for a teenage boy. After all, the most common put-down in American high schools today is "that's so gay," or calling someone a "fag." It refers to anything and everything: what kind of sneakers you have on, what you're eating for lunch, some comment you made in class, who your friends are or what sports team you like. . . . Calling someone gay or a fag has become so universal that it's become synonymous with dumb, stupid, or wrong. (p. 10)

This billboard first appeared along Interstate 15 outside of Orem, Utah in June, 2015. The advertising campaign, in part a reaction to anti-gay billboards, will appear in other states in the coming months.

George Frey/Getty Images

From a conflict perspective, heterosexual men's subordination and devaluation of gay men reinforces gender inequality. "By devaluing gay men . . . heterosexual men devalue the feminine and anything associated with it" (Price and Dalecki 1998, pp. 155–156). Negative views toward lesbians also reinforce the patriarchal system of male dominance. Social disapproval of lesbians is a form of punishment for women who relinquish traditional female sexual and economic dependence on men. Not surprisingly, research findings suggest that individuals with traditional gender role attitudes tend to hold more negative views toward homosexuality than those with more modern perspectives (DeCarlo 2014; Falomir-Pichastor et al. 2010).

Myths and Negative Stereotypes

The stigma associated with homosexuality and bisexuality can also come from myths and negative stereotypes. One negative myth about nonheterosexuals is that they are sexually promiscuous and lack "family values," such as monogamy and commitment to a relationship. Although some gay men and lesbians do engage in casual sex, as do some heterosexuals, many same-sex couples develop and maintain long-term committed relationships. In a survey of LGBT respondents, 66 percent of gay women, 40 percent of gay men, 68 percent of bisexual women, and 40 percent of bisexual men reported that they were in a committed relationship (Pew 2013d).

Another myth is that nonheterosexuals, as a group, are a threat to children—most notably as child molesters. In other words, people confuse homosexuality with pedophilia, or sexual activity with prepubescent children. Having a nonheterosexual orientation is unrelated to pedophilia. Research has *not* demonstrated a connection between an adult's same-sex attraction and an increased likelihood of molesting children and teenagers. In fact, a pedophile does not have an adult sexual orientation because they are fixated on children and teenagers (Schlatter and Steinback 2014).

In 2013, the Boy Scouts announced that they would accept openly gay youth as members and, in 2015, Boy Scouts of America lifted the national ban on gay adult leaders (Toppo 2015). However, the organization also held that local scout troops may continue to exclude gay men as scoutmasters. What do you think? Should local scout groups be permitted to ban gay adult leaders?

Finally, gay men and women are often stereotyped in terms of their appearance and mannerisms—gay men as feminine and gay women as masculine. There are, of course, gay men and women who conform to the stereotype but, statistically, very few. The persistence of the opposite-gender characterization of gay men and women is a result of several sociological processes. First, gay men and women who fit the stereotype are noticed (e.g., gay feminine male hairdresser), while those who don't are not (e.g., gay masculine male construction worker), reinforcing the stereotype that gay men are feminine and work in certain professions. Interestingly, the feminine-looking man may not be gay but the stereotype will be reinforced nonetheless. Second, wanting to avoid that feeling of discomfort when confronted with dissonant information (e.g., gay man are masculine), we reject the possibility with reassuring confidence—"No, he can't be gay—he's too masculine!"

Prejudice against Lesbians, Gays, and Bisexuals

Oppression refers to the use of power to create inequality and limit access to resources, which impedes the physical and/or emotional well-being of individuals or groups of people. A person or group is **privileged** when they have a special advantage or benefits as a result of cultural, economic, societal, legal, and political factors (Guadalupe and Lum 2005). **Heterosexism** is a form of oppression and refers to a belief system that gives power and privilege to heterosexuals, while depriving, oppressing, stigmatizing, and devaluing people who are not heterosexual (Herek 2004; Szymanski et al. 2008).

The belief that heterosexuality is superior to nonheterosexuality results in prejudice and discrimination against gays, lesbians, and bisexuals. **Prejudice** refers to negative attitudes, whereas **discrimination** (discussed in the next section) refers to behavior that denies individuals or groups of individuals equality of treatment (see Chapter 9). In turn, this often leads nonheterosexual individuals to question the legitimacy of their same-sex attractions (Harper et al. 2004). Before reading further, you may wish to complete this chapter's *Self and Society* feature, which assesses your attitudes toward lesbian and gay issues.

Prejudice toward LGB individuals varies significantly by social variables. In general, men are more likely than women to have negative attitudes (Monto and Supinski 2014). In a survey just prior to the legalization of gay marriage in the United States, 60 percent of women compared to 53 percent of men favored marriage equality (Pew 2015a). Attitudes toward same-sex marriage also vary by political party, with 64 percent of Democrats and 30 percent of Republicans in favor of marriage equality (see the "Marriage Inequality" section later in this chapter). Not surprisingly, opposition to same-sex marriages, a proxy for attitudes toward lesbians and gay men, varies by age—as age increases, favorable attitudes toward same-sex marriage decrease. Figure 11.3 graphically portrays attitudes toward gay marriage for all adults surveyed and by respondents' age cohort (Pew 2015a).

oppression The use of power to create inequality and limit access to resources, which impedes the physical and/or emotional well-being of individuals or groups of people.

privilege When a group has a special advantage or benefits as a result of cultural, economic, societal, legal, and political factors.

heterosexism A form of oppression that refers to a belief system that gives power and privilege to heterosexuals, while depriving, oppressing, stigmatizing, and devaluing people who are not heterosexual.

prejudice Negative attitudes and feelings toward or about an entire category of people.

discrimination Actions or practices that result in differential treatment of categories of individuals.

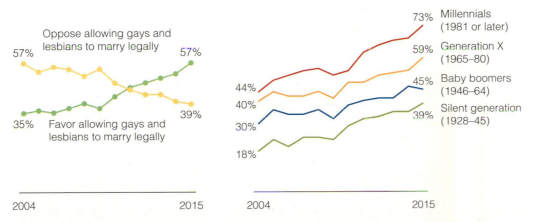

Figure 11.3 Support for Same-Sex Marriage, All Respondents and by Generation, 2004–2015
SOURCE: Pew 2015a.

Attitudes Toward Gay and Lesbian Issues

For each of the questions below, place the number that corresponds to your response on the line following the question. When finished, compare your responses to those from a representative sample of 360 undergraduates at a large midwestern university.

1. What is your view about sexual relations between two consenting adults of the same sex? _____
   0 = Always wrong, almost always wrong   1 = Wrong only sometimes, not wrong at all
2. Sexual relations between two adults of the same sex is unnatural. _____
   0 = Strongly agree, agree   1 = Disagree, strongly disagree
3. If your political party nominated a generally well-qualified person for president who happened to _____
   be homosexual, would you vote for that person?
   0 = No   1 = Yes
4. I could be close friends with a feminine man. _____
   0 = Strongly disagree, disagree   1 = Agree, strongly agree
5. I could be close friends with a masculine woman. _____
   0 = Strongly disagree, disagree   1 = Agree, strongly agree
6. Gay or lesbian people should have the right to marry. _____
   0 = Strongly disagree, disagree   1 = Agree, strongly agree
7. On a scale from 1 to 5 with 1 being *not at all important* and 5 being *very important*, how important _____
   is it for you to be knowledgeable about gay and lesbian persons and issues for your future career?
8. Think of a scale from 1 to 7, with 1 being *not at all comfortable* and 7 being *very comfortable*. _____
   How comfortable would you be with a roommate of the same sex as yourself who is gay or lesbian?

Percent of Students Answering "1"	
1. Consensual sex	60.6
2. Same-sex intimacy unnatural	63.3
3. Gay president	84.7
4. Feminine man	92.2
5. Masculine woman	93.6
6. Right to marry	76.7
Average Response	
7. Future career	3.64
8. Gay roommate	5.16

SOURCE: Sevecke et al. 2015.

The reduction of anti-LGB prejudice is key in the larger fight for LGB and transgender rights. As more LGBT Americans "come out" to their family, friends, and coworkers, heterosexuals and gender-conforming individuals have more personal contact with LGBT individuals (see this chapter's *The Human Side*). Psychologist Gordon Allport (1954) asserted that contact between groups is necessary for the reduction of prejudice—an idea known as the **contact hypothesis**. Research has shown that heterosexuals have more favorable attitudes toward gay men and women if they have had prior contact with gay men and women. For example, the most common reason for switching from an antigay marriage stance to a pro-gay marriage stance is knowing someone who is gay (Pew 2013b).

There are also several indicators of the growing acceptance of gay men and women. The decriminalization of homosexuality in the United States is one, as are opinion polls that document the shift toward greater recognition. The majority of Americans, for example— 63 percent—now believe that gay and lesbian relations are morally acceptable (Jones 2015). A third measure of acceptance is the election of nonheterosexual officials to political offices. All 50 states have been served by openly LGB elected politicians in some capacity, and at least 41 states have elected openly LGB politicians to one or both houses of their state legislatures (Gay and Lesbian Victory Institute 2015).

contact hypothesis The idea that contact between groups is necessary for the reduction of prejudice.

In 2014, Tim Cook, Apple CEO, announced to the world that he was gay. When asked why he decided to make his sexual orientation public, he responded that "It became so clear to me, kids were being bullied in school, kids were getting discriminated against, kids were even being disowned by their own parents...I needed to do something" (Pagliery and Pallotta 2015, p.1).

Throughout my professional life, I've tried to maintain a basic level of privacy. I come from humble roots, and I don't seek to draw attention to myself. Apple is already one of the most closely watched companies in the world, and I like keeping the focus on our products and the incredible things our customers achieve with them.

At the same time, I believe deeply in the words of Dr. Martin Luther King, who said: "Life's most persistent and urgent question is, 'What are you doing for others?'" I often challenge myself with that question, and I've come to realize that my desire for personal privacy has been holding me back from doing something more important. That's what has led me to today.

For years, I've been open with many people about my sexual orientation. Plenty of colleagues at Apple know I'm gay, and it doesn't seem to make a difference in the way they treat me. Of course, I've had the good fortune to work at a company that loves creativity and innovation and knows it can only flourish when you embrace people's differences. Not everyone is so lucky.

While I have never denied my sexuality, I haven't publicly acknowledged it either, until now. So let me be clear: I'm proud to be gay, and I consider being gay among the greatest gifts God has given me.

Being gay has given me a deeper understanding of what it means to be in the minority and provided a window into the challenges that people in other minority groups deal with every day. It's made me more empathetic, which has led to a richer life. It's been tough and uncomfortable at times, but it has given me the confidence to be myself, to follow my own path, and to rise above adversity and bigotry. It's also given me the skin of a rhinoceros, which comes in handy when you're the CEO of Apple.

The world has changed so much since I was a kid. America is moving toward marriage equality, and the public figures who have bravely come out have helped change perceptions and made our culture more tolerant. Still, there are laws on the books in a majority of states that allow employers to fire people based solely on their sexual orientation. There are many places where landlords can evict tenants for being gay, or where we can be barred from visiting sick partners and sharing in their legacies. Countless people, particularly kids, face fear and abuse every day because of their sexual orientation.

I don't consider myself an activist, but I realize how much I've benefited from the sacrifice of others. So if hearing that the CEO of Apple is gay can help someone struggling to come to terms with who he or she is, or bring comfort to anyone who feels alone, or inspire people to insist on their equality, then it's worth the trade-off with my own privacy.

I'll admit that this wasn't an easy choice. Privacy remains important to me, and I'd like to hold on to a small amount of it. I've made Apple my life's work, and I will continue to spend virtually all of my waking time focused on being the best CEO I can be. That's what our employees deserve— and our customers, developers, shareholders, and supplier partners deserve it, too. Part of social progress is understanding that a person is not defined only by one's sexuality, race, or gender. I'm an engineer, an uncle, a nature lover, a fitness nut, a son of the South, a sports fanatic, and many other things. I hope that people will respect my desire to focus on the things I'm best suited for and the work that brings me joy.

The company I am so fortunate to lead has long advocated for human rights and equality for all. We've taken a strong stand in support of a workplace equality bill before Congress, just as we stood for marriage equality in our home state of California. And we spoke up in Arizona when that state's legislature passed a discriminatory bill targeting the gay community. We'll continue to fight for our values, and I believe that any CEO of this incredible company, regardless of race, gender, or sexual orientation, would do the same. And I will personally continue to advocate for equality for all people until my toes point up.

When I arrive in my office each morning, I'm greeted by framed photos of Dr. King and Robert F. Kennedy. I don't pretend that writing this puts me in their league. All it does is allow me to look at those pictures and know that I'm doing my part, however small, to help others. We pave the sunlit path toward justice together, brick by brick. This is my brick.

SOURCE: http://www.bloomberg.com/news/articles/2014-10-30/tim-cook-speaks-up.

Discrimination against Lesbians, Gays, and Bisexuals

In June 2003, a Supreme Court decision in *Lawrence v. Texas* invalidated state laws that criminalize sodomy—oral and anal sexual acts. This historic decision overruled a 1986 Supreme Court case (*Bowers v. Hardwick*), which upheld a Georgia sodomy law as constitutional. The 2003 ruling, which found that sodomy laws were discriminatory and unconstitutional, removed the legal stigma and criminal branding that sodomy laws have long placed on LGB individuals. Nonetheless, sodomy or "unnatural acts" are still illegal in 12 states and are primarily used against gay men and women (Associated Press 2014).

Like other minority groups in American society, LGBT individuals experience various forms of discrimination and, as in the case of other minorities, those opposed to legal protections argue that they are "special rights," unnecessary ("Discrimination against LGBT people isn't a problem!"), and will lead to a rash of frivolous lawsuits claiming differential treatment (Lowndes and Maza 2014). Thus, next, we examine discrimination in the workplace, in marriage and parenting, in the violent expression of hate, and in treatment by the police to assess the extent of LGBT discrimination in the United States.

Workplace Discrimination and Harassment

In 2015, an executive order signed by the president prohibiting discrimination on the basis of sexual orientation or gender identity went into effect. The executive order protects federal workers and the 28 million employees of government contractors, one-fifth of the American workforce (Capehart 2014). There is, however, no federal protection from workplace discrimination for private, state, or local employers, and only 18 states and Washington, DC have laws specifically protecting LGBT workers from discrimination (see Figure 11.4) (White House 2014). Three states—New Hampshire, New York, and Wisconsin—prohibit discrimination on the basis of sexual orientation only.

In addition to state protections, some 200 counties and cities across the country have ordinances prohibiting LGBT discrimination in public and private organizations (HRC 2015a). However, laws don't necessarily prevent discrimination. Mallory and Sears (2014) analyzed complaints of discrimination in 13 states (not all states responded to the survey) and 105 cities or counties where LGBT workers *were* protected by anti-discrimination laws. There were a total of 460 complaints filed with state and local administrative agencies between 1999 and 2007, approximately the same rate of filings on the basis of gender and race.

Whether or not a company has gay-friendly policies, including anti-discrimination provisions, is dependent on the sociopolitical environment of the organization. The probability of having gay-friendly policies increases if (1) the company is in a state with progressive LGBT laws, (2) the company's competitors have gay-friendly policies, or (3) there is a high proportion of women on the board of directors (Everly and Schwarz 2015).

Most U.S. adults agree that LGBT employees should be protected from job discrimination. In fact, in a 2015 survey of 1,000 likely voters in the 2016 election, 51 percent of Republicans, 72 percent of Independents, and 80 percent of Democrats said they were in favor of protections against LGBT discrimination in (Lorenz 2015)

employment, housing, education, credit, jury selection, federal grants and assistance, and access to public places . . . in matters of employment, including the hiring, firing or promoting of employees at all privately run businesses, governments or non-profits . . . [in] housing and credit and . . .buying, selling and renting of houses or apartments . . . in matters of access to public places, including restaurants, restrooms, gyms, retail outlets and other places open to the general public. (p. 1)

LGBT in the Military. For many Americans, their workplace is the military. Since President Clinton signed a 1993 bill instituting a "don't ask, don't tell" (DADT) policy, more than

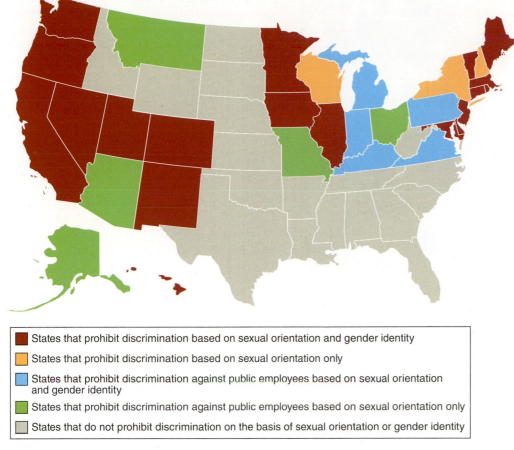

States that prohibit discrimination based on sexual orientation and gender identity

States that prohibit discrimination based on sexual orientation only

States that prohibit discrimination against public employees based on sexual orientation and gender identity

States that prohibit discrimination against public employees based on sexual orientation only

States that do not prohibit discrimination on the basis of sexual orientation or gender identity

Figure 11.4 Statewide LGBT Employment Laws and Policies, 2015
SOURCE: HRC 2015d.

13,000 servicemen and -women have been discharged from the military on the basis of their sexual orientation (Stone 2011), and countless others have decided not to enlist. DADT did not prohibit lesbians, gays, and bisexuals from serving in the military; it prohibited recruiters from asking enlistees their sexual orientation and prohibited LGB servicemen and -women from revealing their sexual orientation.

In 2010, not long after a poll revealed that the majority of Americans supported the repeal of DADT (Morales 2010), the law was repealed (O'Keefe 2010). In July 2011, the president, secretary of defense, and chairman of the Joint Chiefs of Staff certified that the Department of Defense was ready to enact the change. Lesbian, gay, and bisexual troops began to serve openly 60 days later (Singer and Belkin 2012).

Contrary to some people's fears, a report by U.S. military school professors concluded that the repeal of DADT had no overall negative impact on military readiness or its "component dimensions," such as cohesion, recruitment, retention, assaults, harassment, or morale (Belkin et al. 2012). In 2013, the Pentagon extended military benefits (e.g., health care, housing allowances) to spouses of gay military personnel (Ackerman 2013; see also Chapter 15); and, in 2015, 21 U.S. senators signed a letter to the secretary of defense asking the Pentagon to expand its policies to include redress in the case of LGB harassment or discrimination (Reichbach 2015).

Marriage Inequality

In 1996, Congress passed and President Clinton signed the **Defense of Marriage Act (DOMA)**, which (1) stated that marriage is a "legal union between one man and one woman"; (2) denies federal recognition of same-sex marriage; and (3) allows states to

[A] report by U.S. military school professors concluded that the repeal of DADT had no overall negative impact on military readiness or its "component dimensions," such as cohesion, recruitment, retention, assaults, harassment, or morale.

Defense of Marriage Act (DOMA) Federal legislation that states that marriage is a legal union between one man and one woman and denies federal recognition of same-sex marriage.

Two gay Jewish men wearing yarmulkes stand under a chuppah during their outdoor wedding ceremony. The reformed Jewish community was an early supporter of LGBT equality.

either recognize or not recognize same-sex marriages performed in other states. In 2003, Massachusetts became the first state to recognize same-sex marriages; and, over the course of the next 12 years, as a result of legislation, court decisions, or ballot initiatives, 31 additional states had followed suit.

At the federal level, the U.S. Supreme Court issued two key rulings affecting same-sex marriage in the United States. In the first ruling, in 2013, the parts of DOMA that denied equal benefits and recognition to legally married same-sex spouses that married heterosexual spouses have were deemed unconstitutional (*U.S. v. Windsor* 2013). The second ruling, ***Obergefell v. Hodges*** (2015), resulted in marriage equality throughout the United States which, theoretically, assures that all same-sex married couples can now receive the more than 1,000 federal rights, benefits, and responsibilities of marriage heretofore reserved for different-sex married couples (see this chapter's section on "Law and Public Policy").

Nearly 780,000 same-sex couples have been legally married in the United States (Gates and Newport 2015), although that number is likely an underestimate today given the *Obergefell* decision (see Table 11.2). Additionally, there are over 600,000 same-sex domestic partnerships. Over half of gay women (51 percent) report being married or in a co-habituating partnership, as do over half (58 percent) of heterosexual women. The difference is much greater when comparing the percentage of gay men (35 percent) and heterosexual men (63 percent) who are married or in a co-habituating partnership (Gates 2014).

There are also significant differences when comparing married couples, unmarried couples, and those not in a couple on a number of important social dimensions (see Table 11.3). On the average, partners in same-sex marriages are four years younger than partners in different-sex marriages. Individuals in same-sex unmarried couples are significantly less likely to be nonwhite and, as with married couples and couples in general, more likely to be in interracial or interethnic relationships. Significant educational attainment differences also exist. LGB individuals, whether in same-sex married couples, same-sex nonmarried couples, or not part of a couple, are more educated than their different-sex and non-LGB counterparts (Gates 2014).

Arguments against Same-Sex Marriage. Opponents argue that granting legal status to same-sex unions conveys social acceptance of homosexuality and, therefore, teaches youth to view homosexuality as morally acceptable. Opponents are also concerned that the legalization of same-sex marriages will lead to pressure in schools to treat LGB

Obergefell v. Hodges The 2015 U.S. Supreme Court decision that legalized same-sex marriage in the United States.

TABLE 11.2 Pre-Legalization Estimates of Same-Sex Marriages and Domestic Partnerships in the United States, April 2015

	% of adults aged 18 and older	Estimated number
Married with a same-sex spouse	0.3	780,000
Unmarried and living with a same-sex domestic partner	0.5	1,200,000
All others	99.2	–

SOURCE: Gates and Newport 2015.

TABLE 11.3 Mean Age, Race/Ethnicity, Educational Attainment, by Couple Type and Relationship Status, 2014

	Age		Non-white		Inter-racial/ethnic		College (age 25+)	
	Same-sex	Different-sex	Same-sex	Different-sex	Same-sex	Different-sex	Same-sex	Different-sex
All couples	43.9	49.0	24%	29%	19%	9%	49%	34%
Married	46.5	50.5	23%	28%	13%	8%	66%	36%
Unmarried	43.3	37.2	24%	37%	21%	16%	45%	22%
Not in a couple	LGB 37.1	non-LGB 44.3	LGB 35%	non-LGB 38%	LGB N/A	non-LGB N/A	LGB 40%	non-LGB 26%

SOURCE: Gates 2014.

individuals as any other minority group resulting in, for example, classes on gay history and gay literature.

Many opponents of same-sex marriage base their opposition on moral grounds tied to religious beliefs. When a random sample of 1,204 adult Americans were asked, "Regardless of whether or not you think it should be legal, . . . tell me whether you personally believe that in general" gay and lesbian relations are "morally acceptable or morally wrong," 34 percent responded that homosexuality was morally wrong (Jones 2015, p. 2). Eighty-five percent of respondents who reported being "unaffiliated" when asked their religion reported being in favor of marriage equality (Pew 2015a).

The federal government grants tax exempt status to nonprofit organizations and colleges. In 1970, the IRS denied tax-exempt status for private colleges that continued to discriminate on the basis of race, for example, prohibiting interracial dating. Do you think that tax-exempt status should be revoked from private colleges that have anti-gay programs and policies (e.g., expelling students in a same-sex marriage)?

Arguments in Favor of Same-Sex Marriage. In a 2015 poll, 60 percent of Americans agreed with the statement that "same-sex couples should be recognized by the laws as valid, with the same rights as traditional marriages" (McCarthy 2014, p. 1). Advocates of same-sex marriage argue that banning same-sex marriages or refusing to recognize same-sex marriages granted in other states, both held to be unconstitutional in 2015, was a violation of civil rights that denies same-sex couples the countless legal and financial benefits that are granted to heterosexual married couples.

DOMA, before its repeal, prevented many of the more than 1,000 federal rights, benefits, and responsibilities of marriage from being afforded to legally married same-sex couples. As a result of the repeal, however, LGB married same-sex couples now have, for example, the right to assume their spouse's pension, to inherit from a spouse who dies without a will, to avoid inheritance taxes between spouses, to make crucial medical decisions for their spouse, and to take family leave to care for their spouse in the event of critical injury or illness.

Finally, those in favor of marriage equality argue that same-sex marriages promote relationship stability among gay and lesbian couples. Research supports "the conclusion that the extension of marriage rights to same-sex couples has the potential to improve child well-being insofar as the institution of marriage may provide social and legal support to families and enhance family stability, which are key drivers of positive child outcomes" (*Obergefell v. Hodges* 2015, p. 27).

As a result of the legalization of same-sex marriages, children in same-sex families have gained a range of securities and benefits, including the right to get health insurance coverage and Social Security survivor benefits from a nonbiological parent, and the right to continue living with a nonbiological parent should their biological mother or father die (Tobias and Cahill 2003). Ironically, the same pro-marriage groups that stress that children are better off in a married-couple family often disregard the benefits of same-sex marriage for children.

Children and Parental Rights

Children come into a family headed by same-sex partners in a variety of ways—giving birth in the case of lesbians and surrogacy in the case of gay men, children from a previous marriage, adoption, and foster care. In the case of nonheterosexual marriages, both parents *should* be considered the legal parents of a child from birth on as is the case with heterosexual married couples. In the case of adoption, some states allow same-sex married couples to jointly adopt and/or allow the nonbiological parent to adopt the child or children of his or her spouse. In other states, the partner of a biological or adoptive parent is not legally a parent of the child, and has no or few rights to the child or children if the relationship ends (Family Equality Council 2014). There is some speculation, however, that with the legalization of gay marriage, legal restrictions such as these may be abolished (Giambrone 2015).

Several respected national organizations—including the American Academy of Family Physicians, American Academy of Pediatrics, American Bar Association, American Medical Association, American Psychological Association, Child Welfare League of America, National Adoption Center, the National Association of Social Workers, the North American Council on Adoptable Children, and Voice for Adoption—have taken the position that a parent's sexual orientation has nothing to do with his or her ability to be a good parent (American Medical Association 2011; U.S. Court of Appeals 2010; *Position Paper* 2015). Furthermore, research indicates that the majority of Americans, regardless of age or political affiliation, favor adoption rights for same-sex couples (Swift 2014).

That said, a sizeable minority, 35 percent, is opposed. Those opposed traditionally cite concerns that same-sex parents will lead to dysfunctional families and gender role confusion in children. For example, since Mormons hold that gender roles are "specific, complementary, and essential" and determined by God, two women or two men raising children is "unnatural" (Sumerau and Cragun 2014). The results of an investigation into parenting of 41 heterosexual couples (mothers and fathers) and 48 male homosexual couples (half biological fathers, half adoptive fathers) may call into question the assumption of "natural" gender roles (Abraham et al. 2014). All couples were in committed relationships and parents for the first time. Each of the gay couples had used surrogacy to become parents, and the biological mothers were not involved in care of the child.

After videotaping each parent interacting with their child in the home, researchers examined MRI images of the brain when subjects later watched the interactions. When mothers, all primary caregivers, watched their babies, there was increased activity in the *emotion* processing portion of the brain. When heterosexual fathers, none of whom were primary caregivers, watched their babies the *cognitive* portion of their brain, the part that interprets and understands social behavior, increased in activity. When gay primary care-fathers watched their babies, *both* the emotional *and* cognitive portions of their brains were triggered, and the more time spent with his child, the better the connectivity between the two spheres. The authors conclude that the "connectivity between the two networks in primary-caregiving fathers suggests that, although only mothers experience pregnancy, birth, and lactation, . . . [there are] other pathways for adaptation to the parental role in human fathers, and these alternative pathways come with practice, attunement, and day-by-day caregiving" (p. 9795).

In 2015, the U.S. Supreme Court heard arguments both for and against the legalization of gay marriages as part of an appeal by petitioners in Ohio, Tennessee, Michigan, and Kentucky. The petitioners, including James Obergefell, successfully challenged the right of each state to deny legal recognition of their marriage from a different state and/or refusal of the state to issue a marriage license. Within this context, the American Sociological Association (ASA 2015) filed an *amicus* brief outlining social science research on children raised in same-sex families compared to children raised in different-sex families. Child outcome categories included academic performance and cognitive development (e.g., GPA), social development (e.g., number of friends), mental health (e.g., depression), early sexual activity (e.g., age of first intercourse), and substance abuse and behavioral problems (e.g., alcohol abuse). The ASA brief concludes that, consistently and with a consensus, these "studies reveal that children raised by same-sex parents fare just as well as children raised by different-sex parents across a wide spectrum of factors used by social scientists to measure child wellbeing" (ASA 2015, p. 5).

Same-sex couples are three times more likely to be raising adopted or foster children than different-sex families (Gates 2015; Goldberg et al. 2014). Based on the National Health Interview Survey and Gallup data, it is estimated that same-sex couples and single LGB-identified adults are raising somewhere between 1.3 and 2.2 million children (Gates 2014). Nearly 20 percent of married and unmarried same-sex couples are raising children under the age of 18, with an average of 1.7 children per household. Married same-sex couples are more likely to be raising children than unmarried same-sex couples (Gates 2015).

WHAT do you THINK?

In many countries around the world, gay men and women are subjected to torture, imprisonment, corporal punishment, and even death because of their sexual orientation (HRC 2015d). In Nigeria, for example, engaging in same-sex relations is a crime. It is also a crime to "register, operate, associate with, support, or belong to an LGBT group," with penalties as high as 10 years in prison. Do you think that gay men and women who are subjected to violence in their home country should be eligible for political asylum in the United States?

Violence, Hate, and Criminal Victimization

On October 6, 1998, Matthew Shepard, a 21-year-old student at the University of Wyoming, was abducted and brutally beaten. Two motorcyclists who initially thought he was a scarecrow found him tied to a wooden ranch fence. His skull had been smashed, and his head and face had been slashed. Although speculation exists as to the motive of the attack (Jimenez 2013), at the time, the only apparent reason for the attack was that Matthew was gay. On October 12, Matthew died of his injuries. Media coverage of his brutal attack and subsequent death focused nationwide attention on hate crimes against nonheterosexuals. It can be argued that Shepard's death put a face on anti-LGB hate crimes in a way that previously had not been done.

It took 17 years for Laramie, Wyoming, the home of Matthew Shepard and the men who murdered him, to pass an anti-LGBT discrimination law—the first in the state (Sprinkle 2015). However, to date, Wyoming remains one of only five states (Arkansas, Georgia, Indiana, and South Carolina) that does not have hate crime legislation (HRC 2015f). Anti-LGBT hate crimes are crimes against individuals or their property that are based on biases against the victim because of *perceived* sexual orientation or gender identity. Such crimes include verbal threats and intimidation, vandalism, sexual assault and rape, physical assault, and murder.

According to the Federal Bureau of Investigation (FBI), 20.8 percent of reported hate crimes in 2013 were motivated by sexual orientation, involving 1,402 victims. Over half of the incidents, 60.9 percent, were motivated by anti-gay male bias, 22.5 percent were classified as anti-LGBT bias (mixed group), 13.1 percent were motivated by anti-lesbian bias, 1.8 percent were classified as anti-bisexual bias, and 1.6 percent were victims of anti-heterosexual bias (FBI 2014). But, as discussed in Chapter 9, FBI hate

AP Images/Ben Curtis

Protestors in Kenya demonstrate against Uganda's increasingly punitive policies against gay men and women. In 2014, the Ugandan High Commission passed a bill which makes "aggravated" homosexual acts punishable by life in prison.

Anti-LGBT hate crimes are crimes against individuals or their property that are based on biases against the victim because of *perceived* sexual orientation or gender identity.

crime statistics underestimate the incidence of hate crimes. The National Coalition of Anti-Violence Programs (NCAVP) documented 1,359 survivors and victims of violence due to *perceived* nonheterosexuality, gender nonconformity, and HIV-affected status in 2014, including 20 murders. The report found that 80 percent the LGBT homicide victims were people of color, and over half were transgender women (NCAVP 2015).

Anti-LGBT violence is not unique to the United States. Hate crimes based on sexual orientation and gender identity remain a serious problem in many regions of the world. In South Africa, where discrimination against sexual orientation minorities is protected by constitutional provision, the incidence of "corrective rape" of lesbians is estimated to be as high as 10 per week in the capital city of Cape Town (Jaeger 2014). In Jamaica, often called the most anti-gay country in the world, newspapers routinely post incidents of anti-LGBT-motivated hate crimes including assaults ("gay bashing") and murder, all relatively accepted in Jamaica (West and Cowell 2015). In Central America, attacks on LGBT individuals occur daily, with nearly 200 anti-LGBT murders reported between 2000 and 2015 (Ayala 2015).

Anti-LGB Hate and Harassment in Schools and on Campuses. Steven Caruso, a gay 16-year-old high school student, left a suicide note. The note read, in part (Caruso, n.d.):

> I am sorry to the people that I love but I can't . . . take it anymore.
> So I am gay. Why does everyone hate me because of that . . . . I have been punched and spit on and called faggot, queer, loser, pussy, fag boy. Some asshole painted faggot on my locker. Some people do not talk to me. . . . I am so . . . tired of the shit. I have received hate letters telling me to leave school . . . that faggots aren't welcome. (p. 1)

Steven survived his suicide attempt, but many other LGBT youth and young adults do not. In fact, LGBT youth have higher rates of suicide, suicide attempts, and self-harm than their heterosexual and gender-conforming peers (Bostwick et al. 2014).

Hostile school environments, characterized by anti-LGBT attitudes, remarks, and actions by students and teachers, have been documented to exist as early as elementary school. A survey of more than 7,898 LGBT elementary, middle, and high school students documents the prevalence of anti-LGBT sentiment in schools. According to the National School Climate Survey (Kosciw et al. 2014), in the 2013–2014 school year,

- 74.1 percent of LGBT students reported being verbally harassed at school, 36.2 percent reported being physically harassed at school, and 16.5 percent reported being physically assaulted at school because of their sexual orientation.
- 64.5 percent of LGBT students heard homophobic remarks, such as "faggot" or "dyke," frequently or often at school; 56.4 percent heard negative comments about gender expression (e.g., not acting masculine enough) frequently or often.
- Over half of LGBT students, 55.5 percent, reported that they felt unsafe in school because of their sexual orientation; 37.8 percent because of gender expression.
- 30.3 percent of LGBT students missed at least one day of school in the previous month because of safety concerns or feeling uncomfortable.

LGBT bullying is also common among college students. The results of an LGBT campus climate survey at the University of Mississippi indicate that LGBT students are more likely to be assaulted and harassed, to have negative experiences with staff (e.g., hear disparaging comments about sexual orientation), and to feel less comfortable or supported on campus (Strunk et al. 2015). Law enforcement agencies report that nearly 7.2 percent of hate crimes based on sexual orientation occur in schools or on college campuses (FBI 2014).

Police Mistreatment. There is "a significant history of mistreatment of LGBT people by law enforcement in the United States, which included profiling, entrapment, discrimination and harassment" (Mallory et al. 2015, p. 1). Consequently, due to fear of further victimization, many cases of LGBT violence are not reported to the police (Ciarlante and Fountain 2010; NCAVP 2015). In fact, in 2013, only 45 percent of LGBT violence victims reported the incident to the police. Of those who reported their victimization, 32 percent

378 | **CHAPTER 11** Sexual Orientation and the Struggle for Equality reported "hostility and police misconduct" including, in rank order, unfair arrests, excessive force, entrapment, and police raids (NCAVP 2015, p. 11).

As with heterosexuals, intimate partner violence and sexual assault occur in the lives of LGBT individuals. Many cases of LGBT violence are *not* reported to the police out of fear of further victimization and trepidation in formally acknowledging the nature of the relationship (Ciarlante and Fountain 2010; NCAVP 2015). Failure of the police to define LGBT partner violence as *criminal* also contributes to underreporting. Many officers, assuming intimate partner violence is between a man and a woman and that it is the woman who is the victim, simply assign the label of "mutual abuse" and arrest both parties. It is estimated that police are 10 to 15 times more likely to make a dual arrest in cases of same-sex domestic violence than in heterosexual domestic violence (Ciarlante and Fountain 2010).

The Consequences of Anti-LGBT Bias

Sexual minorities, as many other minorities, suffer the negative consequences of harassment, prejudice and discrimination, violence, and isolation. However, unlike, for example, racial and ethnic minorities, gay and transgender men and women are often rejected by their schools and religious institutions, communities, families, and friends. In a qualitative study of LGBTQ youth, when asked about negative factors in their lives, they were most often related to "families, schools, religious institutions, and community and neighborhood" (Higa et al. 2014, p. 663).

Physical and Mental Health

Many of the negative experiences LGBT people have are associated with physical and mental health problems. In general, LGBT status has been associated with a lower sense of well-being (Doyle and Molix 2014). The results of a Gallup-Healthways survey of a representative sample of LGBT and non-LGBT adults in the United States indicate that on every measure of well-being—*financial* (e.g., standard of living), *social* (e.g., relationship with family), *community* (e.g., safety and security), *purpose* (e.g., goal), and *physical* (e.g., diseases) well-being—LGBT adults, and in particular gay and transgender women, scored lower than heterosexual and/or cisgender women (Gates 2014; Kosciw et al. 2012).

LGB men and women who have undergone a prejudice-related stressful event (e.g., discrimination, rejection) are over three times more likely to have a serious physical health problem than LGB men and women who have not had a prejudice-related stressful event (Frost et al. 2011). Youth who report LGBT harassment in school have lower educational outcomes (Aragon et al 2014; Russell et al. 2014) and, later in life, higher rates of depression and suicidal thoughts (Russell et al. 2011). LGBT adults who have been rejected by their families also have higher rates of depression (Ryan et al. 2009). Nationally, gay women attempt suicide at a higher rate than heterosexual women, and the differences are even greater in southern states (Irwin and Austin 2013), where white, conservative, evangelical Protestants disproportionately reside. (See this chapter's *Social Problems Research Up Close.*)

Using the English General Practice Patient Survey, Elliot et al. (2015), after comparing LGBT and non-LGBT adults, conclude that LGB individuals reported significantly worse mental and physical health and more negative healthcare experiences. Furthermore, a survey of nearly 300,000 college students indicates that LGBT students are more likely to have eating disorders when compared to their non-LGBT counterparts, and that transgender college students when compared to sexual minority or cisgender students have the highest rates of bulimia nervosa and anorexia (Diemer et al. 2015).

Finally, for the first time in over 50 years, the annual National Health Interview Survey conducted by the U.S. Department of Health and Human Services included questions on sexual orientation, thereby facilitating comparisons between heterosexuals and nonheterosexuals on a number of health dimensions (Ward et al. 2014). There were no significant differences in self-reported health status in general or for men, although a higher percent

Adolescents encounter an array of challenges in their transition to adulthood; however, LGB adolescents face additional challenges that are related to the social stigma of their sexual orientation. For some lesbian, gay, and bisexual adolescents, this stigma may induce psychosocial stress, leading to increased risky behaviors and poorer health. The following research study sought to determine whether the social environment surrounding lesbian, gay, and bisexual youth contributes to their higher rates of suicide attempts.

Sample and Methods

Nearly 32,000 11th grade students in Oregon completed the Oregon Healthy Teens survey over a two-year period. Students were identified as lesbian, gay, or bisexual (LGB) by asking respondents to indicate "which of the following best describes you"—"heterosexual (straight)," "gay or lesbian," "bisexual," or "not sure." Of the total sample respondents, 90.3 percent identified as heterosexual, 0.9 percent identified as gay or lesbian, and 3.3 percent identified as bisexual. The 1.9 percent of respondents who reported that they were unsure of their sexual orientation was excluded from the analysis.

The extent to which a social environment is supportive, the independent variable, was measured by five indicators across the 34-county area: (1) the percentage of same-sex couples living in a county, (2) the percentage of registered Democrats living in a county, (3) the percentage of schools with Gay-Straight Alliances (GSAs), (4) the percentage of schools with nondiscrimination policies that specifically protect LGB students, and (5) the percentage of schools with anti-bullying policies that specifically protect LGB students. Hatzenbuehler (2011) hypothesizes that the higher each of these indicators and, thus, the higher the social environment measure, the lower LGB student suicide attempts.

The dependent variable was the number of times a student reported attempting suicide in the previous 12 months. The number of suicide attempts was then recoded into 0, no suicide attempts, and 1, the presence of one or more suicide attempt(s). Respondents were also asked about other variables empirically documented to be associated with suicide attempts including depression, binge drinking, harassment at school by other students, and whether or not they had ever been physically abused by an adult. Each of these variables was measured in a yes-or-no format.

of heterosexual women when compared to gay women reported that they were in "excellent or very good" health. Heterosexual men were more likely to report being obese than gay men; bisexual women were significantly more likely to report being obese than heterosexual women.

Substance Abuse

The relationship between LGBT status and increased risk of substance abuse is rooted in sociological variables. Heck et al. (2014) examined the association between school environment and LGBT high school students' use of illicit drugs and misuse of prescription drugs. An online survey was completed by 475 LGBT high school students. School environment was measured by the absence or presence of a high school Gay-Straight Alliance (GSA), a school-based club for all students, and an indicator of how supportive a school is of LGBT students.

The results support previous studies that indicate that the absence of a school GSA is associated with increased risk for drug use. Controlling for such demographic variables as ethnicity, gender, and school type (private versus public), students in schools without a GSA reported higher lifetime use of illegal drugs in general and had higher likelihoods of using cocaine, hallucinogens, and marijuana, and misusing prescription pain killers and ADHD medication. The authors conclude that one possible reason for the relationship between the presence of a school GSA and LGBT students' reduced use of drugs may be the moderating variable victimization. LGBT students in schools without GSAs report higher rates of school-based victimization, and higher rates of victimization are associated with increased drug use. In addition to victimization, a lack of social support, stress, bad coming-out experiences, and a negative housing status have been found to be predictive of LGBT substance abuse (Goldbach et al. 2014).

Findings and Discussion

The analysis took place in several phases, the first answering the question whether or not LGB youth when compared to heterosexual youth are more likely to attempt suicide. Based on the data collected, the answer to this question is yes—LGB youth are more likely to attempt suicide than their heterosexual counterparts, approximately five times more likely. Furthermore, LGB youth have significantly higher rates of depression, binge drinking, peer harassment at school, and adult lifetime physical abuse.

The second question answered is whether or not the supportiveness of the social environment in which a student lives is associated with the presence or absence of a suicide attempt. The answer, once again, is yes. The more supportive the social environment, the lower the odds that a student will attempt suicide. Moreover, when a supportive social environment is divided into a positive social environment and a negative social environment, students residing in negative social environments are 20 percent more likely to attempt suicide. There are also differences in suicide attempts by heterosexuals based on the supportiveness of their social environment. However, the increased risk of a suicide attempt by a heterosexual living in a negative social environment is only 9 percent.

The third question answered is whether or not the relationship between LGB status and social environment is a function of individual-level risk factors. Thus, Hatzenbuehler (2011) examined the relationship between LGB status and social environment adjusting for the presence or absence of depression, binge drinking, peer harassment, and lifetime adult physical abuse. Although LGB status remains a significant predictor of student suicide attempts, its predictive power was reduced in the presence of individual-level risk factors. This means that part of the relationship between LGB status and social environment is explained by individual-level risk variables.

The author concludes that the results of this study can be used in reducing LGB youth suicide attempts. For example, only a 5 percent increase in the supportiveness of the social environment would be associated with a 10 percent decrease in suicide attempts. Hatzenbuehler (2011) is quick to note, however, that the data were collected at one point in time, and therefore, a causal relationship between social environment and suicide attempts cannot be established. Furthermore, in terms of limitations of the study, the sample by virtue of being collected in a school did not include homeless youth, runaways, dropouts, and students absent on the day the survey was administered.

SOURCE: Hatzenbuehler 2011.

Using survey data, Jabson and colleagues (2014) examined substance abuse in a national sample of LGB and heterosexual respondents. The researchers were interested in comparing not only the prevalence of substance abuse between nonheterosexual and heterosexual subsamples, but also the role of stress as a possible mediating variable. The results indicate that gay men and women, and particularly bisexuals, have higher rates of current cigarette smoking, lifetime use of marijuana, and lifetime use of substances other than marijuana (including cocaine, crack cocaine, heroin, or methamphetamine) when compared to heterosexuals. Interestingly, stress did contribute to the relationship between sexual orientation and substance abuse, but only for bisexuals.

Lastly, the National Health Interview Survey indicates that gay and bisexual men and women between the ages of 18 and 64 are more likely to be current cigarette smokers than their heterosexual counterparts (Ward et al. 2014). Gay and bisexual women were also more likely to report binge drinking at least once in the previous year when compared to heterosexual women, although only bisexual men had significantly higher likelihoods of binge drink at least once in the previous year than heterosexual men.

Economic Inequality, Poverty, and Homelessness

There are many stereotypes about sexual minorities, most of them negative. However, there is at least one stereotype that is positive and, as is most often the case, untrue—that gay men and women are disproportionately wealthy (Besen 2014). In fact, as noted earlier, when comparing LGBT and non-LGBT men and women, LGBT men and women have lower standards of living, are less likely to be able to afford basic necessities, and are more likely to worry about finances (Gates 2014).

As Figure 11.5 indicates, poverty rates vary by region of the country and LGBT status (Hasenbush et al. 2014). In each of the five U.S. regions, the percentage of the LGBT

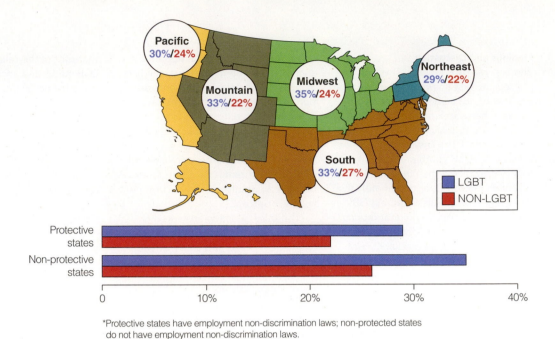

Protective states
Non-protective states

0   10%   20%   30%   40%

■ LGBT
■ NON-LGBT

*Protective states have employment non-discrimination laws; non-protected states do not have employment non-discrimination laws.

Figure 11.5 Percentage of Population with Annual Household Incomes under $24,000 by Region and by Employment Protection, 2014*
SOURCE: Hasenbush et al. 2014.

population with annual household incomes below $24,000 a year is higher than the percentage of the non-LGBT population. Note that in both cases, poverty is highest in the South. Figure 11.5 also indicates that the percentage of the LGBT population with annual household incomes below $24,000 a year is higher than the non-LGBT population in states with (21 protective states) and without LGBT employment protections (29 nonprotective states). Most important, LGBT people in states without employment protections are more likely to report annual household incomes of less than $24,000 than those living in states with employment protections.

Same-sex couple households have a higher average income, likely the result of their higher levels of education, than different-sex couples. However, in states with employment protections, although not in states without employment protections, same-sex couples raising children have a disadvantage of $11,300 a year when compared to their heterosexual counterparts, a statistically significant difference. The income gap between gay and nongay couples with children is highest in the Midwest, where same-sex couples, on the average, earn nearly $20,000 less annually. The relative disadvantage of same-sex versus different-sex couples with children is a function of sociological variables. Same-sex couples raising children, on the average, are younger than their opposite-sex counterparts, disproportionately female, and disproportionately African or Hispanic American. Twenty-four percent of children raised by same-sex couples are living in poverty compared to 11 percent of children raised by different-sex couples (Gates 2015).

Lastly, LGBT youth are disproportionally homeless when compared to their non-LGBT counterparts. A survey of homeless youth providers indicates that 20 percent of youth seeking services are gay and 3 percent are transgender (Choi et al. 2015). The most common reason given by providers as to why LGBT youth were homeless was "being forced out of home or running away from home because of their sexual orientation or gender identity/expression" (p. 5). Of LGBT youth who accessed homeless services, a disproportionate number—31 percent and 14 percent, respectively—reported being African or Hispanic American.

Aging and Retirement

There are approximately 3 million LGBT adults over the age of 55 in the United States, and that number is projected to double by 2030 (Ivers 2015). Aging LGBT individuals are at a higher risk for a range of negative life outcomes. First, LGBT elders are more likely to suffer from a variety of health concerns when compared to non-LGBT elders (Frediksen-Goldsen and Espinoza 2015). The stress associated with being LGBT, often the result of discrimination and victimization, increases the risk of cardiovascular disease and shortens life expectancy. LGBT elders are also more likely to suffer from anxiety, diabetes, HIV/AIDS, and high blood pressure when compared to their non-LGBT counterparts (SAGE 2015).

Second, LGBT elders have higher rates of poverty than their heterosexual and gender-conforming peers. LGBT men and women suffer from "vulnerability to employment discrimination, [a] lack of access to marriage, higher rates of being uninsured, gender and racial inequalities, less family support and family conflict over coming out" (SAGE 2015, p. 1). Eighty percent of long-term health care in the United States is provided by family members. Aging LGBT adults are less likely to have children or a spouse to provide care, and they are more likely to be estranged from other family members and to live alone. The result may be costly hospitals stays or placement in an assisted living facility or nursing home.

Lastly, prior to the *Obergefell* decision, unless married *and* living in a state that recognizes the marriage, the surviving LGBT partner was ineligible for Social Security survivor benefits, or their partners' retirement fund or pension plan (Frediksen-Goldsen and Espinoza 2015; SAGE 2015). However, after the *Windsor* decision in 2013, the Center for Medicare and Medicaid Services proposed a revision to federal requirements for long-term care facilities: "where state law or facility policy provides or allows certain rights or privileges to a patient's opposite-sex spouse under certain provisions, a patient's same-sex spouse must be afforded equal treatment if the marriage is valid in the jurisdiction in which it was celebrated" (Ivers 2015, p. 49).

Strategies for Action: Toward Equality for All

As discussed in this chapter, attitudes toward homosexuality in the United States have become more accepting over the years, and support for protecting civil rights of gays and lesbians is increasing. Many of the efforts to change policies and attitudes regarding non-heterosexuals and gender nonconforming individuals have been spearheaded by organizations that specifically advocate for LGBT rights including the Human Rights Campaign (HRC), the National Gay and Lesbian Task Force (NGLTF), the Gay and Lesbian Alliance Against Defamation (GLAAD), the Gay, Lesbian and Straight Education Network (GLSEN), and Amnesty International. In addition to organizational efforts, the media, local, state, and federal governments, and educational policies offer "strategies for action."

Research by Crosby and Wilson (2015) indicates that when a sample of undergraduates were asked to imagine a scenario where someone used a homophobic slur, about half said they would confront the individual who made the remark. However, when in an experimentally created situation where undergraduates actually heard a homophobic slur, not one confronted the speaker. If you were in the presence of someone who made a homophobic remark, what do you think you would do? What do you think most people would do?

LGBT Status and the Media

The media has been instrumental in the lives of LGBT individuals for several reasons. First, it has provided LGBT individuals with role models. Gomillion and Giuliano (2011) report that media role models are not only a source of pride and comfort, but influenced respondents' self-realization of a gay identity and the coming out process. **Coming out** is a term used to describe an ongoing process whereby an LGBT individual becomes aware of his or her sexual and/or nonconforming gender identity, accepts and incorporates it

coming out The ongoing process whereby a lesbian, gay, or bisexual individual becomes aware of his or her sexuality, accepts and incorporates it into his or her overall sense of self, and shares that information with others such as family, friends, and coworkers.

into his or her overall sense of self, and shares that information with others such as family, friends, and coworkers. The "coming out" process only exists because heterosexuality and gender conformity are considered normative in society.

Second, in addition to providing role models, the media increases LGBT visibility and counteracts stereotypes (GLAAD 2015; Wilcox and Wolpert 2000). Since Ellen DeGeneres's 1998 coming-out episode in her sitcom *Ellen,* many television viewers have watched shows depicting LGBT characters in more realistic ways and that take a supportive stance on LGBT rights issues. Anti LGBT-bias shows include *Buffy the Vampire Slayer, Glee, Queer Eye for the Straight Guy, The Big Gay Sketch Show, Chelsea Lately, True Blood, Will & Grace, Modern Family, Orange Is the New Black, How to Get Away with Murder,* and *Transparent.* Importantly, one study found that college students reported lower levels of anti-gay prejudice after watching television shows with prominent gay characters (Schiappa et al. 2005).

The visibility of famous gay, lesbian, and bisexual individuals also has a societal impact. Many music and television celebrities (e.g., Anderson Cooper, Wanda Sykes, Neil Patrick Harris, Ricky Martin, Jane Lynch), political leaders (e.g., former U.S. representative Barney Frank), and sports figures (e.g., Jason Collins, Robbie Rogers, Abby Wambach) have come out over the years. Similarly, many public figures have actively supported gay rights. For example, former New York Giants defensive end and current Fox NFL analyst Michael Strahan filmed a public service announcement (PSA) supporting the legalization of same-sex marriage.

Finally, social media have been important in addressing the needs of LGBT youth. Following several gay teenage suicides that appeared in the news during September 2010, Dan Savage and his partner created the It Gets Better Project. The It Gets Better Project has turned into a viral cyber-movement, inspiring thousands of user-created videos that instill messages of hope and support for LGBT youth who have been bullied, feel that they must hide in shame, or who experience rejection from family members. The project has received submissions from celebrities, organizations, activists, politicians, and media personalities, ranging widely from President Barack Obama and Lady Gaga to staff members of LinkedIn, Google, and Pixar (It Gets Better Project 2015).

Law and Public Policy

As is often the case with other social problems, federal, state, and local governmental agencies have tackled the issues surrounding LGBT equality. Various policies and laws have been passed or proposed that address LGBT discrimination in the workplace, marriage inequality, restrictive adoption laws, and hate crimes.

Ending Workplace Discrimination. Presently, federal law only protects against discrimination on the bases of race, religion, national origin, sex, age, and disability. When LGBT respondents were asked to prioritize specific policies that affect the LGBT population as "a top priority," "very important but not a top priority," "a somewhat important priority," or "not a priority at all," equal employment rights was the most commonly listed top priority, followed by legally sanctioned marriages for nonheterosexual couples (Pew 2013c, p. 104).

The **Employment Non-Discrimination Act (ENDA)**, a proposed federal bill that would protect LGBT individuals from workplace discrimination, has been debated in Congress since 1994, but it has never been signed into law. It would offer individuals basic protections against workplace discrimination on the basis of sexual orientation or gender identity. It was reintroduced in the U.S. House of Representatives and the U.S. Senate in 2013. To the surprise of some, ENDA passed the U.S. Senate on November 7, 2013, and is now in the House of Representatives awaiting a vote (H. Res. 639).

With the absence of federal legislation prohibiting discrimination based on sexual orientation, some state and local governments, as well as private corporations, prohibit employment discrimination based on sexual orientation, and extend domestic partner benefits to same-sex couples. Twenty-one states and the District of Columbia have laws banning sexual orientation discrimination in the workplace (HRC 2015c).

Employment Non-Discrimination Act (ENDA) This proposed federal bill that would protect LGBTs from workplace discrimination has been up for congressional debate on a number of occasions since 1994 but has never been signed into law.

Additionally, 26 states and the District of Columbia offer benefits for their government employees and their same-sex spouses. Finally, nearly 70 percent of Fortune 500 companies include sexual orientation in their equal employment opportunity or nondiscrimination policies, and more than half offer same-sex domestic partner health benefits to all of their employees (HRC 2015b).

Marriage Equality. Gay marriages are legal in 22 countries (see Figure 11.6) (Pew 2015a), the most recent being the United States and Ireland, which became the first country to legalize same-sex marriage by popular vote. In addition to the 22 countries, jurisdictions in several other countries have legalized gay marriage.

The U.S. Supreme Court decision in *Obergefell v. Hodges* was announced on June 26, 2015. James Obergefell and his partner John Arthur had been in a committed relationship for over 20 years when they received the devastating news that John had ALS, a fatal nerve disorder. With time running out,

Jim Obergefell speaks to the media after the U.S. Supreme Court decision in *Obergefell v. Hodges* which held that denying same-sex couples the right to marry is unconstitutional. Mr. Obergefell, pictured, holds a picture of his late husband, John Arthur.

the couple married on an airport tarmac in Maryland, John too ill to be moved. When they learned that their home state of Ohio would not recognize their marriage and, therefore, would not list James as spouse on John's death certificate, the couple decided to file a federal lawsuit (Rosenwald 2015). After his partner's death, Obergefell continued to fight for equal recognition of his marriage, resulting in the 5–4 U.S. Supreme court decision legalizing same-sex marriage throughout the United States (*Obergefell v. Hodges* 2015).

Although the Supreme Court is considered the "court of last resort," the decision has fueled a backlash from conservatives and faith-based groups. In North Carolina, magistrates and clerks who, "based upon any sincerely held religious exemption," may opt out of performing same-sex marriages or any other kinds of marriages, for that matter (Culhone 2015). The Alabama legislature passed a bill that would bypass the state's involvement in marriages and replaced it with private contracts. Mississippi law, mirroring the repealed religious freedom act in Indiana, states that no one's religious freedom can be "substantially burdened without a 'compelling reason.'" As one political commentator states, "[M]ake no mistake: In certain corners of the country, it's only going to get tougher for gay and lesbian couples to exercise their now-legal right to wed" (p. 1).

Those who oppose the decision include conservative religious and political leaders, as well as the four U.S. Supreme Court justices who dissented from the majority opinion. Religious leaders primarily argue that a Court cannot alter "God's law" (i.e., marriage is between one man and one woman) and replace it with government-made law (i.e., the Fourteenth Amendment guarantees equal protection). Some political leaders, despite the separation of church and state, make the same argument but also question the right of the federal government to impose its will in the face of opposing state law. Finally, despite a long history of what some have disparagingly called "judicial activism" (e.g., *Brown v. Board of Education* [school integration] and *Roe v. Wade* [(abortion)]), one of the primary concerns of the dissenting justices is the majority's use of the Constitution to "invent a new right and impose that right on the rest of the country" (*Obergefell v. Hodges*, p. 102).

Parental Rights. In the United States, the **Every Child Deserves a Family Act (ECDFA)** has been introduced into Congress for several years, most recently in 2015. But, to date, the ECDFA has yet to receive legislative approval. If passed, the act would remove

Every Child Deserves a Family Act (ECDFA) This federal legislation would remove obstacles to nonheterosexual (as well as transgender) individuals from providing homes for adoption or foster care.

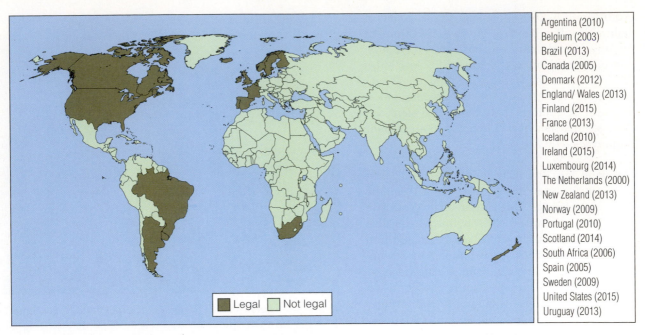

Argentina (2010)
Belgium (2003)
Brazil (2013)
Canada (2005)
Denmark (2012)
England/ Wales (2013)
Finland (2015)
France (2013)
Iceland (2010)
Ireland (2015)
Luxembourg (2014)
The Netherlands (2000)
New Zealand (2013)
Norway (2009)
Portugal (2010)
Scotland (2014)
South Africa (2006)
Spain (2005)
Sweden (2009)
United States (2015)
Uruguay (2013)

Legal   Not legal

Figure 11.6 Countries Where Same-Sex Marriage Is Legal as of August, 2015
SOURCE: Pew 2015b.

obstacles to nonheterosexual individuals or couples providing homes for adoption or foster care by prohibiting public child welfare agencies from discriminating on the basis of sexual orientation, gender identity, or marital status. Agencies that do discriminate would be in danger of losing federal financial assistance (H.R. 2449). Nonetheless, in a peremptory move, Michigan passed a law that allows agencies to refuse to place children in a home if the placement would violate their religious beliefs (Culhane 2015). As of this writing, it is unclear how, or if, the legalization of gay marriage will impact adoption regulations for same-sex couples.

Laws regulating adoption by LGB individuals not only vary by state, as previously discussed, but also vary between nations. Argentina and Uruguay, for example, allow adoption by same-sex married couples. However, in Cuba, adoption by a same-sex couple or a partner of a gay man or woman is illegal. One man expresses his frustrations over his lack of rights in caring for his partner's biological son. He goes on to say, "I've held him in my arms since his birth . . . [but] if the baby has to go to the hospital . . . I have no legal authority to decide anything about his illness" (Gonzalez 2013, p. 1).

Hate Crimes Legislation. In 2009, President Obama signed into law the **Matthew Shepard and James Byrd, Jr. Hate Crimes Prevention Act (HCPA)** (see also Chapter 9). This law expands the original 1969 federal hate crimes law to include hate crimes based on actual or perceived sexual orientation, gender, gender identity, and disability. The law was named after Matthew Shepard who, as discussed earlier, was brutally murdered and James Byrd Jr., an African American man who was attacked, chained to a vehicle, and dragged to his death in Texas. Hate crime laws call for tougher sentencing when prosecutors can prove that the crime committed was a hate crime. According to the most recent data available, 30 states and the District of Columbia have hate crime laws that include sexual orientation, 15 states have hate crime laws that do not include sexual orientation, and 5 states have no hate crime statutes (Waddell 2014).

Educational Strategies and Activism

Educational institutions bear the responsibility of promoting the health and well-being of all students. Thus, they must address the needs and promote acceptance of LGBT youth through certain policies and programs. The strategies for attaining these goals are to include

Matthew Shepard and James Byrd, Jr. Hate Crimes Prevention Act (HCPA) This law expands the original 1969 federal hate crimes law to cover hate crimes based on actual or perceived sexual orientation, gender, gender identity, and disability.

LGBT issues in sex education, include LGBT-affirmative classroom curricula, and promote tolerance in learning environments through policies, education, and activism.

Sex Education and LGBT-Affirmative Classroom Curricula. The censorship of LGBT current issues, historical figures and events, and sexual health in both the classroom and in school libraries is a heated debate across the nation. Whether LGBT themes can be brought into public school classrooms varies considerably between and within states. As of 2015, 22 states require sex education; of those states, 12 require that teachers discuss sexual orientation, and 3 require that only negative information about sexual orientation can be taught (Temblador 2015). For example, in Alabama, sex education teachers are required to tell students that homosexuality "is an unacceptable, criminal lifestyle" (p. 1).

California governor Jerry Brown signed legislation in 2011 "making California the first state to require that textbooks and history lessons include the contributions of gay, lesbian, bisexual, and transgender Americans" (McGreevy 2011, p. 1). In 2012, the guidelines of California's FAIR Education Act were revised in order to prohibit schools from adopting learning materials that portray people with disabilities, lesbians, and/or gay men in a negative light in history and social studies lessons. The act also encourages inclusiveness. For example, when teaching about the Holocaust, teachers are encouraged to mention that in addition to the more than 6 million Jews who were executed, the Nazis targeted and killed people based on sexual orientation and disability status ("FAIR Education Act" 2013).

Promotion of Tolerance in Learning Environments. Concerns over harassment and/or bullying of students have become the focus of many states, often the result of highly publicized student suicides. Nineteen states and Washington, DC, have laws that specifically protect LGBT students from harassment and/or bullying from other students, teachers, and school staff (MAP 2015).

The Student Non-Discrimination Act was introduced into Congress in 2015 (H.R. 846). If passed, the act would prohibit discrimination in schools based on real or perceived sexual orientation or gender identity. It also defines harassment as a form of discrimination, and authorizes the federal government to cut off funds to schools that fail to enforce the new law. Finally, it asserts the rights of victims of discrimination on the basis of sexual orientation or gender identity to seek redress in the courts.

The Gay, Lesbian, and Straight Education Network (GLSEN) is a national organization that collaborates with educators, policy makers, community leaders, and students to, among other things, (1) protect LGBT students from bullying and/or harassment and discrimination, (2) affirm the right of students to learn in a classroom that embraces diversity and to organize and lead Gay Straight Alliances, (3) advance comprehensive safe schools laws and policies, (4) empower principals to make their schools safer, and (5) build the skills of educators to teach respect for all people (GLSEN 2015).

WHAT
do you
THINK?

One strategy to reduce the bullying and harassment that LGBT students endure, is to establish schools which are exclusively for LGBT students such as New York City's Harvey Milk High School and Milwaukee's Alliance High School (Calefati 2015). Do you think that segregation is the answer? What are some of the reasons against such a strategy?

Campus Programs. Many LGBT-affirmative and educational initiatives of middle and high schools also occur on college campuses. Student groups in colleges and universities have been active in the gay liberation movement since the 1960s. In addition to university-wide nondiscrimination policies, other measures to support the LGBT college student population include gay and lesbian studies programs, social centers, and support groups, as well as campus events and activities that celebrate diversity. Some campuses have a "Lavender Graduation" ceremony in which LGBT graduates are honored and receive rainbow tassels for their mortarboards. Many campuses also have Safe Zone programs designed to visibly identify students, staff, and faculty who support the LGBT population.

Understanding Sexual Orientation and the Struggle for Equality

As both structural functionalists and conflict theorists note, nonheterosexuality challenges traditional definitions of family, child rearing, and gender roles. Thus, every victory in achieving legal protection and social recognition for gay men and women fuels the backlash against them by groups who are determined to maintain traditional notions of family and gender, often informed by religious ideology. And, as symbolic interactionists note, the meanings associated with homosexuality are learned, often embedded with stereotypes resulting from myths about sexual orientation. Powerful individuals and groups opposed to gay rights focus their efforts on maintaining these negative meanings of homosexuality to keep the gay, lesbian, and bisexual population marginalized.

Yet, in what can only be called a landmark decision, the Supreme Court held that same-sex marriage is a fundamental right granted by the U.S. Constitution. However, as evidenced by the years of court maneuvering in the struggle against California's Proposition 8, which banned same-sex marriage in a state that had legalized it, gay rights previously won can be taken away. Less than three weeks after the Supreme Court's decision in *Obergefell,* a resolution was introduced into the U.S. House of Representatives asking members to denounce the Court's decision and to encourage states not to issue marriage licenses to same-sex couples or to recognize same-sex marriages performed in other states (Hayworth 2015).

Furthermore, even when laws are passed to protect LGBT individuals, it doesn't ensure behavioral changes. LGBT individuals employed at workplaces with anti-discrimination policies still experience harassment and rejection from their coworkers; students in schools with policies prohibiting bullying are still subjected to antigay taunts; hate crime statutes don't mean LGBT men and women aren't victims of violence; and now, countless LGB couples, many waiting decades, continue to wait for their constitutional right to marry.

> [T]he consensus of scientific research is that LGBT statuses are *ascribed statuses* and can be changed neither by choice nor by the efforts of others.

Embracing the outdated concepts of the binary nature of gender and sexuality, many continue to fight for the subjugation of gay and transgender men and women. But the consensus of scientific research is that LGBT statuses are *ascribed statuses* and can be changed neither by choice nor by the efforts of others. Thus, sexual orientation and gender identity are akin to other ascribed statuses—gender, race, and ethnicity—and, as we are finally beginning to acknowledge, diversity, although difficult, is good.

> Diversity enhances creativity . . . encourages the search for novel information and perspectives, leading to better decision making and problem solving. Diversity can . . . lead to unfettered discoveries and breakthrough innovations. Even simply being exposed to diversity can change the way you think. This is not just wishful thinking: it is the conclusion . . . from decades of research from organizational scientists, psychologists, sociologists, economists and demographers. (Phillips 2014, p. 1)

Imagine the lost contributions of women, racial and ethnic minorities, and gay and transgender men and women who for hundreds of years were unable to reach their full potential as a consequence of stereotyping, prejudice, and discrimination.

Wanting to go to school in a nonhostile environment is not part of a gay agenda. Being evaluated at work on what you do not who you are is not a special right. Getting married, raising children, and settling down with a family are not the actions of a radical. Toobin (2015) notes that the two great causes of the gay rights movement have been the repeal of DADT, and with it the right to serve in the military, and marriage equality. Furthermore, he observes that when the *Obergefell* decision was announced outside the Supreme Court, supporters joined hands and sang "The Star Spangled Banner"—a song "for people who believe in the United States, who want to participate in American society more than they want to transform it" (p. 1).

In these days of divisiveness, hate, and global terror, does it make sense to exclude people who have fought not just to be integrated into American society, but to be able to defend it and fully participate in its institutions?

Chapter Review

- **Are there any countries in which homosexuality is illegal?**
Yes, globally, numerous countries criminalize same-sex relations. Legal penalties vary for violating laws that prohibit homosexual acts. In some countries, homosexuality is punishable by prison sentences and/or corporal punishment, such as whipping or lashing, and in five nations—Iran, Mauritania, Saudi Arabia, Yemen, and Sudan—people found guilty of engaging in same-sex sexual behavior may receive the death penalty.

- **In what ways is the classification of individuals into sexual orientation categories problematic?**
Classifying individuals into sexual orientation categories is problematic for a number of reasons: (1) distinctions among sexual orientation categories are not clear-cut and may better be represented by a continuum; (2) research with same-sex populations has tended to define sexual orientation based on one of three components whereas sexual orientation is complex and multidimensional; and (3) social stigma associated with nonheterosexuality leads people to conceal or falsely portray their sexuality.

- **What is the relationship between beliefs about what "causes" homosexuality and attitudes toward homosexuality?**
Individuals who believe that homosexuality or bisexuality is biologically based or inborn tend to be more accepting of LGB individuals. In contrast, individuals who believe that lesbian, gay, and bisexual people choose their sexual orientation are less tolerant of LGB individuals.

- **What is the official position of numerous respected professional organizations regarding sexual orientation change efforts (SOCE) for gays and lesbians?**
Many professional organizations agree that sexual orientation cannot be changed and that efforts to change sexual orientation (using conversion, reparative, or reorientation therapy) do not work and may, in fact, be harmful.

- **What cultural and structural changes in society have led to a worldwide increase in legal and social support for LGB individuals?**
The increase in worldwide support for LGB individuals is a consequence of (1) the relaxing of traditional gender roles, (2) marriage increasingly seen as a partnership, (3), the greater visibility of LGB individuals, and (4) the emergence of a global society in which nations are influenced by international pressures.

- **What are some of the consequences of anti-LGB bias for individuals?**
Sexual minorities, as many other minorities, suffer the negative consequences of harassment, prejudice and discrimination, violence, and isolation. Among others, the consequences include higher rates of suicide, physical illness, substance abuse, eating disorders, poverty, and homelessness, when compared to their non-LGB counterparts, as well as an uncertain future as they age.

- **Is employment discrimination based on sexual orientation illegal in all 50 states?**
Twenty-one states and the District of Columbia have laws banning sexual orientation discrimination in the workplace (HRC 2015c). Additionally, 26 states and the District of Columbia offer benefits for their government employees and their same-sex spouses. Finally, nearly 70 percent of Fortune 500 companies include sexual orientation in their equal employment opportunity or nondiscrimination policies, and more than half offer same-sex domestic partner health benefits to all of their employees. As of this writing, ENDA which would ban employment discrimination against individuals on the basis of sexual orientation, is pending before the U.S. House of Representatives.

- **What is the Defense of Marriage Act (DOMA)?**
The Defense of Marriage Act is federal legislation that states that marriage is a legal union between one man and one woman and denies federal recognition of same-sex marriage. In a June 2013 5–4 ruling in *United States v. Windsor*, the Supreme Court struck down a provision of the 17-year-old act that denies federal benefits to same-sex couples legally married.

- **What was the decision in *Obergefell v. Hodges*?**
In a 5–4 decision, the Supreme Court of the United States held that it is unconstitutional for a state to ban same-sex marriages or to deny legal recognition of a same-sex marriage from another state.

- **What are some reasons for the underreporting of LGBT intimate partner violence and sexual assault to the police?**
Many cases of LGBT violence are not reported to the police for fear of further victimization by police, including indifference or some form of abuse. Additionally, the failure of police to identify such incidents as occurring in the context of an intimate partnership often erroneously leads to the arrest of both the victim and the perpetrator.

- **Why is the media important in the advancement of LGBT civil rights?**
LGBT visibility in the media counteracts stereotypes of LGBT individuals and allows non-LGBT individuals to see them not as an abstraction, but as real people. This is consistent with Gordon Allport's contact hypothesis that suggests more contact with or exposure to a group results in the reduction of prejudice. The media has also provided LGBT individuals, most importantly youth, with role models for "coming out."

Test Yourself

1. Research has indicated that many individuals are not exclusively heterosexual or homosexual and that sexual orientation can be represented on a continuum.
   a. True
   b. False

2. A national study of U.S. college students found that what percentage identified as gay, lesbian, or bisexual?
   a. 1.2
   b. 3.0
   c. 7.5
   d. 10.1

3. According to estimates, over 30 percent of Americans live with a same-sex partner or spouse.
   a. True
   b. False

4. Individuals who believe that same-sex attraction is a consequence of birth
   a. tend to be in the minority.
   b. are more supportive of gay rights than those who believe it's a "lifestyle."
   c. are more likely to believe that homosexuality is not acceptable.
   d. are correct according to scientific research.

5. Sexual orientation change efforts (SOCE)
   a. have been found to be successful 60 percent of the time.
   b. are rooted in the belief that homosexuality is immoral or pathological.
   c. are supported by many professional associations such as the American Medical Association.
   d. are also suitable for transgender individuals.

6. Heterosexism leads to
   a. prejudice.
   b. the oppression of heterosexuals.
   c. discrimination.
   d. Both a and c are correct.

7. The contact hypotheses suggests that the more gay men and women reveal their sexual orientation, the lower the prejudice against them.
   a. True
   b. False

8. Why was the It Gets Better Project started?
   a. Because same-sex couples continue to encounter obstacles in second-parent and joint adoptions
   b. Because LGBT youth continue to be bullied and harassed by their peers, resulting in shame, isolation, and even suicide
   c. Because LGB servicemen and -women have been waiting a long time to serve openly in the military
   d. Because American society seems to be growing in its acceptance of the legal recognition of same-sex couples

9. The *Obergefell v. Hodges* Supreme Court decision held that states do not have the right to ban same-sex marriages.
   a. True
   b. False

10. The first country to legalize same-sex marriage was
    a. Norway.
    b. England.
    c. The Netherlands.
    d. Argentina.

Answers: 1. A; 2. C; 3. B; 4. B; 5. B; 6. D; 7. A; 8. B; 9. A; 10. C.

Key Terms

bisexuality   358
cisgender   358
coming out   383
contact hypothesis   370
Defense of Marriage Act (DOMA) 373
discrimination   369
Employment Non-Discrimination Act (ENDA)   384
Every Child Deserves a Family Act (ECDFA)   385
First Amendment Defense Act   365

gay   358
gender nonconforming   358
heteronormativity   362
heterosexism   369
heterosexuality   358
homosexuality   358
lesbian   358
LGBT, LGBTQ, and LGBTQI   358
master status   366
Matthew Shepard and James Byrd, Jr. Hate Crimes Prevention Act (HCPA)   386

Obergefell v. Hodges   374
oppression   369
prejudice   369
privilege   369
registered partnerships   360
religious freedom laws   365
sexual orientation   358
sexual orientation change efforts (SOCE)   362

" Population may be the key to all the issues that will shape the future: economic growth; environmental security; and the health and well-being of countries, communities, and families."

NAFIS SADIK,
former executive director, United Nations Population Fund

12

Population Growth and Aging

Learning Objectives

After studying this chapter, you will be able to . . .

1 Describe the history, current trends, and future projections of population growth, and explain the causes of the aging of the world's population.

2 Discuss ways in which structural functionalism, conflict theory, and symbolic interactionism can be applied to the study of population and aging.

3 Explain how population growth is related to environmental problems, poverty, unemployment, global insecurity, and poor maternal and infant/child health, and describe problems related to the aging of the population, including ageism, employment and retirement concerns, and family-provided elder care.

4 Describe strategies to curb global population growth, increase population growth in some countries, combat ageism and age discrimination in the workplace, and reform Social Security.

5 Explain why, given all the other pressing social problems facing the world, controlling population growth should be a priority and why taking care of older populations is everyone's concern.

In this stainless steel bio-cremation container, heat and lye transform human remains into a nontoxic liquid.

Cb2/Zob/Wenn.com/Newscom

generations can continue to be buried before we run out of land? Several cemeteries in New York City have already run out of land and have stopped selling burial plots (Santora 2010).

Japan offers an innovative answer to the problem of not enough grave space: a high-tech graveyard in a multistoried building where the ashes of the dead are stored in urns on shelves. Visitors use plastic swipe cards and a touch screen to identify the remains they wish to "visit," and a robotic arm collects the requested urn and delivers it to a "mourning room" where visitors pay tribute to their deceased loved ones (Here and Now 2009). To maximize grave space, some countries allow the practice of vertical burials in which the deceased are wrapped in a biodegradable shroud and buried in a vertical, standing-up position (Dunn 2008). In another alternative to burials called *alkaline hydrolysis*, heat and lye are used to transform human remains into a nontoxic liquid, leaving behind a small amount of dry bone residue for the family to scatter (Olson 2014). Also known as "liquid cremation" and "bio cremation," this alternative to cremation or burial is legal in at least 10 states and is being considered in other states.

AS PEOPLE AGE, one of the issues they think about is what they would like done with their body after they die. Although traditions, customs, and laws concerning what to do with human remains vary across cultures and across time, some form of cremation and/or burial is common practice around the world. Have you ever driven past a cemetery and wondered how many

Running out of land for grave sites is one of many concerns that stem from the growth in human population over the last century or so. The world's population is not only growing, but it also is aging. In this chapter, we focus on social problems associated with population growth and aging.

The Global Context: A Worldview of Population Growth and Aging

The well-known scientist Jane Goodall remarked, "It's our population growth that underlies just about every single one of the problems that we've inflicted on the planet. If there were just a few of us, then the nasty things we do wouldn't really matter and Mother Nature would take care of it—but there are so many of us" (quoted in Koch 2010, n.p.). In the following sections, we describe how the size and life span of human populations have increased over time.

World Population: History, Current Trends, and Future Projections

Humans have existed on this planet for at least 200,000 years. For 99 percent of human history, population growth was restricted by disease and limited food supplies. Around 8000 B.C., the development of agriculture and the domestication of animals led to increased food supplies and population growth, but even then, harsh living conditions and disease still put limits on the rate of growth. This pattern continued until the mid-18th century, when the Industrial Revolution improved the standard of living for much of the world's population

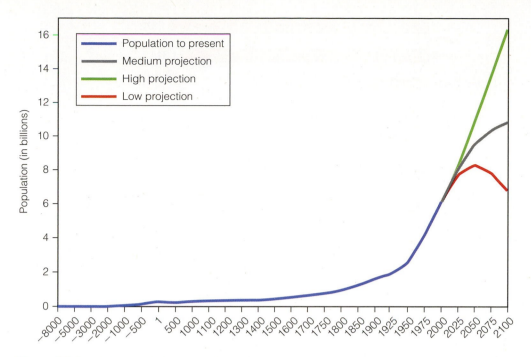

Figure 12.1 World Population through History
SOURCE: Weeks 2015.

through better food, cleaner drinking water, improved housing and sanitation, and advances in medical technology, such as antibiotics and vaccinations against infectious diseases—all of which contributed to rapid increases in population.

Although thousands of years passed before the world's population reached 1 billion in about 1800, the population exploded from 1 billion to 6 billion in less than 200 years (see Figure 12.1). World population was 1.6 billion when we entered the 20th century, and 6.1 billion when we entered the 21st century. World population is projected to grow from 7.2 billion in 2013, to 8.1 billion in 2025, 9.6 billion in 2050, and 10.9 billion by 2100 (United Nations 2013).

Most of the world's population lives in less developed countries, primarily in Asia and Africa (see Figure 12.2). The most populated country in the world is China, where about one in five people on this planet lives (see Figure 12.3).

Nearly all of world population growth from now until 2100 will occur in less developed countries, particularly the least developed countries of the world, and mostly in Africa and Asia. The population of developing countries is expected to increase from 5.9 billion in 2013 to 9.6 billion in 2100. In contrast, the population of more developed countries is expected to grow minimally, from 1.25 billion in 2013 to 1.28 billion in 2100, and would decline if it were not for the migration from developing countries to developed countries (United Nations 2013). Higher population growth in developing countries is largely due to the fact that the least developed countries in the world have the highest **total fertility rates**—the average lifetime number of births per woman in a population. In 22 countries women have an average of 5 or more children in their lifetime (Population Reference Bureau 2015).

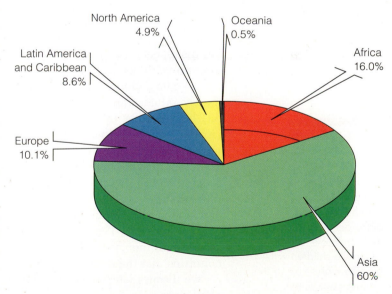

Figure 12.2 Distribution of World Population by Region, 2015
SOURCE: Population Reference Bureau 2015.

total fertility rates The average lifetime number of births per woman in a population.

The Global Context: A Worldview of Population Growth and Aging | **395**

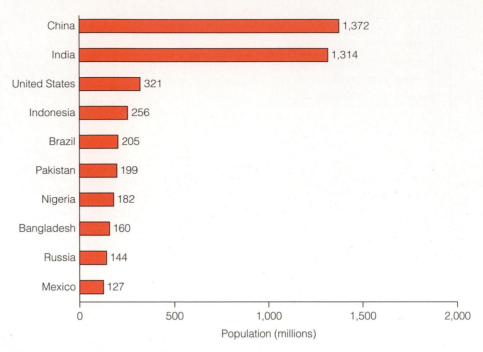

Figure 12.3 World's Ten Largest Countries in Population, 2015
SOURCE: Population Reference Bureau 2015.

Will there be an end to the rapid population growth that has occurred in recent decades? Will the population of the world stabilize? Although some predict that population will stabilize around the middle of the 21st century, no one knows for sure. To reach population stabilization, fertility rates throughout the world would need to achieve what is called "replacement level," whereby births would replace, but not outnumber, deaths. **Replacement-level fertility** is 2.1 births per woman—slightly more than 2 because not all female children will live long enough to reach their reproductive years. In 2015, 85 countries—mostly in Europe and East Asia—had below-replacement fertility (Population Reference Bureau 2015). Weeks (2015) explains that "the motivation to have large families has disappeared, at least for the time being, and has been replaced by a propensity to try to improve the family's standard of living by limiting the number of children" (p. 46).

In some countries with below-replacement fertility, population will continue to grow for several decades because of **population momentum**—population growth due to previous high fertility rates that have resulted in a large number of young women who are currently entering their childbearing years. But the Population Reference Bureau (2015) projects that 34 countries have populations that will decrease over the next few decades. Although the U.S. population has a total fertility rate slightly lower than the replacement level, the U.S. population is expected to continue to increase through 2050 because of immigration.

In sum, two population size trends are occurring simultaneously that appear to be contradictory: (1) The total number of people on this planet is rising and is expected to continue to increase over the coming decades, and (2) fertility rates are so low in some countries that the countries' populations are expected to decline over the coming years. As we discuss later in this chapter, each of these trends presents a set of problems and challenges.

The Aging of the World's Population

Another demographic trend that presents its own set of challenges is the increasing number and proportion of older individuals in the total population. Globally, the population aged 60 and older is the fastest-growing age group (United Nations 2013). Between 2013 and 2050, the percentage of older individuals (aged 60 and over) in the world population is expected

replacement-level fertility The level of fertility at which a population exactly replaces itself from one generation to the next; currently, the number is 2.1 births per woman (slightly more than 2 because not all female children will live long enough to reach their reproductive years).

population momentum Continued population growth as a result of past high fertility rates that have resulted in a large number of young women who are currently entering their childbearing years.

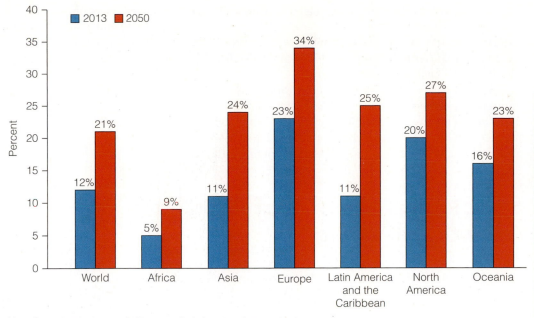

Note: Percentages are rounded to nearest whole percentage.

Figure 12.4 Percent of Population Aged 60+, by Region, 2013–2050 (projected)
SOURCE: United Nations 2013.

In 2050, the global population of people aged 60 years or over will outnumber the population of children (0 to 14 years) for the first time in human history.

to nearly double (see Figure 12.4). In 2050, the global population of people aged 60 years or over will outnumber the population of children (0 to 14 years) for the first time in human history (United Nations 2012). The 80 and older population is also growing throughout the world, increasing from nearly 2 percent of world population in 2013 to 4 percent in 2050 to nearly 8 percent in 2100 (United Nations 2013). In the United States, the 60 and older and 80 and older populations are also growing (see Figure 12.5).

The aging of the population is the result of both increased longevity and declining fertility rates. Global life expectancy increased from 47 years between 1950 and 1955 to 71 years in 2015, and it is expected to increase to 76 years between 2045 and 2050 and to 82 years between 2095 and 2100 (Population Reference Bureau 2015; United Nations 2013). In the United States, population aging is also occurring as the **baby boomers**—the generation of Americans born during a period of high birthrates between 1946 and 1964—are entering their older years.

Fertility rates globally have fallen, from 4.6 in 1970 to 2.5 in 2015 (Population Reference Bureau 2015). In most developed countries, including the United States, fertility rates have fallen to fewer than two children per woman, and they have been below this replacement level for two to three decades. This means that as more of the population is living longer into older ages, fewer children are being born, so an increased percentage of the population is older.

Population aging increases pressure on a society's ability to support its elderly members because as the proportion of older people increases in a population, there are fewer working-age adults to support the elderly population. A commonly used indicator of this pressure is the **elderly support ratio**, calculated as the number of working-age people divided by the number of people 65 or older. Globally, "working age" is considered to be ages 15 to 64; in the United States, "working age" is considered to be ages 20 to 64. Since 1950, the global elderly support ratio has been declining. In 1950, there were 12 working-age people for every older person; in 2012, there were 8 working-age people for every older person, and the elderly support ratio is projected to further decrease to 4:1 by 2050 (United Nations 2012). In developed countries, the elderly support ratio is much

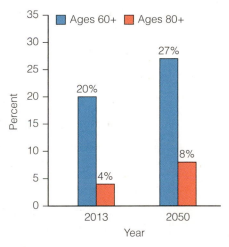

Note: Percentages are rounded to nearest whole percentage.

Figure 12.5 Percent of U.S. Population Age 60+ and 80+, 2013–2050 (projected)
SOURCE: United Nations 2013.

baby boomers The generation of Americans born between 1946 and 1964, a period of high birthrates.

elderly support ratio The ratio of working-age adults (15 to 64) to adults aged 65 and older in a population.

lower than in less developed regions, meaning that there are fewer working-age adults to support each older person in developed countries. Germany, Italy, and Japan have an elderly support ratio of 3:1—the lowest in the world (Population Reference Bureau 2010). By 2050, Japan will have only one working-age adult for every elderly person; Germany and Italy will each have two.

The decrease in the elderly support ratio raises concerns about whether there will be enough workers to take care of the older population. Population aging raises other concerns as well: How will societies provide housing, medical care, transportation, and other needs of the increasing elderly population? Later in this chapter, we look at the problems and challenges of meeting the needs of a growing elderly population.

Sociological Theories of Population Growth and Aging

The three main sociological perspectives—structural functionalism, conflict theory, and symbolic interactionism—can be applied to the study of population and aging.

Structural-Functionalist Perspective

Structural functionalism focuses on how changes in one aspect of the social system affect other aspects of society. For example, the **demographic transition theory** of population describes how industrialization and economic development affect population growth by influencing birth- and death rates. According to this theory, traditional agricultural societies have both high birthrates and high death rates. As a society becomes industrialized and urbanized, improved sanitation, health, and education lead to a decline in mortality. The increased survival rate of infants and children along with the declining economic value of children leads to a decline in birthrates. The demographic transition theory is a generalized model that does not apply to all countries, and it does not account for population change due to HIV/AIDS, war, migration, and changes in gender roles and equality. Many countries with low fertility rates have entered what is known as a "second demographic transition," in which fertility falls below the two-child replacement level. This second demographic transition has been linked to greater educational and job opportunities for women, increased availability of effective contraception, and the rise of individualism and materialism (Population Reference Bureau 2004).

The growing elderly population is a social change that has led to a number of other social changes, such as greater pressure on the workforce and on federal programs to support the aging population (e.g., Medicare and Social Security). Increased longevity also affects families, who often bear the brunt of elder care. The problems and challenges associated with caring for the aging population are discussed later in this chapter.

The structural-functionalist perspective also focuses attention on the unintended, or latent, consequences of social behavior. Although the intended, or manifest, function of modern contraception is to control and limit childbearing, there have also been far-reaching unintended effects of contraception on the social and economic status of women. The development of modern contraceptives—particularly the birth control pill—has led to fewer births to high school and college-aged women, increased age at first marriage, and greater participation by women in the workforce. "The advent of the pill allowed women greater freedom in career decisions, by allowing them to invest in higher education and a career with far less risk of an unplanned pregnancy" (Sonfield 2011, p. 9). Another example of an unintended consequence can be found in the findings of several studies that link social security programs in various countries and lowered fertility (Kornblau 2009). One reason that people have children is to provide security in old age. Social security reduces the need to have children to secure care in old age.

demographic transition theory A theory that attributes population growth patterns to changes in birthrates and death rates associated with the process of industrialization.

Conflict Perspective

The conflict perspective focuses on how wealth and power, or the lack thereof, affect population problems. In 1798, Thomas Malthus predicted that the population would grow faster than the food supply and that masses of people were destined to be poor and hungry. According to Malthusian theory, food shortages would lead to war, disease, and starvation, which would eventually slow population growth. However, conflict theorists argue that food shortages result primarily from inequitable distribution of power and resources (Livernash and Rodenburg 1998).

Conflict theorists also note that population growth results from pervasive poverty and the subordinate position of women in many less developed countries. Poor countries have high infant and child mortality rates. Hence, women in many poor countries feel compelled to have many children to increase the chances that some will survive into adulthood. Their subordinate position prevents many women from limiting their fertility. In many developing countries, a woman must get her husband's consent before she can receive any contraceptive services. Thus, according to conflict theorists, population problems result from continued economic and gender inequality.

Some conflict theorists view the elderly population as a special interest group that competes with younger populations for scarce resources. Debates about funding programs for the elderly (e.g., Social Security and Medicare) versus funding for youth programs (e.g., public schools and child health programs) largely represent conflicting interests of the young versus the old. A growing elderly population means that the elderly have increased power in political issues. In the United States, adults ages 60 and older have the highest rate of voting of any age group (ages 18 to 29 have the lowest voting rate) (U.S Census Bureau 2012).

> In the United States, adults ages 60 and older have the highest rate of voting of any age group.

Symbolic Interactionist Perspective

The symbolic interactionist perspective focuses on how meanings, labels, and definitions learned through interaction affect population problems and issues concerning aging. For example, many societies are characterized by **pronatalism**—a cultural value that promotes having children. Throughout history, many religions have worshiped fertility and recognized it as being necessary for the continuation of the human race. In many countries, religions prohibit or discourage birth control, contraceptives, and abortion. Women in some pronatalistic societies learn through interaction with others that deliberate control of fertility is socially unacceptable. Women who use contraception in communities in which family planning is not socially accepted face ostracism by their community, disdain from relatives and friends, and even divorce and abandonment by their husbands (Women's Studies Project 2003). However, once some women learn new definitions of fertility control, they become role models and influence the attitudes and behaviors of others in their personal networks (Bongaarts and Watkins 1996).

The symbolic interactionist perspective emphasizes the importance of examining social meanings and definitions associated with aging. "Old age" is largely a social construct; there is no biological marker that indicates when a person is "old." Rather, old age is a matter of social definition. In the United States, and much of the world, a person is considered to be "old" (or a "senior citizen") when they reach 65, as this is the age that company pension plans and Social Security have used to define when a person retires and collects benefits. Due to changes in the Social Security system, the retirement age for receiving full Social Security benefits has been increased to 67, yet we continue to use 65 to define "elderly."

WHAT do you THINK?

There are expressions today that suggest that "60 is the new 40" or "70 is the new 50." As life expectancy increases and people redefine 60 as today's 40, do you think that the designation of "senior citizen" or "old" will change from 65 to a higher number? What age do you think of as the beginning of "old age"?

We also learn both positive and negative meanings associated with old age, such as "wise" and "experienced" as well as "frail" and "impaired" (Kornadt and Rothermund 2010). In U.S. culture, negative labels of older people, such as "crone," "old geezer," and

pronatalism A cultural value that promotes having children.

ANIMALS and society

Pet Overpopulation in the United States

More than half of U.S. households include dogs or cats as pets, also referred to as companion animals (PRNewswire 2015). Many pets live out their lives in homes where they are loved and well cared for. But each year, millions of stray and unwanted cats and dogs are taken to animal shelters across the United States, where 2.7 million adoptable dogs and cats are euthanized, most commonly through intravenous injection (Humane Society 2014). In fact, shelter euthanasia is the number one cause of death for both dogs and cats in the United States (American Humane Association 2011).

According to the National Council on Pet Population Study and Policy (2009), the top 10 reasons for taking a companion dog or cat to an animal shelter include moving, cost of pet maintenance, and allergies (see the table within this feature). Many people do not realize that the costs of raising a pet over its lifetime—including initial purchase price, food, treats, vet visits, grooming, toys, medications, training, and other expenses—can be in the tens of thousands of dollars (Buckell 2015). Although it is a crime in all 50 states to abandon a pet, animal shelters in college towns see an increase in abandoned pets at the end of the school year when many students move away and cannot take their pets with them (Kidd 2009).

Samokhin/Shutterstock.com

Not all abandoned pets are taken to shelters; some are left in the street or are dropped off in parking lots or other public places. Abandoned pets left out in the wild struggle with starvation and disease, and they pose a threat to both humans and other animals, such as endangered birds. In the wild, unsterilized cats and dogs can multiply quickly. Consider that if an unsterilized cat gives birth to two litters a year, that can result in 400,000 cats over seven years (Shikina 2011).

The millions of unwanted and feral dogs and cats is part of a larger problem of pet overpopulation. The American Society for the Prevention of Cruelty to Animals (ASPCA) suggests that the best method for reducing pet overpopulation is sterilization, recommending that (1) all companion dogs and cats, except those that are a part of a responsible breeder's program, be spayed or neutered at an early age (2 months); and (2) all communities have accessible and affordable spay/neuter programs (ASPCA 2011). Many vets and animal welfare agencies offer low-cost or free spaying and neutering, and some organizations have mobile clinics that travel to low-income or rural areas where people are not able to transport a pet to a vet.

"old biddy," are predominant and reflect ageism—a topic we discuss later in this chapter. Negative stereotypes of aging have implications for health. One study found that older adults (70 and older) with negative stereotypes of old people were less likely to fully recover from severe disability than those with positive stereotypes of older persons (Levy et al. 2012).

Finally, the symbolic interactionist perspective reminds us that understanding the quality of life among the elderly requires consideration of how the elderly subjectively define their experiences. For example, to assess social isolation among the elderly, researchers often use objective measures such as the frequency of social contacts among the elderly. But it is also important to consider how the elderly subjectively experience their level of social contact, as some older adults have a preference for engaging in solitary activities and experience being alone as enjoyable (Cloutier-Fisher et al. 2011).

Despite the problem of pet overpopulation, about a third of pet owners in the United States do not spay or neuter their pet (American Humane Association 2011). Some pet owners want to breed their pets, either for profit or because they think it would be fun to have a litter of puppies or kittens. Others mistakenly believe that their pet won't become accidentally pregnant.

Some states and municipalities have laws to try to curb pet overpopulation. Rhode Island mandates that dogs and cats must be spayed or neutered before being released from a shelter, a Los Angeles ordinance requires pet owners to sterilize their dogs or cats by age 4 months, and some states have considered laws mandating the spaying and neutering of dogs and cats. Some cities—including Austin, Texas; Albuquerque, New Mexico; and West Hollywood, California—completely ban the retail sales of dogs and cats (Humane Society 2010; Shikina 2011). And some major pet stores, including Petco and PetSmart, refuse to sell puppies and kittens at their stores (although many major pet stores allow local animal and shelter groups to display dogs and cats for adoption).

Finally, people wanting to acquire a dog or cat are encouraged to adopt from an animal shelter or nonprofit animal rescue organization. Dogs and cats that are not adopted or rescued from a shelter are, sadly, destroyed.

The "no kill" movement seeks to end the killing of animals in shelters, and to date at least 160 U.S. cities and towns are designated as "no kill" communities in which shelters save from 90 to 99 percent of their animals

Top 10 Reasons for Pet Relinquishment to U.S. Shelters

Dogs	Cats
1. Moving	1. Too many in house
2. Landlord issues	2. Allergies
3. Cost of pet maintenance	3. Moving
4. No time for pet	4. Cost of pet maintenance
5. Inadequate facilities	5. Landlord issues
6. Too many pets in home	6. No homes for littermates
7. Pet illness	7. House soiling
8. Personal problems	8. Personal problems
9. Biting	9. Inadequate facilities
10. No homes for littermates	10. Doesn't get along with other pets

SOURCE: National Council on Pet Population Study and Policy 2009.

(Sandberg 2013). Nathan Winograd, a leader in the no-kill movement, views pet overpopulation as a myth that is used to justify the unnecessary mass killing of animals. He argues that 3 to 4 million dogs and cats are killed every year in shelters, not because there are too few homes to adopt them but because shelter directors fail to find adoptive homes for the animals. No-kill shelters achieve high-volume adoptions by working with volunteers, foster families, and rescue groups and individuals. They treat medical and behavior problems and neuter and release, rather than kill, feral cats. Although costs associated with pet ownership lead some pet owners to abandon an animal, Winograd argues

that encouraging more adoptions from shelters not only saves animals' lives but also contributes to the local economy:

> Each new pet owner will spend an average of $1,100 a year on his or her pet, and that means more tax revenue for the community. It's an economic boost to local pet stores, groomers, veterinarians, and boarding kennels. No kill is also consistent with public health and safety, because owning pets improves the quality of people's lives and their interactions with one another. We win on so many issues beyond saving the lives of dogs and cats. (quoted in Sandberg 2013, p. 12)

Social Problems Related to Population Growth and Aging

Social problems related to population growth include environmental problems; poverty, unemployment, and global insecurity; and poor maternal and infant/child health. This chapter's *Animals and Society* feature looks at the problem of pet overpopulation and the importance of controlling the fertility of companion animals.

Problems related to the aging of the population include ageism—prejudice and discrimination against older individuals—employment and retirement concerns of older Americans, and the challenge of meeting the various needs of the elderly population, particularly retirement income and health care. Health issues and Medicare are discussed in Chapter 2, and elder abuse is discussed in Chapter 5.

Environmental Problems and Resource Scarcity

According to a survey of faculty at the State University of New York College of Environmental Science and Forestry (2009), overpopulation is the world's top environmental problem, followed closely by climate change and the need to replace fossil fuels with renewable energy sources. As we discuss in Chapter 13, population growth places increased demands on natural resources, such as forests, water, cropland, and oil, and results in increased waste and pollution.

According to the United Nations Environment Programme (UNEP), half of the planet's forests have been cleared for human land use; in 2025, two-thirds of the world's population will be living in countries with water scarcity or stress; and the world's fisheries will be depleted by the middle of this century (Engelman 2011). The countries that suffer most from shortages of water, farmland, and food are developing countries with the highest population growth rates. However, countries with the largest populations do not necessarily have the largest impact on the environment. This is because the demands that humanity makes on the Earth's natural resources—each person's **environmental footprint**—is determined by the patterns of production and consumption in that person's culture. The environmental footprint of an average person in a high-income country is much larger than that of someone in a low-income country. Hence, although population growth is a contributing factor in environmental problems, patterns of production and consumption are at least as important in influencing the effects of population on the environment.

Half of the world's forests have been cleared for human use.

Bob Anderson/Getty Images

WHAT do you THINK?

The growing world population raises questions about how we can provide enough food, water, energy, housing, and so forth, to meet everyone's needs. Gar Smith (2009) argues that "meeting the 'growing demands of a growing population' is not solving the problem: It's perpetuating it" (p. 15). What does this statement mean to you? Do you agree? Why or why not?

Poverty, Unemployment, and Global Insecurity

Poverty and unemployment are problems that plague countries with high population growth. Less developed, poor countries with high birthrates do not have enough jobs for a rapidly growing population, and land for subsistence farming becomes increasingly scarce as populations grow. In some ways, poverty leads to high fertility, because poor women are less likely to have access to contraception and are more likely to have large families in the hope that some children will survive to adulthood and support them in old age. But high fertility also exacerbates poverty, because families have more children to support, and national budgets for education and health care are stretched thin.

Rapid population growth is a contributing factor to global insecurity, including civil unrest, war, and terrorism (Weiland 2005). Although world population is, overall, aging, some countries in Africa and the Middle East are experiencing a "youth bulge"—a high proportion of 15- to 29-year-olds relative to the adult population. Youth bulges result from high fertility rates and declining infant mortality rates, a common pattern in developing countries today. The combination of a youth bulge with other characteristics of rapidly growing populations, such as resource scarcity, high unemployment rates, poverty, and

environmental footprint The demands that humanity makes on the earth's natural resources.

Ageism Survey

Directions: Put a number in the blank that shows how often you have experienced each of the following 10 events. In the following items, "age" means old age.

___ 1. I was told a joke that makes fun of old people.

___ 2. I was given a birthday card that pokes fun at old people.

___ 3. I was ignored or not taken seriously because of my age.

___ 4. I was called an insulting name related to my age.

___ 5. I was patronized or "talked down to" because of my age.

___ 6. I was treated with less dignity and respect because of my age.

___ 7. A doctor or nurse assumed my ailments were caused by my age.

___ 8. Someone assumed I could not hear because of my age.

___ 9. Someone assumed I could not understand because of my age.

___10. Someone told me "You're too old for that."

Comparison data: When 152 U.S. adults over age 60 completed the Ageism Survey, the following percentages indicated that they had experienced the 10 types of ageism in the survey once, twice, or more:

Item	Experienced once	Experienced twice or more
1. Jokes that poke fun	19	49
2. Birthday card that pokes fun	14	23
3. Ignored	16	19
4. Insulting name	8	9
5. Patronized	16	22
6. Treated with less dignity	12	16
7. Assumed ailment caused by age	22	20
8. Assumed deaf	10	20
9. Assumed could not understand	16	17
10. "You're too old"	17	24

SOURCE: Erdman B. Palmore, 2004, "Research Note: Ageism in Canada and the United States," *Journal of Cross-Cultural Gerontology* 19(1):41–46.

rapid urbanization, sets the stage for political unrest. "Large groups of unemployed young people, combined with overcrowded cities and lack of access to farmland and water create a population that is angry and frustrated with the status quo and thus is more likely to resort to violence to bring about change" (Weiland 2005, p. 3).

Poor Maternal, Infant, and Child Health

As noted in Chapter 2, maternal deaths (deaths related to pregnancy and childbirth) are the leading cause of mortality for reproductive-age women in the developing world. Having several children at short intervals increases the chances of premature birth, infectious disease, and death for the mother or the baby. Childbearing at young ages (teens) also increases the risks of health problems and death for both women and infants (United Nations Population Division 2009). In addition, the more children a woman has, the fewer the parental resources (parental income, time, and maternal nutrition) and social resources (health care and education) available to each child.

Ageism: Prejudice and Discrimination toward the Elderly

Ageism refers to negative stereotyping, prejudice, and discrimination based on a person's or group's perceived chronological age. Ageism is reflected in negative stereotypes of the elderly, such as that they are slow, they don't change their ways, they are grumpy, they are poor drivers, they cannot/don't want to learn new things, they are incompetent, or they are physically and/or cognitively impaired (Nelson 2011). Ageism also occurs when older individuals are treated differently because of their age, such as when they are spoken to loudly in simple language, assuming they cannot understand normal speech, or when they are denied employment due to their age. Another form of ageism—**ageism by invisibility**—occurs when older adults are underrepresented in advertising and educational materials. Before reading further, you may want to complete the "Ageism Survey" in this chapter's *Self and Society* feature.

ageism Negative stereotyping, prejudice, and discrimination based on a person's or group's perceived chronological age.

ageism by invisibility The underrepresentation of older adults in advertising and educational materials.

When is the last time you saw a clothing advertisement or catalog that included models in their 50s, 60s, and beyond? The absence of older adults in clothing advertising is an example of ageism by invisibility.

Old age is stereotypically viewed as a negative time during which older individuals suffer a decline in physical and cognitive ability, and have increased dependence on others. Contrary to negative views of aging, people in their later years can be productive and fulfilled (see Table 12.1).

Ageism is embedded in our culture and is much more widely accepted than other "isms" such as racism, sexism, and heterosexism. Nelson (2011) states that "there is no other group like the elderly about which we feel free to openly express stereotypes and even subtle hostility" (p. 40). According to Gullette (2011), "ageism is to the twenty-first century what sexism, racism, homophobia, and ableism were earlier in the twentieth—entrenched and implicit systems of discrimination, without adequate movements of resistance to oppose them" (p. 15). Ageism is different from the other "isms" in that everyone is vulnerable to ageism if they live long enough.

> Ageism is different from the other "isms" in that everyone is vulnerable to ageism if they live long enough.

TABLE 12.1 Accomplishments of Famous Older Individuals

At 70, fitness guru Jack LaLanne towed 70 boats, carrying a total of 70 people, a mile and a half through Long Beach Harbor while handcuffed and shackled.

At 72, feminist author Betty Friedan published *The Fountain of Age*, where she debunks misconceptions about aging.

At 75, Nelson Mandela was elected as the first president of a democratic South Africa.

At 81, Benjamin Franklin facilitated the compromise that led to the adoption of the U.S. Constitution.

At 82, Winston Churchill wrote *A History of the English-Speaking Peoples*.

At 85, actress Mae West starred in the film *Sextette*.

At 85, Coco Chanel was the head of a fashion design firm.

At 87, Mary Baker Eddy created the newspaper *Christian Science Monitor*.

At 88, actress Betty White became the oldest person to ever host *Saturday Night Live*.

At 89, Doris Haddock, also known as "Granny D," began a 3,200-mile walk from Los Angeles to Washington, DC, to raise awareness for the issue of campaign finance reform. She walked 10 miles a day for 14 months, skiing 100 miles when snowfall made walking impossible, and completed her cross-country walk at age 90.

At 90, Pablo Picasso was producing drawings and engravings.

At 93, George Bernard Shaw wrote the play *Farfetched Fables*.

At 94, philosopher Bertrand Russell was active in promoting peace in the Middle East.

At 100, Grandma Moses, noted for her rural American landscapes, was painting. She only started painting at age 78, but by the time she died at 101, she had created over 1,500 works of art.

Ageism is perpetuated in a variety of ways that we commonly accept as harmless fun. For example, birthday greeting cards for aging adults often communicate the message "Sorry to hear you are another year older." Nelson (2011) remarks, "Think about the outrage that would ensue if there was a section of cards that communicated the message 'sorry to hear you're Black' or 'ha ha ha too bad you're Jewish'—yeah, it wouldn't go over so well. So why does society allow, and even condone, the same message directed against older persons?" (p. 41). We give adult birthday gag gifts that make fun of old people with the theme of "over the hill," and we tell or forward jokes that make fun of the elderly, thinking the jokes are humorous rather than offensive. When people are forgetful, they say they are "having a senior moment" without considering that this statement reflects ageism. After considering these points, do you think you will view such acts of making fun of the elderly as harmless fun or as examples of ageism?

WHAT
do you
THINK?

Another indicator of ageism in our society is the negative view of wrinkles, gray hair, and other physical signs of aging. Millions of Americans purchase products or treatments to make them look younger, spending substantial sums of money and undergoing unnecessary and often risky medical procedures.

Unlike more traditional societies that honor and respect their elders, ageism is widespread in modern societies. With the invention of the printing press, elders lost their special status as the keepers of a culture's stories and knowledge (Nelson 2011). Ageism also stems partly from fear and anxiety surrounding aging and death. Many people are uncomfortable around the topic of death and don't like to acknowledge death as a natural part of the life cycle. We use euphemisms to avoid talking about death: We say someone "passed away," "has gone to a better place," "is resting in peace," "kicked the bucket," "has departed," and so on. Old people are reminders of our mortality and, as such, take on negative social meanings.

Employment Age Discrimination. More than one-third of nonretired U.S. adults expect to retire *after* age 65, either out of financial necessity or by choice (Riffkin 2015). One of the obstacles older workers face in finding and keeping employment is age discrimination. In one experimental study, younger job seekers were 40 percent more likely to be offered a job interview than older job seekers with similar résumés (Lahey 2008).

Older workers may be more vulnerable to being "let go" because, although they have seniority on the job, they may also

Allstar Picture Library/Alamy

Nelson Mandela was elected President of South Africa at age 75.

have higher salaries, and businesses that need to cut payroll expenses can save more money by letting higher-salaried personnel go. Prospective employers may view older job applicants as "overqualified" for entry-level positions, less productive than younger workers, and/or more likely to have health problems that could affect not only their productivity but also the cost of employer-based group insurance premiums. Employers may also be concerned about the ability of older workers to learn new skills and adapt to new technology.

Family Caregiving for Our Elders

Many adults have or will provide care and/or financial support for aging spouses, parents, grandparents, and in-laws. Adults who care for their aging parents while also taking care of their own children are referred to as members of the **sandwich generation**—they are "sandwiched" in between taking care of both parents and children. Some adults are caring for not two but *three* generations of family members. The **club sandwich generation** includes adults

sandwich generation A generation of people who care for their aging parents while also taking care of their own children.

club sandwich generation Includes adults (often in their 50s or 60s) who are caring for aging parents, adult children, and grandchildren; or adults (usually in their 30s and 40s) who are caring for their own young children, aging parents, and grandparents.

The Elder Care Study: Everyday Realities and Wishes for Change (Aumann et al. 2010) provides both quantitative and qualitative data about the experience of providing care to elderly family members.

Sample and Methods

The 2008 National Study of the Changing Workforce (NSCW) gathered data through telephone interviews with a nationally representative sample of 3,502 employed people using a random-digit dial procedure. The response rate was 54.6 percent. From this initial study, 1,589 caregivers—both those who were currently providing care and those who had provided care to someone who had died within the past five years—were asked to participate in a telephone interview about their experiences in providing care to an elderly family member. A subsample of 421 family caregivers agreed to participate in a follow-up interview exploring their experiences caring for an elderly relative or in-law; of these, 140 were successfully contacted and interviewed.

Selected Quantitative Findings

In response to the question "Within the past five years, have you provided special attention or care for a relative or in-law 65 years old or older—helping with things that were difficult or impossible for them to do themselves?" nearly half—42 percent—responded yes. The types of care family members provided to elders include direct, in-person care (e.g., preparing meals, performing housework, providing transportation to doctor appointments, bathing, etc.) and indirect care (e.g., shopping, arranging for doctor appointments and other services, managing finances, etc.). Most elders in this study (67 percent) lived in their own homes. Nearly one in four caregivers live with their elderly relative or in-law, either in the caregiver's home (18 percent) or in the elderly person's home (6 percent), and 52 percent live 20 minutes or less from the person for whom they are providing care.

Three-quarters (76 percent) of caregivers relied solely on themselves and their families to care for their elderly relative, with no paid assistance. Elder care providers in this study

provided elder care for an average of 4.1 years; one in four provided care for 5 or more years. Women (20 percent) and men (22 percent) were equally likely to provide care for elders, although on average, women spent more time than men providing care (9.1 hours a week for women versus 5.7 hours for men). Many of these caregivers are in the "sandwich generation"; 46 percent of women caregivers and 40 percent of men caregivers also have children under the age of 18 at home. The majority of caregivers report not having enough time for their children (71 percent), their spouse/partner (63 percent), and themselves (63 percent).

Selected Qualitative Findings

Family caregivers expressed the wish for more support from workplaces in the form of greater job flexibility and more time off for elder care, especially paid time off without having to use vacation time. Caregivers also expressed wanting more active involvement and help from other family members. But despite the challenges

Many adults have or will care for aging parents, grandparents, and/or spouses.

Big Cheese Photo/Getty Images

(often in their 50s or 60s) who are caring for aging parents, adult children, and grandchildren; or adults (usually in their 30s and 40s) who are caring for their own young children, aging parents, and grandparents.

For thousands of years, caring for older adults has been an important function of the family. For example, Chinese culture embraces the tenet of *filial piety,* which entails respecting, obeying, pleasing, and offering support and care to parents. Due to social, economic, and cultural changes, it has become more difficult for families throughout the world to care for aging parents. This chapter's *Social Problems Research Up Close* feature examines the challenges of elder care in U.S. families.

Countries around the world have used various methods to encourage parental support. The United States and Taiwan offer tax deductions and credits to adult children caring for and frustration involved in the caregiving role, many caregivers expressed appreciation for the opportunity to spend time with their elderly relatives and to establish a closer relationship. Interestingly, former caregivers (whose relatives had died) were much more likely to report positive changes in their relationship with their elderly relative during caregiving than were current caregivers. The researchers explained:

> It appears that the death of the elder alters caregivers' perspective on the caregiving experience and its impact on their relationship. It is possible that the demands and challenges of family caregiving may negatively impact the caregiver's perception of the relationship with the elder during the caregiving experience. Quite possibly, caregivers do not have enough time or mental resources to reflect on the caregiving experience and the relationship with the care recipient until after the caregiving experience is over. The grieving process can thus be seen as a healing process. (p. 24)

Many caregivers reported deriving satisfaction from helping their elder avoid

being placed in a nursing home. One caregiver said:

> Knowing that she is not in a nursing home, knowing that there is no one harming her, no one taking advantage of her. . . . It is awful. I work in that industry, and the stories are horrifying, so knowing that she is safe in her own home—those are the comforts. And knowing that I can do everything I can for her. (p. 22)

Many caregivers also described learning valuable lessons from their caregiving experience, including the importance of planning ahead for one's own aging and elder care. When asked what they hoped for in their own aging, caregivers expressed the desire to (1) not "burden" others, especially their children; (2) not burden themselves or others with unaffordable expenses; and (3) not end up in a nursing home.
Finally, researchers noted that

> [a]n alarming theme that emerged from our interviews is that family caregivers overwhelmingly seem to view aging and receiving elder care as profoundly negative, depressing processes to be avoided if at all possible. People seem to dread the

idea of aging and needing care so much they say they would rather be killed in some other way or even commit suicide. (p. 43)

As one caregiver expressed:

> I don't even want to think about it. I want to pass in my sleep of old age. It's an ugly time of life—the last few years of suffering. I would rather die in a car wreck than put anyone through what I had to go through taking care of my mother. (p. 41)

The findings of the Elder Care Study point to the need for better models of elder care that involve more support from the workplace, family/friends, and the health care system. The researchers make this point at the conclusion of their report, where they quote a doctor, who, when standing in a hospital next to an elderly person, said the following:

> Look at that bed and imagine how many more people are going to be in beds just like this in the coming years. You and I will be in those beds someday! We *have* to make things better than they are now. (p. 44)

SOURCE: Aumann et al. 2010.

elderly parents. In China, which has the largest aging population in the world, some parents have taken their adult children to court for failing to support them. Millions of Chinese families have signed a **Family Support Agreement**—a voluntary contract between older parents and adult children that specifies the details of how the adult children will provide parental care (Chou 2011).

In Singapore, the Maintenance of Parents Act of 1995 makes supporting parents a legal obligation, and parents can sue children who fail to support them. And under French law, children are obligated to honor and respect their parents, pay them an allowance, provide or fund a home for them, and stay informed of their parents' state of health and intervene if there are medical problems. France also has a law that obligates parents to leave their estates to their children. What do you think about such policies? Do you think adult children should be legally responsible for providing support for their aging parents? And should parents be legally obligated to leave their estates to their children?

WHAT do you THINK?

Family Support Agreement In China, a voluntary contract between older parents and adult children that specifies the details of how the adult children will provide parental care.

Retirement Concerns of Older Americans and the Role of Social Security

One of the concerns as we age is financial planning for retirement. Four in 10 U.S. workers are "not at all confident" (24 percent) or "not too confident" (17 percent) that they will have enough money to live comfortably in retirement (Helman et al. 2015). These concerns are not unfounded: The Center for Retirement Research estimates that about half of U.S. households will not have enough retirement income to maintain their preretirement standard of living (Munnell et al. 2015).

Types of Retirement Plans. Retirement plans include (1) traditional pensions, which are "defined benefit" plans, in which retirees receive a specified annual amount until their death; and (2) "defined contribution" plans in which workers contribute money to 401k plans or individual retirement accounts (IRAs), without any guarantee of what their future benefits will be.

Pensions, 401ks, and IRAs provide limited financial security for old age. Fewer private employers are offering guaranteed pensions, and budget shortfalls threaten the pensions and other retirement benefits for government workers. Defined contribution retirement plans (401ks and IRAs) are risky because they involve investments in the market and their value fluctuates. In the recent economic crisis, the stock market took a huge hit, producing losses that decimated the IRAs and 401ks of older Americans. Many workers who were planning to retire could no longer afford to stop working. Others who had recently retired and then lost a lot of money in the market felt compelled to reenter the labor force. Although some older workers want to retire but cannot afford to, other older workers want to continue working but are forced to retire due to health problems/disability, job cuts or displacement, or the need to care for parents or spouses (Szinovacz 2011).

The Role of Social Security in Retirement. **Social Security**, actually titled "Old Age, Survivors, Disability, and Health Insurance," is a federal insurance program established in 1935 that protects against loss of income due to retirement, disability, or death. The amount people receive from Social Security is based on how much they earned during their working history—higher lifetime earnings result in higher benefits. Benefit payments also depend on the age at which a person retires. The minimum age for receiving full benefits was 65 for many years, but in 1983, Congress phased in a gradual increase in the full retirement age from 65 to 67. People born in 1960 and later are subject to the new retirement age of 67. Retirees can claim reduced benefits as early as age 62; they receive a larger benefit if they wait until age 70 to claim benefits. In the United States, spouses are entitled to one-half of their deceased partners' benefits regardless of their own work histories and Social Security contributions.

In 2015, the average monthly Social Security benefit to retired workers was $1,328, which totals about $16,000 a year. When Social Security was established in 1935, it was not intended to be a person's sole economic support in old age; rather, it was meant to supplement other savings and assets. But Social Security is a major source of family income for most older Americans: For more than half of Americans over age 65, Social Security provides more than half of their income; and without Social Security income, nearly half of all seniors would be living in poverty (see Figure 12.6). Instead, as noted in Chapter 6, poverty rates for U.S. adults ages 65 and older are lower than for any other age group. Because Social Security payments are based on the number of years of paid work and pre-retirement earnings, women and minorities, who often earn less during their employment years, receive less in retirement benefits.

How Is Social Security Funded? Social Security is funded by workers through a payroll tax called the Federal Insurance Contributions Act (FICA) that comprises 12.4 percent of a worker's wages (6.2 percent is deducted from the worker's paycheck and 6.2 percent is paid by the employer). Self-employed workers pay the entire FICA tax. FICA taxes are paid on wages up to a certain level, or tax cap, which changes each year based on average U.S. wages. In 2015, the tax cap was $118,500, meaning that people whose wages were more than this amount paid FICA tax only on the first $118,500 of their earnings.

Another source of funding for Social Security is a tax on higher-income beneficiaries. For most recipients, Social Security benefits are not taxed, but for recipients who have

> Without Social Security income, nearly half of all seniors would be living in poverty.

Social Security Also called "Old Age, Survivors, Disability, and Health Insurance," a federal program that protects against loss of income due to retirement, disability, or death.

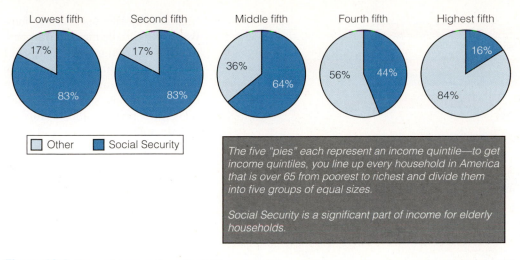

The five "pies" each represent an income quintile—to get income quintiles, you line up every household in America that is over 65 from poorest to richest and divide them into five groups of equal sizes.

Social Security is a significant part of income for elderly households.

Figure 12.6 Share of Income from Social Security of Households 65 or Over by Income Quintile, 2012
SOURCE: Social Security Administration 2014.

income from other sources that is above a certain amount, a portion of Social Security benefits is taxed to help finance Social Security.

Finally, Social Security funds are held in a trust fund and invested in securities guaranteed as to both principal and interest by the federal government. Social Security benefits are paid out of this trust fund, and each year a Board of Trustees issues a report on the financial status of the trust fund.

Is Social Security in Crisis? Many U.S. adults fear that Social Security will run out of money and will not provide them with retirement benefits (Pew Research Center 2015). A number of factors threaten the long-term ability of Social Security to meet its financial obligations to future retirees, including the retirement of the baby boomers, increasing longevity, the declining elderly support ratio (fewer workers per beneficiary), widening wage inequality (which means that more income is not subject to Social Security taxes), high rates of unemployment, and wage stagnation. In addition, the 2013 repeal of the Defense of Marriage Act (in the *United States v. Windsor*) extended federal benefits to same-sex couples. The expansion of Social Security benefits to same-sex couples is projected to have "a small but significant financial impact" on Social Security's budget (Social Security Board of Trustees 2014).

Since 1984, surpluses have been accumulating in the Social Security trust fund, creating significant reserves for the baby boomers' retirement. According to a report of the Social Security Board of Trustees (2014), in the short term, Social Security is not "broke" and there is no immediate crisis. But long-term changes to the Social Security system will be needed to ensure that the program is able to meet its financial obligations for future generations. Options for Social Security reform are discussed in the following "Strategies for Action" section.

Strategies for Action: Responding to Problems of Population Growth and Aging

Next we look at efforts to curb population growth as well as efforts to increase population in countries experiencing population decline. We end the chapter with a discussion of strategies that address social problems associated with aging.

Efforts to Curb Population Growth: Reducing Fertility

Although worldwide fertility rates have fallen significantly since the 1970s, they are still high in many less developed regions (see Figure 12.7). Approaches to reducing fertility in

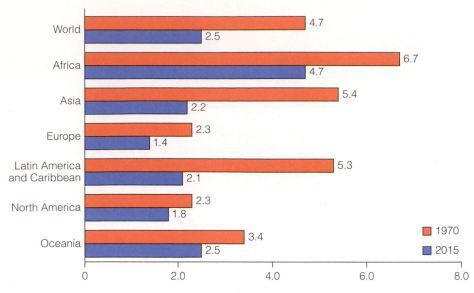

Total Fertility Rates by Region, 1970–2015

Region	1970	2015
World	4.7	2.5
Africa	6.7	4.7
Asia	5.4	2.2
Europe	2.3	1.4
Latin America and Caribbean	5.3	2.1
North America	2.3	1.8
Oceania	3.4	2.5

Figure 12.7 Total Fertility Rates by Region, 1970–2015
SOURCE: Population Reference Bureau 2014; 2015.

high-fertility regions include family planning, economic development, improving the status of women, providing access to safe abortion, government population control policies such as China's one-child policy, and voluntary childlessness.

Family Planning and Contraception. Since the 1950s, governments and nongovernmental organizations such as the International Planned Parenthood Federation have sought to lower fertility through family planning programs that provide reproductive health services and access to contraceptive information and methods. Yet there are still 225 million women in developing countries who want to delay or stop childbearing but are not using effective contraception (World Health Organization 2015). Globally, 40 percent of pregnancies are unintended; half of these pregnancies result in abortion, 13 percent end by miscarriage, and 38 percent result in an unplanned birth (Sedgh et al. 2014).

Population expert William Ryerson (2011) claims that, in most countries, lack of access to family planning is a very minor reason for not using contraception. In Nigeria, less than 1 percent of nonusers who don't want to be pregnant cite lack of access as the reason. Other reasons for not using modern contraception include (1) the belief that modern methods of contraception are dangerous, (2) male partners are opposed to using modern contraception, (3) the belief that one's religion opposes the use of contraception, and (4) the belief that God should decide how many children a woman has (Ryerson 2011). Encouraging women to use modern contraception requires providing access to affordable contraceptive methods, but providing access is not enough. Women and men need education to dispel myths about the dangers of using contraception and to understand the health and economic benefits of delayed, spaced, and limited childbearing. Involving men in family planning services is important because, although men play a central role in family planning decisions, they often do not have access to information and services that would empower them to make informed decisions about contraceptive use (Women's Studies Project 2003).

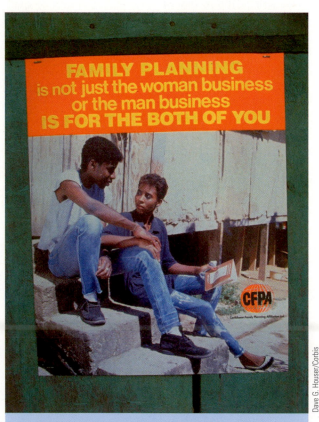

Dave G. Houser/Corbis

Recognizing that men play a crucial role in family planning decisions, family planning programs are making efforts to include men in family planning education and services.

Economic Development. Although fertility reduction can be achieved without industrialization, economic development may play an important role in slowing population growth. Families in poor countries often rely on having many children to provide enough labor and income to support the family. Economic development decreases the economic value of children and is also associated with more education for women and greater gender equality.

Economic development tends to result in improved health status of populations. Reductions in infant and child mortality are important for fertility decline, because couples no longer need to have many pregnancies to ensure that some children survive into adulthood. Finally, the more developed a country is, the more likely women are to be exposed to meanings and values that promote fertility control through their interaction in educational settings and through media and information technologies (Bongaarts and Watkins 1996).

The Status of Women: The Importance of Education and Employment. Throughout the developing world, the primary status of women is that of wife and mother. Women in developing countries traditionally have not been encouraged to seek education or employment; rather, they are encouraged to marry early and have children.

Improving the status of women by providing educational and occupational opportunities is vital to curbing population growth. Educated women are more likely to marry later, want smaller families, and use contraception. Data from many countries have shown that women with at least a secondary-level education eventually give birth to between one-third and one-half as many children as women with no formal education (Population Reference Bureau 2007).

Compared with women with less education, women with more education tend to delay marriage and exercise more control over their reproductive lives, including decisions about childbearing. In addition, "education can result in smaller family size when the education provides access to a job that offers a promising alternative to early marriage and childbearing" (Population Reference Bureau 2004, p. 18). Providing employment opportunities for women is also important to slow population growth, because high levels of female labor force participation and higher wages for women are associated with smaller family size. Primary school enrollment of both young girls and boys is also related to declines in fertility rates. In countries where primary school enrollment is widespread or nearly universal, fertility declines more rapidly because (1) schools help spread attitudes about the benefits of family planning; and (2) universal education increases the cost of having children, because parents sometimes are required to pay school fees for each child and because they lose potential labor that children could provide (Population Reference Bureau 2004). However, providing access to contraception and increasing girls' education are unlikely to slow population growth without changes in male attitudes toward family planning. Indeed, attempts to provide free primary education may increase African men's desire for more children because the costs are decreased (Frost and Dodoo 2009).

Another important component of family planning and reproductive health programs involves changing traditional male attitudes toward women. According to traditional male gender attitudes, (1) a woman's most important role is being a wife and mother, (2) it is a husband's right to have sex with his wife at his demand, and (3) it is a husband's right to refuse to use condoms and to forbid his wife to use any other form of contraception. In parts of Africa, men express their reluctance to use condoms by arguing that "you wouldn't eat a banana with the peel on or candy with the wrapper on" (Frost and Dodoo 2009). A number of programs around the world work with groups of boys and young men to change such traditional male gender attitudes (Schueller 2005).

Access to Safe Abortion. Abortion is a sensitive and controversial issue that has religious, moral, cultural, legal, political, and health implications (see also Chapter 14). Worldwide, 1 in 5 pregnancies ends in abortion, and 1 in 10 pregnancies ends in unsafe abortion, including those performed in unhygienic conditions by unskilled providers (such as traditional or religious healers and herbalists) and those that are self-induced by a woman inserting a foreign object into her uterus, ingesting toxic substances, or inflicting trauma to the abdomen (Mesce and Clifton 2011). Of the estimated 42 million abortions each year, about 47,000 girls and women die. Almost all unsafe abortions take place in developing countries (Mesce and Clifton 2011).

> Families in poor countries often rely on having many children to provide enough labor and income to support the family.

> Women with more education tend to delay marriage and exercise more control over their reproductive lives, including decisions about childbearing.

Some couples in China are eligible for an exemption from the one-child policy and are encouraged to have a second child.

AP Images/Imaginechina

More than one-quarter of the world's population lives in countries where abortion is prohibited or allowed only to save the life of the woman. Where abortion is illegal, it is very often unsafe; and where abortion is legal and widely accessible through formal health systems, it is highly safe (Barot 2011). Because the abortion rate is similar in regions with legal abortion and in regions with restrictive abortion laws, Barot (2011) notes that "legal restrictions on abortion largely do not affect whether women will get an abortion, but they can have a major impact on whether abortion takes place under safe or unsafe conditions and, therefore, whether it jeopardizes women's health and lives" (p. 25).

China's One-Child Policy. In 1979, China initiated a national family planning policy that encourages families to have only one child by imposing a monetary fine on couples that have more than one child, and by conveying public messages through signs and billboards about the benefits of having only one child. The one-child policy has been effective in reducing population growth in China, which has the highest rate of modern contraceptive use in the world (tied with the United Kingdom) (Population Reference Bureau 2015).

In 2013, China relaxed its one-child policy due to growing concerns about the rapid aging of the population, the financial costs of pensions and health care for this aging population, and the shortage of younger workers to fill jobs and help support the older population. Under new rules, couples can have a second child if either parent was an only child, and rural couples can have a second child if their first child is a girl. However, many couples who are eligible to have a second child are reluctant to do so, in part because of the financial costs of having a second child and because decades of propaganda have convinced them of the benefits of having only one child (Denyer 2015).

Voluntary Childlessness. In the United States, as in other countries of the world, the cultural norm is for women and couples to, sooner or later, want to have children. However, a small but growing segment of U.S. women and men does not want children and chooses to be childfree. In the United States, of the 19 percent of women ages 40 to 44 who are childless, half are childless by choice (Notkin 2013). Other women who are childless either are unable to have children or are waiting for love and the right relationship before motherhood. In general, childfree couples are more educated, have higher incomes, live in urban areas, are less religious, and do not adhere to traditional gender ideology (Parks 2005). In a study of 121 childless-by-choice women, the top reason women gave for not wanting children is that they simply love their life as it is (Scott 2009). Other reasons included valuing freedom and independence and not wanting to take on the responsibility. Three-quarters of the women said they "had no desire to have a child, no maternal/paternal instinct." Another reason some individuals choose not to have children is concern for overpopulation and a deep caring for the health of the planet. Yet, ironically, voluntarily childless individuals are often criticized as being selfish and individualistic, as well as less well adjusted and less nurturing (Parks 2005). In this chapter's *The Human Side* feature, one woman explains why she and her husband decided not to have children.

Linda A. Mooney

Dr. Vivian Martin Covington is among a growing segment of people who are choosing to be child-free.

I am 51 years old and my husband, Jeff, is 56 years old. We both came from stable families and have been married over 25 years. We have no children by choice.

I have never had what most of my female friends describe as a desire to have children. I never thought about having children growing up, nor did I think of names for my future children as young girls often do. I do not recall ever having much of a "maternal instinct." I babysat one time, and I called my mother to come over because I was very bored with the children and didn't really know what to do with them, nor did I want to entertain them any longer! My husband and I discussed not having children, and we had a premarital agreement, though not a written one, not to have children. We used birth control strictly to ensure that I would not get pregnant. We did not discuss having children a lot after marriage, though we did not make any decision to permanently prevent it until after I turned 40 and we had been married for about 15 years. I never had a fear of having children; I just had no

desire to have them. Neither of our families gave us any pressure to have children, though we did hear the occasional, "You two would make great parents," and the oft-said, "You'll feel differently when they are your own." Having worked in education for 25 years, I have seen firsthand that for many people, it wasn't different even with their own as many parents of the children I taught didn't seem to give them the care they needed. I trusted my instincts and my feelings, and I have no regrets. It was helpful for both of our families that older siblings did have children, thus there were grandchildren for our parents. I was raised Catholic and my husband was raised Baptist, though neither of us practice these religions. I had no issue reconciling my decision with any religious beliefs or tenets.

When we are asked if we have children, we both respond, "No, we do not and by choice." I learned this after saying no numerous times and seeing people awkwardly change the subject as they always seem to wonder if perhaps we couldn't have children. I do not feel that we have an "empty marriage," as is often said about marriages without children. My husband and I have many friends with whom we socialize. We have worked hard on these relationships over the years. As our friends had children, their lives changed drastically, and we had to respect that. We went to a lot of birthday parties for their small children, bought a lot of 16th birthday presents,

and gave monetary gifts for their high school and college graduations. However, now that their children are grown, we are free to travel and socialize as much with these individuals as we did before they had children. This was important to us and we worked hard to stay connected. If we wanted our friends to respect our choice to not have children, we knew we had to respect their choice to have them as well.

There is one typical societal comment that I have grown to dislike. It is the well-meaning "Who will take care of you when you get older?" Having to care for my mother for two weeks after complications from hip surgery, I couldn't help but wonder what she would do if she had only my older brother to reply upon. Even though he is her child, he would not care for her. There are no guarantees that if you have children, they will care for you. I would take that even further and say they shouldn't have to take care of you. Not all people are comfortable being caregivers, especially to loved ones. And most people have their own families for whom to care, and they have to work to make ends meet. My husband and I have done as much preplanning as we can for our future and our future healthcare needs.

Both my husband and I feel that we have given back to society in many ways and that we have helped other children in our lives. We give to scholarships to help young people get ahead, we donate to organizations that assist young people, and I have worked in education for 25 years, with high school– and college-age students. We live a full life, with friends and family and no regrets.

*Written for this text by Dr. Vivian Martin Covington, College of Education, East Carolina University. Used by permission.

Efforts to Increase Population

Whereas some countries are struggling to slow population growth, others are challenged with maintaining or even increasing their populations. A survey of governmental policies concerning population growth in 176 countries found that 75 countries have policies to lower population growth, 50 have policies to increase population, 29 have policies to maintain current population levels, and 22 have no population policies at all (Weeks 2015).

To increase population, Singapore announced in 2013 that it would offer a package of incentives for having children, including government-paid time off for adoption and paternity leave, funding for fertility treatment, and state-funded assistance with medical care for children. In Australia and Japan, the government has paid women a monetary bonus for having babies. Aside from monetary rewards, many countries encourage childbearing by implementing policies designed to help women combine child rearing with employment. For example, as noted in Chapters 5 and 7, many European countries have generous family leave policies and universal child care. Another way to increase population is to increase immigration. A number of European countries, for example, have eased restrictions on immigration as a way to gain population.

One strategy for encouraging childbearing in European countries with low fertility rates is to provide work–family supports to make it easier for women to combine childbearing with employment. If the United States offered more generous work–family benefits, such as paid parenting leave and government-supported child care, would the U.S. birthrate increase? Would such policies affect the number of children you would want to have?

Combating Ageism and Age Discrimination in the Workplace

One strategy to combat ageism involves incorporating positive views of aging in educational lessons, beginning in preschool, and in other forms of media. Such lessons should teach about aging as a normal part of life, rather than something to fear or be embarrassed about. Younger people should have opportunities to learn from the wisdom, experience, and life perspective of older individuals.

In 1967, Congress passed the Age Discrimination in Employment Act (ADEA), which was designed to ensure continued employment for people between the ages of 40 and 65. In 1986, the upper age limit was removed, making mandatory retirement illegal, but there are exceptions to this rule. In the private sector, certain executives such as partners in accounting firms can be subject to mandatory retirement. Firefighters, law enforcement officers, military personnel, pilots, air traffic controllers, and, in some states, judges are also subject to mandatory retirement. Mandatory retirement is justified by the argument that certain occupations require high levels of physical and/or mental ability, and that age-related decline could jeopardize the safety or well-being of the public and/or the worker. Because mandatory retirement policies are based on a fixed age (which varies according to the profession and by state for judges) and not on an evaluation of the worker's abilities, some say that mandatory retirement is a form of age discrimination.

Under ADEA, it is illegal to discriminate against people because of their age with respect to hiring, firing, promotion, layoff, compensation, benefits, job assignments, and training. Age discrimination is difficult to prove. Nevertheless, thousands of age discrimination cases are filed annually with the Equal Employment Opportunity Commission (EEOC).

A 1994 law ending mandatory retirement for university professors past age 70 has enabled professors to delay retirement past 70 or not retire at all. The number of professors aged 65 and older more than doubled between 2000 and 2011 (Fendrich 2014). The increase in older faculty has led some critics to point out that older faculty should retire because (1) they are not as up-to-date and energetic in their teaching; (2) their higher salaries, health care costs, and retirement benefits contribute to the rising cost of college tuition; and (3) they block opportunities for younger Ph.D.s who aspire to a teaching career (Fendrich 2014). Do you think professors should have to retire after a certain age? Why or why not?

Options for Reforming Social Security

Social Security is the most solvent part of the U.S. government. It is funded through its own separate tax and interest from its trust fund to pay beneficiaries. No other government program or agency is fully funded. Social Security is also a very efficient program. Although it collects taxes from more than 90 percent of the workforce and sends benefits to more than 50 million Americans, Social Security spends less than one cent of every dollar on administration.

Nevertheless, as discussed earlier in this chapter, unless changes are made in years to come, Social Security won't have enough funds to provide payouts to future retirees. Options for reforming Social Security include cutting Social Security benefits, increasing Social Security revenue, and expanding Social Security benefits. The following outlines these options (Edwards et al. 2012; Morrissey 2011).

Cut Social Security Benefits. Some reform proposals call for cuts in Social Security benefits, which would create a significant financial burden on households that depend on Social Security. Aside from simply reducing the amount of benefits paid to recipients, another way to cut benefits is to increase the retirement age. In 1983, the retirement age (the age at which a worker could collect full Social Security retirement benefits) was raised from 65 to 67 to be phased in over 23 years. Some policy makers suggest raising the retirement age even further to 68, 69, or even 70. However, due to the link between higher socioeconomic status and longer life expectancy, raising retirement age imposes the greatest burden on low earners who have lower life expectancies and therefore fewer years to collect Social Security. In addition, older workers already face employment discrimination, and finding or keeping employment is difficult for seniors. Finally, the older workers become, the more likely they will experience health problems and hence file disability claims.

Increase Social Security Revenue. Instead of cutting benefits, many advocates for the elderly suggest raising revenues. One option for increasing Social Security revenue is simply to raise the tax that funds Social Security (the payroll or FICA tax) from its current rate of 12.4 percent (6.2 percent paid by employer; 6.2 percent paid by employee) to a higher rate. To offset gains in life expectancy, Social Security taxes increased 19 times, from 1 percent between 1937 and 1949 to 6.2 percent in 1990 (paid by both employer and employee). As of this writing (August 2015), Social Security taxes have not increased since 1990—the longest period without an increase to the payroll tax. In 2011 and 2012, the workers' share was cut to 4.2 percent under a "payroll tax holiday" economic stimulus measure. The downside to raising the payroll tax is that it could result in fewer jobs, as employers would have a higher financial burden. Paying a higher payroll tax would also disproportionately burden lower-wage earners.

Another option is to raise or even eliminate the tax cap, so that more earnings are taxed, providing more funds to the Social Security program. This would affect higher-income earners, who currently earn more than the tax cap, but only pay payroll taxes on the first $118,500 (the tax cap in 2015) of their earnings.

Finally, policies that increase employment and wages of all workers would lead to more revenue for Social Security. Because health insurance costs are deducted from taxable income, policies to reduce health care costs would also indirectly increase Social Security funding.

Tim Boyle/Getty Images

Many Americans oppose attempts to privatize Social Security, fearing that privatization would lead to lower benefits.

Expand Social Security Benefits. Options for increasing Social Security benefits include raising the minimum benefit amount, offering unemployed parents who are taking care of their children wage credits, and increasing the benefits for the very old (85 years and older). Another proposal involves restoring a student benefit (which existed between 1965 and 1985) so that children of the retired, deceased, or disabled can continue to receive benefits until age 22 if attending college or vocational school.

Not surprisingly, proposals to cut Social Security benefits or raise retirement age for receiving benefits are vehemently opposed by seniors and advocacy groups for older adults. Raising the tax cap would go far to ensure the future financial viability of Social Security, but this option is generally not supported by the wealthy segment of the population who would bear the burden of higher taxes. Until legislators enact changes to prevent the long-term Social Security deficit, there is little chance of seeing Social Security benefits increased.

Understanding Problems of Population Growth and Aging

What can we conclude from our analysis of population growth and aging? First, although fertility rates have declined significantly in recent years and although many countries are experiencing a decline in their fertility rates, world population will continue to grow for several decades. This growth will occur largely in developing regions. Finance columnist Paul Farrell (2009) says that population growth is "the key variable in every economic equation . . . impacting every other major issue facing world economies . . . from peak oil to global warming . . . from foreign policy to nuclear threats . . . from religion to science . . . everything." Given the problems associated with population growth, such as environmental problems and resource depletion, global insecurity, poverty and unemployment, and poor maternal and infant health, most governments recognize the value of controlling population size and supporting family planning programs. However, efforts to control population must go beyond providing safe, effective, and affordable methods of birth control. Slowing population growth necessitates interventions that change the cultural and structural bases for high fertility rates. Reducing fertility necessitates improving the status of women so that women have more power to control their choices regarding contraception and reproduction and have more life options other than being a wife and mother. Addressing problems associated with population growth also requires the willingness of wealthier countries to commit funds to providing reproductive health care to women, improving the health of populations, and providing universal education for people throughout the world.

As fertility rates decline and life expectancy increases, the United States and other countries are experiencing population aging. Many areas of social life are affected by the aging of the population, and more attention is being given to concerns related to older people. What it means to be old is culturally defined. Unfortunately, many cultural meanings associated with the aged are negative stereotypes. Prejudice toward and discrimination against older people constitutes ageism, but unlike other "isms"—racism, sexism, heterosexism—ageism is more socially acceptable and will affect all of us who reach a certain age.

Although older individuals struggle against ageism, families and governments face challenges of meeting the needs of seniors. Caring for elders has traditionally been an important function of the family, but social, cultural, and economic changes have made it more challenging for families to fulfill this function. Social Security plays an important role in supporting many older retirees, but due to projections of a future shortfall, some reforms to Social Security are necessary. What type of reform measures legislators enact continues to be hotly debated. Until such reform measures are in place, the future of Social Security and the well-being of older Americans is uncertain.

Like other social problems, slowing population growth and meeting the needs of aging populations require political will and leadership. Given all the pressing concerns in the world, control of population may not seem like a priority. But as finance columnist Paul Farrell (2009) warns, "Population is the core problem that, unless confronted and dealt with, will render all solutions to all other problems irrelevant. Population is the one variable in an economic equation that impacts, aggravates, irritates, and accelerates all other problems." And if taking care of older populations is not a priority now, those of us who are not yet "old" will suffer the consequences in the future, if—or when—we enter our "golden years."

Chapter Review

- **How long did it take for the world's population to reach 1 billion? How long did it take for it to reach 6 billion?**
  It took thousands of years for the world's population to reach 1 billion, and just another 200 years for the population to grow from 1 billion to 6 billion.

- **Is the world population still increasing? Is it decreasing? Or is it remaining stable?**
  World population is still growing. It is projected to grow from 7.2 billion in 2013, to 8.1 billion in 2025, 9.6 billion in 2050, and 10.9 billion by 2100.

- **Where is most of the world's population growth occurring?**
  Most world population growth is in developing countries, primarily in Africa and Asia.

- **Between 2013 and 2050, how is the percentage of older individuals (ages 60 and over) expected to change globally?**
  Globally, the population aged 60 and older is the fastest growing age group. Between 2013 and 2050, the percentage of older individuals (ages 60 and over) in the world population is expected to nearly double from 12 percent to 21 percent.

- **What factors are contributing to population aging?**
  Population aging is a function of lowered fertility as well as increased longevity. In the United States, population aging is also occurring because the baby boomers are reaching their senior years.

- **What is the demographic transition theory?**
  The demographic transition theory of population describes how industrialization and economic development affect population growth by influencing birth and death rates. According to this theory, traditional agricultural societies have both high birth rates and high death rates. As a society becomes industrialized and urbanized, improved sanitation, health, and education lead to a decline in mortality. The increased survival rate of infants and children along with the declining economic value of children leads to a decline in birthrates. The demographic transition theory is a generalized model that does not apply to all countries and does not account for population change due to HIV/AIDS, war, migration, and changes in gender roles and equality.

- **Many countries are experiencing below-replacement fertility (fewer than 2.1 children born to each woman). Why are some countries concerned about their low fertility?**
  In countries with below-replacement fertility, there are or will be fewer workers to support a growing number of elderly retirees and to maintain a productive economy.

- **What kinds of environmental problems are associated with population growth?**
  Population growth places increased demands on natural resources, such as forests, water, cropland, and oil, and results in increased waste and pollution. Although population growth is a contributing factor in environmental problems, patterns of production and consumption are at least as important in influencing the effects of population on the environment.

- **Why is population growth considered a threat to global security?**
  In developing countries, rapid population growth results in a "youth bulge"—a high proportion of 15- to 29-year-olds relative to the adult population. The combination of a youth bulge with other characteristics of rapidly growing populations—such as resource scarcity, high unemployment rates, poverty, and rapid urbanization—sets the stage for civil unrest, war, and terrorism, because large groups of unemployed young people resort to violence in an attempt to improve their living conditions.

- **How is ageism different from other "isms" (racism, sexism, heterosexism)?**
  Ageism is more widely accepted than other isms and, unlike other isms, everyone is vulnerable to experiencing ageism if they live long enough.

- **How important is Social Security income for senior citizens in the United States?**
  For more than half of Americans over age 65, Social Security provides more than half of their income; and without Social Security income, nearly half of all seniors would be living in poverty.

- **Is Social Security in crisis?**
  In the short term, Social Security is not "broke" and there is no immediate crisis. But long-term changes to the Social Security system will be needed to ensure that the program is able to meet its financial obligations to future retirees.

- **What is the "sandwich generation" and the "club sandwich generation"?**

  The "sandwich generation" refers to adults who care for their aging parents while also taking care of their own children—they are "sandwiched" in between taking care of both parents and children. The "club sandwich generation" includes adults (often in their 50s or 60s) who are caring for aging parents, adult children, and grandchildren or adults (usually in their 30s and 40s) who are caring for their own young children, aging parents, and grandparents.

- **Efforts to curb population growth include what strategies?**

  Efforts to curb population growth include strategies to reduce fertility by providing access to family planning services, involving men in family planning, implementing a one-child policy as in China, and improving the status of women by providing educational and employment opportunities. Achievements in economic development and health are also associated with reductions in fertility.

- **What are three general options for reforming Social Security?**

  Options for reforming Social Security include strategies for increasing Social Security revenue, cutting benefits, and expanding Social Security benefits.

Test Yourself

1. In 2013, world population was 7.2 billion. In 2050, world population is projected to be ___ billion.
   a. 6.5
   b. 7.4
   c. 8.8
   d. 9.6

2. In 2015, how many countries had below-replacement fertility rates?
   a. None
   b. 5
   c. 28
   d. More than 80

3. What would happen if every country in the world achieved below-replacement fertility rates?
   a. Population growth would stop and world population would remain stable.
   b. World population would immediately begin to decline.
   c. World population would continue to grow for several decades.
   d. World population would decline, but then go up again.

4. For more than half of Americans over age 65, Social Security provides more than half of their income.
   a. True
   b. False

5. Conflict theorists argue that food shortages result primarily from overpopulation of the planet.
   a. True
   b. False

6. Pronatalism is a cultural value that promotes which of the following?
   a. Looking young
   b. Abstaining from sex until one is married
   c. Having children
   d. Urban living

7. According to the "Ageism Survey," jokes and birthday cards that poke fun at the elderly represent a form of ageism.
   a. True
   b. False

8. What is the number one cause of death for dogs and cats in the United States?
   a. Getting hit by a car
   b. Shelter euthanasia
   c. Cancer
   d. Eating something poisonous

9. U.S. women with advanced education are more likely than women with less education to voluntarily choose to have no children.
   a. True
   b. False

10. In 1983, the retirement age (the age at which a worker could collect full Social Security retirement benefits) was raised from 65 to 70 to be phased in over 23 years.
    a. True
    b. False

Answers: 1. D; 2. D; 3. C; 4. A; 5. B; 6. C; 7. A; 8. B; 9. A; 10. B

Key Terms

ageism 403
ageism by invisibility 403
baby boomers 397
club sandwich generation 405
demographic transition theory 398

elderly support ratio 397
environmental footprint 402
Family Support Agreement 407
population momentum 396
pronatalism 399

replacement-level fertility 396
sandwich generation 405
Social Security 408
total fertility rates 395

"If grief can be a doorway to love, then let us all weep for the world we are breaking apart so we can love it back to wholeness again."

—ROBIN WALL KIMMERER

13

Environmental Problems

Learning Objectives

After studying this chapter, you will be able to . . .

1 Give an example of an environmental problem that crosses international borders, explain why environmental migrants are an international concern, and discuss how transnational corporations and free trade agreements contribute to environmental problems.

2 Give examples of how structural functionalism, conflict theory, and symbolic interactionism can be applied to our understanding of environmental problems.

3 Identify the sources of the world's energy and briefly describe the following environmental problems: depletion of natural resources; air, land, and water pollution, global warming and climate change; threats to biodiversity; light pollution; and environmental illness.

4 Explain how population growth, industrialization and economic development, individualism, consumerism, and militarism contribute to environmental problems.

5 Describe various strategies and efforts to restore and protect the environment.

6 Summarize the causes of environmental problems, and identify the challenges that must be overcome in order to restore and protect the environment.

The 2015 heat wave in India that took at least 2,500 lives is believed to be related to global warming and climate change.

Anadolu Agency/Getty Images

DURING A HEAT WAVE in May 2015, it got so hot in India, with temperatures rising above 110 degrees Fahrenheit, that pavement melted in the capital city of New Delhi (Di Liberto 2015). The 2015 India heat wave lasted more than two weeks and was responsible for the deaths of at least 2,500 people. Many victims were elders or day laborers who suffered from sunstroke or dehydration. Although hundreds of mainly poor people die from heat in India each year, the number of deaths from the 2015 heat wave was the second highest in India's history, and the fifth highest in recorded global history.

During the heat wave, many Indians did what they could to stay cool and hydrated: they drank a lot of liquids, consumed salted buttermilk and raw onions (both thought to be hydrating), stayed in the shade, and covered their heads. But many farmers and construction workers risked being out in the heat to make their living. "Either we have to work, putting our lives under threat, or we go without food," said one farmer. A construction worker, earning about $3.10 for a day's work, said, "If I don't work due to heat, how will my family survive?" (quoted in Farooq and Daigle 2015, n.p.).

The 2015 Indian heat wave is linked to global warming and climate change—one of the most challenging environmental problems of our time. In this chapter, we discuss the causes and consequences of global warming and climate change and other environmental problems that threaten the lives and well-being of people, plants, and animals all over the world—today and in future generations. After examining how globalization affects environmental problems, we view environmental issues through the lens of structural functionalism, conflict theory, and symbolic interactionism. We then present an overview of major environmental problems, examining their social causes and exploring strategies to reduce or alleviate them.

Toxic chemicals travel thousands of miles from the Southern Hemisphere to the Arctic, where they have been found in the breast milk of Inuit women.

Bryan & Cherry Alexander Photography/Alamy

The Global Context: Globalization and the Environment

In looking at environmental problems from a global perspective, we see that many environmental problems have causes and consequences that cross international borders. For example, global warming and climate change affect the entire planet.

Other environmental problems also can extend far beyond their source to affect distant regions and even the entire planet. For example, toxic chemicals (such as polychlorinated biphenyls [PCBs]) from the Southern Hemisphere have been found in the Arctic. In as few as five days, chemicals from the tropics can evaporate from the soil, ride the winds thousands of miles north, condense in the cold air, and fall on the Arctic in the form of toxic snow or rain (French 2000). This phenomenon was discovered in the mid-1980s, when scientists found high levels of PCBs in the breast milk of Inuit women in the Canadian Arctic region.

Bioinvasion

Another cross-border environmental problem is **bioinvasion**: the intentional or accidental introduction of species in regions where they are not native. Bioinvasion is largely a product of the growth of global trade and tourism (Chafe 2005). Zebra mussels, native to eastern Europe and western Asia, were first discovered in North America in 1988. Zebra mussels were introduced into the Great Lakes in the ballast of a single cargo ship traveling from the Black Sea. This invasive species clogs water supply pipes in industrial facilities, power plants, and public water supply plants and water treatment facilities (U.S. Geological Survey 2015). The invasive kudzu plant, native to east Asia, was brought to the United states in the early to mid-1900s to control soil erosion and feed livestock. Known as the "vine that ate the South," kudzu now covers more than 7 million acres of U.S. land and costs more than $3 million a year in management efforts and in damage to railway lines, telephone poles, and other infrastructure (Grider 2015). You might be surprised to learn that the domestic cat is considered among the world's 100 worst invasive species. Native to northeast Africa, cats have spread to every part of the world and are responsible for the decline and extinction of many species of birds (Global Invasive Species Database 2015).

Environmental Migrants

Environmental migrants—people who flee from their home region due to environmental problems that threaten their survival or livelihood—are also referred to as "environmental refugees," "climate refugees," "environmentally displaced persons," and "climate migrants." Environmental problems that cause people to migrate include floods, hurricanes, droughts, and other natural disasters which displaced an average of 27 million people each year between 2008 and 2013 (Yonetani 2014). Higher levels of displacement are expected in coming decades, in part because of an increase in extreme weather events related to global climate change.

In some cases, environmental migrants find refuge within their home country, but others migrate internationally. Although environmental migrants are sometimes referred to as "refugees," persons forced to leave their countries because of environmental degradation or the effects of climate change are not, under international law, considered refugees. Under current law, governments are under no obligation to treat an "environmental refugee" differently than any other migrant seeking admission into a country. In 2014, a family from the Polynesian island nation of Tuvalu became the world's first environmental migrants to gain residency in another country (New Zealand). The family claimed that the rising sea level resulting from climate change is causing salt water to pollute Tuvala's drinking water (Nuwer 2014). However, this family was not granted residency in New Zealand on the basis of climate change, but rather on humanitarian grounds and, in part, because the family had three generations of relatives living in the country. Another family from the small island nation of Kiribati sought residency in New Zealand on the basis of climate change threatening their island home, but their request was denied (Longeray and Caron 2015). As of this writing, there are no international policies to protect environmental migrants, and countries are facing the need to develop immigration policies dealing with environmental migration. The Nansen Initiative, proposed by Norway and Switzerland in 2011, seeks to develop an international agreement that addresses the needs of environmental migrants (Ferris 2015).

bioinvasion The intentional or accidental introduction of plant, animal, insect, and other species in regions where they are not native.

environmental migrants People who flee from their home region due to environmental problems that threaten their survival or livelihood.

Environmental Problems and the Growth of Transnational Corporations and Free Trade Agreements

As discussed in Chapter 7, the world's economy is dominated by transnational corporations, many of which have established factories and other operations in developing countries where labor and environmental laws are lax. Transnational corporations have been implicated in environmentally destructive activities—from mining and cutting timber to dumping toxic waste.

The World Trade Organization (WTO) and free trade agreements such as the North American Free Trade Agreement (NAFTA) and the Free Trade Area of the Americas (FTAA) have powers that weaken the ability of governments to protect natural resources or to implement environmental legislation (Bruno and Karliner 2002).

Under NAFTA's Chapter 11 provisions, corporations can challenge local and state environmental policies, federal-controlled substances regulations, and court rulings if such regulatory measures and government actions negatively affect the corporation's profits. Any country that decides, for example, to ban the export of raw logs as a means of conserving its forests or, as another example, to ban the use of carcinogenic pesticides, can be charged under the WTO by member states on behalf of their corporations for obstructing the free flow of trade and investment. A secret tribunal of trade officials would then decide whether these laws were "trade restrictive" under the WTO rules and should therefore be struck down. Once the secret tribunal issues its edict, no appeal is possible. The convicted country is obligated to change its laws or face the prospect of perpetual trade sanctions (Clarke 2002, p. 44). For example, in the late 1990s, Ethyl, a U.S. chemical company, used NAFTA rules to challenge Canada's decision to ban the gasoline additive methylcyclopentadienyl manganese tricarbonyl (MMT), which is believed to have harmful effects on human health. Ethyl won the suit, and Canada paid $13 million in damages and legal fees to Ethyl and reversed the ban on MMT (Public Citizen 2005).

As this book goes to press, the United States and the European Union (EU) are negotiating the Trans-Atlantic Trade and Investment Partnership (TTIP)—a free trade agreement that is focused on getting rid of trade barriers by establishing "regulatory cooperation" between the United States and EU member states. The EU has much more stringent laws and regulations regarding the use of pesticides and other chemicals; 82 pesticides are banned in the EU that are used in the United States (Smith et al. 2015). Under the TTIP, such bans could be viewed as "trade irritants," and the EU could be forced to lift or weaken such bans.

Sociological Theories of Environmental Problems

Next we explore how the three main sociological theories—structural functionalism, conflict theory, and symbolic interactionism—can be applied to our understanding of environmental problems.

Structural-Functionalist Perspective

Structural functionalism views social systems (e.g., families, workplaces, societies) as composed of different parts that work together to keep the whole system functioning. Likewise, humans are part of a larger **ecosystem**—which consists of all the organisms living in a particular area, as well as all the nonliving, physical components of the environment—such as air, water, soil, and sunlight—that interact to keep the whole ecosystem functioning. Each living and nonliving part of the ecosystem plays a vital role in maintaining the whole; disrupt or eliminate one element of the ecosystem, and every other part could be affected.

ecosystem A biological environment consisting of all the organisms living in a particular area, as well as all the nonliving, physical components of the environment such as air, water, soil, and sunlight, that interact to keep the whole ecosystem functioning.

WHAT do you THINK?

Pope Francis (2015) said, "There can be no renewal of our relationship with nature without a renewal of humanity itself." What does that statement mean to you?

The ban on trans fats has resulted in higher demand for palm oil, which means more forests are cleared for palm oil plantations.

> The capitalistic pursuit of profit encourages making money from industry regardless of the damage done to the environment.

Structural functionalism focuses on how changes in one aspect of the social system affect other aspects of society. For example, as croplands become scarce or degraded, as forests shrink, and as marine life dwindles, millions of people who make their living from these natural resources must find alternative livelihoods.

The structural-functionalist perspective raises our awareness of latent dysfunctions—negative consequences of social actions that are unintended and not always recognized. For example, while the U.S. ban on trans fats in foods (see Chapter 2) is expected to reduce heart disease and improve health, the ban also means increased demand for palm oil (the leading substitute for trans fat), which means more deforestation as forests are cleared to make way for palm oil plantations (Worland 2015).

Conflict Perspective

The conflict perspective focuses on the role that wealth, power, and the pursuit of profit plays in environmental problems and solutions. The capitalistic pursuit of profit encourages making money from industry regardless of the damage done to the environment. To maximize sales, manufacturers design products intended to become obsolete. As a result of this **planned obsolescence**, consumers continually get rid of used products and purchase replacements. **Perceived obsolescence**—the *perception* that a product is obsolete—is a marketing tool used to convince consumers to replace certain items even though the items are still functional. Fashion is a prime example as consumers are encouraged to buy the latest trends in clothing style every season, even though their current clothing may still be in good condition. Both planned and perceived obsolescence benefit industry profits, but at the expense of the environment, which must sustain the constant production and absorb ever-increasing amounts of waste.

The conflict perspective is also concerned with **environmental injustice** (also known as *environmental racism*)—the tendency for marginalized populations and communities to disproportionately experience adversity because of environmental problems. For example, although developing countries have emitted far more greenhouse gases (which cause global warming and climate change), the effects of global climate change are

planned obsolescence The manufacturing of products that are intended to become inoperative or outdated in a fairly short period of time.

perceived obsolescence The perception that a product is obsolete; used as a marketing tool to convince consumers to replace certain items even though the items are still functional.

environmental injustice Also known as *environmental racism*, the tendency for marginalized populations and communities to disproportionately experience adversity due to environmental problems.

The Climate Deception Dossiers

For nearly three decades, fossil fuel companies have deliberately spread climate misinformation and blocked climate action. A research report called *The Climate Deception Dossiers* documents the fossil fuel's climate deception campaign (Mulvey and Shulman 2015).

Sample and Methods

The research method used involved collecting and analyzing secondary data. The sample consisted of 85 internal company and trade association documents, totaling 336 pages. The documents had been leaked to the public, revealed through lawsuits, or obtained using the federal Freedom of Information Act.

Selected Findings

The *Climate Deception Dossiers* report organizes the findings into seven case studies, five of which we summarize below.

Case Study 1: Dr. Wei-Hock Soon's Research.

Documents show that Dr. Wei-Hock Soon, a scientist (with a degree in aerospace engineering, not climatology) at the Harvard-Smithsonian Center for Astrophysics, received more than $1.2 million in research funding

between 2001 and 2012 from fossil fuel interests including ExxonMobil, the American Petroleum Institute (API), the Charles Koch Foundation, and Southern Company, a large electric utility in Atlanta that generates most of its power from coal. Dr. Soon's industry-funded "research" found that global warming is due to natural variations in the sun's energy, and not due to greenhouse gases caused by burning fossil fuels. "Soon sought to portray his research as independent. . . .The fact that he was paid by fossil fuel interests was never publicly disclosed in his published work or testimony to lawmakers" (p. 7).

Case Study 2: American Petroleum Institute's 1998 Memo.

The American Petroleum Institute (API) is the country's largest oil trade association whose members include ExxonMobil, Shell, BP, ConocoPhillips, and Chevron. A 1998 internal API memo titled "Global Climate Science Communications Plan," described a plan aimed at confusing and misinforming the public about climate change. "According to the memo . . . 'victory' would be achieved for the campaign when

'average citizens' and the media were convinced of 'uncertainties' in climate science" and when "recognition of uncertainties becomes part of the 'conventional wisdom'" (pp. 9–10). The memo outlined tactics to "manufacture uncertainty" in climate science, including (1) recruiting and funding scientists to publish research that creates doubt about prevailing climate science, and (2) disseminating materials to K–12 science teachers that emphasize the "uncertainties" of climate science.

Case Study 3: Western States Petroleum Association's Deception Campaign.

Western States Petroleum Association (WSPA) is a top lobbyist for the oil industry and the oldest U.S. petroleum trade association. In 2014, WSPA president Catherine Reheis-Boyd gave a slide presentation to a business group demonstrating the organization's strategy to orchestrate and fund a network of 16 different fake grassroots groups—also known as front groups or Astroturf groups—to falsely represent citizen opposition to policies and proposals on climate change and renewable energy that threaten fossil fuel industry

expected to be felt most severely by poor, developing nations (Intergovernmental Panel on Climate Change 2014). In the United States, polluting industries and industrial and waste facilities are often located in minority communities. More than half (56 percent) of people living within 1.8 miles of a commercial hazardous waste site are people of color (Bullard et al. 2007). Rates of poverty are also higher among households located near hazardous waste facilities. In North Carolina, hog industries—and the associated environmental and health risks associated with hog waste—tend to be located in communities with large black populations, low voter registration, and low incomes (Edwards and Driscoll 2009). One factor that contributes to environmental injustice is known as **"Not in My Backyard,"** or **NIMBY**, which refers to opposition by local residents to a proposed new development in their community. Residents who have wealth and political clout may be more successful at stopping developments, such as oil or gas drilling sites or hazardous waste storage facilities, "in their backyard," which are then located in more impoverished areas where residents have little political influence.

The conflict perspective is also concerned with how industries use their power and wealth to influence politicians' environmental and energy policies. Editors at *The Nation* explain:

> "Follow the money" is a basic rule of American politics. Find out who funds a given candidate's campaign—and, equally as important, who *isn't* funding

"Not in My Backyard" Opposition by local residents to a proposed new development in their community, also known as NIMBY.

profits. The presentation, which was subsequently leaked to *Bloomberg Businessweek*, referred to front groups as "campaigns and coalitions" with names such as Fed Up at the Pump, Californians for Energy Independence, Californians against Higher Oil Taxes, and Oregonians for Sound Fuel Policy. "Through these groups, the industry attempted to create the impression of a consumer backlash against climate legislation. . . . Concerns are raised by groups of purportedly everyday citizens when, in fact, they are disguised messages from fossil fuel companies seeking to undermine climate legislation" (pp. 13–14).

Case Study 4: Forged Letters from the Coal Industry to Lawmakers.

In 2009, Congress was debating the American Clean Energy and Security Act of 2009 (often known as the Waxman-Markey climate bill), which proposed a federal carbon emissions reduction plan. In an attempt to influence lawmakers to vote against this bill, the American Coalition for Clean Coal Electricity—a front group for the coal industry—hired a public relations firm that sent members of Congress forged letters purporting to be from nonprofit groups including the NAACP, the American Association of University Women, and the American Legion. These forged letters misrepresented the nonprofits' positions on the proposed legislation. For example, Representative Tom Perriello (D-Va.) received a forged letter opposing the climate legislation that was supposedly from Creciendo Juntos, a nonprofit Latino organization based in his district. The letter read:

> My organization, Creciendo Juntos, represents minorities in your district. . . . We ask you to use your important position to help protect minorities and other consumers in your district from higher electricity bills. Please don't vote to force cost increases on us, especially in this volatile economy. (p. 16)

A congressional investigation found that 13 fraudulent letters had been sent to members of Congress, who found out the letters were forged after they voted on the climate bill.

Case Study 5: The Global Climate Coalition's 1995 Primer on Climate Change Science.

In 2009, a memo was leaked to the *New York Times* that "presents the strongest evidence yet that major fossil fuel companies knew the reality of human-caused climate change and its implications even as they continued their deceptive practices" (p. 25). A fossil fuel company scientist wrote the memo, "Predicting Future Climate Change: A Primer," in 1995 for the benefit of the Global Climate Coalition—a fossil fuel industry trade association aimed at opposing mandatory reductions in carbon emissions. Its members included BP, Chevron, ExxonMobil, Shell, and others. The 17-page primer assessed what was known about climate science and unequivocally stated that "the scientific basis for the Greenhouse Effect and the potential impact of human emissions of greenhouse gases such as CO_2 on climate *is well established and cannot be denied*" (p. 25, emphasis in original).

Conclusion

The documents presented in *The Climate Deception Dossiers* show that fossil fuel companies have intentionally spread climate disinformation and blocked actions to reduce greenhouse gases for decades, and they continue to do so today. Furthermore, "fossil fuel company leaders knew that their products were harmful to people and the planet but still chose to actively deceive the public and deny this harm" (p. 2).

SOURCE: Mulvey and Shulman 2015.

---

it—and you can make a pretty good guess as to which interests and policies the candidate will support or oppose once in office. (Editors 2015)

In the 2012 elections, oil, gas, and coal industries contributed at least $91 million to presidential and congressional candidates. In the same year, the federal government used $18.5 billion in tax money to subsidize the fossil fuel industry (Editors 2015). As this book goes to press, *The Nation* has challenged 2016 presidential candidates to neither solicit nor accept campaign contributions from any gas, oil, or coal company.

> Do you think that members of Congress, who make and vote on energy policies, should be allowed to own stocks in oil or gas companies? Would you feel the same way about members of Congress who invested in renewable energy industries, such as solar and wind power?

Just as the tobacco industry in the mid-twentieth century used wealth and power to convince the public that tobacco was harmless, so the fossil fuel industry has created a **climate denial machine**—a well-funded and aggressive misinformation campaign run by the fossil fuel industry and its allies that attacks and discredits climate science. This chapter's *Social Problems Research Up Close* feature documents how the fossil fuel industry has deceived the public on the issue of global warming and climate change.

WHAT do you THINK?

climate denial machine A well-funded and aggressive misinformation campaign run by the fossil fuel industry and its allies that involves attacking and discrediting climate science, scientists, and scientific institutions.

Symbolic Interactionist Perspective

The symbolic interactionist perspective focuses on how meanings, labels, and definitions learned through interaction and through the media affect environmental problems. The words we use reflect and influence how we relate to the environment. Native American Robin Wall Kimmerer (2013) explains that in her native language (Potawatomi), "we speak of the land as *emingoyak*: that which has been given to us. In English, we speak of the land as 'natural resources' or 'ecosystem services,' as if the lives of other beings were our property" (p. 383). The indigenous language that conveys that land is a gift teaches people to be grateful for the Earth and its bounty, to respect and care for the Earth, and to use the Earth's gifts (such as plants that provide food and water that sustains life) for the benefit of all rather than for individual profit or gain.

Large corporations and industries that are environmentally damaging commonly use marketing and public relations strategies to portray their corporation, industry, or products as environmentally friendly—a practice known as **greenwashing**. For example, some brands of household items such as toilet paper and dish soap are advertised as "green," "all natural," or "earth-friendly." Marketing companies know that "green" sells, as consumers are becoming more eco-minded. But how valid are the environmental claims made on the labels of the products we buy? The Environmental Working Group's *Guide to Healthy Cleaning* warns:

> On a cleaning product, the word "natural" can mean anything or nothing at all—there is no regulation of the word's use. Some manufacturers use the term to mean that some or all of the ingredients come from plants or minerals rather than petroleum, but they rarely disclose how much or little of those ingredients is present. The term "natural" can mislead consumers to think that a product is safer or more environmentally friendly than it actually is. ("Decoding the Labels" 2015)

Greenwashing is commonly used by public relations firms that specialize in damage control for clients whose reputations and profits have been hurt by poor environmental practices. For example, coal is associated with the devastation of communities through the mining practice of mountaintop removal, and burning coal is the biggest contributor to pollution that causes global warming. The coal industry has spent enormous sums to convince the public that coal is clean. The "clean coal" campaign has invited widespread criticism from environmentalists: "Saying coal is clean is like talking about healthy cigarettes. There's no such thing as clean coal" (Beinecke 2009).

Corporations also engage in **pinkwashing**—the practice of using the color pink and pink ribbons and other marketing strategies that suggest a company is helping to fight breast cancer, even when the company may be manufacturing or selling products linked to cancer. The cosmetic company Estée Lauder was the first company to use the pink ribbon symbol of fighting breast cancer in its marketing, followed by Avon and Revlon. Yet all three cosmetics companies produce and sell products with ingredients that include hormone disruptors and other suspected carcinogens (Lunden 2013). In another blatant example of pinkwashing, Susan G. Komen, a large nonprofit organization dedicated to fighting breast cancer, teamed up with an oilfield corporation, Baker Hughes, to distribute pink drill bits to oilfields worldwide (Jaggar 2014). The drill bits are shipped in containers along with information packets about breast cancer risk factors and screening tips. Baker Hughes' corporation drills for fossil fuels and conducts fracking—a process for extracting oil and gas using chemicals that include possible carcinogens (discussed later in this chapter).

> These pink drill bits deliver not only barrels of oil, but also good PR and money: Baker Hughes gets to claim it cares about women's health, and Komen will receive a check from the Houston-based company for $100,000. The campaign has even come up with a cute tagline: "Doing their bit for a cure." (Jaggar 2014)

greenwashing The way in which environmentally and socially damaging companies portray their corporate image and products as being "environmentally friendly" or socially responsible.

pinkwashing The practice of using the color pink and pink ribbons to indicate a company is helping to fight breast cancer, even when the company may be using chemicals linked to cancer.

TABLE 13.1 Critical Questions to Ask before You Buy Pink

1.	Does any money from this purchase go to support breast cancer programs? How much?
2.	What organization will get the money? What will they do with the funds, and how do these programs turn the tide of the breast cancer epidemic?
3.	Is there a "cap" on the amount the company will donate? Has this maximum donation already been met? Can you tell?
4.	Does this purchase put you or someone you love at risk for exposure to toxins linked to breast cancer? What is the company doing to ensure that its products are not contributing to the breast cancer epidemic?

SOURCE: Breast Cancer Action, San Francisco.

A wide range of companies has adopted the pink ribbon marketing strategy, including Clorox, Dansko, Evian, Ford, KFC, and American Airlines. Does the presence of a pink ribbon on a product or service influence your purchasing behavior? Why or why not? *Think Before You Pink*, a project of the national organization Breast Cancer Action, encourages consumers to ask critical questions before they buy products with the pink ribbon symbol (see Table 13.1).

The labels used to discuss the environment can be controlled by those in power. In Florida, employees of the Florida Department of Environmental Protection were ordered not to use the terms *global warming* or *climate change* in any official communications, email, or reports. This unwritten policy went into effect after Governor Rick Scott, who has gone on record as being unconvinced that climate change is a scientific fact, went into office in 2011 (Korten 2015). Pennsylvania and North Carolina have also removed the term *climate change* from official state environmental websites (Williams 2015).

Environmental Problems: An Overview

Environmental problems include depletion of natural resources; air, land, and water pollution; problems associated with fracking; global warming and climate change; threats to biodiversity; light pollution; and environmental illness. Because many of these environmental problems are related to the ways that humans produce and consume energy, we begin this section with an overview of global energy use.

Energy Use Worldwide

Most of the world's energy comes from fossil fuels, which include petroleum (or oil), coal, and natural gas (see Figure 13.1). The major environmental problems facing the world today—air, land, and water pollution; destruction of habitats; biodiversity loss; global warming and climate change; and environmental illness—are linked to the production and use of fossil fuels.

For example, the proposed Keystone XL pipeline would carry 900,000 barrels a day of tar sands oil from Canada to the Gulf, crossing Montana, South Dakota, Nebraska, Kansas, Oklahoma, and Texas. **Tar sands** are large, naturally occurring deposits of sand, clay, water, and a dense form of petroleum that looks like tar. **Tar sands oil** has been referred to as the world's dirtiest oil. Converting tar sands into liquid fuel requires energy and generates high levels of greenhouse gases (which cause global warming and climate change), and the process also leaves behind large amounts of toxic waste. The mining of tar sands requires large amounts of water and involves destruction of forests and wetlands, which disrupts wildlife habitats. And transporting the tar sands oil through a pipeline involves risk of leakage that could affect the safety

tar sands Large, naturally occurring deposits of sand, clay, water, and a dense form of petroleum that looks like tar.

tar sands oil Oil that results from converting tar sands into liquid fuel. It is known as the world's dirtiest oil because producing it requires energy, generates high levels of greenhouse gases (which cause global warming and climate change), and leaves behind large amounts of toxic waste.

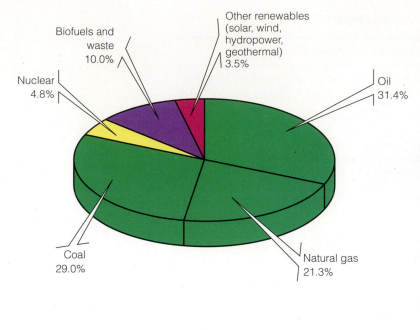

■ Fossil fuels 81.7%

Figure 13.1 World Energy Supply by Source, 2012
SOURCE: International Energy Agency 2014.

of drinking water (Swift et al. 2011). NASA scientist James Hansen said if the Canadian tar sands are fully developed, it is "essentially game over for the climate" (quoted in Elk 2011).

An alternative to fossil fuel energy is renewable energy, also called "green energy" or "clean energy," which includes solar, wind, geothermal (heat from the earth), **biomass** (dung, wood, crop residues, and charcoal), and hydroelectric power, which involves generating electricity from water moving through a turbine in a dam. Although hydroelectric power is nonpolluting and inexpensive and is considered to be a clean and renewable form of energy, it is criticized for disrupting natural habitats and displacing human settlements.

Nuclear Power. As of 2015, there were 437 operable nuclear power reactors in the world, with another 66 under construction (World Nuclear Association 2015). The United States has more nuclear reactors than any other country, with 99 operable U.S. nuclear reactors, most of which are located in nuclear power plants east of the Mississippi River (see Figure 13.2).

Nuclear power is associated with a number of problems related to radioactive nuclear waste (discussed later in this chapter). The safety of nuclear power is highly questionable, as equipment failures and natural disasters can result in leaks of harmful radiation. Following a 9.0 magnitude earthquake and tsunami in 2011, a series of equipment failures and nuclear meltdowns occurred at the Fukushima Daiichi nuclear power plant in

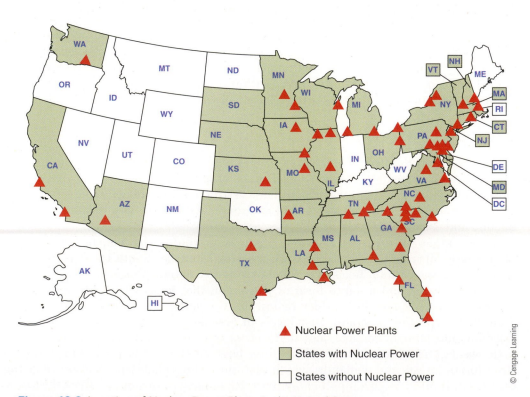

▲ Nuclear Power Plants

■ States with Nuclear Power

□ States without Nuclear Power

© Cengage Learning

biomass Material derived from plants and animals, such as dung, wood, crop residues, and charcoal; used as cooking and heating fuel.

Figure 13.2 Location of Nuclear Power Plants in the United States

Japan, resulting in a mandatory evacuation for the 200,000 people living within 18 miles of the plant. Those living outside this radius were advised to stay indoors, close doors and windows, turn off the air conditioner, cover their mouths with masks, and avoid drinking tap water. The areas surrounding the Fukushima nuclear power plant could remain uninhabitable for decades due to high radiation.

The worst nuclear power plant accident in the United States occurred in 1979, when the cooling system at the Three Mile Island reactor failed, causing a partial meltdown. A small amount of radioactive gas was vented from the building to prevent an explosion. Cleanup of the site cost $1 billion and took 14 years (Clemmitt 2011). Former Nuclear Regulatory Commission chairman Gregory B. Jaczko said he believes that all U.S. nuclear power reactors have safety problems that cannot be fixed and that they should all be phased out of operation (Wald 2013). The Indian Point nuclear power plant, located 30 miles from New York City, has had numerous safety problems and is situated near an earthquake fault line. A nuclear disaster at Indian Point would threaten the entire population of New York City and its surrounding metropolitan area. Emergency evacuation would be impossible (Nader 2013).

Depletion of Natural Resources: Our Growing Environmental Footprint

Humans have used more of the Earth's natural resources since 1950 than in the million years preceding 1950 (Lamm 2006). The demands that humanity makes on the Earth's natural resources are known as the **environmental footprint.** Beginning in about 1970, humanity's environmental footprint has exceeded the Earth's biocapacity—its capacity to produce useful resources such as water and crops, and to absorb waste, such as carbon dioxide (CO_2) emissions. Globally, we exceed the Earth's biocapacity by more than 50 percent, meaning that we need 1.5 planet Earths to support our consumption. The environmental footprint of high-income countries that consume more goods and produce more carbon dioxide is about five times more than that of low-income countries. If everyone in the world lived like an average American, we would need 3.9 planets to support us (World Wildlife Fund [WWF] 2014).

We need 1.5 planet earths to sustain current levels of global consumption.

NPeter/Shutterstock.com

Every year the Global Footprint Network identifies **Earth Overshoot Day**—the approximate date on which humanity's annual demand on the planet's resources exceeds what our planet can renew in a year. In 2015, Earth Overshoot Day was August 13, meaning that in about 7.5 months, we used as much natural resources as our planet can renew in a year.

The depletion of the Earth's resources, such as forests, water, minerals, and fossil fuels, is due to consumption patterns and population growth (see also Chapter 12). More than one-third of the world's population lives in areas that experience severe water scarcity for at least one month each year (WWF 2014). Water supplies around the world are dwindling, while the demand for water continues to increase because of population growth, industrialization, rising living standards, and changing diets that include more food products that require larger amounts of water to produce, such as milk, eggs, chicken, and beef. With 92 percent of water use going to agriculture, water shortages threaten food production and supply (WWF 2014).

The world's forests are also being depleted due to the expansion of agricultural land, human settlements, wood harvesting, and road building. The result is **deforestation**—the

environmental footprint The demands that humanity makes on the earth's natural resources.

Earth Overshoot Day The approximate date on which humanity's annual demand on the planet's resources exceeds what our planet can renew in a year.

deforestation The conversion of forestland to nonforestland.

conversion of forestland to nonforestland. Global forest cover has been reduced by half of what it was 8,000 years ago (Gardner 2005). Between 2000 and 2010, the world's forests shrank by an area roughly the size of France (Normander 2011). Deforestation displaces people and wild species from their habitats; soil erosion caused by deforestation can cause severe flooding; and, as we explain later in this chapter, deforestation contributes to global warming. Deforestation also contributes to **desertification**—the degradation of semiarid land, which results in the expansion of desert land that is unusable for agriculture. As more land turns into desert, populations can no longer sustain a livelihood on the land, and so they migrate to urban areas or other countries, contributing to social and political instability.

> One species of life on Earth goes extinct every three hours.

Threats to Biodiversity

An estimated 8.7 million species of life live on Earth (some scientists believe the number is much higher), 1.4 million of which have been named and cataloged (Staudinger et al. 2012). This enormous diversity of life, known as **biodiversity,** provides food, medicines, fibers, and fuel; purifies air and freshwater; pollinates crops and vegetation; and makes soils fertile.

Between 1970 and 2010, the number of vertebrate species on the earth has declined by about half (WWF 2014). One species of life on Earth goes extinct every three hours (Leahy 2009). As shown in Table 13.2, more than 20,000 species worldwide are threatened with extinction. The primary threats to biodiversity are habitat loss and degradation, exploitation through hunting and fishing, and climate change (WWF 2014).

TABLE 13.2 Threatened Species Worldwide, 2015

Category	Number of Threatened* Species
Mammals	1,200
Birds	1,373
Amphibians	1,961
Reptiles	931
Fishes (bony)	2,248
Insects	1,011
Crustaceans	727
Molluscs	1,949
Plants	10,896
Other	488
Total	22,784

*Threatened species include those classified as critically endangered, endangered, and vulnerable.
SOURCE: IUCN 2015.

Air Pollution

Transportation vehicles, fuel combustion, industrial processes (such as burning coal and processing minerals from mining), and solid waste disposal have contributed to the growing levels of outdoor air pollutants, including carbon monoxide, sulfur dioxide, arsenic, nitrogen dioxide, mercury, dioxins, and lead. Leaded aviation gasoline is one of the few fuels in the United States to still contain lead, and it's the single-largest source of lead emissions in the country (Kessler 2013).

Indoor air pollution is a problem for the approximately 3 billion people, mostly in low- and middle-income countries, which use biomass fuels—wood, charcoal, crop residues, and dung—for cooking and heating their homes (World Health Organization 2014b). Biomass is typically burned on open fires or stoves without chimneys, creating smoke and indoor air pollution. Exposure is particularly high among women and children, who spend the most time near the domestic hearth or stove.

Both outdoor and indoor air pollution are linked to health problems—including heart disease, lung cancer, stroke, chronic obstructive pulmonary emphysema, and lower respiratory infections—and cause about 7 million premature deaths each year, or one in eight of total global deaths. This figure is double that of previous estimates and "confirms that air pollution is now the world's largest single environmental health risk" (World Health Organization 2014b). In the United States, 4 in 10 people live areas with unhealthy levels of either ozone (smog) or particulate pollution (soot) (American Lung Association 2015).

Even in affluent countries, much air pollution is invisible to the eye and exists where we least expect it—in our homes, schools, workplaces, and public buildings. Gas stoves, used by a third of U.S. households, emit nitrogen, carbon monoxide, and formaldehyde, which can exacerbate respiratory problems, especially when no vented hood is used (Nicole 2014).

desertification The degradation of semiarid land, which results in the expansion of desert land that is unusable for agriculture.

biodiversity The diversity of living organisms on Earth.

Other sources of indoor air pollution include lead dust (from old lead-based paint); secondhand tobacco smoke; by-products of combustion (e.g., carbon monoxide) from furnaces, fireplaces, heaters, and dryers; and other common household, personal, and commercial products. Carpeting emits more than a dozen toxic chemicals; mattresses, sofas, and pillows emit formaldehyde and fire retardants; pressed wood found in kitchen cabinets and furniture emits formaldehyde; and dry-cleaned clothing emits perchloroethylene. Air fresheners, deodorizers, and disinfectants emit the pesticide paradichlorobenzene. Potentially harmful organic solvents are present in numerous office supplies, including glue, correction fluid, printing ink, carbonless paper, and felt-tip markers (Rogers 2002).

James Hardy/Altopress/Newscom

Air pollution is linked to a number of health problems, including respiratory problems such as emphysema, and contributes to one in eight deaths worldwide.

Destruction of the Ozone Layer. The ozone layer of the Earth's atmosphere protects life from the sun's harmful ultraviolet rays. Yet the ozone layer has been weakened by the use of certain chemicals, particularly chlorofluorocarbons (CFCs), used in refrigerators, air conditioners, and spray cans. The ozone hole over the Antarctic in 2014 was 9.3 million square miles—about the size of North America (NASA 2014). The depletion of the ozone layer allows hazardous levels of ultraviolet rays to reach the Earth's surface and is linked to increases in skin cancer and cataracts, weakened immune systems, reduced crop yields, damage to ocean ecosystems and reduced fishing yields, and adverse effects on animals. Despite measures that have ended production of CFCs, the ozone is not expected to recover significantly for about another decade because CFCs already in the atmosphere remain for 40 to 100 years.

Acid Rain. Air pollutants, such as sulfur dioxide and nitrogen oxide, mix with precipitation to form **acid rain**. Polluted rain, snow, and fog contaminate crops, forests, lakes, and rivers. As a result of the effects of acid rain, all the fish have died in a third of the lakes in New York's Adirondack Mountains (Blatt 2005). Because winds carry pollutants in the air, industrial pollution in the Midwest falls back to the Earth as acid rain on southeast Canada and the northeast New England states. In China, most of the electricity comes from burning coal, which creates sulfur dioxide pollution and acid rain that falls on one-third of China, damaging lakes, forests, and crops (Woodward 2007). Acid rain also deteriorates the surfaces of buildings, statues, and automotive coatings.

Global Warming and Climate Change

Global warming refers to the increasing average global temperature of the Earth's atmosphere, water, and land. The world's average global temperature in 2014 was the highest on record, marking the 38th consecutive year that the average global temperature was above the 20th-century average (Tompkins and DeConcini 2015).

Causes of Global Warming. Most scientists believe that global warming has resulted from the increase in global atmospheric concentrations of greenhouse gases since industrialization began. **Greenhouse gases**—primarily CO_2, methane, and nitrous oxide—accumulate in the atmosphere and act like the glass in a greenhouse, holding heat from the sun close to the Earth. Global increases in greenhouse gases are caused primarily by humans, particularly the use of fossil fuels, but also population growth, economic activity, lifestyle, technology, and land use patterns. For example, deforestation—the mass

acid rain The mixture of precipitation with air pollutants, such as sulfur dioxide and nitrogen oxide.

global warming The increasing average temperature of the Earth's atmosphere, water, and land, caused mainly by the accumulation of various gases (greenhouse gases) that collect in the atmosphere.

greenhouse gases Gases (primarily carbon dioxide, methane, and nitrous oxide) that accumulate in the atmosphere and act like the glass in a greenhouse, holding heat from the sun close to the Earth.

> Even if greenhouse gases are stabilized, global air temperature and sea level are expected to continue to rise for hundreds of years.

cutting of forests—contributes to increasing levels of carbon dioxide in the atmosphere because trees and other plant life use carbon dioxide and release oxygen into the air. As forests are cut down or burned, there are fewer trees to absorb the carbon dioxide.

China and the United States account for more than 40 percent of the world's carbon emissions from fossil fuel burning. Although China produces 50 percent more total carbon emissions than the United States, the United States has the highest per capita emissions in the world—three times that of China's (Elliot 2015).

Even if greenhouse gases are stabilized, global air temperature and sea level are expected to continue to rise for hundreds of years. That is because global warming that has already occurred contributes to further warming of the planet—a process known as a *positive feedback loop*. For example, the melting of ice and snow due to global warming exposes more land and ocean area, which absorbs more heat than ice and snow, further warming the planet.

Climate Deniers. The Intergovernmental Panel on Climate Change (2014), an international team of scientists from countries around the world, says that the warming of the climate system is "unequivocal:" the atmosphere and ocean have warmed, snow and ice have diminished, sea level has risen, and the concentrations of greenhouse gases have increased. It is often cited that 97 percent of scientists concur that human-caused global warming is a fact. This 97 percent figure may be an understatement: 99.99 percent of scientists who published scientific articles on climate change in peer-reviewed journals in 2013–2014 agree that human-caused global warming is occurring (Hill 2015).

But the fossil fuel industry's aggressive misinformation campaign attacking and discrediting climate science has been effective in swaying public views of climate change: Despite the overwhelming scientific consensus that human activity causes global warming (Cook et al. 2013), 4 in 10 U.S. adults believe that global warming is due more to natural changes in the environment than to human activity (Saad 2015). **Climate deniers**—people who do not accept the scientific consensus that human-caused global warming and climate change are scientific facts—include citizens and lawmakers. Senator James Inhofe (R-Okla.) believes that global warming is a conspiracy among scientists and even wrote a book called *The Greatest Hoax: How the Global Warming Conspiracy Threatens Your Future* (Inhofe 2012).

In 2015, the Senate, including climate denier Inhofe, voted that "climate change is real and is not a hoax" (Barron-Lopez 2015). But about half the Senate, including Inhofe, voted *against* an amendment that stated that "human activity significantly contributes to climate change." Nevertheless, Senator Bernie Sanders (I-Vt.), said the climate votes were a "step forward" for Republicans. He noted, "I think what is exciting is that today we saw for the first time—a number, a minority—but some Republicans going onboard and saying that climate change is real and it's caused by human activity" (quoted in Barron-Lopez 2015).

Climate deniers insist that global warming and climate change are not real.

Jeff Malet Photography/Newscom

climate deniers People who do not accept the scientific consensus that human-caused global warming and climate change are scientific facts.

Effects of Global Warming and Climate Change. Between 2030 and 2050, climate change is expected to cause 250,000 deaths per year, from malnutrition (due to crop failure), malaria (from increased mosquitos), diarrhea (from increased flooding and water-borne diseases), and heat stress (World Health Organization 2014a). The effects of global warming and climate change also include the following:

• *Melting Ice and Sea-Level Rise.* Global warming causes sea level rise because of (1) thermal expansion of sea water (water expands as it warms) and (2) the melting of glaciers and the Greenland and polar ice sheets. Between 1901 and 2010, average global sea level rose by about 7.5 inches (Intergovernmental Panel on Climate Change [IPCC] 2015). Sea level is expected to continue to rise over the 21st century and for many centuries beyond 2100, with the amount of rise dependent on future greenhouse gas emissions. Over the past 3 million years, global sea levels rose about 20 feet on multiple occasions due to small increases in global average temperature (1 to 2 degrees Celsius), which suggests that small amounts of global warming can have significant effects on sea level (Dutton et al. 2015). Research by climate scientist James Hansen and his colleagues suggests that sea level could rise 10 feet in the next 50 to 85 years (Gertz 2015). Rising sea levels could make coastal areas uninhabitable. As sea levels rise, some island countries, as well as some barrier islands off the U.S. coast, are likely to disappear, and low-lying coastal areas will become increasingly vulnerable to storm surges and flooding. Eight of the world's 10 largest cities are near a coast, and nearly 40 percent of the U.S. population lives in coastal areas (Blunden and Arndt 2015).

- *Flooding and Spread of Disease.* Increased heavy rains and flooding caused by global warming contribute to increases in drownings and in human exposure to insect- and water-related diseases, such as malaria and cholera. Flooding, for example, provides fertile breeding grounds for mosquitoes that carry a variety of diseases, including encephalitis, dengue fever, yellow fever, West Nile virus, and malaria (Knoell 2007). With the warming of the planet, mosquitoes are now living in areas in which they previously were not found, placing more people at risk of acquiring one of the diseases carried by the insect.

Anna Henly/Getty Images

Disappearing sea ice makes finding food difficult for polar bears, which are under threat of extinction if greenhouse gases continue at their current rate throughout the 21st century.

- *Threat of Species Extinction.* At least 19 species extinctions have been attributed to climate change (Staudinger et al. 2012). If greenhouse gas emissions continue at the same rate as in 2015, up to one in six species will be under threat of extinction (Urban 2015). For example, disappearing sea ice makes finding food difficult for polar bears, which are under threat of extinction if greenhouse gases continue at their current rate throughout the 21st century (U.S. Fish and Wildlife 2015).

- *Extreme Weather: Hurricanes, Droughts, and Heat Waves.* Rising temperatures are causing drought in some parts of the world and too much rain in other parts. Warmer tropical ocean temperatures can cause more intense hurricanes (Chafe 2006). With rising temperatures, an increase in the number, intensity, and duration of heat waves is expected, with the accompanying adverse health effects (Di Liberto 2015). Droughts, as well as floods, can be devastating to crops and food supplies.

- *Forest Fires.* Another effect of global warming is an increase in the number and size of forest fires (Westerling et al. 2006). Warmer temperatures dry out brush and trees, creating ideal conditions for fires to spread. Warmer weather also allows bark beetles to breed more frequently, which leads to more trees dying from beetle infestation (Staudinger et al. 2012). Dead trees become dry and increase risk of fire. Global warming also means that spring comes earlier, making the fire season longer.

- *Effects on Recreation.* Winter sports and recreation, such as skiing and snowboarding, are threatened by decreased and unreliable snowfall, causing high economic losses for winter recreation businesses, not to speak of frustration for winter sports enthusiasts. In coastal areas, beach recreation is also projected to suffer due to coastal erosion caused by sea-level rise and increased storms association with climate change (Staudinger et al. 2012).

The increase in the number and size of forest fires in recent years has been linked to global warming.

WHAT do you THINK?

A Pew Research (2013) poll of people in 39 countries found that only 4 in 10 Americans view climate change as a major threat, making Americans among the least concerned about global climate change. Why do you think this is so?

Land Pollution

About 30 percent of the world's surface is land, which provides soil to grow the food we eat. Increasingly, humans are polluting the land with nuclear waste, solid waste, and pesticides. In 2015, 1,321 hazardous waste sites in the United States (also called Superfund sites) were on the National Priorities List (Environmental Protection Agency [EPA] 2015b).

Nuclear Waste. Nuclear waste, resulting from both nuclear weapons production and nuclear reactors or power plants, contains radioactive plutonium, a substance linked to cancer and genetic defects. Radioactive wastes and contaminated materials from nuclear power remain potentially harmful to human and other life for 250,000 years (Nader 2013). The question of how to safely dispose of nuclear waste has not been resolved.

Nuclear plants have tons of radioactive spent fuel, with some of it sealed in casks (Vedantam 2005b). Because of inadequate oversight and gaps in safety procedures, radioactive spent fuel is missing or unaccounted for at some U.S. nuclear power plants, which raises serious safety concerns (Vedantam 2005a). Accidents at nuclear power plants, such as the 2011 accident at Fukushima, Japan, and the potential for nuclear reactors to be targeted by terrorists add to the actual and potential dangers of nuclear power plants. Political activist Ralph Nader (2013) commented:

> With all the technological advancements in energy efficiency, solar, wind, and other renewable energy sources, surely there are better and more efficient ways to meet our electricity needs without burdening future generations with deadly waste products and risking the radioactive contamination of entire regions should anything go wrong. . . . It is telling that Wall Street, which rarely considers the consequences of gambling on a risk, will not finance the construction of a nuclear

plant without a full loan guarantee from the U.S. government. Nuclear power is also uninsurable in the private insurance market. The Price-Anderson Act of 1957 requires taxpayers to cover almost all the cost if a meltdown should occur. (n.p.)

Recognizing the hazards of nuclear power plants and their waste, some countries are phasing out or banning nuclear energy. At the 2015 World Uranium Symposium in Quebec, 300 delegates from 20 countries called for stopping all uranium mining (needed for nuclear power), closing down all nuclear power plants, and banning nuclear weapons (Brown 2015). Uranium mining is damaging to indigenous peoples who live near the mines, and harmful to mine workers and the local environment. Delegates at the symposium emphasized that nuclear power is not cost-effective, practical, or safe.

Solid Waste. In the United States, the amount of household garbage each person produced each day increased from 2.7 pounds in 1960 to 4.4 pounds in 2013 (EPA 2015a; see Figure 13.3). This figure does not include mining, agricultural, and industrial waste; demolition and construction wastes; junked autos; or obsolete equipment wastes. About a third of household garbage is composted or recycled; the rest is dumped in landfills.

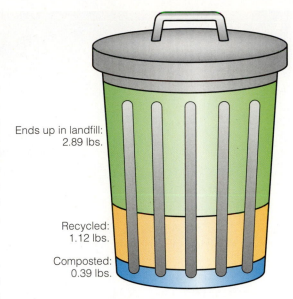

Ends up in landfill: 2.89 lbs.

Recycled: 1.12 lbs.

Composted: 0.39 lbs.

Garbage generated per person: 4.4 lbs.

Figure 13.3 Where Does Our Trash Go?
SOURCE: Data from EPA 2015b.

Most plastic shopping bags, commonly used in grocery and retail stores, are made from petroleum and require a lot of fossil fuel energy to produce. They also contain toxic chemicals and end up in landfills, where it takes 1,000 years for them to degrade, or in oceans, where marine life can choke or starve after swallowing them. Dozens of countries, including the European Union, South Africa, China, and India, have either bans or taxes on plastic shopping bags. In 2007, San Francisco became the first U.S. city to ban plastic bags from supermarkets and chain pharmacies. Since then, a number of plastic bag bans or taxes have been implemented across the country. In 2014, California became the first state to ban plastic shopping bags, followed by Hawaii in 2015 (exceptions include for medical and sanitary purposes). Would you support a ban or a tax on plastic bags in your community or state?

WHAT do you THINK?

Solid waste includes discarded electrical appliances and electronic equipment, known as **e-waste**. Most discarded electronics end up in landfills, incinerators, or hazardous waste exports. The main concern about dumping e-waste in landfills is that hazardous substances, such as lead, cadmium, barium, mercury, PCBs, and polyvinyl chloride, can leach out of e-waste and contaminate the soil and groundwater.

Pesticides. Pesticides are used worldwide for crops and gardens; outdoor mosquito control; the care of lawns, parks, and golf courses; and indoor pest control. Pesticides contaminate food, water, and air and can be absorbed through the skin, swallowed, or inhaled. Many common pesticides are considered potential carcinogens and neurotoxins (Blatt 2005). Even when a pesticide is found to be hazardous and is banned in the United States, other countries from which we import food may continue to use it. In an analysis of more than 5,000 food samples, pesticide residues were detected in 43 percent of the domestic samples and 34 percent of the imported samples (Food and Drug Administration 2015). Pesticides also contaminate our groundwater supplies.

Water Pollution

Water pollution is largely caused by fertilizers and other chemicals used in intensive agriculture, industrial production, mining and untreated urban runoff and wastewater (World Water Assessment Program 2015). Other water pollutants include plastics, pesticides,

e-waste Discarded electrical appliances and electronic equipment.

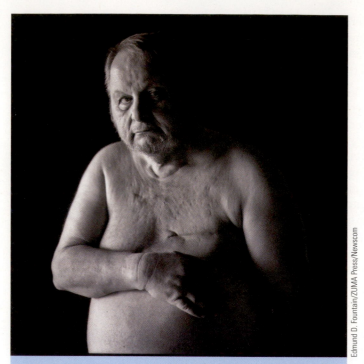

After being exposed to contaminated water at Camp Lejeune, some marines developed a rare form of male breast cancer.

vehicle exhaust, acid rain, oil spills, sewage, and military waste. Water pollution is most severe in developing countries, where untreated sewage is commonly dumped directly into rivers, lakes, and seas that are also used for drinking and bathing.

In 2010, the Deepwater Horizon oil drilling rig exploded, killing 11 and injuring 17 oil rig workers, and opening a gusher that released oil and methane gas into the Gulf of Mexico. By the time the oil well was sealed months later, over 4 million barrels of oil had spilled into the Gulf, creating what may be the worst environmental disaster in U.S. history.

In the United States, one indicator of water pollution is the thousands of fish advisories issued by the EPA that warn against the consumption of certain fish caught in local waters because of contamination with pollutants such as mercury and dioxin. The EPA advises women who may become pregnant, pregnant women, nursing mothers, and young children to avoid eating certain fish altogether (swordfish, shark, king mackerel, and tilefish) because of the high levels of mercury (EPA 2004).

Pollutants in drinking water can cause serious health problems and even death. At Camp Lejeune—a Marine Corps base in Onslow County, North Carolina—as many as 1 million people were exposed to water contaminated with trichloroethylene (TCE), an industrial degreasing solvent, and perchloroethylene (PCE), a dry-cleaning agent from 1957 until 1987 (Nazaryan 2014). Exposure to these chemicals has been linked to a number of health problems, including kidney, liver, and lung damage, as well as cancer, childhood leukemia, miscarriage, and birth defects.

Water pollution also affects the health and survival of fish and other marine life. In the Gulf of Mexico, as well as in the Chesapeake Bay and Lake Erie, there are areas known as "dead zones" that—due to pollution runoff from agricultural uses of fertilizer—have oxygen levels so low they cannot support life (Scavia 2011).

Another growing concern surrounds the increasing amount of plastic pollution found in the world's oceans. There is not a single cubic meter of ocean water that does not contain some plastic (Hill 2014). Much of this plastic is difficult to see because of its small size. Microplastics, which are fragments of plastic that measure less than 5 millimeters, come from the degradation of plastic products and from small pellets that are used to make plastic products such as bottles, bags, and packaging. Some of these pellets are accidentally spilled into the environment and have been found on beaches and in ocean water around the world (Takada 2013). Plastics tend to absorb toxins that are harmful to marine organisms and potentially harmful to humans who eat seafood (Seltenrich 2015).

Problems Associated with Fracking

In recent years, there has been increasing public concern about the effects of hydraulic fracturing, or "**fracking**"—a process used in natural gas production that involves injecting at high pressure a mixture of water, sand, and chemicals into deep underground wells to break apart shale rock and release gas. As of early 2015, tens of thousands of wells were being fracked in 22 U.S. states, with 5 more states likely to soon follow (Hirji and Song 2015).

Fracking poses significant environmental and health problems (Inglis and Rumpler 2015; Ridlington and Rumpler 2013). It involves the use of chemicals that contaminate land, water, and air. Oil and gas companies do not have to disclose to the federal government the chemicals they use in fracking, but scientists have analyzed fracking water and found hundreds of chemicals, including those that are carcinogens and that are toxic to

fracking Hydraulic fracturing, commonly referred to as "fracking," involves injecting a mixture of water, sand, and chemicals into drilled wells to crack shale rock and release natural gas into the well.

human reproductive, neurological, respiratory, and gastrointestinal systems. Fracking extracts materials from underground that can be equally or more toxic than the hydraulic fracturing fluid and include radioactive material, arsenic, and lead. Billions of gallons of toxic fracking wastewater produced each year must be stored, transported, and disposed of. But leaks, overflows, and illegal dumping of wastewater allow chemicals to flow into local land and water supplies.

Fracking requires massive amounts of water, which is problematic in drought-stricken areas such as California, Oregon, and Nevada. Fracking operations have also triggered earthquakes in at least five states (Inglis and Rumpler 2015). Pets and farm animals raised near fracking sites, as well as wildlife, have become sick and died, apparently as a result of contact with and/or consumption of contaminated water from fracking sites (Bamberger and Oswald 2014). Beef cattle and other animals raised for human consumption are not tested for contamination from fracking chemicals, raising concerns about whether humans are eating contaminated animal products. People who live or work near fracking sites are at increased risk for health problems (Inglis and Rumpler 2015). When individuals seek compensation from oil and gas companies for harm done to their health and/or property due to fracking, the company typically settles out of court. As a condition of the

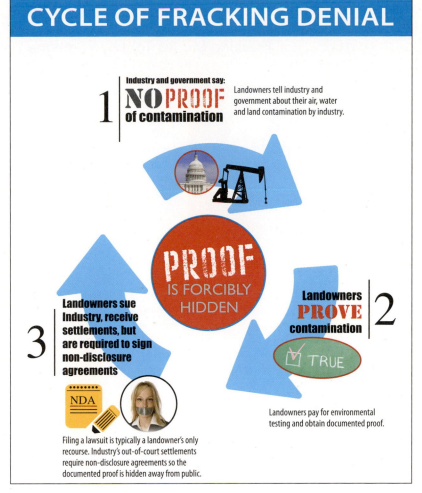

Figure 13.4 Cycle of Fracking Denial
SOURCE: Earthworksaction.org

settlement, companies prohibit the individual from telling their story or sharing any documentation, such as water test results, of the harm they incurred from fracking (see Figure 13.4). In this chapter's *The Human Side* feature, a community organizer, who is not bound by any gag order, shares stories of individuals who are gagged.

A loophole inserted into the 2005 Energy Policy Act (known as the "Halliburton Loophole") exempts fracking from sections of the Safe Drinking Water Act (1974), the Clean Air Act (1970), and the Clean Water Act (1972). The FRAC Act is a bill that, if passed, would repeal the Halliburton Loophole and require the gas industry to disclose the chemicals it uses in fracking (Wassmer 2015).

Because of health and environmental concerns, some countries—including France, Wales, and Scotland—have banned fracking. It has also been banned in New York, Massachusetts, and Vermont. Hundreds of towns, cities, and counties across the United States have also passed bans or moratoriums on fracking. The Permanent Peoples' Tribunal (PPT), an internationally recognized coalition of human rights lawyers and academics, is planning hearings in 2017 to determine whether fracking violates human rights. The PPT will hear testimony to decide whether to indict certain countries for failing to adequately uphold universal human rights by allowing fracking (Wassmer 2015).

Chemicals, Carcinogens, and Health Problems

About 3 million tons of toxic chemicals are released into the environment each year (Pimentel et al. 2007). Chemicals in the environment enter our bodies via the food and

Fracking Stories Told by Someone Who Isn't Gagged

Sharon Wilson

This is the view from the Ruggiero's kitchen window: an oil rig 200 feet from their home.

In 1996, I purchased my dream home: a 42-acre farm in Wise County, Texas. I wanted my sons to grow up wild and free like I did. I wanted horses and to grow my own food, and drink pure well water. At night, we threw lawn chairs in the truck and drove into the pasture to stargaze. Horses, goats, dogs, and cats joined us while the coyotes and owls serenaded in the distance. Paradise found. I didn't know that oilman George Mitchell—known as the "father of fracking"—was experimenting all around me to develop ways to blast oil and gas from the shale formation beneath my paradise.

In the beginning, no one opposed fracking. It seemed like a good idea. Leasing mineral rights could provide

money and leave me with more time to ride my horse. But then my well water turned black and slick, then gray and sandy. When the air turned brown and the noise from the compressors drowned out the birds, we left the farm and moved to the city of Denton, Texas. Soon afterward, Denton City Council issued a permit for the drilling of five gas wells near a neighborhood, park, and hospital. Local residents appealed to both the city and the state for a better city drilling ordinance, but they got nowhere. In 2014, we started a petition to place an initiative to ban drilling in Denton on the November ballot. Citizens of the predominately Republican city of Denton voted 59 to 41 percent to ban fracking in Denton.

The very next day, oil and gas industry and the state of Texas filed lawsuits against the city, calling the ban unconstitutional. To avoid evidence of harm being presented in court, the industry showered the Texas Legislature with $21.3 million in legal bribes to enact HB40—a bill that would remove a 100-year tradition of concurrent regulation by city and state, rendering local governments powerless to stop oil/gas companies from putting facilities near homes or schools. Texas lawmakers passed the bill.

Because I had done anti-fracking organizing work across Texas, Earthworks' Oil & Gas Accountability Project hired me as their full-time Texas organizer. One of my assignments involved compiling stories and data I had collected through the years into case studies, now published in *Flowback: How the Texas Natural Gas Boom Affects Health and Safety* (published by Earthworks, 2011). Many of the people adversely affected by fracking can no longer tell their stories. The only way they could escape their contaminated homes was by filing lawsuits against the industry for damages. But a settlement from the oil and gas industry comes with a nondisclosure or "gag" agreement that prohibits individuals from talking about their damages. All the documentation proving air and water contamination gets sealed away forever. This enables the industry to say there is no documentation of contamination.

But people shared their documentation with me and I am not gagged! So here are three stories:

Tim, Christine, and Reilly Ruggiero
The Ruggieros moved to a 10-acre farm in Wise County, Texas, so their daughter Reilly could grow up in clean country air surrounded by animals. They watched in horror as a rig went up right next to their

water we consume, the air we breathe, and the substances with which we come in contact. For example, bisphenol A (BPA), a chemical that disrupts endocrine function, is commonly found in food packaging, including cans, plastic wraps, and food storage containers (Betts 2011). In a study of umbilical cord blood of 10 newborns, researchers found an average of 200 industrial chemicals, pesticides, and other pollutants (Environmental Working Group 2005).

The National Toxicology Program (2014) lists 243 chemical substances that are "known to be human carcinogens" or "reasonably anticipated to be human carcinogens," meaning that they are linked to cancer. These may constitute only a fraction of actual human carcinogens. When the Toxic Substances Control Act of 1976 was enacted, it "grandfathered" in the 62,000 chemicals then on the market. Since then, the EPA has restricted the uses of only 5 of the 80,000 chemicals used in the United States; 95 percent of the chemicals in use have not been tested for safety (Lunden 2013).

In 2003, the European Union drafted legislation known as Registration, Evaluation, and Authorization and Restriction of Chemicals (REACH) that requires chemical companies to neighbor's home. Looking for information about fracking, they searched the Internet and found me through my blog.* I advised them to get a baseline water test immediately and to document everything. The test showed clean and safe water.

One day a neighbor called Christine to tell her that her horses were running wild, the fence had been cut down, and bulldozers were dozing their property. All this happened without the Ruggieros' permission, which was not required because they only owned the surface of their property. The mineral owner—Aruba Petroleum—had the right to take 4 of their 10 acres and put a rig up 200 feet from their house to extract the minerals it owned under the surface. It's called "split estate."

The family suffered through several spills and constant air pollution. Air testing revealed a host of dangerous pollutants in the air including dangerously high levels of benzene—a known carcinogen for which there is no safe level of exposure. Reilly was diagnosed with asthma. Christine developed a rash and suffered from headaches. Tim had numbness in his extremities. Six weeks after fracking began, a water test revealed high levels of MTBE. Their water was not safe to use for anything, not even watering the grass.

The Ruggieros moved away but they can't tell you why or ever speak about what happened to them. If you ask, they will say, "The matter has been resolved."

Susan

Susan lived on the border of Argyle and Bartonville, Texas, in a gorgeous, custom-built home with a lusciously landscaped yard and a few acres where the vineyard would someday be. Hidden behind a tree line on the hill behind her house was a gas well and a waste pit. More gas wells were tucked in all around her. Argyle Central Facility, a collection facility, compressor station, and processing plant (think mini-refinery), was a short crow-fly away.

Contractors were framing her house during a record 32 inches of rain in June 2007 when there was a "fraccident": The fracking waste pit overflowed, spilling its pollution into her backyard. Subsequently, Susan's well water became foamy, and holding a flame to the water's surface resulted in waxy-plastic bits. Horrible odors killed bugs and frogs and made birds fall from the sky. Her dog developed a rare cancer Perdue had only seen in humans exposed to too much radiation. Geiger counters detected higher than normal radiation in her backyard. Susan started losing her hair.

Susan moved away to a new home. She can't tell you her story. She can only say, "The matter has been resolved."

The Hallowich Family

Stephanie and Chris Hallowich and their two children built a custom dream home on 10 acres in a rural area of Harrisburg, Pennsylvania, on the Marcellus Shale. Soon they were surrounded with natural gas operations and a huge impoundment pit full of smelly fracking waste bordered their property. Their water turned black, and water testing revealed harmful chemicals such as acrylonitrile, toluene, ethyl benzene, tetrachloroethylene, and styrene. They put a big water tank in the garage and had water delivered at great expense. The air pollution made them sick, and the whole family suffered from health impacts. Stephanie's young daughter had frequent and profuse nose bleeds.

The Hallowiches had to find a way to get their children to safety, so they moved to a hotel and filed a lawsuit against Range Resources for damages. They were required to sign a nondisclosure agreement and sign a statement saying that Range Resources' operations was not the cause of their health issues. Later a court made Exhibit B of the nondisclosure agreement public and we learned that Range Resources had demanded that the young Hallowich children also be gagged for the rest of their lives.

*Texas Sharon's Bluedaze at texassharon.com.
SOURCE: Sharon Wilson, 2015. Written for this text.

conduct safety and environmental tests to prove that the chemicals they are producing are safe. If they cannot prove that a chemical is safe, it will be banned from the market (Rifkin 2004). The European Union has become a world leader in environmental stewardship by placing the "precautionary principle" at the center of EU regulatory policy. The precautionary principle requires industry to prove that their products are safe. In contrast, in the United States, chemicals are assumed to be safe unless proven otherwise, and the burden is put on the consumer, the public, or the government to prove that a chemical causes harm.

In general, the European Union seems to be more concerned than the United States about health effects of chemicals—as evidenced by its stricter controls and bans on chemicals. If the United States had a national health insurance plan similar to that of European countries, in which the federal government paid for health care, do you think the U.S. government would enact tougher controls on hazardous chemicals and other environmental issues?

WHAT do you THINK?

This mother puts sunscreen on her child to protect against sunburn. But sunscreen, like other personal care products, may contain harmful chemicals. Learn what chemicals are in the personal care products you use at the Environmental Working Group's website Skin Deep at www.ewg.org/skindeep.

Most cancer researchers believe that the environment in which we live and work may be a major contributor to the development of cancer—a disease that develops in half of U.S. men and a third of U.S. women (National Toxicology Program 2014). Research studies have found a link between breast cancer and environmental pollutants (Brody et al. 2007). Many of the chemicals we are exposed to in our daily lives can cause not only cancer but also other health problems, such as infertility, birth defects, and a number of childhood developmental and learning problems (Fisher 1999; Kaplan and Morris 2000; McGinn 2000; Schapiro 2007). Chemicals found in common household, personal, and commercial products can cause drowsiness, disorientation, headache, dizziness, nausea, fatigue, shortness of breath, cramps, diarrhea, and irritation of the eyes, nose, throat, and lungs. Long-term exposure can affect the nervous system, reproductive system, liver, kidneys, heart, and blood. Fragrances, which are found in perfumes and colognes, shampoos, deodorants, laundry detergents, tampons, air "fresheners," and a host of other consumer products, are known to be respiratory irritants. Some of the 4,000 ingredients that are used in fragrance manufacturing have been linked with cancer, birth defects, neurotoxic effects, and endocrine disruption (Bradshaw 2010).

Multiple chemical sensitivity (MCS), also known as environmental illness, is a condition whereby individuals experience adverse reactions when exposed to low levels of chemicals found in everyday substances (vehicle exhaust, fresh paint, housecleaning products, perfume and other fragrances, synthetic building materials, and numerous other petrochemical-based products). Individuals with MCS often avoid public places and/or wear a protective breathing filter to avoid inhaling the many chemical substances in the environment.

WHAT do you THINK?

Some businesses, universities, hospitals, and local governments are voluntarily limiting fragrances to accommodate employees and consumers who experience ill effects from them. What do you think about banning fragrances in the workplace or other public places? If your college or university were considering instituting a ban on fragrances on campus, would you support the ban? Why or why not?

Although personal care products are regulated by the Food and Drug Administration (FDA), manufacturers are legally responsible for assessing their products as safe, and they are not required to file any safety data or product formulations with the FDA (Kessler 2015). The 2015 Personal Care Products Safety Act, if passed, would require manufacturers to provide the FDA with information about their products' ingredients, would require the FDA to assess the safety of compounds found in personal care products, and would empower the FDA to recall unsafe products.

Light Pollution

The United States, like much of the rest of the world, has become increasingly "lit up" with artificial light. Artificial light has negative impacts on the well-being of both humans

multiple chemical sensitivity Also known as "environmental illness," a condition whereby individuals experience adverse reactions when exposed to low levels of chemicals found in everyday substances.

and wildlife (Bogard 2013). **Light pollution** refers to artificial lighting that is annoying, unnecessary, and/or harmful to life forms on Earth:

> The 24-hour day/night cycle, known as the circadian clock, affects physiologic processes in almost all organisms. These processes include brain wave patterns, hormone production, cell regulation, and other biologic activities. Disruption of the circadian clock is linked to several medical disorders in humans, including depression, insomnia, cardiovascular disease, and cancer. (Chepesiuk 2009, p. A22)

About 30 percent of vertebrates and 60 percent of invertebrates are nocturnal and vulnerable to light pollution disrupting their patterns of mating, migration, feeding, and pollination (Bogard 2013). For example, artificial lighting around beaches where leatherback turtles nest threatens the survival of hatchlings and is a major cause of declining leatherback turtle populations. Hatchlings, which instinctually follow the reflected light of the stars and moon from the beach to the ocean, instead follow the light of hotels and streetlights, with the result that they die of dehydration, are eaten by predators, or are run over by cars.

Social Causes of Environmental Problems

Various structural and cultural factors have contributed to environmental problems. These include population growth, industrialization and economic development, and cultural values and attitudes such as individualism, consumerism, and militarism.

Population Growth

The world's population is growing, exceeding 7 billion in 2013 and projected to grow to more than 9 billion in 2050 (see Chapter 12). Population growth places increased demands on natural resources and results in increased waste. As Hunter (2001) explained:

> Global population size is inherently connected to land, air, and water environments because each and every individual uses environmental resources and contributes to environmental pollution. While the scale of resource use and the level of wastes produced vary across individuals and across cultural contexts, the fact remains that land, water, and air are necessary for human survival. (p. 12)

However, population growth itself is not as critical as the ways in which populations produce, distribute, and consume goods and services.

Industrialization and Economic Development

Many of the environmental problems confronting the world are associated with industrialization, in large part because industrialized countries consume more energy and natural resources and contribute more pollution to the environment than poor countries. The relationship between level of economic development and environmental pollution is curvilinear rather than linear. Industrial emissions are minimal in regions with low levels of economic development and are high in the middle-development range as developing countries move through the early stages of industrialization. However, at more advanced stages of industrialization, industrial emissions ease because heavy-polluting manufacturing industries decline, "cleaner" service industries increase, and rising incomes are associated with a greater demand for environmental quality and cleaner technologies. Although more developed countries, such as the United States, may have fewer and cleaner manufacturing industries compared with developing countries such as China, wealthier countries continue to support pollution-causing manufacturing industries by importing goods from these countries and operating such industries in multinational business operations.

light pollution Artificial lighting that is annoying, unnecessary, and/or harmful to life forms on Earth.

Cultural Values and Attitudes

Cultural values and attitudes that contribute to environmental problems include individualism, consumerism, and militarism.

Individualism. Individualism, which is a characteristic of U.S. culture, puts individual interests over collective welfare. Even though recycling is good for our collective environment, many individuals do not recycle because of the personal inconvenience involved in washing and sorting recyclable items. Similarly, individuals often indulge in countless behaviors that provide enjoyment and convenience for themselves at the expense of the environment: long showers, recreational boating, frequent meat eating, the use of air conditioning, and driving large, gas-guzzling SUVs, to name just a few.

Consumerism. Consumerism—the belief that personal happiness depends on the purchasing of material possessions—also encourages individuals to continually purchase new items and throw away old ones. The media bombard us daily with advertisements that tell us life will be better if we purchase a particular product. Consumerism contributes to pollution and environmental degradation by supporting polluting and resource-depleting industries and by contributing to waste.

Militarism. The cultural value of militarism also contributes to environmental degradation (see also Chapter 15). "It is generally agreed that the number one polluter in the United States is the American military. It is responsible each year for the generation of more than one-third of the nation's toxic waste . . . an amount greater than the five largest international chemical companies combined" (Blatt 2005, p. 25). Toxic substances from military vehicles, weapons materials, and munitions pollute the air, land, and groundwater in and around military bases and training areas. Almost every one of the 4,127 military sites is seriously contaminated (Nazaryan 2014). The Pentagon has asked Congress to loosen environmental laws for the military, and the EPA is forbidden to investigate or sue the military (Blatt 2005; Janofsky 2005).

> The U.S. environmental movement may be the largest single social movement in the United States.

Strategies for Action: Responding to Environmental Problems

Actor Robert Redford (2015) said, "Our planet's resources are limited, but there is no limit to the human imagination and our capacity to solve our biggest problems." Efforts to solve environmental problems include environmental activism, environmental education, reduction of carbon emissions, the use of green energy and energy efficiency, modifications in consumer products and behavior, slowing population growth, and sustainable economic development.

Environmental Activism

With millions of members joining thousands of national and local environmental organizations, the U.S. environmental movement may be the largest single social movement in the United States (Brulle 2009). A Gallup survey found that 16 percent of U.S. adults reports being an active participant in the environmental movement (Gallup Organization 2015) (see Table 13.3).

Environmental organizations exert pressure on government and private industry to initiate or intensify actions related to environmental protection. Environmentalist groups also design and implement their own projects and stage public protests and demonstrations. For example, in September 2014, more

TABLE 13.3 Involvement in the Environmental Movement*

Involvement	Percentage of U.S. Adults
Active participant	16
Sympathetic but not active	41
Neutral	30
Unsympathetic	11
No opinion	1

*In a 2015 Gallup survey, a national sample of U.S. adults was asked, "Do you think of yourself as an active participant in the environmental movement, sympathetic toward the movement, but not active, neutral, or unsympathetic toward the movement?"
SOURCE: Gallup Organization 2015.

than half a million people across the globe participated in demonstrations calling for action on climate. In New York City alone, more than 400,000 people participated in the People's Climate March—the largest environmental demonstration in history.

Environmental organizations also disseminate information to the public about environmental issues and use the Internet and e-mail to send e-mail action alerts to members, informing them when Congress and other decision makers threaten the health of the environment. These members can then send e-mails and faxes to Congress, the president, and business leaders, urging them to support policies that protect the environment.

Wildlife activism efforts are sometimes focused on **charismatic megafauna**—particular species that have popular appeal, such as the panda, polar bear, monarch butterfly, and bald eagle. These species are often used to draw public attention to a larger environmental issue. For example, the image of a stranded polar bear is often used to depict the melting polar region, drawing attention to the larger issue of global warming and climate change.

a katz/Shutterstock.com

In September 2014, more than 400,000 people participated in the People's Climate March in New York City.

Religious Environmentalism. Since the 1990s, religious leaders and scholars have become increasingly concerned with environmental problems and climate change, and every major religion has issued statements about the importance of protecting the environment (Forum on Religion and Ecology at Yale, n.d.). The American Academy of Religion's 2014 annual conference was on "Religion and Climate Change" (Taylor 2015). In response to the growing problem of global warming and climate change, in 2015, the Archbishop of Canterbury, along with religious leaders from the Muslim, Sikh, Jewish, Catholic, and other faiths, signed a declaration urging the global community to transition to a low carbon economy (Archbishop's Council 2015). In the same year, Pope Francis delivered a 184-page encyclical—a letter addressed to the more than 1 billion Catholics around the world—in which he lamented, "Never have we so hurt and mistreated our common home as we have in the last 200 years," and expressed the urgent need to change our lifestyle, production, and consumption (Pope Francis 2015).

Radical Environmentalism. The **radical environmental movement** is a grassroots movement of individuals and groups that employs unconventional and often illegal means of protecting wildlife or the environment. Radical environmentalists believe in what is known as **deep ecology**—the view that maintaining the Earth's natural systems should take precedence over human needs, that nature has a value independent of human existence, and that humans have no right to dominate the Earth and its living inhabitants (Brulle 2009). The best-known radical environmental groups are the Earth Liberation Front (ELF) and the Animal Liberation Front (ALF), which are international underground movements consisting of autonomous individuals and small groups that engage in "direct action" to (1) inflict economic damage on those profiting from the destruction and exploitation of the natural environment, (2) save animals from places of abuse (e.g., laboratories, factory farms, and fur farms), and (3) reveal information and educate the public on atrocities committed against the earth and all the species that populate it. The most common targets of direct action are businesses, such as logging facilities, fur factories, and companies that use animal testing on their products (Yang et al. 2014). University research labs that use animals in their research are also targeted.

charismatic megafauna Particular species that have popular appeal, such as the panda, polar bear, monarch butterfly, and bald eagle, and that are used to draw public attention to larger environmental issues.

radical environmental movement A grassroots movement of individuals and groups that employs unconventional and often illegal means of protecting wildlife or the environment.

deep ecology The view that maintaining the Earth's natural systems should take precedence over human needs, that nature has a value independent of human existence, and that humans have no right to dominate the Earth and its living inhabitants.

Radical environmentalists are sometimes accused of **ecoterrorism**, defined as "the use or threatened use of illegal force by groups or individuals, in order to protect environmental and/or animal rights" (Yang et al. 2014, p. 11). Many environmentalists question whether "terrorist" is an appropriate label and argue that the real terrorists are corporations that exploit and pollute the Earth's natural resources. The FBI once called radical environmentalists the "number one domestic terror threat," but ecoterrorism today is rare (Kirchner 2015).

Should motives be considered in imposing penalties on individuals who are convicted of acts of ecoterrorism? For example, should a person who sets fire to a business to protest that business's environmentally destructive activities receive the same penalty as a person who sets fire to a business in order to collect insurance money?

Hunters/Anglers and Environmentalism. Some people believe that hunting and fishing are incompatible with environmentalism. "For many environmentalists, the word 'hunter' suggests a mindless brute, an enemy of nature who loves guns, kills for fun, and cares nothing for biodiversity or ecological integrity" (Cerulli 2014). Indeed, some commercial hunting and fishing practices contribute to the problem of species depletion. Continued poaching of elephants for their ivory tusks poses a serious threat to the survival of African elephant populations (Mathiesen 2015). And overfishing threatens some fish populations. Yet, many hunters and anglers are avid conservationists who are committed to preserving natural habitats to maintain stable populations of animal and marine life. Hunting has also been used as wildlife management tool to help mitigate overpopulation of species such as deer, which can otherwise populate to unsustainable levels due to the lack of natural predators. Additionally, the excise tax revenue from the purchase of guns, ammunition, angling gear, duck stamps, and hunting and fishing licenses is used to fund various environmental management policies. Although some environmentalists, especially those who live a vegetarian or vegan lifestyle, are philosophically opposed to killing animals, the environmental interests of hunters/anglers and their opponents are often the same.

Environmental Education

One goal of environmental organizations and activists is to educate the public about environmental issues and the seriousness of environmental problems. Being informed about environmental issues is important because people who have higher levels of environmental knowledge tend to engage in higher levels of pro-environment behavior. For example, environmentally knowledgeable people are more likely to save energy in the home, recycle, conserve water, purchase environmentally safe products, avoid using chemicals in yard care, and donate funds to conservation (Coyle 2005).

A main source of information about environmental issues for most Americans is the media. However, because corporations and wealthy individuals with corporate ties own the media, unbiased information about environmental impacts of corporate activities may not readily be found in mainstream media channels, so the public must consider the source in interpreting information about environmental issues. Propaganda by corporations sometimes comes packaged as "environmental education." The American Petroleum Institute (API) distributes curriculum materials that question climate science and promote the value of fossil fuels by offering lesson plans and materials for teachers K–12 and by maintaining a website Classroom Energy! (Mulvey and Shulman 2015).

The education that kindergarten through grade 12 students receive about global warming and climate change varies considerably from school to school and state to state. At least four states—Louisiana, Texas, South Dakota, and Tennessee—have passed laws that weaken teachers' ability to accurately present the science of climate change to their K–12 students. The scientific consensus that human activity is causing global warming and climate change is now, under some state laws, only a "controversial theory among other theories" (Horn 2012).

> Being informed about environmental issues is important because people who have higher levels of environmental knowledge tend to engage in higher levels of pro-environment behavior.

ecoterrorism The use or threatened use of illegal force by groups or individuals, in order to protect environmental and/or animal rights.

The Next Generation Science Standards, proposed in 2013, are the first set of national guidelines to require that K–12 students be taught that climate change is a scientific fact and is mainly caused by burning fossil fuels. These voluntary standards had been accepted in 13 states as of February 2015 but have faced resistance in several states, especially those whose economy is dominated by the fossil fuel industry (Schrank 2015).

Reduction of Carbon Emissions

The IPCC (2014) says that to reduce risks associated with climate change, developed countries must reduce greenhouse gas emissions to nearly zero by 2100 to keep global warming under 2 degrees Celsius—the IPCC's threshold for dangerous climate change. Research conducted by climate scientist James Hansen and colleagues concluded that to avoid sea levels rising by 10 feet over the next 50 to 85 years—which would be catastrophic in that coastal areas, including cities such as New York and Miami, would be uninhabitable—fossil fuel use (the major source of greenhouse gases) must be nearly completely phased out within 30 years (Gertz 2015).

In 1992, countries joined an international treaty, the United Nations Framework Convention on Climate Change, to cooperate in limiting global warming, and coping with climate change and its impacts. In 1997, delegates from 160 nations met in Kyoto, Japan, and forged the **Kyoto Protocol**—the first international agreement to place legally binding limits on greenhouse gas emissions from developed countries. The United States, the world's largest producer of greenhouse gas emissions, rejected the Kyoto Protocol in 2001.

Following Kyoto, a series of international agreements to slow global warming and deal with climate change have been forged, but none have gone far enough to stop global warming. In 2010, world leaders agreed that carbon emissions must be reduced so that global temperature increases are limited to below 2 degrees Celsius (3.6 degrees Fahrenheit) above preindustrial levels in order to avoid catastrophic effects of global warming. A National Academy of Sciences (2015) report concluded that "Although emissions reductions are technologically feasible, they have been difficult to implement for political, economic and social reasons" (n.p.).

One approach to reducing carbon emissions is through **cap and trade programs**—a free-market approach that provides economic incentives to power plants and other industries for reducing carbon emissions. In a cap and trade system, polluters buy credits allowing them to emit a limited amount of carbon dioxide. They can sell leftover credits to other polluters, creating a monetary incentive to reduce emissions. Critics of the cap and trade approach argue that it fails to achieve the lowest possible emissions because it does not require all plants to use the best available technology to reduce emissions. By allowing some plants to have higher emissions, it also exposes populations living near these high-emissions plants to excessive air pollution.

In 2009, the U.S. House passed the American Clean Energy and Security Act, which sought to establish a federal cap and trade system modeled after the European Union Emission Trading Scheme. The bill was defeated in the Senate. In 2010, California adopted the nation's most stringent rules to curb greenhouse gas emissions; the rules reward industries that cut emissions by allowing them to sell carbon credits to other industries. In 2013, nine northeastern states organized in the Regional Greenhouse Gas Initiative (RGGI)—the first mandatory, cap and trade carbon emissions reduction program in the United States. In 2015, RGGI states had cut emissions by 40 percent from 2005 levels, while maintaining robust economies (Cleetus 2015).

In 2015, under President Obama, the EPA established the first-ever federal limits on carbon emissions from power plants—the nation's largest source of carbon pollution. Under the **Clean Power Plan**, the EPA establishes state-by-state targets for carbon emissions reductions, allowing states flexibility in how to meet these targets. The plan's new rules propose a 32 percent cut in carbon emissions from power plants by 2030 (from 2005 levels), and states will have until 2022 to comply (Vaughan 2015). In the Netherlands, a Dutch court issued a landmark ruling, ordering the government to do more to cut greenhouse gas emissions (Schiermeier 2015). As this book goes to press, countries

Kyoto Protocol The first international agreement to place legally binding limits on greenhouse gas emissions from developed countries.

cap and trade programs A free-market approach that provides economic incentives to power plants and other industries for reducing carbon emissions.

Clean Power Plan Establishes the first-ever federal limits on carbon emissions from U.S. power plants, establishing state-by-state targets for carbon emissions reductions and allowing states flexibility in how to meet these targets.

Answer each of the following questions:

1. Some people believe that the United States government should limit the amount of greenhouse gases thought to cause global warming that U.S. businesses can produce. Other people believe that the government should not limit the amount of greenhouse gases that U.S. businesses put out. What do you think?

2. Do you favor or oppose each of the following as a way for the federal government to try to reduce future global warming?

   a. Increased taxes on electricity so people use less of it
   b. Increased taxes on gasoline so people either drive less, or buy cars that use less gas
   c. Tax breaks for companies to produce more electricity from water, wind, and solar power

3. For each of the following, do you think the government should require by law, encourage with tax breaks but not require, or stay out of entirely?

   a. Building cars that use less gasoline
   b. Building air conditioners, refrigerators, and other appliances that use less electricity
   c. Building new homes and offices that use less energy for heating and cooling

4. Do you think that U.S. actions to reduce global warming in the future would hurt the U.S. economy, help the economy, or have no effect on the U.S. economy?

5. Do you think the United States should take action on global warming only if other major industrial countries such as China and India agree to do equally

effective things, that the United States should take action even if these other countries do less, or that the United States should not take action on this at all?

6. Do you think most U.S. business leaders *want* the federal government to do things to stop global warming, or do you think most U.S. business leaders *do not want* the federal government to do things to stop global warming?

7. How important is the issue of global warming to you personally—extremely important, very important, somewhat important, not too important, or not at all important?

Comparison Data

Telephone interviews with a sample of 1,000 U.S. adults from across the nation found that, overall, Americans want government to be involved in reducing global warming:

1. Government should limit greenhouse gases from U.S. businesses (76 percent); government should not limit greenhouse gases from U.S. businesses (20 percent); don't know (3 percent).

2. To reduce future global warming, government should . . .

	Favor	Oppose
a. Increase taxes on electricity	22%	78%
b. Increase taxes on gasoline	28%	71%
c. Give companies tax breaks to produce more electricity from water, wind, and solar power	84%	15%

3. Should the government require by law, encourage with tax breaks but not require, or stay out of entirely?

	Require	Encourage	Stay out of
a. Building cars that use less gasoline	31%	50%	19%
b. Building air conditioners, refrigerators, and other appliances that use less electricity	29%	51%	20%
c. Building new homes and offices that use less energy for heating and cooling	24%	56%	20%

4. U.S. actions to reduce global warming in the future would hurt the U.S. economy (20 percent); help the U.S. economy (56 percent); not affect the economy (23 percent).

5. The United States should take action on global warming only if other countries do (14 percent); even if other countries do less (68 percent); not take action at all (18 percent).

6. Most U.S. business leaders ___ the government to do things to stop global warming: do want (25 percent); do not want (72 percent); don't know (3 percent).

7. How important is the issue of global warming to you personally? Extremely or very important (46 percent); somewhat important (30 percent); not too important (12 percent); not at all important (12 percent).

SOURCE: Adapted from Jon Krosnick, 2010, *Global Warming Poll*, Stanford University. Available at woods.stanford.edu.

are announcing commitments to cut greenhouse gases in preparation for the 2015 global summit on climate change in Paris, where countries will hopefully agree on global emissions reductions. This chapter's *Self and Society* feature looks at U.S. attitudes toward government efforts to reduce global warming.

Green Energy and Energy Efficiency

Other strategies to reduce environmental pollution and degradation are to increase the use of **green energy**—energy that is renewable and nonpolluting—and to increase energy efficiency. In 2012, President Obama issued a new average fuel-efficiency standard for cars and light trucks of 54.5 mpg by 2025, a significant increase from the previous 35.5 mpg. This new standard is expected to reduce U.S. oil consumption by 12 billion barrels a year (White House 2012).

Most states offer incentive programs including tax credits and rebates to encourage energy efficiency and the use of renewable energy sources. For example, some states have offered tax credits for purchases of hybrid or electric cars, housing insulation, or solar panels. In 2004, more than 20 countries committed to specific targets for the renewable share of total energy use (United Nations Environment Programme [UNEP] 2007). A number of U.S. states have set goals of producing a minimum percentage of electricity from wind power, solar power, or other renewable sources.

The World Bank (2014) defines "clean energy" as energy that does not produce carbon dioxide when generated. Under this definition, nuclear energy is considered "clean energy." Should nuclear power be labeled as "clean energy"? Why or why not?

Solar and Wind Energy. Types of solar power include (1) converting sunlight to electricity through the use of photovoltaic cells, (2) using of solar thermal collectors to heat building space and water, and (3) using concentrating solar power plants that use the sun's heat to make steam to turn electricity-producing turbines. Although U.S. solar power production increased 20-fold from 2008 to 2015, it still accounts for less than 1 percent of total U.S. power (Malik 2015).

Wind turbines, which turn wind energy into electricity, are operating in 82 countries. Although wind energy provided less than 5 percent of the nation's electricity in 2015, it could provide 35 percent by 2050 (U.S. Department of Energy 2015). Placement of wind turbines is critical, as wind turbines can impede nearby residents' scenic views and can also create annoying noise pollution that can have adverse health effects, including hypertension, sleep disturbance, and headaches (Seltenrich 2014).

Biofuel. Ethanol and biodiesel are two types of biofuels—fuels derived from agricultural crops. Ethanol is an alcohol-based fuel that is produced by fermenting and distilling corn or sugar. Ethanol is blended with gasoline to create E85 (85 percent ethanol and 15 percent gasoline). Vehicles that run on E85, called *flexible fuel vehicles,* have been used by the government and in private fleets for years and are also available to consumers. Most ethanol is produced in the United States and Brazil (Shrank 2011).

A problem associated with ethanol fuel is that increased demand for corn, which is used to make most ethanol in the United States, has driven up the price of corn, resulting in higher food prices (many processed food items contain corn, and animal feed is largely corn). And as corn prices rise, so too do those of rice and wheat because the crops compete for land. Rising food prices threaten the survival of the world's poorest 2 billion people who depend on grain to survive. The grain it takes to fill a 25-gallon tank with ethanol would feed one person for an entire year (Brown 2007).

Increased corn and/or sugar cane production to meet the demand for ethanol also has adverse environmental effects, including increased use and runoff of fertilizers, pesticides, and herbicides; depletion of water resources; and soil erosion. In addition, tropical forests are being clear-cut to make room for "energy crops," leaving less land for conservation and wildlife (Price 2006); biofuel refineries commonly run on coal and natural gas (which emit greenhouse gases); farm equipment and fertilizer production require fossil fuels; and the use of ethanol involves emissions of several pollutants. Finally, even if 100 percent of the U.S. corn crop were used to produce ethanol, it would only displace less than 15 percent of U.S. gasoline use (Food & Water Watch and Network for New Energy Choices 2007).

green energy Also known as clean energy, energy that is nonpolluting and/or renewable, such as solar power, wind power, biofuel, and hydrogen.

To encourage the use of hydrogen fuel-cell powered cars, California has more than 50 hydrogen fueling stations.

Biodiesel fuel is a cleaner-burning diesel fuel made from vegetable oils and/or animal fats, including recycled cooking oil. Some individuals who make their own biodiesel fuel obtain used cooking oil from restaurants at no charge.

Hydrogen Power. Hydrogen can be used in a battery cell or as a clean-burning fuel for electricity production, heating, cooling, and transportation. Hydrogen fuel from plant waste may one day replace gas for transportation vehicles (Connor 2015). California has more than 50 hydrogen fueling stations, which allows for the first large-scale introduction of hydrogen fuel cell–powered cars in the United States (Reichmuth 2014).

Modifications in Consumer Products and Behavior

In the United States and other industrialized countries, many consumers are making "green" choices in their behavior and purchases that reflect concern for the environment. In some cases, these choices carry a price tag, such as paying more for organically grown food or for clothing made from organic cotton. Consumers are also motivated to make green purchases that save money. Consumers often consider their utility bill when they choose energy-efficient appliances and electrical equipment. Although some eco-minded individuals choose green products and services, others choose to reduce their overall consumption and "buy nothing" rather than "buy green." For example, many consumers are choosing not to buy bottled water and to drink tap water instead. The switch from bottled to tap is partly fueled by the need to cut down on unnecessary spending in hard economic times, but environmental concerns are also a factor. The production and transportation of bottled water uses fossil fuels, and the disposal of plastic water bottles adds to our already overburdened landfills.

Although the average size of new housing in the United States has increased considerably, some homeowners are choosing to downsize their housing. For some, the driving force behind housing downsizing is economic, but others are moving into smaller dwellings out of concern for the environment.

Table 13.4 presents tips for how consumers can reduce the amount of carbon dioxide each of us produces.

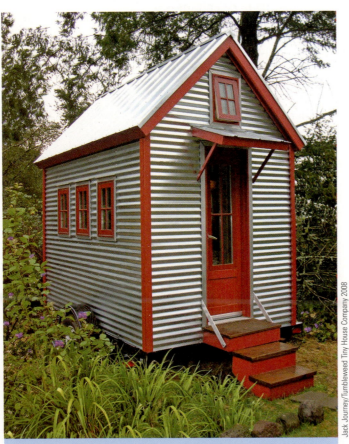

Tiny houses have less impact on the environment.

Slow Population Growth

As discussed in Chapter 12, slowing population growth is an important component of efforts to protect the environment. One study concluded that slowing population growth could make a significant contribution to reducing carbon emissions (O'Neill et al., 2010).

TABLE 13.4 Top 10 Things You Can Do to Fight Global Warming

National Geographic's *Green Guide* lists the following tips for consumers to reduce the amount of carbon dioxide they produce each year. All of the following CO_2 reductions listed are on an annual basis:

1. Replace five incandescent lightbulbs in your home with compact fluorescent bulbs (CFLs): Swapping those 75-watt incandescent bulbs with 19-watt CFLs can cut 275 pounds of CO_2.

2. Instead of short-haul flights of 500 miles or so, take the train and bypass 310 pounds of CO_2.

3. Sure, it may be hot, but get a fan, set your thermostat to 75°F, and blow away 363 pounds of CO_2.

4. Replace refrigerators more than 10 years old with today's more energy-efficient Energy Star models and save more than 500 pounds of CO_2.

5. Shave your eight-minute shower to five minutes for a savings of 513 pounds.

6. Caulk, weatherize, and insulate your home. If you rely on natural gas heating, you'll stop 639 pounds of CO_2 from entering the atmosphere (472 pounds for electric heating). And this summer, you'll save 226 pounds from air conditioner use.

7. Whenever possible, dry your clothes on a line outside or a rack indoors. If you air-dry half your loads, you'll dispense with 723 pounds of CO_2.

8. Trim down on the red meat. Because it takes more fossil fuels to produce red meat than fish, eggs, and poultry, switching to these foods will slim your CO_2 emissions by 950 pounds.

9. Leave the car at home and take public transportation to work. Taking the average U.S. commute of 12 miles by light-rail will leave you 1,366 pounds of CO_2 lighter than driving. The standard, diesel-powered city bus can save 804 pounds, and heavy rail subway users save 288.

10. Finally, support the creation of wind, solar, and other renewable energy facilities by choosing green power if offered by your utility. To find a green power program in your state, call your local utility or visit the U.S. Department of Energy's Green Power Network page at http://apps3.eere.energy.gov/greenpower/.

SOURCE: National Geographic 2007. From National Geographic's *Green Guide*. Copyright © 2007 National Geographic Society.

Although Americans who are concerned about the environment may think about how their home energy use, travel, food choices, and other lifestyle behaviors affect the environment, they rarely consider the environmental impact of their reproductive choices. Researchers estimate that each child born in the United States, and the children that child has as an adult, adds 9,441 metric tons of carbon dioxide to the environment—the equivalent of burning 972,160 gallons of gas (Murtaugh and Schlax 2009).

As shown in Table 13.5, reducing the number of children a woman has by one saves far more carbon dioxide from being released in the environment than do many other environmentally friendly lifestyle choices. Yet policymakers have been reluctant to discuss how family planning can be beneficial for the environment (Plautz 2014).

Sustainable Economic and Human Development

Achieving global cooperation on environmental issues is difficult, in part, because developed countries (primarily in the Northern Hemisphere) have different economic agendas from those of developing countries (primarily in the Southern Hemisphere). The northern agenda emphasizes preserving wealth and affluent lifestyles, whereas the southern agenda focuses on overcoming mass poverty and achieving a higher quality of life. Southern countries are concerned that northern industrialized countries—having already achieved economic wealth—will impose international environmental policies that restrict the economic growth of developing countries.

As discussed in Chapter 6, development involves more than economic growth and the alleviation of poverty. The human development approach views the well-being of

TABLE 13.5 Carbon Reduction Figures for Various Lifestyle Behaviors*

Behavior	Amount of CO_2 Not Released into the Environment
Replace windows with energy-efficient windows	12 metric tons
Recycle newspapers, magazines, glass, plastic, aluminum, and steel cans	17 metric tons
Replace 10 75-watt incandescent bulbs with 25-watt energy-efficient lights	36 metric tons
Increase auto gas mileage from 20 mpg to 30 mpg	148 metric tons
Reduce number of children by one	9,441 metric tons

*Calculated over an 80-year period.
SOURCE: Murtaugh and Schlax 2009.

populations in terms of not only their income but also their access to education and their ability to lead long, healthy lives in societies that respect and value everyone. The 2013 *Human Development Report* adds, "To sustain progress in human development, far more attention needs to be paid to the impact human beings are having on the environment. The goal is high human development and a low ecological footprint per capita" (United Nations Development Programme 2013, p. 94).

The 2013 *Human Development Report* presents data on each country's Human Development Index (based on measures of health and longevity, education, and income) as well as each country's environmental footprint. Although it might seem that carbon productivity—GDP per unit of carbon dioxide emission—would increase with human development, the correlation is weak. This finding supports the idea that progress in human development can be sustainable. **Sustainable development** is development that enables human populations to have fulfilling lives without degrading the planet. "The aim here is for those alive today to meet their own needs without making it impossible for future generations to meet theirs. . . . This in turn calls for an economic structure within which we consume only as much as the natural environment can produce, and make only as much waste as it can absorb" (McMichael et al. 2000, p. 1067). Restoring and protecting the Earth requires that we find alternatives to destructive forms of producing, distributing, and consuming goods and services.

The development and use of clean, renewable energy technologies plays an important role in sustainable human development. Renewable energy projects in developing countries have demonstrated that providing affordable access to green energy helps to alleviate poverty by providing energy for creating business and jobs and by providing power for refrigerating medicine, sterilizing medical equipment, and supplying fresh water and sewer services needed to reduce infectious disease and improve health (Flavin and Aeck 2005).

The Role of Institutions of Higher Education

sustainable development Occurs when human populations can have fulfilling lives without degrading the planet.

Green Revolving Funds (GRFs) College and university funds that are dedicated to financing cost-saving energy-efficiency upgrades and other projects that decrease resource use and minimize environmental impacts.

Colleges and universities can play an important role in efforts to protect the environment by encouraging use of bicycles on campus, using hybrid and electric vehicles, establishing recycling programs, using local and renewable building materials for new buildings, involving students in organic gardening to provide food for the campus, using clean energy, and incorporating environmental education into the curricula. A growing number of colleges and universities are establishing **Green Revolving Funds (GRFs)**, which are funds dedicated to financing cost-saving energy efficiency upgrades and other projects that decrease resource use and minimize environmental impact (Sustainable Endowments Institute 2011). The resulting savings in operating expenses are returned to the fund and then reinvested in additional projects. In some colleges and universities, students play a role in deciding how Green Revolving Funds are spent and in implementing the various energy upgrades and projects. Nevertheless,

David Newport, director of the environmental center at the University of Colorado at Boulder, believes that institutions of higher education are not doing enough to promote sustainability. "We're supposed to be on the leading edge, and we're behind the curve. . . . There are, what 4,500 colleges in the United States, and how many of them are really doing something? Less than 100 or 200?" (quoted by Carlson 2006, p. A10).

Understanding Environmental Problems

Environmental problems are linked to corporate globalization, rapid and dramatic population growth, expanding world industrialization, patterns of excessive consumption, and reliance on fossil fuels for energy. The Global Footprint Network (2010) offers the following analysis of environmental problems:

> Climate change is not the problem. Water shortages, overgrazing, erosion, desertification, and the rapid extinction of species are not the problem. Deforestation, reduced cropland productivity, and the collapse of fisheries are not the problem. Each of these crises, though alarming, is a symptom of a single, overriding issue. Humanity is simply demanding more than the earth can provide.

Whether we understand environmental problems as resulting from a complex set of causes, or from one, simple underlying cause such as consuming more than the Earth can provide, we cannot afford to ignore the growing evidence of the irreversible effects of global warming and loss of biodiversity, and the adverse health effects of toxic waste and other forms of pollution. Although civilizations throughout history have collapsed, these civilizations were limited to a particular region. Never before has civilization on a global scale been threatened, as it is now, by destruction of our ecosystems. To avoid the collapse of the global civilization, it is essential to reduce greenhouse gas emissions to nearly zero by the end of the century. This would require a major shift away from the use of fossil fuels—a shift the fossil fuel industry cannot support without jeopardizing its own profits. "Because the ethics of some businesses include knowingly continuing lethal but profitable activities . . . it is hardly surprising that interests with large financial stakes in fossil fuel burning have launched a gigantic and largely successful disinformation campaign in the USA to confuse people about climate disruption . . . and block attempts to deal with it" (Ehrlich and Ehrlich 2013, p. 3).

Many Americans believe in a "technological fix" for the environment—that science and technology will solve environmental problems. Paradoxically, the same environmental problems that have been caused by technological progress may be solved by technological innovations designed to clean up pollution, preserve natural resources and habitats, and provide clean forms of energy. Technology can be used to help the environment. For example, environmentalists and scientists are using drones to study wildlife in remote regions, measure polar ice melting, collect water samples for water quality monitoring, plant trees to reduce deforestation, and to detect and apprehend illegal poachers (Carroll 2015).

Leaders of government and industry must have the will to finance, develop, and use technologies on a large scale that do not pollute or deplete the environment. When asked how companies can produce products without polluting the environment, Robert Hinkley suggested that, first, it must become a goal to do so:

> I don't have the technological answers for how it can be done, but neither did President John F. Kennedy when he announced a national goal to land a man on the moon by the end of the 1960s. The point is that, to eliminate pollution, we first have to make it our goal. Once we've done that, we will devote the resources necessary to make it happen. We will develop technologies that we never thought possible. But if we don't make it our goal, then we will never devote the resources, never develop the technology, and never solve the problem. (quoted by Cooper 2004, p. 11)

But the direction of technical innovation is largely in the hands of big corporations that place profits over environmental protection. Unless the global community challenges the power of transnational corporations to pursue profits at the expense of environmental and human health, corporate behavior will continue to take a heavy toll on the health of the planet and its inhabitants. Because oil has been implicated in political and military conflicts involving the Middle East (see Chapter 15), such conflicts are likely to continue as long as oil plays the lead role in providing the world's energy.

Global cooperation is also vital to resolving environmental concerns but is difficult to achieve because rich and poor countries have different economic development agendas: Developing poor countries struggle to survive and provide for the basic needs of their citizens; developed wealthy countries struggle to maintain their wealth and relatively high standard of living. Can both agendas be achieved without further pollution and destruction of the environment? Is sustainable economic development an attainable goal? With mounting concern about climate change, the health impacts of air pollution, rising oil prices, and the need to ensure energy access to all, governments worldwide have strengthened their commitment to sustainable, renewable energy policies and projects (UNEP 2007).

In the United States, there is significant opposition to governmental regulations designed to address environmental problems. Opponents of regulation argue that rules and restrictions are harmful to the economy and destroy jobs. Yet, an analysis of the costs and benefits of the Clean Air Act Amendments of 1990 found that value of the benefits ($1.3 trillion) was 25 times the cost ($53 billion). In 2010 alone, the Clean Air Act Amendments of 1990 saved an estimated 160,000 lives (Shapiro and Irons 2011). Lax regulation was a contributing factor in the BP Deepwater Horizon oil spill of 2010—the largest oil spill in U.S. history, which had devastating effects on jobs and the economy in the Gulf states. The claim that environmental regulation hurts jobs is not supported by the evidence, and the costs of regulations are far outweighed by the benefits to health, safety, and well-being (Shapiro and Irons 2011). Protecting the environment does not mean that jobs must be sacrificed in the process. Indeed, transitioning to a low-carbon economy can simultaneously protect the environment, create national energy independence, lower energy prices, *and* create jobs (Brecher 2011).

In 2004, the Nobel Peace Prize was given to Wangari Maathai for leading a grassroots environmental campaign to plant 30 million trees across Kenya. This was the first time ever that the Nobel Peace Prize was awarded to someone for accomplishments in restoring the environment. In her acceptance speech, Maathai explained, "A degraded environment leads to a scramble for scarce resources and may culminate in poverty and even conflict" (quoted by Little 2005, p. 2). For example, increased drought due to climate change can result over conflict over access to water. Conflict in Sudan's Darfur region has been described as the world's first conflict resulting from climate change (Lipott 2015). In a report to Congress, the Department of Defense (2015) described global warming as a security threat: "Global climate change will have wide-ranging implications for U.S. national security interests over the foreseeable future because it will aggravate existing problems—such as poverty, social tensions, environmental degradation, ineffectual leadership, and weak political institutions—that threaten domestic stability in a number of countries" (p. 3). With ongoing conflict around the globe, it is time for world leaders to recognize the importance of a healthy environment for world peace and to prioritize environmental protection in their political agendas.

Robin Wall Kimmerer (2013) says that "it is not just changes in policies that we need but also changes to the heart. . . . Gratitude for all the earth has given us lends us courage to . . . refuse to participate in an economy that destroys the beloved earth to line the pockets of the greedy, to demand an economy that is aligned with life, not stacked against it" (pp. 376, 377). Whether we act out of gratitude, fear, or a sense of responsibility to future generations, we must act quickly if we are to protect the Earth and its living inhabitants. As one sign carried by a participant in the 2014 People's Climate March read, "There is no Planet B."

Chapter Review

- **What are some examples of cross-border environmental problems?**

  Global warming and climate change affect the entire planet. Bioinvasion—the introduction of species in regions where they are not native—can disrupt and damage ecosystems. When environmental migrants seek refuge in another country, there are no international policies to protect them.

- **How do free trade agreements and transnational corporations contribute to environmental problems?**

  Free trade agreements such as NAFTA and the FTAA provide transnational corporations with privileges to pursue profits, expand markets, use natural resources, and exploit cheap labor in developing countries while weakening the ability of governments to protect natural resources or to implement environmental legislation.

- **How does the conflict perspective view environmental problems and solutions?**

  The conflict perspective focuses on the role that wealth, power, and the pursuit of profit plays in environmental problems and solutions. For example, the fossil fuel industry has spent a great deal of money to influence lawmakers to support policies that benefit the fossil fuel industry and to discredit climate science.

- **What does the term environmental injustice refer to?**

  Environmental injustice, also called *environmental racism,* refers to the tendency for marginalized populations and communities to disproportionately experience adversity due to environmental problems. For example, in the United States, polluting industries, industrial and waste facilities, and transportation arteries (which generate vehicle emissions pollution) are often located in minority communities.

- **What is greenwashing?**

  Greenwashing refers to the ways in which environmentally and socially damaging companies portray their corporate image and products as being "environmentally friendly" or socially responsible.

- **Where does most of the world's energy come from?**

  Most of the world's energy comes from fossil fuels, which include oil, coal, and natural gas. This is significant because many of the serious environmental problems in the world today, including global warming and climate change, biodiversity loss, and pollution, stem from the use of fossil fuels.

- **What are some problems associated with nuclear energy?**

  Nuclear waste contains radioactive plutonium, a substance linked to cancer and genetic defects. Nuclear waste in the environment remains potentially harmful to human and other life for thousands of years, and disposing of nuclear waste is problematic. Accidents at nuclear power plants, such as the 2011 Fukushima disaster, and the potential for nuclear reactors to be targeted by terrorists add to the actual and potential dangers of nuclear power plants.

- **What is our environmental footprint?**

  The demands that humanity makes on the Earth's natural resources are known as the environmental footprint. Since 1970, humanity's environmental footprint has exceeded the Earth's capacity to produce useful resources such as water and crops, and to absorb waste, such as CO_2 emissions.

- **How often does a species of life on Earth become extinct?**

  One species of life on Earth goes extinct every three hours.

- **What are the major causes and effects of deforestation?**

  The major causes of deforestation are the expansion of agricultural land, human settlements, wood harvesting, and road building. Deforestation displaces people and wild species from their habitats, contributes to global warming, and contributes to desertification, which results in the expansion of desert land that is unusable for agriculture. Soil erosion caused by deforestation can cause severe flooding.

- **What are the effects of air pollution on human health?**

  Indoor and outdoor air pollution, which is linked to heart disease, lung cancer, and respiratory ailments, kills about 7 million people a year, or one in eight global deaths.

- **What are some examples of common household, personal, and commercial products that contribute to indoor air pollution?**

  Some common indoor air pollutants include carpeting, mattresses, drain cleaners, oven cleaners, spot removers, shoe polish, dry-cleaned clothes, paints, varnishes, furniture polish, potpourri, mothballs, fabric softener, caulking compounds, air fresheners, deodorizers, disinfectants, glue, correction fluid, printing ink, carbonless paper, and felt-tip markers.

- **What is the primary cause of global warming?**

  The prevailing view on what causes global warming is that greenhouse gases—primarily carbon dioxide, methane, and nitrous oxide—accumulate in the atmosphere and act like the glass in a greenhouse, holding heat from the sun close to the Earth. The primary greenhouse gas is carbon dioxide, which is released into the atmosphere by burning fossil fuels.

- **What are "climate deniers"?**

  Climate deniers are people who do not accept the scientific consensus that human-caused global warming and climate change are scientific facts.

- **What are the major sources of water pollution?**

  Water pollution is largely caused by fertilizers and other chemicals used in intensive agriculture, industrial production, mining, and untreated urban runoff and wastewater. Oil spills and plastics also contribute to water pollution.

- **Why is there increasing public concern over fracking?**

  Fracking involves the use of chemicals that contaminate land, water, and air. Oil and gas companies do not have to disclose to the federal government the chemicals they use in fracking, but scientists find carcinogenic and toxic chemicals in fracking fluid and wastewater. Leaks, overflows, and illegal dumping of wastewater allow chemicals to flow into local land and water supplies. Fracking requires massive amounts of water, which is problematic in drought-stricken areas, and fracking operations have triggered earthquakes. People—as well as pets, farm animals, and wildlife near fracking sites—have become sick as a result of contact with and/or consumption of contaminated water from fracking sites.

• **What is the precautionary principle?**
The precautionary principle, adopted by the European Union, requires industries to prove that their products are safe. In contrast, in the United States, chemicals are assumed to be safe unless proven otherwise, and the burden is put on the consumer, the public, or the government to prove that a chemical causes harm.

• **What is the relationship between level of economic development and environmental pollution?**
There is a curvilinear relationship between level of economic development and environmental pollution. In regions with low levels of economic development, industrial emissions are minimal, but emissions rise in countries that are in the middle economic development range as they move through the early stages of industrialization. However, at more advanced stages of industrialization, industrial emissions ease because heavy-polluting manufacturing industries decline, "cleaner" service industries increase, and because rising incomes are associated with a greater demand for environmental quality and cleaner technologies.

• **What social and cultural factors contribute to environmental problems?**
Social and cultural factors that contribute to environmental problems include population growth, industrialization and economic development, and cultural values and attitudes such as individualism, consumerism, and militarism.

• **What are some of the strategies for alleviating environmental problems?**
Strategies for alleviating environmental problems include environmental activism, environmental education, reducing carbon emissions, the use of "green" energy and energy efficiency, modifications in consumer products and behavior, slowing population growth, and sustainable economic development.

• **According to 2004 Nobel Peace Prize winner Wangari Maathai, why is environmental protection important for national and international security?**
In her acceptance speech for the 2004 Nobel Peace Prize, Wangari Maathai explained that "a degraded environment leads to a scramble for scarce resources and may culminate in poverty and even conflict" (quoted by Little 2005, p. 2).

Test Yourself

1. Under international law, environmental migrants are considered refugees.
    a. True
    b. False
2. Kudzu is an example of
    a. a bioinvasion.
    b. an extinct species.
    c. a threatened species.
    d. an alternative fuel.
3. Dr. Wei-Hock Soon published research that found that global warming is due to natural variations in the sun's energy. Who funded Dr. Soon's research?
    a. The Environmental Protection Agency
    b. NASA
    c. University of Florida
    d. Fossil fuel interests
4. If greenhouse gases were to be stabilized today, global air temperature and sea level would be expected to
    a. remain at their current levels.
    b. decrease immediately.
    c. begin to decrease within 20 years.
    d. continue to rise for hundreds of years.
5. The United States has more operating nuclear reactors than any other country.
    a. True
    b. False
6. Most solid waste in the United States is recycled.
    a. True
    b. False
7. Fracking has been banned in which of the following states?
    a. Pennsylvania
    b. Texas
    c. New York
    d. Florida
8. In the United States, the EPA has required testing on all of the more than 80,000 chemicals that have been on the market since 1976.
    a. True
    b. False
9. The Intergovernmental Panel on Climate Change says that the world must keep global warming under ____ degrees C (from 1990 levels) in order to avoid dangerous climate change.
    a. 2
    b. 5
    c. 8
    d. 11
10. In what country did a court ruling order the country to do more to cut greenhouse gases?
    a. The United States
    b. Canada
    c. The Netherlands
    d. Brazil

Answers: 1. B; 2. A; 3. D; 4. D; 5. A; 6. B; 7. C; 8. B; 9. A; 10. C.

Key Terms

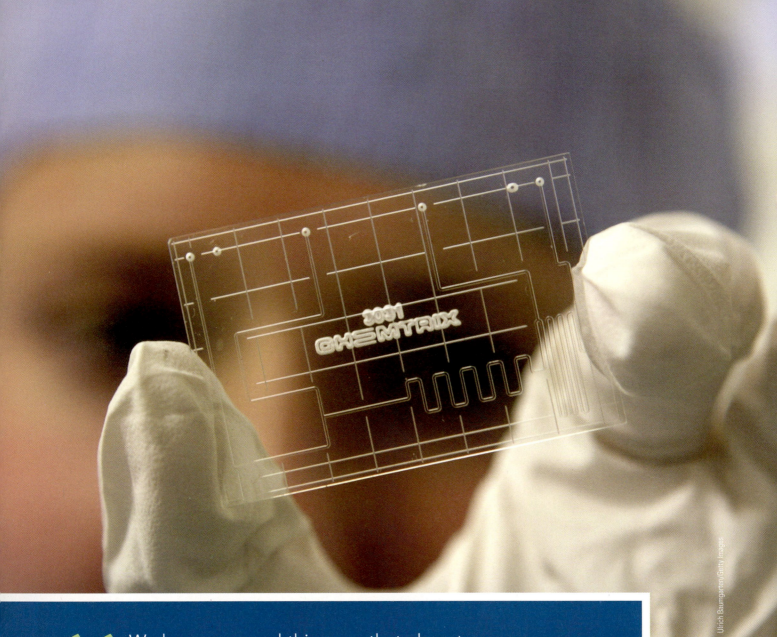

> " We have arranged things so that almost no one understands science and technology. This is a prescription for disaster. We might get away with it for a while, but sooner or later this combustible mixture of ignorance and power is going to blow up in our faces."
>
> **CARL SAGAN**
> Astronomer and astrobiologist

14

Science and Technology

Learning Objectives

After studying this chapter, you will be able to . . .

1 List features of the Technological Revolution.

2 Identify implications for technology use on human society for each sociological theory.

3 Describe and give examples of the ways in which technology transforms aspects of the human experience.

4 Assess the positive and negative effects of science and technology on society.

5 Evaluate social strategies for mitigating the negative effects of technology.

WHEN HANNAH WARREN was born without a windpipe, doctors told her parents there was no hope that she would live (Tanner 2013). Born in Seoul, South Korea, in 2010, Hannah spent the first two years of her life in a hospital, kept alive by machines that allowed her to breathe, eat, drink, and swallow. Hannah's parents read about an experimental treatment that used stem cells to rebuild damaged tracheas in adults. The technology, however, had never been used to rebuild a missing body part. At age 2, Hannah and her parents flew from Seoul to Chicago where doctors at the Children's Hospital of Illinois removed stem cells from her hip and seeded the cells onto a plastic scaffold in a lab. Within a week, the cells had multiplied and formed a new windpipe, which was implanted during a 9-hour surgery in April 2013 (p. 1). Although the trachea successfully functioned, Hannah developed a lung infection as a result of the surgery. She died in July 2013, one month before her third birthday (Moisse 2013). "Our hearts are broken," said Hannah's parents in a statement, "She is a pioneer in stem-cell technology and her impact will reach all corners of our beautiful Earth" (p. 1).

technological dualism The tendency for technology to have both positive and negative consequences.

science The process of discovering, explaining, and predicting natural or social phenomena.

technology Activities that apply the principles of science and mechanics to the solutions of a specific problem.

mechanization Dominant in an agricultural society, the use of tools to accomplish tasks previously done by hand.

automation The replacement of human labor with machinery and equipment.

cybernation Dominant in postindustrial societies, the use of machines to control other machines.

Technological dualism is the tendency for technology to have both positive and negative consequences. Clearly, advances in medical technology have helped people around the world survive illnesses and injuries that would have been deadly only a few decades ago. On the other hand, medical interventions can themselves cause harm. In Hannah's case, the operation giving her the chance to breathe on her own led to a lung infection that caused her death. A recent study of over 1 million Medicare patients found that in a one-year period, more than 40 percent of them received screenings, tests, or treatments that professional medical organizations determined have no benefit or are likely to cause harm (Schwartz et al. 2014). Like medical technology, the 3-D printer is another example of technological dualism. Although efficient and cost-effective for manufacturers, 3-D printers may lead to higher rates of unemployment, particularly among assembly line laborers (see Chapter 7). Similarly, 3-D printers can be used to make a robotic hand for a child with amniotic band syndrome or to make a semiautomatic gun used by a mass murderer.

Science and technology go hand in hand. **Science** is the process of discovering, explaining, and predicting natural or social phenomena. A scientific approach to understanding acquired immunodeficiency syndrome (AIDS), for example, might include investigating the molecular structure of the virus, the means by which it is transmitted, and public attitudes about AIDS. **Technology**, as a form of human cultural activity that applies the principles of science and mechanics to the solution of problems, is intended to accomplish a specific task—in this case, the development of an AIDS vaccine.

Societies differ in their level of technological sophistication and development. In agricultural societies, which emphasize the production of raw materials, the use of tools to accomplish tasks previously done by hand, or **mechanization**, dominates. As societies move toward industrialization and become more concerned with the mass production of goods, automation prevails. **Automation** involves the use of self-operating machines, as in an automated factory where autonomous robots assemble automobiles. Finally, as a society moves toward post-industrialization, it emphasizes service and information professions (Bell 1973). At this stage, technology shifts toward **cybernation**, whereby machines control machines—making production decisions, programming robots, and monitoring assembly performance.

What are the effects of science and technology on humans and their social world? How do science and technology help to remedy social problems, and how do they contribute to social problems? Is technology, as author Neil Postman (1992) suggested, both a friend and a foe to humankind? We address each of these questions in this chapter.

The Global Context: The Technological Revolution

Less than 50 years ago, traveling across state lines was an arduous task, a long-distance phone call was a memorable event, and mail carriers brought belated news of friends and relatives from far away. Today, travelers journey between continents in a matter of hours, and for many, e-mail, faxes, instant messaging, texting, and cell phones have replaced previously conventional means of communication.

The world is a much smaller place than it used to be, and it will become even smaller as the technological revolution continues. In 2014, the Internet had 3 billion users in more than 200 countries, with 277 million users in the United States (Internet Statistics 2015). Of all Internet users, the highest proportion come from Asia (45.7 percent), followed by Europe (19.2 percent), North America (10.2 percent), Latin America and the Caribbean (10.5 percent), Africa (9.8 percent), the Middle East (3.7 percent), and Oceania/Australia (0 percent) (Internet Statistics 2015).

The world was a much smaller place in the 1800s when the Pony Express was the only method to communicate with far-away friends and family. Today, a number of technological innovations make communication between individuals or groups instantaneous. Pictured here is an Apple Watch that acts as a conduit to the Apple iPhone.

Although the **penetration rate**—that is, the percentage of people who have access to and use the Internet in a particular area—is higher in industrialized countries, there is some movement toward the Internet becoming a truly global medium as Africans, Middle Easterners, and Latin Americans increasingly get online. For example, Internet use in the United States grew 182 percent between 2000 and 2013; the number of Internet users in Nigeria increased by 33,560 percent during the same time period (Internet Statistics 2015). In developing nations, at least half of Internet users say they use the Internet daily. The most popular Internet activity for these users was socializing with friends and family (Pew 2015a).

Social media use dominates Internet activity in many parts of the developing world, with users in Saudi Arabia, Vietnam, Malaysia, Mexico, and Argentina spending three hours a day or more on social media (Figure 14.1). Social media use is likely to continue its rapid expansion around the world as Google and Facebook work to bring affordable Internet access and wireless technologies to previously unconnected parts of the developing world (Efrati 2013; Internet.org 2015). The duality of expanding wireless technology use and global social media connections, however, may come at a price to victims of cybercrime. Technology known as "creepware," software that can be purchased for as little as $40, allows cybercriminals to spy on and steal personal information from victims from anywhere in the world (Perez 2014).

The movement toward globalization of technology is, of course, not limited to the use and expansion of the Internet. The world robot market and the U.S. share of it continue to expand, Microsoft's Internet platform and support products are sold all over the world, scientists collect skin and blood samples from remote islanders for genetic research, a global treaty regulating trade of genetically altered products has been signed by more than 100 nations, and South Korea's Samsung leads its competitors in cell phone innovations and sales (Noble 2015).

To achieve such scientific and technological innovations, sometimes called *research and development* (R&D), countries need material and economic resources. *Research*

penetration rate The percentage of people who have access to and use the Internet in a particular area.

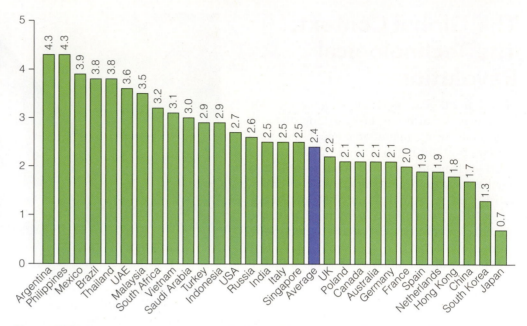

Figure 14.1 Average Hours Spent on Social Media Daily, Selected Countries, 2015
SOURCE: Kemp 2015.

> [D]eveloping countries, most notably China and India, are expanding their scientific and technological capabilities at a faster rate than the United States.

entails the pursuit of knowledge; *development* refers to the production of materials, systems, processes, or devices directed to the solution of practical problems. According to the National Science Foundation (NSF), the United States spends over $400 billion a year in research and development, accounting for about 30 percent of the global total—the largest single-performing country in the world (NSF 2015). As in most other countries, U.S. funding sources are primarily from private industry, 63 percent of the total, followed by the federal government and nonprofit organizations such as research institutes at colleges and universities (NSF 2015).

The United States leads the world in science and technology, although there is some evidence that we are falling behind (Dutta et al. 2015; Information Technology and Innovation Foundation [ITIF] 2012; Price 2008). For example, a report by the World Economic Forum compares information and communication technologies (ICTs) across countries using a Network Readiness Index (NRI) (Figure 14.2). The NRI is composed of four subsections: (1) the quantity and quality of the *environment* for ICTs (e.g., political environment, regulatory environment), (2) ICTs' *readiness* (e.g., affordability), (3) ICTs' *usage* (e.g., individual, business), and (4) the *impact* of ICTs (e.g., social, economic). In 2015, the NRI results indicated that Nordic countries (Finland, Sweden, Netherlands, Norway, and Switzerland) held 5 of the top 10 index positions (Dutta et al. 2015). Asian countries also made a strong showing, with Singapore rated as number 1, and Japan, Taiwan/China, Republic of Korea, and Hong Kong all in the top 18 rankings. The United States ranked number 7 on the overall index but scored higher on the impact subsection and lower on the remaining three subsections.

The decline of U.S. supremacy in science and technology is likely to be the result of several interacting forces (Coleman 2015; ITIF 2009; Lemonick 2006; Price 2008; World Bank 2009). First, the federal government has been scaling back its investment in research and development in response to fiscal deficits (Britt 2015; Plummer 2013). Second, corporations, the largest contributors to research and development, have begun to focus on short-term products and higher profits as pressure from stockholders mounts. Third, developing countries, most notably China and India, are expanding their scientific and technological capabilities at a faster rate than the United States. In 2000, the United States was the leading exporter of ICT goods. By 2012, China had surpassed the United States in the value of ICT exports by

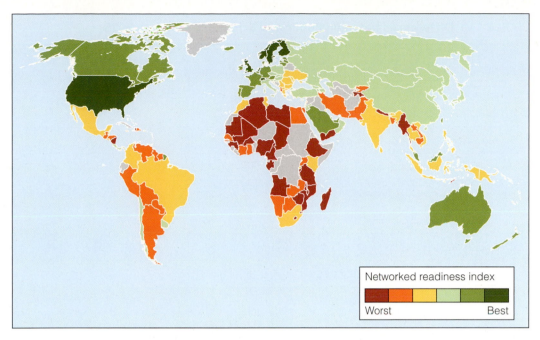

Figure 14.2 Global Network Readiness Index, 2015*
SOURCE: Dutta, Geiger, and Lanvin 2015.
*Networked Readiness Index compiles data on the political and regulatory environment, business and innovation environment, infrastructure, affordability, skills, individual and business usage, and economic impacts.

more than $300 billion (Organisation for Economic Cooperation and Development [OECD] 2014).

Fourth, there has been concern over science and math education in the United States, both in terms of quality and quantity. More than half of students entering college need remedial-level math courses in order to complete college-level work (NSF 2015). Although the United States awards the highest number of science and engineering PhD degrees in the world, about 6 in 10 students receiving degrees in these fields are from foreign nations. The majority of these foreign nationals do not stay in the United States after receiving their degrees, leaving a significant gap between high-skill technology jobs in the United States and American workers able to fill those jobs.

Finally, Mooney and Kirshenbaum (2009) document "unscientific America"—the tremendous disconnect between the citizenry, media, politicians, religious leaders, education, and the entertainment industry (e.g., *CSI, The Big Bang Theory, Grey's Anatomy*), on the one hand, and science and scientists, on the other. Post–World War II America, in part because of the Cold War, invested in R&D, leading to such scientific and technological advances as the space program, the development of the Internet, and the decoding of the genome. Yet, despite these significant contributions and the recognition of the significance of **STEM** disciplines, most Americans know very little about science (NSF 2015) (see this chapter's *Self and Society*). Table 14.1 displays the gaps in opinions about scientific issues between professional scientists and the American public.

One hundred years after the invention of the automobile, a new invention is poised to change society again: the driverless car. Pioneered by Google, early tests indicate that self-driving cars are less prone to accident than those driven by humans (S. Johnson 2015). Convenience for those on long commutes and improved safety, particularly in the age of distracted, cell-phone-wielding drivers, are just a few advantages of the driverless car (*The Economist* 2015). However, technological automation can also have disadvantages. If driverless cars become commonplace, how do you think they will change society?

STEM An acronym for science, technology, engineering, and mathematics.

TABLE 14.1 Differences between Public and Scientists' Opinion, 2015

	U.S. Adults	Professional Scientists	Gap
Percentage of U.S. adults and professional scientists saying each of the following:			
BIOMEDICAL SCIENCES			
Safe to eat genetically modified foods	37	88	51
Favor use of animals in research	47	89	42
Safe to eat foods grown with pesticides	28	68	40
Humans have evolved over time	65	98	33
Childhood vaccines such as MMR should be required	68	86	18
CLIMATE, ENERGY, SPACE SCIENCES			
Climate change is mostly due to human activity	50	87	37
Growing world population will be a major problem	59	82	23
Favor building more nuclear power plants	45	65	20
Favor more offshore drilling	32	52	20
Astronauts essential for future of U.S. space program	47	59	12
Favor increased use of bioengineering fuel	68	78	10
Favor increased use of fracking	31	39	8
Space station has been a good investment for U.S.	64	68	4

Note: Other responses and those saying don't know or no answer are not shown.
SOURCE: Pew 2015b.

Postmodernism and the Technological Fix

Many Americans believe that social problems can be resolved through a **technological fix** (Weinberg 1966) rather than through social engineering. For example, a social engineer might approach the problem of water shortages by persuading people to change their lifestyle: use less water, take shorter showers, and wear clothes more than once before washing. A technologist would avoid the challenge of changing people's habits and motivations and instead concentrate on the development of new technologies that would increase the water supply.

Social problems can be tackled through both social engineering and a technological fix. In recent years, for example, social engineering efforts to address the problem of date rape on college campuses have included sexual assault awareness campaigns and public service announcements about binge drinking and the definition of "consent." An example of a technological fix for the same problem was developed by four students at North Carolina State University who invented a type of nail polish that, when a woman dips her finger into a drink, would change color if common date rape drugs were present (Sullivan 2014).

Not all individuals, however, agree that science and technology are good for society. **Postmodernism** is a worldview that holds that rational thinking and the scientific perspective have fallen short in providing the "truths" they were once presumed to hold. During the industrial era, science, rationality, and technological innovations were thought to hold the promises of a better, safer, and more humane world. Today, postmodernists question the validity of the scientific enterprise, often pointing to the unforeseen

technological fix The use of scientific principles and technology to solve social problems.

postmodernism A worldview that questions the validity of rational thinking and the scientific enterprise.

Answer each of the following questions and then calculate the number that you answered correctly.

1. All radioactivity is man-made.
   a. True
   b. False

2. Electrons are smaller than atoms.
   a. True
   b. False

3. Lasers work by focusing sound waves.
   a. True
   b. False

4. The continents on which we live have been moving their location for millions of years and will continue to move in the future.
   a. True
   b. False

5. Which of the following types of solar radiation does sunscreen protect the skin from?
   a. X-rays
   b. infrared
   c. ultraviolet
   d. microwaves

6. Nanotechnology deals with things that are extremely . . .
   a. small
   b. large
   c. cold
   d. hot

7. Which gas makes up most of the earth's atmosphere?
   a. hydrogen
   b. nitrogen
   c. carbon dioxide
   d. oxygen

8. What is the main function of red blood cells?
   a. fight disease in the body
   b. carry oxygen to all parts of the body
   c. help the blood to clot

9. Which of these is a major concern about the overuse of antibiotics?
   a. it can lead to antibiotic-resistant bacteria
   b. antibiotics are very expensive
   c. people will become addicted to antibiotics

10. Which is an example of a chemical reaction?
    a. water boiling
    b. sugar dissolving
    c. nails rusting

11. Which is the better way to determine whether a new drug is effective in treating diseases? If a scientist has a group of 1,000 volunteers with the disease to study, should she
    a. give the drug to all of them and see how many get better
    b. give the drug to half of them but not to the other half, and compare how many in each group get better

12. Which gas do most scientists believe causes temperatures in the atmosphere to rise?
    a. carbon dioxide
    b. hydrogen
    c. helium
    d. radon

13. Which natural resource is extracted in a process known as "fracking"?
    a. oil
    b. diamonds
    c. natural gas
    d. silicon

Answers: 1. b; 2. a; 3. b; 4. a; 5. c; 6. a; 7. b; 8. b; 9. a; 10. c; 11. b; 12. a; 13. c.

DIRECTIONS: Compare the percentage of questions you answered correctly to the following distribution. For example, if you answered all the questions correctly, you scored better than 93 percent of the public (the sum of all percents below you) and the same as 7 percent of the public. If you answered 11 of the questions correctly, you scored better than 75 percent of the public (the sum of all percents below you), the same as 10 percent of the public, and worse than 15 percent of the public (the sum of all percents above you).

SOURCE: Pew Research Center 2013.

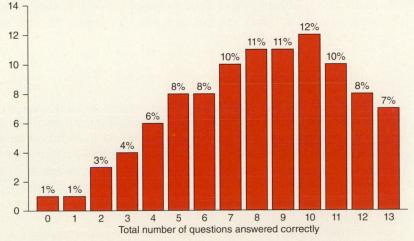

Distribution of the percent of respondents answering 0–13 questions correctly (N=1006)*

Total number of questions answered correctly

*Based on a sample of 1006 randomly selected adults who took part in a national survey in March, 2013.

and unwanted consequences of resulting technologies. Automobiles, for example, began to be mass-produced in the 1930s, in response to consumer demands. But the proliferation of automobiles has also led to increased air pollution and the deterioration of cities as suburbs developed, and, today, traffic fatalities are the number one cause of accident-related deaths.

Sociological Theories of Science and Technology

Each of the three major sociological frameworks helps us to better understand the nature of science and technology in society.

Structural-Functionalist Perspective

Structural functionalists view science and technology as emerging in response to societal needs—that "science was born indicates that society needed it" (Durkheim 1973/1925). As societies become more complex and heterogeneous, finding a common and agreed-on knowledge base becomes more difficult. Science fulfills the need for an assumed objective measure of "truth" and provides a basis for making intelligent and rational decisions. In this regard, science and the resulting technologies are functional for society.

Scientific knowledge has grown at a more rapid rate over time; during each of three historical periods—from about 1650–1750, 1750–1950, and 1950–2012—scientific knowledge grew at triple the rate of the previous phase (Bornmann and Mutz 2015). If society changes too rapidly as a result of science and technology, however, problems may emerge. When the material part of culture (i.e., its physical elements) changes at a faster rate than the nonmaterial part (i.e., its beliefs and values), a **cultural lag** may develop (Ogburn 1957). For example, the typewriter, the conveyor belt, and the computer expanded opportunities for women to work outside the home. With the potential for economic independence, women were able to remain single or to leave unsatisfactory relationships and/or establish careers. But although new technologies have created new opportunities for women, beliefs about women's roles, expectations of female behavior, and values concerning equality, marriage, and divorce have lagged behind.

Robert Merton (1973), a structural functionalist and founder of the subdiscipline sociology of science, also argued that scientific discoveries or technological innovations may be dysfunctional for society and may create instability in the social system. For example, the development of time-saving machines increases production, but it also displaces workers and contributes to higher rates of employee alienation. Defective technology can have disastrous effects on society. For example, defective airbags from Japanese automobile parts provider Takata resulted in deaths and serious injuries, and the recall of more than 34 million cars (BBC 2015).

Conflict Perspective

Conflict theorists, in general, argue that science and technology benefit a select few. For some conflict theorists, technological advances occur primarily as a response to capitalist needs for increased efficiency and productivity and thus are motivated by profit. As McDermott (1993) predicted, most decisions to increase technology are made by "the immediate practitioners of technology, their managerial cronies, and for the profits accruing to their corporations" (p. 93). In the United States, private industry spends more money on research and development than the federal government does. The Dalkon Shield (an intrauterine device, or IUD, a form of contraception) and silicone breast implants are examples of technological advances that promised millions of dollars in profits for their developers. However, the rush to market took precedence over thorough testing of the products' safety. Subsequent lawsuits filed by consumers resulted in large damage awards for the plaintiffs.

cultural lag A condition in which the material part of culture changes at a faster rate than the nonmaterial part.

Science and technology also further the interests of dominant groups to the detriment of others. The need for scientific research on AIDS was evident in the early 1980s, but the required large-scale funding was not made available so long as the virus was thought to be specific to homosexuals and intravenous drug users. Only when the virus became a threat to mainstream Americans were millions of dollars allocated to AIDS research. Hence, conflict theorists argue that granting agencies act as gatekeepers to scientific discoveries and technological innovations. These agencies are influenced by powerful interest groups and the marketability of the product rather than by the needs of society.

When the dominant group feels threatened, it may use technology as a means of social control. For example, the use of the Internet is growing dramatically in China, the world's largest Internet market. Censorship has been consolidated under the State Council Information Office, also known as the "great firewall of China" (Chen 2011). A study by Harvard Law School researchers indicates that, of the 204,000 websites accessed, nearly 20,000 were inaccessible. Top Google search results for such words as *Tibet, equality, Taiwan,* and *democracy China* were consistently blocked (Associated Press 2010). China is not alone, however. The OpenNet Initiative, a collaborative effort of three academic institutions, reports that Uzbekistan, Turkmenistan, Iran, Syria, and Ethiopia also have pervasive political censorship of the Internet (OpenNet 2015).

Finally, conflict theorists as well as feminists argue that technology is an extension of the patriarchal nature of society that promotes the interests of men and ignores the needs and interests of women. As in other aspects of life, women play a subordinate role in reference to technology in terms of both its creation and its use. For example, washing machines, although time-saving devices, disrupted the communal telling of stories and the resulting friendships among women who gathered together to do their chores. Bush (1993) observed that, in a "society characterized by a sex-role division of labor, any tool or technique . . . will have dramatically different effects on men than on women" (p. 204).

Symbolic Interactionist Perspective

Knowledge is relative. It changes over time, over circumstances, and between societies. We no longer believe that the world is flat or that the Earth is the center of the universe, but such beliefs once determined behavior because individuals responded to what they thought to be true. The scientific process is a social process in that "truths"—socially constructed truths—result from the interactions among scientists, researchers, and the lay public.

Kuhn (1973) argued that the process of scientific discovery begins with assumptions about a particular phenomenon (e.g., the world is flat). Because unanswered questions always remain about a topic (e.g., why don't the oceans drain?), science works to fill these gaps. When new information suggests that the initial assumptions were incorrect (e.g., the world is not flat), a new set of assumptions or framework emerges to replace the old one (e.g., the world is round). It then becomes the dominant belief or paradigm.

Symbolic interactionists emphasize the importance of this process and the effect that social forces have on it. Lynch et al. (2008) describe the media's contribution in framing societal beliefs about racial discrimination, racism, and genetic determinism. Social forces also affect technological innovations, and their success depends, in part, on the social meaning assigned to any particular product. As social constructionists argue, individuals socially construct reality as they interpret the social world around them, including the meaning assigned to various technologies. If claims makers can successfully define a product as impractical, cumbersome, inefficient, or immoral, the product is unlikely to gain public acceptance. Such is the case with RU-486, an oral contraceptive known as the "abortion pill" that is widely used in France, Great Britain, China, and the United States but is opposed by many Americans (National Abortion and Reproductive Rights Action League [NARAL] 2013). Similarly, widespread media coverage of the potential risks associated with eating genetically modified (GM) food likely contributes to the disparity between U.S. adults and scientists in the belief that they are unsafe—57 percent versus 11 percent, respectively (Pew 2015b).

When the Internet first came into common usage at the turn of the 21st century, there were significant gender differences in the use of and attitudes toward the Internet. Men used the Internet more than women, were more likely to identify their ability to use the Internet as important to their self-concept, and experienced less anxiety about Internet use than women (Joiner et al. 2005). Ten years later, Internet use has dramatically changed. Dominant trends in Internet use such as social networking, mobile technology, and micro-blogging have led some to argue that the Internet has become "feminized." Has the supposed feminization of the Internet closed the gender gap in Internet use? In 2012, Joiner and colleagues revisited their original study to answer this question.

Sample and Methods

Joiner et al. (2012) researched gender differences in Internet use through survey questionnaires distributed to 501 first-year psychology undergraduate students from six UK universities. The sample included 389 women, 100 men, and 12 individuals who did not specify their gender, with a mean age of 20.1 years. The sample was selected to as closely resemble the 2002 study sample as possible.

Survey questionnaires are a preferred method of collecting quantitative data from a large number of individuals. The survey asked students to identify what types of computer technology they owned and used (e.g., laptop, tablet, smartphone, etc.), at what age they first started using the Internet, what kinds of activities they performed (e.g., social networking, shopping, gaming, banking etc.), and how many hours on average they spent using the Internet every day.

Findings and Conclusions

In many ways, the gender gap in Internet usage has closed. In 2002, researchers found that men had stronger Internet identification and less Internet anxiety than women. Ten years later, there were no gender differences in Internet identification or anxiety. The researchers found no gender differences in the age at which men and women first started using the Internet (mean age = 11 years); their likelihood of owning laptops, smartphones, or tablets; or the amount of time they spend using the Internet per day (mean = 3.4 hours).

However, important gender differences in Internet use remain. First, men used the Internet for a broader range of activities than women. Second, there were differences in the types of activities that men and women engaged in on the Internet. Men and women were equally likely to use the Internet for health information and banking; however, women were more likely than men to use the Internet for communication, social networking, and travel reservations. Men were more likely to use the Internet for entertainment, gaming, gambling, downloading music and videos, and visiting websites with adult content. The authors conclude that gender differences in Internet usage reflect broader sociocultural gender differences. Thus, while young people in general share some commonalities in their relationship to technology, it is not accurate to assume that all young people use Internet technology in the same way or share the same abilities and experiences with Internet use.

SOURCE: Joiner et al. 2012.

Not only are technological innovations subject to social meaning, but who becomes involved in what aspects of science and technology is also socially defined. Men, for example, far outnumber women in earning computer science degrees, as many as 10 to 1 at some schools. Some of the disparity can be explained by variations in interest. In a sample of first-year college students, although fluctuating over time, women consistently expressed less interest in computer science than men (Sax et al. 2015). Societal definitions of men as rational, mathematical, and scientifically minded may have contributed to the interest gender gap. This chapter's *Social Problems Research Up Close* feature highlights the similarities and differences between how men and women use the Internet.

Technology and the Transformation of Society

A number of modern technologies are considerably more sophisticated than technological innovations of the past. Nevertheless, older technologies have influenced the nature of work as profoundly as the most mind-boggling modern inventions. Postman (1992) described how the clock—a relatively simple innovation that is taken for granted in today's world—profoundly influenced the workplace and with it the larger economic institution:

> The clock had its origin in the Benedictine monasteries of the twelfth and thirteenth centuries. . . . The bells of the monastery were to be rung to signal the canonical hours; the mechanical clock was the technology that could provide precision to these rituals of devotion. . . . What the monks did not foresee was that the clock is a means not merely of keeping track of the hours but also of synchronizing and controlling the actions of men. And thus, by the middle of the fourteenth century, the clock had moved outside the walls of the monastery, and brought a new and precise regularity to the life of the workman and the merchant. . . . In short, without the clock, capitalism would have been quite impossible. The paradox . . . is that the clock was invented by men who wanted to devote themselves more rigorously to God; it ended as the technology of greatest use to men who wished to devote themselves to the accumulation of money. (pp. 14–15)

Today, technology continues to have far-reaching effects not only on the economy but also on every aspect of social life. In the following section, we discuss societal transformations resulting from various modern technologies, including workplace technology, computers, the Internet, and science and biotechnology.

> Technology continues to have far-reaching effects not only on the economy but also on every aspect of social life.

Technology and the Workplace

All workplaces—from government offices to factories and from supermarkets to real estate agencies—have felt the impact of technology. A survey of working adults found that more than half say that email and the Internet are "very important" to doing their jobs, and cell phones outrank landline phones in importance at work (Pew 2014e). Some technology lessens the need for supervisors and makes control by employers easier. For example, employees in the Department of Design and Construction in New York City must scan their hands each time they enter or leave the workplace. The use of identifying characteristics such as hands, fingers, and eyes is part of a technology called *biometrics*. Union leaders "called the use of biometrics degrading, intrusive and unnecessary and said that experimenting with the technology could set the stage for a wider use of biometrics to keep tabs on all elements of the workday" (Chan 2007, p. 1).

Technology can also make workers more accountable by gathering information about their performance. In addition, through time-saving devices such as personal digital assistants (PDAs) and battery-powered store-shelf labels, technology can enhance workers' efficiency. Wearable technology is also expected to improve worker productivity. For example, one British grocery store uses digital armbands to track the delivery, transport, and shelving of goods, which makes it unnecessary for workers to mark clipboards. Other companies use wearable technology to provide assembly line workers instructions without having to stop and look at a manual, and to monitor fatigue among backhoe operators to prevent accidents (Boitnott 2015).

Technology, however, can also contribute to worker error. The use of computerized medical records, intended to improve efficiency and reduce medical error, has met with mixed results in the health care industry. In one case, a teenage patient was given a massive overdose of an antibiotic leading to a grand mal seizure that nearly led to his death. The error happened when the doctor ordering the medication through a computer system failed to notice that the drop-down menu on the computer screen was set to "milligrams per kilogram" rather than "milligrams." One doctor who worked on the case suggested that such medical errors are a result of a lag between the advancements of technology and the understanding of that technology among the humans who use it (Wachter 2015).

Technology is also changing the location of work (see Chapter 7). Nearly 3.5 million employees—2.7 percent of the American workforce—telecommute (i.e., consider home their primary place of work) (Global Workforce 2015). Although there is some indication that telecommuting increases worker productivity and satisfaction, the incorporation of advanced technology can also blur the distinction between work and personal life. Unlike structural functionalists who would argue that telecommuting

Automation means that machines can now perform the labor originally provided by human workers, such as robots performing tasks on automobile assembly lines.

Adam Lubroth/The Image Bank/Getty Images

provides a flexible work option for working parents, Noonan and Glass (2012), consistent with a conflict perspective, suggest that "telecommuting appears . . . to have become instrumental in the general expansion of work hours, facilitating workers' needs for additional work time beyond the standard workweek and/or the ability of employers to increase or intensify work demand among their salaried employees" (p. 38).

Robotic technology has also revolutionized work. Although the economic downturn of 2009 and 2010 led to a reduction in the sales of robotic equipment, the record sales and growth from 2011 to 2014 is predicted to continue through 2017 (International Federation of Robotics 2014). Google's intensive, and highly secretive, R&D investments in robotics have fueled wild speculation about the future of robotics in the workplace, but they clearly point to a larger trend toward the incorporation of robots into all occupational sectors (Henn 2014).

Ninety percent of robots work in factories, and more than half of these are used in heavy industry, such as automobile manufacturing. While the most common use for robots has been in manufacturing, a survey of robotics experts indicates that about half believe that robots will play a significant role in both blue-collar and white-collar jobs in the next decade (Pew 2014a). Robotics experts are also divided on whether these "digital agents" will displace existing jobs; 52 percent believe that while some jobs will be eliminated by new technologies, the economy will adapt by inventing new types of work, resulting in a net gain in job creation.

Unoccupied aerial vehicles, also known as drones, are quickly becoming a popular form of robotics technology. The Federal Aviation Administration (FAA) predicts that private drones will be a $90 billion industry within the next decade (Public Broadcasting Service 2015). As the technology becomes more affordable, the use of drones will become more commonplace, for example, in package delivery, photography and filmmaking, agriculture, law enforcement, and recreation. Critics warn, however, that the lag in regulations regarding drone use will soon cause criminal and safety concerns. What restrictions, if any, do you think should be placed on the private use of drones?

The Computer Revolution

Early computers were much larger than the small machines we have today and were thought to have only esoteric uses among members of the scientific and military communities. In 1951, only about a half dozen computers existed (Ceruzzi 1993). The development of the silicon chip and sophisticated microelectronic technology allowed tens of thousands of components to be imprinted on a single chip smaller than a dime. This advancement made computers affordable and led to the development of laptop computers, cellular phones, digital cameras, the iPad, and portable DVDs. Although the first PC was developed over 30 years ago, today 83.8 percent of U.S. households report having a computer in the home compared to 61.8 percent just a decade ago (File and Ryan 2014).

Computer and Internet use is associated with several important demographic variables (see Table 14.2). Most important, when examining the table, note the following points:

- Older Americans and those with a disability are the least likely to live in a house with a computer or to have high-speed Internet access.
- Eighteen- to 34-year-olds are the most likely to live in a house with a computer.
- Asians and white, non-Hispanics are the most likely to live in a house with a computer and to have high-speed Internet access.
- As educational attainment increases, the likelihood of living in a home with a computer and of having access to high-speed Internet increases.

TABLE 14.2. Computer and Internet Use by Individual Characteristics, 2014

Characteristics	LIVES IN A HOUSE WITH A COMPUTER			LIVES IN A HOUSE WITH INTERNET USE	
	Total	Desktop or laptop computer	Handheld computer[1]	With some Internet subscription	With high-speed Internet connection
Total	**88.4**	**83.0**	**71.0**	**79.0**	**78.1**
Age					
0–17 years	92.2	85.1	80.1	81.2	80.6
18–34 years	92.7	85.5	81.6	81.2	80.7
35–44 years	92.5	87.1	79.8	83.3	82.7
45–64 years	88.3	84.5	66.9	80.6	79.6
65 years and older	71.0	68.3	37.8	64.3	62.4
Race and Hispanic origin					
White alone, non-Hispanic	90.1	86.5	71.5	82.5	81.6
Black alone, non-Hispanic	81.9	72.5	65.7	67.3	66.6
Asian alone, non-Hispanic	95.3	93.2	82.5	89.9	89.3
Hispanic (of any race)	84.3	74.6	68.5	71.1	70.2
Sex					
Male	88.7	83.3	71.7	79.4	78.5
Female	88.0	82.6	70.3	78.5	77.6
Disability					
With a disability	73.9	68.7	48.3	63.8	62.5
Without a disability	90.4	85.0	74.2	81.1	80.3
Total civilians 16 years and older	**87.4**	**82.5**	**68.5**	**78.4**	**77.5**
Educational attainment					
Less than high school graduate	66.1	57.6	45.4	53.7	52.6
High school graduate (includes equivalency)	79.9	73.8	55.1	69.7	68.3
Some college or associate's degree	91.2	86.8	70.7	82.4	81.4
Bachelor's degree or higher	96.3	94.6	81.5	91.5	90.8

[1] Handheld computer use includes smart mobile phones and other handheld wireless computers.

SOURCE: File and Ryan 2014.

Although computer and Internet usage has grown rapidly in the past decade, a global divide in computer and Internet access still exists. In the United States, 87 percent of Americans report having at least occasional access to the Internet and 80 percent own a computer, whereas only 9 percent of Pakistanis have occasional access to the Internet and 12 percent own a computer (Pew 2015a). The most common computer Internet activity around the world, as in the United States, is using social media.

Computers are big business, and the United States is one of the most successful producers of computer technology in the world, boasting several of the top companies. The largest PC producer is Lenovo, followed by Hewlett Packard, Dell, Asus, and Acer. Although the sale of desktop and laptop computers has been steadily declining, the sale of mobile devices and tablets has been increasing (Gartner Research 2015).

Computer software is also big business and, in some cases, too big. Since 1999, Microsoft has been battling the U.S. Department of Justice over anti-trust violations. These laws prohibit unreasonable restraint of trade. At issue are Microsoft's Windows operating system and the vast array of Windows-based applications (e.g., spreadsheets, word processors, tax software)—applications that *only* work with Windows. As a result of the appellate process and other delays, Microsoft's compliance with the final judgment remains incomplete and continues to be monitored by the courts (U.S. Department of Justice 2015).

Microsoft is not the only software giant under legal attack. Both Apple and Samsung, in more than 40 lawsuits between 2012 and the present, have claimed patent violations in their mobile devices (Chowdhy 2014). In 2015, the European Union brought antitrust charges against Google, alleging that the company manipulated its search engine to favor its own retail services and, similar to the charges brought against Microsoft, that Google had unfairly compelled manufacturers of Android devices to use only Google-based services. Resolution of the case is expected to take many years (Savel 2015).

Information and Communication Technology and the Internet

Information and communication technology, or ICT, refers to any technology that carries information. Most information technologies were developed within a 100-year span: taking pictures and telegraphy (1830s), rotary power printing (1840s), the typewriter (1860s), transatlantic cable (1866), the telephone (1876), motion pictures (1894), wireless telegraphy (1895), magnetic tape recording (1899), radio (1906), and television (1923) (Beniger 1993). The concept of an "information society" dates back to the 1950s, when an economist identified a work sector he called "the production and distribution of knowledge." In 1958, 31 percent of the labor force was employed in this sector; today, more than 50 percent is. When this figure is combined with those in service occupations, more than 75 percent of the labor force is involved in the information society.

The **Internet** is an international infrastructure—a network of networks—that distributes knowledge and information. In 2014, 87 percent of all Americans used the Internet from some location, although home Internet access is generally lower (File and Ryan 2014; Fox and Rainie 2014). Among those who do not go online, about one-third report that the main reason they don't use the Internet is because they don't think it is relevant to their lives. Another one-third report that it is too difficult to use, and 19 percent say that the expense of owning a computer and subscribing to Internet service is too high. The remaining 7 percent say they lack physical access to a computer or the Internet (Zickuhr 2014). Nearly half of Internet nonusers are over the age of 65.

The most recent advancement in Internet technology is the **Internet of Things (IoT)**, in which nearly every device with an on-off switch can be connected to the Internet. The vast array of network and information sharing made possible by the IoT has the

Internet An international information infrastructure available through universities, research institutes, government agencies, libraries, and businesses.

Internet of Things (IoT) The potential connectivity of all electronic devices; widespread information sharing between people and the objects of everyday life.

potential to radically transform social relations, with both positive and negative consequences (Morgan 2014). It may be convenient for your car to "know" your schedule and automatically redirect you around a traffic jam to make sure you don't miss your appointment, but do you want your wearable device to report to your boss every time you take a coffee break? What are the boundaries of privacy and security when everything you use shares your information with everything else?

IoT is made possible by wireless Internet access. The growth of mobile Internet capabilities has led to "always-present" connectivity as users working on laptops, tablets, and smartphones access the Internet from coffee shops, classrooms, and shopping malls. Approximately one-fourth of teens say that they are online "almost constantly" (Lenhart 2015). The near constant use of the Internet is driven by the high rate of mobile technology usage among teens: Nearly three-quarters have access to a smartphone, with an additional 30 percent using a basic cell phone. Among those with smartphone access, 94 percent go online daily.

High-speed broadband use has also increased over time, from 4 percent of American households in 2001 to 70 percent of American households in 2013. The demographics of broadband access is similar to those of computer and Internet use; that is, those who are younger, more educated, and wealthier are more likely to have home broadband (File and Ryan 2014; Zickuhr and Smith 2013).

E-Commerce. **E-commerce** is the buying and selling of goods and services over the Internet, and primarily includes online shopping and online banking. In OECD countries, the proportion of people who buy goods online rose from 30 percent in 2007 to 47 percent in 2013 (OECD 2014). Online sales represent nearly half of all U.S. sales, with books and magazines being the most common online purchases, followed by clothing, computer hardware, and computer software (U.S. Census Bureau 2013). Motivations to purchase products online include convenience, the ability to compare prices, and less expensive products in some states where there is no Internet sales tax. The Marketplace Fairness Act of 2015, presently being debated in Congress, would "require all sellers not qualifying for a small-seller exception . . . to collect and remit sales and use taxes with respect to remote sales."

Online banking as well as mobile banking has increased significantly over the last decade (Fox 2013). As more financial exchanges occur online, the risks and benefits of e-commerce continue to be debated. Bitcoins provide a recent example of the technological dualism of e-commerce. Invented in 2008, bitcoins are an entirely digital currency that are not backed, and therefore not regulated, by any government. Bitcoins allow for an ease and flexibility in global economic transactions that conventional national currencies do not allow, but critics point out that such unregulated and anonymous exchanges are used to buy and sell illegal goods and services on a digital black market, subject to wild swings in value, and at high risk of theft by hackers (Peek 2015).

Health and Digital Medicine. Many Americans turn to the Internet to address health issues and concerns. In a survey on the Internet and health, over a third of U.S. adults reported going online to research a health condition (Fox and Duggan 2013). Women, youth, those with annual household incomes over a $75,000, and the college educated were the most likely to seek an online diagnosis.

Many other Americans, 72 percent, sought general health or medical information on the Internet, and half of those searches were on behalf of friends or family members. Interestingly, sociologically, there is a "social life" to health information on the Internet. It is not uncommon for people to share their health stories, post symptoms and diagnoses, and otherwise provide peer-to-peer health support (Fox and Duggan 2013). In rural China, where access to the public health care system is limited by geographic distance and long waits for appointments, a mobile app that allows patients to communicate with doctors and receive virtual diagnoses is estimated to have 100 million users in 2015 (Shadbolt 2014).

> The growth of mobile Internet capabilities has led to "always-present" connectivity as users working on laptops, tablets, and smartphones access the Internet from coffee shops, classrooms, and shopping malls.

e-commerce The buying and selling of goods and services over the Internet.

Besides providing medical information, there is considerable evidence that online medical records help improve medical care:

A paper record is a passive, historical document. An electronic health record can be a vibrant tool that reminds and advises doctors. It can hold information on a patient's visits, treatments, and conditions, going back years, even decades. It can be summoned with a mouse click, not hidden in a file drawer in a remote location and thus useless in medical emergencies. (Lohr 2008, p. 1)

Digital patient records are thus the first step in creating **learning health systems**, whereby physicians, looking across patient populations, can identify successful treatments or detect harmful interactions (Lohr 2011). Furthermore, mobile devices such as smartphones and tablets "have the computing capability, display, and battery power to become powerful medical devices that measure vital signs and provide intelligent interpretation or immediate transmission of information" (p. 883).

Lastly, there is evidence that technology can help mediate the soaring cost of health care. For example, high school students in Illinois used a 3-D printer that had been donated to the school to build a prosthetic hand for a 9-year-old girl who had been born without fingers, for a total cost of $5 (Thornhill 2014). Other important sources of cost effectiveness include electronic claims processing and reducing medical errors through more effective diagnostic and treatment interventions. As of July 2012, the federal government spent over $6.6 billion in health information technology incentives (Castro 2013).

The Search for Knowledge and Information. The Internet, perhaps more than any other technology, is the foundation of the information society. Whether reading an online book, taking a MOOC (massive open online course; see Chapter 8), mapping directions, visiting the Louvre virtually, or accessing Wikipedia, the Internet provides millions of surfers with instant answers to questions previously requiring a trip to the library.

There is concern, however, that the very way in which the "Google generation" reads, thinks, and approaches problems has been altered by the new technology. For example, maps on mobile devices change the way we experience the world. Physical barriers and the risks of "getting lost" no longer limit our movements, but the immersive experience of the user-centered map restricts our sense of space and place (McMullan 2014). Journalist Nicholas Carr argues in his 2010 book *The Shallows: What the Internet Is Doing to Our Brains* that the use of the Internet promotes a chronic state of distraction that we are often unaware of:

Our focus on a medium's content can blind us to the deep effects. We're too busy being dazzled or disturbed by the programming to notice what's going on inside our heads. In the end, we come to pretend that the technology itself doesn't matter. It's how we use it that matters, we tell ourselves. The implication, comforting in its hubris, is that we're in control. The technology is a tool, inert until we pick it up and inert again once we set it aside. (p. 3)

Finally, with the lightning growth of the Internet, there are concerns that information is often outdated, difficult to find, and limited in scope (Mateescu 2010). Some are already looking ahead to a time when search engines look for information not syntactically (i.e., based on *combinations* of words and phrases) but semantically (i.e., based on the *meaning* of words and phrases). The **Semantic Web**, sometimes referred to as *Web 3.0,* entails not only pages of information but also pages that describe the interrelationship between the pages of information, resulting in "smart media" (Cambridge 2015).

Politics and e-Government. Technology is changing the world of politics. The number of registered voters who used social media to follow political candidates doubled from 2010 to 2014, and more than 40 percent of registered voters aged 18 to 29 say that they use mobile technology to follow political news (Smith 2014).

learning health systems
The result of electronic records whereby physicians can look across patient populations and identify successful treatments or detect harmful interactions.

Semantic Web Sometimes called Web 3.0, a version of the Internet in which pages not only contain information but also describe the inter-relationship between pages; sometimes called *smart media.*

Social media sites like Twitter and Facebook have been central to political events around the world. For example, Arab Spring protests throughout the Middle East in 2011 were coordinated through social media sites, often in spite of government efforts to restrict access. Although social media can facilitate political change, it can also present risks to the users. Government officials in Tunisia, Iran, Egypt, and elsewhere in the Middle East have used social media postings to track, arrest, and charge political dissidents (Arab Network for Human Rights Information 2015). In the fight against domestic terrorism in the United States, the FBI has been struggling to find a balance between trying to prevent terror groups such as ISIS from using social media to reach potential recruits and using social media as a tool to track the activities of terrorist organizations (K. Johnson 2015).

Social Networking and Blogging. Social networks (e.g., Facebook, Twitter) and blogs comprise a sector of the Internet called **membership communities**. Membership communities have changed in three substantively significant ways in recent years: increasing numbers of people visit membership communities, people are spending substantially more time interacting in these communities, and older people are participating in membership communities. In 2015, over 2 billion people, or about a third of the world's population, had an active social media account; Facebook alone had nearly 1.4 billion active users (Kemp 2015). Globally, nearly a quarter of Facebook users check their account five or more times a day. Among American social media users, 70 percent checked Facebook daily. Although people under the age of 30 still comprise the largest proportion of social media users, of those who use the Internet, 63 percent of 50- to 64-year-olds and 56 percent of those 65 years and older report using Facebook (Duggan et al. 2015).

Science and Biotechnology

Although recent computer innovations and the establishment of the Internet have led to significant cultural and structural changes, science and its resulting biotechnologies have produced not only dramatic changes but also hotly contested issues with public policy implications. In this section, we look at some of the issues raised by developments in genetics, food and biotechnology, and reproductive technologies.

Patents on genes by technology companies led to protests across the United States. In 2013, the U.S. Supreme Court found that a corporation cannot own a gene sequence that, if tested, might reveal that a person has a serious disease.

Genetics. Molecular biology has led to a greater understanding of the genetic material found in all cells—DNA (deoxyribonucleic acid)—and with it the ability for **genetic testing**. Genetic testing "involves examining your DNA, the chemical database that carries instructions for your body's functions," and using that information to identify abnormalities or alterations that may lead to disease or illness (Mayo Clinic 2013, p. 1).

A growing body of research indicates that human characteristics once thought to be largely social or psychological in nature are, at least in part, genetically induced. These include behavioral and psychological traits, such as personality characteristics, addiction, depression, autism, schizophrenia and anorexia, as well as physical illnesses such as sickle-cell disease, breast cancer, Alzheimer's, and cystic fibrosis (De Rubeis et al. 2014; Harmon 2006; Oak Ridge National Laboratory 2011; Picard and Turnbull 2013; Wheelwright 2014).

membership communities Internet sites where participation requires membership and members regularly communicate with one another for personal and/or professional reasons.

genetic testing Examination of DNA in order to identify abnormalities or alterations that may lead to disease or illness.

The U.S. Human Genome Project (HGP), a 13-year effort to decode human DNA, was completed in 2003. Conclusion of the project has transformed medicine, particularly by revealing how the physical and the social interact:

> All diseases have a genetic component whether inherited or resulting from the body's response to environmental stresses like viruses or toxins. The successes of the HGP have . . . enabled researchers to pinpoint errors in genes—the smallest units of heredity—that cause or contribute to disease. (Human Genome Project 2007, p. 1)

The hope is that, if a defective or missing gene can be identified, possibly a healthy duplicate can be acquired and transplanted into the affected cell. This is known as **gene therapy**. Alternatively, viruses have their own genes that can be targeted for removal. Experiments are now under way to accomplish these biotechnological feats.

Food and Biotechnology. **Genetic engineering** is the ability to manipulate the genes of an organism in such a way that the natural outcome is altered. Genetically modified (GM) food, also known as genetically engineered food, and genetically modified organisms involve this process of DNA recombination—scientists transferring genes from one plant into the genetic code of another plant.

The vast majority of soy, corn, canola, and sugar beets grown in the United States are genetically engineered. Because these ingredients are widely used in processed foods, nearly all packaged food sold in the United States and Canada contain genetically modified organisms (GMOs). Most American adults believe that foods grown with pesticides (72 percent) and GM foods are unsafe for human consumption (65 percent), and 92 percent believe the government should require labels that identify food as "genetically modified" or "bio-engineered" (*Consumer Reports* 2015; Pew 2015b).

Sixty countries mandate GMO labeling, but efforts to pass similar regulations in the United States have met with controversy and resistance from the agricultural industry. Biotechnology companies and other supporters of GM foods argue this technology has the potential to alleviate hunger and malnutrition, as it can enable farmers to produce higher-yield crops. In 2015, the Food and Drug Administration approved apples and potatoes that had been genetically modified to not bruise or brown as safe for human consumption (Rossignol 2015). Critics argue that the world already produces enough food for all people to have a healthy diet. If food were distributed equally, every person would be able to consume 2,790 calories a day (World Hunger Education Service 2015). The fundamental causes of hunger, these critics argue, is not lack of technology but rather poverty and unequal access to food and land.

Critics also argue that GMO practices can have wide-ranging effects on human health and the environment. Biotechnology companies claim that crops that are genetically designed to repel insects negate the need for chemical (pesticide) control and thus reduce pesticide poisoning of land, water, animals, foods, and farmworkers. However, critics are concerned that insect populations can build up resistance to GM plants with insect-repelling traits, which would necessitate increased rather than decreased use of pesticides. The inability of humans to control chain reactions in ecosystems also raise concerns. Critics argue that the introduction of genetically modified "frankenfish" into the wild could weaken the genetic pools of wild populations, leading to their collapse and affecting the broader ecosystem (Friends of the Earth [FOE] 2015).

Human health concerns include possible toxicity, carcinogenicity, food intolerance, antibiotic resistance buildup, decreased nutritional value, and food allergens to GM foods—all examples of **technology-induced diseases**. An international study conducted by the World Health Organization of the chemicals in Monsanto's most common weed killer found strong evidence that water contamination from the use of these chemicals in commercial agriculture causes cancer (Guyton et al. 2015).

Nanotechnology. **Nanotechnology**, which refers to the manipulation of materials and creation of structures and systems at the scale of atoms and molecules, is also being used to modify food ingredients (FOE 2014). This technology is used to add nutrition and

---

Human health concerns [of genetically modified foods] include possible toxicity, carcinogenicity, food intolerance, antibiotic resistance buildup, decreased nutritional value, and food allergens.

gene therapy The transplantation of a healthy gene to replace a defective or missing gene.

genetic engineering The manipulation of an organism's genes in such a way that the natural outcome is altered.

technology-induced diseases Diseases that result from the use of technological devices, products, and/or chemicals.

nanotechnology The manipulation of materials and creation of structures and systems at the scale of atoms and molecules.

flavor to foods, as well as to add antibacterial ingredients to food packaging. Between 2008 and 2014, the number of food and beverage products on the market with nano-ingredients present grew more than 10-fold (FOE 2014). Although nanotechnology has the potential to improve nutritional value and shelf life of food, critics argue that there is also a high likelihood that it will have negative consequences for human health and the environment. In particular, nanofoods use particles of silver, titanium dioxide, zinc, and zinc oxide that have been shown to be highly toxic to animals in laboratory testing. Because of the small size of these particles, there is greater risk that these particles will access and alter human cells. The lack of regulation of nanotechnology increases the risk of harmful nanochemicals being released into soil and water, where these particles can alter plant and animal life at the cellular level (FOE 2014).

Activists protesting the agribusiness giant Monsanto and genetically modified organisms (GMOs) wear anonymous masks in front of the White House in 2013.

Reproductive Technologies. The evolution of "reproductive science" has been furthered by scientific developments in biology, medicine, and agriculture. At the same time, at least historically, its development has been hindered by the stigma associated with sexuality and reproduction, its link with unpopular social movements (e.g., contraception), and the feeling that such innovations challenge the natural order (Clarke 1990). Nevertheless, new reproductive technologies have been and continue to be developed.

Although there are other reproductive technologies (e.g., in vitro fertilization), abortion more than any other biotechnology epitomizes the potentially explosive consequences of technological innovation (see also Chapter 13). **Abortion** is the removal of an embryo or fetus from a woman's uterus before it can survive on its own.

In the United States, since the U.S. Supreme Court's ruling in *Roe v. Wade* in 1973, abortion has been legal. However, several Supreme Court decisions have limited the scope of the *Roe v. Wade* decision (e.g., *Planned Parenthood of Southeastern Pennsylvania v. Casey*). Furthermore, several recent legislative initiatives in Congress are intended to restrict or prohibit a woman's access to a lawful abortion. For example, many states have enacted laws that require abortion providers to operate hospital-style surgical centers, regulations that will lead to the closure of many clinics. Clinics in Texas and Mississippi challenged the law, arguing that the new regulations were not necessary to improve the safety of the procedure but rather were designed to place an undue burden on women seeking abortions. The appeals court in Mississippi struck down the law, while the Texas court upheld it (Ludden 2015). However, in 2015, the U.S. Supreme Court granted an emergency appeal from the clinics that, at least for now, halts enforcement of the restrictions (Associated Press 2015).

Historically, abortions are banned when the fetus is considered viable, usually around 22 to 26 weeks from conception. Recently, however, state laws have made legal abortions more restrictive. With an exception for the life or health of the mother, abortions after 20 weeks are prohibited in 13 states (NARAL 2015). Other recently enacted state restrictions include (1) that abortion must take place in a hospital and be performed by a licensed physician, (2) gestational limits, (3) restrictions on state spending for abortions, (4) restrictions on private insurance coverage for abortions, (5) the right of individual or institutional refusal to perform an abortion, (6) state-mandated counseling, (7) waiting periods, and (8) parental involvement if involving a minor (Guttmacher Institute 2015).

abortion The intentional termination of a pregnancy.

the HUMAN side

What Happens to Women Who Are Denied Abortions?

Very little is known about women who seek abortions but are turned away, usually because of being too far along in their pregnancy. Most studies on abortion compare women who have had an abortion to those who have not, which, as researcher Dianna Green Foster notes, is like comparing people who get a divorce with those who remain married rather than to those who wanted a divorce but couldn't get one (Lang 2013). The story of S. is a typical one: young, poor, and alone with an unknown future, like many of the subjects in Foster's study.

S. arrived alone at a Planned Parenthood. . . . As she filled out her paperwork, she looked at the women around her. Nearly all had someone with them; S. wondered if they also felt terrible about themselves. . . . She began to cry quietly. . . .

. . . She texted one of her sisters, "Do you think God would forgive me if I were to murder my unborn child?"

"Where are you?" her sister asked. "Are you O.K.?"

"I'm at Planned Parenthood, about to have an abortion."

"God knows your heart . . . ," her sister texted back. "I think God will understand."

. . . She was a strong believer in birth control—in high school she was selected to help teach sex education. But having been celibate for months and strapped for cash, she stopped taking the pill. Then an ex-boyfriend came around. . . .

S., then 24, took her first pregnancy test . . . and it was late at night when she read the results. She'd never been pregnant before. . . . Her ex had begun a new relationship, and she knew he wouldn't be there to support her or a child. She was working five part-time jobs to keep herself afloat and still didn't always have enough money for proper meals. How could she feed a baby? She . . . made an appointment at Planned Parenthood.

At the clinic, a counselor comforted S. and asked her . . . if anyone had coerced her into making this decision. No, S. explained, she was simply not ready to have a child. . . .S. estimated that she was about three months pregnant.

In the exam room, a technician asked her to lie down. She did an ultrasound, sliding the instrument across S.'s stomach: "Oh . . . it shows here that you are a little further along." She repeated the exam. S., she estimated, was nearly 20 weeks pregnant, too far along for this Planned Parenthood. . . .

Planned Parenthood gave S. a packet of information, including two pieces of paper—one green, for adoption, and one yellow, for other abortion providers. . . She called a clinic in Oakland and took the first available appointment, just after Christmas . . . another of S.'s sisters . . . found a place called First Resort, which provided abortion counseling. S. didn't know that First Resort's president once said that "abortion is never the right answer." . . .

S. went to First Resort the day before her appointment in Oakland. . . . The nurse asked S. if she wanted to see the baby and turned the monitor toward her: "Look! Your baby is smiling at you." S. was shaken, convinced she also saw the baby smiling. The nurse told her that she was at least a week further along than the Planned Parenthood estimate. . . . S. sobbed all the way to her car and called the clinic in Oakland, giving it the First Resort estimate. If it was correct, they told her, she would be past its deadline. . . . She was out of gas money, hadn't eaten a decent meal in weeks and resigned herself to the fact that, no matter what she wanted or how it would affect her life, she was going to have a baby.

SOURCE: Lang 2013.

Abortion is a complex issue for everyone, but especially for women, whose lives are most affected by pregnancy and childbearing. Women who have abortions are disproportionately poor, unmarried minorities who say that they intend to have children in the future. Abortion is also a complex issue for societies, which must respond to the pressures of conflicting attitudes toward abortion and the reality of high rates of unintended and unwanted pregnancy. The debate over abortion has also complicated health care reform as conservatives, citing a 30-year ban on using taxpayer's money to pay for elective abortions, battle the Obama administration and abortion rights supporters.

Attitudes toward abortion tend to be polarized between two opposing groups of abortion activists—pro-choice and pro-life. As recently as 2015, public opinion surveys indicated that 50 percent of Americans identify as pro-choice and 44 percent as pro-life (Gallup 2015a). Advocates of the pro-choice movement hold that freedom of choice is a central human value; that procreation choices must be free of government interference; and that because the woman must bear the burden of moral choices, she should have the right to make such decisions. Alternatively, pro-lifers hold that an unborn fetus has a right to live and be protected, that abortion is immoral, and that alternative means of resolving an unwanted pregnancy should be found.

Attitudes about abortion vary by religion and age. Three-quarters of evangelical Protestants and more than one-half of Catholics say that abortion is morally wrong, while 38 percent of mainline Protestants and 25 percent of religiously unaffiliated Americans say that abortion is morally wrong. Adults under 30 are less likely than adults over 30 to know that *Roe v. Wade* was a decision about abortion (44 percent compared with 62 percent, respectively), and young adults are more likely to say that abortion is "not that important" compared with other issues (62 percent compared with 53 percent, respectively) (Lipka 2015).

Attitudes about abortion are shaped by technological advancements; the likelihood of a premature baby surviving a birth in the second trimester, about 22 to 27 weeks, was unheard of when *Roe v. Wade* was first enacted in 1973. Now the survival rate of the earliest premature births is about 25 percent (Westcott 2015). For the nearly half of Americans who view abortion as morally wrong, the proliferation of reproductive technologies raises ethical dilemmas (Lipka 2015). In a highly controversial decision in 2014, the Supreme Court ruled that some businesses cannot be compelled to provide insurance coverage for their employees for certain reproductive technologies. The majority ruled in a 5–4 decision that coverage of reproductive technologies that violate a business owner's religious beliefs is a substantial burden (Mears and Cohen 2014). In another case, a Christian college was exempted from having to provide full contraception coverage to its employees and students (Holpuch 2014). Both of these cases centered on the issue of whether an employer's religious beliefs are burdened by the provision of coverage for reproductive technologies. However, in her dissent, Justice Sonya Sotomayor argued that "thinking one's religious beliefs are substantially burdened—no matter how sincere or genuine that belief may be—does not make it so" (Holpuch, 2014, p. 1).

> Abortion is . . . a complex issue for societies, which must respond to the pressures of conflicting attitudes toward abortion and the reality of high rates of unintended and unwanted pregnancy.

Cloning, Therapeutic Cloning, and Stem Cells. In July 1996, scientist Ian Wilmut of Scotland successfully cloned an adult sheep named Dolly. To date, cattle, goats, mice, pigs, cats, rabbits, and horses have also been cloned. Although generally legal in the United States, worldwide concerns over human cloning have led to its prohibition in over 60 countries and several international agreements ban the procedure (Rovner 2013).

Human cloning has potential medical value by allowing people to reserve therapeutic cells to treat diseases, and by allowing an alternative reproductive route for infertile couples or those who are at risk of transmitting a genetic disease to their offspring (Kahn 1997). Arguments against cloning are largely based on moral and ethical considerations. Critics of human cloning suggest that, whether used for medical therapeutic purposes or as a means of reproduction, human cloning is a threat to human dignity (Human Cloning Prohibition Act of 2007). Cloned humans would be deprived of their individuality, and as Kahn (1997) pointed out, "creating human life for the sole purpose of preparing therapeutic material would clearly not be for the dignity of the life created" (p. 119). In 2015, just 15 percent of a sample of U.S. adults viewed human cloning as morally acceptable (Gallup 2015b).

There are three different types of cloning. **Genetic cloning**, which is widely accepted and routinely used in laboratory work, copies genes or segments of DNA. This type of cloning is carefully regulated and not considered to have ethical implications. Therapeutic and reproductive cloning are less commonly used and raise ethical debates. **Reproductive cloning** copies the DNA of whole animals to produce an exact genetic copy. This technology has the potential to address infertility issues for sterile couples and allow those who carry deleterious genes to avoid passing genetic illnesses to their offspring. However, this technology raises concerns about human dignity and individual freedom. **Therapeutic cloning** uses stem cells from human embryos; this form of cloning has potential to treat a wide range of illnesses and diseases; however, it requires the destruction of human embryos in order to harvest stem cells (National Human Genome Research Institute 2015). **Stem cells** can produce any type of cell in the human body and thus can be "modeled into replacement parts for people suffering from spinal cord injuries or regenerative diseases, including Parkinson's and diabetes" (Eilperin and Weiss 2003, p. A6).

genetic cloning Producing copies of genes or segments of DNA.

reproductive cloning Producing genetic copies of whole animals.

therapeutic cloning Use of stem cells to produce body cells that can be used to grow needed organs or tissues; regenerative cloning.

stem cells Undifferentiated cells that can produce any type of cell in the human body.

Despite what appears to be a universal race to the future and the indisputable benefits of scientific discoveries such as the workings of DNA and the technology of IVF and stem cells, some people are concerned about the duality of science and technology. Science and the resulting technological innovations are often life-assisting and life-giving; they are also potentially destructive and life-threatening. The same scientific knowledge that led to the discovery of nuclear fission, for example, led to the development of both nuclear power plants and the potential for nuclear destruction. Thus, we now turn our attention to the social problems associated with science and technology.

Societal Consequences of Science and Technology

Scientific discoveries and technological innovations have implications for all social actors and social groups. As such, they also have consequences for society as a whole. Figure 14.3 displays the public's relative assessment of the costs and benefits of scientific research. Although most Americans believe that the benefits of scientific research outweigh the potential for harmful results, there is no denying that science and the resulting technologies have some negative consequences.

Social Relationships, Social Networking, and Social Interaction

Technology affects social relationships and the nature of social interaction. The development of telephones has led to fewer visits with friends and relatives; with the advent of DVRs, cable television, and video streaming, the number of places where social life occurs (e.g., movie theaters) has declined. The average white-collar worker checks his or her email 74 times a day and reports high levels of anxiety when not at work about the need to respond to emails. The potential health risks of "email addiction" have led some companies to install "holiday mode" to prevent workers from receiving emails outside their work hours (Roberts 2014). As technology increases, social relationships and human interaction are transformed.

Students who attend schools where cell phone use is banned showed significantly higher performance on exams than those who attended schools where phones were allowed. This difference was most dramatic for underachieving students, indicating that low-achieving students suffer more from technological distraction than high-achieving students (Beland and Murphy 2015). Technology can also increase the likelihood of people living in a social cocoon, increasing isolation from one another and from diverse viewpoints. A survey researcher at the University of Denver reports that 40 percent of people say they would avoid someone in real life who had "unfriended" them on Facebook (Kelly 2013). Additionally, Internet recommender systems, those systems that personalize search results, contribute to polarization between groups in society by echoing already existing biases (Dandekar et al. 2013).

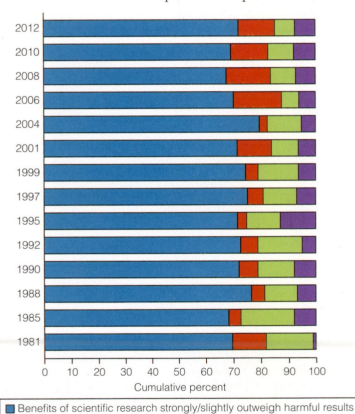

Benefits of scientific research strongly/slightly outweigh harmful results
Benefits of scientific research are about equal to harmful results
Harmful results of scientific research strongly/slightly outweigh benefits
Don't know

Figure 14.3 Public Assessment of Scientific Research, U.S. Adults, 1981–2012
SOURCE: NSF 2014.

Loss of Privacy and Security

In 2014, identity theft was the number one complaint filed with the Federal Trade Commission for the 15th year in a row (FTC 2015; see also Chapter 4). Through computers, individuals can obtain access to someone's phone bills, tax returns, medical reports, credit histories, bank account balances, and driving records. Nearly all Americans are affected by cybersecurity breaches. For example, attacks against large retailers, including Target and The Home Depot, among others, as well as large health insurance providers in 2014 and 2015 compromised the personal information of at least 200 million Americans (Granville 2015). In November 2014, the data centers of Sony Pictures were wiped clean, leading to the cancellation of the theatrical release of the *The Interview,* a comedy about North Korean leader Kim Jong-un. The subsequently leaked emails of celebrities and executives led to several highly publicized scandals and the resignation of the company's CEO (see Chapter 4).

Although just inconvenient for some, unauthorized disclosure of, for example, medical records, is potentially devastating for others. If a person's medical records indicate that he or she is human immunodeficiency virus (HIV)–positive, that person could be in danger of losing his or her job or health benefits. If DNA testing of hair, blood, or skin samples reveals a condition that could make the person a liability in an insurer's or employer's opinion, the individual could be denied insurance benefits, medical care, or even employment. In response to such fears, the **Genetic Information Nondiscrimination Act of 2008** (GINA) was passed. GINA is a federal law that prohibits discrimination in health coverage or employment based on genetic information (Equal Employment Opportunity Commission 2015).

After a five-year court battle, the European Court of Justice ruled in May 2014 that individuals have the right to determine which details should be associated with their names in Internet searches (Powles and Chaparro 2015). Supporters argue that the ruling protects individual freedoms in an age of encroaching and impersonal technology, while critics argue that the ruling opens the door for harmful censorship—for example, should public officials have the right to remove negative press about themselves? So far, the ruling only applies to European citizens. Do you think "right to forget" laws should be enacted in the United States?

Technology has created threats not only to the privacy of individuals but also to the security of entire nations (also see Chapter 15). As revealed by Edward Snowden, the National Security Administration (NSA) computer program XKeyscore allows "analysts to search with no prior authorization through vast databases containing e-mails, online chats, and the browsing histories of millions of individuals" (Greenwald 2013, p. 1). International organizations viewed the NSA program as a violation of human rights; and in 2015, a group of organizations including Wikipedia, Amnesty International, and the Human Rights Watch filed a lawsuit against the NSA (NBC News 2015).

Although the extent to which Snowden is telling the truth is unknown, the Snowden incident led the NSA to reevaluate its policies on communications technologies (Clarke et al. 2013). Not only is the federal government reevaluating how it collects and maintains data for its own uses, but it's also reevaluating its role in addressing the consequences of Internet privacy and security for private citizens:

> Breaches of privacy can cause harm to individuals and groups. It is a role of government to prevent such harm where possible, and to facilitate means of redress when harm occurs. . . . Cameras, sensors, and other observational or mobile technologies raise new privacy concerns. Individuals often do not knowingly consent to providing data. . . . Analysis technology (such as facial, scene, speech, and voice recognition technology) is improving rapidly. Mobile devices provide location information that might not be otherwise volunteered. The combination of data from those sources can yield privacy-threatening information unbeknownst to the affected individuals. (President's Council of Advisors on Science and Technology 2014, p. 1)

Genetic Information Nondiscrimination Act of 2008 A federal law that prohibits discrimination in health coverage or employment based on genetic information.

Ironically, a classified cybersecurity report leaked by Snowden revealed that the U.S. government views the expansion of encryption technology, which would prevent data from being collected from individual users, as the best means of defense against hacking crimes, which are estimated to cost the global economy $400 billion per year (Ball 2015). Cyberattacks are ranked alongside war, water crises, and global unemployment as the highest-impact risks facing global society over the next 10 years (World Economic Forum 2014).

Although newer forms of mobile technology, in particular social media, are being increasingly used by organizations to reach consumers, Americans feel less secure about sharing their information on these platforms compared with older forms of technology. In a nationally representative survey, about two-thirds of Americans felt somewhat or very secure about sharing private information with a trusted person or organization over a landline phone (Pew 2014d). Fewer than half felt secure about sharing that information in text message, email, or chat; and only 16 percent felt at least somewhat secure about sharing information on social media sites (see Figure 14.4).

A survey of cybersecurity experts indicates that Internet vulnerability is likely to increase, with the majority of experts indicating that major cyberattacks causing widespread harm are likely to occur within the next 10 years (Pew 2014b). Opting out of these technologies is not necessarily an option, as journalist Julia Angwin (2014) found when she attempted to avoid sharing all personal information online. Angwin discovered that not only is it impossible to live a normal life in today's modern society without sharing personal data, but the new reality of massive data collection means that people increasingly internalize this sense of insecurity, editing their words and thoughts and ultimately limiting their own freedom.

While Americans are increasingly falling victim to cybercrimes and identity theft, it is important to remember that much of the technology behind these crimes originate in the United States. A study conducted by the computer security company Sophos in 2014 found that the United States produces more malicious spam software than any other nation by a wide margin (Ducklin 2014).

Unemployment, Immigration, and Outsourcing

Some technologies replace human workers, and many of those jobs are unlikely to return to the labor market. A 10-year study of the use of automated technology to replace human labor found that while initial phases of technology adoption in workplaces use automation to replace physical labor, thus displacing manufacturing jobs, newer forms of technology have focused on automation of language and reasoning

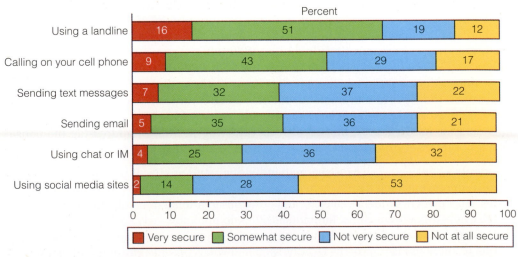

Figure 14.4 Public Perceptions of Online Privacy and Security, 2014*
SOURCE: Pew 2014c.
*Percent of U.S. adults reporting varied levels of security when sharing information with another trusted person or organization.

skills (Elliot 2014). As a result, millions of middle-class jobs in the areas of human services, transportation, education, and health care may soon be displaced. Such a shift has the potential to radically transform our social and economic system in much the same way that the decline of agricultural production and rise of manufacturing transformed 20th-century society:

> Although economists are right in principle that displaced workers should be able to move into new positions . . . it seems unlikely that the structure of the labor market can change as quickly as the technology is advancing. . . . The nation will be forced to reconsider the role that paid employment plays in distributing economic goods and services and in providing a meaningful focus for many people's daily lives. (Elliot 2014, p. 1)

Unemployment rates can also increase when companies **outsource** (sometimes called *off-shore*) jobs to lower-wage countries. Between 2001 and 2013, an estimated 3.2 million American jobs were outsourced to China alone, with three-quarters of those jobs in the manufacturing sector (Peralta 2014). It should be noted, however, that economic trends affect receiving countries as well; in other words, jobs lost in one country impact outsourcing in another. For example, IT outsourcing to South Asia resulted in a loss of jobs in the United States and a gain in India; however, increasing automation in IT services is now resulting in job losses in the Indian IT industry and an increase in jobs in international robotics firms (Overby 2015).

Other employment challenges arise due to the lack of highly skilled American workers in STEM occupations. These jobs are often filled by foreign workers who enter the United States on H1-B visas. These visas allow workers to temporarily hire foreign workers in specialty and high-tech occupations when there are not enough qualified domestic workers available. The number of H-1B visas the federal government issues each year is limited to 65,000 (U.S. Citizenship and Immigration Services 2015). Critics of the H-1B visa limitations are pressuring the federal government to increase the number of visas available annually, arguing that there are more high-tech jobs available than workers to fill them.

The Digital Divide

One of the most significant social problems associated with science and technology is the increased division between the classes. In a now oft-quoted statement, Welter (1997) notes:

> It is a fundamental truth that people who ultimately gain access to, and who can manipulate, the prevalent technology are enfranchised and flourish. Those individuals (or cultures) that are denied access to the new technologies, or cannot master and pass them on to the largest number of their offspring, suffer and perish. (p. 2)

The fear that technology will produce a "virtual elite" is not uncommon. Several theorists hypothesize that, as technology displaces workers—most notably the unskilled and uneducated—certain classes of people will be irreparably disadvantaged: the poor, minorities, and women. There is even concern that biotechnologies will lead to a "genetic stratification," whereby genetic testing, gene therapy, and other types of genetic enhancements are available only to the rich.

Globally, the digital divide reflects the economic and social conditions of a country. In general, wealthier and more educated countries have more technology than poorer countries with a less educated populace, and the gap is growing (see Figure 14.5). Much of the gap can be explained by a lack of access to devices and/or Internet connections at home or at work. Africa, for example, has 15 percent of the world's population, but only 7 percent of households have access to the Internet. Europe, however, has 12 percent of the world's population, but over 75 percent of the households have Internet connections (UN 2014). Facebook CEO Mark Zuckerburg seeks to bridge the digital divide by building infrastructure, expanding broadband access, and innovating affordable mobile technologies in the developing world (Internet.org 2015).

Similarly, in the United States, the wealthier the family, the more likely the family is to have Internet access. Of households with annual incomes of $150,000 or more, 98.1 percent

Several theorists hypothesize that, as technology displaces workers—most notably the unskilled and uneducated—certain classes of people will be irreparably disadvantaged: the poor, minorities, and women.

outsource A practice in which a business subcontracts with a third party to provide business services.

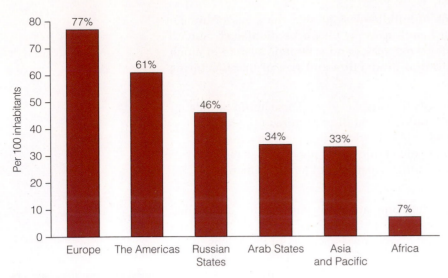

Figure 14.5 Percentage of Households with Access to Internet in 2013, by Region of the World
SOURCE: UN 2014.

have Internet access, and 94.5 percent have a high-speed Internet connection. However, only 62.4 percent of households with income levels below $25,000 a year have an Internet connection, and fewer than half have a high-speed connection (File and Ryan 2014). The digital divide between urban and rural America is also stark. While 17 percent of all Americans lack access to reliable Internet service, more than half of rural Americans lacks access (Federal Communications Commission [FCC] 2015). As a result of this disparity, in 2015, the FCC announced it would spend $7 billion expanding broadband Internet access to poor and rural areas, with the goal of ensuring that at least 98 percent of Americans have access to high-speed Internet service (White House 2015a).

Some of the differences in Internet use and broadband home access are a function of housing patterns. Inner-city neighborhoods are disproportionately populated by racial and ethnic minorities and are simply less likely to have the telecommunications hardware necessary for access to online services. Racial and ethnic minorities' lack of access to computers and the Internet, although signaling a type of digital divide, may be less common than what researchers are now calling the **participation gap.** Black and Latino children, for example, are more likely to access the Internet via mobile devices and, when online, to use Twitter, play games, participate in social networking, and watch video games than their advantaged, white counterparts (McCollum 2011). In fact, black and Latino households are much more likely than white households to have only handheld devices, rather than desktop or laptop computers, to access the Internet (File 2014). Furthermore, some research indicates that having a home computer, although increasing computer skills, actually lowers academic achievement rather than raising it, as one might expect (Vigdor et al. 2014).

There are few gender disparities in computer use and access in developed countries such as the United States and Japan. However, in developing counties, women play a subordinate role in information communication technologies (ICT), which affects their employability. As was the case with the first wave of industrialization in which women were highly concentrated in textile manufacturing, the new wave of ICT production has seen women concentrated largely in tedious, repetitive, and low-skill tasks such as data entry and word processing (Thas et al. 2007).

Concern over accessibility to broadband connectivity has led to a debate over net neutrality. **Net neutrality** advocates hold that Internet users should be able to visit any website and access any content without Internet service providers (ISPs) (e.g., cable or telephone companies) acting as gatekeepers by controlling, for example, the speed of downloads. Why would an ISP do that? Hypothetically, if Internet service provider company X signs an agreement with search engine Y, then it's in the best interest of Internet service provider X to slow down all other search engines' performances so that you will switch to search engine Y. Internet service providers argue that Internet users, be they individuals or corporations, who use more than their "fair share" of the Internet should pay more. Why should you pay the same monthly fee as your neighbor who nightly downloads full-length movie files? Others fear any government regulation of the Internet and/or prefer a strictly market model. The debate is politically charged and partisan: Republican leaders claim that government regulations to enforce net neutrality would stifle business with burdensome bureaucracy, and Democratic leaders argue that the regulations are needed to protect consumers.

In 2015, the FCC invoked its right to regulate net neutrality under Title II of the Federal Communications Act, which gives the agency authority to regulate utilities such as telephone and power companies. FCC chairman Tom Wheeler argued that "the Internet

participation gap The tendency for racial and ethnic minorities to participate in information and communication technologies (e.g., using smartphones to access the Internet rather than a computer) that place them in a disadvantaged position (e.g., difficult to research a term paper on a smartphone).

net neutrality A principle that holds that Internet users should be able to visit any website and access any content without Internet service provider interference.

is simply too important to be left without rules or a referee on the field. . . . The Internet has replaced the function of the telephone and the post office" (Risen 2015, p. 1). The FCC's Open Internet Rules are premised on three principles: *No Blocking* (providers can't block access to otherwise legal content, services, and devices); *No Throttling* (providers can't impede the Internet traffic of otherwise legal content, services, and devices); and *No Paid Prioritization* (providers can't favor some Internet traffic over others in exchange for any other considerations, also known as "no fast lanes") (FCC 2015). The fight over net neutrality is far from over, however, as telecom companies plan to challenge the FCC ruling in court, and the political debate between Republicans and Democrats in Congress will likely continue past the next election cycle (Pagliary 2015).

Problems of Mental and Physical Health

High levels of media use are associated with lower academic achievement, sleep problems, obesity, and a wide range of physical and mental health problems (Department of Health and Human Services 2013). Although the American Academy of Pediatrics recommends that children and adolescents have no more than two hours of "screen time" per day, approximately 73 percent of youth age 12 to 17 exceed this recommendation (Herrick et al. 2014).

> Slade (2012) observes, "Instead of visiting, we phone; instead of phoning, we text; instead of texting, we post 'updates' for our 'friends' on Facebook's wall, and when they don't like these updates, we 'unfriend' them. In public, we catch up on our smartphones with people we no longer have time to visit, or we steal a few moments during our commute to listen to a playlist while reading an e-book on a Kindle. The luminal spaces of our cities are full of people experiencing and practicing uncomfortable 'elevator silences'" (p. 11). Does technology lead to loneliness and separation? How would separating individuals from one another financially benefit corporations?

Technology changes what we do, and when and how we do it. The multitasking that is associated with technology is linked to distraction, a false sense of urgency, and the inability to focus. Not only does the use of technology encourage distraction through multitasking, it also rewires the brain at the cellular level to make prolonged concentration more difficult (Carr 2010). The average attention span dropped from 12 seconds in 2000 to 8 seconds in 2015, contributing to a phenomenon called "slowness rage," in which a decreased attention span combined with the faster pace of technology leads to feelings of anxiety and anger when daily experiences take longer than expected (Wald 2015).

The effects of technology on our brains may have more serious consequences than feeling impatient more often, however. A study of college students from 10 countries found that four out of five exhibited physical and emotional symptoms of addiction withdrawal, including panic attacks, physical discomfort, and depression, when asked to give up cell phone and Internet use for 24 hours (Moeller et al. 2012). Psychologists have also noted a pattern of "Facebook depression" in which increases in social media use are associated with lower levels of self-esteem and life satisfaction, and increased Internet use is correlated with self-harm and suicidal behavior (Kross et al. 2013; Daine et al. 2015).

Malicious Use of the Internet

The Internet may be used for malicious purposes, including but not limited to cybercrime and prostitution (see Chapter 4), hacking, piracy, electronic aggression (e.g., cyberbullying), and "questionable content" sites. **Internet piracy** entails illegally downloading or distributing copyrighted material (e.g., music, games, or software). Court cases indicate that trade organizations (e.g., the Recording Industry Association of America) are pursuing criminal cases against violators, and courts are imposing strict penalties. The founders of The Pirate Bay, a file-sharing website that allowed users free access to popular movies, music, and television shows, were pursued by international authorities for several years and eventually found guilty of hacking and piracy crimes in Denmark and Sweden where they were each sentenced to two- to four-year jail terms (BBC 2014).

Internet piracy Illegally downloading or distributing copyrighted material (e.g., music, games, software).

Malware is a general term that includes any spyware, crimeware, worms, viruses, and adware that is installed on owners' computers without their knowledge. A 2014 study found that crimes committed through the use of malware cost consumers nearly $500 billion annually, or approximately 20 percent of the economic value created by the Internet (Center for Strategic and International Studies 2014). A recent type of malware, ransomware, locks your computer or part of your computer, making it impossible to access your system until you pay a ransom, usually through some online payment method such as PayPal.

In addition to hacking, the Internet also facilitates drug dealing, illegal weapons sales, and human trafficking. Much of this illicit activity occurs on the **deep web**, also known as the *dark web,* the term for the many thousands of illicit websites that can only be accessed through special coding and are not accessible though standard search engines on the conventional, or surface, web. In 2014, the FBI shut down and arrested the leader of Silk Road, a deep website notorious for the black market sales of illegal drugs and weapons using bitcoin currency (Weiser 2015). In a dramatic twist, two FBI agents who worked undercover on the Silk Road investigation were later charged with money laundering and wire fraud when it was discovered that they allegedly converted investigation resources into bitcoins for their own personal use (Weiser and Apuzzo 2015).

Electronic aggression is defined as any kind of aggression that takes place with the use of technology (David-Ferdon and Hertz 2009). For example, **cyberbullying** refers to the use of electronic communication (e.g., websites, e-mail, instant messaging, or text messaging) to send or post negative or hurtful messages or images about an individual or a group (Kharfen 2006). Cyberbullying differs from traditional bullying in several significant ways, including the potential for a larger audience, anonymity, the inability to respond directly and immediately to the bully, and reduced levels of adult or peer supervision (Sticca and Perren 2013). Approximately 40 percent of Internet users report some form of online harassment; men are more likely to report verbal harassment, whereas women are more likely to report stalking and sexual harassment (Duggan 2014) (see Figure 14.6).

Estimates of the frequency of involvement in cyberbullying—as victim, perpetrator, or both—range dramatically, although electronic aggression researchers generally agree that texting is the most common means of cyberbullying (David-Ferdon and Hertz 2009). Because cyberbullying is capable of reaching wider audiences, many states and school districts have begun creating cyberbullying disciplinary policies. As of 2015, 22 states have anti-bullying laws that include cyberbullying, and in 48 states, anti-bullying laws also prohibit electronic harassment (Hinduja and Patchin 2015).

malware A general term that includes any spyware, viruses, and adware that is installed on an owner's computer without their knowledge.

deep web Also known as the *dark web*, the term for the many thousands of illicit websites that can only be accessed through special coding and are not accessible though standard search engines.

cyberbullying The use of electronic communication (e.g., websites, e-mail, instant messaging, text messaging) to send or post negative or hurtful messages or images about an individual or a group.

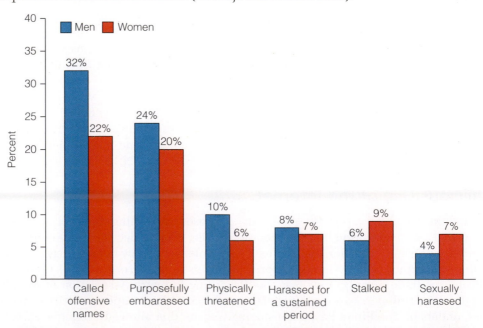

Figure 14.6 Percent of all Internet Users Who Have Experienced Different Types of Online Harassment, by Gender, 2014
SOURCE: Pew 2014e.

The Challenge to Traditional Values and Beliefs

Technological innovations and scientific discoveries often challenge traditionally held values and beliefs, in part because they enable people to achieve goals that were previously unobtainable. For example, technological advances now allow postmenopausal women to give birth, challenging societal beliefs about childbearing and the role of older women. The techniques of egg retrieval, in vitro fertilization, and gamete intrafallopian transfer make it possible for two different women to each make a biological contribution to the creation of a new life. Such technology requires society to reexamine its beliefs about what a family is and what a mother is. Should family be defined by custom, law, or the intentions of the parties involved?

Medical technologies that sustain life lead us to rethink the issue of when life should end. The increasing use of computers throughout society challenges the traditional value of privacy. New weapons systems make questionable the traditional idea of war as something that can be survived and even won. And cloning causes us to wonder about our traditional notions of family, parenthood, and individuality. Toffler (1970) coined the term **future shock** to describe the confusion resulting from rapid scientific and technological changes that unravel our traditional values and beliefs.

Would you volunteer to be part of the first human settlement on another planet? Mars One, a not-for-profit foundation, is currently raising funds and developing technology to establish the first permanent human settlement on Mars. The organization, comprised of an international team of aerospace experts, plans to send the first one-way human settlement mission to Mars in 2026 (Mars One 2015). A call for volunteers brought in 200,000 applications, from which 10 will ultimately be selected to crew the first two missions (Farberov 2015). The mission carries a high level of risk and a guarantee of never returning to Earth. Would you be willing to be part of a one-way trip to Mars?

Strategies for Action: Controlling Science and Technology

As technology increases, so does the need for social responsibility. Nuclear power, genetic engineering, cloning, and computer surveillance all increase the need for social responsibility, creating both "new possibilities for social action as well as new problems that have to be dealt with" (Mesthene 1993, p. 85). In the following sections, we address various aspects of the public debate, including science, ethics and the law, the role of corporate America, and government policy.

Technology and Corporate America

As philosopher Jean-François Lyotard noted, knowledge is increasingly produced to be sold. The development of GMOs, the commodification of women as egg donors, direct-to-consumer genetic testing, and the harvesting of regenerated organ tissues are all examples of market-driven technologies. Like the corporate pursuit of computer technology, profit-motivated biotechnology creates several concerns.

First is the concern that only the rich will have access to life-saving technologies such as genetic testing and cloned organs. Such fears may be justified. Myriad Genetics patented breast and ovarian cancer genes (Bollier 2009). However, because of the resulting **gene monopolies** and the associated astronomical patient costs for genetic testing and treatment, in 2009, the American Civil Liberties Union and several plaintiffs filed a lawsuit against Myriad Genetics, alleging that the patents are "invalid and unconstitutional" (Genomics Law Report 2011). Appeals of several lower court decisions took the case to the U.S. Supreme Court, which ruled in favor of the plaintiffs—human genes cannot be patented (American Civil Liberties Union 2013). As a result of the case, nonprofit genetic testing companies have been working to build a genetic reference database, with the hopes of providing a more cost-effective alternative for patients seeking diagnostic genetic testing (Hayden 2014).

future shock The state of confusion resulting from rapid scientific and technological changes that unravel our traditional values and beliefs.

gene monopolies Exclusive control over a particular gene as a result of government patents.

Annually, "tens of millions of animals are used for biomedical research, chemical testing, and training" (Bowman 2011, p. 1). An estimated 1 million are dogs and cats, monkeys and apes, hamsters and guinea pigs, rabbits, pigs, and sheep, and other farm animals. All else—including birds, fish, rats and mice, frogs, and lizards—number somewhere between 80 to 100 million. Each has or will be subject to **vivisection**, the practice of cutting into or otherwise harming living, nonhuman animals for the purpose of scientific research.

The use of animals for research purposes is not a new phenomenon, but the century-old debate over the morality of using animals in research has recently picked up momentum as animal rights groups such as PETA (People for the Ethical Treatment of Animals) and the Animal Liberation Front make inroads into the American collective consciousness.

The issues surrounding the vivisection of nonhuman animals are complex. Should humans' rights trump those of nonhuman animals? Do nonhuman animals, as human animals, have a right to a clean, safe, and pain-free environment? Should nonhuman animals die in the

hopes of gathering data that might prolong or save human life?

Opinion polls indicate that the number of Americans who believe that medical testing on animals is "morally acceptable" has steadily declined over the years (NSF 2015). Over half of Americans, 56 percent, are in favor of using animals in research although there are striking gender differences. Sixty-two percent of men are in favor of using animals in research compared to 42 percent of women. There are also differences based on the type of animals used—for example, mice versus chimpanzees or dogs (NSF 2014).

The arguments both for and against the use of animals in research generally fall into three camps—yes, no, and sometimes. Advocates argue that any distinction that differentiates dogs and chimpanzees, for example, from birds and fish is an artificial one. The only substantively significant distinction to be made is between human animals and nonhuman animals.

Most scientists, however, make the distinction between life or death issues, arguing that testing on animals is central to providing information on a drug's

safety and effectiveness. If your child were ill, would you choose to give her a drug that hadn't been tested on animals to prove it was safe and effective? Test on animals or your child dies—your choice. In one survey of biomedical scientists, 90 percent responded that the use of animals in research is "essential" (*Nature News* 2011).

Finally, the Food and Drug Administration *requires* that drugs and procedures be tested on nonhuman animals before they can be consumed by or used on human animals. And to opponents who are quick to point out that human and nonhuman animals are too different to result in any meaningful findings, Holland (2010) responds that pig heart valves are used on human patients, and many diseases cross between humans and other mammals and birds.

Hanson (2010), however, convincingly argues that such justifications are an ethical "bait and switch." Few of us could even bear to look at pictures of "monkeys, with their electrode-implanted brains and bolted heads, being put through their paces in a desperate attempt to get a

The commercialization of technology causes several other concerns, including issues of quality control and the tendency for discoveries to remain closely guarded secrets rather than collaborative efforts (Crichton 2007; Lemonick and Thompson 1999; Mayer 2002; Rabino 1998). For example, between 2009 and 2014, the number of jointly written scientific articles steadily declined (Noble 2015). Furthermore, industry involvement has made government control more difficult because researchers depend less and less on federal funding. More than 63 percent of research and development in the United States is supported by private industry using their own company resources (NSF 2015).

Science, Ethics, and the Law

vivisection The practice of cutting into or otherwise harming living, nonhuman animals for the purpose of scientific research.

genetic exception laws Laws that require that genetic information be handled separately from other medical information.

Science and its resulting technologies alter the culture of society through the challenging of traditional values. Public debate and ethical controversies, however, have led to structural alterations in society as the legal system responds to calls for action. For example, several states now have what are called genetic exception laws. **Genetic exception laws** require that genetic information be handled separately from other medical information, leading to what is sometimes called *patient shadow files* (Legay 2001). The logic of such laws rests with the potentially devastating effects of genetic information being revealed to insurance companies, other family members, employers, and the like. Their importance has also grown—there are now genetic tests for more than 1,000 diseases, and life-sustaining sip of water" (p. 3). Yet, researchers argue that such testing techniques may lead to finding the cause or cure for Alzheimer's disease. Hanson, a renowned neuroscientist and PETA member, continues noting that "these experiments . . . are so thoroughly unrelated to the neuro-pathology of Alzheimer's . . . that in more than 28 years of research in the neuroscience of the disease, I have never come across a single reference to them in any scientific literature on neurodegenerative diseases" (p. 3).

Moderates acknowledge that the use of animals in scientific research may be necessary in some cases but is employed more than is essential for life and death issues. For example, most animal testing is conducted not to treat or cure serious illnesses but to predict the toxicity, safety, and effectiveness of chemicals and consumer products. As one leading toxicologist reported, a standard toxicity test is "little more than a ritual mass execution of animals," in which the testing of one substance can involve the use of up to 800 animals at a cost of over $6 million dollars (New England Anti-Vivisection Society 2015). Research on the characteristics that influence attitudes toward the use of animals in research indicate that people are more likely to favor animal testing if lower-level rather than higher-level animals (e.g., mice rather than chimpanzees) are used, if the level of harm is minimal (e.g., discomfort rather than death), and if the disease being studied is severe (e.g., cancer rather than eczema) (Henry and Pulcino 2009). The formal laws and informal cultural norms when it comes to using animals in research echo these preferences: minimize pain and suffering, use as few animals as possible, hire caretakers (e.g., veterinarians), look for alternative means of accomplishing the same ends, and when possible, use lower-level rather than higher-level animals (Research Animal Resources 2003). The Interagency Coordinating Committee on the Validation of Alternative Methods (ICCVAM), established in 2000 under the National Institute of Health, is designed to

promote the development, validation, and regulatory acceptance of new, revised, and alternative regulatory safety testing methods. Emphasis is on alternative methods that will reduce, refine (less pain and distress), and replace the use of animals in testing while maintaining and promoting scientific quality and the protection of human health, animal health, and the environment. (ICCVAM 2011)

Even where scientists attempt to minimize suffering, questions about animal well-being arise. For example, animal welfare advocates have long asked whether mice enjoy or are stressed by running in wheels, an activity used in a wide range of studies. Researchers attempted to answer this question by placing an exercise wheel outdoors and monitoring its use by wild animals. They found that wild mice, rats, shrews, frogs, and even slugs used the wheel voluntarily, and exhibited behavior that indicated recreational use; that is, the animals were not driven by stress or anxiety (Gorman 2014). Although some laboratory animals may enjoy running on a wheel in the same way that humans enjoy going for a jog, the problem is whether the activity is voluntary. There is overwhelming evidence that all animals, even "lower-level" animals such as mice, have emotions; they grimace when experiencing pain and laugh when being tickled (Ferdowsian 2010; Flecknell 2010).

The issue is complex, and there are no easy answers. But asking the right questions is a step in the right direction. Perhaps as philosopher Jeremy Bentham said over 200 years ago, the right question is not "Can they speak?" or "Can they reason?" but "Do they suffer?"

hundreds more are being researched. Examples of protected tests include the BRCA1 and BRCA2 tests for breast cancer and carrier screening for such conditions as sickle-cell anemia and cystic fibrosis (Genetics and Public Policy Center 2010).

Are such regulations necessary? In a society characterized by rapid technological and thus social change—a society in which custody of frozen embryos is part of a divorce agreement—many would say yes. Cloning, for example, is one of the most hotly debated technologies in recent years. Bioethicists and the public vehemently debate the various costs and benefits of this scientific technique. Despite such controversy, however, the chairman of the National Bioethics Advisory Commission warned nearly 20 years ago that human cloning will be "very difficult to stop" (McFarling 1998). Fifteen states have laws pertaining to human cloning, with some prohibiting cloning for reproductive purposes, some prohibiting therapeutic cloning, and still others prohibiting both (National Conference of State Legislatures 2015; Rovner 2013).

Although still in its infancy, direct-to-consumer (DTC) genetic testing is available now. How accurate it is in predicting disease depends on the disease in question. In some cases, the presence of a gene indicates with absolute certainty that you will get the disease; in others, your risk may be much lower. As availability of DTC genetic testing increases and cost decreases, would you consider genetic testing for health purposes?

WHAT
do you
THINK?

Should the choices that we make as a society depend on what we can do or what we should do? Whereas scientists and the agencies and corporations that fund them often determine what we *can* do, who should determine what we *should* do (see this chapter's *Animals and Society* feature)? Although such decisions are likely to have a strong legal component—that is, they must be consistent with the rule of law and the constitutional right of scientific inquiry—legality or the lack thereof often fails to answer the question "What should be done?" *Roe v. Wade* (1973) did little to quash the public debate over abortion and, more specifically, the question of when life begins. Thus, it is likely that the issues surrounding the most controversial of technologies will continue throughout the 21st century with no easy answers.

Runaway Science and Government Policy

Science and technology raise many public policy issues. Policy decisions, for example, address concerns about the safety of nuclear power plants, the privacy of e-mail, the hazards of chemical warfare, and the ethics of cloning. In creating science and technology, have we created a monster that has begun to control us rather than the reverse? What controls, if any, should be placed on science and technology? And are such controls consistent with existing law? Consider the use of the file-sharing network BitTorrent to download music and movie files (the question of intellectual property rights and copyright infringement); laws limiting children's access to material on the Internet (free speech issues); and Acxiom, the "cookie"-collecting company that helps corporations customize advertising on websites by tracing clicks and keystrokes (Fourth Amendment privacy issues).

Concerns over "runaway science" are not uniquely American. Although genetic data may be collected legally—in the course of an arrest, for example—growing evidence indicates that governments around the world are maintaining large genetic databases. In the United States, the Federal Bureau of Investigation's DNA database contains genetic information on more than 11 million suspected or convicted criminals. Is the collection of DNA from a suspected criminal an unreasonable search? In a 5–4 decision, the Supreme Court ruled that it is a reasonable search to take DNA samples from people who have been arrested without warrant, arguing that such a practice is the same as taking fingerprint samples from suspects (Ross 2014). The use of DNA materials in criminal investigations is likely to expand rapidly, as the FBI, Homeland Security, and other government agencies have begun to adopt new DNA scanner technology that can generate a full DNA profile, check it against a national database, and report a match in under 90 minutes (Bauer 2014).

Yet it is the government, often through Congress, regulatory agencies, or departments, that is responsible for controlling technology, prohibiting some (e.g., assisted-suicide devices) and requiring others (e.g., seat belts). A good example is the Stem Cell Research Advancement Act of 2013 that was introduced in the U.S. House of Representatives and remains in committee. As proposed, the act (1) supports the use of embryonic stem cells, including human embryonic stems cells; (2) defines the types of human embryonic stem cells eligible for use in research (e.g., donated from in vitro fertilization clinics);

> Policy decisions . . . address concerns about the safety of nuclear power plants, the privacy of e-mail, the hazards of chemical warfare, and the ethics of cloning.

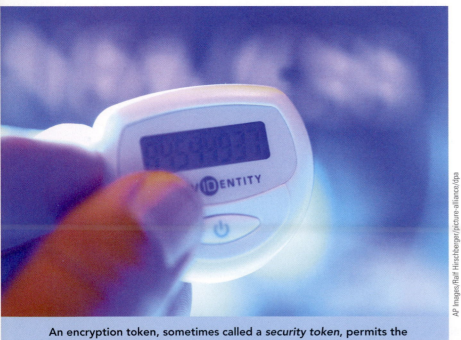

An encryption token, sometimes called a *security token*, permits the holder to access a computer by providing a code sent to the device. Some security tokens act as keys of a sort and can be used to authorize use through Bluetooth or USB connections.

AP Images/Ralf Hirschberger/picture-alliance/dpa

(3) mandates that the Department of Health and Human Services maintain, review, and update guidelines in support of human stem cell research; and (4) prohibits public funds from being used for human cloning (Stem Cell Research Advancement Act 2013).

Of late, two areas of concern have required increased government scrutiny: the National Security Agency (NSA) surveillance programs and international and domestic cyber-threats. After lengthy debate, Congress voted in June 2015 to reform the NSA's surveillance program to prevent the agency from collecting phone records of millions of Americans, while still allowing the agency broad powers to monitor both foreign and domestic cyber-threats (Diamond 2015). Further, confirming the fears of many, in June 2015, the government announced that the records of 4 million federal employees were breached by hackers working for the Chinese government (Nakashima 2015). Not surprisingly, between 2011 and 2014, more than 200 cybersecurity bills were introduced into either the U.S. House of Representatives or the U.S. Senate, five of which were enacted (Fischer 2015).

The government studies and makes recommendations on the use of science and technology through several boards and initiatives, including the National Science and Technology Council, the Office of Science and Technology Policy, the President's Council of Advisors on Science and Technology, and the U.S. National Nanotechnology Initiative. These agencies advise the president on matters of science and technology, including research and development, implementation, national policy, and coordination of different initiatives. Based on the advice of these initiatives, the Obama administration has outlined five key priorities to address cybersecurity (White House 2015b):

1. Protect infrastructure and information systems from cyber threats.
2. Improve ability to identify and report cyber incidents.
3. Build international partnerships to promote Internet freedom and secure, reliable cyberspace.
4. Secure federal networks.
5. Partner with the private sector to build a cyber-savvy workforce.

Understanding Science and Technology

What are we to understand about science and technology from this chapter? As structural functionalists argue, science and technology evolve as a social process and are a natural part of the evolution of society. As society's needs change, scientific discoveries and technological innovations emerge to meet these needs, thereby serving the functions of the whole. Consistent with conflict theory, however, science and technology also meet the needs of select groups and are characterized by political components. As Winner (1993) noted, the structure of science and technology conveys political messages, including "power is centralized," "there are barriers between social classes," "the world is hierarchically structured," and "the good things are distributed unequally" (p. 288).

The demographics of video games is changing, with women, racial and ethnic minorities, and people over age 30 playing more video games than ever before. However, game developers are predominantly young, white, men. Recent threats of violence against female gamers and game reviewers have raised new questions about sexism, racism, and diversity in gaming: "Are games a technology product or a cultural experience?" (Alexander 2014, p. 1). To what extent do you think developers have a social responsibility to be culturally sensitive in their video game designs?

The scientific discoveries and technological innovations that society does or does not embrace are socially determined. Research indicates that science and the resulting technologies have both negative and positive consequences—a *technological dualism*. Technology saves lives, time, and money; it also leads to death, unemployment, alienation, and estrangement. Weighing the costs and benefits of technology poses ethical dilemmas, as does science itself. Ethics, however, "is not only concerned with individual choices and acts. It is also and, perhaps, above all concerned with the cultural shifts and trends of which acts are but the symptoms" (McCormick and Richard 1994, p. 16).

Thus, society makes a choice by the very direction it follows. These choices should be made on the basis of guiding principles that are both fair and just, such as those listed here:

1. Science and technology should be prudent. Adequate testing, safeguards, and impact studies are essential. Impact assessment should include an evaluation of the social, political, environmental, and economic factors.

2. No technology should be developed unless all groups, and particularly those who will be most affected by the technology, have at least some representation "at a very early stage in defining what that technology will be" (Winner 1993, p. 291). Traditionally, the structure of the scientific process and the development of technologies have been centralized (i.e., decisions have been made by a few scientists and engineers); decentralization of the process would increase representation.

3. Means should not exist without ends. Each new innovation should be directed to fulfilling a societal need rather than the more typical pattern in which a technology is developed first (e.g., high-definition television) and then a market is created (e.g., "You'll never watch a regular TV again!"). Indeed, from the space program to research on artificial intelligence, the vested interests of scientists and engineers, whose discoveries and innovations build careers, should be tempered by the demands of society (Buchanan et al. 2000; Eibert 1998; Goodman 1993; Murphie and Potts 2003; Winner 1993).

What the 21st century will hold, as the technological transformation continues, may be beyond the imagination of most of society's members. Technology empowers; it increases efficiency and productivity, extends life, controls the environment, and expands individual capabilities.

Balancing technology with our humanity will continue to be the challenge of the next generation. As humans come to rely on technological automation, the skills, creativity, and innovation that characterize our humanity become dulled. However, it is possible to reconcile technological advancement with the advancement of humanity, as author Nicholas Carr argues, from a technology-centered automation, in which the role of people is to support technology, to a human-centered automation, in which the talents and needs of people direct technological development (Carr 2014). As we proceed further into the first computational millennium, one of the great concerns of civilization will be the attempt to reorder society, culture, and government in a manner that exploits the digital bonanza yet prevents it from running roughshod over the checks and balances so delicately constructed in those simpler pre-computer years.

Chapter Review

- **What are the three types of technology?**
The three types of technology, escalating in sophistication, are mechanization, automation, and cybernation. Mechanization is the use of tools to accomplish tasks previously done by hand. Automation involves the use of self-operating machines, and cybernation is the use of machines to control machines.

- **What are some of the reasons the United States may be "losing its edge" in scientific and technological innovations?**
The decline of U.S. supremacy in science and technology is likely to be the result of five interacting social forces. First, the federal government has been scaling back its investment in research and development. Second, corporations have begun to focus on short-term products and higher profits. Third, there has been a drop in science and math education in U.S. schools in terms of both quality and quantity. Fourth, developing countries, most notably China and India, are expanding their scientific and technological capabilities at a faster rate than the United States. Finally, as documented in the book *Unscientific America* (Mooney and Kirshenbaum 2009), there is a disconnect between American society and the principles of science.

- **What are some Internet global trends?**
In 2014, the Internet had 3 billion users in more than 200 countries, with 277 million users in the United States

(Internet Statistics 2015). Of all Internet users, the highest proportion come from Asia (45.7 percent), followed by Europe (19.2 percent), North America (10.2 percent), Latin America and the Caribbean (10.5 percent), Africa (9.8 percent), the Middle East (3.7 percent), and Oceania/Australia (0 percent) (Internet Statistics 2015).

- **According to Kuhn, what is the scientific process?**
  Kuhn (1973) describes the process of scientific discovery as occurring in three steps. First are assumptions about a particular phenomenon. Next, because unanswered questions always remain about a topic, science works to start filling in the gaps. Then, when new information suggests that the initial assumptions were incorrect, a new set of assumptions or framework emerges to replace the old one. It then becomes the dominant belief or paradigm until it is questioned and the process repeats.

- **What is meant by the computer revolution?**
  The silicon chip made computers affordable. Today, over 75 percent of U.S. households report having a computer in the home compared to 61.8 percent just a decade ago (File 2014).

- **What is the Human Genome Project?**
  The U.S. Human Genome Project is an effort to decode human DNA. The 13-year-old project is now complete, allowing scientists to "transform medicine" through early diagnosis and treatment as well as possibly preventing disease through gene therapy. Gene therapy entails identifying a defective or missing gene and then replacing it with a healthy duplicate that is transplanted to the affected area.

- **How does technology impact laws about abortion?**
  Abortion laws are premised on fetal viability, and technological advancements have made the likelihood of a premature baby surviving a second-trimester birth more likely, leading many states to pass more restrictive abortion laws.

- **How are some of the problems of the Industrial Revolution similar to the problems of the technological revolution?**
  The most obvious example is unemployment. Just as the Industrial Revolution replaced many jobs with technological innovations, so too has the technological revolution. Furthermore, research indicates that many of the jobs created by the Industrial Revolution, such as working on a factory assembly line, were characterized by high rates of alienation. Rising rates of alienation are also a consequence of increased estrangement as high-tech employees work in "white-collar factories."

- **What is meant by *outsourcing,* and why is it important?**
  Outsourcing is the practice of a business subcontracting with a third party, often in low-wage countries such as China and India, for services. The problem with outsourcing is that it tends to lead to higher rates of unemployment in the export countries.

- **What is the digital divide?**
  The digital divide is the tendency for technology to be most accessible to the wealthiest and most educated. For example, some fear that there will be "genetic stratification," whereby the benefits of genetic testing, gene therapy, and other genetic enhancements will be available to only the richest segments of society.

- **What is meant by the commercialization of technology?**
  The commercialization of technology refers to profit-motivated technological innovations. Whether in regard to the isolation of a particular gene, genetically modified organisms, or the regeneration of organ tissues, where there is a possibility for profit, private enterprise will be there.

Test Yourself

1. Which of the following technologies is associated with industrialization?
   a. Mechanization
   b. Cybernation
   c. Hibernation
   d. Automation
2. Analysis of data over time suggests that the science and technology gap between poorer and richer nations is getting greater rather than shrinking.
   a. True
   b. False
3. The U.S. government, as part of the technological revolution, spends more money on research and development than educational institutions and corporations combined.
   a. True
   b. False
4. Which theory argues that technology is often used as a means of social control?
   a. Structural functionalism
   b. Social disorganization
   c. Conflict theory
   d. Symbolic interactionism
5. The global legalization of abortion has all but eliminated unsafe abortion procedures.
   a. True
   b. False
6. The ability to manipulate the genes of an organism to alter the natural outcome is called
   a. gene therapy.
   b. gene splicing.
   c. genetic engineering.
   d. genetic testing.
7. Genetically modified foods have been documented as harmless to humans by the Food and Drug Administration.
   a. True
   b. False
8. In 2007, the U.S. Supreme Court upheld the partial-birth abortion ban in a 5–4 decision.
   a. True
   b. False
9. The practice of outsourcing entails
   a. creating high-tech jobs in the United States for immigrants.
   b. hiring temporary workers to cover for employees who are absent.

c. allowing company workers to work from home.
d. subcontracting jobs often to workers in low-wage countries.

10. The Genetic Information Nondiscrimination Act (GINA)
   a. makes human cloning illegal in the United States.

b. establishes a criminal penalty for human cloning in the United States.
c. is a federal law that prohibits discrimination in health coverage or employment based on genetic information.
d. all of the above

Answers: 1. D; 2. A; 3. B; 4. C; 5. B; 6. C; 7. B; 8. A; 9. D; 10. C.

Key Terms

abortion 477
automation 460
cultural lag 466
cyberbullying 486
cybernation 460
deep web 486
e-commerce 473
future shock 487
gene monopolies 487
gene therapy 476
genetic cloning 479
genetic engineering 476
genetic exception laws 488

Genetic Information Nondiscrimination Act of 2008 481
genetic testing 475
Internet 472
Internet of Things (IoT) 472
Internet piracy 485
learning health systems 474
malware 486
mechanization 460
membership communities 475
nanotechnology 476
net neutrality 484
outsource 483

participation gap 484
penetration rate 461
postmodernism 464
reproductive cloning 479
science 460
Semantic Web 474
STEM 463
stem cells 479
technological dualism 460
technological fix 464
technology 460
technology-induced diseases 476
therapeutic cloning 479
vivisection 488

> 66 "Every gun that is made, every warship launched, every rocket fired signifies, in the final sense, a theft from those who hunger and are not fed, those who are cold and not clothed."
>
> **GENERAL DWIGHT D. EISENHOWER**
> former U.S. president and military leader

15

Conflict, War, and Terrorism

Learning Objectives

After studying this chapter, you will be able to . . .

1 Identify trends in global conflict over the past century.

2 Identify the causes and possible solutions for war from each sociological perspective.

3 Explain how war can be an outcome of other social problems.

4 Assess the pros and cons of various anti-terrorism strategies.

5 Provide examples of other social problems that occur as a consequence of war.

6 Argue for a policy change that would help create a more peaceful world.

FOURTEEN-YEAR-OLD YAHYA, a name given to him by his captors, had been forced to attend an Islamic State training camp after being kidnapped from his home in northern Iraq (Karam and Janssen 2015). Like other young boys in the camp, some captured and others recruited using "gifts, threats, and brainwashing," Yahya had trained 8 to 10 hours a day for nearly 5 months exercising, shouting slogans, reading the Quran, studying jihad, and learning how to fight and how to kill (p. 1). He was forced to attack his younger brother under threat of death and was routinely beaten to "toughen" him up. All of the boys had been shown videos of ISIS militants beheading their enemies, and each was told that someday they would be able to do the same. Practice was essential. Each boy was given a doll, a sword, and instructions on how to cut off the doll's head. Yahya couldn't seem to do it right. He chopped over and over again without success. Seeing his botched attempts, an ISIS trainer showed him the right way to hold the sword and how to cut, telling Yahya that "it was the head of an infidel" (p. 1).

ISIS is not the only radical group that uses children toward their socio-political ends. Here teenagers wearing the headbands of the Sunni Tehreek, a religious political party in Pakistan, take part in an anti-American rally.

War is one of the great paradoxes of human history. It both protects and annihilates. It creates and defends nations but may also destroy them. **War**, the most violent form of conflict, refers to organized armed violence aimed at a social group in pursuit of an objective. Wars have existed throughout human history and continue in the contemporary world. Whether war is just or unjust, defensive or offensive, it involves the most horrendous atrocities known to humankind. This is especially true in the 21st century, when nearly all wars are fought in populated areas rather than on remote battlefields, having deadly consequences for civilians. Thus, war is not only a social problem in and of itself but also contributes to a host of other social problems—death, disease, and disability, crime and immorality, psychological terror, loss of economic resources, and environmental devastation. In this chapter, we discuss each of these issues within the context of conflict, war, and terrorism, the most threatening of all social problems.

The Global Context: Conflict in a Changing World

As societies have evolved and changed throughout history, the nature of war has also changed. Before industrialization and the sophisticated technology that resulted, war occurred primarily between neighboring groups on a relatively small scale. In the modern world, war can be waged between nations that are separated by thousands of miles as well as between neighboring nations. Increasingly, war is a phenomenon internal to states, involving fighting between the government and rebel groups or among rival contenders for state power. Indeed, Figure 15.1 documents that wars between states—that is, interstate wars—recently made up the smallest percentage of armed conflicts. In the following sections, we examine how war has changed our social world and how our changing social world has affected the nature of war in the industrial and postindustrial information age.

war Organized armed violence aimed at a social group in pursuit of an objective.

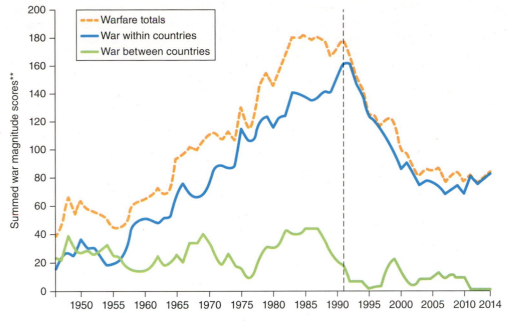

*Broken vertical line represents the collapse of the Soviet Union.
**The higher the magnitude score the greater the destructive impact on the directly-affected society or societies.

Figure 15.1 Global Trends in Armed Conflict, 1946–2014*
SOURCE: Center for Systemic Peace 2015.

War and Social Change

The very act that now threatens modern civilization—war—is largely responsible for creating the advanced civilization in which we live. Before large political states existed, people lived in small groups and villages. War broke the barriers of autonomy between local groups and permitted small villages to be incorporated into larger political units known as chiefdoms. Centuries of warfare between chiefdoms culminated in the development of the state. The **state** is "an apparatus of power, a set of institutions—the central government, the armed forces, the regulatory and police agencies—whose most important functions involve the use of force, the control of territory, and the maintenance of internal order" (Porter 1994, pp. 5–6). Social historian Charles Tilly famously said, "War makes states, and states make war" (1992). The creation of the state in turn led to other profound social and cultural changes:

> And once the state emerged, the gates were flung open to enormous cultural advances, advances undreamed of during—and impossible under—a regimen of small autonomous villages. . . . Only in large political units, far removed in structure from the small autonomous communities from which they sprang, was it possible for great advances to be made in the arts and sciences, in economics and technology, and indeed in every field of culture central to the great industrial civilizations of the world. (Carneiro 1994, pp. 14–15)

Industrialization and technology could not have developed in the small social groups that existed before military action consolidated them into larger states. Thus, war contributed indirectly to the industrialization and technological sophistication that characterize the modern world. Industrialization, in turn, has had two major influences on war. Cohen (1986) calculated the number of wars fought per decade in industrial and preindustrial nations and concluded, "As societies become more industrialized, their proneness to warfare decreases" (p. 265). Thus, for example, in 2015, there were 41 ongoing armed conflicts, all of which were taking place in regions with low economic development, predominantly in the Middle East, South Asia, Africa, and South America (International Institute for Strategic Studies 2015).

Although industrialization may decrease a society's propensity to war, it also increases the potential destruction of war. With industrialization, military technology became more

> The use of new technologies such as high-performance sensors, information processors, directed energy technologies, precision-guided munitions, and computer worms and viruses has changed the very nature of conflict, war, and terrorism.

state The organization of the central government and government agencies such as the military, police, and regulatory agencies.

sophisticated and more lethal. Rifles and cannons replaced the clubs, arrows, and swords used in more primitive warfare and, in turn, were replaced by tanks, bombers, and nuclear warheads. Today, the use of new technologies such as high-performance sensors, information processors, directed energy technologies, precision-guided munitions, and computer worms and viruses has changed the very nature of conflict, war, and terrorism.

The use of unoccupied aerial vehicles (UAVs), also known as drones, has raised national security and ethical concerns. The Obama administration has defended the use of drones as an essential form of defense in the global war on terror in which military ground operations are not always possible to catch suspected terrorists and prevent future violence. Critics argue that the use of drones hurts the national security interests of the United States because the strikes typically violate the sovereign airspace of the countries where the attacks take place, angering the governments of other countries and alienating potential allies. The use of drones in another country's airspace without their permission violates international laws, but defenders suggest such actions are legal when done in self-defense. Critics have also raised ethical concerns. Strikes in Pakistan, Yemen, Somalia, and Afghanistan have resulted in the deaths of over 1,000 unintentional civilian targets, including hundreds of children (Bureau of Investigative Journalism 2015). Defenders of the U.S. drone policy argue, however, that the use of drones is ethically preferable to a large-scale ground war; the civilian casualty rate of conventional ground warfare is estimated to be three times higher than that of drone warfare (Saletan 2015).

WHAT do you THINK?

In April 2015, President Obama gave a public apology to the families of two civilian aid workers who were accidentally killed when a drone strike targeted the al-Qaeda compound in Pakistan where the two aid workers were being held hostage (White House 2015b). The al-Qaeda leader who was the intended target of the attack was also killed. Do you think the risks to civilians associated with drone warfare are worth the potential military gains?

In the postindustrial information age, not only has computer technology revolutionized the nature of warfare, but it has made societies more vulnerable to external attacks. In March 2015, the militant group calling itself the Islamic State of Iraq and Syria (ISIS) released a "kill list" with the names, photos, and home addresses of 100 American soldiers that it claimed it hacked from military databases (Sisk 2015). A 2013 report from the Pentagon also claimed attacks of cyber espionage from the governments of Russia and China (U.S. Department of Defense 2013).

The Economics of Military Spending

The increasing sophistication of military technology has commanded a large share of resources; world military expenditures in 2014 totaled $1.78 trillion (Stockholm International Peace Research Institute [SIPRI] 2015a). This total expenditure represents a decline of about 0.4 percent from 2013, continuing a trend of military spending decline that began in 2011. Although the United States continues to have the highest military expenditure of any country in the world, part of this decline can be attributed to the drawdown of American troops from Iraq and Afghanistan. Other global economic and social factors also played a role. Economic austerity measures in the United States and much of Europe have led to a decline in military spending among many Western nations. For example, between 2013 and 2014, military spending in the United States, and Western and Central Europe declined. At the same time, military spending in Eastern European countries has increased by 8.4 percent, in African countries by 5.9 percent, and in Asian and Middle Eastern countries by around 5 percent (SIPRI 2015a).

Cold War The state of military tension and political rivalry that existed between the United States and the former Soviet Union from the 1950s through the late 1980s.

The **Cold War**, the state of political tension, economic competition, and military rivalry that existed between the United States and the former Soviet Union for nearly 50 years, provided justification for large U.S. expenditures for military preparedness. However, the end of the Cold War, along with the rising national debt, resulted in cutbacks in the U.S. military budget in the 1990s (see this chapter's *Self and Society*). Today, military

National Defense and the U.S. Military

Following each of the questions on national defense and the U.S. military, place a check mark in front of the response that best represents your answer. When complete, compare your responses to those from representative samples of U.S. adults surveyed in 2014 (N = 1,028) and 2015 (N = 837).

1. Do you think we are spending too little, about the right amount, or too much on national defense and military purposes?

   _____too little _____about right _____too much

2. Do you, yourself, feel that our national defense is stronger now than it needs to be, not strong enough, or about right at the present time?

   _____stronger than needs to be _____not strong enough _____about right

3. How much confidence do you yourself have in the military?

   _____a great deal _____quite a lot _____some _____very little

4. Do you think the United States is number one in the world militarily, or that it is only one of several leading military powers?

   _____U.S. is number one _____U.S. is one of several

5. Do you feel that it's important for the United States to be number one in the world militarily or that being number one is not that important, as long as the U.S. is among the leading military powers?

   _____important _____not that important

6. How do you feel about the nation's military strength and preparedness?

   __very satisfied __ somewhat satisfied__ somewhat dissatisfied __ very dissatisfied

7. *Just off the top of your head, which of the five branches of the armed forces in this country would you say is the most prestigious and has the most status in our society?

   ____Air Force ___ Army____Navy ____Marines ____Coast Guard___all the same

8. *Just off the top of your head, which of the five branches of the armed forces in this country would you say is the most important to our national defense today?

   ___Air Force ___ Army ____Navy ____Marines ____Coast Guard___all the same

Responses from a representative sample of Americans; percent selecting response.

1. National defense and military spending?
   34 too little
   29 about right
   32 too much

2. National defense strong?
   13 stronger than needs to be
   44 not strong enough
   42 about right

3. Confidence in military
   42 a great deal
   30 quite a lot
   19 some
   6 very little

4. U. S. number one military power?
   59 U.S. is number one
   38 U.S. is one of several

5. Military number one important?
   68 important
   31 not that important

6. Nation's military strength and preparedness?
   32 very satisfied
   37 somewhat satisfied
   15 somewhat dissatisfied
   11 very dissatisfied

7. *Most prestigious and has the most status branch?
   17 Air Force
   15 Army
   12 Navy
   47 Marines
   2 Coast Guard
   5 they are the same

8. *Most important branch to our national defense today?
   23 Air Force
   26 Army
   17 Navy
   19 Marines
   3 Coast Guard
   9 the same

*2014 survey; all others 2015. Don't know/No opinion responses are not shown.

SOURCES: Gallup 2015; Goldich and Swift 2014.

spending has nearly returned to the levels during the Cold War. In 2014, the United States spent $610 billion on its military, which accounted for 34 percent of the world's military spending and is more than the combined military expenditures of the next eight highest-spending nations, including China ($216 billion) and Russia ($84.5 billion) (SIPRI 2015b).

Weapons Sales. The U.S. government not only spends more money than other countries on its own military and defense but also sells military equipment to other countries, either directly or by helping U.S. companies sell weapons abroad. Although the purchasing countries may use these weapons to defend themselves from hostile attack, foreign military sales may pose a threat to the United States by arming potential antagonists. For example, the United States provided weapons to the Taliban to fight against a Soviet invasion in the 1980s. Years after the Soviets left Afghanistan, rebels continued to fight for control of the country. Using weapons supplied by the United States, the Taliban took over much of Afghanistan and sheltered al Qaeda and Osama bin Laden—also a former recipient of U.S. support—as they planned the attacks on September 11 (Bergen 2002; Rashid 2000).

The United States regularly transfers arms to countries in active conflict, and is the global leader in arms transfers. Between 2001 and 2011, the last period for which the government has released data, 77 percent of the world arms trade was supplied by the United States, with 11 percent supplied by the European Union, 5 percent by Russia, and 2 percent by China (U.S. Department of State 2014). A 2005 report titled *U.S. Weapons at War: Promoting Freedom or Fueling Conflict?* concluded that, far "from serving as a force for security and stability, U.S. weapons sales frequently serve to empower unstable, undemocratic regimes to the detriment of U.S. and global security" (Berrigan and Hartung 2005). For example, the United States provided approximately $1.3 billion to purchase weapons for the Iraq army to support their fight against terrorist insurgents. Many of those weapons were subsequently sold by Iraqi military personnel on the black market and purchased by ISIS militants in Syria (Kirkpatrick 2014).

WHAT
do you
THINK?

Senator John McCain has urged Congress to revoke a decades-old ban on the sale of arms to Vietnam and instead supply Vietnam with maritime weapons in order to curb China's expansion into the South China Sea (Associated Press 2015). McCain and his supporters argue that arming Vietnam will help support the United States' national security interests in the region and deter China from further aggression. Opponents argue that the weapons transfers will only make conflict with China more likely. Do you think America's national interests are ultimately helped or harmed by transferring arms to other countries?

The Costs of War. Historically, wars have been associated with economic growth and technological innovation. During World War II, for example, a surge in government spending and investment in public works projects led to spikes in employment rates and gross domestic product (GDP) that had been at historically low levels during the Great Depression of the 1930s. At the same time, the government raised taxes to fund the war, and U.S. consumption declined as Americans at home sacrificed personal comforts as part of the war effort. The wars in Iraq and Afghanistan, however, represented the first time in modern history that the United States has gone to war without seeing an increase in GDP. It is also the first time that the government has chosen to fund a war through increased deficits rather than increased taxes (Institute for Economics and Peace [IEP] 2013).

The costs of war are difficult to estimate because it is nearly impossible to disentangle direct and indirect costs, as well as the costs associated with the loss of human life, productivity, and the infrastructure. The IEP estimates that in 2012, the most conservative estimate of the direct and indirect costs associated with violence containment to the world economy was $9.46 trillion. The IEP report notes that "were the world to reduce its expenditure on violence by fifteen percent, it would be enough to provide the necessary money for the European stability fund, re-pay Greece's debt, and cover the increase in funding required to achieve the United Nation's Millennium Development Goals" (IEP 2015, p. 4).

Although violence containment efforts include non-war-related expenses, such as the costs of incarceration and violent crime, it is military expenditures that comprise the majority of global spending on violence (see Figure 15.2). The direct costs of war and terrorism to the world economy was approximately $2.5 trillion in 2012 (see Table 15.1). However, direct military expenditures routinely underestimate the economic cost of war as veterans' health care, retirement benefits, and other veterans' services are not included in the defense budget. In 2014, these costs represented an additional $161 billion in spending on top of the nearly $610 billion defense budget (National Center for Veterans Analysis and Statistics 2015).

The wars in Iraq and Afghanistan have also taken a tremendous toll within each country, including the loss of human life and the disruption to social and economic development. Between 2001 and 2014, it is estimated that more than 26,000 Afghan civilians died as a direct result of the war. War-related deaths also occur indirectly, as in Afghanistan, where health-related infrastructure (e.g., water treatment facilities) are destroyed, routine medical care (e.g., vaccines) is unavailable, and refugees are vulnerable to malnutrition and disease (Crawford 2015). Wars are also costly to "populations of concern," which includes refugees, asylum seekers, and other displaced people (United Nations High Commissioner for Refugees 2015). To date, there are 2.7 million Afghan refugees and more than 700,000 displaced persons who remain in Afghanistan.

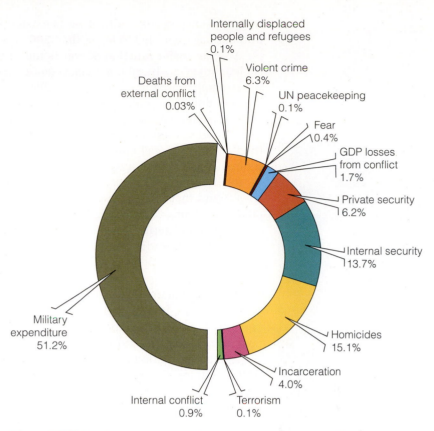

Figure 15.2 The Economic Costs of Violence Containment, 2015
SOURCE: IEP 2015.

Sociological Theories of War

Sociological perspectives can help us understand various aspects of war. In this section, we describe how structural functionalism, conflict theory, and symbolic interactionism can be applied to the study of war.

Structural-Functionalist Perspective

Structural functionalism focuses on the functions that war serves and suggests that war would not exist unless it had positive outcomes for society. We have already noted that war has served to consolidate small autonomous social groups into larger political states. An estimated

TABLE 15.1 Direct Costs of Global War and Terrorism, 2012

Category	Total Direct Cost (US$)
Military expenditure	$2.4 trillion
GDP losses from conflict	$80 billion
Deaths from internal conflict	$40 billion
Terrorism	$5 billion
UN peacekeeping	$5 billion
Internally displaced people and refugees	$3 billion
Deaths from external conflict	$1 billion
Total	**$2.5 trillion**

SOURCE: IEP 2015.

600,000 autonomous political units existed in the world at about 1000 B.C. Today, that number has dwindled to fewer than 200.

Another major function of war is that it produces social cohesion and unity among societal members by giving them a "common cause" and a common enemy. Unless a war is extremely unpopular, military conflict promotes economic and political cooperation. Internal domestic conflicts between political parties, minority groups, and special interest groups often dissolve as they unite to fight the common enemy. During World War II, U.S. citizens worked together as a nation to defeat Germany and Japan. The **rally round the flag effect** refers to the increase in political cohesion that occurs in times of war and international crisis. For example, after the September 11, 2001 attacks, President George W. Bush's approval rating soared to over 90 percent. One study of 51 democratic countries found that voter turnout significantly increased in the months following terrorist attacks (Robbins et al. 2013).

In the short term, war may also increase employment and stimulate the economy. The increased production needed to fight World War II helped pull the United States out of the Great Depression. War can also have the opposite effect, however. In a 2005 restructuring of the military, the Pentagon, seeking a "meaner, leaner fighting machine," recommended shutting down or reconfiguring nearly 180 military installations, "ranging from tiny Army reserve centers to sprawling Air Force bases that have been the economic anchors of their communities for generations" (Schmitt 2005), at a cost of thousands of civilian jobs.

Wars also function to inspire scientific and technological developments that are useful to civilians. For example, innovations in battlefield surgery during World War II and the Korean War resulted in instruments and procedures that later became common practice in civilian hospital emergency wards (Zoroya 2006). Research on laser-based defense systems led to laser surgery, research in nuclear fission facilitated the development of nuclear power, and the Internet evolved from a U.S. Department of Defense research project. Today, **dual-use technologies**, a term referring to defense-funded innovations that also have commercial and civilian applications, are quite common. For example, the fluoride compounds used in toothpaste are also necessary for the manufacturing of sarin, a deadly nerve agent, used by government forces in Syria (Bauer 2013; Deutsch 2015).

War also serves to encourage social reform. After a major war, members of society have a sense of shared sacrifice and a desire to heal wounds and rebuild normal patterns of life. They put political pressure on the state to care for war victims, improve social and political conditions, and reward those who have sacrificed lives, family members, and property in battle. As Porter (1994) explained, "Since . . . the lower economic strata usually contribute more of their blood in battle than the wealthier classes, war often gives impetus to social welfare reforms" (p. 19). For example, much of today's social welfare benefits for impoverished mothers and children grew out of the benefits programs enacted for the widows and wives of deceased and injured soldiers of the U.S. Civil War (Skocpol 1992).

When members of groups who are discriminated against in society are perceived as sacrificing for the benefit of the dominant group, military policy typically precedes the expansion of equal rights laws in the society at large. Thus, the bravery and sacrifice of African American troops in segregated units during World War II led to the racial integration of the military in the early 1950s, more than a decade before the passage of civil rights legislation in U.S. law. Prior to the 2015 Supreme Court ruling that allowed for federal recognition of same-sex marriages, the military formally ended its "don't ask, don't tell" policy to allow LGBT service members to openly serve and live with their partners in 2011, and in 2013 it extended benefits to the same-sex spouses of service members (Hicks 2013; see Chapter 11).

Finally, the U.S. military has historically provided an alternative for the advancement of poor or disadvantaged groups that otherwise face discrimination or limited opportunities in the formal economy. The military's specialized training, tuition assistance programs for college education, and preferential hiring practices improve the prospects of veterans to find a decent job or career after their service (U.S. Department of Veterans Affairs 2015).

> Wars . . . function to inspire scientific and technological developments. . . . Research on laser-based defense systems led to laser surgery, research in nuclear fission facilitated the development of nuclear power, and the Internet evolved from a U.S. Department of Defense research project.

rally round the flag effect Phenomenon of increased political cohesion and civic engagement that commonly occurs during times of war and international crisis.

dual-use technologies Defense-funded technological innovations with commercial and civilian use.

Conflict Perspective

Conflict theorists emphasize that the roots of war are often antagonisms that emerge whenever two or more ethnic groups (e.g., Bosnians and Serbs), countries (United States and Vietnam), or regions within countries (the U.S. North and South) struggle for control of resources or have different political, economic, or religious ideologies. In addition, conflict theory suggests that war benefits the corporate, military, and political elites.

Corporate elites benefit because war often results in the victor taking control of the raw materials of the losing nations, thereby creating a bigger supply of raw materials for its own industries. Furthermore, Pentagon contracts often guarantee a profit to the developing corporations. Even if the project's cost exceeds initial estimates, called a cost overrun, the corporation still receives the agreed-on profit. In the late 1950s, President Dwight D. Eisenhower referred to this close association between the military and the defense industry as the **military-industrial complex**. For example, even as the U.S. government was deciding on the war in Iraq, "many former Republican officials and political associates of the Bush administration [were] associated with the Carlyle Group, an equity investment firm with billions of dollars in military and aerospace assets" (Knickerbocker 2002, p. 2).

In the 20th century, industrialization spurred technological innovations that transformed warfare more rapidly than at any other time in human history. Here, a drone—an unoccupied aerial vehicle—is being prepared for a mission. In 2013, Niger agreed to house U.S. surveillance drones used for watching the actions of Islamic militants in the area. Drones also have nonmilitary uses. In 2015, Amazon proposed drones-only airspace to be used, for example, for robotic delivery of packages.

War benefits the political elite by giving government officials more power. Porter (1994) observed that "throughout modern history, war has been the level by which . . . governments have imposed increasingly larger tax burdens on increasingly broader segments of society, thus enabling ever-higher levels of spending to be sustained, even in peacetime" (p. 14). Political leaders who lead their country to a military victory also benefit from the hero status conferred on them.

The military elite also benefit because war and the preparations for it provide prestige and employment for military officials. Private military and security companies commonly employ retired high-ranking officers into executive leadership positions. Many seats on the boards of directors of security corporations such as DynCorp International, Mission Essential, MPRI, and Kroll Security International, just to name a few, are filled by retired generals and admirals (Spearin 2014). There is, therefore, a strong link between the for-profit private security industry and the highest levels of military leadership.

One private company that provided security personnel to the U.S. military was the North Carolina–based firm Blackwater Worldwide. In September 2007, Blackwater personnel guarding a U.S. diplomatic convoy opened fire at a traffic circle in Baghdad, killing 17 and wounding 24 Iraqi civilians. Corporate officials initially claimed that their contractors responded proportionately to a nearby attack. However, more than seven years after the incident, four of the guards faced trial for the killings. One was found guilty of murder and sentenced to life in prison, the others guilty of manslaughter and sentenced to 30 years each (Hsu and Martin 2015).

Feminist Theories. Feminists, as many other analysts, note the strong association between war and gender. Countries with low levels of gender equality are more likely to experience both civil and interstate war. In fact, much of the recent conflict in the Middle East and Africa has occurred in areas where women are oppressed (SIPRI 2015a). Furthermore, as conflict theorists also note, active combat has historically been carried out by men. Nature-based arguments about gender—that is, that men are innately aggressive or violent and women inherently peaceful—are not generally supported by social science research and do not adequately explain why men are more likely to kill than women. Feminists emphasize the social construction of aggressive masculine identities and their manipulation by elites as important reasons for the association between masculinity and militarized violence (Alexander and Hawkesworth 2008).

military-industrial complex A term first used by Dwight D. Eisenhower to connote the close association between the military and defense industries.

Conflict theorists draw attention to the ways in which war benefits elites and is tied to the economic system. Antiwar demonstrations were common occurrences in the 1960s and 1970s, as students demonstrated against the war in Vietnam (1959 to 1975). They erupted again in opposition to the Gulf War (1990 to 1991), the war in Iraq (2003 to 2011), and the war in Afghanistan (2001 to 2014). There are also those who have protested any U.S. intervention in the Syrian civil war.

> The realities of nonconventional warfare and the operational needs of the military put increasing pressure on the military to revise its policy that excluded women from combat roles.

Although some feminists view women's participation in the military as a matter of equal rights, others see war as an extension of patriarchy and the subordination of women in male-dominated societies. Ironically, because protection of women is perceived as a feature of the masculine identity, feminists also point out that war and other conflicts are often justified using "the language of feminism" (Viner 2002). For example, President George W. Bush used respect for women's rights and protection of women subjugated under the Taliban as a partial justification for the attack on Afghanistan in 2001 (Viner 2002).

Although women have historically been excluded from combat, the unconventional nature of the wars in Iraq and Afghanistan exposed thousands of women and men in occupations defined by the military as noncombat roles into harm's way. Particularly in Iraq, the use of improvised explosive devices (IEDs) turned roads into the front lines of the war, posing new dangers to traditionally noncombat roles such as truck drivers. According to the Department of Defense, 161 women were killed and 1,003 were injured in the wars in Iraq and Afghanistan (Defense Casualty Analysis System [DCAS] 2015).

In both Iraq and Afghanistan, female linguists, intelligence specialists, and military police were needed in combat units to speak with and search Muslim women who traditionally avoid contact with men outside of their families. The realities of nonconventional warfare and the operational needs of the military put increasing pressure on the military to revise its policy that excluded women from combat roles. In January 2013, the Department of Defense announced its plans to rescind the female combat exclusion policy, with full integration implemented by January 1, 2016 (Burrelli 2013). Each branch of service is required to develop "gender-neutral" standards for each occupation; however, it is still not clear how the "gender-neutral" policy will be interpreted by each service branch.

One option is a policy of "gender norming" in which men and woman are measured by equality of effort, rather than equality of output, with the result of men and women having different minimum standards for measures such as push-ups. Past research has found that gender-norming policies result in greater numbers of women being represented in traditionally male dominated occupations; however, these women are typically perceived as incompetent in physically demanding roles, resulting in the reinforcement of gender differences among personnel (Kleykamp and Clever 2015).

A second option is a policy of occupationally specific standards validation, in which each job is assessed for the physical and mental standards necessary to accomplish its requirements, and each individual is assessed on his or her ability to meet those objectives. This approach tends to result in greater gender equality as women are perceived of as highly capable and equally contributing to mission objectives. However, women are less likely than men to meet highly demanding physical standards, resulting in a lower representation of women in physically demanding occupations.

Despite the many advancements women have achieved in the military, many barriers remain. Women make up approximately 15 percent of the military, but only 7 percent of generals and admirals. Because combat experience is a major factor in military promotion decisions, the integration of women into combat roles may help, over time, to bring more women into the highest leadership positions. The lack of women in high levels of leadership and the male-dominated culture of the military have been blamed for the high rate of sexual assault among women in the military, which is about twice the rate among civilian women.

Andrew Harrer/Bloomberg/Getty Images

Tammy Duckworth, a double-amputee veteran of the Iraq War, was elected to the House of Representatives in 2013 and ran for the U.S. Senate in 2016. Duckworth lost her legs in 2004, when insurgents shot down the helicopter she was piloting. Images of women wounded in military service during Operations Iraqi Freedom and Enduring Freedom helped bolster public support for the end of the Combat Exclusion Policy.

General Martin Dempsey, chair of the Joint Chiefs of Staff, stated that one major cause of the sexual assault crisis in the military is a culture of inequality that has stemmed from the military defining men and women as "separate classes of military personnel" (quoted in Portero 2013). In what ways do you think inequality contributes to a "culture of rape," both in the military and in civilian society (see Chapters 4 and 10)?

WHAT
do you
THINK?

Symbolic Interactionist Perspective

The symbolic interactionist perspective focuses on how meanings and definitions influence attitudes and behaviors regarding conflict and war. The development of attitudes and behaviors that support war begins in childhood. American children learn to revere and celebrate the Revolutionary War, which created our nation. Movies romanticize war, children play war games with toy weapons, and various video and computer games glorify heroes conquering villains.

TABLE 15.2 Public Attitudes on the Use of U.S. Ground Troops to Fight Islamic Militants in Iraq and Syria, 2015

	Favor (%)	Oppose (%)
Total	47	49
Men	52	44
Women	41	54
White	49	47
Black	34	61
Hispanic	48	46
18–29	39	59
30–49	52	45
50–64	49	45
65+	45	51
Republican	67	31
Independent	48	48
Democrat	32	63

Notes: "Don't know" responses not shown. Whites and blacks included only those who are "Not Hispanic"; Hispanics are of any race.

SOURCE: Pew Research Center 2015a.

> Governments may use propaganda and appeals to patriotism to generate support for war efforts and to motivate individuals to join armed forces.

Symbolic interactionism helps to explain how military recruits and civilians develop a mind-set for war by defining war and its consequences as acceptable and necessary. The word *war* has achieved a positive connotation through its use in various popular public policies—the war on drugs, the war on poverty, and the war on crime. Positive labels and favorable definitions of military personnel facilitate military recruitment and public support of armed forces. However, because different groups have different experiences in society, not all Americans are equally receptive to messages about the benefits or necessity of war. For example, widespread media coverage of the atrocities committed by the militant group ISIS in Iraq and Syria between 2014 and 2015 coincided with an overall increase in public attitudes in favor of U.S. military intervention in the conflict from 39 percent to 47 percent over a four-month period (Pew Research Center 2015a). However, men, whites, Republicans, and people over age 30 were more likely to express support for armed intervention than women, African Americans, Democrats, and people under age 30 (see Table 15.2).

Many government and military officials convince the public that the way to ensure world peace is to be prepared for war. Governments may use propaganda and appeals to patriotism to generate support for war efforts and to motivate individuals to join armed forces. Both proponents of war and peace employ the language of patriotism—for example, "support the troops" or "peace is patriotic"—to frame their political arguments about war (Salladay 2003; Woehrle et al. 2008). Patriotism is a popular sentiment in American society. Overall, 56 percent of Americans say they often feel proud to be an American, a figure that ranges from 40 percent among liberals to 81 percent among conservatives (Pew Research Center 2014c).

To legitimize war, the act of killing in war is not regarded as "murder." Deaths that result from war are referred to as "casualties." Bombing military and civilian targets appears more acceptable when nuclear missiles are "peacekeepers" that are equipped with multiple "peace heads." Killing the enemy is more acceptable when derogatory and dehumanizing labels such as Gook, Jap, Chink, Kraut, and Haji convey the attitude that the enemy is less than human.

Such labels and their accompanying images are socially constructed, often by the media, and are presented to the public. Social constructionists, like symbolic interactionists in general, emphasize the social aspects of "knowing." Cultural historian John Downer (1986) analyzed propaganda images used by both the American and Japanese governments during World War II, revealing that both sides drew on racist stereotypes about each other to construct dehumanized images of the enemy. Japanese propaganda presented Americans as brutish, ape-like imperialists, while American propaganda presented the Japanese as devious, vermin-like predators.

Not only do media images contribute to political attitudes about war and social attitudes about the enemy, but these images also have a powerful effect on a media consumer's mental and physical health. Psychologists measured study participants' media consumption in the weeks following the 9/11 attacks in 2001 and the invasion of Iraq in 2003, and followed up with their health records annually for the next decade. They found that individuals who watched four or more hours of news coverage per day in the weeks following each event had a significantly higher likelihood than people who had less media exposure of posttraumatic stress symptoms and other health ailments years later (Silver et al. 2013).

In April 2013, after bombs exploded during the Boston Marathon, SWAT teams searched house to house in Boston looking for accused bomber Dzhokhar Tsarnaev. The manhunt led to a "shelter in place" order for the entire city of Boston, forcing the closure of businesses and schools for nearly 24 hours. The lockdown sparked a debate about the boundaries between public safety and American liberty in the wake of an act of domestic terrorism. Representative Dutch Ruppersberger (D-MD) suggested the lockdown plays into the fears that terrorists hope to foster: "We have to stand up as Americans to this. . . . We've got to continue to go to baseball games, continue to go to events. We can't allow these people to shut us down."

Causes of War

The causes of war are numerous and complex. Most wars involve more than one cause. The immediate cause of a war may be a border dispute, for example, but religious tensions that have existed between the two combatant countries for decades may also contribute to the war. The following sections review various causes of war.

Conflict over Land and Other Natural Resources

Nations often go to war in an attempt to acquire or maintain control over natural resources, such as land, water, and oil. Michael Klare, author of *Resource Wars: The New Landscape of Global Conflict* (2001), predicted that wars would increasingly be fought over resources as supplies of the most needed resources diminish. Disputed borders have been common motives for war. Conflicts are most likely to arise when borders are physically easy to cross and are not clearly delineated by natural boundaries, such as major rivers, oceans, or mountain ranges.

In the modern era, oil has been a major resource at the center of many conflicts. Not only do the oil-rich countries in the Middle East present a tempting target in themselves, but war in the region can also threaten other nations that are dependent on Middle Eastern oil. Thus, when Iraq seized Kuwait and threatened the supply of oil from the Persian Gulf, the United States and many other nations reacted militarily in the Gulf War. In the digital era, rare earth minerals needed to make cell phones, computers, and other advanced technologies have come into increasing global demand, leading to fears that control over these resources could be the next major source of conflict. In the Democratic Republic of the Congo, warring militias use the sale of rare earth minerals to Western nations to fund the ongoing civil war, and often use child slaves to work in the mines. The role of these "**conflict minerals**" in perpetuating civil war and human rights abuses associated with the mining prompted the technology firm Intel to announce that beginning in 2016, all of its products would be manufactured using conflict-free minerals (Intel 2015).

Water is another valuable resource that has led to wars. Unlike other resources, water is universally required for survival. At various times, the empires of Egypt, Mesopotamia, India, and China all went to war over irrigation rights. In 1998, five years after Eritrea gained independence from Ethiopia, forces clashed over control of the port city Assab and, with it, access to the Red Sea. In a document prepared for the Center for Strategic and International Studies, Starr and Stoll (1989) warned that soon "water, not oil, will be the dominant resource issue of the Middle East" (p. 1).

Despite such predictions, tensions in the Middle East have erupted into fighting repeatedly in recent years—but not over water. Beginning in 2011, pro-democracy and anti-corruption protests across the Middle East, collectively known as the Arab Spring, led to the outbreak of major civil wars in Egypt, Yemen, Libya, and Syria. In 2015, ongoing tensions in Yemen sparked by Arab Spring protests erupted into a full-scale civil war. Religious, cultural, and political differences (also see "Racial, Ethnic, and Religious Hostilities") between the Sunni-led government in the north and the Shiite rebels seeking greater political autonomy in the south have contributed to the conflict. These differences have escalated tensions between other nations as well. Saudi Arabia began bombing rebel compounds in support of the Yemeni government, leading Iran to provide weapons and tactical support to the rebels' defense (Riedel 2015).

Conflict over Values and Ideologies

Many countries initiate war not over resources but over beliefs. World War II was largely a war over differing political ideologies: democracy versus fascism. The Cold War involved the clash of opposing economic ideologies: capitalism versus communism. Conflicts over values or ideologies are not easily resolved. They are less likely to end in compromise or negotiation because they are fueled by people's convictions. For example, when a representative sample of American Jews was asked, "Do you agree or disagree with the following statement? 'The goal of Arabs is not the return of occupied territories but rather the destruction of Israel,'" 75 percent agreed and 24 percent disagreed (American Jewish Committee 2013).

If ideological differences can contribute to war, do ideological similarities discourage war? The answer seems to be yes; in general, countries with similar ideologies are less likely to engage in war with each other than countries with differing ideological values (Dixon 1994). Referred to as the **democratic peace theory**, research has shown that democratic nations are particularly disinclined to wage war against one another (Brown et al. 1996; Rasler and Thompson 2005).

Racial, Ethnic, and Religious Hostilities

Racial, ethnic, and religious groups vary in their cultural beliefs, values, and traditions. Thus, conflicts between racial, ethnic, and religious groups often stem from conflicting values and ideologies. Such hostilities are also fueled by competition over land and other scarce natural and economic resources. Gioseffi (1993) noted that "experts agree that the depleted world economy, wasted on war efforts, is in great measure the reason for renewed ethnic and religious strife. 'Haves' fight with 'have-nots' for the smaller piece of the pie that must go around" (p. xviii).

Racial, ethnic, and religious hostilities are sometimes perpetuated by a wealthy minority to divert attention away from their exploitations and to maintain their own position of power. Such **constructivist explanations** of ethnic conflict—those that emphasize the role of leaders of ethnic groups in stirring up intercommunal hostility—differ sharply from **primordial explanations**, or those that emphasize the existence of "ancient hatreds" rooted in deep psychological or cultural differences between ethnic groups. For example, most people are familiar with the primordial explanation of the conflict between Israel and

conflict minerals Minerals used in advanced technologies (e.g. cell phones) sold to Western nations by waring groups to fund ongoing civil wars; child slaves often work in the mines, leading to human rights abuses.

democratic peace theory A prevalent theory in international relations suggesting that the ideological similarities between democratic nations make it unlikely that such countries will go to war against each other.

constructivist explanations Those explanations that emphasize the role of leaders of ethnic groups in stirring up hatred toward others external to one's group.

primordial explanations Those explanations that emphasize the existence of "ancient hatreds" rooted in deep psychological or cultural differences between ethnic groups, often involving a history of grievance and victimization, real or imagined, by the enemy group.

Palestine that emphasizes ancient antagonisms between Jewish and Muslim people over the ownership of what both groups believe are their holy sites. Constructivists, however, point out historical evidence that these groups actually coexisted peacefully and shared access to these sites for many centuries. It was only after British and French authorities began dividing the land between the groups in the 1920s, and in many cases promising the same land to both groups simultaneously, that the tensions that characterize the current conflict arose (Kaufman and Hassassian 2009).

As described by Paul (1998), sociologist Daniel Chirot argues that the worldwide increase in ethnic hostilities is a consequence of "retribalization"—that is, the tendency for groups, lost in a globalized culture, to seek solace in the "extended family of an ethnic group" (p. 56). Chirot identified five levels of ethnic conflict: (1) multiethnic societies without serious conflict (e.g., Switzerland); (2) multiethnic societies with controlled conflict (e.g., the United States and Canada); (3) societies with ethnic conflict that has been resolved (e.g., South Africa); (4) societies with serious ethnic conflict leading to warfare (e.g., Sri Lanka); and (5) societies with genocidal ethnic conflict, including "ethnic cleansing" (e.g., Darfur).

Religious differences as a source of conflict have recently come to the forefront. An Islamic jihad, or holy war, has been blamed for the September 11 attacks on the World Trade Center and Pentagon as well as for bombings in Kashmir, Sudan, the Philippines, Indonesia, Kenya, Tanzania, Saudi Arabia, Spain, Great Britain, and France. Some claim that Islamic beliefs in and of themselves have led to recent conflicts (Feder 2003). Others contend that religious fanatics, not the religion itself, are responsible for violent confrontations and emphasize that misunderstandings between cultural groups can further fuel these tensions. For example, most Americans understand the term *jihad* as radical Muslims have used it to justify violent conflict as a holy war, but many moderate Muslims point out that the more conventional understanding of the term is as a faith-based internal struggle to achieve a life of peace (Bonner 2006).

Conflicts between different sects of the same religion can also lead to long-lasting and devastating wars. The civil war that erupted in Syria in 2011 was driven in large part by the perceived injustice of Sunnis representing the largest segment of the population yet holding a minority of leadership positions within the Assad regime. The conflict between these two sects has deep historical roots, and the conflict in Syria has reignited tensions between Sunni and Shiite populations in Iraq, Iran, and Libya (Arango and Barnard 2013). Wars over differing religious beliefs have led to some of the worst episodes of bloodshed in history, in part, because some religions lend themselves to martyrdom—the idea that dying for one's beliefs leads to eternal salvation. For example, Islamic leader Osama bin Laden claimed that unjust U.S.–Middle East policies are responsible for "dividing the whole world into two sides—the side of believers and the side of infidels" (Williams 2003, p. 18).

> Wars over differing religious beliefs have led to some of the worst episodes of bloodshed in history, in part, because some religions lend themselves to martyrdom—the idea that dying for one's beliefs leads to eternal salvation.

Defense against Hostile Attacks

The threat or fear of being attacked may cause leaders of a country to declare war on the nation that poses the threat. This is an example of what experts in international relations refer to as the **security dilemma**:

> The basic premise of the security dilemma is that as one state [nation] takes measures to increase its security. . . . Another state might take similar, reactive measures to make up for the shift in the balance of power. If states perceive the actions of other actors to be offensive in nature, the change in the balance of power is perceived as detrimental to their own security. This creates a cycle in which both states will continually take measures, such as increasing military strength . . . to increase their security. In turn, tensions between the two states can escalate into conflict. (Prasad 2015, p. 1)

Such situations may lead to war inadvertently. The threat may come from a foreign country or from a group within the country. After Germany invaded Poland in 1939, Britain and France declared war on Germany out of fear that they would be Germany's next victims. Germany attacked Russia in World War I, in part out of fear that Russia had entered

security dilemma A characteristic of the international state system that gives rise to unstable relations between states; as State A secures its borders and interests, its behavior may decrease the security of other states and cause them to engage in behavior that decreases A's security.

the arms race and would use its weapons against Germany. Japan bombed Pearl Harbor, hoping to avoid a later confrontation with the U.S. Pacific fleet, which posed a threat to the Japanese military.

In 2001, a U.S.-led coalition bombed Afghanistan in response to the September 11 terrorist attacks. Moreover, in March 2003, the United States, Great Britain, and a loosely coupled "coalition of the willing" invaded Iraq in response to perceived threats of weapons of mass destruction and the reported failure of Saddam Hussein to cooperate with United Nations' weapons inspectors. Yet, in 2005, a presidential commission concluded that the attack on Iraq was based on faulty intelligence and that, in fact, "America's spy agencies were 'dead wrong' in most of their judgments about Iraq's weapons of mass destruction" (Shrader 2005, p. 1). As a result, by 2007, many Americans—more than 60 percent—favored a partial or complete withdrawal from Iraq (CNN/Opinion Research Corporation Poll 2007). Although the Obama administration withdrew all combat operations from Iraq in 2011, the political upheaval caused by the United States–led war contributed to the ongoing civil war in Iraq.

Revolutions and Civil Wars

Revolutions and civil wars involve citizens warring against their own government and often result in significant political, economic, and social change. The difference between a revolution and a civil war is not always easy to determine. Scholars generally agree that revolutions involve sweeping changes that fundamentally alter the distribution of power in society (Skocpol 1994). The American Revolution resulted from colonists revolting against British control. Eventually, they succeeded and established a republic where none existed before. The Russian Revolution involved a revolt against a corrupt, autocratic, and out-of-touch ruler, Czar Nicholas II. Among other changes, the revolution led to wide-scale seizure of land by peasants who formerly were economically dependent on large landowners. North and South Korea, North and South Sudan, Algeria, and Zimbabwe are just a few examples of modern countries that were formed as a result of revolution or civil war.

Civil wars may result in a different government or a new set of leaders but do not necessarily lead to such large-scale social change. Because the distinction between a revolution and a civil war depends on the outcome of the struggle, it may take many years after the fighting before observers agree on how to classify it. Revolutions and civil wars are more likely to occur when a government is weak or divided, when it is not responsive to the concerns and demands of its citizens, and when strong leaders are willing to mount opposition to the government (Barkan and Snowden 2001; Renner 2000).

One of the world's longest-running civil wars came to an end in May 2009. Beginning in 1983, the government of Sri Lanka fought an insurgency led by the Liberation Tigers of Tamil Eelam (LTTE). Also known as the Tamil Tigers, the LTTE were separatist militants who sought to carve an independent state out of the northern and eastern portions of this island country. The Sri Lankan Army defeated the last remnants of the LTTE and killed their leader in May 2009 (Buncombe 2009). The war resulted in upward of 100,000 deaths, including tens of thousands of civilians (Mahr 2013). Like many civil wars, the war in Sri Lanka was also a struggle between a majority community (in this case, Sinhalese Buddhists) and a relatively poor and disadvantaged minority community (Hindu Tamils).

The civil war in Syria has quickly become one of the deadliest in recent decades. Protests against government corruption and the arrests of political dissenters in March 2011 met with a police crackdown that left several protestors dead. Continued protests spurred the Syrian government, led by President Bashar al-Assad, to deploy military force against the protestors. As military and government officials began to defect from the regime and join the side of the rebels, under the banner of the Free Syrian Army, the conflict moved from one of political dissent to a full-scale civil war between two militaries fighting for control of the government.

the HUMAN side — Life and Death in a Refugee Camp

In response to the civil war in Syria, over 4 million refugees have left their homeland and settled in countries throughout the Middle East, North Africa, and Europe. In the excerpts below, refugees who were able to escape to a camp in Jordan describe the brutality of their existence. Note not only the physical impact of living in a refugee camp but the psychological cost experienced by its inhabitants.

A Syrian Mother

I was walking home [and] I came behind two armed men and overheard them taking bets on something. They were planning to use something for target practice. When they then agreed the bets I realised they were talking about an eight-year-old boy who was playing alone on the road. I realised too late—one of them had taken the bet and shot him in the head…

The child was lying on the street…It wasn't a clean shot and he didn't die straight away. It took hours. His mother was inside the house on the same street and she was screaming. She wanted to reach her child, but the men kept firing into the street and taunting this mother: "you can't get to your child, you can't get to your child."

He died alone on the street outside his home.

Wael, 16 year old boy

I knew a boy called Ala'a …He was only six years old. He didn't understand what was happening. His dad was told that his child would die unless he gave himself up. I'd say that this six-year-old boy was tortured more than anyone else…He wasn't given food or water for three days, and he was so weak he used to faint all the time. He was beaten regularly. I watched him die. He only survived for three days and then he simply died…

There's no way I can cope, no way I can turn over a new page. I have seen children slaughtered. I don't think I'll ever be OK again.

Amani, 13 year old girl

…Seven months ago they broke into our uncle's house and starting beating my brother with sticks. Then they took turns jumping on his back. He was beaten so badly that he still can't walk.

…I remember seeing my brother the first time after he was beaten. He was so pale and he couldn't walk. I thought he was about to die. We put him in his bed. He's still there. We had to leave him there when we came here…

When I think about what happened, I can't stop myself crying. I cry all the time.

I don't know how long it will take us to recover—perhaps a lifetime.

Hassan, 14 year old boy

I was at a funeral when I first heard the rocket that caused a massacre. I think it was targeting the funeral. My cousin and my uncle died that day.

Dead bodies along with injured people were scattered on the ground. I found body parts all over each other; and when we reached the mosque we found tens and tens of dead bodies there. We started to rescue people in need.

Dogs were eating the dead bodies for two days after the massacre. There were tons of people in the mosques too. They were dead, all of them. I was afraid, of course I was afraid.

A Syrian Father

Children are on the frontline in this war in many ways. I have seen with my own eyes children used as human shields. When two tanks came into the village I saw children attached to them, tied up by their hands and feet, and by their torsos. The tanks came through the village and no one stood in their way or fought because we knew we would kill the children.

After that happened I cried like a woman. I was close to losing my mind. I have never felt so helpless as the moment I saw those children strapped to those tanks.

The name of the village was Saydeh. Let everyone know this is where this terrible thing happened.

SOURCE: Save the Children 2012.

The ongoing war has devastated Syrian society. At least 220,000 people have been killed, more than 12 million have fled the country, and those who remain live with a decimated infrastructure and lack of public services. There are continued stories of rape, torture, and summary executions, yet no one has been held accountable, leading a coalition of aid organizations to declare that the situation in Syria is a "stain on the conscience of the international community" (Sengupta 2015, p. 1). Furthermore, the secretary general of the United Nations, Ban Ki-moon, said that in five years of civil war, Syria has lost the equivalent of four decades of human development (*The Economist* 2015). See this chapter's *The Human Side*, "Life and Death in a Refugee Camp," for stories of how the Syrian civil war has affected families and children.

The causes of war are often overlapping and complex. ISIS leaders have claimed a wide range of motivations for their wars—which have spread from Syria into Iraq, the Arabian Peninsula, and North Africa—including religious ideology as well as resistance against Western political, economic, and cultural influence. ISIS is known for using both terrorist and conventional military tactics. These ISIS militants pose after a battle in which they took control of a suburban neighborhood of Damascus, Syria in April 2015.

WHAT do you THINK?

After the end of American combat operations in Iraq in 2011, both Iraq and its neighbor Syria fell into civil war, during which time large portions of each country fell under the control of the militant organization ISIS. In June 2015, the Obama administration announced it would begin reestablishing American military bases in Iraq to provide training and support to the Iraqi government in their fight against ISIS (White House 2015a). Under what circumstances, and to what extent, do you think the American military should get involved in civil wars in other nations?

Nationalism

Some countries engage in war in an effort to maintain or restore their national pride. For example, Scheff (1994) argued that "Hitler's rise to power was laid by the treatment Germany received at the end of World War I at the hands of the victors" (p. 121). Excluded from the League of Nations, punished by the Treaty of Versailles, and ostracized by the world community, Germany turned to nationalism as a reaction to material and symbolic exclusion.

In the late 1970s, Iranian militants seized the U.S. Embassy in Tehran and held its occupants hostage for more than a year. President Jimmy Carter's attempt to use military forces to free the hostages was not successful. That failure intensified doubts about the United States' ability to use military power effectively to achieve its goals. The hostages in Iran were eventually released after President Ronald Reagan took office, but doubts about the strength and effectiveness of the U.S. military still called into question the United States' status as a world power. Subsequently, U.S. military forces invaded the small island of Grenada because the government of Grenada was building an airfield large enough to accommodate major military armaments. U.S. officials feared that this airfield would be used by countries in hostile attacks on the United States. From one point of view, the large-scale "successful" attack on Grenada functioned to restore faith in the power and effectiveness of the U.S. military. Many analysts have suggested that Russia's invasions of the Ukraine in 2014, a former Soviet republic, is an example of Russia's efforts to restore its national pride to its former Soviet-era glory (Goode 2014).

Terrorism

Terrorism is the premeditated use, or threatened use, of violence against civilians by an individual or group to gain a political or social objective (Barkan and Snowden 2001; Brauer 2003; Goodwin 2006). Terrorism may be used to publicize a cause, promote an ideology, achieve religious freedom, attain the release of a political prisoner, or rebel against a government. Terrorists use a variety of tactics, including assassinations, skyjackings, suicide bombings, armed attacks, kidnapping and hostage taking, threats, and various forms of bombings. Through such tactics, terrorists struggle to induce fear within a population, create pressure to change policies, or undermine the authority of a government they consider objectionable. Most analysts agree that, unlike war—where a clear winner is more likely—terrorism is unlikely to be completely defeated:

> There can be no final victory in the fight against terrorism, for terrorism (rather than full-scale war) is the contemporary manifestation of conflict, and conflict will not disappear from earth as far as one can look ahead and human nature has not undergone a basic change. But it will be in our power to make life for terrorists and potential terrorists much more difficult. (Laqueur 2006, p. 173)

Types of Terrorism

Terrorism can be either transnational or domestic. **Transnational terrorism** occurs when acts of terrorism are not restricted to or centered within one country. Alternatively, **domestic terrorism** takes place within the territorial jurisdiction of one nation. Sometimes, however, it is difficult to distinguish between the two.

The May 2015 attack in Garland, Texas, followed in the wake of attacks in Paris, in January 2015. Both incidents were motivated by what was perceived by his followers as offensive images of the Islamic prophet Mohammad. The two gunmen in the Texas shooting were both Americans citizens and their actions were limited to the United States. ISIS, however, took responsibility for the attack, and there was some evidence that at least one of the gunmen had planned to join Islamic militants in Africa (Bergen 2015). Similarly, the attacks in Paris have elements of both transnational and domestic terrorism.

The offices of *Charlie Hebdo*, a satirical magazine widely known for publishing cartoons of the Islamic prophet Mohammad, was stormed by two masked gunmen who opened fire at an editorial meeting and fled the scene. During the attack, the shooters were heard to shout "the Prophet Mohammed is avenged!" (BBC 2015). Two other associates stormed a kosher supermarket, holding the customers hostage in a day-long standoff with police. At the end of the attacks, 12 journalists, 4 supermarket customers, 1 police officer, and the 2 shooters were dead. One of the supermarket hostage takers was apprehended by the police, while the other escaped during the police raid and is believed to have fled to Syria. The two attackers of the magazine offices were both French citizens, yet three of the four attackers had belonged to the same jihadist group (Martinez et al. 2015).

Transnational Terrorism. The 2001 attacks on the World Trade Center, the Pentagon, and Flight 93—the most devastating in U.S. history—are also the deadliest examples of transnational terrorism. Al Qaeda, a global alliance of militant Sunni Islamic groups advocating violence against Western targets, was also responsible for attacks on U.S. embassies in Kenya and Tanzania (1998) and the bombing of a naval ship, the USS *Cole*, moored in Aden Harbor, Yemen (2000). Al Qaeda has since been linked to deadly bombings in Bali, Indonesia (2002), Madrid (2004), and London (2005).

After nearly 20 years of effort, President Obama announced on May 2, 2011, that the leader of al Qaeda, Osama bin Laden, had been killed by Navy SEALs and CIA operatives during a raid on a private residential compound in Abbottabad, Pakistan, about 30 miles northeast of Islamabad (Baker et al. 2011). One month after the 9/11 terrorist attacks, 28 percent of Americans polled responded that they were very worried that there would soon be another terrorist attack in the United States; in January 2015, shortly after the attack on the *Charlie Hebdo* offices, 25 percent of Americans responded that they were very worried about another terrorist attack in the United States (see Figure 15.3).

terrorism The premeditated use or threatened use of violence by an individual or group to gain a political objective.

transnational terrorism Transnational terrorism occurs when terrorists' acts are not restricted to or centered within one country.

domestic terrorism Domestic terrorism occurs within the territorial jurisdiction of one nation.

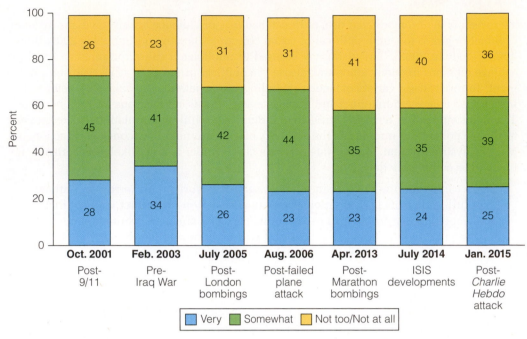

Worried there will soon be another terrorist attack in the United States*

Date	Oct. 2001 Post-9/11	Feb. 2003 Pre-Iraq War	July 2005 Post-London bombings	Aug. 2006 Post-failed plane attack	Apr. 2013 Post-Marathon bombings	July 2014 ISIS developments	Jan. 2015 Post-*Charlie Hebdo* attack
Not too/Not at all	26	23	31	31	41	40	36
Somewhat	45	41	42	44	35	35	39
Very	28	34	26	23	23	24	25

Legend: ■ Very ■ Somewhat ■ Not too/Not at all

*Don't know responses not shown.

Figure 15.3 American Public's Terrorism Worries (N = 1504), 2001–2015
SOURCE: Pew Research Center 2015b.

Many groups other than al Qaeda use terrorism to further their own social and political agendas. In fact, the U.S. Department of State has identified 54 "foreign terrorist organizations . . . [that] threaten the security of U.S. nationals or the national security . . . of the United States" (Office of the Coordinator for Counterterrorism 2014). In 2013, Boko Haram, a militant organization affiliated with the Taliban and al Qaeda and operating in Nigeria, was added to this list after a series of attacks that killed hundreds.

The name *Boko Haram* loosely translates to "Western education is forbidden," and the group is notorious for attacking schools, especially those that educate girls. One assault in 2014 received widespread international attention when Boko Haram attacked a girls' school in northern Nigeria, kidnapping nearly 300 teenage girls and threatening to sell them into slavery (START 2014). The Nigerian army, after widespread condemnation of the kidnappings, conducted a series of raids on Boko Haram camps, rescuing nearly 700 of the estimated 2,000 girls and women who had been kidnapped by the terrorist organization over the previous year (Purefoy and Ellis 2015).

The militant group known as the **Islamic State in Iraq and Syria (ISIS)** is another example of a transnational terrorist organization. The group first emerged as al-Qaeda in Iraq after the 2003 United States led invasion. Their goal was to expel the coalition troops and establish a new nation under a stricter interpretation of Islamic Sharia law. In 2011, after the unrest in Syria, the group expanded its aims to establishing a caliphate, an Islamic-led state, in the Levant, a region that includes Egypt, Iraq, Israel, Jordan, Lebanon, Palestine, Syria, and Turkey. In 2015, ISIS claimed responsibility for the downing of a Russian passenger jet and for the 130 deaths resulting from simultaneous attacks on several Paris venues including a café, soccer stadium, and concert hall.

The group funds its efforts primarily through kidnapping, extortion, smuggling, and other crimes, and it is estimated to net about $8 million a month (Council on Foreign Relations 2014). The group is notorious for its brutal tactics, including its highly produced propaganda videos showing the beheadings of kidnapped journalists and aid workers. According to the CIA, 180 Americans have attempted to travel to Syria to join ISIS, including three teenagers from Colorado who were recruited by ISIS over social media and arrested during a layover in Germany on their way to Syria (Berlinger 2015).

Islamic State in Iraq and Syria (ISIS) Notorious for its brutal tactics, a transnational terrorist organization whose goal is to establish an Islamic-led state in the Levant, a region that includes Egypt, Iraq, Israel, Jordan, Lebanon, Palestine, Syria, and Turkey.

Domestic Terrorism. In 2013, two homemade bombs were detonated near the finish line of the Boston Marathon, killing 3 people and injuring 260. Surveillance footage implicated brothers Dzhokhar and Tamerlan Tsarnaev, U.S. citizens who were born in Dagestan, Russia, of carrying out the attack. After a high-speed chase and gun fight with police, Tamerlan Tsarnaev was killed, and 19-year-old Dzhokhar was ultimately taken into police custody and charged with carrying out the attacks. In May 2015, a jury found him guilty and sentenced him to death (Kamp and Levitz 2015). Other examples of domestic terrorism include the 1995 truck bombing by Americans Timothy McVeigh (who was later executed) and Terry Nichols (who was sentenced to life) of the Alfred P. Murrah Federal Building in Oklahoma City, killing 168 people and injuring more than 680 others; and the 2010 killings of 13 people by an Army psychiatrist at Fort Hood, Texas. Figure 15.4 graphically portrays the various methods used by transnational and domestic terrorists.

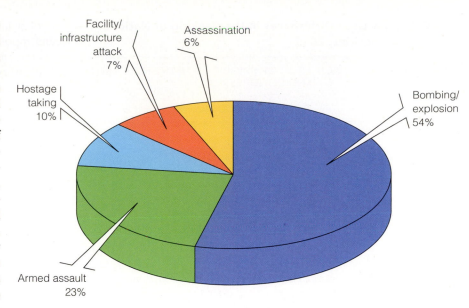

Figure 15.4 Tactics Used in Terrorist Attacks Worldwide, 2014
SOURCE: U.S. Department of State 2015a.

What are the differences among domestic terrorism, hate crimes, and murder? The distinctions are often open to interpretation based on the social meanings attached to the identities of perpetrators and victims. Dylan Roof, a white supremacist, is accused of entering an evening Bible study at the Charleston, South Carolina, African Methodist Emmanuel Church and opening fire, killing nine people. The church was historically associated with the anti-slavery and civil rights movements, and Roof openly declared that his motivations for the attack were to spark a racial conflict in the United States, and he chose the site because of its historical symbolism (Gladstone 2015).

Roof was captured within one day of the attack and charged with nine counts of murder; however, no charges of terrorism were brought against him. Critics were quick to note that the attack fit all the conventional criteria of domestic terrorism, including the ideological and political motivations for the attack, the highly prominent symbolism of the target, and extended planning. However, the statuses of the perpetrator, a young white man, did not fit the conventional understanding of a terrorist—namely, an Islamic insurgent motivated by religious extremism and anti-American ideology.

Patterns of Global Terrorism

A report by the National Counterterrorism Center (U.S. Department of State 2015a) describes patterns of terrorism around the world. In 2014:

- There were 13,463 domestic and transnational terrorist attacks recorded in 95 countries around the world, an average of 809 attacks per month.
- Over 32,700 people lost their lives as a result of these attacks.
- Perpetrators of terrorism kidnapped or held hostage 9,400 people.
- The most commonly used tactic in terrorist attacks was bombings or explosions, followed by armed assault.
- Sixty percent of all attacks took place in five countries: Iraq, Pakistan, India, Nigeria, and Afghanistan.
- The top four perpetrator groups with the highest number of worldwide attacks were ISIS/ISIL, the Taliban, Al-Shabaab, and Boko Haram.

TABLE 15.3 Partisan Differences in the Ranking of National Priorities, 2015

PERCENT RATING EACH AS A TOP PRIORITY FOR THE PRESIDENT AND CONGRESS			
Issue	Republicans	Democrats	Independent
Terrorism	87	71	74
Environmental protection	35	66	48
Education	52	77	67
Military	71	41	51
Poverty	40	70	52
Economy	75	74	75

SOURCE: Pew Research Center 2015b.

In a nationally representative poll, Americans reported that defending against terrorism was the most important political priority, ahead of all other domestic and international issues. Americans were twice as likely to report terrorism as the top political priority as global warming, and terrorism ranked ahead of the economy, jobs, education, and Social Security (Pew Research Center 2015b). Although 76 percent of Americans ranked terrorism as the most important political priority, there were important differences in the ranking of priorities between political parties. Republicans were more likely to rank terrorism as a top priority (87 percent) than Democrats (71 percent) (see Table 15.3).

The Roots of Terrorism

The causes of terrorism are complex. Many scholars have focused on macro-structural factors, such as global economic shifts, foreign occupation, repression of minorities, poverty, and weak governance as creating the underlying conditions that produce terrorism (Bjorgo 2003; Gurr 2015; Piazza 2011). For example, Freytag et al. (2011), studying terrorism in 110 countries over nearly four decades, conclude that the socioeconomic conditions of a country are inversely related to the likelihood of terrorist incident.

These grievance models focus on how social structures and institutions create inequalities that impact the behavior of individuals. When discussing the problem of terrorism from this approach, solutions tend to focus on reforming institutions in targeted countries to address the inequalities that produce the grievances that then, theoretically, lead to terrorism. Because that is an impossible task (Gorka 2014), government-led counterterrorism efforts, no doubt in part as a response to ISIS's success in radicalizing Westerners, increasingly focus on the micro-level psychological processes that make some individuals more prone to radicalization than others (Kruglanski et al. 2014).

How do individuals choose to join terrorist organizations or use terrorist tactics? Borum (2011) reviews several models of how people move from radicalization into violent extremism (RVE) from academic studies, the military, and law enforcement, and he suggests that "understanding the mind-set" of a terrorist can help in the fight against terrorism. The RVE process is very complex, Borum argues, involving social relationship networks, grievances, perceived rewards, experiences with incarceration, and psychosocial vulnerabilities (see Figure 15.5).

Borum points out that the vast majority of people with militant extremist beliefs don't engage in violent action, and that counterviolent extremist (CVE) efforts need to focus on more than the "battle of ideas." Rather, CVE efforts should take a multifaceted approach to understanding and attempting to mitigate the grievances, psychosocial vulnerabilities, and social network influences that interact to produce violent extremism.

> Americans reported that defending against terrorism was the most important political priority, ahead of all other domestic and international issues.

Figure 15.5 The Process of Ideological Development
SOURCE: Borum 2003.

America's Response to Terrorism

A government can use both defensive and offensive strategies to fight terrorism. Defensive strategies include using metal detectors and X-ray machines at airports and strengthening security at potential targets, such as embassies and military command posts. The

Department of Homeland Security (DHS) coordinates such defensive tactics for the U.S. government. As of 2016, DHS has nearly a quarter of a million employees with a budget of $41.2 billion. The objective of the DHS is to provide "a safe and secure homeland . . . in which the liberties of all Americans are assured, privacy is protected, and the means by which we interchange with the world—through travel, lawful immigration, trade, commerce, and exchange—are secured" (U.S. Department of Homeland Security 2015, p. 1). This chapter's *Animals and Society* feature describes some of the ways animals contribute to the fight against terrorism.

Two recent incidents have raised questions about the U.S. government's response to terrorism. In 2010, Private Bradley Manning was arrested for leaking videos and more than 700,000 pages of documents related to the conduct of the U.S. military in Iraq and Afghanistan to the website WikiLeaks. Despite claims of altruistic motives, Manning was convicted in 2013 of 17 of the 22 charges that were brought against him, although he was cleared of the most serious charge, that of "aiding the enemy" (Pilkington 2013). Similarly, Edward Snowden, one of the more than 850,000 civilian contractors with top-secret clearance, leaked information about the National Security Agency's (NSA) domestic surveillance program. Snowden fled to Hong Kong and ultimately Russia where he has been granted asylum. In 2015, Snowden expressed interest in returning to the United States if the government agreed to reduce the charges of espionage and theft of government property (Fuller 2015; Reuters 2013).

Snowden's leak of the NSA surveillance program sparked a public debate about the balance between the rights of American citizens to privacy and the government's obligation to ensure public safety. In 2013, a federal judge ruled that the bulk data collection program was unconstitutional; and, in 2015, after lengthy debate, Congress voted not to renew the portion of the 2001 PATRIOT Act that allowed for the surveillance program. Although many continue to argue that Snowden is a traitor, others suggest that his efforts in addressing an injustice should be commended and that he should be granted a full pardon. According to one commentator:

> After 9/11 . . . the trade-off between security and liberty tipped too far in the direction of intrusion and authoritarianism. Historians will record that Snowden's leaks helped, at least somewhat, to right the balance. At great risk to himself, he stood up to the immensely powerful system for which he worked, and cried foul. (Cassidy 2015, p. 1)

WHAT do you THINK?

Both Bradley Manning and Edward Snowden claimed patriotic motivations for leaking documents about questionable actions of the U.S. government in the fight against terrorism. Some argue that these leaks are acts of whistle-blowing that publicize government corruption and abuse, while others suggest that these leakers are traitors whose actions harm the U.S. fight against terror and endanger the lives of service members. Are the actions of Manning and Snowden those of traitors or heroes? What do you think?

Offensive strategies include retaliatory raids, such as the U.S. bombing of terrorist facilities in Afghanistan, group infiltration, and preemptive strikes. The American public largely supports these offensive strategies; however, many policy analysts are critical about their effectiveness given the high costs of counterterrorism tactics. One study of the costs and benefits of counterterrorism spending found that approximately 300 terrorist attacks on the scale of the Boston Marathon bombing would need to occur in the United States each year to justify the annual cost of counterterrorism (Mueller and Stewart 2014).

Guantánamo Detention Center. Among the most controversial responses to the war on terrorism is the indefinite detention of "enemy combatants" at a military prison and interrogation camp in Guantánamo, Cuba. This is the primary detention center for the Taliban or their allies captured in Afghanistan, as well as suspected terrorists, including al Qaeda members from other regions. Since 2002, as many as 779 detainees have been held at "Gitmo" (*New York Times* 2015). The Bush administration argued that because these detainees were not members of a state's army, they were not covered by the **Geneva Conventions**, the principal international treaties governing the laws of war and in particular the treatment of prisoners

Geneva Conventions A set of international treaties that govern the behavior of states during wartime, including the treatment of prisoners of war.

There are monuments to them all over the world but none quite as spectacular as the Animals in War Memorial in England. Opened in 2004, the memorial honors the millions of animals that valiantly "suffered and served":

To the mules that were silenced for the Burmese jungle by having their vocal cords severed; the donkeys that collapsed under panniers of ammunition; and the dogs that ripped their paws raw digging for survivors, or had half their faces blown off searching for mines but carried on to find more—they are all remembered; the camels and canaries; the elephants and oxen; the messenger pigeons that flew home on one wing; and even the glow worms, by whose gentle light the soldiers read their maps in the first world war. (Price 2004)

Analogous to the Victoria Cross, the most prestigious military award in England, the Dickin's Medal honors animals that have shown "lifesaving bravery in military combat" (Ricks 2014, p. 1). In 2014, after 64 previous winners—including dozens of World War II messenger pigeons, 28 dogs, 3

Pictured, Sergeant Andrew Garrett and K-Dog, a bottlenose dolphin in training. Sea lions "gliding through the water to locate mines, handcuff terrorists and take part in surveillance … are the real Navy Seals" (Dennis, 2014, p.1).

horses, and a cat—a 4-year-old yellow Lab named Sasha was honored. Sasha and her handler, 24-year-old Corporal Ken Rowe, were killed in 2008 in a Taliban ambush. The two were considered the best bomb-sniffing

team in Afghanistan. Rowe and Sasha's deployment had ended and they were scheduled to leave Afghanistan the day before they were killed, but Rowe worried about the safety of his team without Sasha and decided to stay. Sasha saved many military and civilian lives by locating 15 confirmed improvised explosive devices (IEDs) in under one year.

Since 9/11, animals have increasingly been used for guarding U.S. borders and protecting against terrorism. U.S. military forces commonly use dogs because of their loyalty, intelligence, and keen sense of smell. Dogs sniff five times a second, and each nostril can detect different scents (Dean 2013). While dogs are often used to search for illegal drugs by police, military "chem dogs" locate weapons of mass destruction by detecting chemical weapons, or components of chemical weapons, while they are still concealed and before they are released into the atmosphere.

Dogs are not the only animals that serve in the military. Since 1960, the U.S. Navy's Marine Mammal Program has trained bottlenose dolphins and California sea lions to detect underwater mines

of war and civilians during wartime. In 2006, the U.S. Supreme Court rejected this argument, ruling that the detainees were subject to minimal protections under the Conventions. Furthermore, in 2008, the Court ruled that the Constitution guarantees the right of detainees to challenge their detention in a federal court (Greenburg and de Vogue 2008).

The treatment of detainees at Guantánamo and at secret detention centers (so-called black sites) around the world has ignited public debate about interrogation techniques used on suspected terrorists. In 2014, a Senate committee released a report on the CIA's Detention and Interrogation Program (Senate Select Committee on Intelligence 2014). In part, the authors of the report concluded:

- The techniques used by CIA interrogators (e.g., waterboarding, isolation, sleep deprivation, beatings) constitute torture under international law.
- The CIA misrepresented the brutality and the effectiveness of the interrogation techniques to gain approval for and continuation of the program.
- The CIA intentionally impeded oversight of the program, including providing false information to the president, vice president, and the National Security Council.
- The CIA coordinated the release of classified and inaccurate information to the media concerning the effectiveness of the enhanced interrogation program.
- The CIA wrongfully detained people, including an "intellectually challenged" man whose detention was used solely as leverage to get a family member to provide information (p. 12).

and work in underwater object recovery and diving operations (Lee 2014). Both animals have skills that set them apart from other mammals and from each other. When given a cue by a handler, dolphins use *echolocation* or biosonar to search for and find a particular target. The dolphin then uses a series of clicks that bounce off the target and echo back to the dolphin's jawbone, which then transmits the information to the brain, creating a "mental image of the object" (Animals at Arms 2010; Simon 2003; U.S. Navy 2010). The dolphin then returns to the handler and conveys information about the target to him or her through a series of noises and movements. More than two dozen dolphins are working in mine detection, with 50 more dolphins and 30 sea lions working in recovery and diving (*Navy Times* 2012).

Dolphins also protect ships, submarines, harbors, nuclear power plants, and U.S. borders against any kind of waterborne threat. After the attack on the USS *Cole*, there were fears that enemy swimmers might try to attach explosive devices to other U.S. ships. In response to the attack, the U.S. Navy created the anti-swimmer dolphin system (Frey 2003).

Anti-swimmer dolphins are able to detect and mark combat swimmers with "an illuminated floating beacon" and, if necessary, use their rock-hard bottlenoses to delay the combat swimmer's movements until U.S. military personnel arrive (German 2011, p. 1; Larsen 2011). Dolphins, who comfortably work in a variety of environments, have served "tours of duty" in Vietnam, the Gulf War, Bahrain, and Iraq, and can be deployed anywhere in the world within a 72-hour period (Frey 2003; Larsen 2011). Dolphins' unique talents are used throughout the world. In 2014, the Russian Navy captured "a Ukrainian military unit made up of bottlenose dolphins" which they will now give "equipment upgrades" (Lee 2014, p. 1).

Sea lions complete tasks similar to those completed by dolphins but do so in a slightly different manner because of a different skill set. Sea lions can see in the dark as well as underwater, and have underwater directional hearing. Unlike the sheer force that dolphins use to intercept combat swimmers, sea lions use a little finesse:

The sea lions are trained to detect swimmers or divers approaching military ships or piers. The animals carry a clamp in their mouths. They approach the swimmer quietly from behind and attach the clamp, which is connected to a rope, to the swimmer's leg. With the person restrained, sailors aboard ships can pull the swimmer out of the water.... Navy officials say the sea lions, part of the Shallow Water Intruder Detection System program, are so well-trained that the clamp is on the swimmer before he is aware of it. (Leinwand 2003, p. 1)

Dolphins and sea lions may be the U.S. Navy's first choice for some assignments, but "man's best friend" edged out the two sea mammals in the mission to kill bin Laden. In 2011, 79 U.S. Navy SEALs (of the human variety) and a dog entered Abbottabad, Pakistan. However, only 24 commandos slid down the ropes from the hovering Black Hawk helicopter into bin Laden's compound, and only one had a dog strapped to his back (Frankel 2011b; Jeon 2011). Although details about the dog are as top secret as the identity of the Navy SEALs, it is rumored that, soon after the mission, a Belgian Malinois named Cairo, a member of the Navy SEALs, had a closed-door session with President Obama and the rest of the team (Frankel 2011a).

Public opinion on this topic is mixed. In response to a question about whether "torture to gain important information from suspected terrorists is justified," 50 percent of American adults said that they believed torture is "often" or "sometimes" justified, whereas 47 percent believed that torture is "rarely" or "never" justified (Pew Research Center 2014a). A substantial majority of Republicans (71 percent) said that torture could be at least sometimes justified, compared with 51 percent of Independents and 45 percent of Democrats.

Weapons of Mass Destruction. The possibility of terrorists using **weapons of mass destruction (WMDs)**—namely, chemical, biological, and/or nuclear weapons—is a frightening scenario made all the more salient by the Syrian government's 2013 use of chlorine gas against civilians (Human Rights Watch 2014). Banned by international law, the incident led to tensions and near-military conflict between the United States and the Russian-supported Syrian regime. After weeks of tense negotiations, the Russian government brokered an agreement between the Obama administration and the Syrian government to eradicate all chemical weapons stockpiled by the Syrian government in order to prevent a strike by the U.S. military. By October, international mediators assured world leaders that the Syrian stockpile was being eradicated (Rapoza 2013).

Although many praised the diplomatic effort between the United States and Russia for averting military action, others criticized both countries for demanding the eradication of chemical weapons in Syria when both countries have large stockpiles of chemical weapons of their own. Although each signed a treaty in 1993 to destroy their chemical weapons

weapons of mass destruction (WMD) Chemical, biological, and nuclear weapons that have the capacity to kill large numbers of people indiscriminately.

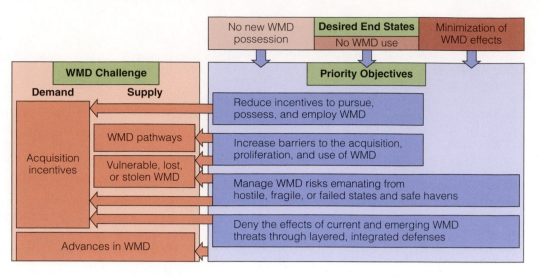

FIGURE 15.6 Desired End States, Priority Objectives, and the WMD Challenge
SOURCE: U.S. Department of Defense 2014.

> Despite the understandable concerns that Americans have about weapons of mass destruction and terrorist attacks in general, it is important to remember that death by either is an extraordinarily rare occurrence...

arsenal by 2012, Russia has an estimated 10,000 metric tons of chemical weapons, while the United States holds an estimated 2,000 metric tons (Arms Control Association 2014). Figure 15.6 displays the challenges of reducing the demand and supply of WMD and the desired end states to meet those challenges (Department of Defense 2014).

Despite the understandable concerns that Americans have about weapons of mass destruction and terrorist attacks in general, it is important to remember that death by either is an extraordinarily rare occurrence worldwide, especially in American society. There has been no major terrorist event in the United States since the attacks of 2001, and as one analyst observed, except for 2001, more Americans have died struck by lightning than by international terrorists (Mueller 2006, p. 13).

Nonetheless, fears about terrorist attacks continue to be widespread. Across seven polls taken between October 2001 and January 2015, more than half of Americans have reported feeling very or somewhat worried that there will soon be another terrorist attack in the United States (Pew Research Center 2015b). These worries were at their highest in February 2003 prior to the invasion of Iraq (75 percent were very or somewhat worried) and at their lowest in April 2013 after the Boston Marathon bombings (58 percent were very or somewhat worried). After the *Charlie Hebdo* shootings in January 2015, 64 percent were very or somewhat worried about a terrorist attack in the United States (see Figure 15.3).

Combating terrorism is difficult, and recent trends will make it increasingly problematic. First, hackers who illegally gain access to classified information can easily acquire data stored on computers. For example, terrorists obtained the fueling and docking schedules of the USS *Cole*, facilitating the attack that killed 17 American sailors and injured many more (CNN 2014). Incidents of cyberwar, attacks on a nation's computer and information networks for malevolent purposes, is increasing as evidenced by the volume of recent data breaches (see Chapter 14). Government officials acknowledge that data, even if collected by state agents, may in turn be sold to others, including terrorists groups (Behm 2015).

Second, social media and the Internet in general have allowed groups with similar interests, once separated by geography and language, to share plans, fund-raising efforts, recruitment strategies, and other coordinated efforts. ISIS in particular has distinguished itself for its sophisticated use of social media to distribute propaganda and recruit followers, particularly young men (Farwell 2014). Even the comments following an online news article may have been deliberately placed there by, for example, Russian employees whose sole task is to post misinformation (Bay and Porche 2015).

Finally, globalization contributes to terrorism by providing international markets where the tools of terrorism—explosives, guns, electronic equipment, and the like—can be purchased. In fact, globalization is a "background factor" for terrorism that provides a "new means of ideological competition and access to new financial resources [e.g. oil] and weapons" (Zimmermann 2011, p. S159).

Social Problems Associated With Conflict, War, and Terrorism

Social problems associated with conflict, war, and terrorism include death and disability; rape, forced prostitution, and displacement of women and children; social-psychological costs; diversion of economic resources; and destruction of the environment.

Death and Disability

Many American lives have been lost in wars, including 116,000 in World War I, 405,000 in World War II, 36,500 in Korea, and 58,000 in Vietnam (DCAS 2015). As of June 2015, 6,828 U.S. military personnel have been killed in the Global War on Terror, and 52,315 have been wounded.

Many civilians and enemy combatants also die or are injured in war. Despite more than 150,000 Iraqi civilian deaths, many Americans are unaware of this tremendous loss of life. In a program the Pentagon developed, American reporters were "embedded" into U.S. military units to provide "journalists with a detailed understanding of military culture and life on the frontlines" (Lindner 2009, p. 21). One of the by-products of this program, however, was that 90 percent of the stories by embedded journalists were written from the perspective of American soldiers, focusing "on the horrors facing the troops, rather than upon the thousands of Iraqis who died" (p. 45).

The impact of war and terrorism extends far beyond those who are killed. Many of those who survive war incur disabling injuries or contract diseases. For example, in South Sudan alone, over 4,200 people were killed or wounded by land mines following the end of civil war with northern Sudan in January 2005 (United National Mine Action Office 2011). In 1997, the Mine Ban Treaty, which requires that governments destroy stockpiles within 4 years and clear land mine fields within 10 years, became international law. To date, 162 countries have signed the agreement; 35 countries remain, including China, India, Israel, Russia, and the United States (International Campaign to Ban Landmines 2015). War-related deaths and disabilities also deplete the labor force, create orphans and single-parent families, and burden taxpayers who must pay for the care of orphans and disabled war veterans (see Chapter 2 for a discussion of military health care).

The killing of unarmed civilians undermines the credibility of armed forces, making their goals more difficult to defend. SEAL Team 6, a highly classified Navy Special Operations force that is most well-known for carrying out the raid that killed Osama Bin Laden and rescuing Captain Richard Phillips from Somali pirates, has been criticized for using excessive force against civilians. A *New York Times* investigation found evidence of widespread indiscriminate killing of civilians, resulting from the limited oversight over Special Forces and the tendency for America to assess success "not by battlefield wins and losses, but by the relentless killing of suspected militants" (Manzetti et al. 2015, p. 1). Excessive use of force by American personnel has repeatedly been used by militant groups to gain support among the civilian population and to justify the use of terrorism.

Rape, Forced Prostitution, and the Displacement of Women and Children

Despite international prohibitions against wartime rape, and "humiliating and degrading treatment, enforced prostitution and any form of indecent assault" (International Committee of the Red Cross 2015), such behaviors have routinely occurred throughout history. Before and during World War II, Japanese officials forced between 100,000 and 200,000 women and teenage girls into prostitution as military "comfort women." These women were forced to have sex with dozens of soldiers every day in "comfort stations." Many of the women died as a result of untreated sexually transmitted diseases, harsh punishment, or indiscriminate acts of torture.

Since 1998, Congolese government forces have fought Ugandan and Rwandan rebels. Women have paid a high price for this civil war, in which gang rape is "so violent, so systematic, so common . . . that thousands of women are suffering from vaginal fistula, leaving them unable to control bodily functions and enduring ostracism and the threat of debilitating health problems" (Wax 2003, p. 1). Though much less common than violence against

women, aid workers also see increasing incidents of rape and sexual violence against men as "yet another way for armed groups to humiliate and demoralize Congolese communities into submission" (Gettleman 2009, p. 1).

Feminist analyses of wartime rape emphasize that the practice reflects not only a military strategy but also ethnic and gender dominance. For example, Refugees International, a humanitarian aid group, reports that rape is "a systematic weapon of ethnic cleansing" against Darfuris and is "linked to the destruction of their communities" (Boustany 2007, p. 9). Under Darfur's traditional law, prosecution of rapists is nearly impossible: Four male witnesses are required to accuse a rapist in court, and single women risk severe corporal punishment for having sex outside marriage.

War, terrorism, and repressive governments have led to 60 million people fleeing to other countries or other regions of their homeland, a quarter of whom live in sub-Saharan Africa (Hadro 2015). Refugee women and female children are particularly vulnerable to sexual abuse and exploitation by locals, members of security forces, border guards, or other refugees. Wars are also dangerous for the very young. As a result of

Nearly 4 million Syrians have become refugees of the ongoing civil war. Here, a Syrian refugee family huddles together under a makeshift shelter during a rainy day in Istanbul, Turkey. As winter approached in 2015, many Eastern European countries, including Croatia, Hungary, and Macedonia, began building walls and closing their borders to refugees. Nearly one million refugees found themselves trapped between the Middle East and Europe without adequate clothing or shelter for the winter weather (Borger et al. 2015).

the civil war in Syria, millions of children have been "killed, maimed, and denied access to food and medicine" (Uenuma 2013). Furthermore, over 4,000 schools have been destroyed since the start of the war in 2011. An estimated 1.6 million Syrian children were not attending school in 2014 (Save the Children 2015).

WHAT do you THINK?

Large numbers of Muslim immigrants and refugees from North Africa and the Middle East seeking safety in Europe have led to anti-immigrant movements in many European countries. Opponents to anti-immigration sentiment point out that policies targeting religious and ethnic minorities in Europe are reminiscent of policies enacted against the Jews just prior to the Holocaust. One analyst argued that in the wake of recent terrorist attacks in Europe, "anti-Semitism, Islamophobia, racism, xenophobia and anti-immigrant sentiment are rising across Europe" (Al-Hussein 2015, p. 1). Do you think European policies limiting immigrants and refugees from entering Europe is an appropriate response to the threat of terrorism?

Social-Psychological Costs

Terrorism, war, and living under the threat of war disrupt social-psychological well-being and family functioning (see this chapter's *The Human Side*). For example, Myers-Brown et al. (2000) report that Yugoslavian children suffered from depression, anxiety, and fear as a response to conflicts in that region, emotional responses not unlike those Americans experienced after the events of 9/11 (National Association of School Psychologists 2003). Furthermore, a study of children in postwar Sierra Leone found that over 70 percent of boys and girls whose parents had been killed were at "serious risk" of suicide (Morgan and Behrendt 2009). A study of 8- to 12-year-old Lebanese children found that children exposed to multiple events of war-related violence became increasingly desensitized to violence; as the number of war-related events children experienced increased, they became more likely to accept violence as normal and to imitate violent acts (Tarabah et al. 2015).

Guerrilla warfare is particularly costly in terms of its psychological toll on soldiers. In Iraq, soldiers were repeatedly traumatized as "guerrilla insurgents attack[ed] with

guerrilla warfare Warfare in which organized groups oppose domestic or foreign governments and their military forces; often involves small groups of individuals who use camouflage and underground tunnels to hide until they are ready to execute a surprise attack.

impunity," and death was as likely to come from "hand grenades thrown by children, [as] earth-rattling bombs in suicide trucks, or snipers hidden in bombed-out buildings" (Waters 2005, p. 1). During the last half of the Iraq War, from 2008 to 2013, the suicide rate for U.S. soldiers surpassed that of the rate for civilians for the first time since the Vietnam War (Haiken 2013). When reservists are included in the totals with active-duty personnel, more military members committed suicide than died in combat in Iraq and Afghanistan (Donnelly 2011).

The suicide rate among young veterans has remained unchanged since combat operations ended in Iraq and Afghanistan: in 2015, nearly 22 veterans a day took their own lives (Shane 2015). The reasons that the suicide rate among veterans is so high is unclear, but may be related to the interactions between an individual's mental health and structural pressures related to combat exposure, wartime stressors, and the social pressures of transitioning to civilian life after service (see this chapter's *Social Problems Research Up Close*).

Military personnel who engage in combat and civilians who are victimized by war may experience a form of psychological distress known as **posttraumatic stress disorder (PTSD)**, a clinical term referring to a set of symptoms that can result from any traumatic experience, including crime victimization, rape, or war. Symptoms of PTSD include sleep disturbances, recurring nightmares, flashbacks, and poor concentration (National Center for Posttraumatic Stress Disorder 2007). For example, Canadian lieutenant-general Roméo Dallaire, head of the United Nations peacekeeping mission in Rwanda, witnessed horrific acts of genocide. For years after his return, he continued to have images of "being in a valley at sunset, waist deep in bodies, covered in blood" (quoted in Rosenberg 2000, p. 14). PTSD is also associated with other personal problems, such as alcoholism, family violence, divorce, and suicide.

Estimates of PTSD vary widely, although they are consistently higher among combat versus noncombat veterans. In a telephone study of 1,965 military personnel who had been deployed to Iraq and Afghanistan, 14 percent reported current symptoms of PTSD, 14 percent reported current symptoms of major depression, and 9 percent reported symptoms consistent with both conditions (Tanielian and Jaycox 2008). The Department of Veterans Affairs estimates that PTSD afflicts 30 percent of Vietnam veterans, 11 percent of Afghan war veterans, and 20 percent of Iraqi war veterans (U.S. Department of Veterans Affairs 2014). According to one PTSD survivor:

> Post-traumatic stress disorder (PTSD) isn't something you just get over. You don't go back to being who you were. It's more like a snow globe. War shakes you up, and suddenly all those pieces of your life—muscles, bones, thoughts, beliefs, relationships, even your dreams—are floating in the air out of your grip. They'll come down. I'm here to tell you that, with hard work, you'll recover. But they'll never come down where they once were. You're a changed person after combat. Not better or worse, just different. (Montalván 2011, p. 50)

The rate of PTSD among soldiers is difficult to measure for several reasons. First, there is a lag, often of several years, between the time of exposure to trauma and the manifestation of symptoms. Second, many are reluctant to seek help because of the social stigma associated with being labeled mentally ill (Mittal et al. 2013). Among those who report PTSD or major depression symptoms, only about half seek treatment (Recovering Warrior Task Force 2014). Evidence indicates that suicide prevention efforts by the Department of Veterans Affairs (VA) do work. A long-term study of suicide trends among veterans found that suicide rates among male veterans who utilized VA Health Services decreased by 30 percent over a 10-year period. Among male veterans who did not use VA Health Services, the suicide rate increased by 60 percent (Kemp 2014).

Diversion of Economic Resources

As discussed earlier, maintaining the military and engaging in warfare require enormous financial capital and human support. However, the decision to spend, for example, $245 million for one F-35 Joint Strike fighter jet, an amount that would fund approximately 31,000

Many are reluctant to seek help because of the social stigma associated with being labeled a mentally ill. Among those who report PTSD or major depression symptoms, only about half seek treatment.

posttraumatic stress disorder (PTSD) A set of symptoms that may result from any traumatic experience, including crime victimization, war, natural disasters, or abuses.

Combat, Mental Illness, and Military Suicides

Not long after the invasion of Iraq in 2003, the suicide rate among military personnel and veterans began to increase dramatically. The wars in Iraq and Afghanistan required the military to deploy its troops to war zones more often and for longer periods of time than ever before. In addition, the nature of the counterinsurgency operations in these war zones introduced new sources of stress, as military personnel had to balance humanitarian outreach with fighting.

The suicide rate among military personnel increased from 2005 to 2009, plateaued in 2010 and 2011, and spiked again in 2012. Although combat operations have ended, the suicide rate among Iraq and Afghanistan veterans has remained steady: approximately 22 veterans commit suicide every day. The unconventional nature of the wars in Iraq and Afghanistan, and the strain associated with repeated and prolonged deployments led many to speculate that the growing suicide rate was linked with the stresses of deployment and combat.

This troubling pattern led Dr. Cynthia LeardMann and colleagues, research scientists at the Naval Health Research Center (NHRC), to launch a long-term study with the goal of understanding whether or not there is a link between combat experience and risk of suicide. The information gained from this research could help military leaders and policy makers understand what kind of organizational changes and outreach efforts might help prevent future suicides.

Sample and Methods
Beginning in July 2001, the NHRC launched the Millennium Cohort Study (MCS), a longitudinal study of more than 150,000 randomly selected military personnel. Every three years, the study participants completed a questionnaire detailing their physical, mental, and functional health. These surveys were

collected among both active-duty personnel and those who had left the service since the study began. Using the names and Social Security numbers of the study participants over an 11-year period, the researchers matched the MCS records with records from the National Death Index and the Defense Medical Mortality Registry. They found that 83 of the 151,560 study participants had committed suicide over the previous decade. The researchers also matched the survey records of the study participants with their Department of Defense service records, indicating the number and length of their deployments and whether they served in combat areas. The MCS survey also asked respondents to report whether they had ever witnessed or been exposed to specific traumatic events, such as witnessing a death, torture, or seeing maimed or decomposing bodies. By matching these records together and employing a statistical hazard ratio model, the researchers were able to determine which factors were associated with risk of suicide.

Findings and Conclusions
Contrary to expectations, the researchers did not find that deployments, combat, or traumatic experiences in war were risk factors for suicide. Instead, the risk factors for suicide among the military population were identical to those among the civilian population: Those who were male, reported alcohol abuse, and/or had a history of mental illness, such as depression or manic-depressive disorder, were at a higher risk of suicide. The frequency, length, and nature of deployment experiences were not associated with suicide. The research team concludes that:

The findings from this study do not support an association between deployment or combat with suicide; rather they are consistent with

previous research indicating that mental health problems increase suicide risk. Therefore, knowing the psychiatric history, screening for mental and substance use disorders, and early recognition of associated suicidal behaviors combined with high-quality treatment are likely to provide the best potential for mitigating suicide risk.

The results of this research help to focus the attention of military leaders on areas where they can help prevent future suicides: by improving substance abuse and mental health resources for soldiers and veterans. However, some questions remain. If the risk of suicide is not linked to deployments and combat, then why did the rate of suicide increase so dramatically during the most intense years of the wars in Iraq and Afghanistan? One of the study investigators, Dr. Nancy Crum-Cianflone suggests that "perhaps it's not being deployed so much as being in a war during a high-stress time period." Other experts also emphasize the link between wartime stressors and mental health problems. Kim Ruocco, the director for TAPS—the military's Tragedy Assistance Program for Survivors—says, "We so often just link military suicide to combat trauma. But there are many others: long hours, separation from support systems, sleeplessness. All are stressors. All add to increases in mental health issues" (Dao 2013, n.p.).

The findings of the study indicate that the increased suicide rate was a result of an increase in mental health and substance abuse problems among military members during the war years, and point to the necessity of improving mental health and substance abuse resources for all service members and veterans, regardless of whether they have served in combat.

SOURCE: LeardMann et al. 2013.

children to attend Head Start, is a choice made by political leaders (Department of Defense 2015; Head Start 2015). Similarly, allocating $8.2 billion to build and repair missiles while schools, bridges, and other infrastructure deteriorate is a political choice.

Between 2001 and 2015, taxpayers paid $1.62 trillion for the cost of the wars in Iraq and Afghanistan, the equivalent of providing (1) one year of low-income health care for 469 million people, or (2) a year's salary for more than 21 million elementary school teachers, or (3) one year of Head Start for 194 million children, or (4) more than 20 million police officers, or (5) one year of scholarships for 198 million university students (National Priorities Project 2015). The requested budget for the Department of Defense for 2016 is two times greater than the combined requested funds for the departments of Education, Justice, Health and Human Services, NASA, the Environmental Protection Agency, and Veterans Affairs (Office of Management and Budget 2015a). (See Figure 15.7.)

Destruction of the Environment

The environmental damage that occurs during war devastates human populations long after war ends. Combatants may intentionally destroy or exploit the environment as a military strategy, a practice known as **ecocide**. For example, during the Vietnam War, the U.S. military dropped millions of gallons of herbicides, including the chemical Agent Orange,

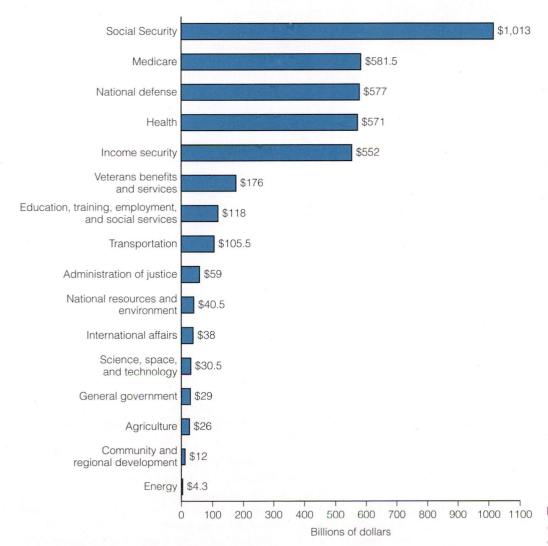

Figure 15.7 Selected Federal U.S. Outlays by Function for 2017 (estimated)
SOURCE: Office of Management and Budget 2015b.

ecocide The intentional destruction of the environment for military strategy.

Oil smoke from the 650 burning oil wells left in the wake of the Gulf War contains soot, sulfur dioxide, and nitrogen oxides, the major components of acid rain, along with a variety of toxic and potentially carcinogenic chemicals and heavy metals.

AP Images/Michel Lipchitz

over the forests of the Mekong Delta in order to prevent the Viet Cong guerillas from finding safe haven in the dense forests (WorldWatch Institute 2015). The use of conflict minerals to fund guerilla groups in many African nations results in mining practices that cause harm to the local communities through forced labor, the loss of economic production, and damage to local water supplies.

Poaching of elephants and rhinos for the sale of ivory to fund militant groups has led to the near elimination of these species in many African nations (Geneva Graduate Institute for International Studies 2015). Sometimes the environmental impact of war is less deliberate. The flood of millions of refugees fleeing from the Rwandan genocide in 1994 brought the Virunga National Park to near endangerment as refugees stripped the forests of wood for fuel and shelter (WorldWatch Institute 2015).

In addition to being affected by war, the environment can also shape how wars are fought. Combat in many parts of the world is a seasonal activity; combatants in tropical climates must stop fighting during the rainy season, in cold climates during the winter, or in agricultural societies during planting and harvesting seasons. These patterns have helped establish some predictability in the tempo and tactics of violence, particularly for communities affected by ongoing civil war. Climate change, however, affects temperature, agriculture, sea levels, and rainfall, thus disrupting local communities' ability to anticipate and plan for fighting seasons (see Chapter 14; Geneva Graduate Institute for International Studies 2015).

The ultimate environmental catastrophe facing the planet is thermonuclear war. Aside from the immediate human casualties, poisoned air, poisoned crops, and radioactive rain, many scientists agree that the dust storms and concentrations of particles created by a massive exchange of nuclear weapons would block vital sunlight and lower temperatures in the Northern Hemisphere, creating a **nuclear winter**. In the event of large-scale nuclear war, most living things on Earth would die. For example, a nuclear blast and the resulting blast wave create overpressure—the amount of pressure in excess of ordinary atmospheric levels as measured by pounds per square inch (psi). As described in a report sponsored by the U.S. Air Force:

- At 20 psi of overpressure, even reinforced-concrete buildings are destroyed.
- Ten psi will collapse most factories and commercial buildings, as well as wood-frame and brick houses.
- Five psi flattens most houses and lightly constructed commercial and industrial structures.
- Three psi suffices to blow away the walls of steel-frame buildings.
- Even 1 psi will produce flying glass and debris sufficient to injure large numbers of people. (Ochmanek and Schwartz 2008, p. 6)

The fear of nuclear war has greatly contributed to the military and arms buildup, which, ironically, also causes environmental destruction even in times of peace. For example, in practicing military maneuvers, the armed forces demolish natural vegetation, disturb wildlife habitats, erode soil, silt up streams, and cause flooding. Bombs exploded during peacetime leak radiation into the atmosphere and groundwater. From 1945 to 1990, 1,908 bombs were tested—that is, exploded—at more than 35 sites around the world. Although underground testing has reduced radiation, some radioactive material still escapes into the atmosphere and is suspected of seeping into groundwater.

nuclear winter The predicted result of a thermonuclear war whereby thick clouds of radioactive dust and particles would block out vital sunlight, lower temperature in the Northern Hemisphere, and lead to the death of most living things on Earth.

Finally, although arms control and disarmament treaties of the last decade have called for the disposal of huge stockpiles of weapons, no completely safe means of disposing of weapons and ammunition exist. Many activist groups have called for placing weapons in storage until safe disposal methods are found. Unfortunately, the longer the weapons are stored, the more they deteriorate, increasing the likelihood of dangerous leakage.

As part of the 1997 Chemical Weapons Convention, the United States agreed to destroy its entire stockpile of chemical weapons within 25 years. By the 2012 deadline, the United States had destroyed 90 percent of its stockpile, mostly mustard gas and nerve agents that were acquired during the Cold War. Destruction of the remaining 10 percent was delayed by local communities who opposed their destruction for fear of the release of harmful vapors. In February 2015, the U.S. Army announced it would begin destroying the remaining weapons using a new and safer technique of chemical neutralization. The last chemical weapons are expected to be destroyed by 2023, at a cost of $11 billion (Elliot 2015).

Strategies in Action: In Search of Global Peace

Various strategies and policies are aimed at creating and maintaining global peace. These include the redistribution of economic resources, the creation of a world government, peacekeeping activities of the United Nations, mediation and arbitration, and arms control.

Redistribution of Economic Resources

Inequality in economic resources contributes to conflict and war because the increasing disparity in wealth and resources between rich and poor nations fuels hostilities and resentment. Therefore, any measures that result in a more equal distribution of economic resources are likely to prevent conflict. John J. Shanahan (1995), retired U.S. Navy vice admiral and former director of the Center for Defense Information, suggested that wealthy nations can help reduce the social and economic roots of conflict by providing economic assistance to poorer countries. Nevertheless, U.S. military expenditures for national defense far outweigh U.S. economic assistance to foreign countries. For instance, the Obama administration's 2017 budget request included $577 billion for national defense, but just $38 billion for international affairs (see Figure 15.7).

As discussed in Chapter 12, strategies that reduce population growth are likely to result in higher levels of economic well-being. Funke (1994) explained that "rapidly increasing populations in poorer countries will lead to environmental overload and resource depletion in the next century, which will most likely result in political upheaval and violence as well as mass starvation" (p. 326). Although achieving worldwide economic well-being is important for minimizing global conflict, it is important that economic development does not occur at the expense of the environment.

Finally, former United Nations secretary general Kofi Annan, in an address to the United Nations, observed that it is not poverty per se that leads to conflict but rather the "inequality among domestic social groups" (Deen 2000). Referencing a research report completed by the Tokyo-based United Nations University, Annan argued that "inequality ... based on ethnicity, religion, national identity, or economic class ... tends to be reflected in unequal access to political power that too often forecloses paths to peaceful change" (Deen 2000).

The United Nations

Founded in 1945 after the devastation of World War II, the United Nations (UN) today includes 193 member states and is the principal organ of world governance. In its early years, the UN's main mission was the elimination of war from society. In fact, the UN charter begins, "We the people of the United Nations—determined to save succeeding generations from the scourge of war." During the past 70 years, the UN has developed major institutions and initiatives in support of international law, economic development, human rights, education, health, and other forms of social progress. The Security Council is the most powerful branch of the United Nations. Comprised of 15 member states, it has the power to impose economic sanctions against states that violate international law. It can also use force, when necessary, to restore international peace and security.

The United Nations has engaged in 71 peacekeeping operations since 1948 (see Table 15.4; United Nations 2015):

> United Nations peacekeepers—military personnel in their distinctive blue helmets or blue berets, civilian police and a range of other civilians—help implement peace agreements, monitor cease-fires, create buffer zones, or support complex military and civilian functions essential to maintain peace and begin reconstruction and institution building in societies devastated by war. (United Nations 2003, p. 1)

In 2015, more than 118,000 UN personnel were involved in overseeing 16 multinational peacekeeping forces in the Western Sahara, the Central African Republic, Mali, Haiti, the Congo, Darfur, Syria, Cyprus, Lebanon, the Sudan, Côte d'Ivoire, Kosovo, Liberia, India, and Pakistan (United Nations 2015).

The UN has recently come under heavy criticism. First, in recent missions, developing nations have supplied more than 75 percent of the troops while developed countries—the United States, Japan, and Europe—have contributed 85 percent of the finances. As one UN official commented, "You can't have a situation where some nations contribute blood and others only money" (quoted by Vesely 2001, p. 8). Second, a review of UN peacekeeping operations noted several failed missions, including an intervention in Somalia in which 44 U.S. marines were killed (Lamont 2001). Third, as typified by the debate over the disarming of Iraq, the UN cannot take sides but must wait for a consensus of its members that, if not forthcoming, undermines the strength of the organization (Goure 2003). Even if a consensus emerges, without a standing army, the UN relies on troop and equipment contributions from member states, and there can be significant delays and logistical problems in assembling a force for intervention.

The consequences of delays can be staggering. Under the Genocide Convention of 1948, the UN is obligated to prevent instances of genocide, defined as "acts committed with the intent to destroy, in whole or in part, a national, racial, ethnical, or religious group" (United Nations 1948). In 2000, the UN Security Council formally acknowledged its failure to prevent the 1994 genocide in Rwanda (BBC 2000). After the death of 10 Belgian soldiers in the days leading up to the genocide, and without a consensus for action among the members, the Security Council ignored warnings from the mission's commander about impending disaster and withdrew its 2,500 peacekeepers.

Finally, the concept of the UN is that its members represent individual nations, not a region or the world. And because nations tend to act in their own best economic and security interests, UN actions performed in the name of world peace may be motivated by nations acting in their own interests.

As a result of such criticisms, outgoing UN secretary general Kofi Annan called on the member states of the UN to approve the most far-reaching changes in the 60-year history of the organization (Lederer 2005). One of the most controversial recommendations concerns the composition of

TABLE 15.4 United Nations Peacekeeping Operations: Summary Data, 2015

Military personnel and civilian police serving in peacekeeping operations	106,595
Countries contributing military personnel and civilian police	120
International civilian personnel	5,2077
Local civilian personnel	11,678
UN volunteers	1,846
Total number of fatalities in peacekeeping operations since 1948	3,348
Approved budgets for July 1, 2014, to June 30, 2015	$8.47 billion

SOURCE: United Nations 2015.

the Security Council, the most important decision-making body of the organization. Annan's recommendation that the 15 members of the Security Council—a body dominated by the United States, Great Britain, France, Russia, and China—be changed to include a more representative number of nations could, if approved, shift the global balance of power. Ban Ki-moon, former foreign minister of South Korea, was elected as the eighth secretary general in October 2006 (MacAskill et al. 2006), after campaigning for the position on a platform that included support for the UN reforms. To date, the expansion of the number of Security Council members is unchanged.

Mediation and Arbitration

Most conflicts are resolved through nonviolent means. Mediation and arbitration are just two of the nonviolent strategies used to resolve conflicts and to stop or prevent war. In **mediation**, a neutral third party intervenes and facilitates negotiation between representatives or leaders of conflicting groups. Good mediators do not impose solutions but rather help disputing parties generate options for resolving the conflict (Conflict Research Consortium 2003). Ideally, a mediated resolution to a conflict meets at least some of the concerns and interests of each party to the conflict. In other words, mediation attempts to find "win-win" solutions in which each side is satisfied with the solution.

Although mediation is used to resolve conflict between individuals, it is also a valuable tool for resolving international conflicts. For example, former U.S. senator George Mitchell successfully mediated talks between parties to the conflict in Northern Ireland in 1998. The resulting political agreement continues to hold today. Also, in May 2008, the government of Qatar mediated talks between Lebanon's political parties that resulted in an agreement that averted civil war after an 18-month political crisis. Using mediation as a means of resolving international conflict is often difficult, given the complexity of the issues. However, research by Bercovitch (2003) shows that there are more mediators available and more instances of mediation in international affairs than ever before. For example, Bercovitch identified more than 250 separate mediation attempts during the Balkan Wars of the early and mid-1990s.

Arbitration also involves a neutral third party who listens to evidence and arguments presented by conflicting groups. Unlike mediation, however, the neutral third party in arbitration arrives at a decision that the two conflicting parties agree in advance to accept. For instance, the Permanent Court of Arbitration—an intergovernmental organization based in The Hague since 1899—arbitrates disputes about territory, treaty compliance, human rights, commerce, and investment among any of its 117 member states who signed

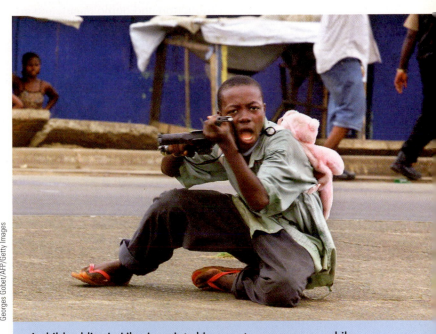

Georges Gobet/AFP/Getty Images

A child soldier in Liberia points his gun at a cameraman while carting a teddy bear on his back. Although reliable figures are hard to obtain, the UN estimates that about 300,000 child soldiers are fighting in wars worldwide.

mediation A neutral third party intervenes and facilitates negotiation between representatives or leaders of conflicting groups.

arbitration Dispute settlement in which a neutral third party listens to evidence and arguments presented by conflicting groups and arrives at a decision that the parties have agreed in advance to accept.

and ratified either of its two founding legal conventions. Recent cases include a dispute between Malaysia and Singapore over the construction of a railroad, between the Philippines and China over maritime rights, as well as between the government of Sudan and the Sudan People's Liberation Army (Permanent Court of Arbitration 2015).

Arms Control and Disarmament

In the 1960s, the United States and the Soviet Union led the world in a nuclear arms race, with each competing to build a larger and more destructive arsenal of nuclear weapons than its adversary. If either superpower were to initiate a full-scale war, the retaliatory powers of the other nation would result in the destruction of both nations as well as much of the planet. Thus, the principle of **mutually assured destruction (MAD)** that developed from nuclear weapons capabilities transformed war from a win-lose proposition to a lose-lose scenario. If both sides would lose in a war, the theory suggested, neither side would initiate war. At its peak year in 1966, the U.S. stockpile of nuclear weapons included more than 32,000 warheads and bombs.

As their arsenals continued to grow at an astronomical cost, both sides recognized the necessity for nuclear arms control, including the reduction of defense spending, weapons production and deployment, and armed forces. Throughout the Cold War and even today, much of the behavior of the United States and the now-former Soviet Union has been governed by major arms control initiatives, including the following:

- The Limited Test Ban Treaty that prohibited testing of nuclear weapons in the atmosphere, underwater, and in outer space
- The Strategic Arms Limitation Treaties (SALT I and II) that limited the development of nuclear missiles and defensive antiballistic missiles
- The Strategic Arms Reduction Treaties (START I and II) that significantly reduced the number of nuclear missiles, warheads, and bombs
- The Strategic Offensive Reduction Treaty (SORT) that required that the United States and the Russian Federation each reduce their number of strategic nuclear warheads to between 1,700 and 2,200 by 2012. This goal was met, with the United States currently holding 1,950 strategic warheads and the Russian Federation holding 1,800 (Center for Arms Control and Non-Proliferation 2013).
- The New Strategic Arms Reduction Treaty (new START) that further reduces the number of deployed strategic nuclear warheads to 1,550 for each country and significantly reduces the number of strategic nuclear missile launchers allowed (Atomic Archive 2011)

With the end of the Cold War came the growing realization that, even as Russia and the United States greatly reduced their arsenals, other countries were poised to acquire nuclear weapons or expand their existing arsenals. Thus, the focus on arms control shifted toward **nuclear nonproliferation**—the prevention of the spread of nuclear technology to nonnuclear states.

Nuclear Nonproliferation Treaty. The Nuclear Nonproliferation Treaty (NPT), signed in 1970, was the first treaty governing the spread of nuclear weapons technology from the original nuclear weapons states (i.e., the United States, the Soviet Union, the United Kingdom, France, and China) to nonnuclear countries. The NPT was renewed in 2000 and is subject to review every five years. Currently, 187 countries have adopted it. The NPT holds that countries without nuclear weapons will not try to get them; in exchange, the countries with nuclear weapons agree that they will not provide nuclear weapons to countries that do not have them. Signatory states without nuclear weapons also agree to allow the International Atomic Energy Agency to verify compliance with the treaty through on-site inspections (Atomic Archive 2015). Only India, Israel, and Pakistan have not signed the agreement, although each of these states is known to possess a nuclear arsenal. Furthermore, many experts suspect that Iran and Syria—both signatories to the NPT—are developing nuclear weapons programs. Both countries claim that their nuclear reactors are for peaceful purposes (e.g., domestic power consumption), not to develop a

mutually assured destruction (MAD) A Cold War doctrine referring to the capacity of two nuclear states to destroy each other, thus reducing the risk that either state will initiate war.

nuclear nonproliferation Efforts to prevent the spread of nuclear weapons, or the materials and technology necessary for the production of nuclear weapons.

nuclear weapons arsenal. However, a report by the International Atomic Energy Agency tentatively concluded that Iran has "sufficient information to be able to design and produce a workable" atomic bomb (Broad and Sanger 2009, p. 1).

Even if military superpowers honor agreements to limit arms, the availability of black-market nuclear weapons and materials presents a threat to global security. One of the most successful nuclear weapons brokers was Pakistan's Abdul Qadeer Khan. Khan, the "father" of Pakistan's nuclear weapons program, sold the technology and equipment required to make nuclear weapons to rogue states such as Iran and North Korea (Powell and McGirk 2005). Further nuclear tests took place after the new Korean leader Kim Jong-un came to power in December 2011. However, after the public failure of several missile tests and intense international pressure, the North Korean leader announced his intentions to renew talks on ending his country's nuclear program (Sang-Hun and Buckley 2013).

In 2003, the United States accused Iran of operating a covert nuclear weapons program. Although Iranian officials responded that their nuclear program was solely for the purpose of generating electricity, concerns escalated with Iran's successful test of a 1,200-mile-range missile in 2009 (Dareini 2009). Hostile public statements about Israel by Iran's president Mahmoud Ahmadinejad deepen already grave concerns about Iran's possible acquisition of nuclear weapons. Analysts fear the development of nuclear weapons by Iran would spur a nuclear competition with Israel, the only other state in the Middle East to possess nuclear weapons, and provoke other states in the region to acquire them (Salem 2007).

Many observers consider South Asia the world's most dangerous nuclear rivalry. India and Pakistan are the only two nuclear powers that share a border and have repeatedly fired upon each other's armies while in possession of nuclear weapons (Stimson Center 2007). India first detonated nuclear weapons in 1974. Pakistan detonated six weapons in 1998, a few weeks after India's second round of nuclear tests. Although precise figures are hard to come by, experts estimate that India has about 110 nuclear bombs, and Pakistan has about 120 nuclear bombs (Ploughshares 2015). Many fear that a conventional military confrontation between these two countries may someday escalate to an exchange of nuclear weapons. Tensions have further escalated in the wake of the U.S. drawdown from Afghanistan as India and Pakistan compete for influence in the region, leading some analysts to speculate that the threat of direct, and possibly nuclear, confrontation will increase over the next decade (Dalrymple 2013).

As states that want to obtain nuclear weapons are quick to point out, nuclear states that advocate for nonproliferation possess well over 25,000 weapons, a huge reduction in the world's arsenal from Cold War days but still a massive potential threat to the earth. President Obama visited Russia in July 2009, to sign the New Strategic Arms Reduction Treaty that reduces the strategic nuclear arsenals of the United States and the Russian Federation by at least 25 percent. The Senate ratified the treaty in December 2010 (Sheridan and Branigin 2010).

This action was reportedly "a first step in a broader effort intended to reduce the threat of such weapons drastically and to prevent their further spread to unstable regions" (Levy and Baker 2009). Still, the United States and Russia hold, by a wide margin, the largest nuclear arsenals in the world. "Do as I say, not as I do" is a weak bargaining position. This is the argument posited by Iran, which for over a decade has been subject to harsh economic sanctions spearheaded by the United States and its allies in an effort to prevent it from developing a nuclear weapons program.

The government of Iran has argued that it has the right, as does every other sovereign nation, to develop nuclear weapons for self-defense and that in fact most of its nuclear development efforts are focused not on weapons technology, but rather on developing nuclear energy. Detractors have argued that Iran poses a substantial threat to United States' interests in the Middle East, particularly to Iran's long-time enemy and U.S. ally, Israel.

In the summer of 2015, after months of negotiations and amid substantial controversy and opposition, the Obama administration reached a historic treaty with Iran, in which the United States agreed to alleviate the crippling economic sanctions against Iran in

> Even if military superpowers honor agreements to limit arms, the availability of black-market nuclear weapons and materials presents a threat to global security.

exchange for Iran reducing its nuclear program and permitting extensive international oversight. The agreement would allow Iran to continue to develop nuclear technology, but only for energy, not for weapons. Conservative hard-liners in both countries opposed the agreement, each arguing that the other side could not be trusted. In particular, the agreement strained relations between the United States and its strongest Middle Eastern ally, Israel, which argued that Iran poses a direct threat to its existence.

The intense political challenges to the agreement within both countries highlight the difficulty of achieving successful strategies of global peace when domestic and international interests conflict. Although the agreement has been approved, it still faces major hurdles. Iran's nuclear program must pass a series of inspections before the first sanctions can be lifted, and experts suggest that it may take up to 10 years before all components of the agreement have been fully enacted (Brumfield 2015).

The Problem of Small Arms

Although the devastation caused by even one nuclear war could affect millions, the easy availability of conventional weapons fuels many active wars around the world. Small arms and light weapons include handguns, submachine guns and automatic weapons, grenades, mortars, land mines, and light missiles. The Small Arms Survey estimated that, in 2013, 875 million firearms were in circulation worldwide, with national armed forces and law enforcement agencies possessing less than one-quarter of these weapons and nonstate armed groups and gangs possessing only 1.3 percent; civilians held the remaining 75 percent of small arms. Half a million people die each year as a result of small arms use, 200,000 in homicides and suicides, and the rest during wars and other armed conflicts (Geneva Graduate Institute for International Studies 2015).

Unlike control of weapons of mass destruction such as chemical and biological weapons, controlling the flow of small arms—especially firearms—is not easy because they have many legitimate uses by the military, by law enforcement officials, and for recreational or sporting activities (Schroeder 2007). Small arms are easy to afford, use, conceal, and transport illegally. The small arms trade is also lucrative. According to official records, the top five exporters of small arms in 2012 were the United States ($1.9 billion), Italy ($544 million), Germany ($472 million), Brazil ($374 million), and Austria ($293 million) (Geneva Graduate Institute for International Studies 2015). From 2001 to 2013, the United States exported over $275 million worth of small arms to Egypt, a country that underwent an extended period of political violence during the Arab Spring.

The U.S. Department of State's Office of Weapons Removal and Abatement (WRA) administers a program that has supported the destruction of over 1.6 million small arms and light weapons and 90,000 tons of ammunition in 38 countries since its founding in 2001 (U.S. Department of State 2014). The availability of these weapons fuels terrorist groups and undermines efforts to promote peace after wars have formally

AP Images/Bullit Marquez

Although the United States is the world's top arms exporter, it is also the world's leader in the destruction of conventional weapons. Since 1992, the Office of Weapons Removal and Abatement has contributed more than $2.4 billion to 90 countries for the safe disposal and destruction of small arms and removal of landmines (U.S. Department of State 2015c).

concluded. "If not expeditiously destroyed or secured, stocks of arms and ammunition left over after the cessation of hostilities frequently recirculate into neighboring regions, exacerbating conflict and crime" (U.S. Department of State 2007, p. 1).

Understanding Conflict, War, and Terrorism

As we come to the close of this chapter, how might we have an informed understanding of conflict, war, and terrorism? Each of the three theoretical positions discussed in this chapter reflects the realities of global conflict. As structural functionalists argue, war offers societal benefits—social cohesion, economic prosperity, scientific and technological developments, and social change. Furthermore, as conflict theorists contend, wars often occur for economic reasons because corporate elites and political leaders benefit from the spoils of war—land and water resources and raw materials. The symbolic interactionist perspective emphasizes the role that meanings, labels, and definitions play in creating conflict and contributing to acts of war.

The September 11 attacks on the World Trade Center and the Pentagon and the aftermath—the battle against terrorism, the wars in Iraq and Afghanistan, and the Arab Spring—changed the world Americans live in. For some theorists, these events were inevitable. Political scientist Samuel P. Huntington argued that such conflict represents a **clash of civilizations**. In *The Clash of Civilizations and the Remaking of World Order* (1996), Huntington argued that in the new world order,

> the most pervasive, important and dangerous conflicts will not be between social classes, rich and poor, or economically defined groups, but between people belonging to different cultural entities . . . . The most dangerous cultural conflicts are those along the fault lines between civilizations . . . the line separating peoples of Western Christianity, on the one hand, from Muslim and Orthodox peoples on the other. (p. 28)

Some public opinion surveys seem to support Huntington's view. In one global survey, at least two-thirds of citizens in Latin America, Europe, Asia, and Africa had a favorable view of the United States, but in the Middle East, two-thirds of citizens had an unfavorable view (Pew Research Center 2014b). Israel was the only Middle Eastern nation in which the public had a majority favorable view of the United States. Conversely, about 50 percent of Americans believe that Islam encourages violence more than other religions (Pew Research Center 2014c).

The clash of civilizations perspective has been vehemently criticized, however, by many scholars who see this view as divisive, overly simplistic, and historically inaccurate. In particular, the clash of civilizations perspective minimizes the vast majority of citizens in predominantly Muslim nations who are religiously and politically moderate, and who share common concerns with Western nations over the dangers of religious extremism and terrorism. For example, a survey of thousands of Muslims in 11 Middle Eastern countries found that the majority were concerned about Islamic extremism in their country; had an unfavorable view of al Qaeda, the Taliban, and Boko Haram; and believed that suicide bombings were never justified (Pew Research Center 2013).

David Brooks summarizes this critique by suggesting that Huntington committed a "Fundamental Attribution Error. That is, he ascribed to traits qualities that are actually determined by context" (2011, p. 1). Huntington suggested that Arab societies are intrinsically opposed to democracy and not nationalistic, but the Arab Spring revolutions highlighted how certain political regimes can effectively, although never permanently, suppress national patriotism and the intrinsic human desire for liberty. Brooks also suggests that Huntington fundamentally misunderstood the nature of culture for, despite our intrinsic and cultural differences, we are all alike: "Huntington minimized the power of universal political values and exaggerated the influence of distinct cultural values . . . underneath cultural differences there are universal aspirations for dignity, for political systems that listen to, respond to and respect the will of the people" (p. 1).

Ultimately, we are all members of one community—Earth—and have a vested interest in staying alive and protecting the resources of our environment for our own and future

clash of civilizations A hypothesis that the primary source of conflict in the 21st century has shifted away from social class and economic issues and toward conflict between religious and cultural groups, especially those between large-scale civilizations such as the peoples of Western Christianity and Muslim and Orthodox peoples.

generations. But, as we have seen, conflict between groups is a feature of social life that is not likely to disappear. What is at stake—human lives and the ability of our planet to sustain life—merits serious attention. World leaders have traditionally followed the advice of philosopher Carl von Clausewitz: "If you want peace, prepare for war." Thus, nations have sought to protect themselves by maintaining large military forces and massive weapons systems. These strategies are associated with serious costs, particularly in hard economic times. In diverting resources away from other social concerns, defense spending undermines a society's ability to improve the overall security and well-being of its citizens. Conversely, defense-spending cutbacks, although unlikely in the present climate, could potentially free up resources for other social agendas, including lowering taxes, reducing the national debt, addressing environmental concerns, eradicating hunger and poverty, improving health care, upgrading educational services, and improving housing and transportation. Therein lies the promise of a "peace dividend." The hope is that future dialogue on the problems of war and terrorism will redefine national and international security to encompass social, economic, and environmental well-being.

Chapter Review

- ## What is the relationship between war and industrialization?
War indirectly affects industrialization and technological sophistication because military research and development advances civilian-used technologies. Industrialization, in turn, has had two major influences on war: The more industrialized a country is, the lower the rate of conflict; and if conflict occurs, the higher the rate of destruction.

- ## What are the latest trends in armed conflicts?
Since World War II, wars between two or more states make up the smallest percentage of armed conflicts. In the contemporary era, the majority of armed conflicts have occurred between groups in a single nation-state, who compete for the power to control the resources of the state or to break away and form their own state.

- ## In general, how do feminists view war?
Feminists are quick to note that wars are part of the patriarchy of society. Although women and children may be used to justify a conflict (e.g., improving women's lives by removing the repressive Taliban in Afghanistan), the basic principles of male dominance and control are realized through war. Feminists also emphasize the social construction of aggressive masculine identities and their manipulation by elites as important reasons for the association between masculinity and militarized violence.

- ## What are some of the causes of war?
The causes of war are numerous and complex. Most wars involve more than one cause. Some of the causes of war are conflict over land and natural resources; values or ideologies; racial, ethnic, and religious hostilities; defense against hostile attacks; revolution; and nationalism.

- ## What is terrorism, and what are the different types of terrorism?
Terrorism is the premeditated use, or threatened use, of violence by an individual or group to gain a political or social objective. Terrorism can be either transnational or domestic. Transnational terrorism occurs when acts of terrorism are not restricted to or centered within one country. Domestic terrorism takes place within the territorial jurisdiction of one nation, such as the 1995 truck bombing of a nine-story federal office building in Oklahoma City.

- ## What are some of the macro-level "roots" of terrorism?
Many scholars have focused on macro-structural factors, such as global economic shifts, foreign occupation, repression of minorities, poverty, and weak governance as creating the underlying conditions that produce terrorism.

- ## How has the United States responded to the threat of terrorism?
The United States has used both defensive and offensive strategies to fight terrorism. Defensive strategies include using metal detectors and X-ray machines at airports and strengthening security at potential targets, such as embassies and military command posts. The Department of Homeland Security coordinates such defensive tactics. Offensive strategies include retaliatory raids such as the U.S. bombing of terrorist facilities in Afghanistan, group infiltration, and preemptive strikes.

- ## What is meant by "diversion of economic resources"?
Worldwide, the billions of dollars used on defense could be channeled into social programs dealing with, for example, education, health, and poverty. Thus, defense monies are economic resources diverted from other needy projects.

- ## What are some of the criticisms of the United Nations?
First, in recent missions, developing nations have supplied more than 75 percent of the troops. Second, several recent UN peacekeeping operations have failed. Third, the UN cannot take sides but must wait for a consensus of its members that, if not forthcoming, undermines the strength of the organization. Fourth, the concept of the UN is that its members represent individual nations, not a region or the world. Because nations tend to act in their own best economic and security interests, UN actions performed in the name of world peace may be motivated by nations acting in their own interests. Finally, the Security Council limits power to a small number of states.

- **What problems do small arms pose?**

Even after a conflict ends, these weapons circulate in society, making crime worse or falling into the hands of terrorists. Trade in small arms is legal because they have many legitimate uses—for example, by the military, police, and hunters. Because they are small and simple to handle, these weapons are easily concealed and transported, making it difficult to control them.

Test Yourself

1. War between states is still the most common form of warfare.
   a. True
   b. False
2. The rise of the modern state is most directly a result of
   a. industrialization and the creation of national markets.
   b. innovations in communications technology.
   c. the development of armies to control territory.
   d. the development of police to control a population.
3. Structural-functionalist explanations about war emphasize
   a. that war is a biological necessity.
   b. that despite its destructive power, war persists because it fulfills social needs.
   c. that war is an anachronism that will eventually disappear.
   d. that war is necessary because it benefits political and military elites.
4. On the whole, conflicts over values and ideologies are more difficult to resolve than those over material resources.
   a. True
   b. False
5. All wars are a result of unequal distribution of wealth.
   a. True
   b. False
6. Next to defense spending, transfers to foreign governments are the most expensive item in the U.S. government budget.
   a. True
   b. False
7. Which of the following factors is a likely cause of revolutions or civil wars?
   a. A weak or failed state
   b. An authoritarian government that ignores major demands from citizens
   c. The availability of strong opposition leaders
   d. All of the above
8. Primordial explanations of ethnic conflict suggest that
   a. ethnic leaders instigate hostilities to serve their own interests.
   b. people become hostile when they blame their frustration with economic hardship on competing ethnic groups.
   c. ancient hatreds compel ethnic groups to continue fighting.
   d. None of the above
9. The consequences of war and the military on the environment
   a. are prevalent only during wartime.
   b. persist in peacetime or for many years after a war is over.
   c. are negligible.
   d. have been mostly reduced by technological innovations.
10. Advocates of nuclear nonproliferation seek to
    a. ban all nuclear weapons.
    b. prevent construction of nuclear power plants.
    c. prevent new states from acquiring nuclear weapons.
    d. None of the above

Answers: 1. B; 2. C; 3. B; 4. A; 5. B; 6. B; 7. D; 8. C; 9. B; 10. C.

Key Terms

arbitration 531
clash of civilizations 535
Cold War 500
conflict minerals 510
constructivist explanations 510
democratic peace theory 510
domestic terrorism 515
dual-use technologies 504
ecocide 527
Geneva Conventions 519
guerrilla warfare 524

Islamic State in Iraq and Syria (ISIS) 516
mediation 531
military-industrial complex 505
mutually assured destruction (MAD) 532
nuclear nonproliferation 532
nuclear winter 528
posttraumatic stress disorder (PTSD) 525
primordial explanations 510

rally round the flag effect 504
security dilemma 511
state 499
terrorism 515
transnational terrorism 515
war 498
weapons of mass destruction (WMD) 521

Appendix
Methods of Data Analysis

Description, Correlation, Causation, Reliability and Validity, and Ethical Guidelines in Social Problems Research

There are three levels of data analysis: description, correlation, and causation. Data analysis also involves assessing reliability and validity.

Description

Qualitative research involves verbal descriptions of social phenomena. Having a homeless and single pregnant teenager describe her situation is an example of qualitative research.

Quantitative research often involves numerical descriptions of social phenomena. Quantitative descriptive analysis may involve computing the following: (1) means (averages), (2) frequencies, (3) mode (the most frequently occurring observation in the data), (4) median (the middle point in the data; half of the data points are above the median and half are below), and (5) range (the highest and lowest values in a set of data).

Correlation

Researchers are often interested in the relationship between variables. *Correlation* refers to a relationship between or among two or more variables. The following are examples of correlational research questions: What is the relationship between poverty and educational achievement? What is the relationship between race and crime victimization? What is the relationship between religious affiliation and divorce?

If there is a correlation or relationship between two variables, then a change in one variable is associated with a change in the other variable. When both variables change in the same direction, the correlation is positive. For example, in general, the more sexual partners a person has, the greater the risk of contracting a sexually transmissible disease (STD). As variable A (number of sexual partners) increases, variable B (chance of contracting an STD) also increases. Similarly, as the number of sexual partners decreases, the chance of contracting an STD decreases. Notice that in both cases, the variables change in the same direction, suggesting a positive correlation (see Figure A.1).

When two variables change in opposite directions, the correlation is negative. For example, there is a negative correlation between condom use and contracting STDs. In other words, as condom use increases, the chance of contracting an STD decreases (see Figure A.2).

The relationship between two variables may also be curvilinear, which means that it varies in both the same and opposite directions. For example, suppose a researcher finds that after drinking one alcoholic beverage, research participants are more prone to violent behavior. After two drinks, violent behavior is even more likely, and this trend continues for three and four drinks. So far, the correlation between alcohol consumption and violent behavior is positive. After the research participants have five alcoholic drinks, however, they become less prone to violent behavior. After six and seven drinks, the likelihood of engaging in violent behavior decreases further. Now the correlation between alcohol consumption and violent behavior is negative. Because the correlation changed from positive to negative, we say that the correlation is curvilinear (the correlation may also change from negative to positive) (see Figure A.3).

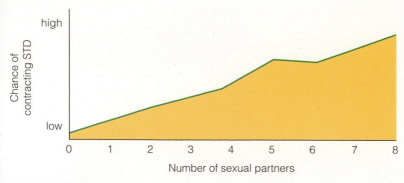

FIGURE A.1

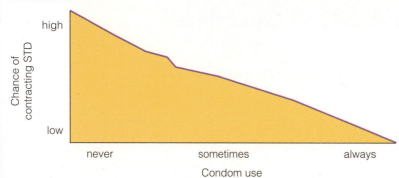

FIGURE A.2

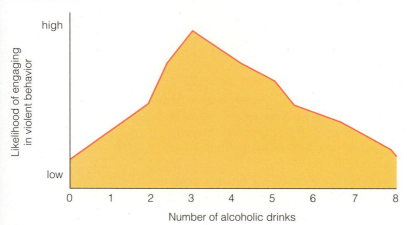

FIGURE A.3

A fourth type of correlation is called a spurious correlation. Such a correlation exists when two variables appear to be related, but the apparent relationship occurs only because each variable is related to a third variable. When the third variable is controlled through a statistical method in which the variable is held constant, the apparent relationship between the first two variables disappears. For example, blacks have a lower average life expectancy than whites do. Thus, race and life expectancy appear to be related. This apparent correlation exists, however, because both race and life expectancy are related to socioeconomic status. Because blacks are more likely than whites to be impoverished, they are less likely to have adequate nutrition and medical care.

Causation

If the data analysis reveals that two variables are correlated, we know only that a change in one variable is associated with a change in another variable. We cannot assume, however, that a change in one variable causes a change in the other variable unless our data collection and analysis are specifically designed to assess causation. The research method that best assesses causality is the experimental method (discussed in Chapter 1).

To demonstrate causality, three conditions must be met. First, the data analysis must demonstrate that variable A is correlated with variable B. Second, the data analysis must demonstrate that the observed correlation is not spurious. Third, the analysis must demonstrate that the presumed cause (variable A) occurs or changes before the presumed effect (variable B). In other words, the cause must precede the effect.

It is extremely difficult to establish causality in social science research. Therefore, much social research is descriptive or correlative, rather than causative. Nevertheless, many people make the mistake of interpreting a correlation as a statement of causation. As you read correlative research findings, remember the following adage: "Correlation does not equal causation."

Reliability and Validity

Assessing reliability and validity is an important aspect of data analysis. *Reliability* refers to the consistency of the measuring instrument or technique; that is, the degree to which the way information is obtained produces the same results if repeated. Measures of reliability are made on scales and indexes (such as those in the *Self and Society* features in this text) and on information-gathering techniques, such as the survey methods described in Chapter 1.

Various statistical methods are used to determine reliability. A frequently used method is called the "test-retest method." The researcher gathers data on the same sample of people twice (usually 1 or 2 weeks apart) using a particular instrument or method and then correlates the results. To the degree that the results of the two tests are the same (or highly correlated), the instrument or method is considered reliable.

Measures that are perfectly reliable may be absolutely useless unless they also have a high validity. *Validity* refers to the extent to which an instrument or device measures what it intends to measure. For example, police officers administer "Breathalyzer" tests to determine the level of alcohol in a person's system. The Breathalyzer is a valid test for measuring alcohol consumption.

Validity measures are important in research that uses scales or indexes as measuring instruments. Validity measures are also important in assessing the accuracy of self-report data that are obtained in survey research. For example, survey research on high-risk sexual behaviors associated with the spread of HIV relies heavily on self-report data on topics such as number of sexual partners, types of sexual activities, and condom use. Yet how valid are these data? Do survey respondents underreport the number of their sexual partners? Do people who say they use a condom every time they engage in intercourse really use a condom every time? Because of the difficulties in validating self-reports of number of sexual partners and condom use, we may not be able to answer these questions.

Ethical Guidelines in Social Problems Research

Social scientists are responsible for following ethical standards designed to protect the dignity and welfare of people who participate in research. These ethical guidelines include the following:

1. Freedom from coercion to participate. Research participants have the right to decline to participate in a research study or to discontinue participation at any time during the study. For example, professors who are conducting research using college students should not require their students to participate in their research.
2. Informed consent. Researchers are required to inform potential participants of any aspect of the research that might influence a subject's willingness to participate. After informing potential participants about the nature of the research, researchers typically ask participants to sign a consent form indicating that the participants are informed about the research and agree to participate in it.
3. Deception and debriefing. Sometimes the researcher must disguise the purpose of the research to obtain valid data. Researchers may deceive participants as to the purpose or nature of a study only if there is no other way to study the problem. When deceit is used, participants should be informed of this deception (debriefed) as soon as possible. Participants should be given a complete and honest description of the study and why deception was necessary.
4. Protection from harm. Researchers must protect participants from any physical and psychological harm that might result from participating in a research study. This is both a moral and a legal obligation. It would not be ethical, for example, for a researcher studying drinking and driving behavior to observe an intoxicated individual leaving a bar, getting into the driver's seat of a car, and driving away.

    Researchers are also obligated to respect the privacy rights of research participants. If anonymity is promised, it should be kept. Anonymity is maintained in mail surveys by identifying questionnaires with a number coding system rather than with the participants' names. When such anonymity is not possible, as is the case with face-to-face interviews, researchers should tell participants that the information they provide will be treated as confidential. Although interviews may be summarized and excerpts quoted in published material, the identity of the individual participants is not revealed. If a research participant experiences either physical or psychological harm as a result of participation in a research study, the researcher is ethically obligated to provide remediation for the harm.
5. Reporting of research. Ethical guidelines also govern the reporting of research results. Researchers must make research reports freely available to the public. In these reports, a researcher should fully describe all evidence obtained in the study, regardless of whether the evidence supports the researcher's hypothesis. The raw data collected by the researcher should be made available to other researchers who might request it for purposes of analysis. Finally, published research reports should include a description of the sponsorship of the research study, its purpose, and all sources of financial support.

Glossary

3-D printing A revolutionary manufacturing technology that involves downloading a digital file containing a design for a product. A printer reads the file and then shoots out the product (made of specialized plastic or other raw materials) through a heated nozzle.

abortion The intentional termination of a pregnancy.

absolute poverty The lack of resources necessary for well-being—most importantly, food and water, but also housing, sanitation, education, and health care.

abusive head trauma A form of inflicted brain injury resulting from violent shaking or blunt impact; a leading cause of death in children under 1 year (see also **shaken baby syndrome**).

acculturation The process of adopting the culture of a group different from the one in which a person was originally raised.

achieved status A status that society assigns to an individual on the basis of factors over which the individual has some control (e.g., high school graduate).

acid rain The mixture of precipitation with air pollutants, such as sulfur dioxide and nitrogen oxide.

acquaintance rape Rape committed by someone known to the victim.

adaptive discrimination Discrimination that is based on the prejudice of others.

affirmative action A broad range of policies and practices in the workplace and educational institutions to promote equal opportunity as well as diversity.

Affordable Care Act (ACA) Health care reform legislation that President Obama signed into law in 2010, with the goal of expanding health insurance coverage to more Americans; also known as the Patient Protection and Affordable Care Act, or "Obamacare."

ageism Negative stereotyping, prejudice, and discrimination based on a person's or group's perceived chronological age.

ageism by invisibility The underrepresentation of older adults in advertising and educational materials.

alienation A sense of powerlessness and meaninglessness in people's lives.

allopathic medicine The conventional or mainstream practice of medicine; also known as Western medicine.

alternative certification programs Programs whereby college graduates with degrees in fields other than education can become certified if they have "life experience" in industry, the military, or other relevant jobs.

androgyny Having both traditionally defined feminine and masculine characteristics.

anomie A state of normlessness in which norms and values are weak or unclear.

anti-miscegenation laws Laws banning interracial marriage until 1967, when the Supreme Court (in *Loving v. Virginia*) declared these laws unconstitutional.

arbitration Dispute settlement in which a neutral third party listens to evidence and arguments presented by conflicting groups and arrives at a decision that the parties have agreed in advance to accept.

ascribed status A status that society assigns to an individual on the basis of factors over which the individual has no control (e.g., race).

assimilation The process by which formerly distinct and separate groups merge and become integrated as one.

attributional gender bias The practice of explaining the same behavior(s) of females and males using different explanations.

automation The replacement of human labor with machinery and equipment.

aversive racism A subtle form of prejudice that involves feelings of discomfort, uneasiness, disgust, fear, and pro-white attitudes.

baby boomers The generation of Americans born between 1946 and 1964, a period of high birthrates.

behavior-based safety programs A strategy used by business management that attributes health and safety problems in the workplace to workers' behavior, rather than to work processes and conditions.

beliefs Definitions and explanations about what is assumed to be true.

bigamy The criminal offense in the United States of marrying one person while still legally married to another.

bilingual education In the United States, teaching children in both English and their non-English native language.

binge drinking As defined by the U.S. Department of Health and Human Services, drinking five or more drinks on the same occasion on at least one day in the past 30 days prior to the National Survey on Drug Use and Health.

biodiversity The diversity of living organisms on earth.

bioinvasion The intentional or accidental introduction of plant, animal, insect, and other species in regions where they are not native.

biomass Material derived from plants and animals, such as dung, wood, crop residues, and charcoal; used as cooking and heating fuel.

biphobia Prejudice toward bisexual individuals.

bisexuality The emotional, cognitive, and sexual attraction to members of both sexes.

blended families Also known as *stepfamilies*, families involving children from a previous relationship.

boy code A set of societal expectations that discourages males from expressing emotion, weakness, or vulnerability, or asking for help.

breeding ground hypothesis A hypothesis that argues that incarceration serves to increase criminal behavior through the transmission of criminal skills, techniques, and motivations.

bullying Bullying "entails an imbalance of power that exists over a long period of time in which the more powerful intimidate or belittle others" (Hurst 2005, p. 1).

capital punishment The state (the federal government or a state) takes the life of a person as punishment for a crime.

capitalism An economic system characterized by private ownership of the means of production and distribution of goods and services for profit in a competitive market.

character education Education that emphasizes the moral and ethical aspects of an individual.

charismatic megafauna Particular species that have popular appeal, such as the panda, polar bear, monarch butterfly, and bald eagle, and that are used to draw public attention to larger environmental issues.

charter schools Schools that originate in contracts, or charters, which articulate a plan of instruction that local or state authorities must approve.

chattel slavery A form of slavery in which slaves are considered property that can be bought and sold.

chemical dependency A condition in which drug use is compulsive and users are unable to stop because of physical and/or psychological dependency.

child abuse The physical or mental injury, sexual abuse, negligent treatment, or maltreatment of a child younger than age 18 by a person who is responsible for the child's welfare.

child labor Involves a child performing work that is hazardous, that interferes with a child's education, or that harms a child's health or physical, mental, social, or moral development.

cisgender Refers to a person whose gender identity is consistent with his or her birth sex.

civil union A legal status that entitles same-sex couples who apply for and receive a civil union certificate to nearly all of the benefits available to married couples.

clash of civilizations A hypothesis that the primary source of conflict in the 21st century has shifted away from social class and economic issues and toward conflict between religious and cultural groups, especially those between large-scale civilizations such as the peoples of Western Christianity and Muslim and Orthodox peoples.

classic rape Rape committed by a stranger, with the use of a weapon, resulting in serious bodily injury to the victim.

Clean Power Plan Establishes the first-ever federal limits on carbon emissions from U.S. power plants, establishing state-by-state targets for carbon emissions reductions and allowing states flexibility in how to meet these targets.

clearance rate The percentage of crimes in which an arrest and official charge have been made and the case has been turned over to the courts.

climate denial machine A well-funded and aggressive misinformation campaign run by the fossil fuel industry and its allies that involves attacking and discrediting climate science, scientists, and scientific institutions.

climate deniers People who do not accept the scientific consensus that human-caused global warming and climate change are scientific facts.

club sandwich generation Includes adults (often in their 50s or 60s) who are caring for aging parents, adult children, and grandchildren, or adults (usually in their 30s and 40s) who are caring for their own young children, aging parents, and grandparents.

Cold War The state of military tension and political rivalry that existed between the United States and the former Soviet Union from the 1950s through the late 1980s.

color-blind racism A form of racism that is based on the idea that overcoming racism means ignoring race, but color-blindness is, in itself, a form of racism because it prevents acknowledgment of privilege and disadvantage associated with race, and therefore allows the continuation of institutional forms of racial bias.

colorism Prejudice or discrimination based on skin tone.

coming out The ongoing process whereby a lesbian, gay, or bisexual individual becomes aware of his or her sexuality, accepts and incorporates it into his or her overall sense of self, and shares that information with others such as family, friends, and coworkers.

communism A political system that usually takes on the form of totalitarianism—a one party system where the state controls all aspects of public and private life, permitting no individual freedom.

complementary and alternative medicine (CAM) A broad range of health care approaches, practices, and products that are not considered part of conventional medicine.

computer crime Any violation of the law in which a computer is the target or means of criminal activity.

constructivist explanations Those explanations that emphasize the role of leaders of ethnic groups in stirring up hatred toward others external to one's group.

contact hypothesis The idea that contact between groups is necessary for the reduction of prejudice.

corporal punishment The intentional infliction of pain intended to change or control behavior.

corporate tax inversion The practice in which a company lowers its taxes by merging with a foreign company and changing to an offshore address.

corporate violence The production of unsafe products and the failure of corporations to provide a safe working environment for their employees.

corporate welfare Laws and policies that benefit corporations.

corporatocracy A system of government that serves the interests of corporations and that involves ties between government and business.

covenant marriage A type of marriage (offered in a few states) that requires premarital counseling and that permits divorce only under condition of fault or after a marital separation of more than two years.

crack A crystallized illegal drug product produced by boiling a mixture of baking soda, water, and cocaine.

crime An act, or the omission of an act, that is a violation of a federal, state, or local criminal law for which the state can apply sanctions.

crime rate The number of crimes committed per 100,000 population.

cultural competence In reference to education, refers to having an awareness of and appreciation for different cultural groups in order to inform student–teacher and student–student interaction.

cultural imperialism The indoctrination into the dominant culture of a society.

cultural lag A condition in which the material part of culture changes at a faster rate than the nonmaterial part.

cultural sexism The ways in which the culture of society perpetuates the subordination of an individual or group based on the sex classification of that individual or group.

culture The meanings and ways of life that characterize a society, including beliefs, values, norms, sanctions, and symbols.

cyber-bullying The use of electronic communication (e.g., websites, e-mail, instant messaging, text messaging) to send or post negative or hurtful messages or images about an individual or a group.

cybernation Dominant in postindustrial societies, the use of machines to control other machines.

cycle of abuse A pattern of abuse in which a violent or abusive episode is followed by a makeup period when the abuser expresses sorrow and asks for forgiveness and "one more chance," before another instance of abuse occurs.

decriminalization Entails removing state penalties for certain drugs, and promotes a medical rather than criminal approach to drug use that encourages users to seek treatment and adopt preventive practices.

deep ecology The view that maintaining the earth's natural systems should take precedence over human needs, that nature has a value independent of human existence, and that humans have no right to dominate the earth and its living inhabitants.

Defense of Marriage Act (DOMA) Federal legislation that states that marriage is a legal union between one man and one woman and denies federal recognition of same-sex marriage.

deforestation The conversion of forestland to nonforestland.

deinstitutionalization In the U.S. model for psychiatric care, the transformation from long-term inpatient care in institutions to drug therapy and community-based mental health centers.

demand reduction One of two strategies in the U.S. war on drugs (the other is supply reduction), demand reduction focuses on reducing the demand for drugs through treatment, prevention, and research.

democratic peace theory A prevalent theory in international relations suggesting that the ideological similarities between democratic nations make it unlikely that such countries will go to war against each other.

demographic transition theory A theory that attributes population growth patterns to changes in birth rates and death rates associated with the process of industrialization.

dependent variable The variable that the researcher wants to explain; the variable of interest.

deregulation The reduction of government control over, for example, certain drugs.

desertification The degradation of semiarid land, which results in the expansion of desert land that is unusable for agriculture.

deterrence The use of harm or the threat of harm to prevent unwanted behaviors.

devaluation hypothesis The hypothesis that women are paid less because the work they perform is socially defined as less valuable than the work men perform.

developed countries Countries that have relatively high gross national income per capita, also known as high-income countries.

developing countries Countries that have relatively low gross national income per capita, also known as less-developed or middle-income countries.

discrimination Actions or practices that result in differential treatment of categories of individuals.

diversity gap The difference between the racial, ethnic, and socio-economic statuses of public school teachers versus public school students.

divorce mediation A process in which divorcing couples meet with a neutral third party (mediator) who assists the individuals in resolving issues such as property division, child custody, child support, and spousal support in a way that minimizes conflict and encourages cooperation.

domestic partnership A status that some states, counties, cities, and workplaces grant to unmarried couples, including gay and lesbian couples, which conveys various rights and responsibilities.

domestic terrorism Domestic terrorism occurs within the territorial jurisdiction of one nation.

drug Any substance other than food that alters the structure or functioning of a living organism when it enters the bloodstream.

drug abuse The violation of social standards of acceptable drug use, resulting in adverse physiological, psychological, and/or social consequences.

drug courts Special courts that divert drug offenders to treatment programs in lieu of probation or incarceration.

dual-use technologies Defense-funded technological innovations with commercial and civilian use.

earned income tax credit (EITC) A refundable tax credit based on a working family's income and number of children.

earnings premium The benefits of having a college degree far outweighs the cost of getting one.

Earth Overshoot Day The approximate date on which humanity's annual demand on the planet's resources exceeds what our planet can renew in a year.

ecocide The intentional destruction of the environment for military strategy.

e-commerce The buying and selling of goods and services over the Internet.

economic institution The structure and means by which a society produces, distributes, and consumes goods and services.

ecosystem A biological environment consisting of all the organisms living in a particular area, as well as all the nonliving, physical components of the environment such as air, water, soil, and sunlight, that interact to keep the whole ecosystem functioning.

ecoterrorism The use or threatened use of illegal force by groups or individuals, in order to protect environmental and/or animal rights.

education dividend The associated benefits of educating women including, for example, reducing the mortality rate of children under 5 years old.

elder abuse The physical or psychological abuse, financial exploitation, or medical abuse or neglect of the elderly.

elderly support ratio The ratio of working-age adults (15 to 64) to adults ages 65 and older in a population.

emotion work Work that involves caring for, negotiating, and empathizing with people.

Employment Non-Discrimination Act (ENDA) This proposed federal bill that would protect LGBTs from workplace discrimination has been up for congressional debate on a number of occasions since 1994, but has never been signed into law.

environmental footprint The demands that humanity makes on the Earth's natural resources.

environmental injustice Also known as **environmental racism**, the tendency for marginalized populations and communities to disproportionately experience adversity due to environmental problems.

environmental migrants People who flee from their home region due to environmental problems that threaten their survival or livelihood.

equal rights amendment (ERA) The proposed 28th amendment to the Constitution, which states that "equality of rights under the law shall not be denied or abridged by the United States, or by any state, on account of sex."

ethnicity A shared cultural heritage or nationality.

Every Child Deserves a Family Act This piece of federal legislation would remove obstacles to non-heterosexual (as well as transgender) individuals providing homes for adoption or foster care.

e-waste Discarded electrical appliances and electronic equipment.

experiments Research methods that involve manipulating the independent variable to determine how it affects the dependent variable.

expressive roles Roles into which women are traditionally socialized (i.e., nurturing and emotionally supportive roles).

expulsion Occurs when a dominant group forces a subordinate group to leave the country or to live only in designated areas of the country.

extreme poverty Living on less than $1.25 a day.

family A kinship system of all relatives living together or recognized as a social unit, including adopted people, foster families, unmarried same-sex and opposite-sex couples with or without children, and any relationships that function and feel like a family.

Family and Medical Leave Act (FMLA) A federal law that requires public agencies and companies with 50 or more employees to provide eligible workers with up to 12 weeks of job-protected, unpaid leave so that they can care for an ill child, spouse, or parent; stay home to care for their newborn, newly adopted, or newly placed foster child; or take time off when they are seriously ill, and up to 26 weeks of unpaid leave to care for a seriously ill or injured family member who is in the armed forces, including the National Guard or Reserves.

Family Support Agreement In China, a voluntary contract between older parents and adult children that specifies the details of how the adult children will provide parental care.

feminism The belief that men and women should have equal rights and responsibilities.

feminist criminology An approach that focuses on how the subordinate position of women in society affects their criminal behavior and victimization.

feminization of poverty The disproportionate distribution of poverty among women.

fetal alcohol syndrome A syndrome characterized by serious physical and mental handicaps as a result of maternal drinking during pregnancy.

field research Research that involves observing and studying social behavior in settings in which it occurs naturally.

First Amendment Defense Act An act, if passed, that would prohibit the government from imposing penalties for discriminating against married same-sex couples if it is demonstrated that the individual or organization held a "religious belief or moral conviction" against such marriages.

food deserts Areas where residents lack access to grocery stores that sell fresh fruits and vegetables, and instead rely on convenience stores and fast-food chains that sell mostly high-calorie processed food.

forced labor Also known as slavery, any work that is performed under the threat of punishment and is undertaken involuntarily.

fracking Hydraulic fracturing, commonly referred to as "fracking," involves injecting a mixture of water, sand, and chemicals into drilled wells to crack shale rock and release natural gas into the well.

free trade agreement (FTA) A pact between two or more countries that makes it easier to trade goods across national boundaries and that protects intellectual property rights.

full employment Exists when there are jobs available for all workers; usually when unemployment rates are below 5 percent.

future shock The state of confusion resulting from rapid scientific and technological changes that unravel our traditional values and beliefs.

gateway drug A drug (e.g., marijuana) that is believed to lead to the use of other drugs (e.g., cocaine).

gay A term that can refer to either women or men who are attracted to same-sex partners.

gender The social definitions and expectations associated with being female or male.

gender deviance hypothesis The tendency to overconform to gender norms after an act(s) of gender deviance; a method of neutralization.

gender expression The way in which a person presents her- or himself as a gendered individual (i.e., masculine, feminine, or androgynous) in society. A person could, for example, have a gender identity as male but nonetheless present their gender as female.

gender non-conforming Often used synonymously with transgender, gender non-conforming (sometimes called gender variant) refers to displays of gender that are inconsistent with society's expectations.

gender roles Patterns of socially defined behaviors and expectations associated with being female or male.

gene monopolies Exclusive control over a particular gene as a result of government patents.

gene therapy The transplantation of a healthy gene to replace a defective or missing gene.

genetic engineering The manipulation of an organism's genes in such a way that the natural outcome is altered.

genetic exception laws Laws that require that genetic information be handled separately from other medical information.

Genetic Information Nondiscrimination Act of 2008 A federal law that prohibits discrimination in health coverage or employment based on genetic information.

genetic testing Examination of DNA in order to identify abnormalities or alterations that may lead to disease or illness.

Geneva Conventions A set of international treaties that govern the behavior of states during wartime, including the treatment of prisoners of war.

genocide The deliberate, systematic annihilation of an entire nation or people.

glass ceiling An invisible barrier that prevents women and other minorities from moving into top corporate positions.

glass escalator effect The tendency for men seeking or working in traditionally female occupations to benefit from their minority status.

GLBT See *LGBT*.

global economy An interconnected network of economic activity that transcends national borders and spans the world.

global warming The increasing average global temperature of Earth's atmosphere, water, and land, caused mainly by the accumulation of various gases (greenhouse gases) that collect in the atmosphere.

globalization The growing economic, political, and social interconnectedness among societies throughout the world.

globesity The high prevalence of obesity around the world.

grandfamilies Families in which children reside with and are being raised by grandparents; parents may or may not also live in the household.

gray divorces Divorces among people ages 50 and older.

green energy Also known as clean energy, energy that is nonpolluting and/or renewable, such as solar power, wind power, biofuel, and hydrogen.

green growth Economic growth that is environmentally sustainable.

greenhouse gases Gases (primarily carbon dioxide, methane, and nitrous oxide) that accumulate in the atmosphere and act like the glass in a greenhouse, holding heat from the sun close to the earth.

Green Revolving Funds (GRFs) College and university funds that are dedicated to financing cost-saving energy-efficiency upgrades and other projects that decrease resource use and minimize environmental impacts.

greenwashing The ways in which environmentally and socially damaging companies portray their corporate image and products as being "environmentally friendly" or socially responsible.

harm reduction A public health position that advocates reducing the harmful consequences of drug use for the user as well as for society as a whole.

hate crime An unlawful act of violence motivated by prejudice or bias.

Head Start Begun in 1965 to help preschool children from the most disadvantaged homes, Head Start provides an integrated program of health care, parental involvement, education, and social services for qualifying children.

health According to the World Health Organization, "a state of complete physical, mental, and social well-being."

heavy drinking As defined by the U.S. Department of Health and Human Services, five or more drinks on the same occasion on each of five or more days in the past 30 days prior to the National Survey on Drug Use and Health.

heteronormativity The cultural presumption that heterosexuality is the norm whereby other orientations, by default, become abnormal and something to change.

heterosexism A form of oppression that refers to a belief system that gives power and privilege to heterosexuals, while depriving, oppressing, stigmatizing, and devaluing people who are not heterosexual.

heterosexuality The predominance of emotional, cognitive, and sexual attraction to individuals of the other sex.

homophobia Negative or hostile attitudes directed toward non-heterosexual sexual behavior, a non-heterosexually identified individual, and communities of non-heterosexuals.

homosexuality The predominance of emotional, cognitive, and sexual attraction to individuals of the same sex.

honor killings Murders, often public, as a result of a female dishonoring, or being perceived to have dishonored, her family or community.

human capital hypothesis The hypothesis that pay differences between females and males are a function of differences in women's and men's levels of education, skills, training, and work experience.

hypothesis A prediction or educated guess about how one variable is related to another variable.

identity theft The use of someone else's identification (e.g., Social Security number, birth date) to obtain credit or other economic rewards.

implicit prejudice A prevalent form of racial bias over which a person has little or no conscious awareness or control.

incapacitated assault An assault that takes place when the victim is unable to make an informed decision because of alcohol or other drugs, being unconscious, and the like.

incapacitation A criminal justice philosophy that argues that recidivism can be reduced by placing offenders in prison so that they are unable to commit further crimes against the general public.

independent variable The variable that is expected to explain change in the dependent variable.

index offenses Crimes identified by the FBI as the most serious, including personal or violent crimes (homicide, assault, rape, and robbery) and property crimes (larceny, motor vehicle theft, burglary, and arson).

Indian Health Service A federal agency that provides health services to members of 566 federally recognized American Indian and Alaska Native tribes and their descendants.

individual discrimination The unfair or unequal treatment of individuals because of their group membership.

individualism The tendency to focus on one's individual self-interests and personal happiness rather than on the interests of one's family and community.

infant mortality Deaths of live-born infants under 1 year of age.

insider trading The use of privileged (i.e., nonpublic) information by an employee of an organization that gives that employee an unfair advantage in buying, selling, and trading stocks and securities.

institution An established and enduring pattern of social relationships.

institutional discrimination Discrimination in which institutional policies and procedures result in unequal treatment of and opportunities for minorities.

institutional racism The systematic distribution of power, resources, and opportunity in ways that benefit whites and disadvantage minorities.

instrumental roles Roles into which men are traditionally socialized (i.e., task-oriented roles).

integration hypothesis A theory that the only way to achieve quality education for all racial and ethnic groups is to desegregate the schools.

intentional communities Communities where people live together based on shared values and goals.

intergenerational poverty Poverty that is transmitted from one generation to the next.

internalized homophobia (or internalized heterosexism) The internalization of negative messages about homosexuality by lesbian, gay, and bisexual individuals as a result of direct or indirect social rejection and stigmatization.

Internet An international information infrastructure available through universities, research institutes, government agencies, libraries, and businesses.

Internet of Things (IoT) The potential connectivity of all electronic devices; widespread information sharing between people and the objects of everyday life.

Internet piracy Illegally downloading or distributing copyrighted material (e.g., music, games, software).

Interpol The largest international police organization in the world.

intimate partner violence (IPV) Actual or threatened violent crimes committed against individuals by their current or former spouses, cohabiting partners, boyfriends, or girlfriends.

Islamic State in Iraq and Syria (ISIS) Notorious for its brutal tactics, a transnational terrorist organization whose goal is to establish an Islamic-led state in the Levant, a region that includes Egypt, Iraq, Israel, Jordan, Lebanon, Palestine, Syria, and Turkey.

Islamophobia Anti-Muslim and anti-Islam bias.

job burnout Prolonged job stress that can cause or contribute to high blood pressure, ulcers, headaches, anxiety, depression, and other health problems.

job exportation The relocation of jobs to other countries where products can be produced more cheaply.

Kyoto Protocol The first international agreement to place legally binding limits on greenhouse gas emissions from developed countries.

labor unions Worker advocacy organizations that developed to protect workers and represent them at negotiations between management and labor.

larceny Larceny is simple theft; it does not entail force or the use of force, or breaking and entering.

latent functions Consequences that are unintended and often hidden.

learning health systems The result of electronic records whereby physicians can look across patient populations and identify successful treatments or detect harmful interactions.

least developed countries The poorest countries of the world.

legalization Making prohibited behaviors legal; for example, legalizing drug use or prostitution.

lesbian A woman who is attracted to same-sex partners.

LGBT, LGBTQ, and LGBTQI Terms used to refer collectively to lesbian, gay, bisexual, transgender, questioning or "queer," and/or intersexed individuals.

life expectancy The average number of years that individuals born in a given year can expect to live.

light pollution Artificial lighting that is annoying, unnecessary, and/or harmful to life forms on earth.

living apart together (LAT) relationships An emerging family form in which couples—married or unmarried—live apart in separate residences.

living wage laws Laws that require state or municipal contractors, recipients of public subsidies or tax breaks, or, in some cases, all businesses to pay employees wages that are significantly above the federal minimum, enabling families to live above the poverty line.

long-term unemployment rate The share of the unemployed who have been out of work for 27 weeks or more.

malware A general term that includes any spyware, viruses, and adware that is installed on an owner's computer without their knowledge.

managed care Any medical insurance plan that controls costs through monitoring and controlling the decisions of health care providers.

manifest functions Consequences that are intended and commonly recognized.

marital decline perspective A view of the current state of marriage that includes the beliefs that (1) personal happiness has become more important than marital commitment and family obligations, and (2) the decline in lifelong marriage and the increase in single-parent families have contributed to a variety of social problems.

marital resiliency perspective A view of the current state of marriage that includes the beliefs that (1) poverty, unemployment, poorly funded schools, discrimination, and the lack of basic services (such as health insurance and child care) represent more serious threats to the well-being of children and adults than does the decline in married two-parent families, and (2) divorce provides a second chance for happiness for adults and an escape from dysfunctional environments.

masculine overcompensation thesis The thesis that men have a tendency to act out in an exaggerated male role when believing their masculinity to be threatened.

master status The status that is considered the most significant in a person's social identity.

maternal mortality Deaths that result from complications associated with pregnancy, childbirth, and unsafe abortion.

Matthew Shepard and James Byrd, Jr. Hate Crimes Prevention Act (HCPA) This law expands the original 1969 federal hate crimes law to cover hate crimes based on actual or perceived sexual orientation, gender, gender identity, and disability.

McDonaldization The process by which principles of the fast-food industry (efficiency, calculability, predictability, and control through technology) are being applied to more sectors of society, particularly the workplace.

means-tested programs Assistance programs that have eligibility requirements based on income and/or assets.

mechanization Dominant in an agricultural society, the use of tools to accomplish tasks previously done by hand.

Medicaid A public health insurance program, jointly funded by the federal and state governments, that provides health insurance coverage for the poor who meet eligibility requirements.

medicalization Defining or labeling behaviors and conditions as medical problems.

medical tourism A global industry that involves traveling, primarily across international borders, for the purpose of obtaining medical care.

Medicare A federally funded program that provides health insurance benefits to the elderly, disabled, and those with advanced kidney disease.

membership communities Internet sites where participation requires membership and members regularly communicate with one another for personal and/or professional reasons.

mental health Psychological, emotional, and social well-being.

mental illness Refers collectively to all mental disorders, which are characterized by sustained patterns of abnormal thinking, mood (emotion), or behaviors that are accompanied by significant distress and/or impairment in daily functioning.

meritocracy A social system in which individuals get ahead and earn rewards based on their individual efforts and abilities.

meta-analysis Meta-analysis combines the results of several studies addressing a research question; i.e., it is the analysis of analyses.

microcredit programs The provision of loans to people who are generally excluded from traditional credit services because of their low socioeconomic status.

Military Health System (MHS) The federal entity that provides medical care in military hospitals and clinics, and in combat zones and at bases overseas and on ships, and that provides health insurance known as Tricare to active duty service members, military retirees, their eligible family members, and their survivors.

military-industrial complex A term first used by Dwight D. Eisenhower to connote the close association between the military and defense industries.

minority group A category of people who have unequal access to positions of power, prestige, and wealth in a society and who tend to be targets of prejudice and discrimination.

modern racism A subtle form of racism that involves the belief that serious discrimination in the United States no longer exists, that any continuing racial inequality is the fault of minority group members, and that the demands for affirmative action for minorities are unfair and unjustified.

monogamy Marriage between two partners; the only legal form of marriage in the United States.

mortality Death.

motherhood penalty The tendency for women with children, particularly young children, to be disadvantaged in hiring, wages, and the like compared to women without children.

multicultural education Education that includes all racial and ethnic groups in the school curriculum, thereby promoting awareness and appreciation for cultural diversity.

Multidimensional Poverty Index A measure of serious deprivation in the dimensions of health, education, and living standards that combines the number of deprived and the intensity of their deprivation.

multiple chemical sensitivity Also known as "environmental illness," a condition whereby individuals experience adverse reactions when exposed to low levels of chemicals found in everyday substances.

mutually assured destruction (MAD) A Cold War doctrine referring to the capacity of two nuclear states to destroy each other, thus reducing the risk that either state will initiate war.

nativist extremist groups Organizations that not only advocate restrictive immigration policy, but also encourage their members to use vigilante tactics to confront or harass suspected undocumented immigrants.

naturalized citizens Immigrants who apply for and meet the requirements for U.S. citizenship.

neglect A form of abuse involving the failure to provide adequate attention, supervision, nutrition, hygiene, health care, and a safe and clean living environment for a minor child or a dependent elderly individual.

neonatal abstinence syndrome (NAS) A condition in which a child, at birth, goes through withdrawal as a consequence of maternal drug use.

net neutrality A principle that holds that Internet users should be able to visit any website and access any content without Internet service provider interference.

no-fault divorce A divorce that is granted based on the claim that there are irreconcilable differences within a marriage (as opposed to one spouse being legally at fault for the marital breakup).

norms Socially defined rules of behavior, including folkways, mores, and laws.

"Not in My Backyard" Opposition by local residents to a proposed new development in their community; also known as *NIMBY*.

nuclear nonproliferation Efforts to prevent the spread of nuclear weapons, or the materials and technology necessary for the production of nuclear weapons.

nuclear winter The predicted result of a thermonuclear war whereby thick clouds of radioactive dust and particles would block out vital sunlight, lower temperature in the northern hemisphere, and lead to the death of most living things on earth.

objective element of a social problem Awareness of social conditions through one's own life experiences and through reports in the media.

occupational sex segregation The concentration of women in certain occupations and men in other occupations.

Occupy Wall Street A protest movement that began in 2011, and is concerned with economic inequality, greed, corruption, and the influence of corporations on government.

offshoring The relocation of jobs to other countries.

oppression The use of power to create inequality and limit access to resources, which impedes the physical and/or emotional well-being of individuals or groups of people.

organized crime Criminal activity conducted by members of a hierarchically arranged structure devoted primarily to making money through illegal means.

outsource See *outsourcing*.

outsourcing A practice in which a business subcontracts with a third party to provide business services.

overt discrimination Discrimination that occurs because of an individual's own prejudicial attitudes.

parental alienation The intentional efforts of one parent to turn a child against the other parent and essentially destroy any positive relationship a child has with the other parent.

parent trigger laws State legislation that allows parents to intervene in their children's education and schools.

parity In health care, a concept requiring equality between mental health care insurance coverage and other health care coverage.

parole Release from prison, for a specific time period and subject to certain conditions, before an inmate's sentence is finished.

participation gap The tendency for racial and ethnic minorities to participate in information and communication technologies (e.g., using smartphones to access the Internet rather than a computer) that place them in a disadvantaged position (e.g., difficult to research a term paper on a smartphone).

patriarchy A male-dominated family system that is reflected in the tradition of wives taking their husband's last name and children taking their father's name.

penetration rate The percentage of people who have access to and use the Internet in a particular area.

perceived obsolescence The perception that a product is obsolete; used as a marketing tool to convince consumers to replace certain items even though the items are still functional.

period of PURPLE crying The phase at 2 weeks to 3 to 4 months of age when infants' crying is characterized by the acronym PURPLE: **P**eaks in crying that are **U**nexplained, **R**esists soothing, are accompanied by a **P**ain-like face, are **L**ong-lasting, and occur in the **E**vening and late afternoon.

pink-collar jobs Jobs that offer few benefits, often have low prestige, and are disproportionately held by women.

pinkwashing The practice of using the color pink and pink ribbons to indicate a company is helping to fight breast cancer, even when the company may be using chemicals linked to cancer.

planned obsolescence The manufacturing of products that are intended to become inoperative or outdated in a fairly short period of time.

pluralism A state in which racial and ethnic groups maintain their distinctness but respect each other and have equal access to social resources.

plutocracy A country governed by the wealthy.

political alienation A rejection of or estrangement from the political system accompanied by a sense of powerlessness in influencing government.

polyamory Multiple intimate sexual and/or loving relationships with the knowledge and consent of all partners involved.

polyandry The concurrent marriage of one woman to two or more men.

poly families Multi-partner relationships that raise children and function as families.

polygamy A form of marriage in which one person may have two or more spouses.

polygyny A form of marriage in which one husband has more than one wife.

population momentum Continued population growth as a result of past high fertility rates that have resulted in a large number of young women who are currently entering their childbearing years.

postmodernism A worldview that questions the validity of rational thinking and the scientific enterprise.

post-traumatic stress disorder (PTSD) A set of symptoms that may result from any traumatic experience, including crime victimization, war, natural disasters, or abuse.

prejudice Negative attitudes and feelings toward or about an entire category of people.

primary deviance Deviant behavior committed before a person is caught and labeled an offender.

primary groups Usually small numbers of individuals characterized by intimate and informal interaction.

primordial explanations Those explanations that emphasize the existence of "ancient hatreds" rooted in deep psychological or cultural differences between ethnic groups, often involving a history of grievance and victimization, real or imagined, by the enemy group.

privilege When a group has a special advantage or benefits as a result of cultural, economic, societal, legal, and political factors.

probation The conditional release of an offender who, for a specific time period and subject to certain conditions, remains under court supervision in the community.

progressive taxes Taxes in which the tax rate increases as income increases, so that those who have higher incomes are taxed at higher rates.

pronatalism A cultural value that promotes having children.

psychotherapeutic drugs The nonmedical use of any prescription pain reliever, stimulant, sedative, or tranquilizer.

public housing Federally subsidized housing that is owned and operated by local public housing authorities (PHAs).

race A category of people who are perceived to share distinct physical characteristics that are deemed socially significant.

racial microaggressions Brief and commonplace daily verbal, behavioral, or environmental indignities, whether intentional or unintentional, that communicate hostile, derogatory, or negative racial slights and insults toward the target person or group.

racial profiling The law enforcement practice of targeting suspects on the basis of race.

racism The belief that race accounts for differences in human character and ability and that a particular race is superior to others.

radical environmental movement A grassroots movement of individuals and groups that employs unconventional and often illegal means of protecting wildlife or the environment.

recession A significant decline in economic activity spread across the economy and lasting for at least six months.

recidivism A return to criminal behavior by a former inmate, most often measured by rearrest, reconviction, or reincarceration.

refined divorce rate The number of divorces per 1,000 married women.

registered partnerships Federally recognized relationships that convey most but not all the rights of marriage.

rehabilitation A criminal justice philosophy that argues that recidivism can be reduced by changing the criminal through such programs as substance abuse counseling, job training, education, and so on.

rehabilitative alimony Alimony that is paid to an ex-spouse for a specified length of time to allow the recipient time to find a job or to complete education or job training.

relationship literacy education Voluntary, community-based educational programs that teach youth, young adults, cohabiting couples, and engaged and married couples what healthy relationships look like and the skills and attitudes that make them work.

relative poverty The lack of material and economic resources compared with some other population.

religious freedom laws Laws that protect business owners who discriminate against customers (e.g., gay men and women) based on religious grounds.

replacement-level fertility The level of fertility at which a population exactly replaces itself from one generation to the next; currently, the number is 2.1 births per woman (slightly more than 2 because not all female children will live long enough to reach their reproductive years).

restorative justice A philosophy primarily concerned with reconciling conflict between the victim, the offender, and the community.

roles The set of rights, obligations, and expectations associated with a status.

sample A portion of the population, selected to be representative so that the information from the sample can be generalized to a larger population.

sanctions Social consequences for conforming to or violating norms.

sandwich generation A generation of people who care for their aging parents while also taking care of their own children.

school-to-prison pipeline The established relationship between severe disciplinary practices, increased rates of dropping out of school, lowered academic achievement, and court or juvenile detention involvement.

school vouchers Tax credits that are transferred to the public or private school that parents select for their child.

science The process of discovering, explaining, and predicting natural or social phenomena.

second shift The household work and child care that employed parents (usually women) do when they return home from their jobs.

secondary deviance Deviant behavior that results from being caught and labeled as an offender.

secondary groups Involving small or large numbers of individuals, groups that are task-oriented and are characterized by impersonal and formal interaction.

Section 8 housing A housing assistance program in which federal rent subsidies are provided either to tenants (in the form of certificates and vouchers) or to private landlords.

security dilemma A characteristic of the international state system that gives rise to unstable relations between states; as State A secures its borders and interests, its behavior may decrease the security of other states and cause them to engage in behavior that decreases A's security.

segregation The physical separation of two groups in residence, workplace, and social functions.

self-fulfilling prophecy A concept referring to the tendency for people to act in a manner consistent with the expectations of others.

Semantic Web Sometimes called Web 3.0, a version of the Internet in which pages not only contain information but also describe the interrelationship between pages; sometimes called smart media.

serial monogamy A succession of marriages in which a person has more than one spouse over a lifetime but is legally married to only one person at a time.

sex A person's biological classification as male or female.

sexism The belief that innate psychological, behavioral, and/or intellectual differences exist between women and men and that these differences connote the superiority of one group and the inferiority of the other.

sexual harassment In reference to workplace harassment, when an employer requires sexual favors in exchange for a promotion, salary increase, or any other employee benefit and/or the existence of a hostile environment that unreasonably interferes with job performance.

sexual orientation A person's emotional and sexual attractions, relationships, self-identity, and behavior.

sexual orientation change efforts (SOCE) Collectively refers to reparative, conversion, and reorientation therapies, according to the APA.

shaken baby syndrome A form of child abuse whereby a caretaker shakes a baby to the point of causing the child to experience brain or retinal hemorrhage.

single-payer health care A health care system in which a single tax-financed public insurance program replaces private insurance companies.

slums Concentrated areas of poverty and poor housing in urban areas.

social group Two or more people who have a common identity, interact, and form a social relationship.

social movement An organized group of individuals with a common purpose to either promote or resist social change through collective action.

social problem A social condition that a segment of society views as harmful to members of society and in need of remedy.

Social Security Also called "Old Age, Survivors, Disability, and Health Insurance," a federal program that protects against loss of income due to retirement, disability, or death.

socialism An economic system characterized by state ownership of the means of production and distribution of goods and services.

sociological imagination The ability to see the connections between our personal lives and the social world in which we live.

state The organization of the central government and government agencies such as the military, police, and regulatory agencies.

State Children's Health Insurance Program (SCHIP) A public health insurance program, jointly funded by the federal and state governments, that provides health insurance coverage for children whose families meet income eligibility standards.

status A position that a person occupies within a social group.

STEM An acronym for science, technology, engineering, and mathematics.

stem cells Undifferentiated cells that can produce any type of cell in the human body.

stereotype threat The tendency of minorities and women to perform poorly on high-stakes tests because of the anxiety created by the fear that a negative performance will validate societal stereotypes about one's member group.

stereotypes Exaggerations or generalizations about the characteristics and behavior of a particular group.

stigma A discrediting label that affects an individual's self-concept and disqualifies that person from full social acceptance.

structural sexism The ways in which the organization of society, and specifically its institutions, subordinate individuals and groups based on their sex classification.

structure The way society is organized including institutions, social groups, statuses, and roles.

structured choice Choices that are limited by the structure of society.

subjective element of a social problem The belief that a particular social condition is harmful to society, or to a segment of society, and that it should and can be changed.

subprime mortgages High-interest or adjustable-rate mortgages that require little money down and are issued to borrowers with poor credit ratings or limited credit history.

sundown towns Communities that are purposely "all-white" and that have used various means to deliberately keep racial and ethnic minorities out.

Supplemental Nutrition Assistance Program (SNAP) The largest U.S. food assistance program.

supply reduction One of two strategies in the U.S. war on drugs (the other is demand reduction), supply reduction concentrates on reducing the supply of drugs available on the streets through international efforts, interdiction, and domestic law enforcement.

survey research A research method that involves eliciting information from respondents through questions.

sustainable development Occurs when human populations can have fulfilling lives without degrading the planet.

Sustainable Development Goals A set of 17 goals that comprise an international agenda for reducing poverty and economic inequality and improving lives.

sweatshops Work environments that are characterized by less-than-minimum wage pay, excessively long hours of work (often without overtime pay), unsafe or inhumane working conditions, abusive treatment of workers by employers, and/or the lack of worker organizations aimed to negotiate better working conditions.

symbol Something that represents something else.

synthetic drugs A category of drugs that are "designed" in laboratories rather than naturally occurring in plant material.

tar sands Large, naturally occurring deposits of sand, clay, water, and a dense form of petroleum (that looks like tar).

tar sands oil Oil that results from converting tar sands into liquid fuel. It is known as the world's dirtiest oil because producing it requires energy and generates high levels of greenhouse gases (that cause global warming and climate change), and also leaves behind large amounts of toxic waste.

technological dualism The tendency for technology to have both positive and negative consequences.

technological fix The use of scientific principles and technology to solve social problems.

technology Activities that apply the principles of science and mechanics to the solutions of a specific problem.

technology-induced diseases Diseases that result from the use of technological devices, products, and/or chemicals.

Temporary Assistance for Needy Families (TANF) A federal cash welfare program that involves work requirements and a five-year lifetime limit.

terrorism The premeditated use or threatened use of violence by an individual or group to gain a political objective.

theory A set of interrelated propositions or principles designed to answer a question or explain a particular phenomenon.

therapeutic cloning Use of stem cells to produce body cells that can be used to grow needed organs or tissues; regenerative cloning.

therapeutic communities Organizations in which approximately 35 to 500 individuals reside for up to 15 months to abstain from drugs, develop marketable skills, and receive counseling.

total fertility rates The average lifetime number of births per woman in a population.

transgender individual A person whose sense of gender identity is inconsistent with their birth (sometimes called chromosomal) sex (male or female).

transnational corporations Also known as multinational corporations, corporations that have their home base in one country and branches, or affiliates, in other countries.

transnational crime Criminal activity that occurs across one or more national borders.

transnational terrorism Terrorism that occurs when a terrorist act in one country involves victims, targets, institutions, governments, or citizens of another country.

under-5 mortality Deaths of children under age 5.

underemployment Unemployed workers as well as (1) those working part-time but who wish to work full-time, (2) those who want to work but have been discouraged from searching by their lack of success, and (3) others who are neither working nor seeking work but who want and are available to work and have looked for employment in the last year. Also refers to the employment of workers with high skills and/or educational attainment working in low-skill or low-wage jobs.

unemployment To be currently without employment, actively seeking employment, and available for employment, according to U.S. measures of unemployment.

union density The percentage of workers who belong to unions.

universal health care A system of health care, typically financed by the government, that ensures health care coverage for all citizens.

value-added measurement (VAM) VAM is the use of student achievement data to assess teacher effectiveness.

values Social agreements about what is considered good and bad, right and wrong, desirable and undesirable.

variable Any measurable event, characteristic, or property that varies or is subject to change.

Veterans Health Administration (VHA) A system of hospitals, clinics, counseling centers, and long-term care facilities that provides care to military veterans.

victimless crimes Illegal activities that have no complaining participant(s) and are often thought of as crimes against morality, such as prostitution.

vivisection The practice of cutting into or otherwise harming living, nonhuman animals for the purpose of scientific research.

vulnerable employment Employment that is characterized by informal working arrangements, little job security, few benefits, and little recourse in the face of an unreasonable demand.

wage theft Occurs when employers "steal" workers' wages by requiring them to work off the clock or refusing to pay them for overtime.

war Organized armed violence aimed at a social group in pursuit of an objective.

wealth The total assets of an individual or household minus liabilities.

wealthfare Laws and policies that benefit the rich.

weapons of mass destruction (WMD) Chemical, biological, and nuclear weapons that have the capacity to kill large numbers of people indiscriminately.

white-collar crime Includes both *occupational crime*, in which individuals commit crimes in the course of their employment, and *corporate crime*, in which corporations violate the law in the interest of maximizing profit.

worker cooperatives Democratic business organizations controlled by their members, who actively participate in setting their policies and making decisions; also known as *workers' self-directed enterprises*.

workers' self-directed enterprises See *worker cooperatives*.

Workforce Investment Act Legislation passed in 1998 that provides a wide array of programs and services designed to assist individuals to prepare for and find employment.

working poor Individuals who spend at least 27 weeks per year in the labor force (working or looking for work) but whose income falls below the official poverty level.

work/life conflict The day-to-day struggle to simultaneously meet the demands of work and other life responsibilities and goals, including family, education, exercise, and recreation.

References

Chapter 1

Berger, Peter. 1963. *Invitation to Sociology*. Garden City, NJ: Doubleday.

Blumer, Herbert. 1971. "Social Problems as Collective Behavior." *Social Problems* 8(3):298–306.

Boak, Josh. 2014. "The CEO Got a Huge Raised. You Didn't. Here's Why." The Big Story. *Associated Press*. Available at www.bigstory.ap.org

Basu, Moni. 2014. "Unprecedented Verdict: Peanut Executive Guilty in Deadly Salmonella Outbreak." September 19. Available at www.cnn.com

British Petroleum (BP). 2015. "Gulf of Mexico: Progress of Restoration Efforts." Available at www.bp.com

Career Cast. 2013. "Jobs Rated 2013: Ranking 200 Jobs from Best to Worse." Available at www.careercast.com

Centers for Disease Control and Prevention. 2008. (August 1). "Trends In HIV- and STD-Related Risk Behaviors among High School Students—United States, 1991–2007." *Morbidity and Mortality Weekly Report* 57(30):817–822.

Centers for Disease Control and Prevention. 2010 (June 4). "Youth Risk Behavior Surveillance, United States, 2009." *Morbidity and Mortality Weekly Report* 59 No. SS-5.

Centers for Disease Control and Prevention. 2012 (July 27). "Trends in HIV-Related Risk Behaviors among High School Students—United States, 1991–2011." *Morbidity and Mortality Weekly Report* 61(29):556–560.

Centers for Disease Control and Prevention. 2014 (June 13). "Youth Risk Behavior Surveillance, 2013." *Morbidity and Mortality Weekly Report* 63(4):1–5.

Eagan, K., Lozano, J. B., Hurtado, S., and Case, M. H. 2014. *The American Freshman: National Norms: Fall 2013*. Los Angeles: Higher Education Research Institute, UCLA.

Farrell, Dan, and James C. Petersen. 2010. "The Growth of Internet Research Methods and the Reluctant Sociologist." *Sociological Inquiry* 80(1):114–125.

Fleming, Zachary. 2003. "The Thrill of It All." In *In Their Own Words*, ed. Paul Cromwell, 99–107. Los Angeles: Roxbury.

Gallup Poll. 2015a (June). "Most Important Problem." Historical Trends. Available at www.gallup.com

Gallup Poll. 2015b (June). "Satisfaction with the United States." Historical Trends. Available at www.gallup.com

Herd, D. 2011. "Voices from the Field: The Social Construction of Alcohol Problems in Inner-City Communities." *Contemporary Drug Problems*, 38(1):7–39.

Holland, Steve, and Roberta Rampton. 2014 (November 21). "Obama Unveils U.S. Immigration Reform, Setting Up Fight with Republicans." Available at www.reuters.com

Indiana State Government. 2013. "General Information." Department of Toxicology. Available at www.in.gov/isdt/2340.htm

Internet Live. 2015 (June 22). "United States Internet Users." *Internet Live Statistics*. Available at www.internetlivestats.com

Kmec, Julie A. 2003. "Minority Job Concentration and Wages." *Social Problems* 50:38–59.

Matza, David. 1990. *Delinquency and Drift*. Brunswick, NJ: Transaction Publishers.

Merton, Robert K. 1968. *Social Theory and Social Structure*. New York: Free Press.

Mills, C. Wright. 1959. *The Sociological Imagination*. London: Oxford University Press.

Mooney, Linda A. 2015. *An Author's "Human Side."* Personal essay.

National Highway Traffic Safety Administration. 2014. *Distracted Driving: Key Facts and Statistics*. Available at www.distraction.gov

Newport, Frank. 2012 (November 29). "Democrats, Republicans Diverge on Capitalism, Federal Gov't." Available at www.gallup.com

News Service, April 5. Available at www.reuters.com

Obama, Barack. 2013 (January 21). "Inaugural Address by President Barack Obama." Available at www.whitehouse.gov

Palacios, Wilson R., and Melissa E. Fenwick. 2003. "'E' Is for Ecstasy." In *In Their Own Words*, ed. Paul Cromwell, 277–283. Los Angeles Roxbury.

Pew. 2014 (October 16). "Middle Easterners See Religious and Ethnic Hatred as Top Global Threat." Available at www.pewglobal.org

Pope, Carl. 2011. "No, BP Won't Make It Right." *Huffington Post*, February 25. Available at www.huffingtonpost.com

Recovery.gov. 2015. "Overview of Funding." Available at www.recovery.gov

Reiman, Jeffrey, and Paul Leighton. 2013. *The Rich Get Richer and the Poor Get Prison*, 10th ed. New York: Pearson.

Shortell, David. 2014. "Seek Student Sues U.S. Army, Says ROTC Enlistment Locked Because of His Faith." *CNN News*, November 14. Available at www.CNN.com

Simi, Pete, and Robert Futrell. 2009. "Negotiating White Power." *Social Problems* 56(1):98–110.

Smith, Jacquelyn. 2013. "The Best and Worst Jobs for 2013." *Forbes*, April 23. Available at www.forbes.com

Times, June 27. Available at www.nytimes.com

Thomas, W. I. 1931/1966. "The Relation of Research to the Social Process." In *W. I. Thomas on Social Organization and Social Personality*, ed. Morris Janowitz, 289–305. Chicago: University of Chicago Press.

Ungar, Sanford J. 2015. "American Ignorance." March 23. *Inside Higher Education*. Available at www.insidehighered.com

U.S. Department of State. 2013. "USA Jobs." Available at www.usajobs.gov

Wall Street Journal. 2013. "Best and Worst Jobs of 2013." April 22. Available at www.wsj.com

Weir, Sara, and Constance Faulkner. 2004. *Voices of a New Generation: A Feminist Anthology*. Boston: Pearson Education.

Wilhelm, Ian. 2012. "Northern Arizona U. Overhauls Curriculum to Focus on 'Global Competence.'" *Chronicle of Higher Education*, May 20. Available at chronicle.com

Wilson, John. 1983. *Social Theory*. Englewood Cliffs, NJ: Prentice Hall.

Chapter 2

AKA-NAMI Partnership. 2015. *Mental Health: Know the Warning Signs*. Available at www.nami.org.

Allen, P. L. 2000. *The Wages of Sin: Sex and Disease, Past and Present*. Chicago: University of Chicago Press.

American College Health Association. 2014. *American College Health Association National College Health Assessment II: Reference Group Executive Summary Spring 2014*. Hanover, MD: American College Health Association.

Arehart-Treichel, Joan. 2008. "Psychiatrists and Farmers: Alliance in the Making?" *Psychiatric News* 43(10):15.

Artiga, Samantha, Rachel Arguello, and Philethea Duckett. 2013. "Health Coverage and Care for American Indians and Alaska Natives." Available at www.kff.org

*The authors and Wadsworth acknowledge that some of the Internet sources may have become unstable; that is, they are no longer hot links to the intended reference. In that case, the reader may want to access the article through the search engine or archives of the homepage cited (e.g., fbi.gov, cbsnews.com).

Bafana, Busani. 2013 (January 11). "Morphine Kills Pain but Its Price Kills Patients." InterPress Service. Available at ipsnews.net

Baker, Peter, Shari L. Dwarkin, Sengfah Tong, Ian Banks, Time Shand, and Gavin Yamen. 2014. "The Men's Health Gap: Men Must Be Included in the Global Health Equity Agenda." *Bulletin of the World Health Organization* 92:618–670.

Barker, Kristin. 2002. "Self-Help Literature and the Making of an Illness Identity: The Case of Fibromyalgia Syndrome (FMS)." *Social Problems* 49(3):279–300.

Berman, Mark, Lena H. Sun, and Ashley Halsey III. 2014. "Travelers from West African Countries Will Face Stronger Ebola Screening at U.S. Airports." *Washington Post*, October 8. Available at www.washingtonpost.com.

Beronio, K., R. Po, L. Skopec, and S. Glied. 2013 (February 20). "Affordable Care Act Expands Mental Health and Substance Use Disorder Benefits and Federal Parity Protections for 62 Million Americans." *ASPE Issue Brief.* U.S. Department of Health and Human Services. Available at aspe.hhs.gov

Biset, Blain. 2013 (January 31). "No Woman Should Die Giving Life." Inter Press Service. Available at www.ipsnews.net

Blumenthal, David, Melinda Abrams, and Rache Nuzum. 2015 (June 18). "The Affordable Care Act at 5 Years." *New England Journal of Medicine* 372:2451–2458.

Borgelt, L. M., K. L. Franson, A. M. Nussbaum, and G. S. Wang. 2013. "The Pharmacologic and Clinical Effects of Medical Cannabis." *Pharmacotherapy* 33(2):195–209.

Braine, Theresa. 2011. "Race Against Time to Develop New Antibiotics." *Bulletin of the World Health Organization* 89:88–89.

Brill, Steven (2013, March 4). "Bitter Pill," *Time* 181(8):16–55.

Carolla, Bob. 2013 (March). "Entry on Mental Illness Added to AP Stylebook." *NAMI NOW.* Available at www.nami.org

Centers for Disease Control and Prevention. 2013 (November 22). *CDC Health Disparities and Inequalities Report—United States, 2013.* MMWR 62(3):all.

Centers for Disease Control and Prevention. 2015. "Deaths: Final Data for 2013." *National Vital Statistics Report* 64(2).

Chandler, C. K. 2005. *Animal Assisted Therapy in Counseling.* New York: Routledge.

Cockerham, William C. 2007. *Medical Sociology,* 10th ed. Upper Saddle River, NJ: Prentice Hall.

Cockerham, William C. 2013. "Sociological Theory in Medical Sociology in the Early Twenty-first Century." *Social Theory & Health. Suppl. Ten Year Anniversary Issue* 11(3):241–255.

Cohen, Elizabeth, and John Bonifield. 2014 (September 11). "The Reason Ebola Isn't Being Stopped." CNN. Available at www.cnn.com

Collins, S. R., P. W. Rasmussen, M. M. Doty, and S. Beutel. 2015. "The Rise in Health Care Coverage and Affordability Since Health Reform Took Effect." Commonwealth Fund. Available at www.commonwealth.org

Crudo, Dana. 2013 (January). "New Semester, New NAMI on Campus Clubs." *NAMI Now.* Available at www.nami.org

Davis, K., K. Stremikis, D. Squires, and C. Schoen. 2014. *Mirror, Mirror on the Wall, 2014 Update: How the Performance of the U.S. Health Care System Compares Internationally.* Commonwealth Fund. Available at commonwealthfund.org

Davis, Mathew A., Brook I. Martin, Ian D. Coulter, and William B. Weeks. 2013. "U.S. Spending on Complementary and Alternative Medicine During 2002–2008 Plateaued, Suggesting Role in Reformed Health System." *Health Affairs* 32(1):45–52.

Delany, Bill. 2012 (June 26). "New Hampshire Native: Allow Compassionate Use in the Granite State." *Medical Cannabis: Voices from the Frontlines. Blog Archive.* Safe Access Now. Available at safeaccessnow.org

Devries, K. M., J. Y. T. Mak, C. García-Moreno, M. Petzold, J. C. Child, G. Falder, S. Lim, L. J. Bacchus, R. E. Engell, L. Rosenfeld, C. Pallitto, T. Vos, N. Abrahams, and C. H. Watts. 2013 (June 28). "The Global Prevalence of Intimate Partner Violence Against Women." *Science*:1527–1528.

Dickman, Sam, David Himmelstein, Danny McCormick, and Steffie Woolhandler. 2014 (January 30). "Opting Out of Medicaid Expansion: The Health and Financial Impacts." *Health Affairs Blog.* Available at www.healthaffairs.org

Dingfelder, Sadie F. 2009 (June). "The Military's War on Stigma." *Monitor on Psychology* 40(6):52.

Families USA. 2007 (January 9). "No Bargain: Medicare Drug Plans Deliver Higher Prices." Available at www.familiesusa.org

Farmer, Paul, Julio Frenk, Felicia M. Knaul, Lawrence N. Shulman, George Alleyne, Lance Armstrong, Rifat Atun, Douglas Blayney, Lincoln Chen, Richard Feachem, Mary Gospodarowicz, Julie Gralow, Sanjay Gupta, Ana Langer, Julian Lob-Levyt, Claire Neal, Anthony Mbewu, Dina Mired, Peter Piot, K. Srinath Reddy, Jeffrey D. Sachs, Mahmoud Sarhan, and John R. Seffrin. 2010 (October 2). "Expansion of Cancer Care and Control in Countries of Low and Middle Income: A Call to Action." *Lancet* 376(9747):1186–1193.

Federal Trade Commission. 2012. "A Review of Food Marketing to Children and Adolescents." Available at www.ftc.gov

Fine, Aubrey H. 2010. "Forward." In *Handbook on Animal-Assisted Therapy,* 3rd ed., ed. Aubrey H. Fine, xix–xxi. Burlington, MA: Academic Press.

Fischman, J. 2010. "The Pressure of Race." *Chronicle of Higher Education,* September 12. Available at chronicle.com

Freeman, Daniel, and Jason Freeman. 2013a. "Let's Talk about the Gender Differences That Really Matter—in Mental Health." *The Guardian* (December 13). Available at www.theguardian.com

Freeman, Daniel, and Jason Freeman. 2013b. *The Stressed Sex: Uncovering the Truth about Men, Women, and Mental Health.* Oxford: Oxford University Press.

Gebelhoff, Robert. 2015. "Sugary Drinks Linked to 180,000 Deaths a Year, Study Says." *The Washington Post* (June 29). Available at www.washingtonpost.com

Goldstein, Michael S. 1999. "The Origins of the Health Movement." In *Health, Illness, and Healing: Society, Social Context, and Self,* ed. Kathy Charmaz and Debora A. Paterniti, 31–41. Los Angeles: Roxbury.

Goode, Erica, and Jack Healy. 2013. "Focus on Mental Health Laws to Curb Violence Is Unfair, Some Say." *New York Times,* January 31. Available at www.nytimes.com

Goodnough, Abby, Reed Abelson, and Anemona Hartocollis. 2014. "Has Insurance under the Law Been Affordable?" *New York Times,* October 6. Available at www.nytimes.com

Grossman, Amy. 2009. "A Birth Pill." *New York Times,* May 9. Available at www.nytimes.com

Grow, Brian, P. J. Huffstutter, and Michael Erman. 2014 (September 15). "Farmaceuticals: The Drugs Fed to Farm Animals and the Risks to Humans." *Reuters Investigates.* Available at www.reuters.com

Hamel, Liz, Jamie Firth, and Mollyann Brodie. 2014 (August 1). *Kaiser Health Tracking Poll: July 2014.* Kaiser Family Foundation. Available at www.kff.org

Harrison, Joel A. 2008 (May/June). "How Much Is the Sick U.S. Health Care System Costing You?" *Dollars and Sense.* Available at dollarsandsense.com

Harvard School of Public Health. 2013. "The Obesity Prevention Source. Obesity Causes; Globalization." Available at www.hsph.harvard.edu

Hawkes, Corinna. 2006. "Uneven Dietary Development: Linking the Policies and Processes of Globalization with the Nutrition Transition, Obesity, and Diet-Related Chronic Diseases." *Globalization and Health* 2:4.

Himmelstein, D. U., and S. Woolhandler. 2014. "High Administrative Costs: The Authors Reply." *Health Affairs* 33(1):2081.

Holmes, Lindsay. 2014. "These Wildly Successful People Will Prompt You to Rethink What It Means to Have a Mental Illness." *Huffington Post,* August 4. Available at www.huffingtonpost.com

Hummer, Robert A., and Elaine M. Hernandez. 2013. "The Effects of Educational Attainment on Adult Mortality in the United States." *Population Bulletin* 68(1).

Indian Health Service. 2014. "Fact Sheet: Quick Look." Available at www.ihs.gov

Jones, Jared. 2014. "Ebola, Emerging: The Limitations of Cultural Discourse in Epidemiology." *Journal of Global Health* (August 9). Available at www.ghjournal.org

Kaiser Family Foundation. 2007. "How Changes in Medical Technology Affect Health Care Costs." Available at www.kff.org

Kaiser Family Foundation. 2012 (May). *Kaiser Health Tracking Poll.* Available at www.kff.org

Kaiser Family Foundation/HRET. 2014. "Employer Health Benefits 2014 Annual Survey." Available at www.kff.org

Katel, Peter. 2010. "Food Safety." *CQ Researcher* 20(44).

Kindig, David A., and Erika R. Cheng. 2013. "Even as Mortality Fell in Most U.S. Counties, Female Mortality Nonetheless Rose in 42.8 Percent of Counties from 1992 to 2006." *Health Affairs* 32(3):451–458.

Kirsch, D. J., S. L. Pinder-Amaker, C. Morse, M. L. Ellison, L. A. Doerfler, and M. B. Rib. 2014. "Population-Based Initiatives in College Mental Health: Students Helping Students to Overcome Obstacles." *Current Psychiatry Reports* 16:524–532.

Kolappa, K., D. C. Henderson, and S. P. Kishore. 2013. "No Physical Health without Mental Health: Lessons Unlearned?" *Bulletin of the World Health Organization* 91(3):3–3A.

Kruger, K. A., and J. A. Serpell. 2010. "Animal-Assisted Interventions in Mental Health: Definitions and Theoretical Foundations." In *Handbook on Animal-Assisted Therapy*, 3rd ed., ed. Aubrey H. Fine, 33–48. San Diego: Academic Press.

LaMontagne, Christina. 2014 (October 8). "Nerd-Wallet Health Study: Medical Debt Crisis Worsening Despite Policy Advances." NerdWallet. Available at www.nerdwallet.com

Lee, Kelley. 2003. "Introduction." In *Health Impacts of Globalization*, ed. Kelley Lee, 1–10. New York: Palgrave Macmillan.

Lewis, Tene T., David R. Williams, Mahader Tamene, and Cheryl R. Clark. 2014. "Self-Reported Experiences of Discrimination and Cardiovascular Disease." *Current Cardiovascular Risk Reports* 8(1):365–380.

Light, Donald W., and Rebecca Warburton. 2011 (February 7). "Demythologizing the High Costs of Pharmaceutical Research." *Biosocieties*, pp. 1–17.

Link, Bruce G., and Jo Phelan. 2001. "Social Conditions as Fundamental Causes of Disease." In *Readings in Medical Sociology*, 2nd ed., ed. William C. Cockerham, Michael Glasser, and Linda S. Heuser, 3–17. Upper Saddle River, NJ: Prentice Hall.

Mahar, Maggie. 2006. *Money-Driven Medicine*. New York: HarperCollins.

Mannix, Jeff. 2009. "The Patients' Perspective: Locals Find Relief with Medical Marijuana." *Durango Telegraph*, November 19. Available at www.durangotelegraph.com

Mayer, Lindsay Renick. 2009. "Insurers Fight Public Health Plan." Capitol Eye Blog, June 18. Center for Responsive Politics. Available at www.opensecrets.org

Mental Health Advocacy, Inc. n.d. "Role Models: Successful People with Mental Illness." Available at www.mentalhealthadvocacy.org

Musumeci, Mary Beth. 2012 (July). "A Guide to the Supreme Court's Affordable Care Act Decision." Kaiser Family Foundation. Available at www.kff.org

Nader, Ralph. 2009 (July 25). "Health Care Hypocrisy." *Common Dreams*. Available at www.commondreams.org

National Alliance on Mental Illness. 2012. *College Students Speak: A Survey Report on Mental Health*. Available at www.nami.org

National Center for Complementary and Alternative Medicine (NCCAM). 2015. "Complementary, Alternative, or Integrative Health: What's in a Name?" Available at www. nccam.nih.gov

National Center for Health Statistics. 2015. *Health, United States, 2014*. Hyattsville, MD: U.S. Government Printing Office.

National Coalition on Health Care. 2009. "Health Insurance Costs." Available at www.nchc.org

National Conference of State Legislators. 2015a. "Indoor Tanning Restrictions for Minors—A State-by-State Comparison." Available at www.ncsl.org

National Conference of State Legislators. 2015b. "State Medical Marijuana Laws." Available at www.ncsl.org

National Research Council and Institute of Medicine. 2013. *U.S. Health in International Perspective: Shorter Lives, Poorer Health*. Washington D.C.: The National Academies Press.

Park, Madison. 2009 (September 18). "45,000 American Deaths Associated with Lack of Insurance." CNN. Available at articles.cnn.com

Peters, Sharon. 2011. "Animals Can Assist in Psychotherapy." *USA Today*, January 17. Available at www.usatoday.com

Pollitz, Karen, Cynthia Cox, Kevin Lucia, and Katie Keith. 2014 (January 7). "Medical Debt Among People with Health Insurance." Kaiser Family Foundation. Available at www.kff.org

Population Reference Bureau. 2013. *Family Planning Worldwide 2013 Data Sheet*. Available at www.prb.org

Potter, Wendell. 2010. *Deadly Spin: An Insurance Company Insider Speaks Out on How Corporate PR Is Killing Health Care and Deceiving Americans*. New York: Bloomsbury Press.

Quadagno, Jill. 2004. "Why the United States Has No National Health Insurance: Stakeholder Mobilization against the Welfare State 1945–1996." *Journal of Health and Social Behavior* 45:25–44.

Reinberg, Steven. 2009. "Tanning Beds Get Highest Carcinogen Rating." *U.S. News & World Report*, July 28. Available at health .usnews.com

Ruiz, John M., Patrick Steffen, and Timothy B. Smith. 2013. "Hispanic Mortality Paradox: A Systematic Review and Meta-Analysis of the Longitudinal Literature." *American Journal of Public Health* 103(3):e52–e60.

Sanger-Katz, Margot. 2014. "Is the Affordable Care Act Working?" *New York Times*, October 6. Available at www.nytimes.com

Scal, Peter, and Robert Town. 2007. "Losing Insurance and Using the Emergency Department: Critical Effect of Transition to Adulthood for Youth with Chronic Conditions." *Journal of Adolescent Health* 40 (2 Suppl. 1):S4.

Sered, Susan Starr, and Rushika Fernandopulle. 2005. *Uninsured in America: Life and Death in the Land of Opportunity*. Berkeley: University of California Press.

Shally-Jensen, Michael. 2013. "Introduction." In *Mental Health Care Issues in America: An Encyclopedia*, ed. M. Shally-Jensen, i–xxix. Santa Barbara, CA: ABC-CLIO, LLC.

Shern, David, and Wane Lindstrom. 2013. "After Newtown: Mental Illness and Violence." *Health Affairs* 32(3):447–450.

Sidel, Victor W., and Barry S. Levy. 2002. "The Health and Social Consequences of Diversion of Economic Resources to War and Preparation for War." In *War or Health: A Reader*, ed. Ilkka Taipale, P. Helena Makela, Kati Juva, and Vappu Taipale, 208–221. New York: Palgrave Macmillan.

Smedley, B.D., A. Y. Stitth, and A. R. Nelson. 2003. *Unequal Treatment: Confronting Racial and Ethnic Disparities in Health Care*. Washington, DC: National Academy Press.

Smith, Jessica C., and Carla Medalia. 2015. *Health Insurance Coverage in the United States: 2014*. U.S. Census Bureau. Current Population Reports P60-253. Available at www.census.gov

Smith-McDowell, Keiana. 2013 (March). "Executive Order Calls for New Recommendations on Mental Health Issues." *NAMI Now*. Available at www.nami.org

Squires, David A. 2012 (May). "Explaining High Health Care Spending in the United States: An International Comparison of Supply, Utilization, Prices, and Quality." *Issues in International Health Policy*. Commonwealth Fund. Available at www.commonwealthfund.org

Substance Abuse and Mental Health Services Administration. 2014 (September 4). "Results from the 2013 National Survey on Drug Use and Health: Mental Health Findings." Available at www.samhsa.gov

Szasz, Thomas. 1961/1970. *The Myth of Mental Illness: Foundations of a Theory of Personal Conduct*. New York: Harper & Row.

"Testimonials." 2013. NORML. Available at norml.org/about/item/testimonials

Thomson, George, and Nick Wilson. 2005. "Policy Lessons from Comparing Mortality from Two Global Forces: International Terrorism and Tobacco." *Globalization and Health* 1:18.

Trust for America's Health. 2012. "F as in Fat: How Obesity Threatens America's Future." Available at healthyamericans.org

Turner, Leigh, and Jill R. Hodges. 2012. "Introduction: Health Care Goes Global." In *Risks and Challenges in Medical Tourism*, ed. J. R. Hodges, L. Turner, and A. M. Kimball, 1–18. Santa Barbara, CA: Praeger.

UNICEF. 2014. *Ending Child Marriage*. Available at data.unicef.org

Urichuk, L., with D. Anderson. 2003. *Improving Mental Health through Animal-Assisted Therapy*. Edmonton, Alberta: Chimo Project.

U.S. Department of Defense. 2014 (July 29). "Post Traumatic Stress Disorder (PTSD) Not Qualified for Purple Heart Award." Available at kb.defense.gov

Weitz, Rose. 2013. *The Sociology of Health, Illness, and Health Care: A Critical Approach*. 6th ed. Belmont, CA: Wadsworth/Cengage.

White, Frank. 2003. "Can International Public Health Law Help to Prevent War?" *Bulletin of the World Health Organization* 81(3):228.

The White House. 2014 (August 26). "Fact Sheet: President Obama Announces New Executive Actions to Fulfill Our Promises to Service Members, Veterans, and Their Families." Available at www.thewhitehouse.gov

Williams, David R. 2012. "Miles to Go before We Sleep: Racial Inequities in Health." *Journal of Health and Social Behavior* 53(3):279–295.

Williams, D. R., M. B. McClellan, and A. M. Rivlin. 2010 (August). "Beyond the Affordable Care Act: Achieving Real Improvements in Americans' Health." *Health Affairs* 29(8):1481–1488.

Willingham, Val. 2014 (May 29). "Sunlamps and Tanning Beds Get FDA Warning." CNN. Available at www.CNN.com

Wisner, K. L., D. K. Y. Sit, M. C. McShea, et al. 2013. "Onset Timing, Thoughts of Self-Harm, and Diagnoses in Postpartum Women with Screen-Positive Depression Findings." *JAMA Psychiatry* (Online March 13). Available at archpsych.jamanetwork.com

Wootan, Margo. 2012 (December 21). "Little Improvement Seen in Food Marketing." www.cspinet.org

World Health Organization. 1946. *Constitution of the World Health Organization*. New York: World Health Organization Interim Commission.

World Health Organization. 2010. "Mental Health and Development: Targeting People with Mental Health Conditions as a Vulnerable Group." Available at www.who.org

World Health Organization. 2012. *World Health Statistics 2012*. Available at www.who.int

World Health Organization. 2014a. "Maternal Mortality." *Fact Sheet No. 348*. Available at www.who.int

World Health Organization. 2014b. "The Top 10 Causes of Death." *Fact Sheet No. 310*. Available at www.who.int

World Health Organization. 2015a. "Ebola Virus Disease." *Fact Sheet No. 103*. Available at www.who.int

World Health Organization. 2015b. "Family Planning Contraception." *Fact Sheet No. 351*. Available at www.who.int

World Health Organization. 2015c. "Obesity and Overweight." *Fact Sheet No. 311*. Available at www.who.int

World Health Organization. 2015d. *World Health Statistics 2015*. Available at www.who.int

Chapter 3

Abadinsky, Howard. 2013. *Drug Use and Abuse: A Comprehensive Introduction*. Belmont, CA: Wadsworth.

Alabama Code. 2015. *ALA Code 26-15-32.2*. Available at codes.lp.findlaw.com

Aleksander, Irina. 2013. "Molly: Pure, but Not So Simple." *New York Times,* June 21. Available at www.nytimes.com

Alexander, Michelle. 2010. *The New Jim Crow: Mass Incarceration in the Age of Colorblindness*. New York: New York Press.

Almasy, Steve. 2013. "FDA Changes Course on Graphic Warning Labels for Cigarettes." *CNN News,* March 20. Available at www.cnn.com

Allsop, D. J., J. Copeland, M. M. Norberg, S. Fu, A. Molnar, J. Lewis, and A. J. Budney. 2012. "Quantifying the Clinical Significance of Cannabis Withdrawal." *Public Library of Science* 7(9):44864–44864.

American College Health Association. 2013. *American College Health Association National College Health Assessment II*. Reference Group Executive Summary Fall 2012. Hanover, MD: American College Health Association.

American College Health Association. 2014. *American College Health Association National College Health Assessment II*. Reference Group Executive Summary Spring 2014. Hanover, MD: American College Health Association.

American Legacy Foundation (ALF). 2014. *Vaporized: E Cigarettes, Advertising, and Youth*. May. Available at legacyforhealth.org (no www)

Arnowitz, Leora. 2013. "Miley Cyrus Sings about Molly Again; Experts Warn of Its Dangers." Fox News, October 15. Available at www.foxnews.com

Ausness, Richard C. 2014. "The Role of Litigation in the Fight against Prescription Drug Abuse." *West Virginia Law Review* 116: 1117–1165.

Balsa, A. I., J. F. Homer, and M. T. French. 2009. "The Health Effects of Parental Problem Drinking on Adult Children." *Journal of Mental Health Policy and Economics* 12(2):55–66.

Becker, H. S. 1966. *Outsiders: Studies in the Sociology of Deviance*. New York: Free Press.

Behrendt, S., H.-U. Wittchen, M. Höfler, R. Lieb, and K. Beesdo. 2009. "Transitions from First Substance Use to Substance Use Disorders in Adolescence: Is Early Onset Associated with a Rapid Escalation?" *Drug and Alcohol Dependence* 99:68–78.

Berends, Lynda, Jason Ferris, and Anne-Marie Laslett. 2014. "On the Nature of Harms Reported by Those Identifying a Problem Drinker in the Family, an Exploratory Study." *Journal of Family Violence* 29:197–204.

Blomeyer, Dorothea, Chris M. Friemel, Arlette F. Buchmann, Tobias Banaschewski, Manfred Laucht, and Miriam Schneider. 2013. "Impact of Pubertal Stage at First Drink on Adult Drinking Behavior." *Alcoholism, Clinical and Experimental Research* 37 (10):1804–1811.

Botticelli, Michael. 2015 (June 26). "Drug Policy Reform at Home and Abroad." Office of National Drug Control Policy. Available at www.whitehouse.gov

Boyette, Chris, and Jacque Wilson. 2015 (January 7). "It's 2015: Is Weed Legal in Your State?" *CNN News*. Available at www.cnn.com

Branson, Richard. 2012 (December 7). "War on Drugs Trillion-Dollar Failure." Available at www.cnn.com

British Medical Association. 2013. *Drugs of Dependence: The Role of Medical Professionals*. BMA Board of Science. January. Available at bma.org.uk

Bunnell, Rebecca E., Israel T. Agaku, Renee A. Arrazola, Benjamin J. Apelberg, Ralph S. Caraballo, Catherine G. Corey, Blair N. Coleman, Shanta R. Dube, and Brian A. King. 2014. "Intentions to Smoke Cigarettes among Never Smoking U.S. Middle and High School Electronic Cigarette Users, National Youth Tobacco Survey, 2011–2013." *Nicotine and Tobacco Research* (August).

Campaign for Tobacco Free Kids. 2013 (March 13). *Not Your Grandfather's Cigars*. Available at www.tobaccofreekids.org

Carrigan, Matthew A., Oleg Uryasev, Carol B. Frye, Blair L. Eckman, Candace R. Myers, Thomas D. Hurley, and Stephen A. Benner. 2014 (December 1). "Hominids Adapted to Metabolize Ethanol Long Before Human-Directed Fermentation." *Proceedings of the National Academy of Sciences (PNAS) of the United States of America*.

Center (for Court Innovation). 2015. "Drug Courts." Available at www.courtinnovation.net

Center for Disease Control and Prevention (CDC). 2011 (September 22). "Fact about FASDs." *Fetal Alcohol Spectrum Disorders (FASDs)*. Available at: www.cdc.gov

Center for Disease Control and Prevention (CDC). 2012 (January 10). "Health Effects of Cigarette Smoking." *Smoking and Tobacco Use*. Available at www.cdc.gov

Center for Disease Control and Prevention (CDC). 2013a. *Addressing Prescription Drug Abuse in the United States: Current Activities and Future Opportunities*. Available at www.cdc.gov

Center for Disease Control and Prevention (CDC). 2013b. "Binge Drinking: A Serious, Underrecognized Problem among Women and Girls." *Vital Signs*, January. Available at www.cdc.gov

Center for Disease Control and Prevention (CDC). 2014a. "Fact Sheet: Alcohol Use and your Health." Available at www.cdc.gov

Center for Disease Control and Prevention (CDC). 2014b. "Smokefree Policies Reduce Smoking." Available at www.cdc.gov

Champion, Katrina E., Nicola C. Newton, Emma L. Barrett, and Maree Teesson. 2013. "A Systematic Review of School-Based Alcohol and Other Drug Prevention Programs Facilitated by Computers or the Internet." *Drug and Alcohol Review* 32(2):115–123.

Chauvin, Chantel D. 2012. "Social Norms and Motivations Associated with College Binge Drinking." *Sociological Inquiry* 82(2):257–281.

Chouvy, Piere-Arnaud. 2013. "A Typology of Unintended Consequences of Drug Crop Reduction." *Journal of Drug Issues* 43(2):216–230.

Christensen, Jen, and Jacque Wilson. 2014 (January 20). "Is Marijuana as Safe as—or Safer Than—Alcohol?" CNN. Available at www.cnn.com

Cohen, Jodi S. 2014. "Judge Dismisses Lawsuit against NIU Fraternity Tied to Party Death." *Chicago Tribune* (December 12). Available at www.chicagotribune.com

Copes, Heigh, Andy Hochstetler, and J. Patrick Williams. 2008. "'We Weren't Like No Regular Dope Fiends': Negotiating Hustler and Crack-head Identities." *Social Problems* 55(2):254–270.

Count the Costs. 2015. "The Seven Costs." Available at www.countthecosts.org

Crisp, Elizabeth. 2012. "Mo. Legislature Approves Change in Crack Cocaine Sentencing." *St. Louis Post-Dispatch,* May 18.

Crowley, D. Max, Damon E. Jones, Donna L. Coffman, and Mark T. Greenberg. 2014. "Can We Build an Efficient Response to the Prescription Drug Abuse Epidemic? Assessing the Cost Effectiveness of Universal Prevention in the PROSPER Trial." *Preventative Medicine* 62:71–77.

Csomor, Marina. 2012 (August 16). "There's Something (Potentially Dangerous) about Molly." CNN Health. Available at www.cnn.com

Degenhardt, Louisa, Wai-Tat Chiu, Nancy Sampson, Ronald C. Kessler, James C. Anthony, Matthias Angermeyer, Ronny Bruffaerts, Giovanni de Girolamo, Oye Gureje, Yueqin Huang, Aimee Karam, Stanislav Kostyuchenko, Jean Pierre Lepine, Maria Elena Medina Mora, Yehuda Neumark, J. Hans Ormel, Alejandra Pinto-Meza, José Posada-Villa, Dan J. Stein, Tadashi Takeshima, and J. Elisabeth Wells. 2008. "Toward a Global View of Alcohol, Tobacco, Cannabis, and Cocaine Use: Findings from the WHO World Mental Health Surveys." *PLoS Medicine* 5(1):1053–1077.

DeJong, William, and Jason Blanchette. 2014. "Case Closed: Research Evidence on the Positive Public Health Impact of the 21 Minimum Legal Drinking Age in the United States." *Journal of Studies on Alcohol and Drugs* 75:108–115.

Dinan, Stephen, and Ben Conery. 2009. "DEA Pot Raids Go On; Obama Opposes." *The Washington Times,* February 5. Available at www.washingtontimes.com

Dinno, Alexis, and Stanton Glantz. 2009. "Tobacco Control Policies Are Egalitarian: A Vulnerabilities Perspective on Clean Indoor Air Laws, Cigarette Prices, and Tobacco Use Disparities." *Social Science & Medicine* 68:1439–1447.

Dodes, Lance, and Zachary Dodes. 2014. *The Sober Truth: Debunking the Bad Science behind 12-Step Programs and the Rehab Industry.* Boston: Beacon.

Doward, Jamie. 2013. "Western Leaders Study 'Gamechanging' Report on Global Drugs Trade." *The Guardian,* May 18. Available at www.guardian.co.uk

Downey, P. Mitchell, and John K. Roman. 2014 (June). "Cost-Benefit Analysis: A Guide for Drug Courts and Other Criminal Justice Programs." *Research in Brief.* U.S. Department of Justice, Office of Justice Programs, National Institute of Justice.

Driving Under the Influence of Drugs, Alcohol, and Medicines (DRUID). 2012. *Final Report: Work Performed, Main Results and Recommendations.* January 8. Available at www.druid-project.eu

Drug Enforcement Administration (DEA). 2010. "Fiction: Drug Production Does Not Damage the Environment." Just Think Twice: Facts and Fiction. Available at www.justthinktwice.com

Drug Enforcement Administration (DEA). 2014 (May). *The Dangers and Consequences of Marijuana Abuse.* U.S. Department of Justice. Available at www.dea.gov

Drug Policy Alliance (DPA). 2015a. "Drug War Statistics." Available at www.drugpolicy.org/

Drug Policy Alliance (DPA). 2015b. *Fact Sheet: Women, Prison, and the Drug War.* Available at www.drugpolicy.org/

Duchnowski, Julian. 2014. "Family Unsure of Potential NIU Hazing Plea Deals." *Daily Chronicle,* August 12. Available at www.dailychronicle.com

Duke, Jennifer C., Youn O. Lee, Annice E. Kim, Kimberly A. Watson, Kristin Y. Arnold, James M. Nonnemaker, and Lauren Porter. 2014. "Exposure to Electronic Cigarette Television Advertising among Youth and Young Adults." *Pediatrics* 134(1): 1–8.

Dutch Drug Policy. 2014. "Facts." Available at www.holland.com

The Economist. 2013 (February 23). "Winding Down the War on Drugs towards a Ceasefire." Available at www.economist.com

Elder, Jeff. 2014. "Icann, Regulators Clash over Illegal Online Drug Sales; FDA, Interpol Want Internet Gatekeeper to Take Action against Suspicious Websites, but Icann Says Its Powers Are Limited." *Wall Street Journal,* October 27. Available at www.wsj.com

Elliott, Tim. 2013. "Online Synthetic Cocaine Selling for $80 a Gram." *Sydney Morning Herald,* October 21. Available at www.smh.com

Eriksen, M., J. Mackay, and H. Ross. 2012. *The Tobacco Atlas,* 4th ed. Atlanta, GA: American Cancer Society; New York: World Lung Foundation. Available at www.TobaccoAtlas.org

European Monitoring Centre for Drugs and Drug Addiction (EMCDDA). 2013 (June 6). "Perspectives on Drugs: The New EU Drugs Strategy (2013–2020)." Available at www.emcdda.org

European Monitoring Centre for Drugs and Drug Addiction (EMCDDA). 2014a. *European Drug Report: Trends and Developments.* Available at www.emcdda.europa.eu/publications/edr

European Monitoring Centre for Drugs and Drug Addiction (EMCDDA). 2014b. *Perspectives on Drugs: The EU Drugs Strategy (2013–2020).* Available at www.emcdda.europa.eu

Fang, Lee. 2014. "The Real Reason Pot Is Still Illegal." *The Nation,* July 1. Available at www.thenation.com

Feagin, Joe R., and C. B. Feagin. 1994. *Social Problems.* Englewood Cliffs, NJ: Prentice Hall.

Feilding, Amanda. 2013. "At Last, the Edifice of Drugs Prohibition Starts to Crumble." *The Guardian,* June 14. Available at www.guardian.co.uk

Feliz, Josie. 2014 (July 22). "National Study: Teens Report Higher Use of Performance-Enhancing Substances." Partnership for Drug-Free Kids. Available at www.drugfree.org

Filbey, Francesca, Sina Asian, Vince D. Calhoun, Jeffrey S. Spence, Eswar Damaraju, Arvind Caprihan, and Judith Segal. 2014. "Long-Term Effects of Marijuana Use on the Brain." *Proceedings of the National Academy of Science* 111(47):16913–6918.

Flanagin, Jake. 2014. "The Surprising Failures of 12 Steps." *The Atlantic,* March 25. Available at www.theatlantic.com

Food and Drug Administration (FDA). 2013b. "Preliminary Scientific Evaluation of the Possible Public Health Effects of Menthol versus Non-Menthol Cigarettes." Available at www.fda.gov

Food and Drug Administration (FDA). 2014a. "Deeming- Extending Authorities to Additional Tobacco Products." Available at www.fda.gov

Food and Drug Administration (FDA). 2014b. "Issues Snapshot on Deeming: Regulating Additional Tobacco Products." Available at www.fda.gov

Food and Drug Administration (FDA). 2014c. "The Real Cost Campaign." Available at www.fda.gov

Forliti, Amy, Dan Sewall, and Nigel Duara. 2014 (April 5). "We're All Paying: Heroin Spreads Misery in US." Available at www.bigstory.com

Foundation for a Drug Free World. 2013. "Real Life Stories about Drug Abuse." Available at www.drugfreeworld.org

Freeman, Dan, Merrie Brucks, and Melanie Wallendorf. 2005. "Young Children's Understanding of Cigarette Smoking." *Addiction* 100(10):1537–1545.

Friedman-Rudovsky, Jean. 2009. "Red Bull's New Cola: A Kick from Cocaine?" *Time/CNN,* May 25. Available at www.time.com

Fryer, Ronald G, Paul S. Heaton, Stephen D. Levitt and Kevin M. Murphy. 2014. "Measuring Crack Cocaine and Its Impact." *Economic Inquiry* 51(3):1651–1681.

Gaffney, Alexander. 2015 (June 16). "DEA to Allow Huge Increase in Marijuana Production to Meet Research Requirements." Regulatory Affairs Professional Society. Available at www.raps.org

Gallahue, Patrick, Ricky Gunawan, Fifa Rahman, Karim El Mufti, Najam U Din, and Rita Felten. 2012. "The Death Penalty for Drug Offences: Global Overview 2012, Tipping the Scales for Abolition." London: International Harm Reduction Association. Available at www.ihra.net

Ghosh, Palash. 2014. "Off with Their Heads: Saudi Arabia Executes Two Pakistani Smugglers." *International Business Times,* January 14. Available at www.ibtimes.com

Gilbert, R., C. S. Widom, K. Browne, D. Fergusson, E. Webb, and S. Janson. 2009. "Burden and Consequence of Child Maltreatment in High-Income Countries." *The Lancet* 73(9657):68–81.

Government Accounting Office (GAO). 2013 (January). "State Approaches Taken to Control Access to Key Methamphetamine Ingredient Show Varied Impact on Domestic Drug Labs." GAO-13-204. Available at www.gao.gov

Gusfield, Joseph. 1963. *Symbolic Crusade: Status Politics and the American Temperance Movement.* Urbana: University of Illinois Press.

Halper, Evan. 2014. "Congress Quietly Ends Federal Government's Ban on Medical Marijuana." *Los Angeles Times,* December 16. Available at latimes.com

Hansen, Chris. 2013. "Following DOJ Decision, FDA Should Quickly Create New Cigarette Warning Labels." Media Center, March 19. American Cancer Society Cancer Action Network.

Hanson, David J. 1997. "History of Alcohol and Drinking around the World." *Alcohol Problems and Solutions.* Available at www.2.potsdam.edu/alcohol

Hanson, David J. 2013. "World Alcohol and Drinking History Timeline." *Alcohol Problems and Solutions.* Available at www.2.potsdam.edu/alcohol

Heinrich, Henry. 2009. "Obama Drug Policy to Do More to Ease Health Risks." Reuters, March 16. Available at: www.reuters.com

Henrisken, L., N. C. Schleicher, A. L. Dauphinee, and S. P. Fortmann. 2011. "Targeted Advertising, Promotion, and Price for Menthol Cigarettes in California High School Neighborhoods." *Nicotine and Tobacco Research* 14(1):116–121.

Hingson, Ralph W., Timothy Heeren, and Michael R. Winter. 2006. "Age at Drinking Onset and Alcohol Dependence." *Archives of Pediatrics & Adolescent Medicine* 160:739–746.

Hollersen, Wiebke. 2013. "'This Is Working': Portugal, 12 Years after Decriminalizing Drugs." *Spiegel International,* March 27 Available at www.spiegel.de/international

Human Rights Watch (HRW). 2014. "UN: Freeze Funding of Iranian Counter-Narcotics Efforts." December 17. Available at www.hrw.org

Human Rights Watch (HRW). 2007. "Reforming the Rockefeller Drug Laws. " Available at www.hrw.org

Jarecki, Eugene. 2014. "As the Marijuana Economy Takes Off, Let's Not Forget the Casualties of the U.S. War on Drugs." *The Guardian,* August 2.

Jargin, Sergei V. 2012. "Social Aspects of Alcohol Consumption in Russia." *South African Medical Journal* 102(9):719.

Kelly, J. F., and B. B. Hoeppner. 2013. "Does Alcoholics Anonymous Work Differently for Men and Women? A Moderated Multiple-Mediation Analysis in a Large Clinical Sample." *Drug and Alcohol Dependence* 130(1–3):186–193.

Kelly, Adrian B., Gary C. K. Chan, John W. Toumbourou, Martin O'Flaherty, Ross Homel, George C. Patton, and Joanne Williams. 2012. "Very Young Adolescents and Alcohol: Evidence of a Unique Susceptibility to Peer Alcohol Use." *Addiction* 37(4):414–419.

Kirschbaum, Katrina M., Lisa Grigoleit, Cornelius Hess, Burkhard Madea, and Frank Musshoff. 2013. "Illegal Drugs and Delinquency." *Forensic Science International* 226(1):230–234.

Knapp, Clifford M., Matthew O'Malley, Subimal Datta, and Domenic A. Ciraulo. 2014. "The K Lee 7 Potassium Channel Activator Retigabine Decreases Alcohol Consumption in Rats." *American Journal of Drug and Alcohol Abuse* 40 (3): 244–250.

Knox, Becca. 2015 (June 24). "Increasing the Minimum Legal Sale Age for Tobacco Products to 21." Campaign for Tobacco Free Kids. Available at www.tobaccofreekids.org

Kravchenko, Stepan. 2014. "Russia Rolls Out Bar Smoking Ban as Cigarette Producers Protest." Bloomberg, June 2. Available at www.bloomberg.com

Lee, Yon, and M. Abdel-Ghany. 2004. "American Youth Consumption of Licit and Illicit Substances." *International Journal of Consumer Studies* 28(5):454–465.

LSE Expert Group. 2014. *Report of the LSE Expert Group on the Economics of Drug Policy.* London School of Economics and Political Science. Available at www.lse.ac.uk

Lurie, Peter. 2015 (June 30). "Naloxone-FDA Hosts Meeting to Discuss Expanded Use of Overdose Medicine." *FDA Voices.* Available at blog.fda.gov

Margolis, Robert D., and Joan E. Zweben. 2011. *Treating Patients with Alcohol and Other Drug Problems: An Integrated Approach.* Chapter 3: "Models and Theories of Addiction." Washington DC: American Psychological Association.

Marine-Street, Natalie. 2012. "Stanford Researchers' Cigarette Ad Collection Reveals How Big Tobacco Targets Women and Adolescent Girls." *Gender News,* April 26. Available at gender.stanford.edu

Martin, David. S. 2012. "Vets Feel Abandoned after Secret Drug Experiments." *CNN Health,* March 1. Available at www.cnn.com

May, Philip A., Amy Baete, Jaymi Russo, Amy J. Elliott, Jason Blankenship, Wendy O. Kalberg, David Buckley, Marita Brooks, Julie Hasken, Omar Abdul-Rahman, Margaret P. Adam, Luther K. Robinson, Melanie Manning, and H. Eugene Hoyme. 2014. "Prevalence and Characteristics of Fetal Alcohol Spectrum Disorders." *Pediatrics* 134(5): 855–866.

McCoy, Kevin. 2015. "Silk Road Mastermind Appeals Life Sentence." *USA Today,* June 5. Available at www.usatoday.com

McLaughlin, Michael. 2012. "Bath Salt Incidents Down since DEA Banned Synthetic Drug." *Huffington Post,* September 4. www.huffingtonpost.com

Mears, Bill. 2012 (November 26). "Tobacco Companies Ordered to Publicly Admit Deception on Smoking Dangers." CNN. Available at www.cnn.com

Medicine Abuse Project (MAP). 2014. *The Medicine Abuse Project.* Partnership for Drug-Free Kids. Available at medicineabuseproject.org

Merolla, David. 2008. "The War on Drugs and the Gender Gap in Arrests: A Critical Perspective." *Critical Sociology* 34 (March):255–270.

Monitoring the Future (MTF). 2014. *Key Findings on Adolescent Drug Use: 2013 Overview.* Ann Arbor: Institute for Social Research, the University of Michigan.

Monitoring the Future (MTF). 2015. *Monitoring the Future National Results on Adolescent Drug Use: Overview of Key Findings, 2014.* Ann Arbor: Institute for Social Research, University of Michigan.

Morgan, Patricia A. 1978. "The Legislation of Drug Law: Economic Crisis and Social Control." *Journal of Drug Issues* 8:53–62.

Myers, Matthew. 2011 (April 25). "FDA Acts to Protect Public Health by Extending Authority over Tobacco Products, Including E-Cigarettes." Available at www.tobaccofreekids.org

Myers, Matthew L. 2009. "U.S. Court of Appeals Affirms 2006 Lower Court Ruling That Tobacco Companies Committed Fraud for Five Decades and Lied about the Dangers of Smoking." Press Office Release, May 22. Available at www.tobaccofreekids.org

National Center on Addiction and Substance Abuse (CASA). 2009. "The Impact of Substance Abuse on Federal, State, and Local Budgets." New York: Columbia University Press.

National Center on Addiction and Substance Abuse (CASA). 2012. "National Survey of American Attitudes on Substance Abuse XVII: Teens." Available at www.casacolumbia.org

National Highway Traffic Safety Administration. 2014. "2013 Motor Vehicle Crashes: Overview." Available at www-nrd.nhsta.dot.gov

National Institute on Alcohol Abuse and Alcoholism (NIAAA). 2013. "The Genetics of Alcoholism." *Alcohol Alert* 84:1–6.

National Institute on Drug Abuse (NIDA). 2006. *Principles of Drug Addiction Treatment: A Research-Based Guide.* Available at drugabuse.gov

National Institute on Drug Abuse (NIDA). 2013. "Monitoring the Future 2012 Survey Results." Available at www.drugabuse.gov

National Institute on Drug Abuse (NIDA). 2014. *Teen Drug Use Infographic.* Available at www.drugabuse.gov

National Survey on Drug Use and Health (NSDUH). 2014. "Results from the 2013 National Survey on Drug Use and Health: Summary of National Findings." Substance Abuse and Mental Health Services Administration. Office of Applied Statistics, NSDUH Series H-48, HHS Publication No. (SMA) 14-4863. Rockville, MD.

Nelson, Steven. 2014. "House Leaders Rush to Defend E-Cigarettes from Possible FDA Ban." *U.S. News & World Report,* December 3. Available at www.usnews.com

Ng, Marie, Michael K. Freeman, Thomas D. Fleming, Margaret Robinson, Laura Dwyer-Lindgren, Blake Thomson, Alexandra Wollum, Ella Sanman, Sarah Wulf, Alan D. Lopez, Christopher J. L. Murray, and Emmanuela Gakidou. 2014. "Prevalence and Cigarette Consumption in 187 Countries, 1980–2012." *Journal of the American Medical Association* 31 (2):183–192.

Nordrum, Amy. 2014. "The New D.A.R.E. Program—This One Works." *Scientific American,* September 10. Available at www.scientificamerican.com

Nordqvist, Christian. 2012. "Teen Cannabis Use Linked to Lower IQ." *Medical News Today,* August 28. Available at www.medicalnewstoday.com

Office of National Drug Policy (ONDCP). 2014. "Living Above." Available at abovetheinfluence,com

Office of National Drug Control Policy (ONDCP). 2013. "Above the Influence." *Fact Sheet.* Available at www.whitehoused.gov

Office of National Drug Control Policy (ONDCP). 2006. *Methamphetamine.* Available at www.whitehousedrugpolicy.gov/drugfact/methamphetamine

Office of National Drug Control Policy (ONDCP). 2009. *National Youth Anti-Drug Media Campaign.* Retrieved from www.theantidrug.com

Organization of American States. 2013 (May 17). "Highlights." *OAS Report on the Drug Problem in the Americas.* Available at www.oas.org/documents/eng/press/highlights.pdf

Palazzolo, Joe. 2014. "R.J. Reynolds Loses $23.6 Billion Verdict." *Wall Street Journal,* July 10. Available at www.wsj.com

Partnership Attitude Tracking Study. 2013 (April 23). *PATS Key Findings: Partnership Attitude Tracking Study.* MetLife Foundation in conjunction with the Partnership at Drugfree.org. Available at www.drugfree.org

Paynter, Ben. 2011. "The Big Business of Synthetic Highs." *Businessweek,* June 16. Available at www.businessweek.com

Peralta, Robert L. Jennifer L. Steele, Stacey Nofziger, and Michael Rickles. 2010. "The Impact of Gender on Binge Drinking Behavior among U.S. College Students Attending a Midwestern University: An Analysis of Two Gender Measures." *Feminist Criminology* 10(5):355–379.

Peters, Jeremy W. 2009. "Albany Takes Step to Repeal '70s-Era Drug Laws." *New York Times,* March 5. Available at www.nytimes.com

Pew. 2014. "America's New Drug Policy Landscape." April 2. Pew Research Center. Available at www.people-press.org

Pew. 2015 (April 14). "In Debate over Legalizing Marijuana, Disagreement over Drugs Danger." Pew Research Center. Available at www.people-press.org

Pierre, Joseph M. 2011. "Cannabis, Synthetic Cannabinoids, and Psychosis Risk: What the Evidence Says." *Current Psychiatry* 10(9): 49–58.

Placko, Dane. 2014 (March 24). "Parents of NIU Freshman Who Died of Alcohol, Hazing Speak Out." Fox 32. Available at www.myfoxchicago.com

Polansky, Jonathan R., Kori Titus, Natalie Lanning, Stanton A. Glantz. 2014 (June 12). "Smoking in Top Grossing U.S. Movies, 2013." San Francisco: Center for Tobacco Control Research and Education, University of California.

Porter, Nicole D. 2015. *The State of Sentencing 2014: Development in Policy and Practice.* The Sentencing Project. Available at www.sentencingproject.org

Primack, Brian A., James E. Bost, Stephanie R. Land, and Michael J. Fine. 2007. "Volume of Tobacco Advertising in African American Markets: Systematic Review and Meta-Analysis." *Public Health Reports* 122(5):607–615.

Rabinowitz, Mikaela, and Arthur Lurigio. 2009. "A Century of Losing Battles: The Costly and Ill-Advised War on Drugs." Conference Paper. American Sociological Association. San Francisco, August 8–11.

Research Society on Alcoholism. 2011. "Impact of Alcoholism and Alcohol-Induced Disease on Americans." April 20. Available at www.rosa.org

Rorabaugh, W. J. 1979. *The Alcoholic Republic: An American Tradition*. New York: Oxford University Press.

Saad, Lydia. 2014. "Beer is American's Adult Beverage of Choice This Year." Available at www.gallup.com

Saloner, B., and B. LeCook. 2013. "Blacks and Hispanics Are Less Likely Than Whites to Complete Addiction Treatment, Largely Due to Socioeconomic Factors." *Health Affairs,* January 7. Available at www.rwjf.org

Sample, Ian. 2014. "Surge in Illegal Sales of Drugs as Gangs Exploit 'Phenomenal Market' Online." *The Guardian*, December 28. Available at www.theguardian.com

Sánchez-Moreno, Maria McFarland. 2012 (September 4). "A Discussion about Drug Policy Is Long Overdue." Human Rights Watch. Available at www.hrw.org

Schmidt, Lorna. 2013 (June 18). "Tobacco Company Marketing to Kids." Campaign for Tobacco Free Kids. Available at www.tobaccofreekids.org

Schmidt, Lorna. 2014. "Tobacco Industry Targeting of Women and Girls" Campaign for Tobacco Free Kids. (April 23). Available at www.tobaccofreekids.org

The Sentencing Project. 2013. "Trends in U.S. Corrections." Fact Sheet. Available at sentencingproject.org

Shuman, Phil. 2014 (November 4). "Fox 11 Investigates: Synthetic Marijuana." *Fox News.* Available at www.myfoxla.com

Sifferlin, Alexandra. 2015 (January 9). "Why you've Never Heard of the Vaccine for Heroin Addiction." *Time.* Available at time.com

Sifferlin, Alexandra. 2013. "FDA Approves New Cigarettes in First Use of New Regulatory Power over Tobacco." *Time,* June 26. Available at: healthland.time.com

Smith, Chad L., Gregory Hooks, and Michael Lengefeld. 2014. "The War on Drugs in Columbia: The Environment, the Treadmill of Destruction and Risk Transfer Militarism." *Journal of World Systems Research* 20(2):185–206.

Substance Abuse and Mental Health Services Administration (SAMHSA). 2007. "Parental Substance Abuse Raises Children's Risk." *Practice What You Preach,* February 20. Retrieved from www.family.samhsa.gov

Substance Abuse and Mental Health Services Administration (SAMHSA). 2009. "Children Living with Substance-Dependent or Substance-Abusing Parents: 2002–2007." *The NSDUH Report,* April 16. Available at oas.samhsa.gov

Substance Abuse and Mental Health Services Administration (SAMHSA). 2012. "More Than 7 Million Children Live with a Parent with Alcohol Problems." *Data Spotlight,* February 16.

Szalavitz, Maia. 2009. "Drugs in Portugal: Did Decriminalization Work?" *Time*, April 26. Available at www.time.com

Taifia, Nkechi. 2006 (May). "The 'Crack/Powder Disparity': Can the International Race Convention Provide a Basis for Relief?" American Constitution Society for Law and Policy white paper. Available at acslaw.org

Taylor, Matthew. 2014. "Health Warnings on Alcohol Bottles Should Be Compulsory—MPs." *The Guardian*, August 10. Available at www.theguardian.com

Thio, Alex. 2007. *Deviant Behavior*. Boston: Allyn and Bacon.

Thomas, Gerald, Ginny Gonneau, Nancy Poole, and Jaclyn Cook. 2014. "The Effectiveness of Alcohol Warning Labels in the Prevention of Fetal Alcohol Spectrum Disorder: A Brief Review." *International Journal of Alcohol and Drug Research* 3(1):91–103.

Timeline. 2001 (March 8). "Timeline of Tobacco Litigation." Fox News. Available at www.foxnews.com

U.S. Department of Justice. 2010. "The Impact of Drugs on Society." *National Drug Threat Assessment 2010*. Available at www.justice.gov

U.S. Surgeon General. 2012. *Preventing Tobacco Use among Youth and Young Adults*. Executive Summary. U.S. Department of Health and Human Services. Available at www.surgeongeneral.gov

U.S. Surgeon General. 2014. *2014 Surgeon General's Report: The Health Consequences of Smoking—50 Years of Progress*. U.S. Department of Health and Human Services. Available at www.cdc.gov

Van Dyck, C., and R. Byck. 1982. "Cocaine." *Scientific American* 246:128–141.

Volpp, K. G., A. B. Troxel, M. V. Pauly, H. A. Glick, A. Puig, D. A. Asch, R. Galvin, J. Zhu, F. Wan, J. DeGuzman, E. Corbett, J. Weiner, J. Audrain-McGovern. 2009. "A Randomized, Controlled Trial of Financial Incentives for Smoking Cessation." *New England Journal of Medicine* 360(7):699–709.

Wadley, Jared. 2012. "American Teens Are Less Likely Than European Teens to Use Cigarettes and Alcohol, but More Likely to Use Illicit Drugs." University of Michigan News Service, June 1. Available at www.ns.umich.edu

Walberg, Matthew. 2013. "Parents Sue NIU Fraternity, Members in Son's Hazing Death." *Chicago Tribune,* February 15. Available at articles.Chicagotribune.com

Ward, Clifford. 2015. "22 Former NIU Frat Members Guilty of Misdemeanors in Death of Pledge." *Chicago Tribune*, May 8. Available at www.chicagotribune

Wechsler, William, and Toben F. Nelson. 2008. "What We Have Learned from the Harvard School of Public Health College Alcohol Study: Focusing Attention on College Student Alcohol Consumption and the Environmental Conditions That Promote It." *Journal of Alcohol Studies* (July):1–9.

White House. 2014. *National Drug Control Strategy*. Available at www.whitehouse.gov

White House. 2015. Drug Control Funding Priorities in the FY 2016 President's Budget. Available at www.whitehouse.gov

Williams, Jenny, Frank J. Chaloupka, and Henry Wechsler. 2005. "Are There Differential Effects of Price and Policy on College Students' Drinking Intensity?" *Contemporary Economic Policy* 23(1):78–90.

Willing, Richard. 2014. "Lawsuits Target Alcohol Industry" *USA Today*, May 13. Available usatoday30.usatoday.com

Wiltz, Teresa. 2015. "You'll Never Believe Who's Abusing Heroin Most." *Fiscal Times*, February 4. Available at www.thefiscaltimes.com

Witters, Weldon, Peter Venturelli, and Glen Hanson. 1992. *Drugs and Society*, 3rd ed. Boston: Jones & Bartlett.

World Drug Report (WDR). 2013. *World Drug Report 2013*. United Nations Office on Drugs and Crime (UNODC). New York: United Nations.

World Drug Report (WDR). 2014. *World Drug Report 2014*. United Nations Office on Drugs and Crime (UNODC). New York: United Nations.

World Health Organization (WHO). 2008. *WHO Report on the Global Tobacco Epidemic, 2008: The MPOWER Package*. Geneva: World Health Organization.

World Health Organization (WHO). 2011a. *Global Status Report on Alcohol and Health 2011*. Geneva: World Health Organization. Available at www.who.int

World Health Organization (WHO). 2011b (February). "Tobacco." Geneva: World Health Organization. Available at www.who.int

World Health Organization (WHO). 2013a. *MPOWER in Action: Defeating the Global Tobacco Epidemic*. December 23. Available at www.who.int

World Health Organization (WHO). 2013b. "Tobacco." Geneva: World Health Organization.

World Health Organization (WHO). 2014a. *Global Status Report on Alcohol and Health 2011*. Geneva: World Health Organization. Available at www.who.int

World Health Organization (WHO). 2014b. "Raising Tax on Tobacco: What You Need to Know." May. Available at www.who.int

World Health Organization (WHO). 2014c. "Tobacco: Fact Sheet." May. Available at www.who.int

World Health Organization (WHO). 2015. "Alcohol: Fact Sheet." January. Available at www.who.int

Wu, Li-Tzy, George E. Woody, Chongming Yang, Jeng-Jong, and Dan G. Blazer. 2011. "Racial/Ethnic Variations in Substance-Related Disorders among Adolescents in the United States." *Archives of General Psychiatry* 68(11):1176–1185.

Wysong, Earl, Richard Aniskiewicz, and David Wright. 1994. "Truth and Dare: Tracking Drug Education to Graduation and as Symbolic Politics." *Social Problems* 41:448–468.

Zehe, Jennifer M., and Craig R. Colder. 2015. "A Latent Growth Curve Analysis of Alcohol-Use Specific Parenting and Adolescent Alcohol Use." *Addictive Behaviors* 39(12):1701.

Zickler, Patrick. 2003. "Study Demonstrates That Marijuana Smokers Experience Significant Withdrawal." *NIDA Notes* 17:7, 0.

Chapter 4

Afterschool Alliance. 2013. "After School Programs: Making a Difference in America's Communities by Improving Academic Achievement, Keeping Kids Safe and Healthy Working Families." Available at www.afterschoolalliance.org

Afterschool Alliance. 2014. "Programs Keep Kids Safe, Engage Kids in Learning and Help Working Families." Available at www.afterschoolalliance.org

Agnew, Robert. 2013. "When Criminal Coping Is Likely: An Extension of General Strain Theory." *Deviant Behavior* 34(8):653–670.

Albanese, Jay S. 2012. "Deciphering the Linkages between Organized Crime and Transnational Crime." *Journal of International Affairs* 66 (1): 1-11.

Alsup, David. 2015 (February 10). "Jesse Matthew Charge in Slaying of UVA Student Hannah Graham." CNN News. Available at www.cnn.com

American Civil Liberties Union (ACLU). 2011. *Chicago's Video Surveillance Cameras: A Pervasive and Unregulated Threat to Our Privacy.* February. A report from the ACLU of Illinois. Available at www.aclu-il.org

American Civil Liberties Union (ACLU). 2013. "The War on Marijuana in Black and White." June. Available at www.aclu.org

American Medical Association (AMA). 2015. "Code of Ethics." Section (Opinion) 2.06. Available at www.ama-org

Amnesty International. 2015. *Death Sentences and Executions 2014.* Available at amnestyusa.org

Anderson, David. 2012. "The Cost of Crime." *Foundations and Trends in Microeconomics* 7(3):209–265.

Anega, Abhat, John Donohue, and Alexandria Zhang. 2014. "The Impact of Right to Carry Laws and the NRC Report: The Latest Lessons for the Empirical Evaluation of Law and Policy." September 4. Stanford Law and Economics Olin Working Paper No. 461 Available at ss.com

Arnold, Tim. 2013. "The Real Weapons of Mass Destruction: America's 300 Million Guns." *Huffington Post,* May 17. Available at www. huffingtonpost.com

Ascani, Nathaniel. 2012. "Labeling Theory and the Effects of Sanctioning on a Staff Delinquent Peer Association: A New Approach to Sentencing Juveniles." *Perspectives* (Spring):80–84.

Austin, Andrew D. 2015 (February 5). "Discretionary Budget Authority by Subfunction: An Overview." Washington, DC: Congressional Research Service.

Barrett, Ted, and Tom Cohen. 2013 (April 18). "Senate Rejects Expanded Gun Background Checks." CNN. Available at www.cnn.com

Becker, Howard S. 1963. *Outsiders: Studies in the Sociology of Deviance.* New York: Free Press.

Bell, Kerryn E. 2009. "Gender and Gangs: A Quantitative Comparison." *Crime and Delinquency* 55(3):363–387.

Berman, Dennis. 2014. "The Recent History of States Scrambling to Keep Using Lethal Injections." *Washington Post,* February 19. Available at www.washingtonpost.com

Berlow, Alan, and Gordon Witkin. 2013 (May 1). "Gun Lobby's Money and Power Still Holds Sway over Congress." Available at www .publicintegrity.org

Bonn, Scott A. 2014. "Serial Killer Myth Number 3: They Are All Men." *Psychology Today,* August 18. Available at www.psychologytoday.com

Boyle, Louise. (April 27) 2015. "Cleveland 'House of Horrors' Survivors Detail Their Decade of Abuse and How Ariel Castro Raped Them Up To 5 Times A Day - The Only Time They Were Released From Their Chains." *Daily Mail.* Available at www.dailymail.co.uk

Burdine, Nikki and Meta Pettus. 2015. "Accused Hannah Graham Killer on Trial for 2005 Attack." *USA Today* (June 8). Available at www .usatoday.com

Bureau of Justice Statistics (BJS). 2013a (March 7). "Female Victims of Sexual Violence, 1994–2010." Available at www.bjs.gov

Bureau of Justice Statistics (BJS). 2013b (June 28). "NCVS Redesign." Available at www.bjs.gov

Bureau of Justice Statistics (BJS). 2013c (September 24). "Study Finds Some Racial Differences in Perceptions of Police Behavior during Contact with Police." Available at www.bjs.gov

Bureau of Justice Statistics (BJS). 2014a. "Data Collection: National Crime Victimization Survey." Available at www.bjs.gov

Bureau of Justice Statistics (BJS). 2014b *Probation and Parole in the United States, 2013.* U.S. Department of Justice. Available at www.bjs.gov

Bureau of Prisons. 2014 (March 18). "Annual Determination of Average Cost of Incarceration." Available at www.gpo.gov

Button, Mark, Chris Lewis, and Jacki Tapley. 2014. "Not a Victimless Crime: The Impact of Fraud on Individual Victims and Their Families." *Security Journal* 27(1):36–54.

Byrne, James, and Gary Marx. 2011. "Technological Innovations in Crime Preventions and Policing: A Review of the Research on Implementation and Impact." *Journal of Police Studies* 20(3):17–38.

Canadian Press. 2014. "More Than 60 Organizations and Agencies Call for Repeal of New Prostitution Law." *Huffington Post, Canada,* December 6. Available at www.huffingtonpost.ca

Carson, E. Ann. 2014. *Prisoners in 2013.* Bureau of Justice Statistics. September. Available at www .bjs.gov

Carter, Zach, and Lauren Berlin. 2012. "State of the Union: Pres. Obama's Financial Fraud Team Tied to Banks." *Huffington Post,* January 24. Available at www.huffingtonhpost.com

Center for Disease Control and Prevention. 2014. "Deaths, Percent of Total Deaths, and Death Rates for the 15 Leading Causes of Death in 5 Year Age Groups, by Race and Sex: United States, 1999–2013." Available at www.cdc.gov

Chapman University. 2014. *The Chapman Survey of American fears, Wave 1.* Orange, CA: Earl Babbie Research Center, Wilkinson College of Humanities and Social Sciences.

Chesney-Lind, Meda, and Randall G. Shelden. 2004. *Girls, Delinquency, and Juvenile Justice.* Belmont, CA: Wadsworth.

Cho, Seo-Young, Axel Dreher, and Eric Neumayer. 2013. "Does Legalized Prostitution Increase Human Trafficking?" *World Development* 41:67–82.

Civitas. 2012. "Comparisons of Crime in OECD Countries." *Crime Briefing,* January. Available at www.civitas.org.uk

Cohn, Scott. 2014 (March 5). "DOJ Official Says Groundwork Laid for More White-Collar Crime Enforcement." CNBC. Available at www.cnbc.com

Commercial Crime Services. 2015. "SE Asia Tanker Hijacks Rose in 2014 Despite Global Drop in Sea Piracy, IMB Report Reveals." January 12. Available at www.icc-ccs.org

Community Oriented Policing Services (COPS). 2014. "About." Available at www.cops.usdoj.gov

Consumer Reports. 2015. "Everything You Need to Know about the Takata Airbag Recall." July 16. Available at consumerreports.org

Conklin, John E. 2007. *Criminology,* 9th ed. Boston: Allyn and Bacon.

Cox, Robynn J. A. 2015. *Where Do We Go from Here? Mass Incarceration in the Struggle for Civil Rights.* January 16. Economic Policy Institute.

D'Alessio, David, and Lisa Stolzenberg. 2002. "A Multilevel Analysis of the Relationship between Labor Surplus and Pretrial Incarceration." *Social Problems* 49:178–193.

Daugherty, Scott. 2013. "Somali Pirates Receive Life Sentences from Federal Jury." *Virginian-Pilot,* August 3. Available at hamptonroads.com

Death Penalty Information Center (DPIC). 2014. *The Death Penalty in 2014: Year End Report.* Available at deathpenaltyinfo.org

Debies-Carl, Jeffrey S. 2013. "Are the Kids All Right? It Critique an Agenda for Taking Youth Cultures Seriously." *Social Science Information* 52(1):110–133.

Devaney, Tim. 2015. "Online Gambling, the Target of New Legislation." *The Hill,* February 6. Available at thehill.com

Diamond, Milton, Eva Jozifkova, and Petr Weiss. 2011. "Pornography and Sex Crimes in the Czech Republic." *Archives of Sexual Behavior* 40:1037–1043.

Dilanian, Ken. 2015 (June 29). "Union Sues Feds over Hack, Says Agency Had Ample Warning." Associated Press. Available at hosted.ap.org

Do the Right Thing (DTRT). 2015. "Our Mission." available at dotherightthinginc.org

Doherty, Carroll. 2015. "A Public Opinion Trend That Matters: Priorities for Gun Policy." January 9. Available at www.pewresearch.org

Dolak, Kevin. 2013 (September 17). "Green River Killer Claims He Murdered Dozens More Women." ABC News. Available at abcnews .go.com

Drakulich, Kevin M. 2013. "Strangers, Neighbors, and Race: A Content Model of Stereotypes and Racial Anxieties about Crime." *Race and Justice* 2(4):322–355.

Dugan, Andrew. 2014 (November 24). "In U.S., 37 Percent Do Not Feel Safe Walking at Night Near Home." Gallup Research. Available at www .gallup.poll

Eisen, Lauren-Brooke. 2013 (May 22). "Should Judges Consider the Cost of Sentences?" Brennan Center for Justice. New York University School of Law. Available at www.brennancenter.org

Eng, James. 2012 (April 20). "Judge Rules Race Tainted North Carolina Death Penalty Case; Inmate Marcus Robinson Spared from Death Row." NBC News. Available at usnews .nbcnews.com

Erikson, Kai T. 1966. *Wayward Puritans.* New York: Wiley.

Europol. 2015. "About Us." Available at www .europol.europa.eu

Fact Sheet. 2012 (April). "Major Federal Legislation Concerned with Child Protection, Child Welfare, and Adoption." Child Welfare Information Gateway. Available at www.childwelfare.gov

Federal Bureau of Investigation (FBI). 2014a. "Crime in the United States, 2013." *Annual*

Uniform Crime Report. Washington, DC: U.S. Government Printing Office.

Federal Bureau of Investigation (FBI). 2014b. "Data Tables." *2013 National Incident-Based Reporting System.* Available at www.fbi.gov

Federal Bureau of Investigation (FBI). 2014c. "Innocence Lost National Initiative." May 29. *Podcasts and Radio: FBI This Week.* Available at www.fbi.gov

Federal Bureau of Investigation (FBI). 2014d. *2013 National Incident-Based Reporting System.* Available at www.fbi.gov

Federal Trade Commission. 2015 (February). *Consumer Sentinel Network Data Book.* Available at www.ftc.gov

Fellner, Jamie. 2015. "Dispatches: Evermore US Prisoners Growing Old behind Bars." [Human Rights Watch] *Dispatches,* February 9. Available at www.hrw.org

Fight Crime: Invest in Kids. 2014. "Members in Action Brochure." Available at www.fightcrime.org

French Press Agency (Agence France-Presse). 2014. "Membership of Japan's Yakuza Crime Gangs Falls to All-Time Low." *The Guardian,* March 6. Available at www.theguardian.com

Gallup Poll. 2011 "Gallup's Pulse of Democracy: Crime." Historic Trends. Available at www.gallup.com

Gault-Sherman, Martha. 2012. "It's a Two-Way Street: The Bi-directional Relationship between Parenting and Delinquency." *Journal of Youth and Adolescence* 41(2):121–145.

Geier, Ben. 2014. "GM Starts 2015 with Three More Recalls." *Fortune,* January 2. Available at fortune.com

Glossip v. Gross. 2015. No. 14-7955.

Goad, Benjamin. 2014. "Obama Renews Gun Control Push." *The Hill,* March 4. Available at thehill.com

Gorzelany, Jim. 2014. "The Most Stolen New and Used Cars in America." *Forbes,* August 18. Available at www.forbes.com

Greenblatt, Alan. 2008. "Second Chance Programs Quietly Gain Acceptance." *Congressional Quarterly Weekly,* September 15. Available at www.cq.com

Greene, Richard Allen. 2012. "Norway Massacre Could Have Been Avoided, Report Finds." *CNN News,* August 13. Available at www.cnn.com

Greenwood, Chris. 2015. "Crime Victims Don't Trust the Police and Thousands of Offenses Are Being Unrecorded Because People Feel 'Nothing Will Be Done,' Says Federation Chairman." *Daily Mail,* January 5. Available at www.dailymail.co.uk

Hartney, Christopher, and Linh Vuong. 2009. *Created Equal: Racial and Ethnic Disparities in the U.S. Criminal Justice System.* Oakland, CA: National Council on Crime and Delinquency.

Heath, Brad. 2014. "Racial Gap in U.S. Arrests Rates: 'Staggering Disparity'." *USA Today,* November 19. Available at www.usatoday.com

Heckman, James J., Seong Hyeok, Rodrigo Pinto, Pater A. Savelyve, and Adam Yavitz. 2010. "The Rate of Return to the HighScope Perry Preschool Program." *Journal of Public Economics* 94:114–128.

Hefling, Kimberly. 2014 (December 11). "Justice Department: Majority of Campus Sexual Assault

Goes Unreported to Police." *PBS NewsHour.* Available at www.pbs.org

Heimer, Karen, Stacy Wittrock, and Halime Unal. 2005. "Economic Marginalization and the Gender Gap in Crime." In *Gender and Crime: Patterns of Victimization and Offending,* ed. Karen Heimer and Candace Kruttschnitt, 115–136. New York: New York University Press.

Herald Sun. 2008 (January 17). "Jessica's Victim Impact Statement." Available at www.news.com.au/heraldsun

Hetey, Rebecca C., and Jennifer L. Eberhardt. 2014. "Racial Disparities in Incarceration Increase Acceptance of Punitive Policies." *Psychological Science* 25(10):1949–1954.

Hilzenrath, David. 2011. "Goldman Sachs Subpoenaed." *Washington Post,* June 2. Available at www.washingtonpost.com

Hirschi, Travis. 1969. *Causes of Delinquency.* Berkeley: University of California Press.

Holtfreter, Kristy, Shanna Van Slyke, Jason Bratton, and Marc Gertz. 2008. "Public Perceptions of White-Collar Crime and Punishment." *Bureau of Criminal Justice* 36:50–60.

Human Rights Watch. 2014 (May 6). *Nation behind Bars: A Human Rights Solution.* Available at www.hrw.org

The Innocence Project. 2015 (June 15). "DNA Exonerations Nationwide." Available at www.innocenceproject.org

International Center for Prison Studies (ICPS). 2015. "Highest to Lowest Prison Population Rate." Available at www.prison studies.org

International Centre for the Prevention of Crime (ICPC). 2014. "Missions and Activities." Available at www.crime-prevention-intl.org/

International Labour Office (ILO). 2014. *Profits and Poverty: The Economics of Forced Labour.* Available at www.ilo.org

Internet Crime Complaint Center (ICCC). 2014a. "Internet Crime Schemes." Available at www.ic3.gov

Internet Crime Complaint Center (ICCC). 2014b. *2013 Internet Crime Report.* Federal Bureau of Investigation. Available at www.ic3.gov

Interpol. 2014. "About Interpol." Available at www.interpol.int

Isidore, Chris. 2014 (November 20). "Takata Airbag Victims Looked Like They Had Been Shot or Stabbed." CNN Money. Available at money.cnn.com

Johnson, Brian D., and Jacqueline G. Lee. 2013. "Racial Disparity under Sentencing Guidelines: A Survey of Recent Research and Emerging Perspectives." *Sociology Compass* 7:50 –514.

Jones, Jeffrey M. 2013a (April 5). "Minneapolis–St. Paul Area Residents Most Likely to Feel Safe." Available at www.gallup.com

Jones, Jeffrey M. 2013b (January 29). "Party Views Diverge Most on U.S. Gun Policies." Available at www.gallup.com

Jones, Jeffrey M. 2014 (October 23). "Americans' Support for Death Penalty Stable." Available at www.gallup.com

Kaba, Fatos, Pamela Diamond, Alpha Haque, Ross McDonald, and Homer Venters. 2014. "Traumatic Brain Injury among Newly Admitted Adolescents in the New York City Jail System." *Journal of Adolescent Health* 54(5):61–617.

Katz, Rebecca. 2012. "Environmental Pollution: Corporate Crime and Cancer Mortality." *Contemporary Justice Review: Issues in Criminal, Social, and Restorative Justice* 15(1):97–125.

Kearney, Melissa S., Benjamin H. Harris, Elisa Jacome, and Lucie Parker. (May) 2015. *Ten Economic Facts about Crime and Incarceration in the United States.* The Hamilton Project. Available at www.hamiltonproject.org

Kelly, Heather. 2014 (May 26). "Police Embracing Content That Predicts Crime." CNN News. Available at www.cnn.com

Keneally, Meghan. 2014 (October 24). "Remains of Missing UVA Student Hannah Graham Identified." ABC News. Available at abcnews.go.com

Kubrin, Charis E. 2005. "Gangsters, Thugs, and Hustlers: Identity and the Code of the Street in Rap Music." *Social Problems* 52(3):360–378.

Kubrin, Charis, and Ronald Weitzer. 2003. "Retaliatory Homicide: Concentrated Disadvantage and Neighborhood Culture." *Social Problems* 50:157–180.

Langton, Lynn, Marcus Berzofsky, Christopher Krebs, and Hope Smiley-McDonald. 2012 (August). *Victimization Is Not Reported to the Police, 2006–2010.* Capital NCJ 238536. Washington, DC: U.S. Department of Justice. Available at www.bjs.gov

Langton, Lynn, and Jennifer Truman. 2014. *Special Report: Social-Emotional Impact of Violent Crime.* September. NCJ247076. Washington, DC: U.S. Department of Justice, Office of Justice Programs. Available at www.bjs.gov

Laughland, Oliver, Jon Swaine, and Jamiles Lartey. 2015. "U.S. Police Killings Headed for 1,100 This Year, with Black Americans Twice as Likely to Die." *The Guardian,* July 1. Available at www.theguardian.com

Leigh, David, James Ball, Juliette Garside, and David Pegg. 2015. "HSBC Files Show How Swiss Bank Helped Clients Dodge Taxes and Hide Millions." *The Guardian,* February 8. Available at www.theguardian.com

Leverentz, Andrea. 2012. "Narratives of Crime and Criminals: How Places Socially Construct the Crime Problem." *Sociological Forum* 27(2):348–371.

Lichtblau, Eric, David Johnston, and Ron Nixon. 2008. "F.B.I. Struggles to Handle Financial Fraud Cases." *New York Times,* October 19. Available at www.nytimes.com

Lott, John R., Jr. 2003. "Guns Are an Effective Means of Self-Defense." In *Gun Control,* ed. Helen Cothran, 86–93. Farmington Hills, MI: Greenhaven Press.

MAD DADS. 2014. "Who Are Mad Dads?" Available at maddads.com

Malamuth, Neil M., Gert Martin Hald, and Mary Koss. 2012. "Pornography, Individual Differences in Risk of Men's Acceptance of Violence against Women in a Representative Sample." *Sex Roles* 66 (7/8):427–439.

Marks, Alexandria. 2006. "Prosecutions Drop for US White Collar Crime." *Christian Science Monitor,* August 31. Available at www.csmonitor.com

Martinez, Michael. 2014 (October 29). "Remains Found Last Week Are Those of UVA Student Hannah Graham." CNN News. Available at www.cnn.com

Masters, Jonathan. 2013 (July 15). "U.S. Gun Policy: Global Comparisons." Council on Foreign Relations. Available at www.cfr.org

Maynard, Micheline. 2010. "Toyota Cited $100 Million Savings After Limiting Recall." *New York Times,* February 21. Available at www.nytimes.com

McCoy, Kevin. 2014. "Ex-JPMorgan Lawyer Tells Tale of Wrongdoing." *USA Today,* November 8. Available at www.usatoday.com

McCurry, Justin. 2013. "Yakuza Magazine—for the Well-Read Gangster." *The Guardian,* July 11. Available at www.guardian.co.uk

Mellman Group. 2012. "Public Opinion on Sentencing and Corrections Policy in America." *Public Opinion Strategies,* March. Available at www.prisonpolicy.org

Merton, Robert. 1957. *Social Theory and Social Structure.* Glencoe, IL: Free Press.

Michaels, Daniel. 2013. "Arrests Made in Diamond Heist." *Wall Street Journal,* May 8. Available at www.wsj.com

Mokhiber, Russell. 2007 (June 15). "Twenty Things You Should Know about Corporate Crime." Alternet. Available at www.alternet.org

Morin, Rich. 2015 (July 15). "The Demographics of Politics of Gun-Owning Households." July 15. Available at www.pewresearch.org

Mouilso, Emily R., and Karen S. Calhoun. 2013. "The Role of Rape Myth Acceptance and Psychopathy in Sexual Assault Perpetration." *Journal of Aggression, Maltreatment and Trauma* 22(2):159–174.

Murphy, Shelley, and Milton J. Valencia. 2013. "Confessed Murderer Ties Bulger to 6 Killings." *Boston Globe,* June 17. Available at www.bostonglobe.com

National Association of State Budget Officers. 2013 (September 11). "State Spending for Corrections: Long-Term Trends and Recent Criminal Justice Policy Reforms." Available at www.nasbo.org

National Association of Town Watch (NATW). 2014. "About Us." Available at www.natw.org

National Center on Missing and Exploited Children (NCEMC). 2014. *National Center for Missing and Exploited Children, Annual Report 2013.* Available at www.missingkids.com

National Center on Missing and Exploited Children (NCEMC). 2015. "Child Sex Trafficking: Infographic." Available at www.missingkids.com

National Gang Report (NGR). 2014. *National Gang Report 2013.* FBI: National Gang Intelligence Center. Available at www.fbi.gov

National Research Council. 1994. *Violence in Urban America: Mobilizing a Response.* Washington, DC: National Academy Press.

National White Collar Crime Center (NWCCC). 2010. "National Public Survey on White Collar Crime." Available at crimesurvey.nw3c.org

NBC News. 2014a (December 16). "Death Toll from GM Ignition Switches Rises to 42." NBC News. Available www.nbcnews.com

NBC News. 2014b (December 8). "Two More Deaths Linked to Faulty GM Ignition Switch." NBC News. Available www.nbcnews.com

National Crime Prevention Council. 2005. "Preventing Crime Saves Money." Available at www.ncpc.org

Newport, Frank. 2014 (August 20). "A Gallup Review: Black-and-White Attitudes towards the Police." Gallup. Available at www.gallup.com

O'Toole, James. 2013 (June 13). "Smith and Wesson Will Record Sales as Gun Debate Raged." CNN Money. Available at money.cnn.com

Ouimet, Marc. 2012. "The Effect of Economic Development, Income Inequality, and Excess Infant Mortality on the Homicide Rate for 165 Countries in 2010." *Homicide Studies* 16(3):238–258.

Patterson, Thom. 2014 (May 26). "Data Surveillance Centers: Crime Fighters or 'Spy Machines'?" CNN News. Available at www.cnn.com

Pearson, Jake. 2014. "Nearly Half of All Jailed Youths in New York City Have Brain Injury." *Huffington Post,* June 18. Available at www.huffingtonpost.com

Pertossi, Mayra. 2000 (September 27). "Analysis: Argentine Crime Rate Soars." Available at news.excite.com

Pew. (June 6) 2012. "Time Served: The High Cost, Low Return of Longer Prison Terms." Available at pewstates.org

Pew. 2013 (January 14). "In Gun Control Debate, Several Options Draw Majority Support." Available at www.people-press.org

Pew. 2014. (September 11). "Prison and Crime: A Complex Link." Pew Charitable Trust. Available at pewtrusts.org

Pickett, Justin T., Ted Chiricos, Kristin M. Golden, and Marc Gertz. 2012. "Reconsidering the Relationship between Perceived Neighborhood Racial Composition in Whites' Perceptions of Victimization Risk: Do Racial Stereotypes Matter?" *Criminology* 50(1):14–186.

Polaris Project. 2013. "Sex Trafficking in the U.S." Available at www.polarisproject.org

Ponemon Institute. 2014. *2014 Global Report on the Cost of Cyber Crime.* Hewlett Packard. Available at ssl.www8.hp.com

Prichard, Jeremy, Caroline Spiranovic, Paul Watters, and Christopher Lueg. 2013. "Young People, Child Pornography, and Subcultural Norms on the Internet." *Journal of the American Society for Information Science and Technology* 65(5):992–1000.

Pridemore, William Alex, and Sang-Weon Kim. 2007. "Socioeconomic Change and Homicide in a Transitional Society." *Sociological Quarterly* 48:229–251.

Reiman, Jeffrey, and Paul Leighton. 2013. *Rich Get Richer and the Poor Get Prison.* Boston: Pearson.

Resource Center (Racial Profiling Data Collection Resource Center). 2013. "Legislation and Litigation." Northeastern University. Available at racialprofilinganalysis.neu.edu

Richinick, Michelle. 2014 (July 1). "Georgia Guns Everywhere, Bill Takes Effect." *MSNBC News.* Available at www.msnbc.com

Riffkin, Rebecca. 2014 (October 27). "Hacking Tops List of Crimes Americans Worry about the Most." Gallup. Available at www.gallup.com

Roeder, Oliver, Lauren-Brooke Eisen, and Julia Bowling. 2015. *What Caused the Crime Decline?* New York University School of Law, Brennan Center for Justice. Available at www.brennancenter.org

Roche. 2015 (January 16). "Roche Statement." Available at deathpenaltyinfo.org

Rossi, Rosalind. 2015. "Serious Crime on CTA Falls to 4- Year Low, Emmanuel Says." *Chicago Sun-Times,* February 17. Available at chicago.suntimes.com

Rubin, Paul H. 2002. "The Death Penalty and Deterrence." *Forum* (Winter):10–12.

Sampson, Robert J., Jeffrey D. Morenoff, and Stephen W. Raudenbush. 2005. "Social Anatomy of Racial and Ethnic Disparities in Violence." *American Journal of Public Health* 95(2):224–232.

Sanders, Sam. 2015 (January 8). "Honda Fined 70 Million for Underreporting Deaths and Injuries." NPR News. Available at www.npr.org

Sanger, David E., and Nicole Perlroth. 2015. "Bank Hackers Steal Millions via Malware." *New York Times,* February 14. Available at nytimes.com

Schelzig, Erik. 2007 (September 19). "Court Ruling Halts Tennessee Executions." Available at www.wral.com

Schiesel, Seth. 2011. "Supreme Court Has Ruled: Now Games Have a Duty." *New York Times,* June 28. Available at www.nytimes.com

Schweinhart, Lawrence J. 2007. "Crime Prevention by the High/Scope Perry Preschool Program." *Victims and Offenders* 2:141–160.

Seelye, Katherine Q. 2013. "Bulger Guilty of Gangland Crimes, Including Murder." *New York Times,* August 12. Available at www.nytimes.com

The Sentencing Project. 2013 (August). *Report of The Sentencing Project to the United Nations Human Rights Committee Regarding Racial Disparities in the United States Criminal Justice System.* Available at sentencing project.org

Shapiro, Robert J., and Kevin Hassett. 2012 (June). *The Economic Benefits of Reducing Violent Crime.* Center for American Progress. Available at cdn.americanprogress.org

Sheldon, Randall. 2013 (February 12). "Part I: Trends in Girls' Crime." Center on Juvenile and Criminal Justice. Available at wwwcjcj.org

Shelley, Louise. 2007. "Terrorism, Transnational Crime and Corruption Center." American University. Available at traccc.gmu.edu/

Sherman, Lawrence. 2003. "Reasons for Emotions." *Criminology* 42:1–37.

Siegel, Larry. 2006. C*riminology,* 9th ed. Belmont, CA: Wadsworth.

Smyth, Julie. 2012. "Dual Punishment: Incarcerated Mothers and Their Children." *Columbia Social Work Review* 3:33–45.

Statista. 2015. "Average Cost of Cyber Crime in Selected Countries as of June 2014 (in Million U.S. Dollars)." Available at www.statista.com

Stuart, Keith. 2015. "Hatred: Gaming's Most Contrived Controversy." *The Guardian,* May 29 Available at theguardian.com

Sutherland, Edwin H. 1939. *Criminology.* Philadelphia: Lippincott.

Sweeten, Gary, Alex R. Piquero, and Laurence Steinberg. 2013. "Age and the Explanation of Crime, Revisited." *Journal of Youth and Adolescence* 42:921–938.

Tabuchi, Hiroko. 2014. "Takata Saw and Hid Risk in Airbags 2004, Former Workers Say." *New York Times,* November 7. Available at www.nytimes.com

Thomas, Pierre, Jack Date, Jason Ryan, and Jack Cloherty. 2013 (January 25). "Hot Guns Fueling Crime, US Study Says." ABC News. Available at abcnews.go.com

Thompson, Mark, and Evan Perez. 2015 (June 30). "BNP Paribas to Pay Nearly $9 Billion Penalty." CNN Money. Available at money.cnn.com

Tides Center. 2015 (March). "The Opportunity Survey: Criminal Justice Findings." Prison Policy Initiative. Available at prisonpolicy.org

Tobin, Mike. 2015 (May 12). "Security Camera Surge in Chicago Sparks Concerns of 'Mass Surveillance System.'" Fox News. Available at www.foxnews.com

Travis, Jeremy, Bruce Western, and Steve Redburn (Eds.). 2014. *The Growth of Incarceration in the United States: Exploring Causes and Consequences.* Committee on Law and Justice, Division of Behavioral and Social Sciences and Education. National Research Council. Washington, DC: National Academies Press.

Truman, Jennifer, and Lynn Langton. 2014 (September 19). *Criminal Victimization, 2013.* U.S. Department of Justice, Bureau of Justice Statistics. Available at www.bjs.gov

Turney, Kristin. 2014. "The Consequences of Paternal Incarceration for Maternal Neglect and Harsh Parenting." *Social Forces* 92(4):160–1636.

Turney, Kristin, and Anna R. Haskins. 2014. "Falling Behind? Children and Early Grade Retention after Paternal Incarceration." *Sociology of Education* 87(4):241–258.

Turney, Kristin, and Christopher Wildman. 2013 (April). "Redefining Relationships: Explaining the Countervailing Consequences of Paternal Incarceration for Parenting." Working Paper Series. National Center for Family and Marriage Research, Bowling Green State University. Available at www.firelands.bgsu.edu

United Nations (UN). 2014 (December 14). "Adopting Resolution, Security Council Urges Fight against Nexus of Transnational Crime, Terrorism." UN News Centre. Available at www.un.org

United Nations Office on Drug and Crimes (UNODC). 2012. "Hidden Cost to Society." In *Transnational Organized Crime: The Globalized Illegal Economy.* Available at www.unodc.org

United Nations Office on Drug and Crimes (UNODC). 2014a. *Global Study on Homicide, 2013.* Available at www.unodc.org

United Nations Office on Drug and Crimes (UNODC). 2014b (April 12–16). *World Crime Trends and Emerging Issues and Responses in the Field of Crime Prevention and Criminal Justice.* Available at www.unodc.org

U.S. Congress. 2015–2016a. "Bill Summary and Status." Available at thomas.loc.gov

U.S. Congress. (2015–2016b). *National Criminal Justice Commission Act of 2015.* 114th Congress, S. 1119. Available at www.congress.gov

U.S. Department of Commerce. 2011. *Service Annual Survey: 2009.* Washington, D.C. Government Printing Office.

U.S. Department of Justice. 2008. "Serial Murder: Multi-Disciplinary Perspectives for Investigators." Washington, DC: Behavioral Analysis Unit, National Center for the Analysis of Violent Crime.

U.S. Department of Justice. 2012a. "The Prostitution of Children in the United States." *Child Exploitation and Obscenity Section.* Available at www.justice.gov

U.S. Department of Justice. 2012b. "Transnational Organized Crime." Available at justice.gov

U.S. Department of Justice. 2014 (April 11). "The Attorney General's 'Smart on Crime' Initiative." Available at www.justice.gov

U.S. Department of State. 2014. *Trafficking in Persons Report.* June. Available at www.state.gov

Van Slyke, Shanna, and William D. Bales. 2012. "A Contemporary Study of the Decision to Incarcerate White-Collar and Street Property Offenders." *Punishment and Society* 14(2):217–246.

Verizon. 2014 (April). *2014 Data Breach Investigations Report.* Available at rp-Verizon-DBIR-2014X

Victim Statements. 2009. *U.S. v. Bernard L. Madoff.* 2009. U.S. Department of Justice. Available at www.pbs.org

Violence Policy Center. 2015 (January 29). "States with Weak Gun Laws and Higher Gun Ownership Lead Nation in Gun Deaths, New Data for 2013 Confirms." Available at www.vps.org

Wagner, Peter, and Leah Sakala. 2014 (March 12). "Mass Incarceration: The Whole Pie." Prison Policy Initiative. Available at www.prisonpolicy.org

Walmsley, Roy. 2012. *World Prison Population List,* 9th ed. London: International Centre for Prison Studies. Available at www.prisonstudies.org

Welsh-Huggins, Andrew. 2013. "Hundreds of New Charges Filed in U.S. Kidnap Case." *Associated Press,* July 13. Available at apnews.com

Wildeman, Christopher, Jason Schnittker, and Kristen Turney. 2014. "Despair by Association? The Mental Health of Mothers with Children by Recently Incarcerated Fathers." *American Sociological Review* 77(2):216–243.

Williams, Linda. 1984. "The Classic Rape: When Do Victims Report?" *Social Problems* 31:459–467.

Wilson, Reid. 2014. "Prison in These 17 States Are over Capacity." *Washington Post,* September 20. Available at www.washington.com

Winslow, Robert W., and Sheldon Zhang. 2008. *Criminology: A Global Perspective.* Englewood Cliffs, NJ: Prentice Hall.

Yost, Pete. 2013. "Nearly 2 in 3 Hate Crimes Unreported, Study Finds." *Washington Times,* March 21. Available at www.washingtontimes.com

Young, Angelo. 2015. "Airbag Recall List Grows: 2015 Has Already Seen Safety Recalls Covering Millions of Vehicles." *International Business Times,* February 9. Available at www.ibtimes.com

Young, Jeffrey. 2013. "Gun Violence Costs U.S. Healthcare System, Taxpayers Billions Each Year." *Huffington Post,* March 28. Available at www.huffingtonpost.com

Zahn, Margaret A., Robert Agnew, Diana Fishbein, Shari Miller, Donna-Marie Winn, Gayle Dakoff, Candace Kruttschnitt, Peggy Giordano, Denise C. Gottfredson, Allison A. Payne, Barry C. Feld, and Meda Chesney-Lind. 2010 (April). "Girls Study Group: Causes and Correlates of Girls' Delinquency." U.S. Department of Justice. Office of Juvenile Justice and Delinquency Prevention. Available at www.ncjrs.gov

Chapter 5

Adams, Rebecca. 2014. "Here's How People Really Feel about a Woman Taking Her Husband's Last Name." Huffington Post, October 20. Available at www.huffingtonpost.com

Ahrons, C. 2004. *We're Still Family: What Grown Children Have to Say about Their Parents' Divorce.* New York: HarperCollins.

Allendorf, Keera. 2013. "Schemas of Marital Change: From Arranged Marriages to Eloping for Love." *Journal of Marriage and Family* (April): 453–464.

Amato, Paul. 2003. "The Consequences of Divorce for Adults and Children." In *Family in Transition,* 12th ed., ed. Arlene S. Skolnick and Jerome H. Skolnick, 190–213. Boston: Allyn and Bacon.

Amato, Paul. 2004. "Tension between Institutional and Individual Views of Marriage." *Journal of Marriage and Family* 66:959–965.

Amato, P. R., A. Booth, D. R. Johnson, and S. J. Rogers. 2007. *Alone Together: How Marriage in America Is Changing.* Cambridge MA: Harvard University Press.

Amato, P. R., and J. Cheadle. 2005. "The Long Reach of Divorce: Divorce and Child Well-Being across Three Generations." *Journal of Marriage and the Family* 67:191–206.

American Humane Association. 2010 (April 20). "Orange County Animal Services and Harbor House Create First Pets and Women's Shelter (PAWS) Program in Central Florida." News Release. Available at www.americanhumane.org

Amundsen, Amy J., and Mike Kelly. 2014. Trends in Alimony Modifications. Chicago: American Bar Association. Available at www.americanbar.org

Anderson, Kristin L. 2013. "Why Do We Fail to Ask 'Why' about Gender and Intimate Partner Violence?" *Journal of Marriage and Family* 75(April):314–318.

Applewhite, Ashton. 2003. "Covenant Marriage Would Not Benefit the Family." In *The Family: Opposing Viewpoints,* ed. Auriana Ojeda, 189–195. Farmington Hill, MI: Greenhaven Press.

Ascione, F. R. 2007. "Emerging Research on Animal Abuse as a Risk Factor for Intimate Partner Violence." In *Intimate Partner Violence,* ed. K. Kendall-Tackett and S. Giacomoni, 3.1–3.17. Kingston, NJ: Civic Research Institute.

Baker, Amy J. L. 2006. "The Power of Stories/Stories about Power: Why Therapists and Clients Should Read Stories about Parental Alienation Syndrome." *American Journal of Family Therapy* 34:191–203.

Baker, Amy J. L. 2007. *Adult Children of Parental Alienation Syndrome: Breaking the Ties That Bind.* New York: Norton.

Baker, Amy J. L., and Jaclyn Chambers. 2011. "Adult Recall of Childhood Exposure to Parental Conflict: Unpacking the Black Box of Parental Alienation." *Journal of Divorce & Remarriage* 52(1):55–76.

Beckmeyer, Jonathon J., Marilyn Coleman, and Lawrence H. Ganong. 2014. "Postdivorce Coparenting Typologies and Children's Adjustment." Family Relations 63(October):526–537.

Bernet, William, and Amy J. L. Baker. 2013. "Parental Alienation, DSM-5, and ICD-11: Response to Critics." *Journal of the American Academy of Psychiatric Law* 41(1):98–104.

Brown, Susan L., and I-Fen Lin. 2014 (October 8). "Gray Divorce: A Growing Risk Regardless of Class or Education." Council on Contemporary Families. Available at www.contemporaryfamilies.org

Bureau of Justice Statistics. 2011. "Intimate Partner Violence in the U.S.: Victim Characteristics." Available at bjs.ofp.usdoj.gov

Bureau of Labor Statistics. 2015a. *Employment Characteristics of Families: 2014.* Available at www.bls.gov

Bureau of Labor Statistics. 2015b. "Wives Who Earn More Than Their Husbands, 1987–2013." Available at www.bls.gov

Bureau of Labor Statistics. 2015c. "American Time Use Survey—2014 Results." Available at www.bls.gov

Bureau of Labor Statistics. 2010 (May 7). "Labor Force Participation among Mothers." Available at www.bls.gov

Carrington, Victoria. 2002. *New Times: New Families*. Dordrecht, The Netherlands: Kluwer Academic.

Carter, Lucy S. 2010. *Batterer Intervention: Doing the Work and Measuring the Progress*. Family Violence and Prevention Fund. Available at www.endabuse.org

Catalano, Shannon. 2012. *Intimate Partner Violence, 1993–2010*. Bureau of Justice Statistics. Available at www.ojp.usdoj.gov

Centers for Disease Control and Prevention. 2014. *Understanding Child Maltreatment*. Fact Sheet. Available at www.cdc.gov

Cherlin, Andrew J. 2009. *The Marriage-Go-Round: The State of Marriage and Family in America Today*. New York: Knopf.

Child Trends. 2013. Attitudes Towards Spanking. Available at www.childtrendsdatabank.org

Child Trends. 2014. World Family Map 2014. Available at www.worldfamilymap.org

Cohn, Jonathan. 2014 (September 14). "Five Things We Can Do to Reduce Domestic Violence." New Republic. Available at www.newrepublic.com

Coontz, Stephanie. 2000. "Marriage: Then and Now." *Phi Kappa Phi Journal* 80:10–15.

Coontz, Stephanie. 2004. "The World Historical Transformation of Marriage." *Journal of Marriage and Family* 66(4):974–979.

Coontz, Stephanie. 2005. *Marriage, a History*. New York: Penguin Books.

Cross, Terry L. 2014. "Child Welfare in Indian Country: A Story of Painful Removals." Health Affairs 33(12):2256–2259.

Cui, M., K. Ueno, M. Gordon, and F. D. Fincham. 2013. "The Continuation of Intimate Partner Violence from Adolescence to Young Adulthood." *Journal of Marriage and Family* 75(April):300–313.

Daniel, Elycia. 2005. "Sexual Abuse of Males." In *Sexual Assault: The Victims, the Perpetrators, and the Criminal Justice System*, ed. Frances P. Reddington and Betsy Wright Kreisel, 133–140. Durham, NC: Carolina Academic Press.

Davis, John W. 2014 (July 15). "Rosemary Pate Remembered as Lawmaker Works on 'Parent Abuse' Bill." News 13. Available at www.mynews.com

Decuzzi, A., D. Knox, and M. Zusman. 2004. "The Effect of Parental Divorce on Relationships with Parents and Romantic Partners of College Students." Roundtable Discussion, Southern Sociological Society, Atlanta, April 17.

DeMaris, Alfred, Laura A. Sanchez, and Kristi Krivickas. 2012. "Developmental Patterns in Marital Satisfaction: Another Look at Covenant Marriage." *Journal of Marriage and Family* 74(October):989–1004.

Demo, David H., Mark A. Fine, and Lawrence H. Ganong. 2000. "Divorce as a Family Stressor." In *Families and Change: Coping with Stressful Events and Transitions*, 2nd ed., ed. P. C. McKenry and S. J. Price, 279–302. Thousand Oaks, CA: Sage.

Dennison, R. P., and S. Koerner. 2008. "A Look at Hopes and Worries about Marriage: The Views of Adolescents Following a Parental Divorce." *Journal of Divorce & Remarriage* 48:91–107.

Dugan, Andrew. 2015 (June 19). "Men, Women Differ on Morals of Sex, Relationships." Gallup Organization. Available at www.gallup.org

Dyson, Michael Eric. 2014. "Punishment? Or Child Abuse?" New York Times, September 18, p. A33.

Edin, Kathryn. 2000. "What Do Low-Income Single Mothers Say about Marriage?" *Social Problems* 47(1):112–133.

Ellis, Renee R., and Tavia Simmons. 2014. "Coresident Grandparents and Their Grandchildren: 2012." U.S. Census Bureau. Available at www.census.gov

Emery, Robert E. 2014. "How Divorced Parents Lost Their Rights." New York Times, September 6. Available at www.nytimes.com

Emery, Robert E., David Sbarra, and Tara Grover. 2005. "Divorce Mediation: Research and Reflections." *Family Court Review* 43(1):22–37.

Fincham, F., M. Cui, M. Gordon, and K. Ueno. 2013. "What Comes before Why: Specifying the Phenomenon of Intimate Partner Violence." *Journal of Marriage and Family* 75(April):319–324.

Finkelhor, D., R. Ormrod, H. A. Turner, and S. L. Hamby. 2005. "The Victimization of Children and Youth: A Comprehensive National Survey." *Child Maltreatment* 10:5–25.

Fogle, Jean M. 2003. "Domestic Violence Hurts Dogs, Too." *Dog Fancy*, April, p. 12.

Follingstad, D. R., and M. Edmundson. 2010. "Is Psychological Abuse Reciprocal in Intimate Relationships? Data from a National Sample of American Adults." *Journal of Family Violence* 25:495–508.

Foubert, J. D., E. E. Godin, and J. L. Tatum. 2010. "In Their Own Words: Sophomore College Men Describe Attitude and Behavior Changes Resulting from a Rape Prevention Program 2 Years after Their Participation." *Journal of Interpersonal Violence* 25:2237–2257.

Fowler, K. A., and D. Westen. 2011. "Subtyping Male Perpetrators of Intimate Partner Violence." *Journal of Interpersonal Violence* 26(4):607–639.

Gadalla, Tahany M. 2009. "Impact of Marital Dissolution on Men's and Women's Income: A Longitudinal Study." *Journal of Divorce & Remarriage* 50(1):55–65.

Gartrell, Nanette K., Henny M. W. Bos, and Naomi G. Goldberg. 2010. "Adolescents of the U.S. National Longitudinal Lesbian Family Study: Sexual Orientation, Sexual Behavior, and Sexual Risk Exposure." *Archives of Sexual Behavior*, online November 6.

Generations United. 2014. The State of Grandfamilies in America, 2014. Available at www.gu.org

Global Initiative to End All Corporal Punishment of Children. 2015. Countdown to Universal Prohibition. Available at www.endcorporalpunishment.org

Greenwood, Joleen Loucks. 2014. "Effects of a Mid- to Late-Life Parental Divorce on Adult Children." Journal of Divorce & Remarriage 55(7): 539–556.

Gromoske, Andrea N., and Kathryn Maguire-Jack. 2012. "Transactional and Cascading Relations between Early Spanking and Children's Social-Emotional Development." *Journal of Marriage and Family* 74:1054–1068.

Gustafsson, Hanna C., and Martha J. Cox. 2012. "Relations among Intimate Partner Violence, Maternal Depressive Symptoms, and Maternal Parenting Behaviors." *Journal of Marriage and Family* 74(October):1005–1020.

Halligan, C., D. Knox, and J. Brinkley. 2013 (February 22). "TRAPPED: Technology as a Barrier to Leaving an Abusive Relationship." Poster, Southeastern Council on Family Relations, Birmingham, Alabama.

Halpern-Meekin, Sarah, Wendy D. Manning, Peggy C. Giordana, and Monica A. Longmore. 2013. "Relationship Churning, Physical Violence, and Verbal Abuse in Young Adult Relationships." *Journal of Marriage and Family* (February):2–12.

Hamilton, Brady E., Joyce A. Martin, Michelle J. K. Osterman, and Sally C. Curtin. 2015. "Births: Preliminary Data for 2014." *National Vital Statistics Reports* 64(6). Available at www.cdc.gov/nchs

Hardesty, J. L., L. Khaw, M. D. Ridgway, C. Weber, and T. Miles. 2013 (May 13). "Coercive Control and Abused Women's Decisions about Their Pets When Seeking Shelter." *Journal of Interpersonal Violence* (May 13):1–24.

Hawkins, Alan J., and Betsy VanDenBerghe. 2014. Facilitating Forever. The National Marriage Project. Available at www.nationalmarriageproject.org

Hewlett, Sylvia Ann, and Cornel West. 1998. *The War against Parents: What We Can Do for Beleaguered Moms and Dads*. Boston: Houghton Mifflin.

Hochschild, Arlie Russell. 1989. *The Second Shift: Working Parents and the Revolution at Home*. New York: Viking.

Hymowitz, Kay, Jason S. Carroll, W. Bradford Wilcox, and Kelleen Kaye. 2013. *Knot Yet: The Benefits and Costs of Delayed Marriage in America*. National Marriage Project, National Campaign to Prevent Teen and Unplanned Pregnancy, and the Relate Institute. Available at nationalmarriageproject.org

Jackson, Shelly, Lynette Feder, David R. Forde, Robert C. Davis, Christopher D. Maxwell, and Bruce G. Taylor. 2003 (June). *Batterer Intervention Programs: Where Do We Go from Here?* U.S. Department of Justice. Available at www.usdoj.gov

Jalovaara, M. 2003. "The Joint Effects of Marriage Partners' Socioeconomic Positions on the Risk of Divorce." *Demography* 40:67–81.

James, Susan Donaldson. 2014 (September 12). "'Shaken Baby Syndrome': Those Who Survive are Horribly Afflicted." NBC News. Available at www.nbcnews.com

Jasinski, J. L., L. M. Williams, and J. Siegel. 2000. "Childhood Physical and Sexual Abuse as Risk Factors for Heavy Drinking among African-American Women: A Prospective Study." *Child Abuse and Neglect* 24:1061–1071.

Jekielek, Susan M. 1998. "Parental Conflict, Marital Disruption, and Children's Emotional Well-Being." *Social Forces* 76:905–935.

Johnson, Michael P. 2001. "Patriarchal Terrorism and Common Couple Violence: Two Forms of Violence against Women." In *Men and Masculinity: A Text Reader*, ed. T. F. Cohen, 248–260. Belmont, CA: Wadsworth.

Johnson, Michael P., and Kathleen Ferraro. 2003. "Research on Domestic Violence in the 1990s: Making Distinctions." In *Family in Transition*,

12th ed., ed. A. S. Skolnick and J. H. Skolnick493–514. Boston: Allyn and Bacon.

Jones, Jeffrey. 2015 (May 28). "Approval of Out-of-Wedlock Births Growing in U.S." Gallup Organization. Available at www.gallup.com

Kalmijn, Matthijs, and Christiaan W. S. Monden. 2006. "Are the Negative Effects of Divorce on Well-Being Dependent on Marital Quality?" *Journal of Marriage and the Family* 68:1197–1213.

Kamp Dush, Claire M. 2013. "Marital and Cohabitation Dissolution and Parental Depressive Symptoms in Fragile Families." *Journal of Marriage and Family* 75(February):91–109.

Kaufman, Joan, and Edward Zigler. 1992. "The Prevention of Child Maltreatment: Programming, Research, and Policy." In *Prevention of Child Maltreatment: Developmental and Ecological Perspectives*, ed. Diane J. Willis, E. Wayne Holden, and Mindy Rosenberg, 269–295. New York: Wiley.

Khan, Roxanne, and Paul Rogers. 2015. "The Normalization of Sibling Violence: Does Gender and Personal Experience of Violence Influence Perceptions of Physical Assault against Siblings?" Journal of Interpersonal Violence 30(3):437–458.

Kitzmann, K. M., N. K. Gaylord, A. R. Holt, and E. D. Kenny. 2003. "Child Witnesses to Domestic Violence: A Meta-Analytic Review." *Journal of Clinical and Consulting Psychology* 71:339–352.

Klesse, Christian. 2014 (January). "Polyamory: Intimate Practice, Identity, or Sexual Orientation?" Sexualities 17(1–2):81–99.

Knox, David (with Kermit Leggett). 1998. *The Divorced Dad's Survival Book: How to Stay Connected with Your Kids*. New York: Insight Books.

Koch, Wendy. 2009. "Fees Cut Down Private Adoptions." *USA Today*, April 27, p. 1A.

Lacey, K.K., D. G. Saunders, and L. Zhang. 2011. "A Comparison of Women of Color and Non-Hispanic White Women on Factors Related to Leaving a Violent Relationship." *Journal of Interpersonal Violence* 26:1036–1055.

LaFraniere, Sharon. 2005. "Entrenched Epidemic: Wife-Beatings in Africa." *New York Times*, August 11, pp. A1, A8.

Lara, Adair. 2005. "One for the Price of Two: Some Couples Find Their Marriages Thrive When They Share Separate Quarters." *San Francisco Chronicle*, June 29. Available at www.sfgate.com

Levin, Irene. 2004. "Living Apart Together: A New Family Form." *Current Sociology* 52(2):223–240.

Levtov, R., N. van der Gaag, M. Greene, M. Kaufman, and G. Barker. 2015. State of the World's Fathers: Executive Summary: A MenCare Advocacy Publication. Available at www.sowf.men-care.org

Lewin, Tamar. 2000. "Fears for Children's Well-Being Complicates a Debate over Marriage." *New York Times*, November 4. Available at www.nytimes.com

Livingston, Gretchen. 2014 (December 22). "Less Than Half of U.S. Kids Today Live in a 'Traditional' Family." Pew Research Center. Available at www.pewresearch.org

Lloyd, Sally A. 2000. "Intimate Violence: Paradoxes of Romance, Conflict, and Control." *National Forum* 80(4):19–22.

Lloyd, Sally A., and Beth C. Emery. 2000. *The Dark Side of Courtship: Physical and Sexual Aggression*. Thousand Oaks, CA: Sage.

Mason, Mary Ann, Arlene Skolnick, and Stephen D. Sugarman. 2003. "Introduction." In *All Our Families*, 2nd ed., ed. Mary Ann Mason, Arlene Skolnick, and Stephen D. Sugarman, 1–13. New York: Oxford University Press.

Mathews, T. J., and Brady Hamilton. 2014 (May). "First Births to Older Women Continue to Rise." NCHS Data Brief #152. National Center for Health Statistics. Available at www.cdc.gov

Mohr, Holbrook, and Garance Burke. 2014. "786 Abused Kids Died in Plain View of Authorities: Report." Huffington Post, December 18. Available at www.huffingtonpost.com

Morgan, Rachel E., and Jennifer L. Truman. 2014 (April 17). "Nonfatal Domestic Violence, 2003–2012." Bureau of Justice Statistics. Available at www.bjs.gov

National Center for Injury Prevention and Control. 2014a. Child Maltreatment: Facts at a Glance. Available at www.cdc.gov

National Center for Injury Prevention and Control. 2014b. *Understanding Intimate Partner Violence*. Available at www.cdc.gov

National Conference of State Legislators. 2014a (October 16). Domestic Violence Legislation 2013–2014. Available at www.ncsl.org

National Conference of State Legislators. 2014b (September 23). "Mandatory Reporting of Child Abuse and Neglect 2013 Introduced State Legislation." Available at www.ncsl.org

National Marriage Project and the Institute for American Values. 2012. *The State of Our Unions: Marriage in America 2012*. Available at www.stateofourunions.org

National Network to End Domestic Violence. 2014. Domestic Violence Counts 2013. Available at www.nnedv.org

Nelson, B. S., and K. S. Wampler. 2000. "Systemic Effects of Trauma in Clinic Couples: An Exploratory Study of Secondary Trauma Resulting from Childhood Abuse." *Journal of Marriage and Family Counseling* 26:171–184.

NESN Staff. 2015 (August 6). "Janay Rice: Ray Rice Made a 'Mistake,' Should Be Allowed to Play in NFL (Video)." New England Sports Network. Available at www.nesn.com

Nielsen, Linda. 2014. "Shared Physical Custody: Summary of 40 Studies on Outcomes for Children." Journal of Divorce & Remarriage 55(8):613–635.

Ogolsky, Brian G., James K. Monk, and Renee P. Dennison. 2014. "The Role of Couple Discrepancies in Cognitive and Behavioral Egalitarianism in Marital Quality." Sex Roles 70(7–8):329–342.

Parker, K. 2011. "A Portrait of Stepfamilies." Pew Research Center. Available at www.pewsocialtrends.org

Parker, K., and W. Wang. 2013 (March 14). "Modern Parenthood." Pew Research Center. Available at www.pewsocialtrends.org

Parker, Marcie R., Edward Bergmark, Mark Attridge, and Jude Miller-Burke. 2000. "Domestic Violence and Its Effect on Children." *National Council on Family Relations Report* 45(4):F6–F7.

Pasley, Kay, and Carmelle Minton. 2001. "Generative Fathering after Divorce and Remarriage: Beyond the 'Disappearing Dad.'" In *Men and Masculinity: A Text Reader*, ed. T. F. Cohen, 239–248. Belmont CA: Wadsworth.

Pew Research Center. 2008. "Women Call the Shots at Home: Public Mixed on Gender Roles in Jobs." Available at pewresearch.org

Phillips, Allie. 2014. "Understanding the Link between Violence to Animals and People." National District Attorneys Association. Available at www.ndaa.org

Planty, Michael G. 2014 (October 10). "Domestic Violence." U.S. Department of Justice, Bureau of Justice Statistics. Available at www.bjs.gov

Planty, M., L. Langton, C. Krebs, M. Berzofsky, and H. Smiley-McDonald. 2013. "Female Victims of Sexual Violence, 1994–2010." U.S. Department of Justice, Bureau of Justice Statistics. Available at www.bjs.gov

Population Reference Bureau. 2011. *The World's Women and Girls 2011 Data Sheet*. Available at www.prb.org

Reese, Laura Schwab, Erin O. Heiden, Kimberly Q. Kim, and Jingzhen Yang. 2014. "Evaluation of Period of PURPLE Crying, an Abusive Head Trauma Prevention Program." Journal of Obstetric, Gynecologic, & Neonatal Nursing 43(6):752–761.

Russell, D. E. 1990. *Rape in Marriage*. Bloomington: Indiana University Press.

Santich, Kate. 2014. "Parent Abuse Would Be a Crime under Proposed Law." Orlando Sentinel, February 24. Available at www.articlesorlandosentinel.com.

Sawhill, Isabel V. 2014. Generation Unbound: Drifting into Sex and Parenthood without Marriage. Washington, DC: Brookings Institution Press.

Schwartz, Christine R., and Hongyun Han. 2014. "The Reversal of the Gender Gap in Education and Trends in Marital Dissolution." American Sociological Review 79(4):605–629.

Shah, Neil. 2014. "For More American Moms, Kids Are a Late-30s Thing." Wall Street Journal, May 9. Available at www.blogs.wsj.com

Sheff, Elisabeth. 2014. *The Polyamorists Next Door: Inside Multiple-Partner Relationships and Families*. Lanham, MD.: Rowman and Littlefield.

Shepard, Melanie F., and James A. Campbell. 1992. "The Abusive Behavior Inventory: A Measure of Psychological and Physical Abuse." *Journal of Interpersonal Violence* 7(3):291–305.

Simiao, L., A. Levick, A. Eichman, and J. Chang. 2015. "Women's Perspectives on the Context of Violence and Role of Police in Their Intimate Partner Arrest Experiences." Journal of Interpersonal Violence 30(3):400–419.

Skinner, Jessica A., and Robin M. Kowalski. 2013. "Profiles of Sibling Bullying." *Journal of Interpersonal Violence* 28(8):1726–1736.

Starkweather, Katherine E., and Raymond Hames. 2012. "A Survey of Non-classical Polyandry." Human Nature 23:149–172.

Straus, Murray. 2000. "Corporal Punishment and Primary Prevention of Physical Abuse." *Child Abuse and Neglect* 24:1109–1114.

Straus, Murray. 2010. "Prevalence, Societal Causes, and Trends in Corporal Punishment by Parents in World Perspective." *Law and Contemporary Problems* 73(1):1–30.

Sullivan, Erin. 2010. "Abused Pasco Dog Taken in by Victim Advocate Now Pays It Forward." *St. Petersburg Times,* January 23. Available at www.tampabay.com

Swan, S. C., L. J. Gambone, J. E. Caldwell, T. P. Sullivan, and D. L Snow. 2008. "A Review of Research on Women's Use of Violence with Male Intimate Partners." *Violence and Victims* 23:301–315.

Sweeney, M. M. 2010. "Remarriage and Stepfamilies: Strategic Sites for Family Scholarship in the 21st Century." *Journal of Marriage and the Family* 72:667–684.

Swiss, Liam, and Celine Le Bourdais. 2009. "Father–Child Contact after Separation: The Influence of Living Arrangements." *Journal of Family Issues* 30(5):623–652.

Teaster, Pamela B., Tyler A. Dugar, Marta S. Mendiondo, Erin L. Abner, Kara A. Cecil, and Joanne M. Otto. 2006 (February). *The 2004 Survey of State Adult Protective Services: Abuse of Adults 60 Years of Age and Older*. National Center on Elder Abuse. Washington, DC.

Trail, Thomas E., and Benjamin R. Karney. 2012. "What's (Not) Wrong with Low-Income Marriages." Journal of Marriage and Family (June):413–427.

Trinder, L. 2008. "Maternal Gate Closing and Gate Opening in Postdivorce Families." *Journal of Family Issues* 29:1298–1298.

Tucker, Corinna Jenkins, David Finkelhor, Heather Turner, and Anne M. Shattuck. 2014. "Sibling and Peer Victimization in Childhood and Adolescence." Child Abuse & Neglect 38:1599–1606.

Umberson, D., K. L. Anderson, K. Williams, and M. D. Chen. 2003. "Relationship Dynamics, Emotion State, and Domestic Violence: A Stress and Masculine Perspective." *Journal of Marriage and the Family* 65:233–247.

UNICEF. 2014. Ending Child Marriage: Progress and Prospects. Available at www.unicef.org

United Nations Development Programme. 2009 (November 23). "Ending Violence against Women Helps Achieve Development Goals." Available at www.beta.undp.org

U.S. Census Bureau. 2012 (April 25). "Interethnic Married Couples Grew by 28 Percent over Decade." Available at www.census.gov

U.S. Census Bureau. 2014. *Families and Living Arrangements: 2014*. Available at www.census.gov

U.S. Department of Health and Human Services. 2015. Child Maltreatment 2013. Available at www.acf.hhs.gov

Walby, S. 2013. "Violence and Society: Introduction to an Emerging Field of Sociology." *Current Sociology* 61:95–111.

Walker, Alexis J. 2001. "Refracted Knowledge: Viewing Families through the Prism of Social Science." In *Understanding Families into the New Millennium: A Decade in Review*, ed. Robert M. Milardo, 52–65. Minneapolis, MN: National Council on Family Relations.

Wallerstein, Judith S. 2003. "Children of Divorce: A Society in Search of Policy." In *All Our Families*, 2nd ed., ed. Mary Ann Mason, Arlene Skolnick, and Stephen D. Sugarman, 66–95. New York: Oxford University Press.

Wang, Wendy, and Kim Parker. 2014. "Record Share of Americans Have Never Married: As Values, Economics and Gender Patterns Change." Pew Research Center. Available at www.pewsocialtrends.org

Wang, Wendy, and Paul Taylor. 2011 (March 19). "For Millennials, Parenthood Trumps Marriage." Pew Research Center. Available at www.pewsocialtrends.org

Whiffen, V. E., J. M. Thompson, and J. A. Aube. 2000. "Mediators of the Link between Childhood Sexual Abuse and Adult Depressive Symptoms." *Journal of Interpersonal Violence* 15:1100–1120.

Williams, K., and A. Dunne-Bryant. 2006. "Divorce and Adult Psychological Well-Being: Clarifying the Role of Gender and Child Age." *Journal of Marriage and the Family* 68:1178–1196.

World Health Organization. 2014a. Child Maltreatment. Fact Sheet No. 150. Available at www.who.int

World Health Organization. 2014b. Violence against Women. Fact Sheet No. 239. Available at www.who.int

Zeitzen, Miriam K. 2008. *Polygamy: A Cross-Cultural Analysis*. Oxford: Berg.

Chapter 6

Administration for Children and Families. 2002. *Early Head Start Benefits Children and Families*. U.S. Department of Health and Human Services. Available at www.acf.hhs.gov

Agazzi, Isoida. 2012 (November 26). "Fixing the 'Silent' Sanitation Crisis." Inter Press Service. Available at www.ipsnews.net

Alex-Assensoh, Yvette. 1995. "Myths about Race and the Underclass." *Urban Affairs Review* 31:3–19.

Americans for Tax Justice. 2014 (August). *Tax Fairness Briefing Booklet*. Available at www.americansfortaxjustice.org

Badgett, M. V. Lee, Laura E. Durso, and Alyssa Schneebaum. 2013. "New Patterns of Poverty in the Lesbian, Gay, and Bisexual Community." Williams Institute. Available at williamsinstitute.law.ucla.edu

Bajak, Frank. 2010. "Chile-Haiti Earthquake Comparison: Chile Was More Prepared." *Huffington Post,* February 27. Available at www.huffingtonpost.com

Bickel, G., M. Nord, C. Price, W. Hamilton, and J. Cook. 2000. *United States Department of Agriculture Guide to Measuring Household Food Security*. Alexandria, VA: U.S. Department of Agriculture, Food and Nutrition Service.

Bishaw, Alemayehu and Kayla Fontenot. 2014 (September). "Poverty: 2012 and 2013." *American Community Briefs*. U.S. Census Bureau. Available at www.census.gov

Blankenhorn, David, William Galston, Jonathan Rauch, and Barbara Dafoe Whitehead. 2015 (March/April/May). "Can Gay Wedlock Break Political Gridlock?" *Washington Monthly*. Available at www.washingtonmonthly.com

Blumenthal, Susan. 2012 (March 12). "Debunking Myths about Food Stamps." SNAP to Health. Available at www.snaptohealth.org

Bolen, Ed. 2015 (January 5). "Approximately 1 Million Unemployed Childless Adults Will Lose SNAP Benefits in 2016 as State Waivers Expire." Center on Budget and Policy Priorities. Available at www.cbpp.org

Callahan, David, and J. Mijin Cha. 2013. *Stacked Deck: How the Dominance of Politics by the Affluent and Business Undermines Economic Mobility*. Demos. Available at www.demos.org

Carey, Kevin. 2015. "A Quiet Revolution in Helping Lift the Burden of Student Debt." *New York Times*, January 24. Available at www.nytimes.com

Center on Budget and Policy Priorities. 2014. "Chart Book: TANF at 18." Available at www.cbpp.org

Center on Budget and Policy Priorities. 2015 (January 16). "ChartBook: The Earned Income Tax Credit and Child Tax Credit." Available at www.cbpp.org

Child Care Aware of America. 2014. *Parents and the High Cost of Child Care: 2014 Report*. Available at www.usa.childcareaware.org

Coleman-Jensen, Alisha, Christian Gregory, and Anita Singh. 2014. *Household Food Security in the United States, 2013*. USDA Economic Research Service. Available at www.ers.usda.gov

Credit Suisse Research Institute. 2014 (October). *Global Wealth Report 2014*. Available at www.publications.credit-suisse.com

Davis, Kingsley, and Wilbert Moore. 1945. "Some Principles of Stratification." *American Sociological Review* 10:242–249.

Davis, Susan. 2015. "An Approach to Ending Poverty That Works." *Harvard Business Review,* January 22. Available at www.hbr.org

DeNavas-Walt, Carmen, and Bernadette D. Proctor. 2015. *Income and Poverty in the United States: 2014*. U.S. Census Bureau, Current Population Reports P60-249. Washington, DC: U.S. Government Printing Office.

Dolan, Kerry A., and Luisa Kroll. 2014. "Inside the 2014 Forbes Billionaires List: Facts and Figures." *Forbes,* March 3. Available at www.Forbes.com

Dvorak, Petula. 2009. "Increase Seen in Attacks on Homeless." *Washington Post*, February 5, p. DZ01.

Easley, Jason. 2015 (January 21). "Bernie Sanders Files a New Constitutional Amendment to Overturn Citizens United." *Politicususa*. Available at www.politicususa.org

Epstein, William M. 2004. "Cleavage in American Attitudes toward Social Welfare." *Journal of Sociology and Social Welfare* 31(4):177–201.

Food and Agriculture Organization. 2014. *The State of Food Insecurity in the World: 2014*. Available at www.fao.org

Gabe, Thomas. 2015. *Poverty in the United States: 2013*. Congressional Research Office. Available at www.crs.gov

Gans, Herbert. 1972. "The Positive Functions of Poverty." *American Journal of Sociology* 78:275–289.

Garfinkel, Irwin. 2013. "The Welfare State: Myths & Measurement." *Spectrum* (Winter):6–7.

Gilens, Martin and Benjamin I. Page. 2014. "Testing Theories of American Politics: Elites, Interest Group, and Average Citizens." *Perspectives on Politics* 12(3): 564–581.

Giovanni, Thomas, and Roopal Patel. 2013. *Gideon at 50: Three Reforms to Revive the Right to Counsel*. Brennan Center for Justice. Available at www.brennancenter.org

Golden, Olivia. 2013 (May 9). "Poverty in America: How We Can Help Families." Urban Institute. Available at www.urban.org

Gould, Elise. 2014 (August 27). "Why America's Workers Need Faster Wage Growth—and What We Can Do About It." Economic Policy Institute. Available at www.epi.org

Gould, Elise, Hilary Wething, Natalie Sabadish, and Nicholas Finio. 2013 (July 3). "What Families Need to Get By: The 2013 Update of EPI's Family Budget Calculator." Economic Policy Institute. Available at www.epi.org

Graham, Carol. 2015 (February 19). "The High Costs of Being Poor in America: Stress, Pain, and Worry." *Social Mobility Memos.* Brookings Institution. Available at www.brookings.edu

Green, Autumn R. 2013. "Patchwork: Poor Women's Stories of Resewing the Shredded Safety Net." *Affilia* 28:51–64.

Grunwald, Michael. 2006. "The Housing Crisis Goes Suburban." *Washington Post,* August 27. Available at www.washingtonpost.com

Hardoon, Deborah. 2015 (January 19). "Wealth: Having It All and Wanting More." Oxfam International. Available at www.Oxfam.org

Harell, Allison, Stuart Soroka, and Adam Mahon. 2008. "Is Welfare a Dirty Word?: Canadian Public Opinion on Social Assistance Policies." *Options Politiques* (September):53–56.

Hoback, Alan, and Scott Anderson. 2007. "Proposed Method for Estimating Local Population of Precariously Housed." National Coalition for the Homeless. Available at www.nationalhomeless.org

Human Rights Watch. 2015. *World Report 2015.* Available at www.hrw.org

Institute for Economics & Peace. 2013. *Pillars of Peace.* Available at economicsandpeace.org

Irvine, Leslie. 2013. "Animals as Lifechangers and Lifesavers: Pets in the Redemption Narratives of Homeless People." *Journal of Contemporary Ethnography* 42(1):3–36.

Katz, Michael B. 2013. *The Undeserving Poor: America's Enduring Confrontation with Poverty: Fully Updated and Revised.* New York: Oxford University Press.

Kochhar, Rakesh, and Richard Fry. 2014 (December 12). "Wealth Inequality Has Widened along Racial, Ethnic Lines since the End of Great Recession." Pew Research Center. Available at www.pewresearch.org

Kondo, Naoki, Grace Sembajwe, Ichiro Kawachi, Rob M. van Dam, S. V. Subramanian, and Zentaro Yamagata. 2009. "Income Inequality, Mortality, and Self-Rated Health: Meta-analysis of Multilevel Studies." *British Medical Journal* 339(7731):1178–1181.

Kraut, Karen, Scott Klinger, and Chuck Collins. 2000. *Choosing the High Road: Businesses That Pay a Living Wage and Prosper.* Boston: United for a Fair Economy.

Ku, Leighton, and Brian Bruen. 2013 (February 19). "The Use of Public Assistance Benefits by Citizens and Non-Citizen Immigrants in the United States." CATO Working Paper. Washington, DC: Cato Institute.

Ladd, Helen F. 2012. "Education and Poverty: Confronting the Evidence." *Journal of Policy Analysis and Management* 31(2):203–227.

Lanier, Paul, Katie Maguire-Jack, Tova Walsh, Brett Drake, and Grace Hubel. 2014 (September). "Race and Ethnic Differences in Early Childhood Maltreatment in the United States." *Journal of Developmental & Behavioral Pediatrics* 35(7):419–426.

Leigh, J. Paul. 2013 (March 6). "Raising the Minimum Wage Could Improve Public Health." Economic Policy Institute Blog. Available at www.epi.org

Luker, Kristin. 1996. *Dubious Conceptions: The Politics of Teenage Pregnancy.* Cambridge, MA: Harvard University Press.

Massey, D. S. 1991. "American Apartheid: Segregation and the Making of the American Underclass." *American Journal of Sociology* 96:329–357.

McDonagh, Thomas. 2013. *Unfair, Unsustainable, and under the Radar: How Corporations Use Global Investment Rules to Undermine a Sustainable Future.* Democracy Center. Available at www.democracyctr.org

McNamee, Stephen J., and Robert K. Miller Jr. 2009. *The Meritocracy Myth,* 2nd ed. Lanham, MD: Rowman and Littlefield.

Meixell, Braky and Ross Eisenbrey. 2014 (September 11). "An Epidemic of Wage Theft Is Costing Workers Hundreds of Millions of Dollars a Year." Economic Policy Institute. Available at www.epi.org

Mencimer, Stephanie. 2013. "Inadequate Diaper Supply Linked to Child Abuse, Depression." *Mother Jones,* July 30. Available at www.motherjones.com

Mishel, Lawrence. 2015 (February 4). "Congressional Testimony: Policies That Do and Do Not Address the Challenges of Raising Wages and Creating Jobs." Economic Policy Institute. Available at www.epi.org

Mishel, Lawrence, and Alyssa Davis. 2015 (June 21). "Top CEOs Make 300 Times More Than Typical Workers." Economic Policy Institute. Available at www.epi.org

National Alliance to End Homelessness. 2015. *The State of Homelessness in America 2054.* Washington, DC: National Alliance to End Homelessness.

National Coalition for the Homeless. 2014a. *Share No More: The Criminalization of Efforts to Feed People in Need.* Available at www.nationalhomeless.org

National Coalition for the Homeless. 2014b. *Vulnerable to Hate: A Survey of Hate Crimes and Violence Committed Against Homeless People in 2013.* Available at www.nationalhomeless.org

National Diaper Bank Network. N.d. "What Is Diaper Need?" Available at www.nationaldiaperbanknetwork.org

National Research Council and Institute of Medicine. 2013. *U.S. Health in International Perspective: Shorter Lives, Poorer Health.* Washington, DC: National Academies Press.

"Nipped in the Bud." 2015. *The Economist,* June 6–12, pp. 24–25.

Norton, Michael I. 2014 (October). "Unequality: Who Gets What and Why It Matters." *Policy Insights from the Behavioral and Brain Sciences* 1:151–155.

Office of Family Assistance. 2014. *Tenth Report to Congress.* Available at www.acf.hhs.gov

Oxfam. 2014. *Even It Up: Time to End Extreme Inequality.* Available at www.oxfam.org

Oxfam International. 2013 (January 19). "Annual Income of Richest 100 People Enough to End Global Poverty Four Times Over." Available at www.oxfam.org

Piketty, Thomas. 2014. *Capital in the Twenty-first Century.* Cambridge, MA: Harvard University Press.

Plumer, Brad. 2012. "Who Receives Government Benefits, in Six Charts." *The Washington Post,* September 18. Available at www.washingtonpost.com

Pugh, Tony. 2007. "U.S. Economy Leaving Record Numbers in Severe Poverty." *McClatchy Newspapers,* February 22. Available at www.mcclatchydc.com

Ramos, Alcida Rita, Rafael Guerreiro Osorio, and Jose Pimenta. 2009. "Indigenising Development." *Poverty in Focus* 17(May):pp. 3–5. International Policy Centre for Inclusive Growth.

Ratcliffe, Caroline, and Mary-Signe McKernan. 2010 (June 10). "Childhood Poverty Persistence: Facts and Consequences." Urban Institute. Available at www.urban.org

Reeves, Richard R., and Joanna Venator. 2015 (February). "Sex, Contraception, or Abortion? Explaining Class Gaps in Unintended Childbearing." Brookings Center on Children and Family. Available at www.brookings.edu

Rohde, David. 2012. "The Hideous Inequality Exposed by Hurricane Sandy." *The Atlantic,* October 12. Available at www.theatlantic.com

Roseland, Mark, and Lena Soots. 2007. "Strengthening Local Economies." In *2007 State of the World,* ed. Linda Starke, 152–169. New York: Norton.

Saez, Emmanuel, and Gabriel Zucman. 2014 (October). "Wealth Inequality in the United States since 1913: Evidence from Capitalized Income Data." NBER Working Paper No. 20625. National Bureau of Economic Research. Available at www.nber.org

Semuels, Alana. 2015. "Suburbs and the New American Poverty." *The Atlantic,* January 7 Available at www.theatlantic.com

Sobolewski, Juliana M., and Paul R. Amato. 2005. "Economic Hardship in the Family of Origin and Children's Psychological Well-Being in Adulthood." *Journal of Marriage and Family* 67(1):141–156.

Sommeiller, Estelle, and Mark Price. 2015 (January 26). "The Increasingly Unequal States of America." Economic Policy Institute. Available at www.epi.org

Straus, Rebecca. 2013. "Schooling Ourselves in an Unequal America." *New York Times,* June 16. Available at www.nytimes.com

Susskind, Yifat. 2005 (May). *Ending Poverty, Promoting Development: MADRE Criticizes the United Nations Millennium Development Goals.* Available at www.madre.org

Talberth, John, Daphne Wysham, and Karen Dolan. 2013. "Closing the Inequality Divide: A Strategy for Fostering Genuine Progress in Maryland." Center for Sustainable Economy and Institute for Policy Studies. Available at www.sustainable-economy.org

Turner, Margery Austin, Susan J. Popkin, G. Thomas Kingsley, and Deborah Kaye. 2005 (April). *Distressed Public Housing: What It Costs to Do Nothing.* Urban Institute. Available at www.urban.org

UN-Habitat. 2010. *State of the World's Cities 2010/2011.* Available at www.unhabitat.org

United for a Fair Economy. 2012. *Born on Third Base: What the Forbes 400 Really Says About Economic Equality & Opportunity in America.* Available at www.faireconomy.org

United Nations. 2005. *Report on the World Social Situation 2005.* New York: United Nations.

United Nations. 2013. *Inequality Matters: Report on the World Social Situation 2013.* New York: United Nations.

United Nations Development Programme (UNDP). 2010. *Human Development Report 2010.* Available at hdr.undp.org

United Nations Development Programme (UNDP). 2014. *Human Development Report 2014.* Available at hdr.undp.org

Urban Institute. 2015 (February). "Nine Charts about Wealth Inequality in America." Available at www.datatools.urban.org

U.S. Census Bureau. 2014. *2013 American Community Survey.* Available at www.census.gov

U.S. Census Bureau. 2015a. "One in Five Children Receive Food Stamps, Census Bureau Reports." Available at www.census.gov

U.S. Census Bureau. 2015b. *Poverty Thresholds for 2014.* Available at www.census.gov

USDA Food and Nutrition Service. 2015. *Supplemental Nutrition Assistance Program Participation and Costs.* Available at www.fns.usda.gov

vanden Heuvel, Katrina. 2011. "Putting Poverty on the Agenda." *The Nation,* January 17. Available at www.thenation.com

Weeks, Daniel. 2014. "Why Are the Poor and Minorities Less Likely to Vote?" *The Atlantic,* January 10. Available at www.theatlantic.com

Wider Opportunities for Women. 2010. *The Basic Economic Security Tables for the United States.* Washington DC: Wider Opportunities for Women.

Wilson, William J. 1987. *The Truly Disadvantaged: The Inner City, the Underclass, and Public Policy.* Chicago: University of Chicago Press.

Wilson, William J. 1996. *When Work Disappears: The World of the New Urban Poor.* New York: Knopf.

World Bank. 2005. *Global Monitoring Report 2005.* Available at www.worldbank.org

World Bank. 2015. "Poverty Data." Available at www.data.worldbank.org.

World Bank Group. International Monetary Fund. 2015. *Global Monitoring Report 2014/2015: Ending Poverty and Sharing Prosperity.* Washington, DC: World Bank Group.

World Health Organization and UNICEF. 2014. *Progress on Sanitation and Drinking-Water 2014 Update.* Geneva: WHO Press.

Wright, Erik Olin, and Joel Rogers. 2011. *American Society: How It Really Works.* New York: Norton.

Zedlewski, Sheila R. 2003. *Work and Barriers to Work among Welfare Recipients in 2002.* Urban Institute. Available at www.urban.org

Chapter 7

AFL-CIO. 2015. *Death on the Job: The Toll of Neglect,* 24th ed. Available at www.aflcio.org

Austin, Colin. 2002. "The Struggle for Health in Times of Plenty." In *The Human Cost of Food: Farmworkers' Lives, Labor, and Advocacy,* ed. C. D. Thompson Jr. and M. F. Wiggins, 198–217. Austin: University of Texas Press.

Barsamian, David. 2012. "Capitalism and Its Discontents: Richard Wolff on What Went Wrong." *The Sun* 434:4–13.

Bassi, Laurie J., and Jens Ludwig. 2000. "School-to-Work Programs in the United States: A Multi-Firm Case Study of Training, Benefits, and Costs." *Industrial and Labor Relations Review* 53(2):219–239.

Birth, Allyssa. 2014 (December 18). "Americans' Spending/Savings Plans Suggest Cautious Optimism." The Harris Poll. Available at www.harrisinteractive.com

Brand, Jennie E., and Sarah A. Burgard. 2008. "Job Displacement and Social Participation over the Lifecourse: Findings for a Cohort of Joiners." *Social Forces* 87(1):211–242.

Bureau of Labor Statistics. 2015a. "Census of Fatal Occupational Injuries: 2014." Available at www.bls.gov

Bureau of Labor Statistics. 2015b. "The Employment Situation—March 2015." *Employment Situation Summary.* Available at www.bls.gov

Bureau of Labor Statistics. 2015c (April 7). "Job Openings and Labor Turnover Survey Highlights February 2015." Available at www.bls.gov

Bureau of Labor Statistics. 2015d. *Union Members—2014.* Available at www.bls.gov

Butterworth, P., L. S. Leach, L. Strazdins, S. C. Olesen, B. Rodgers, and D. H. Broom. 2011. "The Psychosocial Quality of Work Determines Whether Employment Has Benefits for Mental Health: Results from a Longitudinal National Household Survey." *Occupational and Environmental Medicine.* Advance online publication. doi:10.1136/oem.2010.059030

Carney, Eliza Newlin. 2012. "Rules of the Game: Workplace Intimidation Becomes Murky in Post-Citizens United Era." *Roll Call,* October 12. Available at www.rollcall.com

Cockburn, Andrew. 2003. "21st Century Slaves." *National Geographic,* September, pp. 2–11, 18–24.

Davidson, Paul. 2012. "More U.S. Service Jobs Heading Offshore." *USA Today,* December 6. Available at www.usatoday.com

Davis, Alyssa, Will Kimball, and Elise Gould. 2015 (May 27). "The Class of 2015." Economic Policy Institute. Available at www.epi.org

Davis, Kingsley, and Wilbert Moore. 1945. "Some Principles of Stratification." *American Sociological Review* 10:242–249.

Ebeling, Richard M. 2009 (February 12). "Capitalism the Solution, Not Cause of the Current Economic Crisis." American Institute for Economic Research. Available at www.aier.org

The Editors. 2013 (May 12). "Your Future Will Be Manufactured on a 3-D Printer." Bloomberg. Available at bloomberg.com

Flavin, Patrick, and Gregory Shufeldt. 2014 (October 27). "Labor Union Membership and Life Satisfaction in the United States." Working Paper, Baylor University. Available at www.blogs.baylor.edu

Flounders, Sara. 2013 (February 4). "The Pentagon and Slave Labor in U.S. Prisons." Global Research. Available at www.globalresearch.ca

Frederick, James, and Nancy Lessin. 2000. "Blame the Worker: The Rise of Behavior-Based Safety Programs." *Multinational Monitor* 21(11). Available at www.essential.org/monitor

The Global Slavery Index. 2014. Available at www.globalslaveryindex.org

Goh, Joel, Jeffrey Pfeffer, and Stefanos A. Zenios. 2015 (April). "The Relationship between Workplace Stressors and Mortality and Healthcare Costs in the United States." Working Paper. Harvard Business School. Available at www.hbs.edu

Greenhouse, Steven. 2008. *The Big Squeeze.* New York: Knopf.

Heymann, Jody, Alison Earle, and Jeffrey Hayes. 2007. *The Work, Family and Equity Index.*

Montreal, QC: The Project on Global Working Families and The Institute for Health and Social Policy.

Human Rights Watch. 2014. *Tobacco's Hidden Children.* Available at www.hrw.org.

Human Rights Watch. 2015a (January). "Failing Its Families: Lack of Paid Leave and Work–Family Supports in the U.S." Available at www.hrw.org

Human Rights Watch. 2015b (March 12). "Work Faster or Get Out: Labor Rights Abuses in Cambodia's Garment Industry." Available at www.hrw.org

International Federation of Social Workers. 2015 (February 3). "Bangladeshi Social Workers Supporting Survivors at the Rana Plaza Building Collapse." Available at www.ifsw.org

International Labour Organization (ILO). 2015a. "Child Labour." Available at www.ilo.org

International Labour Organization (ILO). 2015b. *World Employment Social Outlook.* Available at www.ilo.org

Institute for Global Labour and Human Rights. 2011 (March 23). "Triangle Returns: Young Women Continue to Die in Locked Sweatshops." Available at www.globallabourrights.org

International Trade Union Confederation. 2014. *ITUC Global Rights Index.* Available at www.ituc-csi.org

Jensen, Derrick. 2002 (June). "The Disenchanted Kingdom: George Ritzer on the Disappearance of Authentic Culture." *The Sun,* pp. 38–53.

Jones, Jeffrey M. 2014a (August 28). "Americans Approve of Unions but Support 'Right to Work.'" Gallup Organization. Available at www.gallup.com

Jones, Jeffrey M. 2014b (February 13). "In U.S., 14% of Those Aged 24 to 34 are Living with Parents." Gallup Organization. Available at www.gallup.com

Jones, Jeff, and Lydia Saad. 2014 (August 7-10). Gallup News Service: Gallup Poll Social Series: Work and Education. Available at www.galluppoll.com

Korten, David. 2015 (March 9). "Do Corporations Really Need More Rights? Why Fast Track for the TPP Is a Bad Idea." *Yes! Magazine.* Available at www.yesmagazine.org

Lenski, Gerard, and J. Lenski. 1987. *Human Societies: An Introduction to Macrosociology,* 5th ed. New York: McGraw-Hill.

Leonard, Bill. 1996 (July). "From School to Work: Partnerships Smooth the Transition." *HR Magazine* (Society for Human Resource Management). Available at www.shrm.org

Lockard, C. Brett, and Michael Wolf. 2012. "Education and Training Outlook for Occupations, 2010–2020." Bureau of Labor Statistics. Available at www.bls.gov

Luhby, Tami. 2013. "Recent College Grads Face 36% 'Mal-Employment' Rate." *CNNMoney,* June 25. Available at money.cnn.com

MacEnulty, Pat. 2005 (September). "An Offer They Can't Refuse: John Perkins on His Former Life as an Economic Hit Man." *The Sun* 357:4–13.

Martinson, Karin, and Pamela Holcomb. 2007. *Innovative Employment Approaches and Programs for Low-Income Families.* Washington, DC: Urban Institute.

Matos, Kenneth. 2014. *Highlights from the 2014 Older Adult Caregiver Study.* Families and Work Institute. Available at www.familiesandwork.org

Matos, Kenneth, and Ellen Galinsky. 2014. *2014 National Study of Employers.* Families and Work Institute. Available at familiesandwork.org

Mendenhall, Ruby, Ariel Kalil, Laurel J. Spindel, and Cassandra M. D. Hart. 2008. "Job Loss at Mid-Life: Managers and Executives Face the 'New-Risk Economy.'" *Social Forces* 87(1): 185–209.

Miles, Kathleen. 2012. "'Sweatshop Conditions' Found in LA Fashion District at Contractors for Urban Outfitters, Aldo, Forever 21." *Huffington Post,* December 14. Available at www.huffingtonpost.com

Miller, Claire Cain. 2015. "The Economic Benefits of Paid Parental Leave." *New York Times,* January 30. Available at www.nytimes.com

Motlagh, Jason. 2014. "What Hasn't Changed since the Bangladesh Factory Collapse." *Washington Post,* April 18. Available at www.washingtonpost.com

National Chicken Council. 2012. "Broiler Chicken Industry Key Facts." Available at www.nationalchickencouncil.org

National Labor Committee. 2007. "Senate Minority Leader Senator Harry Reid, Congressman Bernie Sanders, AFL-CIO and Others Endorse Anti-Sweatshop Bill." Available at www.nlcnet.org

Newport, Frank. 2011 (March 31). "Americans' Top Job-Creation Idea: Stop Sending Work Overseas." Gallup. Available at www.gallup.com

Newport, Frank. 2012 (November 29). "Democrats, Republicans, Diverge on Capitalism, Federal Gov't." Gallup. Available at www.gallup.com

Palley, Thomas. 2015 (February 9). "The Federal Reserve and Shared Prosperity Why Working Families Need a Fed That Works for Them." Economic Policy Institute. Available at www.epi.org

Parenti, Michael. 2007 (February 16). "Mystery: How Wealth Creates Poverty in the World." Common Dreams NewsCenter. Available at www.commondreams.org

Parker, Kim, and Wendy Wang. 2013 (March 14). *Modern Parenthood.* Pew Research Center. Available at www.pewresearch.org

Perkins, John. 2004. *Confessions of an Economic Hit Man.* San Francisco: Berrett-Koehler.

Pew Research Center. 2014 (October 9). "Emerging and Developing Economies Much More Optimistic Than Rich Countries about the Future." Available at www.pewglobal.org

Rampell, Ed. 2013 (April 16). "An Interview with Richard Wolff." Counterpunch. Available at www.counterpunch.org

Ritzer, George. 1995. *The McDonaldization of Society: An Investigation into the Changing Character of Contemporary Social Life.* Thousand Oaks, CA: Pine Forge Press.

Scott, Robert. E. 2015 (April 8). "Misleading Math on the Korea Free Trade Agreement." Economic Policy Institute blog. Available at www.epi.org/blog

Scott, Robert E., and David Ratner. 2005 (July 20). *NAFTA's Cautionary Tale.* Economic Policy Institute Briefing Paper 214. Available at www.epi.org

"Sex Trade Enslaves Millions of Women, Youth." 2003. *Popline* 25:6.

Shierholz, Heidi. 2011 (May 11). "Continuing Dearth of Job Opportunities Leaves Many Workers Still Sidelined." Economic Policy Institute. Available at www.epi.org

Shierholz, Heidi. 2013 (June 11). "Unemployed Workers Still Far Outnumber Job Openings in Every Major Sector." The Economic Policy Institute. Available at www.epi.org

Shierholz, Heidi, Alyssa Davis, and Will Kimball. 2014. "The Weak Economy Is Idling Too Many Young Graduates." Briefing Paper #377. Economic Policy Institute. Available at www.epi.org

Shipler, David K. 2005. *The Working Poor.* New York: Vintage Books.

Skinner, E. Benjamin. 2008. *A Crime So Monstrous: Face-to-Face with Modern-Day Slavery.* New York: Free Press.

Southern Poverty Law Center and Alabama Appleseed. 2013. *Unsafe at These Speeds: Alabama's Poultry Industry and Its Disposable Workers.* Available at www.splc.org

Strully, Kate W. 2009. "Job Loss and Health in the U.S. Labor Market." *Demography* 46(2): 221–247.

SweatFree Communities. n.d. "Adopted Policies." Available at www.sweatfree.org

Tate, Deborah. 2007 (February 14). "U.S. Lawmakers Seek to Crack Down on Foreign Sweatshops." National Labor Committee. Available at www.nlcnet.org

Thompson, Charles D., Jr. 2002. "Introduction." In *The Human Cost of Food: Farmworkers' Lives, Labor, and Advocacy,* ed. C. D. Thompson, Jr., and M. F. Wiggins, 2–19. Austin: University of Texas Press.

Turner, Anna, and John Irons. 2009 (July). "Mass Layoffs at Highest Level since at Least 1995." Economic Policy Institute. Available at www.epi.org

Uchitelle, Louis. 2006. *The Disposable American: Layoffs and Their Consequences.* New York: Knopf.

United Nations. 2005. *The Millennium Development Goals Report.* New York: United Nations.

Walsh, Janet. 2015 (March 19). "Dispatches: U.S. May Join Rest of the World in Offering Paid Family Leave." Human Rights Watch. Available at www.hrw.org

Williams, Timothy. 2014. "Seeking New Start, Finding Steep Cost." *New York Times,* August 17. Available at www.nytimes.com

Wilson, Jill H. 2015 (February 26). "Improving Workers' Skills Must Include Immigrants." Brookings Institution. Available at www.brookings.edu

Wolff, Richard D. 2013a. "Alternatives to Capitalism." *Critical Sociology* 39(4):487–490.

Wolff, Richard D. 2013b (June 13). "Capitalism, Democracy, and Elections." *Democracy at Work* (blog). Available at www.democracyatwork.info

Wright, Erik Olin. 2013. "Transforming Capitalism through Real Utopias." *American Sociological Review* 78(1):1–25.

Chapter 8

Abel, Jaison R., and Richard Deitz. 2014 (September 2). "The Value of a College Degree." Liberty Street Economics. Available at libertystreeteconomics.newyorkfed.org

Adams, Caralee. 2011. "Obama Calls Community Colleges 'Key to the Future.'" *Education Week,* October 5. Available at blogs.edweek.org

Advisory Committee on Student Financial Assistance. 2010 (June). *The Rising Price of Inequality.* Washington, DC. Available at www.ed.gov/acsfa

Alexander, Debbie, Laurie Lewis, and John Ralph. 2014 (March). *Condition of America's Public School Facilities: 2012–13.* National Center for Education Statistics. Available at nces.ed.gov

Alexander, Karl, Doris Entwisle, and Linda Olson. 2014. *The Long Shadow: Family Background, Disadvantaged Urban Youth, and the Transition to Adulthood.* Russell Sage Foundation.

Allen, I. Elaine, and Jeff Seaman. 2014. *Great Change: Tracking Online Education in the United States, 2014.* January. Bobson Survey Research Group. Available at online www.learningsurvey.com

Alliance for Excellent Education (AEE). 2014a (May 5). "The Economic Case for Reducing the High School Dropout Rate." Available at all4ed.org.

Alliance for Excellent Education (AEE). 2014b (July). "On the Path to Equity: Improving the Effectiveness of Beginning Teachers." Available at all4ed.org.

American Federation of Teachers. 2009. *Building Minds, Minding Buildings: A Union's Roadmap to Green and Sustainable Schools.* Available at www.aft.org

American Society of Civil Engineers. 2014. *2013 Report Card for America's Infrastructure: Schools.* Available at www.infrastructurereportcard.org

American Statistical Association (ASA). 2014 (April 8). "ASA Statement on Using Value-Added Models for Educational Assessment." Available at www.amstat.org

Amurao, Carla. 2013 "Fact Sheet: How Bad Is the School-to-Prison Pipeline?" PBS. Available at www.pbs.org

Anderson, Melinda D. 2014 (November 24). "When Educators Understand Race and Racism." Teaching Tolerance. Available at www.tolerance.org

Avery, Chris. 2014. "Meet the 2014 Teaching Tolerance Award Winners." *Teaching Tolerance,* July 14. Available at www.tolerance.org

Avvisati, Francesco. 2014. "Are Disadvantaged Students More Likely to Repeat Grades?" *PISA in Focus,* September 9. Organisation for Economic Cooperation and Development. Available at www.oecd.org

Barton, Paul E. 2004. "Why Does the Gap Persist?" *Educational Leadership* 62(3):9–13.

Beard, Aaron. 2014. "Fake Classes, Inflated Grade: Massive UNC Scandal Included Athletes over 2 Decades." *Star Tribune,* October 23. Available at www.startribune.com

Biecek, Przemyslaw, and Francesca Borgonovi. 2014. "Do Parents' Occupations Have an Impact on Student Performance?" *PISA in Focus.* February 2. Organisation for Economic Cooperation and Development. Available at www.oecd.org

Borgonovi, Francesca, and Marilyn Achiron. 2015. "What Lies behind Gender Inequality in Education?" *PISA in Focus*, March 3. Available at www.oecd-ililibrary.org

Boser, Ulrich. 2014. "Teacher Diversity Revisited: A New State-by-State Analysis." Center for American Progress. May. Available at cdn.americanprogress.org

Brammer, John Paul. 2014. "Wheels: How We Teach Latino Inferiority in Schools." *Huffington Post*, September 4. Available at www.huffingtonpost.com

Buchmann, Bryce. 2014. "Cheating in College: Where It Happens, Why Students Do It and How to Stop It." *Huffington Post*, April 22. Available at www.huffingtonpost.com

Bureau of Labor Statistics (BLS). 2015. "Employment Projections." U.S. Department of Labor, Current Population Survey. Available at www.bls.gov

Bushaw, William J., and Valerie Calderon. 2014. "America Puts Teacher Quality on Center Stage." 46th Annual PDK/Gallup Poll of the Public's Attitudes towards the Public Schools: Part II. Available at pdkintl.org

Carnevale, Anthony P., and Jeff Strohl. 2013 (July). *Separate and Unequal: How Higher Education Reinforces the Intergenerational Reproduction of White Racial Privilege.* Georgetown Public Policy Institute, Center on Education and the Workforce, Georgetown University.

Caryl, D. (2013, September 13). "Why I Stopped Following My Dream and Gave Up Teaching: Your Letters." Syracuse.com. Available at www.syracuse.com

Center for Promise (CFP). 2014. "Don't Call Them Dropouts." A Report from America's Promise Alliance and Its Center for Promise at Tufts University. Available at gradnation.org

Center for Research on Educational Outcomes (CREDO). 2015. *Urban Charter School Study.* Stanford, CA: Stanford University.

Chen, Grace. 2015 (March 3). "Public School vs. Private School." Public School Review. Available at www.publicschoolreview.com

Coleman, James S., J. E. Campbell, L. Hobson, J. McPartland, A. Mood, F. Weinfield, and R. York. 1966. *Equality of Educational Opportunity.* Washington, DC: U.S. Government Printing Office.

Collman, Ashley. 2014. "College Professor Brutally Beaten Up in Campus Parking Garage Believes a Vengeful Student Hired Attack over Bad Grade." *Daily Mail,* April 22. Available at www.dailymail.co.uk

Corbett, Christianne, Catherine Hill, and Andresse St. Rose. 2008. *Where the Girls Are.* Washington, DC: American Association of University Women.

Cordell, LaDoris. 2013. "Hate Crimes Alleged at San Jose State Could Be Charged as Felonies." *Mercury News,* November 28. Available at www.mercurynews.com

Cowen, Joshua M., David J. Fleming, John F. Witte, Patrick J. Wolf, and Brian Kisida. 2013. "School Vouchers and Student Attainment: Evidence from a State-Mandated Study of Milwaukee's Parental Choice Program." *Policy Studies Journal* 44(1):147–168.

Davis, Michele R. 2013. "Education Industry Players Exert Public Policy Influence." *When Public Mission Meets Private Opportunity*, supplement to *Education Week* 32(29):S52–S16.

DeNeui, Daniel, and Tiffany Dodge. 2006. "Asynchronous Learning Networks and Student Outcomes." *Journal of Instructional Psychology* 33(4):256–259.

Dobbs, Michael. 2005. "Youngest Students Most Likely to Be Expelled." *Washington Post,* May 16. Available at www.washingtonpost.com

Dockterman, Eliana. 2015. "The Hunting Ground Reignites the Debate over Campus Rape." *Time,* March 5. Available at time.com

Dorff, Victor. 2014. "A World without Integrity." *Huffington Post,* April 16. Available at www.huffingtonpost.com

Dynarski, Susan. 2014. "Finding Shock Absorbers for Student Debt." *New York Times,* June 14. Available at www.nytimes.com

Education Justice. 2015. "North Carolina Litigation." *Hoke County Board of Education v. State of North Carolina.* Available at www.educationjustice.org

Eliot, Lise. 2010. *The Myth of PINK & BLUE Brains. Educational Leadership* 68(3):32–36.

Ertas, Nevbahar, and Christine H. Roch. 2014. "Charter Schools, Equity, and Student Enrollments: The Role of For-Profit Educational Management Organizations." *Education and Urban Society* 46(5):548–579.

Ewert, Stephanie. 2013 (January). "The Decline in Private School Enrollment." Working Paper FY 12-117. Social, Economic, and Housing Statistics Division.

Federal Education Budget Project (FEBP). 2014 (April 21). "School Finance: Federal, State, and Local K–12 School Finance Overview." Available at febp.newamerica.net

Fletcher, Robert S. 1943. *History of Oberlin College to the Civil War.* Oberlin, OH: Oberlin College Press.

Flexner, Eleanor. 1972. *Century of Struggle: The Women's Rights Movement in the United States.* New York: Atheneum.

Fox News. 2015 (February 2). "Proposed Tax Hike to Fund Education Dies in House Committee." Fox 13 News. Available at fox13now.com

Fraise, Nicole Jaenee, and Jeffrey Brooks. 2015. "Toward a Theory of Culturally Relevant Leadership for School Community Culture." *International Journal of Multicultural Education* 1 (1):6–21.

Friedlander, Brett. 2014. "Timeline of Events in UNC Athletic/Academic Scandal." *Star News,* October 22. Available at acc.blogs.starnewsonline.com

Gambino, Lauren. 2015. "Oklahoma Educators Fear High School History Bill Will Have Devastating Impact." *The Guardian,* February 20. Available at www.theguardian.com

Gardner, Walter. 2013. "Push Back on Standardized Testing." *Education Week,* January 14. Available at blogs.edweek.org

Ginder, Scott A., Janice E. Kelly-Reid, and Farah B. Mann. 2014. *Postsecondary Institutions and Cost of Attendance in 2013–14; Degrees and Other Awards Conferred, 2012–13; and Twelve-Month Enrollment, 2012–13.* Available at nces.ed.gov

Goldenberg, Claude. 2008. "Teaching English Language Learners." *American Educator* (Summer):8–11, 14–19, 22–23, 42–44.

Goldhaber, Dan, and Lesley Lavery, and Roddy Theobald. (June/July) 2015. "Uneven Playing Field: Assessing the Teacher Quality Gap between Advantaged and Disadvantaged Students." *Educational Researcher* 44:293–307.

Greenhouse, Linda. 2007. "Supreme Court Votes to Limit the Use of Race in Integration Plans." *New York Times,* June 29. Available at www.nytimes.com

Gross-Loh, Christine. 2014. "The Never Ending Controversy over All Girls Education." *The Atlantic,* March 20. Available at www.theatlantic.com

Haberman, Martin. 1991. "The Pedagogy of Poverty versus Good Teaching." *Phi Delta Kappan* 73(4):290–294. Available at www.det.nsw.edu.au

Harris, Philip, Bruce M. Smith, and Joan Harris. 2011. *The Myths of Standardized Tests: Why They Don't Tell You What You Think They Do.* Lanham, MD: Rowman and Littlefield.

Head Start. 2015. "Head Start Program Facts Fiscal Year 2014." Available at eclkc.ohs.acf.gov

Hefling, Kimberly. 2014 (May 7). "No Gains for 12th Graders on National Exam." Associated Press: The Big Story. Available at bigstory.ap.org

Henderson, Michael B., Paul A. Peterson, and Martin R. West. 2015. "No Comment Opinion on the Common Core." *Education Next* 15(1):8–12.

Herrera, Guillermo. 2014. "Professor Hospitalized after Being Attacked in Parking Garage at Kendall Campus." *The Reporter,* April 22. Available at www.mdc.edu

Hewitt, Damon, Rachel Kleinman, Lazar Treschan, and Apurva Mehrotra. 2013. *The Meaning of Merit: Alternatives for Determining Admission to New York City's Specialized High Schools.* October. A Report by the Community Service Society of New York and the NAACP Legal Defense and Education Fund. Available at www.naacpldf.org

Hightower, Amy M. 2013 (January 4). "States Show Spotty Progress on Education Gauges." Available at www.edweek.org

Hope, Elan C., Alexandria B. Skoog, and Robert J. Jagers. 2015. "'It'll Never Be the White Kids, It'll Always Be Us'": Black High School Students Evolving Critical Analysis of Racial Discrimination and Inequity in Schools." *Journal of Adolescent Research* 30(1):83–112.

Hopkinson, Natalie. 2011. "The McEducation of the Negro." *The Root,* January 3. Available at www.theroot.com

Hurdle, Jon. 2013. "Philadelphia Officials to Close 23 Schools." *New York Times,* March 7. Available at www.nytimes.com

Juvonen, Jaana, and Sandra Graham. 2014. "Bullying in Schools: The Power of Bullies and the Plight of Victims." *Annual Review of Psychology* 65:159–185.

Kahlenberg, Richard D. 2006. "A New Way of School Integration." *Issue Brief.* Century Foundation. Available at www.equaleducation.org

Kahlenberg, Richard D. 2013. "From All Walks of Life: New Hope for School Integration." *American Educator* (Winter):1–40.

Kanter, Rosabeth Moss. 1972. "The Organization Child: Experience Management in a Nursery School." *Sociology of Education* 45:186–211.

Kaplan, Tracey, and Katy Murphy. 2014. "San Jose State Hate Crime Case: Black Student Files $5 Million Claim." *Mercury News,* March 20. Available at www.mercurynews.com

Kastberg, David, David Ferraro, Nita Lemanski, Stephen Roey, and Frank Jenkins. 2013. *Highlights from TIMSS 2011.* NCES 2013-009. Revised. Available at nces.ed.gov

Khan, Z. R., and S. Balasubramanian. 2012. "Students Go Click, Click, and Cheat: E-Cheating, Technologies, and More." *Journal of Academic and Business Ethics* 6:1–26.

Kids Count. 2014. *2014 Kids Count Data Book.* Annie E. Casey Foundation. Available at www.aecf

King, Darwin L., and Carl J. Case. 2014. "Each Cheating: Incidents and Trends among College Students." *Issues in Information Systems* 15(1):20–27.

Klein, Rebecca. 2014. "American Teachers Spend More Time in the Classroom Than World Peers, Says Report." *Huffington Post,* September 9. Available at www.huffingtonpost.com

Klein, Rebecca. 2015. "Here Are the States That Spend the Most on Public School Students." *Huffington Post,* January 29. Available at www.huffingtonpost.com

Kohn, Alfie. 2011. "How Education Reform Traps Poor Children." *Education Week* 30(29):32–33.

Kozol, Jonathan. 1991. *Savage Inequalities: Children in America's Schools.* New York: Crown.

Lahey, Jessica. 2013. "The Benefits of Character Education." *The Atlantic,* May 6. Available at www.theatlantic.com

Leandro v. State, 488 S.E.2d 249 (N.C. 1997).

Leachman, Michael, and Chris Mai. 2014 (May 20). "Most States Funding Schools Less Than before the Recession." Center on Budget and Policy Priorities. Available at www.cbpp.org

Lederman, Doug. 2014. "The Enrollment Slow Down." February 28. *Inside Higher Education.* Available at insidehighered.com

Lickona, Thomas, and Matthew Davidson. 2005. *A Report to the Nation: Smart and Good High Schools.* Available at www.cortland.edu

Lindstrom, Natosha. 2015 (March 20). "Parent Trigger Laws Spreading from California to Other States." KPCC Public Radio. Available at www.scpr.org

Loveless, Tom. 2013. "The Resurgence of Ability Grouping in Persistence of Tracking." Brown Center Report on American Education. Available at www.brookings.edu

Lu, Adrienne. 2013 (July 1). "Parents Revolt against Failing Schools." Pew Charitable Trust. Available at www.pewstates.org

Lubienski, Christopher, Janelle T. Scott, John Rogers, and Kevin G. Welner. 2012 (September 5). "Missing the Target? The Parent Trigger as a Strategy for Parental Engagement and School Reform." National Education Policy Center, School of Education, University of Colorado at Boulder.

Lumina Foundation. 2013 (February 5). *Americans Call for Higher Education Redesign.* Available at luminafoundation.org

Majority Staff. 2014 (July 9). "Sexual Violence on Campus." A Report Prepared by the U.S. Senate Subcommittee on Financial and Contracting Oversight, Senator Claire McCaskill. Available at www.mccaskill.senate.gov

Marcus, John, and Holly K. Hacker. 2014. "Poorer Families Are Bearing the Brunt of College Price Hikes, Data Show." *Hechinger Report,* March 9. Available at hechingerr.org

Mares, Marie-Louise, and Zhongdang Pan. 2013. "Effects of *Sesame Street:* A Meta-Analysis of Children's Learning in 15 Countries." *Journal of Applied Developmental Psychology* 34(3):140–151.

Martinez, Michael. 2014 (June 11). "U.S. Schools Chief Calls California Ruling 'A Mandate' to Fix Tenure, Firing Laws." CNN News. Available at www.cnn.com

Maxwell, Leslie A. 2013. "Head Start Gains Found to Wash Out by Third Grade." *Education Week,* January 9. Available at www.edweek.org

McDonnell, Sanford. 2009. "America's Crisis of Character—and What to Do about It." *Education Week,* October 3. Available at www.edweek.org

Mead, Sara. 2006 (June). "The Truth about Boys and Girls." Education Sector. Available at www.educationsector.org

Merton, Robert K. 1968. *Social Theory and Social Structure.* New York: Free Press.

MetLife. 2013. *MetLife Survey of the American Teacher: Past, Present and Future.* Available at www.metlife.com

Mettler, Suzanne. 2014. *Degrees of Inequality: How the Politics of Higher Education Sabotaged the American Dream.* New York: Basic Books.

Milkie, Melissa A., and Catherine H. Warner. 2011. "Classroom Learning Environments and Mental Health of First Grade Children." *Journal of Health and Social Behavior* 42:4–22.

Mitchell, Corey. 2015. "Boys-Only Programs Raise Legal Concerns." *Education Week* 34(23):1–12.

Muller, Chandra, and Katherine Schiller. 2000. "Leveling the Playing Field?" *Sociology of Education* 73:196–218.

Mullis, I. V. S., M. O. Martin, P. Foy, and A. Arora. 2012. *The TIMSS 2011 International Results in Mathematics.* Chestnut Hill, MA: International Study Center.

National Assessment of Educational Progress (NAEP). 2014. "Results for 2013 NAEP Mathematics and Reading Assessments Are In." *The Nation's Report Card.* Available at www.nationsreportcard.gov

National Board for Professional Teaching Standards. 2015. "Value for Students." Available at boardcertifiedteachers.org

National Center for Educational Statistics (NCES). 2011. *Digest of Education Statistics, 2010.* Available at nces.ed.gov

National Center for Educational Statistics (NCES). 2014a. *The Condition of Education.* Washington, DC: U.S. Department of Education. Available at nces.ed.gov

National Center for Educational Statistics (NCES). 2014b (June). "Enrollment in Distance Education Courses, by State: Fall 2012." Web Tables. NCES 2014-023. Washington, DC: U.S. Department of Education.

National Center for Educational Statistics (NCES). 2014c. *Fast Facts: Back to School Statistics.* U.S. Department of Education. Available at nces.ed.gov

National Center for Educational Statistics (NCES). 2015a. *The Condition of Education.* Washington, DC: U.S. Department of Education. Available at nces.ed.gov

National Center for Educational Statistics (NCES). 2015b. *Digest of Education Statistics, 2013.* Washington, DC: U.S. Department of Education. Available at nces.ed.gov

National Center for Educational Statistics (NCES). 2015c. "Early High School Dropouts: What Are Their Characteristics?" *DataPoint,* February. NCES 2015-066. Washington, DC: U.S. Department of Education.

National Center for Fair and Open Testing (NCFOT). 2012 (May 22). "What's Wrong with Standardized Tests?" Available at fairtest.org

National Conference of State Legislatures (NCSL). 2013a. "Parent Trigger Laws in the States." Available at www.ncsl.org

National Conference of State Legislatures (NCSL). 2013b. "School Vouchers." Available at www.ncsl.org

National Education Policy Center (NEPC). 2013. "Promoting Consumption at School." Available at nepc.colorado.edu

National Public Radio. 2013 (June 2). "Why Some Schools Want to Expel Suspensions." Available at www.npr.org

Network for Public Education (NPE). 2015. "About the Network for Public Education." Available at networkforpubliceducation.org

North Carolina Justice Center. 2012. "*Leandro* and the 2012 Court of Appeals Decision on NC Pre-K." *Education and Law Project.* Available at www.ncjustice.org

Office of Management and Budget. 2015. "Fiscal Year 2016 Budget Overview." Available at www.whitehouse.gov

Orfield, Gary, Erica Frankenberg, Jongyeon Ee, and John Kuscera. 2014 (May 15). "Brown at 60: Great Progress, a Long Retreat and an Uncertain Future." Civil Rights Project, UCLA. Available at civilrightsproject.ucla.edu

Organisation for Economic Cooperation and Development (OECD). 2013. "About PISA." Available at www.oecd.org

Organisation for Economic Cooperation and Development (OECD). 2014. *Education at a Glance 2014.* Available at www.oecd.org

Parent Revolution. 2014. "About Us" Available at parentrevolution.org

Perry, Mark J. 2014 (July 9). "Recent Parents Revolution National Test Data (NAEP, SAT) Refute the Claim That 'There Just Aren't Gender Differences in Math Performance.'" American Enterprise Institute. Available at www.aei.org

Pew Research Center. 2015 (January 15). "Public's Policy Priorities Reflect Changing Conditions at Home and Abroad." Available at www.people-press.org

Picciano, Anthony G., and Jeff Seaman. 2010 (August). *Class Connections: High School Reform in the Role of Online Learning.* Available at www.babson.edu

Poliakoff, Anne Rogers. 2006. "Closing the Gap: An Overview." *ASCD InfoBrief* 44(January):1–10. Alexandria, VA: Association for Supervision and Curricular Development.

Primary Sources. 2014. *Teachers Bring Passion and Commitment to Their Challenging Work."* Scholastic and the Bill and Melinda Gates Foundation. Available at www.scholastic.com

Program for International Student Assessment (PISA). 2014. "PISA 2012 Results in Focus." Organisation for Economic Cooperation and Development. Available at www.oecd.org

Prothero, Arianna. 2014. "Twenty Columbus Schools Qualified for Overhauls under Parent-Trigger Law." *Education Week*, September 15. Available at blogs.edweek.org

Public Agenda. 2013 (September). "Not Yet Sold: What Employers, and Community College Students Think about Online Education." September. Available at www.publicagenda.org

Quality Counts. 2015. "Quality Counts 2015: State Report Cards Map." *Education Week* 34(16):1. Available at www.edweek.org

Quinn, David M. 2015. "Kindergarten, Black–White Test Score Gaps: Re-examining the Roles of Social and Economic Status and School Quality with New Data." *Sociology of Education* 88(2):120–139.

Ramey, Garey, and Valerie A. Ramey. 2010. "The Rug Rat Race." *Brookings Paper on Economic Activity* (Spring):129–176.

Ravitch, Diane. 2010. *The Death and Life of the Great American School System: How Testing and Choice Are Undermining Education.* New York: Basic Books.

Ray, Paul Ryan D. 2015 (July 6). "Research Facts on Homeschooling." National Home Education Research Institute. Available at www.nheri.org

Reardon, Sean F. 2013. "No Rich Child Left Behind." *New York Times,* April 27. Available at www.nytimes.com

Riehl, Carolyn. 2004. "Bridges to the Future: Contributions of Qualitative Research to the Sociology of Education." In *Schools and Society*, ed. Jeanne Ballantine and Joan Spade, 56–72. Belmont, CA: Thomson Wadsworth.

Robers, S., A. Zhang, R. E. Morgan, and L. Musu-Gillette. (May) 2015. *Indicators of School Crime and Safety: 2014.* NCES 2015-072/NCJ 248036. Washington, DC: National Center for Education Statistics, U.S. Department of Education, and Bureau of Justice Statistics, Office of Justice Programs, U.S. Department of Justice.

Rogers, John, and Nicole Mirra. 2014 (November). "It's about Time: Learning Time and Educational Opportunity in California High Schools." Institute for Democracy, Education, and Access. Available at idea.gseis.ucla.edu

Rooks, Noliew. 2012. "Why It's Time to Get Rid of Standardized Tests." *Time,* October 11. Available at ideas.time.com

Rosen, Jill. 2014. "Study: Children's Life Trajectories Largely Determined by Family They Are Born Into." *Johns Hopkins News Network,* June 2. Available at hub.jhu.edu

Rosenthal, Robert, and Lenore Jacobson. 1968. *Pygmalion in the Classroom: Teacher Expectations and Pupils' Intellectual Development.* New York: Holt, Rinehart, and Winston.

Rothstein, Richard, Helen F. Ladd, Diane Ravitch, Eva L. Baker, Paul E. Barton, Linda Darling-Hammond, Edward Haertel, Robert L. Linn, Richard J. Shavelson, and Lorrie A. Shepard. 2010 (August 29). *Problems with the Use of Student Test Scores to Evaluate Teachers.* Educational Policy Institute. Available at www.epi.org

Russo, Ralph D. 2015. "Academic Misconduct Is on the Rise in College Athletics, Capital NCAA Says." *Huffington Post,* January 29. Available at huffingtonpost.com

Sadovnik, Alan. 2004. "Theories in the Sociology of Education." In *Schools and Society*, ed. Jeanne Ballantine and Joan Spade, 7–26. Belmont, CA: Thomson Wadsworth.

Sawchuk, Stephen. 2011. "New Teacher Distribution Methods Hold Promise." *Education Week* 29(35):16–17.

Schott Report. 2015. *Black Lives Matter: The 2012 Schott 50 State Report on Public Education and Black Males.* Schott Foundation for Public Education. Available at blackboysreport.org

Segal, Tom. 2013. "The Impact of Investing in Education." *Education Week*, March 26. blogs.edweek.org

Service Employees International Union. 2013. *Falling Further Apart: Decaying Schools in New York City's Poorest Neighborhoods.* Available at www.seiu32bj.org

Shah, Nirvi. 2013 (January). "Discipline Policy Shift with Views on What Works." *Education Week* 32(16):12.

Smyth, Julie Carr. 2013. "Longer School Year: Will It Help or Hurt US Students?" *Huffington Post,* January 13. Available at www.huffingtonpost.com

Sparks, Sarah D. 2013. "Student Social, Emotional Needs Intertwined with Learning, Security." *Education Week* 32(16)16–21.

Stephens, Jason M., and David B. Wangaard. 2013. "Using the Epidemic of Academic Dishonesty as an Opportunity for Character Education: A Three-Year Mixed Methods Study (with Mixed Results)." *Peabody Journal of Education* 88(2):159–179.

Tanner, C. K. 2008. "Explaining Relationships among Student Outcomes and the School's Physical Environment." *Journal of Advanced Academics* 19:444–471.

Toppo, Greg. 2013. "Growing Number of Educators Boycott Standardized Test." *USA Today,* February 1. Available at www.usatoday.com

Tyre, Peg. 2013 (February 14). "It's Time to Worry: Boys Are Rapidly Falling behind Girls in School." Take Part News. Available at www.takepart.com

UNESCO. 2014a (June). "Progress in Getting All Children to School Stalls but Some Countries Show the Way Forward." *Policy Paper 14/Fact Sheet 28.* Available at www.uis.unesco.org

UNESCO. 2014b. "Statistics on Literacy." Available at www.uis.unesco.org

UNESCO. 2014c. *Teaching and Learning: Achieving Equality for All.* EFA Global Monitoring Report. Available www.unescdoc.unesco.org

U.S. Census Bureau. 2015. "Educational Attainment in the United States: 2014—Detailed Tables." Educational Attainment: Table 2. Available at www.census.gov

U.S. Department of Education. 2012. "ESEA Flexibility." Available at www.ed.gov

U.S. Department of Education. 2014a (March). "Data Snapshot: School Discipline." Office of Civil Rights. Available at www.ed.gov

U.S. Department of Education. 2014b. "Improving Teacher Preparation: Building on Innovation." *Great Teachers Matter.* Available at www.ed.gov

Venator, Joanna, and Richard V. Reeves. 2015 (March 18). "Building the Soft Skills for Success." Brookings Institute. Available at www.brookings.edu

Waggoner, Martha. 2009. "Judge: 'Academic Genocide' in Halifax Schools." *News & Observer,* March 19. Available at www.newsobserver.com

Watson, John, Larry Pate, Amy Murin, Butch Gemin, and Lauren Vashaw. 2014. *Keeping Pace with K–12 Digital Learning: An Annual Review of Policy and Practice,* 11th ed. Available at www.kpk12.com

Welner, Kevin. 2014. "Poverty and the Education Opportunity Gap: Will Obama Step Up to SOTU?" Valerie Strauss, The Answer Sheet Blog. *Washington Post,* January 26. Available at www.washingtonpost.com

White House. 2014a (June 9). "Fact Sheet: Making Student Loans More Affordable." Office of the Vice President, Statements and Releases. Available at www.whitehouse.gov

White House. 2014b (April). *Not Alone: The First Report of the White House Task Force to Protect Students from Sexual Assault.* Available at www.notalone.gov

White House. 2015 (January 9). "White House Unveils America's College Promise Proposal: Tuition-Free Community College for Responsible Students." Office of the Press Secretary. Available at www.whitehouse.gov

Williams, Corey, and Kimberly Hefling. 2014 (May 2). "55 Colleges Face Federal Sex Assault Investigations." Associated Press: The Big Story. Available at bigstory.ap.org

Witt, Ariana. 2014 (July 21). *On the Path to Equity: Improving the Effectiveness of Beginning Teachers.* Alliance for Excellent Education. Available at all4ed.org

Chapter 9

Alexander, Michelle. 2010. *The New Jim Crow: Mass Incarceration in the Age of Colorblindness.* New York: New Press.

Allport, G. W. 1954. *The Nature of Prejudice.* Cambridge, MA: Addison-Wesley.

American Civil Liberties Union. 2014 (July 9). "United States' Compliance with the International Convention on the Elimination of All Forms of Racial Discrimination." Available at www.aclu.org

American Council on Education and American Association of University Professors. 2000. *Does Diversity Make a Difference? Three Research Studies on Diversity in College Classrooms.* Washington, DC: American Council on Education and American Association of University Professors.

Apfelbaum, Evan. 2011. "Prof. Evan Apfelbaum: A Blind Pursuit of Racial Colorblindness—Research Has Implications for How Companies Manage Multicultural Teams." MIT *Sloan Experts*, MIT Sloan Management blog. Available at mitsloanexperts.mit.edu

Armario, Christine. 2011. "Feds: All Kids, Legal or Not, Deserve K–12 Education." *Chron*, May 7. Available at www.chron.com

Associated Press. 2011 (August 17). "FBI Investigate Fatal Rundown of Black Miss. Man." National Public Radio. Available at www.npr.org

Ayala, Elaine, and Ellen Huet. 2013. "Hispanic May Be a Race on 2020 Census." *San Francisco Chronicle*, February 4. Available at www.sfgate .com

Balko, Radley. 2009 (July 6). "The El Paso Miracle." Reasononline. Available at www.reason.com

Bauer, Mary, and Sarah Reynolds. 2009. *Under Siege: Life for Low-Income Latinos in the South*. Montgomery, AL: Southern Poverty Law Center.

Beirich, Heidi. 2007 (Summer). "Getting Immigration Facts Straight." *Intelligence Report*. Available at www.splcenter.org

Beirich, Heidi. 2013 (Spring). "The Year in Nativism." *Intelligence Report* (149). Available at www.splcenter.org

Bonilla-Silva, Eduardo. 2012. "The Invisible Weight of Whiteness: The Racial Grammar of Everyday Life in Contemporary America." *Ethnic and Racial Studies* 35(2):173–194.

Bonilla-Silva, Eduardo. 2013. *Racism without Racists: Color-Blind Racism and the Persistence of Racial Inequality*, 4th ed. Lanham, MD: Rowan and Littlefield.

Brace, C. Loring. 2005. *"Race" Is a Four-Letter Word*. New York: Oxford University Press.

Bradner, Eric. 2015 (August 18). "Huckabee: MLK Would be 'Appalled' by Black Lives Matter Movement." *CNN*. Available at www.cnn.com

Brown, Anna, and Eileen Patten. 2014 (April 29). "Statistical Portrait of the Foreign-Born Population in the United States, 2012." Pew Research Center. Available at www.pewhispanic.org

Brown University Steering Committee on *Slavery and Justice*. 2007. Slavery and Justice. Providence, RI: Brown University.

CNN. 2009 (June 18). "Senate Approves Resolution Apology for Slavery." Available at www.cnn.com

Colby, Sandra L., and Jennifer M. Ortman. 2015 (March). "Projections of the Size and Composition of the U.S. Population: 2014 to 2060." *Current Population Reports* P25-1143. U.S. Census Bureau. Available at www.census.gov

Colford, Paul. 2013 (April 2). "'Illegal Immigrant' No More." *The Definitive Source*, AP Blog. Available at blog.ap.org

Conley, Dalton. 1999. *Being Black, Living in the Red: Race, Wealth, and Social Policy in America*. Berkeley: University of California Press.

Cooper, David. 2015 (May 13). "The Policy Failures Exposed by the *New York Times'* Nail Salon Investigation." Economic Policy Institute. Available at www.epi.org

Croll, Paul R. 2013. "Explanations for Racial Disadvantage and Racial Advantage: Beliefs about Both Sides of Inequality in America." *Ethnic and Racial Studies* 36(1):47–74.

"DOJ Study: Hate Crimes More Prevalent Than Previously Known." 2013 (Fall). *Intelligence Report* 151:7.

Dudziak, Mary. 2000. Cold War Civil Rights: *Race and the Image of American Democracy*. Princeton, NJ: Princeton University Press.

Dwyer, Devin. 2011 (January 5). "Opponents of Illegal Immigration Target Birthright Citizenship." ABC News. Available at www.abcnews.go.com

Equal Employment Opportunity Commission (EEOC). 2011 (June 22). "A. C. Widenhouse Sued by EEOC for Racial Harassment." Press Release. Available at www.eeoc.gov

Ennis, Sharon R., Merarys Rios-Vargas, and Nora G. Albert. 2011 (May). "The Hispanic Population: 2010." *2010 Census Briefs*. U.S. Census Bureau. Available at www.census.gov

Esposito, John L. 2011. "Getting It Right about Islam and American Muslims." *Huffington Post*, May 25. Available at www.huffingtonpost.com

Federal Bureau of Investigation (FBI). 2014. *Hate Crime Statistics 2013*. Available at www.fbi.gov

Frey, William H. 2015 (February 23). "Our Rising Black-White Multiracial Population." Brookings Institution. Available at www.brookings.edu

Frieden, Bonnie. 2013 (April 3). "'I Don't See Race': The Pitfalls of the Colorblind Mindset." *Washington University Political Review*. Available at www.wupr.org

Fry, Richard. 2009. "Sharp Growth in Suburban Minority Enrollment Yields Modest Gains in School Diversity." Pew Hispanic Center. Available at www.pewhispanic.org

Futon, Deirdre. 2014 (October 13). "In Rejecting Columbus, Cities Forge Path toward System Alternative." Common Dream. Available at www .commondream.org

Gaertner, Samuel L., and John F. Dovidio. 2000. *Reducing Intergroup Bias: The Common In-Group Identity Model*. Philadelphia: Taylor & Francis.

Gallup Organization. 2013. Race Relations. Available at www.gallup.com

Gambino, C. P., Y. D. Acosta, and E. M. Grieco. 2014 (June). "English-Speaking Ability of the Foreign-Born Population in the United States: 2012." *American Community Survey Reports*. U.S. Census Bureau. Available at www.census.gov

Gayle, Damien. 2015 (May 11). "Michelle Obama: I was 'Knocked Back' by Race Perceptions." *The Guardian*. Available at www.theguardian.com

Gebreyes, Rahel. 2015. "The Unfortunate Consequences of Banning Sharia Law." *Huffington Post*, March 3. Available at www.huffingpost.com

Glaser, Jack, Jay Dixit, and Donald P. Green. 2002. "Studying Hate Crime with the Internet: What Makes Racists Advocate Racial Violence?" *Journal of Social Issues* 58(1):177–193.

Grieco, Elizabeth M., Yesenia D. Acosta, G. Patricia de la Cruz, Christine Gambino, Luke J. Larsen, Edward N. Trevalyan, and Nathan P. Walters. 2012. "The Foreign-Born Population in the United States: 2010." *American Community Survey Reports*. U.S. Census Bureau. Available at www.census.gov

Grinberg, Emanuella. 2014 (October 13). "Instead of Columbus Day, Some U.S. Cities Celebrate Indigenous People's Day." CNN. Available at www.cnn.com

Halper, Katie. 2014 (October 13). "Five Scary Christopher Columbus Quotes That Let You Celebrate the Holiday the Right Way." Raw Story. Available at www.rawstory.com

Hankes, Keegan. 2014 (Winter). "Music & Money & Hate." *Intelligence Report,* Issue 156, pp. 34–36.

Hannon, Lance. 2015. "White Colorism." *Social Currents* 2(1):13–21.

Harwood, S. A., S. Choi, M. Orozco, M. Browne-Huntt, and R. Mendenhall. 2015. *Racial Microaggressions at the University of Illinois at Urbana-Champaign: Voices of Students of Color in the Classroom*. Urbana: University of Illinois.

Harwood, Stacy A., Margaret Browne Huntt, Ruby Mendenhall, and Jioni A. Lewis. 2012. "Racial Microaggressions in the Residence Halls: Experiences of Students of Color at a Predominantly White University." *Journal of Diversity in Higher Education* 5(3):159–173.

"Hear and Now." 2000 (Fall). *Teaching Tolerance*, p. 5.

Higginbotham, Elizabeth, and Margaret L. Andersen. 2012. "The Social Construction of Race and Ethnicity." In *Race and Ethnicity in Society*, 3rd ed., ed. E. Higginbotham and M. L. Anderson, 3–6. Belmont, CA: Wadsworth, Cengage Learning.

Hodgkinson, Harold L. 1995. "What Should We Call People? Race, Class, and the Census for 2000." *Phi Delta Kappa*, October, pp. 173–179.

hooks, bell. 2000. *Where We Stand: Class Matters*. New York: Routledge.

Humes, Karen R., Nicholas A. Jones, and Roberto R. Ramirez. 2011 (March). "Overview of Race and Hispanic Origin: 2010." *2010 Census Briefs*. Available at www.census.gov

Humphreys, Debra. 1999. "Diversity and the College Curriculum: How Colleges and Universities Are Preparing Students for a Changing World." *Diversity-Web*. Available at www .inform.umd.edu

"Ivy League Legacy Admissions." 2015 (May 12). Available at www.ivycoach.com

Jones, Jeffrey M. 2013 (July 24). "In U.S., Most Reject Considering Race in College Admissions." Gallup Organization. Available at www.gallup.org

Jordan, Reed. 2014 (August 27). "America's Public Schools Remain Highly Segregated." Urban Institute. Available at www.urban.org

Kahlenberg, Richard D. 2013. "How to Fight Growing Economic and Racial Segregation in Higher Education." *Chronicle of Higher Education*, August 7. Available at www.chronicle.com

Keenan, Jillian. 2014. "Kick Andrew Jackson off the $20 Bill!" *Slate*, March 3. Available at www .slate.com

King, Joyce E. 2000. "A Moral Choice." *Teaching Tolerance* 18 (Fall):14–15.

Kozol, Jonathan. 1991. *Savage Inequalities: Children in America's Schools*. New York: Crown.

Krogstad, Jens Manuel. 2015 (April 8). "Reflecting a Racial Shift, 78 Counties Turned Majority-Minority since 2000." Pew Research Center. Available at www.pewresearch.org

Krogstad, Jens Manuel, and Jeffrey S. Passel. 2014 (November 14). "Obama's Expected Immigration Action: How Many Would Be Affected." Pew Research Center. Available at www.pewresearch .org

Kwok, Irene, and Yuzhou Wang. 2013. "Locate the Hate: Detecting Tweets against Blacks." Proceedings of the Twenty-seventh AAAI Conference on Artificial Intelligence. Available at www.aaai.org

Leadership Council on Civil Rights Education Fund. 2009. *Confronting the New Faces of Hate: Hate Crimes in America*. Available at www.civilrights.org

Levin, Jack, and Jack McDevitt. 1995. "Landmark Study Reveals Hate Crimes Vary Significantly by Offender Motivation." *Klanwatch Intelligence Report*, August, pp. 7–9.

Loewen, James. 2006. *Sundown Towns*. New York: Touchstone.

Logan, Trevon, and John Parman. 2015. "The National Rise in Residential Segregation." National Bureau of Economic Research, Working Paper 20934. Available at www.econ.ucsb.edu

Ly, Laura. 2013 (March 4). "Oberlin Cancels Classes to Address Racial Incidents." CNN. Available at www.cnn.com

Marger, Martin N. 2012. *Race & Ethnic Relations: American and Global Perspectives*, 9th ed. Belmont CA: Wadsworth, Cengage Learning.

Maril, Robert Lee. 2004. *Patrolling Chaos: The U.S. Border Patrol in Deep South Texas*. Lubbock: Texas Tech University Press.

Maril, Robert Lee. 2011. *The Fence: National Security, Public Safety, and Illegal Immigration along the U.S.-Mexico Border*. Lubbock: Texas Tech University Press.

Martin, N. D., W. Tobin, and K. I. Spenner. 2014. "Interracial Friendships across the College Years: Evidence from a Longitudinal Case Study." *Journal of College Student Development* 55(7):720–725.

Matthews, Cate. 2014. "He Dropped One Letter in His Name While Applying for Jobs, and the Responses Rolled In." *Huffington Post*, September 2. Available at www.huffingtonpost.com

McIntosh, Peggy. 1990 (Winter). "White Privilege: Unpacking the Invisible Knapsack." *Independent School* 49(2):31–35.

Mooney, Chris. 2014 (December 1). "The Science of Why Cops Shoot Young Black Men." *Mother Jones*. Available at www.motherjones.com

Mukhopadhyay, Carol C., Rosemary Henze, and Yolanda T. Moses. 2007. *How Real Is Race?* Lanham, MD: Rowman & Littlefield Education.

Nadal, Kevin L., Yinglee Wong, Katie E. Griffin, Kristin Davidoff, and Julie Sriken. 2014. "The Adverse Impact of Racial Microaggressions on College Students' Self-Esteem." *Journal of College Student Development* 55(5):461–474.

Nagle, Mary Kathryn. 2015 (June 22). "Take Andrew Jackson off the $20 Bill." MSNBC. Available at www.msnbc.com

National Conference of State Legislatures (NCSL). 2013 (June). "Affirmative Action: An Overview." Available at www.ncsl.org

National Immigration Law Center. 2015 (April 6). "Table: Laws & Policies Improving Access to Higher Education for Immigrants." Available at www.nilc.org

Ossorio, Pilar, and Troy Duster. 2005. "Race and Genetics." *American Psychologist* 60(1):115–128.

Pager, Devah. 2003. "The Mark of a Criminal Record." *American Journal of Sociology* 108(5):937–975.

Passel, Jeffrey S., and D'Vera Cohn. 2009. "A Portrait of Unauthorized Immigrants in the United States." Pew Hispanic Research Center. Available at www.pewhispanic.org

Passel, Jeffrey S., and D'Vera Cohn. 2011. "Unauthorized Immigrant Population: National and State Trends." Pew Hispanic Research Center. Available at www.pewhispanic.org

Passel, Jeffrey S., and Paul Taylor. 2009. "Who's Hispanic?" Pew Hispanic Research Center. Available at www.pewhispanic.org

Pérez-Peña, Richard. 2015. "Black or White? Woman's Story Stirs Up a Furor." *New York Times*, June 12. Available at www.nytimes.com

Pew Hispanic Center. 2013 (January 29). *A Nation of Immigrants*. Available at www.pewhispanic.org

Pew Research Center. 2010 (January 12). "Blacks Upbeat about Black Progress, Prospects." Available at www.pewsocialtrends.org

Pew Research Center. 2015 (January 15). "Unauthorized Immigrants: Who They Are and What the Public Thinks." Available at www.pewresearch.org

Picca, Leslie, and Joe R. Feagin. 2007. *Two-Faced Racism*. New York: Routledge.

Pollin, Robert. 2011. "Economic Prospects: Can We Stop Blaming Immigrants?" *New Labor Forum* 20(1):86–89.

Potok, Mark. 2015 (Spring). "The Year in Hate and Extremism." *Intelligence Report,* No. 157. Available at www.splc.org

Riffkin, Rebecca. 2015 (August 26). "HIgher Support for Gender Affirmative Action Than Race." Gallup Organization. Available at www.gallup.com

Schiller, Bradley R. 2004. *The Economics of Poverty and Discrimination*, 9th ed. Upper Saddle River, NJ: Pearson Education.

Schuman, Howard, and Maria Krysan. 1999. "A Historical Note on Whites' Beliefs about Racial Inequality." *American Sociological Review* 64:847–855.

Shierholz, Heidi. 2010 (February 4). "Immigration and Wages—Methodological Advancements Confirm Modest Gains for Native Workers." EPI Briefing Paper #255. Available at www.epi.org

Shierholz, Heidi, Natalie Sabadish, and Nicholas Finio. 2013. "Young Graduates Still Face Dim Job Prospects." Briefing Paper #360. Economic Policy Institute. Available at www.epi.org

Shipler, David K. 1998. "Subtle vs. Overt Racism." *Washington Spectator* 24(6):1–3.

Sidanius, Jim, Shana Levin, Colette Van Laar, and David O. Sears. 2010. *The Diversity Challenge: Social Identity and Intergroup Relations on the College Campus*. New York: Russell Sage Foundation.

Sinclair, Stacey, Andreana C. Kenrick, and Drew S. Jacoby-Senghor. 2014 (October). "Whites' Interpersonal Interactions Shape, and Are Shaped by, Implicit Prejudice." *Policy Insights from the Behavioral and Brain Sciences* 1(1):81–87.

Southern Poverty Law Center (SPLC). 2013. *Close to Slavery: Guestworker Programs in the United States (2013 edition)*. Montgomery, AL: Southern Poverty Law Center.

Southern Poverty Law Center (SPLC). 2015 (Spring). "iTunes Dumps Hate Music but Spotify and Amazon Still Selling It." *Intelligence Report*, No. 157. Available at www.splcenter.org

Stepler, Renee, and Anna Brown. 2015 (May 12). "Statistical Portrait of Hispanics in the United States, 1980–2013." Pew Research Center. Available at www.pewhispanic.org

Sue, D. W., C. M. Capodilup, G. C. Torino, M. Bucceri, A. M. B. Holder, K. L. Nadal, et al. 2007. "Racial Microaggressions in Everyday Life: Implications for Clinical Practice." *American Psychology* 62(4):271–286.

Tanneeru, Manav. 2007 (May 11). "Asian-Americans' Diverse Voices Share Similar Stories." CNN.com. Available at www.cnn.com

Tavernise, Sabrina. 2011. "In Census, Young Americans Increasingly Diverse." *New York Times*, February 4. Available at www.nytimes.com

Teaching Tolerance. 2011 (Spring). "10 Myths about Immigration." Available at www.tolerance.org

Tolbert, Caroline J., and John A. Grummel. 2003. "Revisiting the Racial Threat Hypothesis: White Voter Support for California's Proposition 209." *State Politics and Policy Quarterly* 3(2):183–202, 215–216.

Turn It Down. 2009 (May 3). "Social Networking: A Place for Hate?" Available at turnitdown.newcomm.org

Turner, Margery Austin, Stephen L. Ross, George Galster, and John Yinger. 2002. *Discrimination in Metropolitan Housing Markets*. Washington, DC: Urban Institute.

Urban Dictionary. Available at www.urbandictionary.com

U.S. Census Bureau (2011; generated by C. Schacht, July 3). *2010 Census Data*. Available at 2010.census.gov/2010census/data/index.php

U.S. Census Bureau. 2013. Current Population Survey. Table A-3. "Mean Earnings of Workers 18 Years and Over, by Educational Attainment, Race, and Hispanic Origin: 1975–2011." Available at www.census.gov

U.S. Citizenship and Immigration Services. 2011. *A Guide to Naturalization*. Available at www.uscis.gov

U.S. Customs and Border Protection. 2013. "U.S. Border Patrol Fiscal Year 2012 Statistics." Available at www.cbp.gov

U.S. Department of Labor. 2002. *Facts on Executive Order 11246 Affirmative Action*. Available at www.dol.gov

Wang, Wendy. 2012. "The Rise of Intermarriage." Pew Research Center. Available at www.pewresearch.org

Washington Post Staff. 2013. "President Obama's Remarks on Trayvon Martin (full transcript)." *Washington Post*, July 19. Available at www.washingtonpost.com

Williams, Eddie N., and Milton D. Morris. 1993. "Racism and Our Future." In *Race in America: The Struggle for Equality*, ed. Herbert Hill and James E. Jones Jr., 417–424. Madison: University of Wisconsin Press.

Williams, Richard, Reynold Nesiba, and Eileen Diaz McConnell. 2005. "The Changing Face of Inequality in Home Mortgage Lending." *Social Problems* 52(2):181–208.

Wilson, Jill H. 2014 (September 24). "Investing in English Skills: The Limited English Proficient Workforce in U.S. Metropolitan Areas." Metropolitan Policy Program. Available at www.brookings.edu

Winfrey, Oprah. 2009. "Oprah Talks to Jay-Z." *O, The Oprah Magazine*, October. Available at www.oprah.com

Winter, Greg. 2003. "Schools Resegregate, Study Finds." *New York Times,* January 21. Available at www.nytimes.com

Wise, Tim. 2009. *Between Barack and a Hard Place: Racism and White Denial in the Age of Obama.* San Francisco: City Light Books.

Yeung, Jeffrey G., Lisa B. Spanierman, and Jocelyn Landrum-Brown. 2013. "'Being White in a Multicultural Society': Critical Whiteness Pedagogy in a Dialogue Course." *Journal of Diversity in Higher Education* 6(1):17–32.

Chapter 10

Abernathy, Michael. 2003. *Male Bashing on TV.* Tolerance in the News. Available at www.tolerance.org

Almed, Shumaila, and Juliana Abdul Wahab. 2014. "Animation and Socialization Process: Gender Role Portrayal in Cartoon Network." *Asian Social Science* 10(3):44–53.

American Association of University Women (AAUW). 2015. *The Simple Truth about the Gender Pay Gap.* Available at www.aauw.org

American Women. 2015 (January 20). "National Survey: Women Continue to Support Paycheck Fairness Act, and Economic Proposals to Give Women and Families of Fair Shot." Available at www.americanwomen.org

Amnesty International. 2015. "The International Violence against Women Act (IVAWA)." Spring. *Issue Brief,* No. 4. Available at amnestyusa.org

Andersen, Margaret L. 1997. *Thinking about Women,* 4th ed. New York: Macmillan.

Arrindell, W. A., Sonja Van Well, Annemarie M. Kolk, Dick P. H. Barelds, Tian P. S. Oei, Pui Yi Lau, and the Cultural Clinical Psychology Study Group. 2013. "Higher Levels of Masculine Gender Role Stress in Masculine Than in Feminine Nations: A 13-Nations Study." *Cross-Cultural Research* 47(1):51–67.

Askari, Sabrina F., Mirian Liss, Mindy J. Erchull, Samantha E. Staebell, and Sarah J. Axelson. 2010. "Men Want Equality, but Women Don't Expect It: Young Adults' Expectations for Participation in Household and Child Care Chores." *Psychology of Women Quarterly* 34(2): 243–252.

Bajekal, Naina. 2015. "Pope Francis Says Church Must Listen to Women." *Time,* April 15. Available at time.com

Bassett, Laura. (April 14) 2015. "Republicans Float Stripped-Down Equal Pay Bill." *The Huffington Post.* Available at www.huffingtonpost.com

Bedi, Rahul. 2013. "Indian Dowry Deaths on the Rise." *The Telegraph,* February 28. Available at www.telegraph.co.uk

Begley, Sharon. 2000. "The Stereotype Trap." *Newsweek,* November 6, pp. 66–68.

Benenson Strategy Group. 2015. *January 2015 Survey of Millennials.* Report to Fusion.net. Available at fusion.net

Bertrand, Marianne, Claudia Goldin, and Lawrence F. Katz. 2009 (January). "Dynamics of the Gender Gap for Young Professionals in the Financial and Corporate Sectors." Working Paper. Available at www.economics.harvard.edu

Blau, Francine D., and Lawrence M. Kahn. 2013. "Female Labor Supply: Why Is the US Falling Behind?" January. NBER Working Paper No.18702. National Bureau of Economic Research. Available at www.nber.org

Bly, Robert. 1990. *Iron John: A Book about Men.* Boston: Addison-Wesley.

Boorstein, Michelle. 2014. "US Evangelicals Headed for Showdown over Gender Roles." *Washington Post,* September 3Available at www.washingtonpost.com

Boylan, Jennifer Finney. 2015. "How to Save Your Life: A Response to Leelah Alcorn's Suicide Note." *New York Times,* January 6. Available at www.nytimes.com

British Broadcasting Corporation. "Libby Lane: First Female Church of England. Bishop Consecrated." *BBC News.* January 26. Available at www.bbc.co

Bruce, Adrienne, N., Alexis Battista, Michael W. Plankey, Lynt B. Johnson, and M. Blair Marshall. 2015. "Perceptions of Gender-Based Discrimination during Surgical Training and Practice." *Medical Education Online* 20 (February):1–10.

Bureau of Labor Statistics (BLS). 2013. "Characteristics of Minimum Wage Workers: 2012." Labor Force Statistics from the Current Population Survey. Available at www.bls.gov

Bureau of Labor Statistics (BLS). 2015. "Employed Persons by Detailed Occupation, Sex, Race, and Hispanic or Latino Ethnicity." Labor Force Statistics from the Current Population Survey. Available at www.bls.gov

Bush, Bianca, and Adrian Furnham 2013 "Gender Jenga: The Role of Advertising in Gender Stereotypes within Educational and Non-Educational Games" *Young Consumers* 14(3):216–229.

Callaway, Chris. 2015 (January 28). "What Is the Average Salary of a Basketball Player?" Available at www.livestrong.com

Ceci, Stephen J., Wendy M. Williams, and Susan M. Barnett. 2009. "Women's Underrepresentation in Science: Socio-cultural and Biological Considerations." *Psychological Bulletin* 135(2):218–261.

Center for American Women and Politics (CAWP). CAWP. 2015a. "Women of Color in Elective Offices, 2015." Available at www.cawp.rutgers.edu

Center for American Women and Politics (CAWP). 2015b. "Women in Elected Office, 2015." *Fact Sheet.* Available at www.cawp.rutgers.edu

Center for American Women and Politics (CAWP). CAWP. 2015c. "Women in State Legislatures, 2015." Available at www.cawp.rutgers.edu

Cohen, Philip. 2004. "The Gender Division of Labor: 'Keeping House' and Occupational Sex Segregations in the United States." *Gender and Society* 18 (2) 239–252.

Cohen, Theodore. 2001. *Men and Masculinity.* Belmont, CA: Wadsworth.

Congress.gov. 2015. "S. 84—Paycheck Fairness Act." Available at www.congress.gov

Cook, Carolyn. 2009. "ERA Would End Women's Second-Class Citizenship: Only Three More States Are Needed to Declare Gender Bias Unconstitutional." *Philadelphia Inquirer,* April 12. Available at www.philly.com

Corbett, Christianne, and Catherine Hill. 2012 (October 24). *Graduating to a Pay Gap: The Earnings of Women and Men One Year after College Graduation.* American Association of University Women. Available at www.aauw.org

Correll, Shelly J., Stephen Benard, and In Paik. 2007. "Getting a Job: Is There a Motherhood Penalty?" *American Journal of Sociology* 112(5):1297–1338.

Dallesasse, Starla L., and Annette S. Kluck. 2013. "Reality Television and the Muscular Male Ideal." *Body Image* 10:309–315.

DeNavas-Walt, Carmen, and Bernadette D. Proctor. 2014. "Income and Poverty in the United States 2013." *Current Population Reports,* September. Available at www.census.gov

DeSilver, Drew. 2015. "Despite Progress, U.S. Still Lags Many Nations in Women Leaders." Pew Research Center. January 26. Available at www.pewresearch.org

Dewan, Shaila, and Robert Gebeloff. 2012. "More Men Enter Fields Dominated by Women." *New York Times,* May 20. Available at www.nytimes.com

Dunn, James. 2015."Sportswear Company Apologises after Washing Instructions to 'Give It to a Woman. It's Her Job.'" *The Independent,* March 8. Available at www.independent.co.uk

Dunn, Marianne G, Aaron B. Rochlen, and Karen M. O'Brien. 2013. "Employee, Mother, and Partner: An Exploratory Investigation of Working Women with Stay-at-Home Fathers." *Journal of Career Development* 40(1):3–22.

Eckler, Petya, Yusuf Kalyango, and Ellen Paasch. 2014 (April). "Increased Time on Facebook could Lead to Negative Body Images." Paper presented at the 64th Annual meetings of the International Communication Association, Seattle.

Edmeades, J. R. Hayes, and G. Gaynair. 2014. "Improving the Lives of Married Adolescent Girls in Amhara, Ethiopia." Washington, DC: International Research Center on Women.

Eisner, Manuel, and Lana Ghuneim. 2013. "Honor Killing Attitudes amongst Adolescents in Amman, Jordan." *Aggressive Behavior* 39(5):405–417.

Equal Employment Opportunity Commission (EEOC). 2012 "Facts about Sexual Harassment." Available at www.eeoc.gov

Equal Employment Opportunity Commission (EEOC). 2013. *Charge Statistics: FY 1997 through FY 2012.* Available at www.eeoc.gov

Equal Employment Opportunity Commission (EEOC). 2015. "Sex-Based Charges" FY 1997 through FY 2014. Available at www.eeoc.gov

Espinoza, Penelope, Ann B. Areas du Luz Fontes, and Clarissa J. Arms-Chavez. 2014. "Attributional Gender Bias: Teacher's Ability and Effort Explanations for Student's Math Performance." *Social Psychology Education* 17:105–126.

European Social Survey. 2013. *Attitudes and Behaviors in a Changing Europe.* July. Available at www.esrc.ac.uk

Faludi, Susan. 1991. *Backlash: The Undeclared War against American Women.* New York: Crown.

Fischer, Jocelyn, and Jeff Hayes. 2013 (August 13). "The Importance of Social Security in the Incomes of Older Americans." Institute for Women's Policy Research. Available www.iwpr.org

Freeman, Daniel, and Jason Freeman. 2013a. "Let's Talk about the Gender Differences That Really Matter—in Mental Health." *The Guardian,* December 13. Available at www.theguardian .com

Freeman, Daniel, and Jason Freeman 2013b. *Stressed Sex: Uncovering the Truth about Men, Women, and Mental Health.* Oxford: Oxford University Press.

Gallagher, Sally K. 2004. "The Marginalization of Evangelical Feminism." *Sociology of Religion* 65:215–237.

Girls' Attitudes Survey. 2013. "What Girls Say about Equality for Girls." Available at girlsat-titudes.girlguiding.org

Goffman, Erving. 1963. *Stigma.* Englewood Cliffs, NJ: Prentice Hall.

Goodkind, Sara, Lisa Schelbe, Andrea A. Joseph, Daphne E. Beers, and Stephanie L. Pinsky. 2013. "Providing New Opportunities or Reinforcing Old Stereotypes? Perceptions and Experiences of Single-Sex Public Education." *Children and Youth Services Review* 35:1174–1181.

Goodstein, Laurie. 2009. "U.S. Nuns Facing Vatican Scrutiny." *New York Times,* July 2. Available at www.nytimes.com

Goodstein, Laurie. 2012. "Vatican Reprimands a Group of U.S. Nuns and Plans Changes." *New York Times,* April 18. Available at www.nytimes.com

Grant, Jaime M., Lisa A. Mottet, Justine Tants, Jack Harrison, Jody L. Herman, and Mara Ketsling. 2011. *Injustice at Every Turn: A Report of the National TransGender Discrimination Survey.* Washington, DC: National Center for Transgender Equality and the National Gay and Lesbian Task Force.

Grogan, Sarah. 2008. *Body Image: Understanding Body Dissatisfaction in Men, Women and Children.* New York: Routledge.

Gupta, Sanjay. 2003. "Why Men Die Young." *Time,* May 12, p. 84.

Guy, Mary Ellen, and Meredith A. Newman. 2004. "Women's Jobs, Men's Jobs: Sex Segregation and Emotional Labor." *Public Administration Review* 64:289–299.

Hamel, Liz, Jamie Firth, and Mollyann Brodie. 2014 (December 11). "Kaiser Family Foundation/ New York Times/CBS News, Non-Employed Whole." Available at kff.org

Hamilton, Mykol C., David Anderson, Michelle Broaddus, and Kate Young. 2006. "Gender Stereotyping and Under-representation of Female Characters in 200 Popular Children's Picture Books: A Twenty-first Century Update." *Sex Roles* 55:757–765.

Hass, Ann P., Philip L. Rodgers, and Jody L. Herman. 2014. "Suicide Attempts among Transgender and Gender-Non-conforming Adults: Findings of the National Transgender Discrimination Survey." Williams Institute. Available at williamsinstitute.law.ucla.edu

Hausmann, Ricardo, Laura D. Tyson, Yasmina Bekhouche, and Saadia Zahidi. 2014. *The Global Gender Gap Report 2014.* Geneva: World Economic Forum.

Head, Sarah K., Sally Zweimueller, Claudia Marchena, and Elliott, Hoel. 2014. *Women's Lives and Challenges: Equality and Empowerment since 2000.* United States Agency for

International Development. Rockville, MD: ICF International.

Hegewisch, Ariane, and Emily Ellis. 2015 (April). "The Gender Wage Gap by Occupation 2014 and by Race and Ethnicity." Institute for Women's Policy Research. Available at www.iwpr.org

Hegewisch, Ariane, Emily Ellis, and Heidi Hartmann. 2015. *The Gender Wage Gap 2014: Earning Differences by Race and Ethnicity.* Institute for Women's Policy Research. September. Available at www.iwpr.org

Hersch, Joni. 2013. "Opting Out among Women with Elite Education." Vanderbilt Law and Economics Research Paper No. 13-05. Nashville, TN: Vanderbilt University.

Heyder, Anke, and Ursula Kessels. 2013. "Is School Feminine? Implicit Gender Stereotyping of School as a Predictor pf Academic Achievement." *Sex Roles* 69:605–617.

Hochschild, Arlie. 1989. *The Second Shift.* London: Penguin.

Human Rights Campaign. 2015. "Sexual Orientation and Gender Identity Definitions." Resources. Available at www.hrc.org

Hvistendahl, Mara. 2011. *Unnatural Selection: Choosing Boys over Girls, and the Consequences of a World Full of Men.* Philadelphia: Public Affairs.

Institute for Women's Policy Research (IWPR). 2014 (September). "Women's Median Earnings as a Percent of Men's Median Earnings, 1960–2013 (Full-time, Year-round Workers) with Projection for Pay Equity in 2058." Quick Figures. Available at www.iwpr.org

Inter-Agency Standing Committee. 2009. "IASC Policy Statement Gender Equality in Humanitarian Action." Available at www.humanitarian.org

International Labour Organization (ILO). 2012 (December). *Global Employment Trends for Women.* Available at www.ilo.org

International Labour Organization (ILO). 2014 (March 10). "Trafficking in Human Beings: A Severe Form of Violence against Women and Girls, and a Flagrant Violation of Human Rights." *Speeches and Statements.* Available at www.ilo.org

International Labour Organization (ILO). 2015a (January). "Global Momentum Means More Women Move into Management." Available at www.ilo.org

International Labour Organization (ILO). 2015b. *Global Wage Report 2014/15.* Available at www .ilo.org

International Labour Organization (ILO). 2015c. *World Employment Social Outlook.* Available at www.ilo.org

International Violence Against Women Act. 2013 (June). "Issue Brief: The International Violence against Women Act." Amnesty International. Available www.amnestyusa.org

International Trade Union Commission. 2011 (March 8). Decisions for Work: An Examination of the Factors Influencing Women's Decisions for Work. Available www.ituc-csi.org

Jackson, Janna. 2010. "'Dangerous Presumptions': How Single-Sex Schooling Reifies False Notions of Sex, Gender, and Sexuality." *Gender and Education* 22(2):227–238.

Jitha, T. J. 2013. "Mediating Production, Re-Powering Patriarchy: The Case of Micro Credit." *India Journal of Gender Studies* 20(2):253–278.

Jose, Paul, and Isobel Brown. 2009. "When Does the Gender Difference in Rumination Begin? Gender and Age Differences in the Use of Rumination by Adolescents." *Journal of Youth and Adolescence* 37:180–192.

Kimmel, Michael. 2011 (Winter). "Gay Bashing Is about Masculinity." *Voice Male.* Available at www.voicemalemagazine.org

Kimmel, Michael. 2012 (December 19). "Masculinity, Mental Illness and Guns: A Lethal Equation?" *CNN News.* Available at www.cnn.com

Klein, Sue, Jennifer Lee, Page McKinsey, and Charmaine Archer. 2014 (December). *Identifying U.S. K–12 Public Schools with Delivered Sex Segregation.* Arlington, Virginia: Feminist Majority Foundation. Available at www .feminist.org/education

Kostakis, Adam. 2014 (May 26). "Lecture 2: The Same Old Gynocentric Story." A Voice for Men. Available at www.avoiceformen.com

Largent, Emily. 2015. *M.C. v. Aaronson* March 5. Bill of Health. Harvard Law. Available at blogs. law.harvard.edu

Leadership Conference on Civil Rights Education Fund. 2009. "Confronting the New Faces of Hate: Hate Crimes in America." Available at www.civilrights.org

Leopold, Thomas, and Jan Skopek, 2014. "Gender and the Division of Labor in Older Couples: How European Grandparents Share Market Work and Childcare." *Social Forces* 93 (1):63–91.

Lepkowska, Dorothy. 2008. "Playing Fair?" *The Guardian,* December 16. Available at www.guardian.co.uk

Leslie, David W. 2007 (March). "The Reshaping of America's Academic Workforce." *Research Dialogue* 87. New York: TIAA-CREF Institute. Available at www.tiaa-crefinstitute.org

Levin, Diane E., and Jean Kilbourne. 2009. *So Sexy So Soon: The New Sexualized Childhood and What Parents Can Do to Protect Their Kids.* New York: Random House.

Lopez-Claros, Augusto, and Saadia Zahidi. 2005. *Women's Empowerment: Measuring the Global Gender Gap.* World Economic Forum.

Machitt, Nico. 2014. "The Next Civil Rights Frontier: How the Transgender Movement is Taking Over." *Huffington Post,* October 14. Available at www.huffington.post

Madera, Juan M., Michelle M. Hebl, and Randi C. Martin. 2009. "Gender and Letters of Recommendation for Academia: Agentic and Communal Differences." *Journal of Applied Psychology* 94(6):1591–1599.

Maliniak, Daniel, Ryan M. Powers, and Barbara F. Walter. 2013. "The Gender Citation Gap." Paper presented at the 2013 meeting of the American Political Science Association, Chicago.

Mason, Katherine. 2012. "The Unequal Weight of Discrimination: Gender, Body Size, and Income Inequality." *Social Problems* 59(3):411–435.

M.C. v. Medical University of South Carolina. U.S. District Court. Filed May 14, 2013.

Mears, Bill. 2014 (April 23). "Michigan's Ban on Affirmative Action Upheld by Supreme Court." CNN News. Available at www.cnn.com

MediaSmarts. 2012. "Body Image—Girls" Available at mediasmarts.ca

Mehmood, Isha. 2009 (January 29). "Lilly Ledbetter Fair Pay Act Becomes Law." Available at www.civilrights.org

Misra, Joya, Jennifer Hickes Lundquist, Elissa Holmes, and Stephanie Agiomavritis. 2011. "The Ivory Ceiling of Service Work." *Academe* 97:22–26.

Messner, Michael A., and Jeffrey Montez de Oca. 2005. "The Male Consumer as Loser: Beer and Liquor Ads in Mega Sports Media Events." *Signs* 30:1879–1909.

Moen, Phyllis, and Yan Yu. 2000. "Effective Work/Life Strategies: Working Couples, Working Conditions, Gender, and Life Quality." *Social Problems* 47:291–326.

Morgan, Whitney Botsford, Katherine B. Elder, and Eden B. King. 2013. "The Emergence and Reduction of Bias in Letters of Recommendation." *Journal of Applied Social Psychology* 43:2297–2306.

Morin, Rich, and Paul Taylor. 2008 (September 15). *Revisiting the Mommy Wars: Politics, Gender and Parenthood*. Available at www.pewsocialtrends.org

Nandita, B., P. Achyut, N. Khan, and S. Walia. 2014. Are Schools Safe and Gender Equal Spaces? Findings from a Baseline Study of School Related Gender-Based Violence in Five Countries in Asia. Washington, DC, and Woking, UK: UK International Centre for Research on Women and Plan International. National Center for Educational Statistics (NCES). 2014. Digest of Education, 2013. U.S. Department of Education. Available at nces.ed.gov

National Center for Educational Statistics (NCES). 2014. *The Condition of Education, 2013*. U.S. Department of Education. Available at nces.ed.gov

National Coalition for Men. 2015. "Philosophy." Available at ncfm.org

National Coalition for Women and Girls in Education. 2012. *Title IX Working to Ensure Gender Equity in Education*. Washington, DC: National Coalition for Women and Girls in Education.

National Conference of State Legislators (NCSL). 2014 (December 8). "Equal Pay 2014 Year-End Legislation: Summary." Available at www.ncsl.org

National Organization for Men against Sexism. 2015. "Principles." Available at www.site.nomas.org

National Organization for Women (NOW). 2015. "About." Available at now.org

National Public Radio. 2011. "Two Spirits: A Map of Gender—Diverse Cultures." Public Broadcasting System. Available at www.pbs.org

National Women's Law Center (NWLC). 2014 (October 1). "Fair Pay for Women Requires Increasing the Minimum Wage and Tipped Minimum Wage." Available at www.nwlc.org

National Women's Law Center (NWLC). 2015a (April). "How the Wage Gap Hurts Women and Families." *Fact Sheet*. Available at www.nwlc.org

National Women's Law Center (NWLC). 2015b. "President Obama's Fiscal Year 2016 Budget: Investing in Women and Their Families." *Issue Brief*. March. Available at www.nwlc.org

National Women's Law Center (NWLC). 2015c (October). "The Wage Gap over Time." *Fact Sheet*. Available at www.nwlc.org

Nordberg, Jenny. 2010. "Afghan Boys Are Prized, So Girls Live the Part." *New York Times*, September 20. Available at www.nytimes.com

Norman, Moss E. 2011. "Embodying the Double-Bind of Masculinity: Young Men and Discourses of Normalcy, Health, Heterosexuality, and Individualism." *Men and Masculinities* 14(4):430–449.

Office of Civil Rights. 2014 (December 31). "Overview of Title IX of the Education Amendments of 1972." Available at justice.gov

Pedulla, David S., and Sarah Thebaud. 2013. "Can We Finish the Revolution? Gender, Work–Family Ideals, and Institutional Constraint." *American Sociological Review* 80(1):116–139.

Pew. 2014a. "The Divide over Ordaining Women." September 9. Pew Research Center. Available at www.pewresearch.org

Pew. 2014b. "Size of Gender Pay Gap Varies by State, Job." January 23. The Pew Charitable Trust. Available at www.pewtrusts.org

Pew. 2015 (January 14). "Women and Leadership." *Social and Demographic Trends*. Pew Research Center. Available at www.pewsocialtrends.org

Polavieja, Javier G., and Lucinda Platt. 2014. "Nurse or Mechanic? The Role of Parental Socialization and Children's Personality in the Formation of Sex-Typed Occupational Aspirations." *Social Forces* 93(1):31–61.

Political Parity. 2014 (July). *Shifting Gears: How Women Navigate the Road to Higher Office*. Available at www.politicalparity.org

Pollack, William. 2000. *Real Boys' Voices*. New York: Random House.

Portnoy, Jenna. 2015. "Debate Rages On over Transgender Elementary School Student in Stafford." *Washington Post*, April 6. Available at www.washingtonpost.com

Quist-Areton, Ofeibea. 2003. "Fighting Prejudice and Sexual Harassment of Girls in Schools." *All Africa*, June 12. Available at www.globalpolicy.org

Recknagel, Charles. 2014 (April 27). "Explainer: Why Is It So Hard to Stop Honor Killings?" *Radio Free Europe Radio Liberty*. Available at www.rferl.org

Rehel, Erin. M. 2014. "When Dad Stays Home Too: Paternity Leave, Gender, and Parenting." *Gender and Society* 28 (1):110–132.

Reskin, Barbara, and Debra McBrier. 2000. "Why Not Ascription? Organizations' Employment of Male and Female Managers." *American Sociological Review* 65:210–233.

Ridgeway, Sicilia L. 2011. Framed by Gender: How Gender Inequality Persists in the Modern World. New York: Oxford University Press.

Sadker, David, and Karen Zittleman. 2009. Still Failing at Fairness: How Gender Bias Cheats Boys and Girls in Schools. New York: Simon and Schuster.

Sanchez, Diana T., and Jennifer Crocker. 2005. "How Investment in Gender Ideals Affects Well-Being: The Role of External Contingencies of Self-Worth." *Psychology of Women Quarterly* 29:63–77.

Sayman, Donna M. 2007. "The Elimination of Sexism and Stereotyping in Occupational Education." *Journal of Men's Studies* 15(1):19–30.

Schneider, Daniel. 2012. "Gender Deviance and Household Work: The Role of Occupation." *American Journal of Sociology* 117(4):1029–1072.

"See Jane." 2013. PSA. Geena Davis Institute on Gender in Media. Available at seejane.org

Simister, John. 2013. "Is Men's Share of House Work Reduced by Gender Deviance Neutralization? Evidence from Seven Countries." *Journal of Comparative Family Studies* 44(3):311–325.

Simpson, Ruth. 2005. "Men in Non-traditional Occupations: Career Entry, Career Orientation, and Experience of Role Strain." *Gender Work and Organization* 12(4):363–380.

Sloan, Colleen A., Danielle S. Berke, and Amos Zeichner. 2015. "Bias-Motivation against Men: Gender Expressions and Sexual Orientation as Risk Factors for Victimization." *Sex Roles* 72:140–149.

Smith, Stacy L., Marc Choueiti, Ashley Prescott, and Catherine Pieper. 2015. *Gender Bias without Borders*. Annenberg School for Communication and Journalism. University of Southern California. Available at www.seejane.org

Snyder, Karrie Ann, and Adam Isaiah Green. 2008. "Revisiting the Glass Escalator: The Case of Gender Segregation in a Female Dominated Occupation." *Social Problems* 55(2):271–299.

Southern Poverty Law Center. 2013 (May 14). "Groundbreaking SPLC Lawsuit Accuses South Carolina, Doctors and Hospitals of Unnecessary Surgery on Infant." Available at www.splcenter.org

Stancill, Jane. 2013. "UNC-CH Women Wage National Campaign against Sexual Assault." *The News and Observer*, June 1. Available at www.newsobserver.com

Stohr, Greg. 2011. "Wal-Mart Million-Worker Bias Suit Thrown Out by High Court." *Bloomberg*, June 20. Available at www.bloomberg.com

Strain, Michael R. 2013. "Single-Sex Class and Student Outcomes: Evidence from North Carolina." *Economics of Education Review* 36:73–87.

Tenenbaum, Harriet R. 2009. "You'd Be Good at That: Gender Patterns in Parent–Child Talk about Courses." *Social Development* 18(2):447–463.

Uggen, C., and A. Blackstone. 2004. "Sexual Harassment as a Gendered Expression of Power." *American Sociological Review* 69:64–92.

UNESCO. 2014a. "Statistics on Literacy." Available at www.uis.unesco.org

UNESCO. 2014b. *Teaching and Learning: Achieving Equality For All*. EFA Global Monitoring Report. Available www.unescdoc.unesco.org

Ungar-Sargon, Batya. 2013. "Orthodox Yeshiva Set to Ordain 3 Women; Just Don't Call Them Rabbi." *Tablet*, June 1. Available at tabletmag.com

United Equality. 2015. *Equal Rights Amendment 2015*. Available at www.united4equality.com

United Nations (UN). 2014. *The Millennium Development Goals Report 2014*. Available at www.un.org

United Nations (UN). 2015a. "Education for All 2000–2015: Achievements and Challenges." *EFA Global Monitoring Report 2015*. Available at unesdoc.unesco.org

United Nations (UN). 2015b. "Facts and Figures: Leadership and Political Participation." *UN Women*. Available at www.unwomen.org

United Nations (UN). 2015c. "2015: Empowering Women—Empowering Humanity: Picture It." March 8. International Women's Day. Available at www.un.org

United Nations Development Fund for Women. 2007. "Harmful Traditional Practices." Available at www.unifem.org

"The Upshot." 2014. "Why US Women Are Leaving Jobs Behind." *New York Times,* December 12. Available at www.nytimes.com

USAID. 2014 United States Strategy to Prevent and Respond to Gender-Based Violence. Available at pdf.usaid.gov.

U.S. Census Bureau. 2009. *Statistical Abstract of the United States: 2008,* 128th ed. Washington, DC: U.S. Government Printing Office.

U.S. Census Bureau. 2013. *Statistical Abstract of the United States: 2012,* 130th ed. Washington, DC: U.S. Government Printing Office.

U.S. Department of State. 2012 (May 30). "The U.S. Response to Global Maternal Mortality: Saving Mothers, Giving Life." Available at www.state.gov

Valiente, Christian, and Xeno Rasmusson. 2015. "Bucking the Stereotypes: *My Little Pony* and the Challenges to Traditional Gender Roles. *Journal of Psychological Issues in Organizational Culture* 5(4):88–97.

Vandello, Joseph A., Jennifer K. Bosson, Dov Cohen, Rochelle M. Burnaford, and Jonathan R. Weaver. 2008. "Precarious Manhood." *Journal of Personality and Social Psychology* 95 (6):1325–1339.

Van der Gaag, Nikki. 2014. "Women Are Better Off Today, but Still Far from the People with Men." *The Guardian,* December 29. Available at www.theguardian.com

Wang, Wendy, Kim Parker, and Paul Taylor. 2013 (May 29). *Breadwinner Moms.* Pew Research Social and Demographic Trends. Available at www.pewsocialtrends.org

Warner, Ann, Kirsten Stoebenau, and Allison M. Glinski. 2014. *More Power to Her: How Empowering Girls Can End Child Marriage.* International Center for Research on Women. Available at www.icrw.org

Weeks, Linton. 2011 (June 23). "The End of Gender?" National Public Radio. Available at www.npr.org

Westbrook, Laurel, and Kristen Schilt. 2014. "Doing Gender, Determining Gender: Transgender People, Gender Panics, and the Maintenance of the Sex/Gender/Sexuality System." *Gender and Society* 28(1):32–57.

White, Alan, and Karl Witty. 2009. "Men's Under-use of Health Services—Finding Alternative Approaches." *Journal of Men's Health* 6(2):95–97.

Willer, Robb, Christabel L. Rogalin, Bridget Conlon, and Michael T. Wojnowicz. 2013. "Overdoing Gender: A Test of the Masculine Overcompensation Thesis." *American Journal of Sociology* 118(4): 980–1022.

Wikle, Jocelyn. 2014. "Patterns in Housework and Childcare among Girls and Boys." *Journal of Research on Women and Gender* 5:17–29.

Williams, Christine L. 2007. "The Glass Escalator: Hidden Advantages for Men in the 'Female' Occupations." In *Men's Lives,* 7th ed., ed. Michael S. Kimmel and Michael Messner, 242–255. Boston: Allyn and Bacon.

Williams, Joan. 2000. *Unbending Gender: Why Family and Work Conflict and What to Do about It.* Oxford: Oxford University Press.

Williams, Kristi, and Debra Umberson. 2004. "Marital Status, Marital Transitions, and Health: A Gendered Life Course Perspective." *Journal of Health and Social Behavior* 45:81–98.

Women in Higher Education. 2013. "Faculty Sex Bias Costs, N.J. Med U $4.6 Million." January. *Women in Higher Education* 2 (1):4.

Wood, Wendy, and Alice H. Eagly. 2002. "A Cross-Cultural Analysis of the Behavior of Women and Men: Implications for the Origins of Sex Differences." *Psychological Bulletin* 128(5):699–727.

World Health Organization (WHO). 2008. *Eliminating Female Genital Mutilation: An Interagency Statement.* Available at wholibdoc.who.int

World Health Organization (WHO). 2009. "Ten Facts about Women's Health." Available at www.who.int

World Health Organization (WHO). 2010. "Gender, Women and Primary Heal Care Renewal." Discussion paper. Available at from wholibdoc.who.int

World Health Organization (WHO). 2011. Global Sector Strategy on HIV/AIDS 2011–2015. Available at wholibdoc.who.ints

World Health Organization (WHO). 2014a (February). "Female Genital Mutilation." Available at www.who.int

World Health Organization (WHO). 2014b. "Maternal Mortality." *Fact Sheet.* May. Available at www.who.int

World Health Organization (WHO). 2014c. *Preventing Suicide: A Global Imperative.* Available at www.who.int

World Health Organization (WHO). 2014d. *World Health Statistics 2014.* Available at apps.who.int

Xinran. 2012. Message from an Unknown Chinese Mother: Stories of Loss and Love. New York: Scribner.

Yoder, P. Stanley, N. Abderrahim, and A. Zhuzhuni. 2004. *Female Genital Cutting in the Demographic and Health Surveys: A Critical and Comparative Analysis.* DHS Comparative Reports 7. Calverton, MD: ORC Macro.

YouGov. 2014 (August 1). "Feminism Today: What Does It Mean?" Available at today.yougov.com

Zakrzewski, Paul. 2005. "Daddy, What Did You Do in the Men's Movement?" *Boston Globe,* June 19. Available at www.bostonglobe.com

Chapter 11

Abraham, Eyal, Talma Hendler, Irit Shapira-Lichter, Yaniv Kanat-Maymon, Oma Zagoory-Sharon, and Ruth Feldman. "Father's Brain Is Sensitive to Childcare Experiences." *Proceedings of the National Academy of Sciences of the United States of America: Current Issues* 111(27):9792–9797.

Ackerman, Spencer. 2013. "Gay Military Couples Welcome Pentagon Decision to Extend Benefits." *The Guardian,* August 14. Available at www.theguardian.com

Allport, G. W. 1954. *The Nature of Prejudice.* Cambridge, MA: Addison-Wesley.

American College Health Association. 2014. "National College Health Assessment II: Reference Group Executive Summary Spring 2014." Available at www.acha-ncha.org

American Medical Association. 2011. "AMA Policy Regarding Sexual Orientation." *GLBT Advisory Committee.* Available at www.ama-assn.org

American Sociological Association (ASA). 2015 (March). "Brief of *Amicus Curiae* American Sociological Association In Support of Petitioners." Available at www.asanet.org

Aragon, S. R., V. Paul Poteat, and Dorothy L. Espelage. 2014. "The Influence of Peer Victimization on Educational Outcomes for LGBTQ and Non-LGBTQ High School Students." *Journal of LGBT Youth* 11(1):1–19.

Associated Press. 2014. "Tell 12 States Still Banned Sodomy a Decade after Court Ruling." *USA Today,* April 21. Available at will www.usatoday.com

Ayala, Edgardo. 2015 (February 18). "LGBTI Community in Central America Fights Stigma and Abuse." Available at www.ipsnews.net

Bariso, the Hon. Peter F. 2015 (February 10). "Order Granting Plaintiff's Motion for Partial Summary Judgment." Docket No. L-5473-12. Available at www.splcenter.org

Bayer, Ronald. 1987. *Homosexuality and American Psychiatry: The Politics of Diagnosis,* 2nd ed. Princeton, NJ: Princeton University Press.

Belkin, A., M. Ender, N. Frank, S. Furia, G. R. Lucas, G. Packard, T. S. Schultz, S. M. Samuels, and D. R. Segal. 2012. *One Year Out: An Assessment of DADT Repeal's Impact on Military Readiness.* Palm Center: University of California, Los Angeles.

Besen, Wayne. 2010. "Ex-Gay Group Should Repent, Not Revel." *Huffington Post,* June 17. Available at www.huffingtonpost.com

Bostwick, W. B., I. Meyer, F. Aranda, S. Russell, T. Hughes, M. Birkett, and B. Mustanski. 2014. "Mental Health and Suicidality among Racially/Ethnically Diverse Sexual Minority Youths." *American Journal of Public Health* 104(6): 1129–1136.

Brown, Michael J., and Ernesto Henriquez. 2008. "Socio-Demographic Predictors of Attitudes towards Gays and Lesbians." *Individual Differences Research* 6:193–202.

Calamur, Krishnadev. 2015 (May 21). "Head of Boy Scouts Says Group's Ban on Gay Adults 'Unsustainable.'" National Public Radio. Available at www.npr.org

Calefati, Jessica. 2015 (June 10). "Gay High Schools Offer a Haven from Bullies." *U.S. News & World Report.* Available at www.usnews.com

Capehart, Jonathan. 2014. "Obama Moves to Protect LGBT Federal Contractors and Employees." *Washington Post,* July 21. Available at www.washingtonpost.com

Carroll, Aengus, and Lucas Paoli Itaborahy. 2015. *State Sponsored Homophobia 2015:* World Survey of Laws: Criminalization, Protection and Recognition of Same-Sex Love. International Lesbian, Gay, Bisexual, Trans and Intersex Association. Geneva ILGA.

Caruso, Kevin. n.d. "Suicide Note of a Gay Teen." Suicide Survivors Forum. Available at www.suicide.org

Choi, Soon Kyu, Bianca D. M. Wilson, Jama Shelton, and Gary Gates. 2015. *Serving Our Youth 2015: The Needs and Experiences of Lesbian, Gay, Bisexual, Transgender, and Questioning Youth Experiencing Homelessness*. Williams Institute. Available at www.williamsinstitute.law.ucla.edu

Cianciotto, Jason, and Sean Cahill. 2007. "Anatomy of a Pseudo-Science." *The Gay and Lesbian Review,* July–August, pp. 22–24.

Ciarlante, Mitru, and Kim Fountain. 2010. "Why It Matters: Rethinking Victim Assistance for Lesbian, Gay, Bisexual, Transgender, and Queer Victims of Hate Violence and Intimate Partner Violence." National Center for Victims of Crime and the New York City Anti-Violence Project. Available at www.avp.org

Cook, Tony, Tom LoBianco, and Doug Stranglin. 2015. "Indiana Gov. Signs Amended Religious Freedom Law." *USA Today,* April 2. Available at www.usatoday.com

Crosby, Jennifer Randall, and Johannes Wilson. 2015. "Let's Not, and Say We Would: Imagined and Actual Responses to Witnessing Homophobia." *Journal of Homosexuality* 62(7):957–970.

Culhone, John. 2015. "The Gay Marriage Fight Isn't Over." *Politico,* June 26. Available at www.politico.com

DeCarlo, Aubrey. 2014. "The Relationship between Traditional Gender Roles and Negative Attitudes towards Lesbians and Gay Men in Greek-Affiliated and Independent Male College Students." Dissertation, Lehigh University.

Diemer, Elizabeth W., Julia D. Grant, Melissa A. Munn-Chernoff, David A. Patterson, and Alexis E. Duncan. 2015. "Gender Identity, Sexual Orientation, and Eating-Related Pathology in a National Sample of College Students." *Journal of Adolescent Health* 57:144–149.

Doyle, David M., and Lisa Molix, 2014. "Perceived Discrimination and Well-Being In Gay Men: The Protective Role of Behavioral Identification." *Psychology and Sexuality* 5(2):117–130.

Drake, Bruce. 2013 (June 18). "As More Americans Have Contacts with Gays and Lesbians, Social Acceptance Rises." Available at www.pewresearch.org

Durkheim, Emile. 1993 [1938]. "The Normal and the Pathological." In *Social Deviance,* ed. Henry N. Pontell, 33–63. Englewood Cliffs, NJ: Prentice Hall. (Originally published in *The Rules of Sociological Method.*)

The Economist. 2015. "How to Count How Many People Are Gay." May 5. Available at www.economist.com

Elliott, M. N., D. E. Kanouse, Q., Burkhart, G. A. Abel, G. Lyratzopoulos, M. K. Beckett, and M. Roland. 2015. "Sexual Minorities in England Have Poorer Health and Worse Health Care Experiences: A National Survey." *Journal of General Internal Medicine* 30(1):9–16.

Everly, Benjamin A., and Joshua Schwarz. 2015. "Predictors of the Adoption of LGBT-Friendly HR Policies." *Human Resource Management* 54(2):367–384.

"FAIR Education Act." (2013). Available at www.faireducationact.com

Falomir-Pichastor, Juan Manual, Carmen Martinez, and Consuelo Paterna. 2010. "Gender Role's Attitude, Perceived Similarity, and Sexual Prejudice against Gay Men." *Spanish Journal of Psychology* 13(2):841–848.

Family Equality Council. 2014. "Joint Adoption Laws." Available at www.familyequality.org

FBI. 2014 (November). *Hate Crime Statistics, 2013.* U.S. Department of Justice. Available at www.fbi.gov

Fredriksen-Goldsen K. I., and R. Espinoza. (in press). "Transforming Public Policies to Achieve Health Equity for LGBT Older Adults." *Generations.*

Frost, David M., Keren Levahot, and Ilan H. Meyer. (2011). "Minority Stress and Physical Health among Sexual Minorities." Williams Institute. Available at www.law.ucla.edu

Gates, Gary J. 2011. "How Many People Are Lesbian, Gay, Bisexual, and Transgender?" Williams Institute. Available at www.law.ucla.edu

Gates, Gary J. 2014 (October). "LGB Families and Relationships: Analysis of the 2013 National Health Interview Survey." Executive Summary. Available at williamsinstitute.law.ucla.edu

Gates, Gary J. 2015. *Amicus Curiae Brief in Support of Appellees and Affirmance.* Court of Appeals for the Eighth Circuit. Available at williamsinstitute.law.ucla.edu

Gates, Gary J., and Frank Newport. 2012 (October 18). "Special Report: 3.4 percent of U.S. Adults Identify as LGBT." Available at www.gallup.com

Gates, Gary J., and Frank Newport. 2013 (February 15). "LGBT Percentage Highest in DC, Lowest in North Dakota." Available at www.gallup.com

Gates, Gary J., and Frank Newport. 2015 (April 24). "An Estimated 780,000 Americans in Same-Sex Marriages." Available at www.gallup.com

Gay and Lesbian Victory Institute. 2015. "Out Officials." Available at victoryinstitute.org

Geyer, Thomas. 2014. "Two Q-C Women Marry after 72 Years Together." *Quad-City Times,* September 8. Available at qctimes.com

Giambrone, Andrew. 2015. "Equality in Marriage May Not Bring Equality in Adoption." *The Atlantic,* May 26. Available at www.theatlantic.com

GLAAD. 2015. 2015 *Studio Responsibility Index.* Available at www.glaad.org

GLSEN. 2015. "Value Statement." Available at www.glsen.org

Goldbach, Jeremy T., Emily E. Tanner-Smith, Meredith. Bagwell, and Shannon. Dunlap. 2014. "Minority Stress and Substance Use in Sexual Minority Adolescents: A Meta-Analysis." *Prevention Science* 15:350–363.

Goldberg, Arthur. 2015. "A Message from the Co-director of Jonah: Arthur Goldberg." Available at jonahwhen.org

Goldberg, Arthur, Nanette K. Gartrell, and Gary Gates. 2014 (July). "Research Report on LGB-Parent Families." Williams Institute. Available at williamsinstitutelaw.ucla.edu

Gomillion, Sarah C., and Traci A. Guiliano. 2015. "The Influence of Media on Gay, Lesbian and Bisexual Identity." *Journal of Homosexuality* 58(3):330–354.

Gonzalez, Ivet. 2013. "Gay Parents in Cuba Demand Legal Right to Adopt." *Global Issues,* June 4. Available at www.globalissues.org

Guadalupe, Krishna L., and Doman Lum. 2005. *Multidimensional Contextual Practice: Diversity and Transcendence.* Belmont, CA: Thomson Brooks/Cole.

Guillory, Sean. 2013. "Repression and Gay Rights in Russia." *The Nation,* September 26. Available at www.thenation.com

Haider-Markel, Donald P., and Mark R. Joslyn. 2008. "Beliefs about the Origins of Homosexuality and Support for Gay Rights: An Empirical Test of Attribution Theory." *Public Opinion Quarterly* 72:291–310.

Harper, Gary W., Nadine Jernewall, and Maria C. Zea. 2004. "Giving Voice to Emerging Science and Theory for Lesbian, Gay, and Bisexual People of Color." *Cultural Diversity and Ethnic Minority Psychology* 10:187–199.

Hasenbush, Amira, Andres R. Flores, Angeliki Kastanis, Brad Sears, Gary J. Gates. 2014 (December). *The LGBT Divide: A Data Portrait of the South, Midwest and Mountain States.* Available at williamsinstitute.law.ucla.edu

Hatzenbuehler, M. L. 2011. "The Social Environment and Suicide Attempts in Lesbian, Gay, and Bisexual Youth." *Pediatrics* 127:896–904.

Hayworth, Brett. 2015. "Steve King Continues Pushback on Gay Marriage." *Sioux City Journal.* Available at siouxcityjournal.com

Heck, N. C., N. A. Livingston, A. Flentje, K. Oost, B. T. Stewart, and B. N. Cochran. 2014. "Reducing Risk for Illicit Drug Use and Prescription Drug Misuse: High School Gay-Straight Alliances and Lesbian, Gay, Bisexual, and Transgender Youth." *Addictive Behaviors* 39(4):824–828.

Herek, Gregory M. 2004. "Beyond 'Homophobia': Thinking about Sexual Prejudice and Stigma in the Twenty-First Century." *Sexuality Research and Social Policy: A Journal of the NSRC* 1:6–24.

Herek, Gregory M., and Linda D. Garnets. 2007. "Sexual Orientation and Mental Health." *Annual Review of Clinical Psychology* 3:353–375.

Higa, Darrel, and Marilyn J. Hoppe, Taryn Lindhorst, Shaw Mincer, Blair Beadnell, Diane M. Morrion, Elizabeth A. Wells, Avry Todd, and Sarah Mountz. "Negative and Positive Factors Associated with the Well-Being of Lesbian, Gay, Bisexual, Transgender, Queer, and Questioning (LGBTQ) Youth." *Youth and Society* 46(5):663–687.

Hines, Judith M. 2014. *Internalized Heterosexism, Outness, Relationship Satisfaction, and Violence in Lesbian Relationships.* Dissertation, University of Illinois, Chicago.

H.R. 1755. 2013. *Employment Non-Discrimination Act of 2013.* Introduced in House April 25. Available at www.congress.gov

H.R. 3135. 2013. *Domestic Partnership Benefits and Obligations Act of 2013.* Introduced in House September 19. Available at www.govtrack.us

H.R. 846. 2015. *The Student Non-Discrimination Act of 2015.* Introduced in House February 10. Available at govyrack.us/congress

H.R. 2449. 2015. *Every Child Deserves a Family Act of 2015.* Introduced in House May 19. Available at www.govtrack.us/congress

H.R. 2802. 2015. *First Amendment Defense Act of 2015.* Introduced in House June 17. Available at www.congress.gov

Human Right Campaign (HRC). 2014. *Equality Rising: HRC Global Equality Report.* Available at www.hrc.org

Human Rights Campaign (HRC). 2015a. *Beyond Marriage Equality: A Blueprint for Federal Nondiscrimination Protections.* Available at www.hrc.org

Human Rights Campaign (HRC). 2015b. *The Domestic Partnership Benefits and Obligations Act.* Available at www.hrc.org

Human Rights Campaign (HRC). 2015c (March 9). *Employment Non-Discrimination Act.* Available at www.hrc.org

Human Rights Campaign (HRC). 2015d. *Equality Rising: HRC Global Equality Report 2014.* Available at Available at www.hrc.org

Human Rights Campaign (HRC). 2015e. "The Lies and Dangers of Efforts to Change Sexual Orientation or Gender Identity." Available at www.hrc.org

Human Rights Campaign (HRC). 2015f. *Maps of State Laws and Policies.* Available at www.hrc.org

It Gets Better Project. 2015. "What Is the It Gets Better Project?" Available at www.itgetsbetter.org

Irwin, J. A., and E. L. Austin. 2013. "Suicide Ideation and Suicide Attempts among White Southern Lesbians." *Journal of Gay and Lesbian Mental Health* 17(1):4–20.

Ivers, Tracy C. (March) 2015. "Legal Developments and Practical Considerations: Improving the Quality of Care for Older LGBT Adults in the Long Term Care Setting." *AHLA Connections.* Available at issuu.com

Jabson, Jennifer M., Grant W. Farmer, and Deborah J. Bowen. 2014. "Stress Mediates the Relationship between Sexual Orientation and Behavioral Risk Disparities." *Biomedical Center Public Health* 14:401–407.

Jaeger, Joel. 2014 (September 9). "LGBT Visibility in Africa Also Brings Backlash." *Inter Press Service.* Available at www.ipsnews.net

Jimenez, Stephen. 2014. *The Book of Matt: Hidden Truths about the Murder of Matthew Shepard.* Hanover, NH: Steerforth Press.

Jones, Jeffrey M. 2015 (May 20). "Majority in US Now Say Gays and Lesbians Born, Not Made." Available at www.gallup.com

Jones, Robert J., Daniel Cox, and Juhem Navarro-Rivera. 2014 (February 26). *A Shifting Landscape.* Available at publicreligion.org.

Kane, Melinda D. 2013. "LGBT Religious Activism: Predicting State Variations in the Number of Metropolitan Community Churches, 1774–2000." *Sociological Forum* 28(1):135–158.

Kastanis, Angeliki, Matt Strieker, and Archipelago Web. 2015 (December). "The Business Impact of Opening Marriage to Same-Sex Couples." Williams Institute. Available at www.law.ucla.edu

Kaufman, Alexander C. 2015. "More Than 70 Tech Execs Sign Historic Statement against LGBT Discrimination." *Huffington Post,* April 1. Available at www.huffingtonhpost.com

Kimmel, Michael. 2011 (Winter). "Gay Bashing Is about Masculinity." *Voice Male.* Available at www.voicemalemagazine.org

Kinsey, A. C., W. B. Pomeroy, and C. E. Martin. 1948. *Sexual Behavior in the Human Male.* Philadelphia: Saunders.

Kinsey, A. C., W. B. Pomeroy, C. E. Martin, and P. H. Gebhard. 1953. *Sexual Behavior in the Human Female.* Philadelphia: Saunders.

Kosciw, J. G., E. A. Greytak, M. J. Bartkiewicz, M. J. Boesen, and N. A. Palmer. (2012). *The 2011 National School Climate Survey: The Experiences of Lesbian, Gay, Bisexual and Transgender Youth in Our Nation's Schools.* New York: GLSEN.

Kosciw, J. G., E. A. Greytak, N. A. Palmer, and M. J. Boesen. 2014. *The 2013 National School Climate Survey: The Experiences of Lesbian, Gay, Bisexual and Transgender Youth in Our Nation's Schools.* New York: GLSEN.

Kutner, Jenny. 2015. "Transgender Team Leelah Alcorn's Death Ruled a Suicide—Mother Threw Away Handwritten Note." *Salon,* April 30. Available at www.salon.com

Lipka, Michael. 2014 (October 16). "Young U.S. Catholics Overwhelmingly Accepting of Homosexuality." Fact Tank. Available at www.prewresarch.org.

Lorenz, Brandon. 2015 (March 17). "New HRC Poll Shows Overwhelming Support for Federal LGBT Non-Discrimination Bill." *Human Rights Campaign.* Available at www.hrc.org

Lowndes, Coleman, and Carlos Maza. 2014 (September 23). "The Top 5 Myths about LGBT Nondiscrimination Laws Debunked." *Media Matters.* Available at mediamatters.org

Mallory, Christie, Amira Hassenbush, and Brad Sears. 2015. "Harassment by law enforcement officers and LGBT community." Williams Institute. Available at williamsinstitute.law.ucla.edu

Mallory, Christy and Brad Sears. 2014. "Discrimination against State and Local Government LGBT Employees." *LGBTQ Policy Journal at the Harvard Kennedy School* 4:37–54.

MAP. 2015. "Safe Schools Laws." Available at www.lgbtmap.org

Masci, David. 2015 (March 18). "Where Christian Churches, Other Religions Stand on Gay Marriage." Available at www.pewresearch.org

McCarthy, Justin. 2014 (May 28). "Americans' Views on Origins of Homosexuality Remain Split." Available at www.gallup.com

McGreevy, Patrick. 2011. "New State Law Requires Textbooks to Include Gays' Achievements." *Los Angeles Times,* July 15. Available at articles.latimes.com

Mercer, Phil. 2015 (May 8). "Gay Marriage is Economic Sense, Say Australian Firms." BBC News. Available at www.bbb.com

Metropolitan Community Church (MCC). 2014. "Our Churches." Available at mcchurch.org

Metropolitan Community Church (MCC). 2015 (February). "Philosophy." Available at mcchurch.org

Meyer, Ilian. 2003. "Prejudice, Social Stress, and Mental Health in Lesbian, Gay, and Bisexual Populations: Conceptual Issues and Research Evidence." *Psychological Bulletin* 129:674–697.

Monto, Marin A. and Jessica Supinski. 2014. "Discomfort with Homosexuality: A New Measure Captures Differences in Attitudes toward Gay Men and Lesbians." *Journal of Homosexuality* 61:899–916.

Morales, Lymari. 2010. (December 9). "In U.S., 67 Percent Support Repealing 'Don't Ask, Don't Tell.'" Gallup Organization. Available at www.gallup.com

Morgan, E. M. and E. M. Thompson. 2011. "Processes of Sexual Orientation Questioning among Heterosexual Women." *Journal of Sex Research* 48:16–28.

National Coalition of Anti-Violence Programs (NCAVP). 2015. "A Report from the National Coalition of Anti-Violence Programs." Available at www.avp.org

National Conference of State Legislatures. 2015 (June 4). "2015 State Religious Freedom Restoration Legislation." Available at www.ncsl.org

Obama, Barack. 2015. "Petition Response: On Conversion Therapy." April 8. Available at www.whitehouse.gov

Obergefell v. Hodges. 2015. Available at www.supremecourt.gov

O'Keefe, Ed. 2010. "'Don't Ask, Don't Tell' Is Repealed by Senate; Bill Awaits Obama's Signing." *Washington Post,* December 19. Available at www.washingtonpost.com

Olson, Laura R., Cadge, Wendy, and Harrison, James T. 2006. "Religion and Public Opinion about Same-Sex Marriage." *Social Science Quarterly* 87:340–360.

Pagliery, Jose and Frank Pallotta. 2015. "Tim Cook Talks Charity, Coming Out as Gay on Late Show." CNN Money. September 15, 2015. Available at money.CNN.com

Pew. 2013a (June 6). "In Gay Marriage Debate, Both Supporters and Opponents See Legal Recognition as 'Inevitable.'" Available at www.people-press.org

Pew. 2013b (June 4). "The Global Divide on Homosexuality." Available at www.pewglobal.org

Pew. 2013c (March 20). "Growing Support for Same-Sex Marriage: Changed Minds and Changing Demographics." Available at www.people-press.org

Pew. 2013d (June 13). "A Survey of LGBT Americans: Attitudes, Experiences and Values in Changing Times." Available at www.pewsocialtrends.org

Pew. 2015a (June 8). "Changing Attitudes on Gay Marriage." Available at www.pewforum.org

Pew. 2015b (June 26). "Gay Marriage around the World." Available at www.pewforum.org

Phillips, Katherine W. 2014. "How Diversity Makes Us Smarter." *Scientific American,* September 16. Available at www.scientificamerican.com

Position Paper. 2015 (April). "Position Statements on Parenting of Children By Lesbian, Gay, Bisexual, and Transgender Adults." *Child Welfare League of America.* Available at www.cwla.org

Price, Jammie, and Michael G. Dalecki. 1998. "The Social Basis of Homophobia: An Empirical Illustration." *Sociological Spectrum* 18:143–159.

Reichbach, Matthew. 2015 (May 22). "Heinrich: More Protections Needed Following DADT Appeal." *NM Political Report.* Available at nmpoliticalreport.com

Reyna, Christine, Geoffrey Wetherell, Caitlyn Yantis, and Mark J. Brandt. 2014. "Attributions for Sexual Orientation vs. Stereotypes: How Beliefs about Value Violations Account for Attribution Effects on Anti-Gay Discrimination." *Journal of Applied Social Psychology* 4(4):289–302.

Röndahl, Gerd, and Sune Innala. 2008. "To Hide or Not to Hide, That Is the Question!" *Journal of Homosexuality* 52:211–233.

Rosenwald, Michael S. 2015. "How Jim Obergefell Became the Face of the Supreme Court Gay Marriage Case." *Washington Post,* April 6. Available at www.washingtonpost.com

Rothblum, Esther D. 2000. "Sexual Orientation and Sex in Women's Lives: Conceptual and Methodological Issues." *Journal of Social Issues* 56:193–204.

Russell, Stephen, C. Ryan, Russell B. Toomey, Rafael Diaz, and J. Sanchez. 2011. "Lesbian, Gay, Bisexual, and Transgender Adolescent School Victimization: Implications for Young Adult Health and Adjustment." *Journal of School Health* 81(5):223–230.

Ryan, Caitlin, David Huebner, Rafael M. Diaz, and Jorge Sanchez. 2009. "Family Rejection as a Predictor of Negative Health Outcomes in White and Latino Lesbian, Gay, and Bisexual Young Adults." *Pediatrics* 123(1):346–352.

SAGE (Services and Advocacy for Gay, Lesbian, Bisexual, and Transgender Elders). 2015. "Economic Security." Available at sageusa.org

Sasnett Sherri. 2015. "Are the Kids All Right? A Qualitative Study of Adults with Gay and Lesbian Parents." *Journal of Contemporary Ethnography* 44 (2):196–222.

Savin-Williams, R. C. 2006. "Who's Gay? Does It Matter?" *Current Directions in Psychological Science* 15:40–44.

Savin-Williams, R.C., and Z. Vrangalova. 2013. "Mostly Heterosexual as a Distinct Sexual Orientation Group: A Systematic Review of the Empirical Evidence." *Developmental Review* 33:58–88.

Schiappa, E., P. B. Gregg, and D. E. Hewes. 2005. "The Parasocial Contact Hypothesis." *Communication Monographs* 72(1):92–115.

Schlatter, Evelyn, and Robert Steinback. 2014 (Update). "10 Anti-Gay Myths Debunked." *Intelligence Report* 140(Winter).

Servick, Kelly. 2014. "Study of Gay Brothers May Confirm X Chromosome Link to Homosexuality." *Science*, November 17. Available at news.sciencemag.org

Sevecke, Jessica R., Katrina N. Rhymer, Elbert P. Almazan, and Susan Jacob. 2015. "Effects of Interaction Experiences and Undergraduate Coursework on Attitudes toward Gay and Lesbian Issues." *Journal of Homosexuality* 62(6):821–840.

Shackelford, Todd K., and Avi Besser. 2007. "Predicting Attitudes toward Homosexuality: Insights from Personality Psychology." *Individual Differences Research* 5:106–114.

Singer, P., and Belkin, A. 2012 (September 20). "A Year after DADT Repeal, No Harm Done." Available at www.cnn.com/

Sink, Justin. 2014. "U.S. Slaps Sanctions on Uganda's Anti-Gay Law." *The Hill*, June 19. Available at thehill.com

Slaatten, Hilda, and Leena Gabrys, L. (2014). "Gay-Related Name-Calling as a Response to the Violation of Gender Norms." *Journal of Men's Studies* 22(1):28–33.

Socarides, Richard. 2015. "Corporate America's Evolution on LGBT Rights." *The New Yorker*, April 27. Available at www.newyorker.com

Sprinkle, Stephen V. 2015. "How Laramie's LGBT Decision Awakens Us." *Huffington Post*, May 17. Available at www.huffingtonpost.com

Stone, Andrea. 2011. "Pentagon Discharged Hundreds of Service Members under 'Don't Ask, Don't Tell' in Fiscal 2010: Report." *Huffington Post*, March 24. Available at www.huffingtonpost.com

Strunk, Kamden K., Joy R. Suggs, and Kenneth Thompson, Eds. 2015. *Campus Climate Survey: Findings and Conclusions*. Research Initiative on Social Justice and Equity, University of Southern Mississippi. Available at aquila.usm.edu

Sumerau, J. Edward, and Ryan T. Cragun. 2014. "'Why Would Our Heavenly Father Do That to Anyone?': Oppressive Othering through Sexual Classification Schemes in the Church of Jesus Christ of Latter-Day Saints." *Symbolic Interactionism* 37(3):331–352.

Summers, Bryce B. 2010. "Factor Structure and Validity of the Lesbian, Gay, and Bisexual Knowledge and Attitude Scale for Heterosexuals (LGB-KASH)." Proquest Dissertations and Theses (UMI No. 3425038).

Swift, Art. 2014 (May 30). "Most Americans Say Same-Sex Couples Entitled to Adopt." Available at www.gallup.com

Szymanski, Dawn M., Susan Kashubeck-West, and Jill Meyer. 2008. "Internalized Heterosexism: Measurement, Psychosocial Correlates, and Research Directions." *The Counseling Psychologist*, 36:525–574.

Tobias, Sarah, and Sean Cahill. 2003. "School Lunches, the Wright Brothers and Gay Families." National Gay and Lesbian Task Force. Available at www thetaskforce.org

Toobin, Jeffrey. 2015. "Why Gay Marriage Victory Anthem Was 'Star- Spangled Banner.'" June 30. *CNN News*. Available at www.cnn.com

Toppo, Greg. 2015. "Boy Scouts of America Ends Gay Ban on Scout Leaders." *USA Today*, July 27. Available at www.usatoday.com

Temblador, Alexandra. 2015. "Why Is LGBT-Inclusive Sex Education Still So Taboo?" *Huffington Post*, March 7. Available at www.huffingtonpost.com

U.S. Court of Appeals, District 3. 2010 (September 22). *In re: Matter of Adoption of X.X.G. and N.R.G. the state of Florida*. No. 3D08-3044. Available at www.3dca.flcourts.org

U.S. v. Windsor. 2013. U.S. Supreme Court. Available at www.supremecourt.giv

Valenza, Alessia. 2014 (September 26). "Top UN Human Rights Body Condemns Violence and Discrimination on the Basis of Sexual Orientation, Gender Identity." International Lesbian, Gay, Bisexual, Trans and Intersex Rights Association. Available at igla.org

Varjas, Kris, Brian Dew, Megan Marshall, Emily Graybill, Anneliese Singh, Joel Meyers, and Lamar Birckbichler. 2008. "Bullying in Schools towards Sexual Minority Youth." *Journal of School Violence* 7:59–86.

Vrangalova, Zhana, and Savin-Williams, Ritch C. 2010. "Correlates of Same-Sex Sexuality in Heterosexually Identified Young Adults." *Journal of Sex Research* 47:92–102.

Vrangalova, Zhana, and Savin-Williams, Ritch C. 2012. "Mostly Heterosexual and Mostly Gay/Lesbian: Evidence for New Sexual Orientation Identities." *Archives of Sexual Behavior* 41:85–101.

Waddell, Kaveh. 2014. "The Fight for LGBT Hate Crimes Legislation Is Moving to the States." *National Journal*, October 28. Available at www.nationaljournal.com

Ward, Brian W., James M. Dahlhamer, Adena M. Galinsky, and Sarah S. Joestel. 2014 (July 15). "Sexual Orientation and Health among US Adults: National Health Interview Survey, 2013." *National Health Statistics Statistics Report*, Number 77.

West, Keon, and Noel M. Cowell. 2015. "Predictors of Prejudice against Lesbians and Gay Men in Jamaica." *Journal of Sex Research* 52(3):296–305.

White House. 2014 (July 21). "Taking Action to Support LGBT Workplace Equality Is Good for Business." Press Office. Available at www.whitehouse.gov

Wilcox, Clyde, and Robin Wolpert. 2000. "Gay Rights in the Public Sphere: Public Opinion on Gay and Lesbian Equality." In *The Politics of Gay Rights*, ed. Craig A. Rimmerman, Kenneth D. Wald, and Clyde Wilcox, 409–432. Chicago: University of Chicago Press.

Woodford, Michael R., Michael L. Howell, Perry Silverschanz, and Lotus Yu. 2012. "'That's So Gay!': Examining the Covariates of Hearing this Expression among Gay, Lesbian, and Bisexual Students." *Journal of American College Health* 60 (6): 429-434.

Chapter 12

American Humane Association. 2011. "Pet Overpopulation." Available at www.americanhumane.org

American Society for the Prevention of Cruelty to Animals (ASPCA). 2011. "Position Statement on Mandatory Spay/Neuter Laws." Available at www.aspca.org

Aumann, Kerstin, Ellen Galinsky, Kelly Sakai, Melissa Brown, and James T. Bond. 2010. *The Elder Care Study: Everyday Realities and Wishes for Change.* Families and Work Institute. Available at www.familiesandwork.org

Barot, Sneha. 2011 (Spring). "Unsafe Abortion: The Missing Link in Global Efforts to Improve Maternal Health." *Guttmacher Policy Review* 14(2):24–28.

Bongaarts, John, and Susan Cotts Watkins. 1996. "Social Interactions and Contemporary Fertility Transitions." *Population and Development Review* 22(4):639–682.

Buckell, Tobias. 2015 (April 21). "Your New Pet Dog Could Cost You $50,000." *Money Nation*. Available at www.moneynation.com

Chou, Rita Jing-Ann. 2011. "Filial Piety by Contract? The Emergence, Implementation, and Implications of the 'Family Support Agreement' in China." *The Gerontologist* 51(1):3–16.

Cloutier-Fisher, Denise, Karen Kobayashi, and Andre Smith. 2011. "The Subjective Dimension of Social Isolation: A Qualitative Investigation of Older Adults' Experiences in Small Social Support Networks." *Journal of Aging Studies*, doi:10.1016/j.jaging.2011.03.012

Denyer, Simon. 2015. "'One Is Enough': Chinese Families Lukewarm over Easing of One-Child Policy." *Washington Post*, January 25. Available at www.washingtonpost.com

Dunn, Mark. 2008. "Darlington Will Have Australia's First Vertical Cemetery." *Herald Sun*, November 21. Available at www.heraldsun.com

Edwards, Kathryn A., Anna Turner, and Alexander Hertel-Fernandez. 2012. *A Young Person's Guide to Social Security*. Economic Policy Institute. Available at www.epi.org

Engelman, Robert. 2011 (July 18). "The World at 7 Billion: Can We Stop Growing Now?" *Yale Environment 360*. Available at e360.yale.edu

Farrell, Paul. 2009 (January 26). "Peak Oil? Global Warming? No, It's 'Boomsday'!" MarketWatch. Available at www.marketwatch.com

Fendrich, Laurie. 2014. "The Forever Professors." *Chronicle of Higher Education*, November 14. Available at www.chronicle.com

Frost, Ashley E., and F. Nii-Amoo Dodoo. 2009. "Men Are Missing from African Family Planning." *Contexts: Understanding People in Their Social Worlds* 8(1):44–49.

Gullette, Margaret M. 2011. *Agewise: Fighting the New Ageism in America*. Chicago: University of Chicago Press.

Helman, Ruth, Greenwald & Associates, Craig Copeland, Jack Van Derhei, and Employee Benefit Research Institute. 2015 (April). "The 2015 Retirement Confidence Survey: Having a Retirement Savings Plan a Key Factor in Americans' Retirement Confidence." EBRI Issue Brief, no. 413. Available at www.ebri.org

Here and Now. 2009 (October 18). "Japan's High-Tech Graveyard in the Sky." Public Radio International. Available at www.pri.org

Humane Society. 2009 (November 23). "HSUS Pet Overpopulation Estimates." Available at www.humanesociety.org

Humane Society. 2010 (December 16). "Austin City Council Prohibits Retail Sales of Dogs and Cats." Available at www.humansociety.org

Humane Society. 2014. "Pets by the Numbers." Available at www.humanesociety.org

Kidd, Andrew. 2009 (May 9). "Shelters See Rise in Abandoned Pets as College Students' Year Ends." Fox News. Available at www.foxnews.com

Koch, Wendy. 2010. "Curb Population Growth to Fight Climate Change?" *USA Today*, May 25. Available at content.usatoday.com

Kornadt, Anna E., and Klaus Rothermund. 2010. "Constructs of Aging: Assessing Evaluative Age Stereotypes in Different Life Domains." *Educational Gerontology* 36(6).

Kornblau, Melissa. 2009. "Social Security Systems around the World." *Today's Research on Aging* 15. Population Reference Bureau. Available at www.prb.org

Lahey, Johanna. 2008. "Age, Women, and Hiring: An Experimental Study." *Journal of Human Resources* 43:30–56.

Levy, B. R., M. D. Slade, T. E. Murphy, and T. M. Gill. 2012. "Association between Positive Age Stereotypes and Recovery from Disability in Older Persons." *JAMA* 308(19):1972–1973.

Livernash, Robert, and Eric Rodenburg. 1998. "Population Change, Resources, and the Environment." *Population Bulletin* 53(1):1–36.

Mesce, Deborah, and Donna Clifton. 2011. *Abortion: Facts and Figures 2011*. Washington DC: Population Reference Bureau.

Morrissey, Monique. 2011 (January 26). "Beyond 'Normal': Raising the Retirement Age Is the Wrong Approach for Social Security." EPI Briefing Paper #287. Available at www.epi.org

Munnell, Alicia H., Matthew S. Rutledge, and Anthony Webb. 2015 (March). "Are Retirees Falling Short? Reconciling Conflicting Evidence." Center for Retirement Research. Available at crr.bc.edu

National Council on Pet Population Study and Policy. 2009. "The Top Ten Reasons for Pet Relinquishment to Shelters in the United States." Available at petpopulation.org

Nelson, Todd D. 2011. "Ageism: The Strange Case of Prejudice against the Older You." In *Disability and Aging Discrimination*, ed. R. L. Wiener and S. L. Willborn, 37. New York: Springer Science + Business Media.

Notkin, Melanie. 2013. "The Truth about the Childless Life." *Huffington Post*, August 1. Available at www.huffingtonpost.com

Olson, Philip R. 2014. "Flush and Bone: Funeralizing Alkaline Hydrolysis in the United States." *Science, Technology, & Human Values* 39(5):666–693.

Palmore, Erdman B. 2004. "Research Note: Ageism in Canada and the United States." *Journal of Cross-Cultural Gerontology* 19(1):41–46.

Parks, Kristin. 2005. "Choosing Childlessness: Weber's Typology of Action and Motives of the Voluntarily Childless." *Sociological Inquiry* 75(3):372–402.

Pew Research Center. 2015. *Family Support in Graying Societies: How Americans, Germans, and Italians Are Coping with an Aging Population*. Available at www.pewresearch.org

Population Reference Bureau. 2004. "Transitions in World Population." *Population Bulletin* 59(1).

Population Reference Bureau. 2007. *World Population Data Sheet*. Washington, DC: Population Reference Bureau. Available at www.prb.org

Population Reference Bureau. 2010. *World Population Data Sheet*. Washington, DC: Population Reference Bureau. Available at prb.org

Population Reference Bureau. 2014. *2014 World Population Data Sheet*. Washington, DC: Population Reference Bureau. Available at www.prb.org

Population Reference Bureau. 2015. *2015 World Population Data Sheet*. Washington, DC: Population Reference Bureau. Available at www.prb.org

PRNewswire. 2015 (April 27). "Packaged Facts: Pet Ownership at 54% of U.S. Households." Available at www.PRNewswire.com

Riffkin, Rebecca. 2015 (April 29). "Americans Settling on Older Retirement Age." Gallup Organization. Available at www.gallup.com

Ryerson, William N. 2011 (March 11). "Family Planning: Looking Beyond Access." *Science* 331:1265.

Sandberg, Lisa. 2013. "Inhumane: Nathan J. Winograd on Reforming Animal Shelters." *The Sun* 453 (September):4–13.

Santora, Marc. 2010. "City Cemeteries Face Gridlock." *New York Times*, August 13. Available at www.nytimes.com

Schueller, Jane. 2005 (August). "Boys and Changing Gender Roles." YouthNet. *YouthLens* 16. Available at www.fhi.org

Scott, Laura. 2009. *Two Is Enough: A Couple's Guide to Living Childless by Choice*. Berkeley, CA: Seal Press.

Sedgh, Gilda, Susheela Sing, and Rubina Hussain. 2014. "Intended and Unintended Pregnancies Worldwide in 2012 and Recent Trends." *Studies in Family Planning* 45(3):301–314.

Shikina, Rob. 2011. "Bill Mandates 'Fixing' of Cats, Dogs before Sale." *Star Advertiser*, April 17. Available at www.staradvertiser.com

Smith, Gar. 2009. "Planet Girth." *Earth Island Journal* 24(2):15.

Social Security Administration. 2013 (July). "Monthly Statistical Snapshot." Available at www.ssa.gov

Social Security Administration. 2014. *Income of the Population 55 or Older, 2012*. Available at www.ssa.gov

Social Security Trustees. 2013. *The 2013 Annual Report of the Board of Trustees of the Federal Old Age and Survivors Insurance and Federal Disability Insurance Trust Funds*. Washington, DC: U.S. Government Printing Office.

Sonfield, Adam. 2011 (Winter). "The Case for Insurance Coverage of Contraceptive Services and Supplies without Cost-Sharing." *Guttmacher Policy Review* 14(11):7–15.

State University of New York College of Environmental Science and Forestry. 2009. "Worst Environmental Problem? Overpopulation, Experts Say." *Science Daily*, April 20. Available at www.sciencedaily.com

Szinovacz, Maximiliane E. 2011. "Introduction: The Aging Workforce: Challenges for Societies, Employers, and Older Workers." *Journal of Aging & Social Policy* 23(2):95–100.

United Nations. 2012. *Population Ageing and Development 2012*. Available at www.un.org

United Nations. 2013. *World Population Prospects: The 2012 Revision*. Available at www.un.org/esa/population/publications/publications.htm

United Nations Population Division. 2009. "What Would It Take to Accelerate Fertility Decline in the Least Developed Countries?" United Nations Population Division Policy Brief No. 2009/1. New York: United Nations.

U.S. Census Bureau. 2012. "Voting and Registration." Available at www.census.gov

Weeks, John R. 2015. *Population: An Introduction to Concepts and Issues*, 12th ed. Belmont, CA: Wadsworth, Cengage Learning.

Weiland, Katherine. 2005. *Breeding Insecurity: Global Security Implications of Rapid Population Growth*. Washington, DC: Population Institute.

Women's Studies Project. 2003. *Women's Voices, Women's Lives: The Impact of Family Planning*. Family Health International. Available at www.fhi.org

World Health Organization. 2015 (May). "Family Planning/Contraception." Fact Sheet No. 351. Available at www.who.int

Chapter 13

American Lung Association. 2015. *State of the Air: 2015*. Available at lungaction.org

Archbishop's Council. 2015 (June 16). "Archbishop of Canterbury Join Faith Leaders in Call for Urgent Action to Tackle Climate Change." Available at www.churchofengland.org

Bamberger, Michelle, and Robert Oswald. 2014. *The Real Cost of Fracking*. Boston: Beacon Press.

Barron-Lopez, Laura. 2015. "Senate Votes That Climate Change is Real." *The Hill*, January 21. Available at www.thehill.com

Beinecke, Frances. 2009 (Spring). "Debunking the Myth of Clean Coal." OnEarth. Available at www.onearth.org

Betts, Kellyn S. 2011. "Plastics and Food Sources: Dietary Intervention to Reduce BPA and DEHP." *Environmental Health Perspectives* 119(7): A306.

Blatt, Harvey. 2005. *America's Environmental Report Card: Are We Making the Grade?* Cambridge, MA: MIT Press.

Blunden, Jessica, and Derek S. Arndt. 2015. "State of the Climate in 2014." *Bulletin of the American Meteorological Society* 96(7).

Bogard, Paul. 2013 (August 19). "Bringing Back the Night: A Fight against Light Pollution." *Yale Environment 360*. Available at e360.yale.edu

Bradshaw, Nancy. 2010 (March). *Fragrance-Free Policy Management Presentation.* Women's College Hospital, University of Toronto.

Brecher, Jeremy. 2011 (January 4). "Climate Protection Strategy: Beyond Business as Usual." Labor Network for Sustainability. Available at www.labor4sustainability.org

Broder, John M. 2013 (February 13). "Keystone XL Protesters Seized at White House." *Green* (*New York Times* blog). Available at green.blogs.nytimes.com

Brody, Julia Gree, Kirsten B. Moysich, Olivier Humblet, Kathleen R. Attfield, Gregory P. Beehler, and Ruthann A. Rudel. 2007. "Environmental Pollutants and Breast Cancer." *Cancer* 109(S12):2667–2711.

Brown, Paul. 2015 (April 24). "World Group Seeks Ban on Uranium and Nuclear Power." *Climate News Network.* Available at www.climatenewsnetwork.net

Brown, Lester R. 2007. "Distillery Demand for Grain to Fuel Cars Vastly Understated: World May Be Facing Highest Grain Prices in History." *Earth Policy News,* January 4. Available at www.earthpolicy.org

Brulle, Robert J. 2009. "U.S. Environmental Movements." In *Twenty Lessons in Environmental Sociology,* ed. Kenneth A. Gould and Tammy L. Lewis, 211–227. New York: Oxford University Press.

Bruno, Kenny, and Joshua Karliner. 2002. *Earthsummit.biz: The Corporate Takeover of Sustainable Development.* CorpWatch and Food First Books. Available at www.corpwatch.org

Bullard, Robert D., Paul Mohai, Robin Saha, and Beverly Wright. 2007 (March). *Toxic Wastes and Race at Twenty 1987–2007.* Cleveland, OH: United Church of Christ.

Carlson, Scott. 2006. "In Search of the Sustainable Campus." *Chronicle of Higher Education* 53(9):A10–A12, A14.

Carroll, Linda. 2015 (February 20). "Eco-Drones Aid Researchers in Fight to Save the Environment." NBC News. Available at www.NBCNews.com

Cerulli, Tovar. 2014 (March 14). "A Caretaker and a Killer: How Hunters Can Save the Wilderness." *The Atlantic.* Available at www.theatlantic.com

Chafe, Zoe. 2005. "Bioinvasions." In *State of the World 2005,* ed. L. Starke, 60–61. New York: Norton.

Chafe, Zoe. 2006. "Weather-Related Disasters Affect Millions." In *Vital Signs,* ed. L. Starke 44–45. New York: Norton.

Chepesiuk, Ron. 2009. "Missing the Dark: Health Effects of Light Pollution." *Environmental Health Perspectives* 117(1):A20–A27.

Clarke, Tony. 2002. "Twilight of the Corporation." In *Social Problems, Annual Editions 02/03,* 30th ed., ed. Kurt Finster-Busch, 41–45. Guilford, CT: McGraw-Hill/Dushkin.

Cleetus, Rachel. 2015 (April 22). "How to Cut Carbon and Save Money: RGI Delivers Yet Again." Union of Concerned Scientists. Available at www.blog.ucsusa.org

Clemmitt, Marcia. 2011. "Nuclear Power." *CQ Researcher* 21(22):all.

Connor, Steve. 2015. "Breakthrough in Hydrogen-Powered Cars May Spell End for Petrol Stations." *The Independent,* April 7. Available at www.independent.co.uk

Cook, John, Dana Nuccitelli, Sarah A. Green, Mark Richardson, Barbel Winkler, Rob Painting, Robert Way, Peter Jacobs, and Andrew Skuce. 2013. "Quantifying the Consensus on Anthropogenic Global Warming in the Scientific Literature." *Environmental Research Letters* 8(2)1–7.

Cooper, Arnie. 2004. "Twenty-Eight Words that Could Change the World: Robert Hinkley's Plan to Tame Corporate Power." *The Sun* 345(September):4–11.

Coyle, Kevin. 2005. *Environmental Literacy in America.* Washington, DC: National Environmental Education and Training Foundation.

"Decoding the Labels." 2015. *Guide to Healthy Cleaning.* Environmental Working Group. Available at www.ewg.org/guides/cleaners/content/decoding_labels

Department of Defense. 2015 (July 23). "National Security Implications of Climate-Related Risks and a Changing Climate." Available at www.defense.gov

Di Liberto, Tom. 2015 (June 9). "India Heat Wave Kills Thousands." NOAA. Available at www.climate.gov

Dutton, A., A. E. Carlson, A. J. Long, G. A. Milne, P. U. Clark, R. DeConto, B. P. Horton, S. Rahmstorf, and M. E. Rayo. 2015 (July 10). "Sea-Level Rise Due to Polar Ice-Sheet Mass Loss during Past Warm Periods." *Science* 349(6244).

Editors. 2015. "Why the Presidential Candidates Should Reject Donations from Fossil-Fuel." *The Nation,* July 20–27. Available at www.thenation.com

Edwards, Bob, and Adam Driscoll. 2009. "From Farms to Factories: The Environmental Consequences of Swine Industrialization in North Carolina." In *Twenty Lessons in Environmental Sociology,* ed. Kenneth A. Gould and Tammy L. Lewis, 153–175. New York: Oxford University Press.

Ehrlich, Paul R., and Anne H. Ehrlich. 2013. "Can a Collapse of Global Civilization Be Avoided?" *Proceedings of the Royal Society B* 280 (January): 20122845. Available at dx.doi.org/10.1098/rspb.2012.2845

Elk, Mike. 2011 (September 5). "Which Is More Likely to Rebuild the Labor Market: Environmental Allies or 6,000 Temp Jobs?" *Working in These Times.* Available at www.inthesetimes.com

Elliot, Kennedy. 2015. "Where Carbon Emissions Are Greatest." *Washington Post,* March 31. Available at www.washingtonpost.com

Environmental Protection Agency (EPA). 2004. *What You Need to Know about Mercury in Fish and Shellfish.* Available at www.epa.gov

Environmental Protection Agency (EPA). 2015a. *Advancing Sustainable Materials Management: 2013 Fact Sheet.* Available at www.epa.gov

Environmental Protection Agency (EPA). 2015b. *National Priorities List (NPL).* Available at www.epa.gov/superfund/sites

Environmental Working Group. 2005 (July 14). *Body Burden: The Pollution in Newborns.* Available at www.ewg.org

Farooq, Omer, and Katy Daigle. 2015. "India Heat Wave Persists, Claiming over 1,400 Lives." *Huffington Post,* May 28. Available at www.huffington.com

Ferris, Elizabeth. 2015 (April 22). "Earth Day: Climate Change and Displacement across Borders." Brookings Institution. Available at www.brookings.edu

Fisher, Brandy E. 1999. "Focus: Most Unwanted." *Environmental Health Perspectives* 107(1). Available at ehpnet1.niehs.nih.gov/docs/1999/107-1/focus-abs.html

Flavin, Chris, and Molly Hull Aeck. 2005 (September 15). "Cleaner, Greener, and Richer." Tom.Paine.com. Available at www.tompaine.com/articles/2005/09/15/cleaner_greener_and_richer.php

Food and Drug Administration. 2015. *Pesticide Residue Monitoring Program Results and Discussion FY 2012.* Available at www.fda.gov

Food & Water Watch and Network for New Energy Choices. 2007. *The Rush to Ethanol: Not All Biofuels Are Created Equal.* Available at www.newenergychoices.org

Forum on Religion and Ecology at Yale. N.d. Available at www.fore.yale.edu

French, Hilary. 2000. *Vanishing Borders: Protecting the Planet in the Age of Globalization.* New York: Norton.

Gallup Organization. 2015. *Environment.* Available at www.gallup.com/poll

Gardner, Gary. 2005. "Forest Loss Continues." In *Vital Signs 2005,* ed. Linda Starke, 92–93. New York: Norton.

Gertz, Emily. 2015 (July 22). "Three Things to Know about the Terrifying New Climate Study." *Takepart.* Available at www.takepart.com

Global Footprint Network. 2010. *2010 Annual Report.* Oakland, CA: Global Footprint Network. Available at www.footprintnetwork.org

Global Invasive Species Database. 2015. "*Felis catus.*" Available at www.issg.org

Grider, Jeanette. 2015 (June 3). "Biology Graduate Student Publishes Research on Invasive Kudzu Plant." Saint Louis University. Available at www.slu.edu

Hill, Taylor. 2014. "Plastics in Paradise: Scientists Collect 60 Tons of Junk Surrounding Remote Hawaiian Islands." *Takepart.* Available at go.takepart.com

Hill, Taylor. 2015 (July 9). "The 97 Percent Scientific Consensus Is Wrong—It's Even Higher." *Takepart.* Available at www.takepart.com

Hirji, Zahra and Lisa Song. 2015 (January 20). "Map: The Fracking Boom, State by State." *Inside Climate News.* Available at www.insideclimatenews.org

Horn, Steve. 2012 (March 21). "ALEC Climate Change Denial Model Bill Passes in Tennessee." Desmogblog.com. Available at www.Desmogblog.com

Hunter, Lori M. 2001. *The Environmental Implications of Population Dynamics.* Santa Monica, CA: Rand Corporation.

Inglis, Jeff, and John Rumpler. 2015. *Fracking Failures: Oil and Gas Industry Environmental Violations in Pennsylvania and What They Mean for the United States.* Environment North Carolina Research Policy Center. Available at www.environmentnorthcarolinacenter.org

Inhofe, James. 2012. *The Greatest Hoax: How the Global Warming Conspiracy Threatens Your Future.* Washington, DC: WND Books.

Intergovernmental Panel on Climate Change. 2015. *Climate Change 2014: Synthesis Report.* United Nations Environmental Programme and the World Meteorological Organization. Available at www.ipcc

International Energy Agency. 2014. *2014 Key World Energy Statistics.* Available at www.iea.org

IUCN. 2015. *The IUCN Red List of Threatened Species, 2015.* Available at www.iucnredlist.org

Jaggar, Karuna. 2014. "Komen Is Supposed to be Curing Breast Cancer. So Why Is Its Pink Ribbon on So Many Carcinogenic Products?" *Washington Post,* October 21. Available at www.washingtonpost.com

Janofsky, Michael. 2005. "Pentagon Is Asking Congress to Loosen Environmental Laws." *New York Times,* May 11. Available at www.nytimes.com

Kaplan, Sheila, and Jim Morris. 2000. "Kids at Risk." *U.S. News & World Report,* June 19, pp. 47–53.

Kessler, Rebecca. 2015 (May). "More Than Cosmetic Changes: Taking Stock of Personal Care Product Safety." *Environmental Health Perspectives* 123(5):A120–A127.

Kessler, Rebecca. 2013. "Sunset for Leaded Aviation Gasoline?" *Environmental Health Perspectives* 121(2):A54–A57.

Kimmerer, Robin Wall. 2013. *Braiding Sweetgrass: Indigenous Wisdom, Scientific Knowledge, and the Teachings of Plants.* Minneapolis: Milkweed Editions.

Kirchner, Lauren. 2015. "Whatever Happened to 'Eco-Terrorism'?" *Pacific Standard,* January 26. Available at www.psmag.com

Knoell, Carly. 2007 (August 9). "Malaria: Climbing in Elevation as Temperature Rises." *Population Connection.* Available at www.populationconnection.org

Korten, Tristram. 2015. "In Florida, Officials Ban Term 'Climate Change.'" *Miami Herald,* March 8. Available at www.miamiherald.com

Lamm, Richard. 2006. "The Culture of Growth and the Culture of Limits." *Conservation Biology* 20(2):269–271.

Leahy, Stephen. 2009 (May 21). "Alien Species Eroding Ecosystems and Livelihoods." Interpress Service News Agency. Available at www.ipsnews.net

Lipott, Sigrid. 2015 (May 19). "Emerging Climate Change Cross-Border Conflicts among the Kenyan Borderlands." Aid & International Development Forum. Available at www.aidforum.org

Little, Amanda Griscom. 2005. "Maathai on the Prize: An Interview with Nobel Peace Prize Winner Wangari Maathai." *Grist Magazine,* February 15. Available at www.grist.org

Longeray, Pierre and Pierre-Louis Caron. 2015. "New Zealand Supreme Court Rejects Kiribati Man's Bid to Become a 'Climate Refugee.'" *Vice News,* July 21. Available at www.news.vice.com

Lunden, Jennifer. 2013. "Exposed." *Orion Magazine* (September/October). Available at www.orionmagazine.com.

Malik, Naureen. 2015. "Solar Shines as Sellers Sometimes Pay Buyers to Use Power." *Bloomberg,* May 26. Available at www.bloomberg.com

Mathiesen, Karl. 2015. "Elephant Poaching Crisis Unchanged a Year after Global Pledge." *The Guardian,* March 23. Available at www.theguardian.com

McAllister, Lucy. 2013. "The Human and Environmental Effects of E-Waste." Population Reference Bureau. Available at www.prb.org

McGinn, Anne Platt. 2000. "Endocrine Disrupters Raise Concern." In *Vital Signs 2000,* ed. Lester R. Brown, Michael Renner, and Brian Halweil, 130–131. New York: Norton.

McMichael, Anthony J., Kirk R. Smith, and Carlos F. Corvalan. 2000. "The Sustainability Transition: A New Challenge." *Bulletin of the World Health Organization* 78(9):1067.

Mulvey, Kathy, and Seth Shulman. 2015. *The Climate Deception Dossiers: Internal Fossil Fuel Industry Memos Reveal Decades of Corporate Disinformation.* Union of Concerned Scientists. Available at www.ucsusa.org

Murtaugh, Paul, and Michael Schlax. 2009. "Reproduction and the Carbon Legacies of Individuals." *Global Environmental Change* 19:14–20.

Nader, Ralph. 2013 (October 14). "Why Atomic Energy Stinks Worse Than You Thought." Counterpunch. Available at www.counterpunch.org

NASA. 2014 (October 30). "2014 Antarctic Ozone Hole Holds Steady." Available at www.nasa.gov

National Academy of Science. 2015. "Climate Intervention." *Report in Brief.* Available at www.dels.nas.edu

National Geographic. 2007 (June 5). "Top Ten Tips to Fight Global Warming." *Green Guide.* Available at www.thegreenguide.com

National Toxicology Program. 2014. *Report on Carcinogens,* 13th ed. Research Triangle Park, NC: U.S. Department of Health and Human Services, Public Health Service.

Nazaryan, Alexander. 2014. "Camp Lejeune and the U.S. Military's Polluted Legacy." *Newsweek,* July 16. Available at www.Newsweek.com

Nicole, Wendee. 2014. "Cooking Up Indoor Air Pollution." *Environmental Health Perspectives* 122(1): A27.

Normander, Bo. 2011 (February 23). "World's Forests Continue to Shrink." Vital Signs, World Watch Institute. Available at vitalsigns.worldwatch.org

Nuwer, Rachel. 2014 (August 7). "The World's First Climate Change Refugees Were Granted Residency in New Zealand." *Smithsonian.* Available at www.Smithsonian.com

O'Neill, B. C., M. Dalton, R. Fuchs, L. Jiang, S. Pachauri, and K. Zigova. 2010. "Global Demographic Trends and Future Carbon Reductions." *PNAS* 107(41):17521–17526.

Pew Research. 2013 (June 24). "Climate Change and Financial Instability Seen as Top Global Threats." Available at www.pewglobal.org

Pimentel, D., S. Cooperstein, H. Randell, D. Filiberto, S. Sorrentino, B. Kaye, C. Nicklin, J. Yagi, J. Brian, J. O'Hern, A. Habas, and C. Weinstein. 2007. "Ecology of Increasing Diseases: Population Growth and Environmental Degradation." *Human Ecology* 35(6):653–668.

Plautz, Jason. 2014. "The Climate-Change Solution No One Will Talk About." *The Atlantic,* November 1. Available at www.theatlantic.com

Pope Francis. 2015 (May 24). Encyclical Letter Laudato Si' of the Holy Father Francis.

On Care for Our Common Home. Vatican Press. Available at www.w2.vatican.va

Price, Tom. 2006. "The New Environmentalism." *CQ Researcher* 16(42):987–1007.

Public Citizen. 2005 (February). *NAFTA Chapter 11 Investor-to-State Cases: Lessons for the Central America Free Trade Agreement.* Public Citizens Global Trade Watch Publication E9014. Available at www.citizen.org

Redford, Robert. 2015 (July 10). "Our Last Chance." Natural Resources Defense Council. Available at www.nrdconline.org

Reichmuth, David. 2014 (May 2). "California Is Filling Up the Hydrogen Station Map." Union of Concerned Scientists. Available at www.blog.ucsusa.org

Ridlington, Elizabeth, and John Rumpler. 2013. *Fracking by the Numbers: Key Impacts of Dirty Drilling at the State and National Level.* Environment America Research & Policy Center. Available at www.environmentamerica.org

Rifkin, Jeremy. 2004. *The European Dream: How Europe's Vision of the Future Is Quietly Eclipsing the American Dream.* New York: Tarcher/Penguin.

Rogers, Sherry A. 2002. *Detoxify or Die.* Sarasota, FL: Sand Key.

Saad, Lydia. 2015 (March 25). "U.S. Views on Climate Change Stable after Extreme Winter." Gallup. Available at www.gallup.com

Scavia, Donald. 2011 (September 2). "Dead Zones in Gulf of Mexico and Other Waters Require a Tougher Approach: Donald Scavia." *Nola.com.* Available at www.nola.com

Schapiro, Mark. 2007. *Exposed: The Toxic Chemistry of Everyday Products and What's at Stake for American Power.* White River Junction, VT: Chelsea Green Publishing.

Schiermeier, Quirin. 2015. "Landmark Court Ruling Tells Dutch Government to Do More on Climate Change." *Nature,* June 24. Available at www.nature.com

Schrank, Aaron. 2015 (February 16). "Science Standards Draw Climate Change Debate Back into Wyo. Classrooms." NPR. Available at www.npr.org

Seltenrich, Nate. 2014. "Wind Turbines: A Different Breed of Noise?" *Environmental Health Perspectives* 122(1):A21–A25.

Seltenrich, Nate. 2015. "New Link in the Food Chain? Marine Plastic Pollution and Seafood Safety." *Environmental Health Perspectives* 123(2):A35–A41.

Shapiro, Isaac, and John Irons. 2011. *Regulation, Employment, and the Economy.* EPI Briefing Paper #305. Washington, DC: Economic Policy Institute.

Shrank, Samuel. 2011. "Growth of Biofuel Production Slows." In *Vital Signs,* ed. Linda Starke, 16–18. Washington DC: Worldwatch Institute.

Smith, Erica, David Azoulay, and Baskut Tuncak. 2015. *Lowest Common Denominator: How the Proposed EU-US Trade Deal Threatens to Lower Standards of Protection from Toxic Pesticides.* Center for International Environmental Law. Available at www.ciel.org

Staudinger, Michelle D., Nancy B. Grimm, Amanda Staudt, Shawn L. Carter, F. Stuart Chapin III, Peter Kareiva, Mary Ruckelshaus, and Bruce A. Stein. 2012. *Impacts of Climate*

Change on Biodiversity, Ecosystems, and Ecosystem Services: Technical Input to the 2013 National Climate Assessment. Available at assessment.globalchange.gov

Sustainable Endowments Institute. 2011. *Greening the Bottom Line: The Trend toward Green Revolving Funds on Campus.* Cambridge MA: Sustainable Endowments Institute.

Swift, Anthony, Susan Casey-Lefkowitz, and Elizabeth Shope. 2011. *Tar Sands Pipelines Safety Risks.* Natural Resources Defense Council. Available at www.nrdc.org

Takada, Hideshige. 2013 (May 10). *Microplastics and the Threat to Our Seafood.* Ocean Health Index. Available at www.oceanhealthindex.org

Taylor, Bron. 2015. "Religion to the Rescue (?) in an Age of Climate Disruption." *Journal for the Study of Religion, Nature and Culture* 9(1):7–18.

Tompkins, Forbes and Christina DeConcini. 2015. "Fact Sheet: A Year of Temperature Records and Landmark Climate Findings." World Resources Institute. Available at www.wri.org

United Nations Development Programme. 2013. *Human Development Report 2013.* Available at hdr.undp.org

United Nations Environment Programme (UNEP). 2013. *UNEP Yearbook 2013: Emerging Issues in Our Global Environment.* Available at www.unep.org

Urban, Mark C. 2015 (May 1). "Accelerating Extinction Risk from Climate Change." *Science* 348(6234):571-573.

U.S. Department of Energy. 2015. "Wind Vision: A New Era for Wind Power in the United States." Available at www.energy.gov

U.S. Fish and Wildlife. 2015. *Polar Bear* (Ursus maritimus) *Conservation Management Plan, Draft.* Anchorage: U.S. Fish and Wildlife.

U.S. Geological Survey. 2015. "Fact Sheet: Zebra Mussel." Available at www.nas.er.usgs.gov

Vaughan, Adam. 2015. "Obama's Clean Power Plan Hailed as US's Strongest Climate Action Ever." *The Guardian,* August 3. Available at www.theguardian.com

Vedantam, Shankar. 2005a. "Nuclear Plants Not Keeping Track of Waste." *Washington Post,* April 19. Available at www.washingtonpost.com

Vedantam, Shankar. 2005b. "Storage Plan Approved for Nuclear Waste." *Washington Post,* September 10. Available at www.washingtonpost.com

Wald, Matthew L. 2013. "Ex-Regulator Says Reactors are Flawed." *New York Times,* April 8. Available at www.nytimes.com

Wassmer, Julie. 2015 (July 23). "Fracking Goes on Trial." *New Internationalist* (blog). Available at www.newint.org/blog

Westerling, A. L., H. G. Hidalgo, D. R. Cayan, and T. W. Swetnam. 2006. "Warming and Earlier Spring Increase Western U.S. Forest Wildfire Activity." *Science* 313 (5789):940–943.

White House. 2012 (August 28). "Obama Administration Finalizes Historic 54.5 MPG Fuel Efficiency Standard." Office of the Press Secretary. Available at www.whitehouse.gov

Williams, Carolyn. 2015 (March 11). "Florida Isn't Alone: North Carolina, Pennsylvania Ban 'Climate Change.'" Weather Channel. Available at www.weather.com

Woodward, Colin. 2007. "Curbing Climate Change." *CQ Global Researcher* 1(2):27–50. Available at www.globalresearcher.com

Worland, Justin. 2015 (June 19). "The Surprising Link between Trans Fat and Deforestation." *Time.* Available at www.time.com

World Bank. 2014 (May 28). "State & Trends Report Charts Global Growth of Carbon Pricing." Available at www.theworldbank.org

World Health Organization. 2014a. "Climate Change and Health." Fact Sheet No. 266. Available at www.who.int

World Health Organization. 2014b (March 25). "7 Million Premature Deaths Annually Linked to Air Pollution." News Release. Available at www.who.int

World Nuclear Association. 2015. "World Nuclear Power Reactors & Uranium Requirements." Available at world-nuclear.org

World Water Assessment Program. 2015. *The United Nations World Water Development Report: Water for a Sustainable Future.* Paris: UNESCO.

World Wildlife Fund (WWF). 2014. *Living Planet Report.* World Wildlife Fund, Zoological Society of London, Global Footprint Network, and Water Footprint Network. Available at www.assetsworldwildlife.org

Yang, Sue-Ming, Yi-Yuan Su, and Jennifer V. Carson. 2014. "Eco-Terrorism and the Corresponding Legislative Efforts to Intervene and Prevent Future Attacks." TSAS Working Paper # 14-04. Canadian Network for Research on Terrorism, Security, and Society. Available at www.library.tsas.ca/tsas-working-papers

Yonetani, Michelle. 2015. *Global Estimates 2014: People Displaced by Disasters.* Internal Displacement Monitoring Centre and Norwegian Refugee Council." Available at www.internal.displacement.org

Chapter 14

Alexander, Leigh. 2014. "Sexism, Lies and Video Games: The Culture War Nobody is Winning." *Time,* September 5. Available at www.time.com

American Civil Liberties Union. 2013 (June 13). "2013 Supreme Court Invalidates Breast and Ovarian Cancer Genes." Available at www.aclu.org

Angwin, Julia. 2014. *Dragnet Nation.* New York: Times Books.

Arabic Network for Human Rights Information. 2015 (May 4). "#Turn_Around_and_Go_Back: Internet in the Arab World." Available at www.anhri.net

Associated Press. 2010 (January 11). "China's Internet Censorship." CBS News. Available at www.cbsnews.com

Associated Press. 2015. "U.S. Supreme Court Refuses to Let Texas Close 10 Abortion Clinics." *The Guardian,* June 29. Available at www.theguardian.com

Ball, James. 2015. "Secret US Cybersecurity Report: Encryption Vital to Protect Private Data." *The Guardian,* January 15. Available www.guardian.com

Bauer, Shane. 2014. "The FBI Is Very Excited about This Machine That Can Scan Your DNA in 90 Minutes." *Mother Jones,* November 20. Available at www.motherjones.com

BBC. 2014 (October 24). "Pirate Bay Founder Gottfrid Warg Gets Lengthy Jail Term." *BBC News.* Available at www.bbcnews.com

BBC. 2015 (June 26). "Japanese Carmakers Expand Takata Airbag Recall." June 26. Available at www.bbc.com

Beland, Louis-Philippe and Richard Murphy. 2015 (May). "Communication: Technology, Distraction and Student Performance." *London School and Economics Centre for Economic Performance,* Discussion Paper No. 1350. Available at www.cep.lse.ac.uk

Bell, Daniel. 1973. *The Coming of Post-Industrial Society: A Venture in Social Forecasting.* New York: Basic Books.

Beniger, James R. 1993. "The Control Revolution." In *Technology and the Future,* ed. Albert H. Teich, 40–65. New York: St. Martin's Press.

Boitnott, John. 2015. "Wearable Tech Is Improving Worker Productivity and Happiness." *Entrepreneur,* April 28. Available at www.entrepreneur.com

Bollier, David. 2009 (July 22). "Deadly Medical Monopolies." *On the Commons.* Available at onthecommons.org

Bornmann, Lutz, and Ruediger Mutz. 2015. "Growth Rates of Modern Science: A Bibliometric Analysis Based on the Number of Publications and Cited References." *Journal of the Association for Information Science and Technology.* Forthcoming.

Bowman, Lee. 2011 (May 7). "Animal Testing: Biomedical Researchers Using Millions of Animals Yearly." Scripps Howard News Service.

Britt, Ronda. 2015 (April). "Majority of Federally Funded R&D Centers Report Declines in R&D Spending In FY 2013." *InfoBrief.* National Center for Science and Engineering Statistics. Available at www.nsf.gov

Buchanan, Allen, Dan Brock, Norman Daniels, and Daniel Wikler. 2000. *From Chance to Choice: Genetics and Justice.* New York: Cambridge University Press.

Bush, Corlann G. 1993. "Women and the Assessment of Technology." In *Technology and the Future,* ed. Albert H. Teich, 192–214. New York: St. Martin's Press.

Cambridge. 2015. "Introduction to the Semantic Web." *Cambridge Semantics.* Available at www.cambridgesemantics.com

Carr, Nicholas. 2010. *The Shallows: What the Internet Is Doing to Our Brains.* New York: Norton.

Carr, Nicholas. 2014. "Automation Makes Us Dumb." *Wall Street Journal,* November 21. Available at www.wsj.com

Castro, Daniel. 2013. "Health IT 2013: A Renewed Focus on Efficiency and Effectiveness." *Electronic Health Reporter* (February 4). Available electronichealthreporter.com

Center for Strategic and International Studies. 2014. "Estimating the Global Cost of Cybercrime." Available at www.mcafee.com/us

Ceruzzi, Paul. 1993. "An Unforeseen Revolution." In *Technology and the Future,* ed. Albert H. Teich, 160–174. New York: St. Martin's Press.

Chan, Sewell. 2007. "New Scanners for Tracking City Workers." *New York Times,* January 23. Available at www.nytimes.com

Chen, Shirong. 2011. "China Tightens Internet Censorship Controls." BBC. Available at www.bbc.co.uk

Chowdhy, Amit. 2014. "Apple and Samsung Drop Patent Disputes against Each Other Outside of U.S." *Forbes,* August 6. Available at www.forbes.com

Clarke, Adele E. 1990. "Controversy and the Development of Reproductive Sciences." *Social Problems* 37(1):18–37.

Clarke, Richard A., Michael J. Morell, Geoffrey R. Stone, Cass R. Sunstein, and Peter Swire. 2013 (December 12). "Liberty and Security in a Changing World." National Security Administration. Available at www.nsa.gov

Coleman, Kevin G. 2015 (May 15). "Has the U.S. Lost Technological Supremacy?" *C4ISR Networks.* Available at wwwc4isrnet.com

Consumer Reports. 2015. "GMO Foods: What You Need to Know." February 26. Available at www.consumerreports.org

Crichton, Michael. 2007. "Patenting Life." *New York Times,* February 13. Available at www.nytimes.com

Daine, Kate, Keith Hawton, Vinod Singaravelu, Anne Stewart, Sue Simkin, and Paul Montgomery. 2013. "The Power of the Web: A Systematic Review of Studies of the Influence of the Internet on Self-Harm and Suicide in Young People." *PLoS ONE* 8(10):e77555.

Dandekar, Pranav, Ashish Goel, and David T. Lee. 2013. "Biased Assimilation, Homophily, and the Dynamics of Polarization." *Proceedings of the National Academy of Sciences of the United States of America,* February 28. Available at www.pnas.org

David-Ferdon, Corrine, and Marci Feldman Hertz. 2009. *Electronic Media and Youth Violence: A CDC Issue Brief for Researchers.* Atlanta: Centers for Disease Control.

Department of Health and Human Services. 2013. "Teen Media Use." Available at www.hhs.gov

De Rubeis, Silvia, et al. 2014. "Synaptic, Transcriptional, and Chromatin Genes Disrupted in Autism." *Nature* 515 (November):209–215.

Diamond, Jeremy. 2015 (June 2). "NSA Surveillance Bill Passes after Weeks-long Showdown." CNN. Available at www.cnn.com

Ducklin, Paul. 2014 (July 22). "Dirty Dozen Spampionship – Which Country Is Spewing the Most Spam?" *Sophos: Naked Security.* Available at www.nakedsecurity.sophos.com

Duggan, Maeve. 2014 (October 22). "Experiencing Online Harassment." Pew Research Center. Available at www.pewinternet.org

Duggan, Maeve, Nicole B. Ellison, Cliff Lampe, Amanda Lenhart, and Mary Madden. 2015 (January 9). "Social Media Update 2014." Pew Research Center. Available at www.pewresearch.org

Durkheim, Emile. 1973/1925. *Moral Education.* New York: Free Press.

Dutta, Soumitra, Thierry Geiger, and Bruno Lanvin. 2015. *The Global Information Technology Report 2015.* World Economic Forum. Available at reports.weforum.org

The Economist. 2015. "Driven from Distraction." April 25. Available at www.economist.com

Efrati, Amir. 2013. "Google to Find, Develop Wireless Networks in Emerging Markets." *Wall Street Journal,* May 24. Available at online.wsj.com

Eibert, Mark D. 1998. "Clone Wars." *Reason* 30(2):52–54.

Eilperin, Juliet, and Rick Weiss. 2003. "House Votes to Prohibit All Human Cloning." *Washington Post,* February 28. Available at www.washingtonpost.com

Elliot, Stuart W. 2014. "Anticipating a Luddite Revival." *Issues in Science and Technology* 30(3) (Spring). Available at www.issues.org

Equal Employment Opportunity Commission. 2015. "Facts about the Genetic Information Nondisclosure Act." Available at www.eeoc.gov

Farberov, Snejana. 2015. "Meet the 33 Americans Who Could Live on Mars—and Never Return." *Daily Mail,* February 17. Available at www.dailymail.co.uk

Federal Communications Commission (FCC). 2015. "Open Internet Rules." Available at www.fcc.gov

Federal Trade Commission (FTC). 2015 (February 27). "Consumers Told It to the FTC: Top 10 Complaints for 2014." Available at www.consumer.ftc.gov

Ferdowsian, Hope. 2010. "Animal Research: Why We Need Alternatives." *Chronicle of Higher Education,* November 7. Available at chronicle.com

File, Thom. 2014. *Demographics of Handheld -Only Households.* November 13. Available at blogs.census.gov

File, Thom, and Camille Ryan. 2014 (November). *Computer and Internet use in the United States.* U.S. Bureau of the Census. Publication No. ACS-28. Washington, DC: U.S. Government Printing Office.

Fischer, Eric A. 2015 (January 27). *Statement before U.S. House of Representatives on the Expanding Cyber Threat.* Congressional Research Service. Available at docs.house.gov

Flecknell, Paul A. 2010. "Do Mice Have a Pain Face?" *Nature Methods* 7:437–438.

Friends of the Earth (FOE). 2014. "Tiny Ingredients: Big Risks." Available at www.foe.org

Friends of the Earth (FOE). 2015. "Genetically Engineered Fish: An Unnecessary Risk to the Environment, Public Health and Fishing Communities." *Friends of the Earth Issue Brief,* N.D. Available at www.foe.org

Fox, Susannah. 2013 (August 7). "51% of U.S. Adults Bank Online." Pew Internet and the American Life Project. Available at www.pewinternet.org

Fox, Susannah, and Maeve Duggan. 2013 (January 15). "Online Health 2013." Pew Internet and the American Life Project. Available at www.pewinternet.org

Fox, Susannah, and Lee Rainie. 2014 (February 27). "How the Internet Has Woven Itself into American Life." Pew Research Center. Available at pewinternet.org

Gallup. 2015a (May 6). "Abortion." Gallup Poll: Abortion. Available at www.gallup.com

Gallup. 2015b (May 6). "Cloning." Gallup Poll: Moral Issues. Available at www.gallup.com

Gartner Research. 2015 (April 9). "Gartner Says Declining Worldwide PC Shipments Declined 5.2 Percent in First Quarter of 2015." Available at www.gartner.com

Genetics and Public Policy Center. 2010. "Frequently Asked Questions." Johns Hopkins University, Berman Institute of Bioethics, Washington, DC.

Genomics Law Report. 2011. "Myriad Gene Patent Litigation." A Publication of Robinson Bradshaw and Hinson. Available at www.genomicslawreport.com

Global Workforce. 2015. "Latest Telecommuting Statistics." Available at globalworkplaceanalytics.com

Goodman, Paul. 1993. "Can Technology Be Humane?" In *Technology and the Future,* ed. Albert H. Teich, 239–255. New York: St. Martin's Press.

Gorman, James. 2014. "Mice Run for Fun, Not Just Work, Research Shows." *New York Times,* May 20. Available at www.nytimes.com

Government Tracks. 2013. "Abortion." Available at www.govtrack.us

Granville, Kevin. 2015. "9 Recent Cyberattacks against Big Businesses." *New York Times,* February 5. Available at www.nytimes.com

Greenwald, Glenn. 2013. "XKeyscore: NSA Tool Collects 'Nearly Everything a User Does on the Internet.'" *The Guardian,* July 31. Available at www.theguardian.com

Guttmacher Institute. 2015. (May 1). "State Policies in Brief: An Overview of Abortion Laws." Available at www.guttmacher.org

Guyton, Kathryn et al. 2015. "Carcinogenicity of Ttrachlorvinphos, Parathion, Malathion, Diazinon, and Glyphosate." *The Lancet* 16(5):490–491.

Hanson, Lawrence A. 2010 (November 7). "Animal Research: Groupthink in Both Camps." *Chronicle of Higher Education.* Available at chronicle.com

Harmon, Amy. 2006. "That Wild Streak? Maybe It Runs in the Family." *New York Times,* June 15. Available at www.nytimes.com

Hayden, Erika Check. 2014. "Cancer-Gene Data Sharing Boosted." *Nature,* June 10. Available at www.nature.com

Henn, Steve. 2014 (March 17). "With Google's Robot-Buying Bing, a Hat Tip to the Future." *All Tech Considered.* Available at www.npr.org

Henry, Bill, and Roarke Pulcino. 2009. "Individual Differences and Study-Specific Characteristics Influencing Attitudes about the Use of Animals in Medical Research." *Society and Animals* 17:305–324.

Herrick, Kristin A., Tala H. I. Fakhouri, Susan A. Carlson, and Janet E. Fulton. 2014. "TV Watching and Computer Use in U.S. Youth Aged 12–15, 2012." NCHS data brief, no. 157. Hyattsville, MD: National Center for Health Statistics.

Hinduja, Sameer, and Justin W. Patchin. 2015 (April). "State Cyberbullying Laws." Cyberbullying Research Center. Available at www.cyberbullying.us

Holland, Earle. 2010 (November 7). "Animal Research: Activists' Wishful Thinking, Primitive Reasoning." *Chronicle of Higher Education.* Available at chronicle.com

Holpuch, Amanda. 2015. "U.S. High Court's Female Justices Issue Fierce Dissent on Contraception Ruling." The Guardian, July 4. Available at www.theguardian.com

Human Cloning Prohibition Act of 2007. U.S. House of Representatives, Washington, DC. Available at www.govtrack.us

Human Genome Project. 2007. *Medicine and the New Genetics.* Available at www.ornl.gov

Information Technology and Innovation Foundation (ITIF). 2009. "Benchmarking EU & U.S. Innovation and Competitiveness." European-American Business Council (February). Available at www.itif.org

Information Technology and Innovation Foundation (ITIF). 2011 (July). "The Atlantic Century II." European-American Business Council (February). Available at www.itif.org

Information Technology and Innovation Foundation (ITIF). 2012 (December). The 2012 State New Economy Index. Available at www.itif.org

Interagency Coordinating Committee on the Validation of Alternative Methods (ICCVAM). 2011. "Since You Asked: Alternatives to Animal Testing." Available at www.niehs.nih.gov

International Federation of Robotics. 2013. "World Robotics 2012 Industrial Robots." Available at www.ifr.org

Internet.org. 2015. "Who We Are." Available at www.internet.org

Internet Statistics. 2015. Internet Usage Statistics. Available at www.internetworldstats.com

Johnson, Kevin. 2015. "FBI Director Says Islamic State Influence Growing in the U.S." USA Today, May 7. Available at www.usatoday.com

Johnson, Steven. 2015 (January 6). "The Revolution in the Driver's Seat." CNN. Available at www.cnn.com

Joiner, Richard, Jeff Gavin, Mark Brosnan, John Cromby, Helen Gregory, Jane Guiller, Pam Maras, and Amy Moon. 2012. "Gender, Internet Experience, Internet Identification, and Internet Anxiety: A Ten-Year Followup." Cyberpsychology, Behavior, and Social Networking 15(7):370–372.

Joiner, Richard, Jeff Gavin, Jill Duffield, Mark Brosnan, Charles Crook, Alan Durndell, Pam Maras, Jane Miller, Adrian J. Scott, and Peter Lovatt. 2005. "Gender, Internet Identification, and Internet Anxiety: Correlates of Internet Use." CyberPsychology & Behavior 8(4):371–378.

Kahn, A. 1997. "Clone Mammals . . . Clone Man?" Nature 386:119.

Kaplan, Karen. 2009. "Corn Fortified with Vitamins Devised by Scientists." Los Angeles Times, April 29. Available at www.latimes.com

Kelly, David. 2013 (February 4). "Study Shows Facebook Unfriending Has Real Off-Line Consequences." University Communications. Available at www.ucdenver.edu

Kemp, Simon. 2015 (January 21). "Digital, Social & Mobile Worldwide in 2015." Available at www.wearesocial.net

Kharfen, Michael. 2006. "1 of 3 and 1 in 6 Pre-Teens Are Victims of Cyber-Bullying." Available at www.fightcrime.org

Kross, Ethan et al. 2013. "Facebook Use Predicts Declines in Subjective Well-Being in Young Adults." PLoS ONE 8(8): e69841

Kuhn, Thomas. 1973. The Structure of Scientific Revolutions. Chicago: University of Chicago Press.

Lang, Joshua. 2013. "What Happens to Women Who Are Denied Abortions?" New York Times, June 12. Available at www.nytimes.com

Legay, F. 2001 (January). "Genetics: Should Genetic Information Be Treated Separately?" Virtual Mentor, p. E5. Available at virtualmentor.ama-assn.org

Lemonick, Michael. 2006. "Are We Losing Our Edge?" Time, February 13, pp. 22–33.

Lemonick, Michael, and Dick Thompson. 1999. "Racing to Map Our DNA." Time Daily 153:1–6. Available at www.time.com

Lenhart, Amanda. 2015 (April 9). "Teens, Social Media & Technology Overview 2015." Pew Research Center. Available at www.pewinternet.org

Lipka, Michael. 2015 (January 21). "5 Facts about Abortion." Pew Research Fact Tank. Available at www.pewresearch.org

Lohr, Steve. 2008. "Health Care That Puts a Computer on the Team." New York Times, December 27. Available at www.nytimes.com

Lohr, Steve. 2011. "Carrots, Sticks and Digital Health Records." New York Times, February 26. Available at www.nytimes.com

Ludden, Jennifer. 2015 (June 10). "Court Decision on Texas Abortion Law Could Hasten Clinic Closures." NPR. Available at www.npr.org

Lynch, John, Jennifer Bevan, Paul Achter, Tim Harris, and Celeste M. Condit. 2008. "A Preliminary Study of How Multiple Exposures to Messages about Genetics Impact Lay Attitudes towards Racial and Genetic Discrimination." New Genetics and Society 27(1):43–56.

Mars One. 2015. "Mission." Available at www.mars-one.com

Marketplace Fairness Act. 2015. "Summary." Available at www.congress.gov

Mateescu, Oana. 2010. "Introduction: Life in the Web." Journal of Comparative Research in Anthropology and Sociology 1(2):1–21.

Mayer, Sue. 2002. "Are Gene Patents in the Public Interest?" BIO-IT World, November 12. Available at www.bio-itworld.com

Mayo Clinic. 2013. "Genetic Testing: Definition." Available at www.mayoclinic.com

McCollum, Sean. 2011. "Getting Past the 'Digital Divide.'" Teaching Tolerance 39. Available at www.tolerance.org

McCormick, S. J., and A. Richard. 1994. "Blastomere Separation." Hastings Center Report, March–April, pp. 14–16.

McDermott, John. 1993. "Technology: The Opiate of the Intellectuals." In Technology and the Future, ed. Albert H. Teich, pp. 89–107. New York: St. Martin's Press.

McFarling, Usha L. 1998. "Bioethicists Warn Human Cloning Will Be Difficult to Stop." Raleigh News and Observer, November 18, p. A5.

McMullan, Thomas. 2014. "How Digital Maps Are Changing the Way We Understand Our World." The Guardian, December 2. Available at www.theguardian.com

Mears, Bill and Tom Cohen. 2014 (June 30). "Supreme Court Rules against Obama in Contraception Case." CNN. Available at www.edition.cnn.com

Merton, Robert K. 1973. "The Normative Structure of Science." In The Sociology of Science, ed. Robert K. Merton. Chicago: University of Chicago Press.

Mesthene, Emmanuel G. 1993. "The Role of Technology in Society." In Technology and the Future, ed. Albert H. Teich, pp. 73–88. New York: St. Martin's Press.

Moeller, Susan, Elia Powers and Jessica Roberts. 2012. "The World Unplugged and 24 Hours Without Media: Media Literacy to Develop Self-Awareness Regarding Media." Scientific Journal of Media Education 39(20):45–52.

Moisse, Katie. 2013 (July 8). "Girl Dies after Groundbreaking Trachea Transplant." ABC News. Available at www.abcnews.go.com

Mooney, Chris, and Sheril Kirshenbaum. 2009. Unscientific America: How Scientific Literacy Threatens Our Future. Philadelphia: Basic Books.

Morgan, Jacob. 2014. "A Simple Explanation of 'The Internet of Things.'" Forbes, May 13. Available at www.forbes.com

Murphie, Andrew, and John Potts. 2003. Culture and Technology. New York: Palgrave Macmillan.

Nakashima, Ellen. 2015. "Chinese Breach Data of 4 Million Federal Workers." Washington Post, June 4. Available at www.washingtonpost.com

National Abortion and Reproductive Rights Action League (NARAL). 2013. Mifepristine: The Impact of Abortion Politics on Women's Health in Scientific Research. February 28. Available at www.prochoiceamerica.org

National Abortion and Reproductive Rights Action League (NARAL). 2015. Abortion Bans at 20 Weeks: A Dangerous Restriction for Women. January 1. Available at www.prochoiceamerica.org

Nature News. 2011 (February 23). "Animal Research: Battle Scars." Nature 470 (7335): 452–453.

NBC News. 2015 (March 10). "Wikipedia Sues NSA over Mass Surveillance." Available at www.getbriefme.com

National Conference of State Legislatures. 2015. "Human Cloning Laws." Available at www.ncsl.org

National Human Genome Research Institute. 2015. "Education." Available at www.genome.gov

National Science Foundation (NSF). 2015. "Science and Engineering Indicators, 2014." Available at www.nsf.gov

New England Anti-Vivisection Society. 2015. "Product Development and Drug Testing." Available at www.neavs.org

Near, Christopher. 2013. "Selling Gender: Associations of Box Art Representation of Female Characters with Sales for Teen and Mature Rated Video Games." Sex Roles 68(3/4):252–269.

Noble, Phil. 2015. The Future is Open: 2015 State of Innovation. Thomson Reuters. Available at images.info.science.thomsonreuters

Noonan, Mary C., and Jennifer L. Glass. 2012. "The Hard Truth about Telecommuting." Monthly Labor Review (June):38–45.

Oak Ridge National Laboratory. 2011. "Medicine and the New Genetics." Office of Biological and Environmental Research. Available at www.ornl.gov

Ogburn, William F. 1957. "Cultural Lag as Theory." Sociology and Social Research 41:167–174.

OpenNet. 2015 OpenNet Initiative: Global Internet Filtering Map. Available at map.opennet.net

Organisation for Economic Cooperation and Development (OECD). 2014. Measuring the Digital Economy: A New Perspective, OECD Publishing. Available at dx.doi.org/10.1787/97889264221796-en

Overby, Stephanie. 2015. "IT Outsourcing Deal Values Hit 10-Year Low." CIO. Available at www.cio.com

Peek, Liz. 2015. "Regulating Bitcoin Legitimizes a Treacherous Business." Fiscal Times, May 13. Available at www.thefiscaltimes.com

Peralta, Katherine. 2014. "Outsourcing to China Cost U.S. 3.2 Million Jobs since 2001." *U.S. News & World Report*, December 11. Available at www.usnews.com

Perez, Evan. 2014 (May 19). "More Than 90 People Nabbed in Global Hacker Crackdown." CNN. Available at www.cnn.com

Pew. 2014a (August 6). "AI, Robotics, and the Future of Jobs." Available at www.pewinternet.com

Pew. 2014b (October 29). "Cyber Attacks Likely to Increase." Available at www.pewinternet.org

Pew. 2014c (June 1). "The Darkest Side of Online Harassment: Menacing Behavior." Available at www.pewresearch.org

Pew. 2014d (November 12). "Public Perceptions of Privacy and Security in the Post-Snowden Era." Available at www.pewinternet.org

Pew. 2014e (December 30). "Technology's Impact on Workers." Available at www.pewinternet.com

Pew. 2015a (March 19). "Online Activities in Emerging and Developing Nations." Available at www.pewglobal.com

Pew. 2015b (January 29). "Public and Scientists' Views on Science and Society." Available at www.pewinternet.org

Picard, Martin, and Doug M. Turnbull. 2013. "Linking the Metabolic State and Mitochondrial DNA in Chronic Disease, Health, and Aging." *Diabetes* 62 (March):672–678.

Plummer, Brad. 2013. "The Coming R&D Crash." *Washington Post,* February 26. Available at www.washingtonpost.com

Postman, Neil. 1992. *Technopoly: The Surrender of Culture to Technology.* New York: Knopf.

Powles, Julia and Enrique Chaparro. 2015. "How Google Determined Our Right to Be Forgotten." *The Guardian,* February 18. Available at www.theguardian.com

President's Council of Advisors on Science and Technology. 2014 (April). Report to the *President on Big Data and Privacy: A Technological Perspective.* Washington, DC. Available at www.whitehouse.gov

Price, Tom. 2008. "Science in America: Are We Falling Behind in Science and Technology?" *CQ Researcher* 18(2):24–48.

Public Broadcasting Service. 2015 (January 7). "While the Drone Industry Grows Faster Than a Flick, of a Joystick, Regulations Lag." *PBS Newshour.* Available at www.pbs.org

Rabino, Isaac. 1998. "The Biotech Future." *American Scientist* 86(2):110–112.

Research Animal Resources. 2003. "Ethics and Alternatives." University of Minnesota. Available at www.ahc.umn.edu

Risen, Tom. 2015. "FCC Enacts Title II Net Neutrality Rules with Partisan Vote." *US News & World Report,* February 26. Available at www.usnews.com

Roberts, Yvonne. 2014. "Addicted to Email? The Germans Have an Answer." *The Guardian,* August 31. Available at www.theguardian.com

Roe v. Wade. 1973. 410 U.S. 113.

Ross, Valerie. 2014. "Forget Fingerprints: Law Enforcement DNA Databases Poised to Expand." *PBS NOVA,* January 2. Available at www.pbs.org

Rossignol, Pascal. 2015. "FDA approves GMO Apples, Potatoes That Don't Bruise or Brown as Being Safe to Eat." Reuters, March 22. Available at www.rt.com/usa

Rovner, Julie. 2013 (May 16). "Cloning, Stem Cells Long Mired in Legislative Gridlock." National Public Radio. Available at www.npr.org

Savel, Maria. 2015. "EU Strikes Twice Against Google with Antitrust Charges." *World Politics Review,* April 28. Available at www.worldpoliticsreview.com

Sax, Linda, Kathleen Lehman, Allison Kanny, Gloria Lim, Laura Paulson, Hilary Zimmerman, and Jerry Jacobs. 2015 (April 20). "Anatomy of an Enduring Gender Gap: The Evolution of Women's Participation in Computer Science." Paper presented at the American Educational Research Association, Chicago.

Schwartz, Aaron L., Bruce E. Landon, Adam G. Elshaug, Michael E. Chernew, and J. Michael McWilliams. 2014. "Measuring Low-Value Care in Medicare." *JAMA Internal Medicine* 174(7):1067–1076.

Shadboldt, Peter. 2014 (October 2). "Feeling Sick? Cyber Doctor Is Just a Click Away." CNN. Available at www.cnn.com

Slade, Giles. 2012. *The Big Disconnect: The Story of Technology and Loneliness.* Amherst, NY: Prometheus Books.

Smith, Aaron. 2014 (November 3). "Cell Phones, Social Media and Campaign 2014." Available at www.pewinternet.org

Stem Cell Research Advancement Act. 2013. H.R. 2433. Available at www.govtrack.us

Sticca, Fabio, and Sonja Perren. 2013. "Is Cyberbullying Worse Than Traditional Bullying? Examining the Differential Roles of Media, Publicity, and Anonymity for the Perceived Severity of Bullying." *Journal of Youth and Adolescence* 42(5):739–750.

Sullivan, Gail. 2014. "Students Develop Nail Polish to Detect Date-Rape Drugs." *Washington Post,* August 26. Available at www.washingtonpost.com

Tanner, Lindsey. 2013. "2-year-old Girl Gets Windpipe Made from Stem Cells." *USA Today,* April 30. Available at www.usatoday.com/

Thas, Angela, Chat Garcia Ramilo, and Cheekay Garcia Cinco. 2007. *Gender and ICT.* United Nations Development Programme. Bangkok, Thailand: Pacific Development Information Programme. Available at www.undp.org

Thornhill, Ted. 2014. "'I Can Grab My Handlebars!' Girl Born without Fingers Delighted after High School Builds Her a Prosthetic Hand for Just Five dollars." *Daily Mail,* May 5. Available at: www.dailymail.co.uk Toffler, Alvin. 1970. *Future Shock.* New York: Random House.

United Nations. 2014. "United Nations E-Government Survey 2014." Available at unpan1.un.org

U.S. Census Bureau. 2013. *Statistical Abstract of the United States, 2012,* 128 ed. Washington, DC: U.S. Government Printing Office.

U.S. Citizenship and Immigration Services. 2015 (May 4). "H-1B Fiscal Year (FY) 2016 Season." Available at www.uscis.gov

U.S. Department of Justice. 2015. "Anti-Trust Case Filings: *United States v. Microsoft Corporation.*" Available at www.justice.gov

Vigdor, Jacob L., Helen F. Ladd, and Erika Martinez. 2014. "Scaling the Digital Divide: Home Computer Technology and Student Achievement." *Economic Inquiry* 52(3)(July):1103–1119.

Wachter, Robert M. 2015. "Why Health Care Tech Is Still So Bad." *New York Times*, March 21. Available at www.nytimes.com

Wald, Chelsea. 2015. "Why Your Brain Hates Slowpokes." *Nautilus,* March 5. Available at www.nautil.us

Weinberg, Alvin. 1966. "Can Technology Replace Social Engineering?" *University of Chicago Magazine* 59(October):6–10.

Weiser, Benjamin. 2015. "Man behind Silk Road Website is Convicted on All Counts." *New York Times.* February 4. Available at www.nytimes.com

Weiser, Benjamin, and Matt Apuzzo. 2015. "Inquiry of Silk Road Website Spurred Agents Own Illegal Acts, Officials Say." *New York Times,* March 30. Available at www.nytimes.com

Welter, Cole H. 1997. "Technological Segregation: A Peek through the Looking Glass at the Rich and Poor in an Information Age." *Arts Education Policy Review* 99(2):1–6.

Westcott, Lucy. 2015. "Finding That Babies Born at 22 Weeks Can Survive Could Change Abortion Debate." *Newsweek,* May 7. Available at www.newsweek.com

Wheelwright, Jeff. 2014. "Schizophrenia Study Finds New Genetic Links." *Discover*, November 26. Available at www.discovermagazine.com

White House. 2015a (March 23). *Fact Sheet: Next Steps in Delivering Fast, Affordable Broadband.* Available at www.whitehouse.gov

White House. 2015b (February 13). *Summit on Cybersecurity and Consumer Protection.* Available at www.whitehouse.gov

Winner, Langdon. 1993. "Artifact/Ideas as Political Culture." In *Technology and the Future,* ed. Albert H. Teich, 283–294. New York: St. Martin's Press.

World Bank. 2009. *Global Economic Prospects: Technology Diffusion in the Developing World.* Washington, DC: World Bank.

World Economic Forum. 2015. "Global Risks Landscape 2015." *World Economic Forum.* Available at reports.weforum.org

World Hunger Education Service. 2015. "2015 World Hunger and Poverty Facts and Statistics." Available at www.worldhunger.org

Zickuhr, Kathryn. 2013 (September 25). *Who's Not Online and Why.* Pew Research Center's Internet and American Life. Available at pewinternet.org

Zickuhr, Kathryn, and Aaron Smith. 2013 (August 26). *Home Broadband 2013.* Pew Research Center's Internet and American Life. Available at pewinternet.org

Chapter 15

Al-Hussein, Zaid Raad. 2015. "Europe Should Remember Its Past When Rejecting Migrants." *Daily Star,* June 6. Available at www.dailystar.com.lb

Alexander, Karen, and Mary E. Hawkesworth, eds. 2008. *War and Terror: Feminist Perspectives.* Chicago: University of Chicago Press.

American Jewish Committee. 2013 (October 28). *Annual Survey of American Jewish Opinion.* Available at www.ajc.org

Animals at Arms. 2010. "Animals at Arms." *Army-Technology,* December 22. Available at www.army-technology.com

Arango, Tim, and Anne Barnard. 2013. "As Syrians Fight, Sectarian Strife Infects Mideast." *New York Times,* June 1. Available at www.nytimes.com/

Arms Control Association. 2014 (February). "Chemical and Biological Weapons Status at a Glance." Available at armscontrol.org

Associated Press. 2015. "John McCain Urges US to Provide Defensive Weapons to Vietnam." *The Guardian,* May 30. Available at www.theguardian.com

Atomic Archive. 2015. *Arms Control Treaties.* Available at www.atomicarchive.com

Baker, Peter, Helene Cooper, and Mark Mazzetti. 2011. "Bin Laden Is Dead, Obama Says." *New York Times,* May 1. Available at www.nytimes.com

Barkan, Steven, and Lynne Snowden. 2001. *Collective Violence.* Boston: Allyn and Bacon.

Bauer, Sibylle. 2013. "For the Bathroom or the Missile Factory: Why Dual Use Trade Controls Matter." December 12. Available at www.sipri.org

Bay, Mrten, and Isaac R. Porche III. 2015. "War on the Web." World Report. *U.S. News & World Report,* July 23. Available at www.usnews.com

BBC. 2000 (April 15). "UN Admits Rwanda Genocide Failure." *BBC News.* Available at news.bbc.co.uk

BBC. 2015 (January 15). "Charlie Hebdo Attack: Three Days of Terror." *BBC News Europe.* Available at www.bbc.com

Behm, Sharon. 2015 (July 22). "Cyber War: Sex, Fingerprints, and Spearfishing." *Voice of America.* Available at www.voanews.com

Bello, Hakim and Daniel Trilling. 2015. "I Was a Lampedusa Refugee. Here's My Story of Fleeing Libya—and Surviving." *The Guardian,* April 20. Available at www.theguardian.com

Bercovitch, Jacob, ed. 2003. *Studies in International Mediation: Advances in Foreign Policy Analysis.* New York: Palgrave Macmillan.

Bergen, Peter. 2002. *Holy War, Inc.: Inside the Secret World of Osama bin Laden.* New York: Free Press.

Bergen, Peter. 2015 (June 26). "Terror on Three Continents," CNN. Available at www.cnn.com

Berlinger, Joshua. 2015 (February 26). "The Names: Who Has Been Recruited to ISIS from the West." CNN. Available at www.cnn.com

Berrigan, Frida, and William Hartung. 2005. *U.S. Weapons at War: Promoting Freedom or Fueling Conflict?* World Policy Institute Report. Available at www.worldpolicy.org/projects/arms/reports/wawjune2005.html

Bjorgo, Tore. 2003. *Root Causes of Terrorism.* Paper presented at the International Expert Meeting, June 9–11. Oslo: Norwegian Institute of International Affairs.

Bonner, Michael. 2006. *Jihad in Islamic History: Doctrines and Practice.* Princeton, NJ: Princeton University Press.

Borger, Julian, Andrew MacDowell, Amelia Gentleman, Kate Connelly, David Crouch, and Frances Paraudin. 2015. "Winter is Coming: The New Crisis for Refugees in Europe." *The Guardian,* November 2. Available at www.theguardian.com

Borum, Randy. 2011. "Radicalization into Violent Extremism II: A Review of Conceptual Models and Empirical Research." *Journal of Strategic Security* 4(4): 37–62. Available at scholarcommons.usf.edu/jss/vol4/iss4/3

Boustany, Nora. 2007. "Janjaweed Using Rape as 'Integral' Weapon in Darfur, Aid Group Says." *Washington Post,* July 3. Available at www.washingtonpost.com

Brauer, Jurgen. 2003. "On the Economics of Terrorism." *Phi Kappa Phi Forum,* Spring, pp. 38–41.

Broad, William J., and David E. Sanger. 2009. "Report Says Iran Has Data to Make a Nuclear Bomb." *New York Times,* October 3. Available at www.nytimes.com

Brooks, David. 2011. "Huntington's Clash Revisited." *New York Times,* March 3. Available at www.nytimes.com

Brown, Michael E., Sean M. Lynn-Jones, and Steven E. Miller, eds. 1996. *Debating the Democratic Peace.* Cambridge, MA: MIT Press.

Brumfield, Ben. 2015. "It's 'Adoption Day' - Launch Time for the Iran Nuclear Deal." CNN, October 18. Available at www.cnn.com

Bumiller, Elisabeth. 2011. "The Dogs of War: Beloved Comrades in Afghanistan." *New York Times,* May 11. Available at www.nytimes.com

Buncombe, Andrew. 2009. "End of Sri Lanka's Civil War Brings Back Tourists." *The Independent,* August 16. Available at www.independent.co.uk

Bureau of Investigative Journalism. 2015. "Casualty Estimates." Available at www.thebureauinvestigates.com

Burns, Robert. 2011. "Panetta: U.S. within Reach of Defeating Al Qaeda." *Washington Times,* July 9. Available at www.washingtontimes.com

Burrelli, David F. 2013 (May 9). "Women in Combat: Issues for Congress." *Congressional Research Service.* Available at www.fas.org/sgp/crs/natsec/R42075.pdf

Carneiro, Robert L. 1994. "War and Peace: Alternating Realities in Human History." In *Studying War: Anthropological Perspectives,* ed. S. P. Reyna and R. E. Downs, 3–27. Langhorne, PA: Gordon & Breach.

Cassidy, John. 2015. "It's Time to Let Edward Snowden Come Home." *New Yorker,* June 3. Available at www.newyorker.com

Center for Arms Control and Non-Proliferation. 2013. "Fact Sheet: Global Nuclear Weapons Inventories in 2013." Available at www.armscontrolcenter.org.

Center for Systemic Peace. 2015. "Assessing the Qualities of Systemic Peace." Available at www.systemicpeace.org

Chandler, Adam. 2015. "ISIS Takes Credit for the Texas Attack." *The Atlantic,* May 5. Available at www.theatlantic.com

CNN. (October 8) 2014. "USS *Cole* Bombing Fast Facts." Available at www.cnn.com

CNN/Opinion Research Corporation Poll. 2007 (June 22–24). *Iraq.* Available at www.pollingreport.com/iraq3.htm

Cohen, Ronald. 1986. "War and Peace Proneness in Pre- and Post-industrial States." In *Peace and War: Cross-Cultural Perspectives,* ed. M. L. Foster and R. A. Rubinstein, 253–267. New Brunswick, NJ: Transaction Books.

Conflict Research Consortium. 2003. *Mediation.* Available at www.colorado.edu/conflict/peace

Council on Foreign Relations. 2014 (June 24). "Background Briefing: What is ISIL?" Available at www.pbs.org

Crawford, Neta C. 2015 (May 22). "War-Related Death, Injury, and Displacement in Afghanistan and Pakistan 2001–2014." *Costs of War.* Watson Institute for International Studies. Available at watson-brown.edu

Dalrymple, William. 2013 (June 25). "A Deadly Triangle: Afghanistan, Pakistan, & India." The Brookings Essay. Available at www.brookings.edu/series/the-brookings-essay

Dao, James. 2013. "Deployment Factors Are Not Related to Rise in Military Suicides, Study Says." *New York Times,* August 6. Available at www.nytimes.com

Dareini, Ali Akbar. 2009. "Iran Missile Test: Ahmadinejad Says It's within Israel's Range." *Huffington Post,* May 20. Available at www.huffingtonpost.com

Dean, Josh. 2013 (October 22). "In the Bomb-Detecting Dogs Business, iK9 Sniffs Out an Empire." Bloomberg Business. Available at www.bloomberg.com

Deen, Thalif. 2000 (September 9). *Inequality Primary Cause of Wars, Says Annan.* Available at www.hartford-hwp.com/archives

Defense Casualty Analysis System (DCAS). 2015. "Conflict Casualties." U.S. Department of Defense. Available at www.dmdc.osd.mil

Dennis, Vanessa. 2014. "Photos: U.S. Navy's Dolphin and Seal Program." March 26. Available at www.pbs.org

Deutsch, Anthony. 2015 (May 8). "Exclusive: Weapons Inspectors Find Undeclared Sarin and VX Traces in Syria." Available at www.reuters.com

Dixon, William J. 1994. "Democracy and the Peaceful Settlement of International Conflict." *American Political Science Review* 88(1): 14–32.

Donnelly, John. 2011 (January 24). "More Troops Lost to Suicide." Available at www.congress.org

Dower, John. 1986. *War without Mercy.* New York: Pantheon Books.

The Economist. 2015. "Syria's Humanitarian Crisis." *The Economist,* June 8. Available at www.economist.com

Elliot, Dan. 2015. "U.S. to Destroy Largest Remaining Chemical Weapons Cache." *USA Today,* February 4. Available at www.usatoday.com

Farwell, James P. 2014 (January). "The Media Strategy of ISIS." *Survival: Global Politics and Strategy* 56(6):49–55.

Feder, Don. 2003. "Islamic Beliefs Led to the Attack on America." In *The Terrorist Attack on America,* ed. Mary E. Williams, 20–23. Farmington Hills, MA: Greenhaven Press.

Frankel, Rebecca. 2011a. "War Dogs: The Legend of the Bin Laden Hunter Continues." *Foreign Policy,* May 12. Available at www.foreignpolicy.com

Frankel, Rebecca. 2011b. "War Dogs: There's a Reason Why They Brought One to Get Osama Bin Laden." *Foreign Policy,* May 4. Available at www.foreignpolicy.com/articles

Frey, Josh. 2003 (August 12). "Anti-Swimmer Dolphins Ready to Defend Gulf." Available at www.navy.mil

Freytag, Andreas, Jens J. Kruger, Daniel Meierrieks, and Friedrich Schneider. 2011. "The Origins of Terrorism: Cross-country Estimates of Socio-economic Determinants of Terrorism." *European Journal of Political Economy* 27(1):S5–S16.

Fuller, Jaime. 2015. "Snowden Would Come Home If Politicians Promise Not to Hang Him." *New York Magazine,* July 22. Available at nymag.com

Funke, Odelia. 1994. "National Security and the Environment." *In Environmental Policy in the 1990s: Toward a New Agenda,* 2nd ed., Norman J. Vig and Michael E. Kraft, eds. (pp. 323–345). Washington, DC: Congressional Quarterly.

Gallup. 2013 (April 25). "Terrorism in the United States." Available at www.gallup.com

Gallup. 2015. "Military and National Defense." Available at www.gallup.com

Garrett, Laurie. 2001. "The Nightmare of Bioterrorism." *Foreign Affairs* 80:76.

Geneva Graduate Institute for International Studies. 2015. *Small Arms Survey, 2015.* New York: Oxford University Press.

German, Erik. 2011. "Flipper Goes to War." *The Daily,* July 18. Available at www.thedaily.com

Gettleman, Jeffrey. 2008. "Rape Victims' Words Help Jolt Congo into Change." *New York Times,* October 18. Available at www.nytimes.com

Gettleman, Jeffrey. 2009. "Symbol of Unhealed Congo—Male Rape Victims." *New York Times,* August 4. Available at www.nytimes.com

Gioseffi, Daniela. 1993. "Introduction." In *On Prejudice: A Global Perspective,* ed. Daniela Gioseffi, xi–l. New York: Anchor Books, Doubleday.

Gladstone, Rick. 2015. "Many Ask, Why Not Call Church Shooting Terrorism?" *New York Times,* June 18. Available at www.nytimes.com

Goldich, Dave, and Art Swift. 2014 (May 23). "Americans Say Army Most Important Branch to U.S. Defense." Available at www.gallup.com

Goode, Paul. 2014. "How Russian Nationalism Explains—and Does Not Explain—the Crimean Crisis." *Washington Post,* March 3. Available at www.washingtonpost.com

Goodwin, Jeff. 2006. "A Theory of Categorical Terrorism." *Social Forces* 84(4):2027–2046.

Gorka, Katharine C. 2014. "The Flawed Science behind America's Counter-Terrorism Strategy." Council on Global Security. Available at councilonglobalsecurity.org

Goure, Don. 2003 (March 20). *First Casualties? NATO, the U.N.* MSNBC News. Available at www.msnbc.com/news

Greenburg, Jan Crawford, and Ariane de Vogue. 2008 (June 12). "Supreme Court: Guantanamo Detainees Have Rights in Court." *ABC News.* Available at abcnews.go.com

Gurr, Ted Robert. 2015. *Political Rebellion: Causes, Outcomes and Alternatives.* New York: Routledge.

Hadro, Matt. 2015. "Catholic News Agency: The Cost of Violence: 15 Million Africans Displaced, with Little Hope for the Future." Refugees International. July 14. Available refugeesinternational.org

Haiken, Melanie. 2013. "Suicide Rate among Vets and Active Duty Military Jumps—Now 22 a Day." *Forbes,* February 5. Available at www.forbes.com

Head Start. 2015. *Head Start Program Facts: FY2014.* Available at eclkc.ohs.acf.hhs.gov/

Hicks, Josh. 2013. "Pentagon Extends Benefits to Same-Sex Military Spouses." *Washington Post,* August 14. Available at www.washingtonpost.com

Hsu, Spencer S. and Victoria St. Martin. 2015. "Four Blackwater Guards Sentenced in Iraq Shootings of 31 Unarmed Civilians." *Washington Post,* April 13. Available at www.washingtonpost.com

Human Rights Watch (HRW). 2014 (May 13). "Syria: Strong Evidence Government Used Chemicals as Weapons." Available at www.hrw.org

Huntington, Samuel. 1996. *The Clash of Civilizations and the Remaking of World Order.* New York: Simon and Schuster.

Institute for Economics and Peace (IEP). 2013. The Economic Consequences of War on the U.S. Economy. Available at economicsandpeace.org

Institute for Economics and Peace (IEP). 2015. The Economic Cost of Violence Containment. Available at economicsandpeace.org

Intel. 2015. "In Pursuit of Conflict Free Minerals." Available at www.intel.com

International Campaign to Ban Landmines. 2015. *States Not Party.* Available at www.icbl.org

International Committee of the Red Cross. (2015). "Rule 93. Rape and Other Forms of Sexual Violence." *Customary International Humanitarian Law.* Available at www.icrc.org

International Institute for Strategic Studies. 2015. "Armed Conflict Database: All Conflicts." Available at acd.iiss.org

Jeon, Arthur. 2011 (May 5). "German Shepherd? Belgian Malinois? Navy SEAL Hero Dog Is Top Secret." *Global Animal.* Available at www.globalanimal.org

Kamp, Jon, and Jennifer Levitz. 2015. "Boston Marathon Bombing Trial Verdict: Jury Sentences Dzhokhar Tsarnaev to Death." *Wall Street Journal,* May 15. Available at www.wsj.com

Karam, Zeina, and Bram Janssen. 2015. "Islamic State Recruits, Indoctrinates Children in Iraq, Syria." *Columbus Dispatch,* July 20. Available at www.dispatch.com

Karnitschnig, Matthew, William Horobin, and Anton Troianovski. 2015. "A Backlash Swells in Europe after Charlie Hebdo Attack." *Wall Street Journal,* January 8. Available at www.wsj.com

Kaufman, Edward, and Manuel Hassassian. 2009. "Understanding Our Israeli-Palestinian Conflict and Searching for Its Resolution." In *Regional and Ethnic Conflicts,* ed. Judy Carter, George Irani, and Vamik D. Volkan, 90–129. Upper Saddle River, NJ: Pearson.

Kemp, Janet E. 2014 (January 24). "Suicide Rates in VHA Patients through 2011 with Comparisons with Other Americans and Veterans through 2010." Veterans Health Administration. Available at www.mentalhealth.va.gov

Kirkpatrick, David D. 2014. "Graft Hobbles Iraq's Military in Fighting ISIS." *New York Times,* November 23. Available at www.nytimes.com

Klare, Michael. 2001. *Resource Wars: The New Landscape of Global Conflict.* New York: Metropolitan Books.

Kleykamp, Meredith and Molly Clever. 2015. "Women in Combat: The Quest for Full Inclusion." In *Women, War, and Violence,* vol. 2, ed. Mariam Kurtz and Lester Kurtz. Santa Barbara, CA: Praeger.

Knickerbocker, Brad. 2002. "Return of the Military-Industrial Complex?" *Christian Science Monitor,* February 13. Available at www.csmonitor.com

Kruglanski, Arie W., Michelle J. Gelfand, Jocelyn J. Belandr, Anna Sheveland, Malkanthi Hetiarachchi, and Rohen Gunaratna. 2014. "The Psychology of Radicalization and Deradicalization: How Significance Quest Impacts Violent Extremism." *Political Sociology* 35(1):69–93.

Lamont, Beth. 2001. "The New Mandate for UN Peacekeeping." *The Humanist* 61:39–41.

Langley, Robert. n.d. "Chemical Sniffing Dogs Deployed along Borders." Available at usgovinfo.about.com

Laqueur, Walter. 2006. "The Terrorism to Come." In *Annual Editions 05–06,* ed. Kurt Finsterbusch, 169–176. Dubuque, IA: McGraw-Hill/Dushkin.

Larrabee, F. Stephen, Stuart E. Johnson, John Gordon, Peter A. Wilson, Caroline Baxter, Deborah Lai, and Calin Trenkov-Wermuth. (2012). *NATO and the Challenges of Austerity.* Santa Monica, CA: RAND Corporation. Available at www.rand.org/pubs/monographs/MG1196

Larsen, Kaj. 2011 (July 31). "Harnessing the Military Power of Animal Intelligence." CNN. Available at articles.cnn.com

LeardMann, Cynthia A., Teresa M. Powell, Tyler C. Smith, Michael R. Bell, Besa Smith, Edward J. Boyko, Tomoko I. Hooper, Gary D. Gackstetter, Mark Ghamsary, and Charles W. Hoge. 2013. "Risk Factors Associated with Suicide in Current and Former U.S. Military Personnel." *Journal of the American Medical Association* 310(5):496–506.

Lederer, Edith. 2005 (May 20). "Annan Lays Out Sweeping Changes to U.N." Associated Press. Available at www.apnews.com

Lee, Jane J. 2014. "Military Dolphins and Sea Lions: What Do They Do and Who Uses Them?" *National Geographic* (March 29). Available at news.nationalgeographic.com

Leinwand, Donna. 2003. "Sea Lions Called to Duty in Persian Gulf." *USA Today,* February 16. Available at www.usatoday.com

Levy, Clifford J., and Peter Baker. 2009. "U.S.-Russian Nuclear Agreement Is First Step in Broad Effort." *Washington Post,* July 6. Available at www.washingtonpost.com

Levy, Jack S. 2001. "Theories of Interstate and Intrastate War: A Levels of Analysis Approach." In *Turbulent Peace: The Challenges of Managing International Conflict,* ed. Chester A. Crocker, Fen Osler Hampson, and Pamela Aall, 3–27. Washington, DC: U.S. Institute of Peace.

Liguori, Mike. 2015. "What It's Like to Live as a Veteran with a Mental Health Stigma." *Huffington Post,* March 6. Available at www.huffingtonpost.com

Lindner, Andrew M. 2009. "Among the Troops: Seeing the Iraq War through Three Journalistic Vantage Points." *Social Problems* 56(1):21–48.

MacAskill, Ewen, Ed Pilkington, and Jon Watts. 2006. "Despair at UN over Selection of 'Faceless' Ban Ki-moon as General Secretary." *The Guardian,* October 7. Available at www.guardian.co.uk

Mahr, Krista. 2013. "Sri Lanka to Start Tally of Civil-War Dead." *Time,* November 28. Available at world.time.com

Manzetti, Mark et al. 2015. "The Secret History of SEAL Team 6." *New York Times,* June 7. Available at www.nytimes.com

Martinez, Michael, Jethro Mullen, and Josh Levs. 2015 (January 10). "Who Are Suspects in Two Violent French Standoffs?" CNN. Available at www.cnn.com

Mazzetti, Mark, and Scott Shane. 2009. "Interrogation Memos Detail Harsh Tactics by the C.I.A."

New York Times, April 16. Available at www .nytimes.com

Miller, Erin, and Gary LaFree. 2014 (April). "Country Reports on Terrorism 2013: Annex of Statistical Information." National Consortium for the Study of Terrorism and Responses to Terrorism (START). Available at www.state .gov/j/ct/rls/crt/2013/224831.htm

Millman, Jason. 2008. "Industry Applauds New Dual-Use Rule." *Hartford Business Journal*, September 29. Available at www.hartfordbusiness.com

Mittal, Dinesh, K.L. Drummond, D. Belvins, G. Curran, P. Corrigan, and G. Sullivan. 2013 (June). "Stigma Associated with PTSD: Perceptions of Treatment Seeking Combat Veterans." *Psychiatric Rehabilitation Journal* 36(2):86–92.

Montalván, Luis Carlos. 2011. *Until Tuesday: A Wounded Warrior and the Golden Retriever Who Saved Him*. New York: Hyperion.

Morgan, Jenny, and Alic Behrendt. 2009. *Silent Suffering: The Psychological Impact of War, HIV, and Other High-Risk Situations on Girls and Boys in West and Central Africa*. Plan International. Available at plan-international.org

Mueller, John. 2006. *Overblown: How Politicians and the Terrorism Industry Inflate National Security Threats, and Why We Believe Them*. New York: Free Press.

Mueller, John, and Mark G. Stewart. 2014 (Summer). "Evaluating Counterterrorism Spending." *Journal of Economic Perspectives* 28(3):237–247.

Myers-Brown, Karen, Kathleen Walker, and Judith A. Myers-Walls. 2000 (November 20). "Children's Reactions to International Conflict: A Cross-Cultural Analysis." Paper presented at the National Council of Family Relations, Minneapolis.

National Association of School Psychologists. 2003. *Children and Fear of War and Terrorism*. Available at www.nasponline.org

National Center for Posttraumatic Stress Disorder. 2007. "What Is Post Traumatic Stress Disorder?" Available at www.ncptsd.va.gov

National Center for Veterans Analysis and Statistics. 2015. "Expenditures." U.S. Department of Veterans' Affairs. Available at www.va.gov /vetdata/Expenditures.asp

National Counterterrorism Center. 2012 (March 12). *2011 Report on Terrorism*. Available at www.nctc.gov

National Priorities Project. 2015. *Cost of War: Trade-Offs*. Available at costofwar.com

Navy Times. 2012 (November 23). "Navy Mine-Detecting Dolphins to Retire by 2017." Available at www.navytimes.com

New York Times. 2015 (June 13). "The Guantánamo Docket." Available at projects.nytimes.com

Norman, Laurence, and Jay Solomon. 2015. "Iran, World Powers Reach Nuclear Deal." *Wall Street Journal*, July 14. Available at www.wsj.com

Ochmanek, David, and Lowell H. Schwartz. 2008. "The Challenge of Nuclear-Armed Regional Adversaries." Rand Project Air Force. Available at www.rand.org

Office of the Coordinator for Counterterrorism. 2014 (January 14). "Foreign Terrorist Organizations." Available at www.state.gov/j/ct

Office of Management and Budget. 2015a. "*Table S-11: Funding Levels for Appropriated ("Discretionary") Programs by Agency.*" Available at www.whitehouse.gov

Office of Management and Budget. 2015b. "*Table 28-1: Policy Budget Authority and Outlays by Function, Category, and Program.*" Available at www.whitehouse.gov/sites/default/files/omb/ budget/fy2015/assets/28_1.pdf

Office of Weapons Removal and Abatement. 2013. *To Walk the Earth in Safety: The United States' Commitment to Humanitarian Mine Action and Conventional Weapons Destruction*. Available at www.state.gov

Oliver, Amy. 2011. "Mammals with a Porpoise . . . Meet the Dolphins and Sea Lions Who Go to War with the U.S. Navy." *Daily Mail*, July 6. Available at www.dailymail.co.uk

Paul, Annie Murphy. 1998. "Psychology's Own Peace Corps." *Psychology Today* 31:56–60.

Permanent Court of Arbitration. 2015. *About Us and Cases*. Available at www.pca-cpa.org

Perry, Tony. 2015. "Dolphins, Sea Lions Train for Navy Deployment to Overseas Trouble Spots." *Los Angeles Times*, March 29. Available at www.latimes.com

Pew Research Center. 2013 (September 10). "Muslim Publics Share Concerns about Extremist Groups." Available at www.pewglobal.org

Pew Research Center. 2014a (December 9). "Americans' Views on Use of Torture in Fighting Terrorism Have Been Mixed." Available at www .pewresearch.org

Pew Research Center. 2014b (July 14). "Global Opposition to U.S. Surveillance and Drones, but Limited Harm to America's Image." Available at www.pewglobal.com

Pew Research Center. 2014c (June 26). "Patriotism, Personal Traits, Lifestyles, and Demographics." Available at www.pewresearch.org

Pew Research Center. 2015a (February 24). "Growing Support for Campaign against ISIS— and Possible Use of U.S. Ground Troops." Available at www.pewresearch.org

Pew Research Center. 2015b (January 12). "Terrorism Worries Little Changed; Most Give Government Good Marks for Reducing Threat." Available at www.pewresearch.org

Piazza, James A. 2011 (March). "Poverty, Minority Economic Discrimination, and Domestic Terrorism." *Journal of Peace Research* 28(3):339–353.

Pilkington, Ed. 2013. "Bradley Manning Verdict: Cleared of 'Aiding the Enemy' but Guilty of Other Charges." *The Guardian*, July 30. Available at www.theguardian.com

Ploughshares. 2015. *World Nuclear Weapon Stockpiles*. Available at www.ploughshares.org

Plumer, Brad. 2013. "The U.S. Gives Egypt $1.5 Billion a Year in Aid. Here's What It Does." *Washington Post*, July 9. Available at www .washingtonpost.com

Porter, Bruce D. 1994. *War and the Rise of the State: The Military Foundations of Modern Politics*. New York: Free Press.

Portero, Ashley. 2013. "Women in Combat Units Could Help Reduce Sexual Assaults: US Joint Chiefs Chairman." *International Business Times*, January 25. Available at www.ibtimes.com

Powell, Bill, and Tim McGirk. 2005. "The Man Who Sold the Bomb." *Time*, February 14, pp. 22–31.

Prasad, Hari. (May 29) 2015. "The Security Dilemma and ISIS." *International Affairs Review*. Available at www.iar-gwu.org

Price, Eluned. 2004. "The Served and Suffered for Us." *The Telegraph*, November 1. Available at www.telegraph.co.uk

Purefoy, Christian, and Ralph Ellis. 2015 (May 4). "Nigeria: 234 More Women, Children Rescued." CNN. Available at www.cnn.com

Rapoza, Kenneth. 2013. "Syria's Chemical Weapons Slowly Being Eradicated." *Forbes*, October 16. Available at www.forbes.com

Rashid, Ahmed. 2000. *Taliban: Militant Islam, Oil, and Fundamentalism in Central Asia*. New Haven, CT: Yale University Press.

Rasler, Karen, and William R. Thompson. 2005. *Puzzles of the Democratic Peace: Theory, Geopolitics, and the Transformation of World Politics*. New York: Palgrave Macmillan.

Recovering Warrior Task Force. 2014. *Annual Report*. Department of Defense. Washington, D.C. Available at rwtf.defense.gov

Renner, Michael. 2000. "Number of Wars on Upswing." In *Vital Signs: The Environmental Trends That Are Shaping Our Future*, ed. Linda Starke, 110–111. New York: Norton.

Reuters. 2013 (June 21). "U.S. Files Criminal Charges against Snowden over Leaks: Sources." Available at www.reuters.com

Ricks, Thomas E. 2014. "Rebecca's War Dog of the Week: Sasha Awarded Dickin Medal for Gallantry in Afghanistan." *Foreign Policy*, May 2. Available at www.foreignpolicy .com

Riedel, Bruce. 2015. "Why Saudi Arabia's Yemen War Is Not Producing Victory." *U.S. News & World Report*, May 26. Available at www .usnews.com

Robbins, Joseph, Lance Hunter, and Gregg R. Murray. 2013 (July). "Voters versus Terrorists: Analyzing the Effect of Terrorist Events on Voter Turnout." *Journal of Peace Research* 50(4):495–508.

Romero, Anthony. 2003. "Civil Liberties Should Not Be Restricted during Wartime." In *The Terrorist Attack on America*, ed. Mary Williams, 27–34. Farmington Hills, MA: Greenhaven Press.

Rosenberg, Tina. 2000. "The Unbearable Memories of a U.N. Peacekeeper." *New York Times*, October 8, pp. 4, 14.

Roughton, Randy. 2011 (February 3). "Fallen Marine's Family Adopts His Best Friend." U.S. Air Force Official Website, Available at www.af.mil

Salem, Paul. 2007. "Dealing with Iran's Rapid Rise in Regional Influence." *Japan Times*, February 22. Available at www.carnegieendowment.org

Saletan, William. 2015. "Don't Blame Drones." *Slate*, April 24. Available at www.slate.com

Salladay, Robert. 2003 (April 7). *Anti-War Patriots Find They Need to Reclaim Words, Symbols, Even U.S. Flag from Conservatives*. Available at www.commondreams.org

Sang-Hun, Choe, and Chris Buckley. 2013. "North Korean Leader Supports Resumption of Nuclear Talks, State Media Say." *New York Times*, July 26. Available at www.nytimes.com

Save the Children. 2012. *Untold Atrocities: The Stories of Syria's Children*. October. Available at www.savethechildren.org

Save the Children. 2015. *Failing Syria: Assessing the Impact of UN Security Council Resolutions in Protecting and Assisting Civilians in Syria*. Available at www.savethechildren.org

Scheff, Thomas. 1994. *Bloody Revenge*. Boulder, CO: Westview Press.

Schmitt, Eric. 2005. "Pentagon Seeks to Shut Down Bases across Nation." *New York Times*, May 14. Available at www.nytimes.com

Schroeder, Matt. 2007. *The Illicit Arms Trade*. Washington, DC: Federation of American Scientists. Available at www.fas.org

Schultz, George P., William J. Perry, Henry A. Kissinger, and Sam Nunn. 2007. "A World Free of Nuclear Weapons." *Wall Street Journal*, January 4. Available at www.wsj.com

Senate Select Committee on Intelligence. 2014. Committee Study of the Central Intelligence Agency's Detention and Interrogation Program. Available at www.documentcloud.org

Sengupta, Somini. 2015. "United Nations' Reputation Slips as Four-Year War in Syria Drags On." *New York Times*, March 12. Available at www.nytimes.om

Shanahan, John J. 1995. "Director's Letter." *Defense Monitor* 24(6):8.

Shane, Leo III. 2015. "Report: Suicide Rate Spikes among Young Veterans." *Stars and Stripes*, January 9. Available at www.stripes.com

Sheridan, Mary Beth, and William Branigin. 2010. "Senate Ratifies New U.S.-Russia Nuclear Weapons Treaty." *Washington Post*, December 22. Available at washingtonpost.com

Shrader, Katherine. 2005. "WMD Commission Releases Scathing Report." *Washington Post*, March 31. Available at www.washingtonpost.com

Silver, Roxane Cohen, E. Alison Holman, Judith Pizarro Andersen, Michael Poulin, Daniel N. McIntosh, and Virginia Gil-Rivas. 2013 (September). "Mental- and Physical-Health Effects of Acute Exposure to Media Images of the September 11, 2011, Attacks and the Iraq War." *Psychological Science* 24(9):1623–1634.

Simon, Scott. 2003 (March 29). "Marine Mammals on Active Duty: Navy Uses Dolphins, Sea Lions to Patrol Waters in Persian Gulf." Available at www.npr.org

Sisk, Richard. 2015. "ISIS 'Hacking Division' Claims It Posted US Troops' Names, Addresses." *Military News*, March 22. Available at www.military.com

Skocpol, Theda. 1992. *Protecting Soldiers and Mothers: The Political Origins of Social Policy in the United States*. Cambridge, MA: Belknap Press of Harvard University Press.

Skocpol, Theda. 1994. *Social Revolutions in the Modern World*. Cambridge: Cambridge University Press.

Smith, Lamar. 2003. "Restricting Civil Liberties during Wartime Is Justifiable." In *The Terrorist Attack on America*, ed. Mary Williams, 23–26. Farmington Hills, MA: Greenhaven Press.

Smith-Spark, Laura, Jim Sciutto, and Elise Labbott. 2013 (October 16). "Iran Nuclear Talks Start in Geneva amid 'Cautious Optimism.'" *CNN*. Available at www.cnn.com

Spearin, Christopher. 2014 (Summer). "Special Operations Forces and Private Security Companies." *Parameters* 44(2):61–73.

Starr, J. R., and D. C. Stoll. 1989. *U.S. Foreign Policy on Water Resources in the Middle East*. Washington, DC: Center for Strategic and International Studies.

START (National Consortium for the Study of Terrorism and Responses to Terrorism). 2014 (May). *Background Report: Boko Haram Recent Attacks*. Available at www.start.umd.edu

Stimson Center. 2007. *Reducing Nuclear Dangers in South Asia*. Available at www.stimson.org

Stockholm International Peace Research Institute (SIPRI). 2014. *Recent Trends in Military Expenditure*. Available at www.sipri.org

Stockholm International Peace Research Institute (SIPRI). 2015a. *SIPRI Yearbook, 2015*. Available at www.sipri.org

Stockholm International Peace Research Institute (SIPRI). 2015b (April). *Trends in World Military Expenditures, 2014*. Available at www.sipri.org

Tanielian, Terri and Lisa H. 2008. Jaycox. *Invisible Wounds of War*. Santa Monica, CA: RAND.

Tarabah, Asma, Lina Kurdahi Badr, Jinan Usta, and John Doyle. 2015. "Exposure to Children's Desensitization Attitudes in Lebanon." *Journal of Interpersonal Violence*, May 26.

Tavernise, Sabrina. 2007. "U.S. Contractor Banned by Iraq over Shootings." *New York Times*, September 18. Available at www.nytimes.com

Tilly, Charles. 1992. *Coercion, Capital and European States: AD 990–1992*. Cambridge, MA: Basil Blackwell.

Uenuma, Francine. 2013 (August 23). "Number of Children Who Have Fled Reaches One Million." Save the Children Media Release. Available at www.savethechildren.org

United National Mine Action Office. 2011 (June). "UNMAO Regional Fact Sheet." *Southern Sudan*. Available at reliefweb.int/report/sudan/unmao-regional-fact-sheet-southern-sudan-updated-june-201

United Nations. 1948. *Convention on the Prevention and Punishment of the Crime of Genocide*. Available at www.un.org

United Nations. 2003. *Some Questions and Answers*. Available at www.unicef.org

United Nations. 2015. *United Nations Peacekeeping Operations*. Available at www.un.org/en/peacekeeping/resources/statistics/factsheet.shtml

United Nations High Commissioner for Refugees. 2015. *UNHCR Population Statistics*. Available at www.unhcr.org

United Nations Security Council. 2011 (March 17). *Resolution 1973.S/RES/1973*. Available at www.un.org

U.S. Department of Defense. 2013. *Annual Report to Congress: Military and Security Developments Involving the People's Republic of China*. Available at www.defense.gov/pubs/2013_China_Report_FINAL.pdf

U.S. Department of Defense. 2014 (June). *Strategy for Countering Weapons of Mass Destruction*. Available at www.defense.gov

U.S. Department of Defense. 2015 (March). *Program Acquisition Cost by Weapon System*. Available at www.defense.gov

U.S. Department of Homeland Security. 2015. *FY 2016 Budget-in Brief*. Available at www.dhs.gov

U.S. Department of State. 2007. "Small Arms/Light Weapons Destruction." Available at www.state.gov

U.S. Department of State. 2014. "World Military Expenditures and Arms Transfers." Available at www.state.gov

U.S. Department of State. 2015a. "Country Reports on Terrorism 2014: Annex of Statistical Information." National Consortium for the Study of Terrorism and Responses to Terrorism (START). Available at www.state.gov

U.S. Department of State. 2015b. Parameters for a Joint Comprehensive Plan of Action Regarding the Islamic Republic of Iran's Nuclear Program. Available at www.state.gov

U.S. Department of State. 2015c (April 3). "U.S. Global Leadership in Landmine Clearance and Conventional Weapons Destruction." *Fact Sheet: Bureau of Political-Military Affairs*. Available at www.state.gov

U.S. Department of Veterans Affairs. 2014. *PTSD: National Center for PTSD*. Available at www.ptsd.va.gov

U.S. Department of Veterans Affairs. 2015. *Education and Training*. Available at www.benefits.va.gov

U.S. Navy. 2010. "Navy Marine Mammal Program Excels during Frontier Sentinel." Available at www.navy.mil/search/display.asp?story_id=53979

Vesely, Milan. 2001. "UN Peacekeepers: Warriors or Victims?" *African Business* 261:8–10.

Viner, Katharine. 2002. "Feminism as Imperialism." *The Guardian*, September 21. Available at www.guardian.co.uk

Walker, Peter. 2013. "The Bradley Manning Trial: What We Know from the Leaked WikiLeaks Documents." *The Guardian*, July 30. Available at www.theguardian.com

Ward, Benjamin. 2015 (February 25). "The EU Stands By as Thousands of Migrants Drown in the Mediterranean." *Human Rights Watch*. Available at www.hrw.org

Waters, Rob. 2005. "The Psychic Costs of War." *Psychotherapy Networker*, March–April, pp. 1–3.

Wax, Emily. 2003. "War Horror: Rape Ruining Women's Health." *Miami Herald*, November 3. Available at www.miami.com

Weisman, Jonathan. 2013. "Iran Talks Face Resistance in U.S. Congress." *New York Times*, November 12. Available at www.nytimes.com

White House. 2015a (June 10). "Press Conference Call on Additional Steps to Counter ISIL." Office of the Press Secretary. Available at www.whitehouse.gov

White House. 2015b (April 23). "Statement by the President on the Deaths of Warren Weinstein and Giovanni Lo Porto." Office of the Press Secretary. Available at www.whitehouse.gov

Williams, Mary E., ed. 2003. *The Terrorist Attack on America*. Farmington Hills, MA: Greenhaven Press.

Woehrle, Lynne M., Patrick G. Coy, and Gregory M. Maney. 2008. *Contesting Patriotism: Culture, Power, and Strategy in the Peace Movement*. Lanham, MD: Rowman & Littlefield.

WorldWatch Institute. 2015. "Modern Warfare Causes Unprecedented Environmental Damage." *Vital Signs*, June 15. Available at www.worldwatch.org

Zimmermann, Ekkart. 2011. "Globalization and Terrorism." *European Journal of Political Economy* 27(1):S5–S16.

Zoroya, Gregg. 2006. "Lifesaving Knowledge, Innovation Emerge in War Clinic." *USA Today*, March 27. Available at www.usatoday.com

Name index

Page numbers in *italics* denote figures, illustrations, and captions.

Subject Index

Note: Page numbers in **boldface** type denote definitions. Page numbers in *italics* denote tables, figures, illustrations, and captions.

Violent and property crimes, 101, 108–11, *108*
Violent crime, 101, 108–9, *108*
Violent Crime Control and Law Enforcement Act, 129, 136
Vivisection, **488**
A Voice for Men, 350–51
Voting restrictions, 303
Voting rights, institutional discrimination in, 303
Vulnerable employment, **329**

Wage gap, *330*, 331–33, *332*, 351; explanations for, **333**; legislation on, 351–52
Wages: inequality in, 218; living wage laws, **198**; minimum wage, **198**; wage gap, *330*, 331–33, *332*, 351; wage theft, **198**
War, 496–537, **498**; arms control and, 532–34; casualty numbers, 523; causes of, 509–14, *514*, 529; chemical weapons, 521–22, 529; civilian casualties, 523; clash of civilizations and, **535**; constructivist explanations, **510**–11; costs of, 500–503, *503*; deaths and disabilities, 523; economic growth and, 502, 504, *506*; economic inequality and, 192–93; economic resources and, 525–27, *527*, 529; environmental costs, 527–29; euphemisms about, 508; fear of and defense against, 511–12; female combat exclusion policy, 506, *507*; Geneva Conventions on, **519**–20; global context of, 498–503, *499*; guerrilla warfare, **520**, 524–25; language of, 508; language of feminism and, 506; military spending, 527, *527*; nuclear war, 528–29; population growth and, 402; poverty and, 192–93; primordial explanations, **510**–11; rally round the flag effect, **504**; rape and, 523–24; refugees and, 503, 513, *513*, 523–24; revolutions and civil wars, 512–13; security dilemma and, **511**–12; social problems associated with, 523–29; social-psychological costs,

524–25; sociological theories of, 524–25; strategies for action, 529–35; terrorism and, **515**–22; understanding, 535–36; weapons of mass destruction and, **521**–22, *522. See also* Military; Terrorism; *specific wars*
War on drugs, 66, 92–94, *93*
Wars, specific: Afghanistan, 503, 506, 525; Cold War, **500**–502, 510, 529; Global War on Terror, 523; Gulf War, *528*; Iraq, 503, 506, 512; Korea, 523; Vietnam, 525, 527–28; World War I, 523; World War II, 504, 510, 523
Waste. *See* E-waste; Nuclear waste; Solid waste
Water: fracking and, 438–39; as resource leading to wars, 510; safe drinking water, 189
Waterboarding, 520
Water contamination, 476
Water pollution, 189, 437–38
Wealth, 176–211, **178**; capitalism and, 216; global context, 178–79; inequality in, 178–79, 184–85, *185*; laws and policies favoring, 182; political influence and, 182; sociological theories of, 181–83; taxes on, 195; understanding, 208–9; wealthiest Americans, 185–86
Wealthfare, **182**
Weapons: arms control, 532–34; sale of, 502; small arms, 534–35
Weapons of mass destruction (WMD), **521**–22, *522*
Web 3.0, 474
Web-based surveys, 21
Welfare, 198–208; as a "dirty word", 183; corporate welfare, **182**–83; means-tested programs, **198**–203, *205*; myths and realities, 203–8; Romney on, 203; wealthfare, **182**
Whistle-blowers, 113
White-collar crime, *113*, **113**–14, 124–26
Whiteness Studies, 312
White racial advantage, 312, *313*
Wife beating, 152, *152*
WikiLeaks, 519
Wind energy, 449

Women: body image and, 345, 347; child brides, *336*, 337; crime and, 116–17, *117*, 121; division of labor in a marriage, 152, 154–55, 337–39; economic autonomy of, 154; economic marginalization of, 117; education and, 51, 146–47, 257–59, *258*, 327–29, *328*, 411; equal rights amendment, **350**; family planning, access to, 50; female genital mutilation/cutting, 323; film characters of, 342, *342*; honor killings, **348**–49; intimate partner violence and, 41, **159**–62; motherhood penalty, **331**, 332–33; politics and, 333–34, *335*, *336*; poverty and, 186, 344, *344*; status of, 411; tobacco marketing to, 74–75; on U.S. Supreme Court, 333; victimization of, 348–49, *348*, 353; violence against, 41, 117, 348–49, *348*; violence against, preventing, 353; wage gap and, *330*, 331–33, *332*, 351; war, combat roles in, 506, *507*; war, effects on, 523–24; wife beating, 152, *152. See also gender entries;* Rape; Sexism
Women's movement, 349–50
Work, 212–42; alienation and, 230–31; automation and, **221**, **460**, 470, 482; behavior-based safety programs, **238**; child labor, **225**–26; full employment, **233**; garment industry, 214, 225; gender and, 329–31; global context, 214–17, 234; health and safety issues, 226–27, *227*, *228–29*, 237–38; job burnout, **229**; job creation and preservation, 234; job stress, 228–30, *230*; labor unions and, **231**–33, 239–40; leave and vacation policies, 153, 238–39; minimum wage/living wage, **198**; offshoring of jobs, 217; by prison inmates, 224; problems of, 219–33; self-concept and, *220*; slavery (forced labor), **224**–25, 235–36; sociological theories of, 218–19; strategies for action, 233–40; structural sexism

and, 329–31; sweatshops, **225**, 236–37, 291; technology and, 469–70; telecommuting, 469–70; unauthorized immigrants in workforce, 291; understanding, 240; unemployment and underemployment, 219–22, *220*; vulnerable employment, **329**; wage gap, *330*, 331–33, *332*, 351; work/life conflict, **230**. *See also* Employment; Unemployment; Workplace
Worker cooperatives, 234–35
Worker Rights Consortium (WRC), 236–37
Workers' self-directed enterprises, **235**
Workforce, unauthorized immigrants in, 291
Workforce Investment Act (WIA), **233**–34
Working poor, **187**, 233
Work/life conflict, **230**
Work-life policies and programs, 238–39, *239*
Workplace: family support in, 168; health and safety issues, 226–27, *227*, *228–29*, 237–38; McDonaldization of, **231**; responses to health and safety issues, 237–38; technology and, 469–70
Workplace discrimination, 304–5, *305*; based on age, 405; based on sexual orientation, 372–73, *373*, 384–85; Employment Non-Discrimination Act, **384**–85; ending, 384–85, 414
World Bank, 219
World population, 394–96, *395*, *396*; aging of, 396–98, *397*
World Trade Center. *See* September 11, 2001, attacks
World War I, 523
World War II, 504, 510, 523

Yakuza, 112
Youth bulge, 402–3
Youth programs, 127–28
YouTube, 91

Zero-tolerance policies, 263